Pavement Analysis and Design

YANG H. HUANG

University of Kentucky

PRENTICE HALL
Englewood Cliffs, New Jersey 07632

Library of Congress Cataloging-in-Publication Data

HUANG, YANG H. (YANG HSIEN)
   Pavement analysis and design / Yang H. Huang.
      p.   cm.
   Includes bibliographical references and index.
   ISBN 0-13-655275-7
   1. Pavements—Design and construction.   2. Structural analysis
(Engineering)   I. Title.
TE251.H77   1993
625.8—dc20   92-26720
                                                        CIP

Acquisitions editor: Doug Humphrey
Production editor: Merrill Peterson
Editorial/production supervision and interior
design: Joan Stone
Copy editor: Virginia Dunn
Cover designer: Karen Marsilio
Prepress buyer: Linda Behrens
Manufacturing buyer: Dave Dickey
Editorial assistant: Jaime Zampino

 © 1993 by Prentice-Hall, Inc.
A Paramount Communications Company
Englewood Cliffs, New Jersey 07632

The author and publisher of this book have used their best efforts in preparing this book. These efforts include the development, research, and testing of the theories and programs to determine their effectiveness. The author and publisher make no warranty of any kind, expressed or implied, with regard to these programs or the documentation contained in this book. The author and publisher shall not be liable in any event for incidental or consequential damages in connection with, or arising out of, the furnishing, performance, or use of these programs.

Printed in the United States of America

10 9 8 7 6

ISBN 0-13-655275-7

PRENTICE-HALL INTERNATIONAL (UK) LIMITED, *London*
PRENTICE-HALL OF AUSTRALIA PTY. LIMITED, *Sydney*
PRENTICE-HALL CANADA INC., *Toronto*
PRENTICE-HALL HISPANOAMERICANA, S.A., *Mexico*
PRENTICE-HALL OF INDIA PRIVATE LIMITED, *New Delhi*
PRENTICE-HALL OF JAPAN, INC., *Tokyo*
SIMON & SCHUSTER ASIA PTE. LTD., *Singapore*
EDITORA PRENTICE-HALL DO BRASIL, LTDA., *Rio de Janeiro*

Contents

Contents   **v**

Preface

During the past two decades, I have been teaching a course on Pavement Analysis and Design to both seniors and graduate students at the University of Kentucky. I had difficulty finding a suitable textbook for the course because very few are available. There are at least two reasons that a professor does not like to write a textbook on pavement analysis and design. First, the subject is very broad. It covers both highway and airport pavements and involves analysis, design, performance, evaluation, maintenance, rehabilitation, and management. It is difficult, if not impossible, to cover all these topics in sufficient detail to serve as a textbook with enough illustrative examples and homework problems for the students. Second, empirical methods have been used most frequently for pavement analysis and design, and a book based on empirical procedures becomes out of date within a short time. No one is willing to write a book with such a short life. Because of the above difficulties, I have written this book as an alternative. I have limited the content to the structural analysis and design of highway pavements and covered essentially the mechanistic–empirical design procedures rather than the purely empirical methods.

To facilitate the teaching of mechanistic–empirical methods, I have included two computer programs that I developed for pavement analysis and design. These programs have been used by my students for more than ten years and have been constantly updated and improved. They are original and contain salient features not available elsewhere. For example, the KENLAYER program for flexible pavements can be applied to a multilayered system under stationary or moving multiple wheel loads with each layer being either linear elastic, nonlinear elastic, or viscoelastic. The KENSLABS program for rigid pavements can be applied to multiple slabs fully or partially supported on a liquid, solid, or layered foundation with moment or shear transfer across the joints. Both programs can perform damage analysis by dividing a year into a number of periods, each having a different set of material properties and subjected to varying repetitions of different axle loads. These programs were originally written for an IBM mainframe but were later adapted to an IBM PC and can be run using the diskette provided with this book.

In addition to the documentation of the computer programs, this book presents the theory of pavement design and reviews the methods developed by several organizations, such as the American Association of State Highway and Transportation Officials (AASHTO), the Asphalt Institute (AI), and the Portland Cement Association (PCA). Because most of the advanced theory and detailed information are presented in the appendices, the book can be used either as a text for an undergraduate course by skipping the appendices or as an advanced graduate course by including them. Although this book covers only the analysis and design of highway pavements, the same principles can be applied to airport pavements and railroad trackbeds.

This book is divided into 13 chapters. Chapter 1 introduces the historical development of pavement design, the major road tests, the various design factors, and the differences in design concepts among highway pavements, airport pavements, and railroad trackbeds. Chapter 2 discusses stresses and strains in flexible pavements, including the analysis of homogeneous mass and layered systems composed of linear elastic, nonlinear elastic, and linear viscoelastic materials. Simplified charts and tables for determining stresses and strains are also presented. Chapter 3 presents the KENLAYER computer program, based on Burmister's layered theory, including theoretical developments, program description, comparison with available solutions, and sensitivity analysis on the effect of various factors on pavement responses. Chapter 4 discusses stresses and deflections in rigid pavements due to curling, loading, and friction, as well as the design of dowels and joints. Influence charts for determining stresses and deflections are also presented. Chapter 5 presents the KENSLABS computer program, based on the finite element method, including theoretical developments, program description, comparison with available solutions, and sensitivity analysis. Chapter 6 discusses the concept of equivalent single-wheel and single-axle loads and the prediction of traffic. Chapter 7 describes the material characterization for mechanistic–empirical methods of pavement design including the determination of resilient modulus, fatigue and permanent deformation properties, and the modulus of subgrade reaction. Their correlations with other empirical tests are also presented. Chapter 8 outlines the subdrainage design including general principles, drainage materials, and design procedures. Chapter 9 discusses pavement performance including distress, serviceability, skid resistance, nondestructive testing, and the evaluation of pavement performance. Chapter 10 illustrates the reliability concept of pavement design in which the variabilities of traffic, material, and geometric parameters are all taken into consideration. A simple and powerful probabilistic procedure, originally developed by Rosenblueth, is described and two probabilistic computer programs including VESYS for flexible pavements and PMRPD for rigid pavements are discussed. Chapter 11 outlines an idealistic mechanistic method of flexible pavement design and presents in detail the Asphalt Institute method and the AASHTO method, as well as the design of flexible pavement shoulders. Chapter 12 outlines an idealistic mechanistic method of rigid pavement design and presents in detail the Portland Cement Association method and the AASHTO method. The design of continuous reinforced concrete pavements and rigid pavement shoulders is also included. Chapter 13 outlines the design of overlay on both flexible and rigid pavements including the AASHTO,

AI, and PCA procedures. More advanced theory and detailed information related to some of the chapters as well as a list of symbols and references are included in the appendices.

Other than the empirical AASHTO methods used by many state highway departments, this book emphasizes principally the mechanistic–empirical method of design. With the availability of personal computers and the sophisticated methods of material testing, the trend toward mechanistic–empirical methods is quite apparent. It is believed that a book based on mechanistic–empirical methods is more interesting and challenging than one based on empirical methods. This book, with the accompanying computer programs and the large number of illustrative examples, will serve as a classroom text and useful reference for people interested in learning about the structural analysis and design of highway pavements.

Although considerable portions of the materials presented in this book were developed by myself through years of research, teaching, and engineering practice, much information was obtained from the published literature. Grateful acknowledgment is offered to AASHTO, the Asphalt Institute, the Federal Highway Administration, the Portland Cement Association, the Transportation Research Board, and many other organizations and individuals that have permitted me to use the information they developed. The helpful comments by James Lai, Professor of Civil Engineering, Georgia Institute of Technology, are highly appreciated.

Yang H. Huang, P.E.

Professor of Civil Engineering
University of Kentucky

1

Introduction

1.1 HISTORICAL DEVELOPMENTS

Although pavement design has gradually evolved from art to science, empiricism still plays an important role even up to the present day. Prior to the early 1920s, the thickness of pavement was based purely on experience. The same thickness was used for a section of highway even though widely different soils were encountered. As experience was gained throughout the years, various methods were developed by different agencies for determining the thickness of pavement required. It is neither feasible nor desirable to document all the methods that have been used so far. Only a few typical methods will be cited to indicate the trend.

Some technical terms will be used in this introductory and review chapter. It is presumed that the students using this book as a text are seniors or graduate students who have courses on transportation engineering, civil engineering materials, and soil mechanics and are familiar with these terms. In case this is not true, these terms can be ignored for the time being because most are explained and clarified in later chapters.

1.1.1 Flexible Pavements

Flexible pavements are constructed of bituminous and granular materials. The first asphalt roadway in the United States was constructed in 1870 at Newark, New Jersey. The first sheet asphalt pavement, which is a hot mixture of asphalt cement with clean, angular, graded sand and mineral filler, was laid in 1876 on Pennsylvania Avenue in Washington, D.C., with imported asphalt from Trinidad Lake. As of 1990 (FHWA, 1990), there are about 2.2 million miles of paved roads in the United States, of which 94% are asphalt surfaced.

Design Methods

Methods of flexible pavement design can be classified into five categories: empirical method with or without a soil strength test, limiting shear failure method, limiting deflection method, regression method based on pavement performance or road test, and mechanistic–empirical method.

Empirical Methods

The use of empirical method without a strength test dates back to the development of the Public Roads (PR) soil classification system (Hogentogler and Terzaghi, 1929), in which the subgrade was classified as uniform from A-1 to A-8 and nonuniform from B-1 to B-3. The PR system was later modified by the Highway Research Board (HRB, 1945), in which soils were grouped from A-1 to A-7 and a group index was added to differentiate the soil within each group. Steele (1945) discussed the application of HRB classification and group index in estimating the subbase and total pavement thickness without a strength test. The empirical method with a strength test was first used by the California Highway Department in 1929 (Porter, 1950). The thickness of pavements was related to the California Bearing Ratio (CBR), defined as the penetration resistance of a subgrade soil relative to a standard crushed rock. The CBR method of design was studied extensively by the U.S. Corps of Engineers during World War II and became a very popular method after the war.

The disadvantage of empirical method is that it can be applied only to a given set of environmental, material, and loading conditions. If these conditions are changed, the design is no longer valid and a new method must be developed through trial and error to be commensurate with the new conditions.

Limiting Shear Failure Methods

The limiting shear failure method is used to determine the thickness of pavements so that shear failures will not occur. The major properties of pavement components and subgrade soils to be considered are their cohesion and angle of internal friction. Barber (1946) applied Terzaghi's bearing capacity formula (Terzaghi, 1943) to determine pavement thickness. McLeod (1953) advocated the use of logarithmic spirals to determine the bearing capacity of pavements. These methods were reviewed by Yoder (1959) in his book *Principles of Pavement Design* but were not even mentioned in the second edition (Yoder and Witczak, 1975). This is not surprising because, with the ever increasing speed and volume of traffic, pavements should be designed for riding comfort rather than barely preventing shear failures.

Limiting Deflection Methods

The limiting deflection method is used to determine the thickness of pavements so that the vertical deflection will not exceed the allowable limit. The Kansas State Highway Commission (1947) modified Boussinesq's equation (Boussinesq, 1885) and limited the deflection of subgrade to 0.1 in. (2.54 mm). The U.S. Navy (1953) applied Burmister's two-layer theory (Burmister, 1943) and limited the surface deflection to 0.25 in. (6.35 mm). The use of deflection as a design criterion has the apparent advantage that it can be easily measured in the field. Unfortunately, pavement failures are caused by excessive stresses and strains instead of deflections.

Regression Methods Based on Pavement Performance or Road Tests

A good example of the use of regression equations for pavement design is the AASHTO method based on the results of road tests. The disadvantage of the method is that the design equations can be applied only to the conditions at the road test site. For conditions other than those under which the equations were developed, extensive modifications based on theory or experience are needed. Regression equations can also be developed from the performance of existing pavements such as those used in the pavement evaluation systems COPES (Darter et al., 1985) and EXPEAR (Hall et al., 1989). Unlike road tests, the materials and construction of these pavements were not well controlled, so a wide scatter of data and a large standard error are expected. Although these equations can illustrate the effect of various factors on pavement performance, their usefulness in pavement design is limited because of the many uncertainties involved.

Mechanistic–Empirical Methods

The mechanistic–empirical method of design is based on the mechanics of materials that relates an input, such as a wheel load, to an output or pavement response, such as stress or strain. The response values are used to predict distress based on laboratory test and field performance data. Dependence on observed performance is necessary because theory alone has not proven sufficient to design pavements realistically.

Kerkhoven and Dormon (1953) first suggested the use of vertical compressive strain on the surface of subgrade as a failure criterion to reduce permanent deformation, while Saal and Pell (1960) recommended the use of horizontal tensile strain at the bottom of asphalt layer to minimize fatigue cracking, as shown in Figure 1.1. The use of the above concepts for pavement design was first presented in the United States by Dormon and Metcalf (1965).

The use of vertical compressive strain to control permanent deformation is based on the fact that plastic strains are proportional to elastic strains in paving materials. Thus, by limiting the elastic strains on the subgrade, the elastic strains

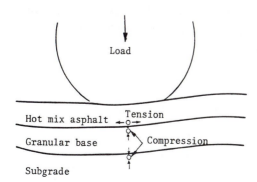

Figure 1.1 Tensile and compressive strains in flexible pavements.

in other components above the subgrade will also be controlled; hence, the magnitude of permanent deformation on the pavement surface will be controlled as well. These two criteria have since been adopted by Shell Petroleum International (Claussen et al., 1977) and the Asphalt Institute (Shook et al., 1982) in their mechanistic–empirical methods of design. The advantages of mechanistic methods are the improvement in the reliability of a design, the ability to predict the types of distress, and the feasibility to extrapolate from limited field and laboratory data.

The term "hot mix asphalt" in Figure 1.1 is synonymous to the commonly used "asphalt concrete." It is an asphalt aggregate mixture produced at a batch or drum mixing facility that must be mixed, spread, and compacted at an elevated temperature. To avoid the confusion between portland cement concrete (PCC) and asphalt concrete (AC), the term hot mix concrete (HMA) will be used frequently throughout this book in place of asphalt concrete.

Other Developments

Other developments in flexible pavement design include the application of computer programs, the incorporation of serviceability and reliability, and the consideration of dynamic loading.

Computer Programs

Various computer programs based on Burmister's layered theory have been developed. The earliest and the most well known is the CHEV program developed by the Chevron Research Company (Warren and Dieckmann, 1963). The program can be applied only to linear elastic materials but was modified by the Asphalt Institute in the DAMA program to account for nonlinear elastic granular materials (Hwang and Witczak, 1979). Another well-publicized program is BISAR developed by Shell, which considers not only vertical loads but also horizontal loads (De Jong et al., 1973). Another program, originally developed at the University of California, Berkeley, and later adapted to microcomputers, is ELSYM5, for elastic five-layer systems under multiple wheel loads (Kopperman et al., 1986). Based on the layered theory with stress-dependent material properties, Finn et al. (1986) developed a computer program named PDMAP (Probabilistic Distress Models for Asphalt Pavements) for predicting the fatigue cracking and rutting in asphalt pavements. They found that the critical responses obtained from PDMAP compared favorably with SAPIV, which is a finite element stress analysis program developed at the University of California, Berkeley.

A major disadvantage of the layered theory is the assumption that each layer is homogeneous with the same properties throughout the layer. This assumption makes it difficult to analyze layered systems composed of nonlinear materials, such as untreated granular bases and subbases. The elastic modulus of these materials is stress dependent and varies throughout the layer, so a question immediately arises: Which point in the nonlinear layer should be selected to represent the entire layer? If only the most critical stress, strain, or deflection is desired, as is usually the case in pavement design, a point near to the applied load can be reasonably selected. However, if the stresses, strains, or deflections at different points, some near to and some far away from the load, are desired, it will be difficult to use the layered theory for analyzing nonlinear materials. This difficulty can be overcome by using the finite element method.

Duncan et al. (1968) first applied the finite element method for the analysis of flexible pavements. The method was later incorporated in the ILLI-PAVE computer program (Raad and Figueroa, 1980). Due to the large amount of computer time and storage required, the program has not been used for routine design purposes. However, a number of regression equations, based on the responses obtained by ILLI-PAVE, were developed for use in design (Thompson and Elliot, 1985; Gomez-Achecar and Thompson, 1986). The nonlinear finite element method was also used in the MICH-PAVE computer program developed at Michigan State University (Harichandran et al., 1989). More information about ILLI-PAVE and MICH-PAVE and their deficiencies is presented in Section 3.3.2.

Serviceability and Reliability

As a result of the AASHO Road Test, Carey and Irick (1960) developed the pavement serviceability performance concept and indicated that pavement thickness should also depend on the terminal serviceability index required. Lemer and Moavenzadeh (1971) presented the concept of reliability as a pavement design factor, and a probabilistic computer program for analyzing a three-layer viscoelastic pavement system was developed (Moavenzadeh et al., 1974). This program, which incorporated the concepts of serviceability and reliability, was modified by the Federal Highway Administration (FHWA, 1978) and renamed VESYS II. Several versions of the VESYS program were developed; the one described in Section 10.5.1 is called VESYS IV-B (Jordahl and Rauhut, 1983). The reliability concept was also incorporated in the Texas flexible pavement design system (Darter et al., 1973b) and in the AASHTO Design Guide (AASHTO, 1986). Although the AASHTO procedures are basically empirical, the replacements of the empirical soil support value by the subgrade resilient modulus and the empirical layer coefficients by the resilient modulus of each layer clearly indicate the trend toward mechanistic methods. The resilient modulus is the elastic modulus under repeated loads and can be determined by laboratory tests. Details about resilient modulus are presented in Section 7.1, serviceability in Section 9.2, and reliability in Chapter 10.

Dynamic Loads

All the methods discussed so far are based on static or moving loads without considering the inertia effects due to dynamic loads. Mamlouk (1987) described a computer program capable of considering the inertial effect and indicated that the effect is most pronounced when shallow bedrock or frozen subgrade is encountered and becomes more important for vibratory than for impulse loading. The program requires a large amount of computer time to run and is limited to the analysis of linear elastic materials. The inclusion of inertial effect for routine pavement design involving nonlinear elastic and viscoelastic materials is still a dream to be realized in the future.

Recent research by Monismith et al. (1988) has shown that for asphalt concrete pavements it is unnecessary to perform a complete dynamic analysis.

Inertia effects can be ignored and the local dynamic response can thus be determined by an essentially static method using material properties compatible with the rate of loading. However, due to the impulse loading, a dynamic problem of more immediate interest is the effect of vehicle dynamics on pavement design. Current design procedures do not consider the damage caused by pavement roughness. As trucks become larger and heavier, some benefits can be gained by designing proper suspension systems to minimize the damage effect.

1.1.2 Rigid Pavements

Rigid pavements are constructed of portland cement concrete. The first concrete road was built in Detroit, Michigan, in 1908. As of 1990, there were about 130,000 miles (200,000 km) of rigid pavements in the United States. The development of design methods for rigid pavements is not as dramatic as that of flexible pavements because the flexural stress in concrete has long been considered as a major, or even the only, design factor.

Analytical Solutions

Analytical solutions ranging from simple closed form formulas to complex derivations are available for determining the stresses and deflections in concrete pavements.

Goldbeck's Formula

By assuming the pavement as a cantilever beam with a load concentrated at the corner, Goldbeck (1919) developed a simple equation for the design of rigid pavements, as indicated by Eq. 4.12 in Chapter 4. The same equation was applied by Older (1924) in the Bates Road Test.

Westergaard's Analysis Based on Liquid Foundations

The most extensive theoretical studies on the stresses and deflections in concrete pavements were made by Westergaard (1926a, 1926b, 1927, 1933, 1939, 1943, 1948), who developed equations due to temperature curling as well as three cases of loading: load applied near the corner of a large slab, load applied near the edge of a large slab but at a considerable distance from any corner, and load applied at the interior of a large slab at a considerable distance from any edge. The analysis was based on the simplifying assumption that the reactive pressure between the slab and the subgrade at any given point is proportional to the deflection at that point, independent of the deflections at any other points. This type of foundation is called a liquid or Winkler foundation. Westergaard also assumed that the slab and subgrade were in full contact.

In conjunction with Westergaard's investigation, the U.S. Bureau of Public Roads conducted at the Arlington Experimental Farm, Virginia, an extensive investigation on the structural behavior of concrete pavements. The results were published in *Public Roads* from 1935 to 1943 as a series of six papers and reprinted as a single volume for wider distribution (Teller and Sutherland, 1935–1943).

In comparing the critical corner stress obtained from Westergaard's corner formula with that from field measurements, Pickett found that Westergaard's corner formula, based on the assumption that the slab and subgrade were in full contact, always yielded a stress that was too small. By assuming that part of the slab was not in contact with the subgrade, he developed a semiempirical formula that was in good agreement with experimental results. Pickett's corner formula (PCA, 1951), with a 20% allowance for load transfer, was used by the Portland Cement Association until 1966 when a new method based on the stress at the transverse joint was developed (PCA, 1966).

The PCA method was based on Westergaard's analysis. However, in determining the stress at the joint, the PCA assumed that there was no load transfer across the joint, so the stress thus determined was similar to the case of edge loading with a different tire orientation. The replacement of corner loading by the stress at the joint was due to the use of a 12-ft-wide (3.63-m) lane with most traffic moving away from the corner. The PCA method was revised again in 1984, in which an erosion criterion based on the corner deflection, in addition to the fatigue criterion based on the edge stress, was employed (PCA, 1984).

Pickett's Analysis Based on Solid Foundations

In view of the fact that the actual subgrade behaved more like an elastic solid than a dense liquid, Pickett et al. (1951) developed theoretical solutions for concrete slabs on an elastic half space. Due to the complexities of the mathematics involved, this refined method has not received the attention it merits. However, a simple influence chart based on solid foundations was developed by Pickett and Badaruddin (1956) for determining the edge stress, which is presented in Section 5.3.2.

Numerical Solutions

All the analytical solutions mentioned above were based on the assumption that the slab and the subgrade are in full contact. It is well known that, due to pumping, temperature curling, and moisture warping, the slab and subgrade are usually not in contact. With the advent of computers and numerical methods, some analyses based on partial contact were developed.

Discrete Element Methods

Hudson and Matlock (1966) applied the discrete element method by assuming the subgrade to be a dense liquid. The discrete element method is more or less similar to the finite difference method in that the slab is seen as an assemblage of elastic joints, rigid bars, and torsional bars. The method was later extended by Saxena (1973) for analyzing slabs on an elastic solid foundation.

Finite Element Methods

With the development of the powerful finite element method, a breakthrough was made in the analysis of rigid pavements. Cheung and Zienkiewicz (1965) developed finite element methods for analyzing slabs on elastic foundations of both liquid and solid types. The methods were applied to jointed slabs on liquid foundations by Huang and Wang (1973, 1974) and on solid foundations by Huang (1974a). In collaboration with Huang (Chou and Huang, 1979, 1981, 1982), Chou (1981) developed finite element computer programs named WESLIQID and WESLAYER for the analysis of liquid and layered foundations, respectively. The consideration of foundation as a layered system is more realistic when layers of base and subbase exist above the subgrade. Other finite element computer programs available include ILLI-SLAB developed at the University of Illinois (Tabatabaie and Barenberg, 1979, 1980), JSLAB developed by the Portland Cement Association (Tayabji and Colley, 1986), and RISC developed by Resource International, Inc. (Majidzadeh et al., 1984).

Other Developments

Even though theoretical methods are helpful in improving and extrapolating design procedures, the knowledge gained from pavement performance is the most important. Westergaard (1927), who contributed so much to the theory of concrete pavement design, was one of the first to recognize that theoretical results need to be checked against pavement performance. In addition to the above theoretical developments, the following events are worthy of note.

Fatigue of Concrete

An extensive study was made by the Illinois Division of Highways during the Bates Road Test on the fatigue properties of concrete (Clemmer, 1923). It was found that an induced flexural stress could be repeated indefinitely without causing rupture, provided the intensity of extreme fiber stress did not exceed approximately 50% of the modulus of rupture, and that, if the stress ratio was above 50%, the allowable number of stress repetitions to cause failures decreased drastically as the stress ratio increased. Although the arbitrary use of 50% stress ratio as a dividing line was not actually proved, this assumption has been used most frequently as a basis for rigid pavement design. To obtain a smoother fatigue curve, the current PCA method assumes a stress ratio of 0.45, below which no fatigue damage need be considered.

Pumping

With increasing truck traffic, particularly just before World War II, it became evident that subgrade type played an important role in pavement performance. The phenomenon of pumping, which is the ejection of water and subgrade soils through joints and cracks and along the pavement edge, was first described by Gage (1932). After pavement pumping became critical during the war, rigid pavements were constructed on granular base courses of varying thickness to protect loss of subgrade support due to pumping. Many studies were made on the design of base courses for the correction of pumping.

Probabilistic Methods

The application of probabilistic concepts to rigid pavement design was presented by Kher and Darter (1973) and the concepts were incorporated in the

AASHTO design guide (AASHTO, 1986). Huang and Sharpe (1989) developed a finite element probabilistic computer program for the design of rigid pavements and showed that the use of a cracking index in a reliability context was far superior than the current deterministic approach.

1.2 PAVEMENT TYPES

There are three major types of pavements: flexible or asphalt pavements, rigid or concrete pavements, and composite pavements.

1.2.1 Flexible Pavements

Flexible pavements can be analyzed by Burmister's layered theory, which is discussed in Chapters 2 and 3. A major limitation of the theory is the assumption of a layered system infinite in areal extent. This assumption makes the theory inapplicable to rigid pavements with transverse joints. Nor can the layered theory be applied to rigid pavements when the wheel loads are less than 2 or 3 ft (0.6 or 0.9 m) from the pavement edge because discontinuity causes a large stress at the edge. Its application to flexible pavements is valid due to the limited area of stress distribution through flexible materials. As long as the wheel load is more than 2 ft (0.61 m) from the edge, the discontinuity at the edge has very little effect on the critical stresses and strains obtained.

Three types of construction have been used for flexible pavements: conventional flexible pavement, full-depth asphalt pavement, and contained rock asphalt mat (CRAM). The CRAM construction is still in the experimental stage and has not been widely accepted for practical use.

Conventional Flexible Pavements

Conventional flexible pavements are layered systems with better materials on top where the intensity of stress is high and inferior materials at the bottom where the intensity is low. Adherence to this design principle makes possible the use of local materials and usually results in a most economical design. This is particularly true in regions where high-quality materials are expensive but local materials of inferior quality are readily available.

Figure 1.2 shows the cross section of a conventional flexible pavement. Starting from the top, the pavement consists of seal coat, surface course, tack coat, binder course, prime coat, base course, subbase course, compacted subgrade, and natural subgrade. The use of the various courses is based on either necessity or economy and some of the courses may be omitted.

Seal Coat

Seal coat is a thin asphalt surface treatment used to waterproof the surface or to provide skid resistance where the aggregates in the surface course may be polished by traffic and become slippery. Depending on the purpose, seal coats

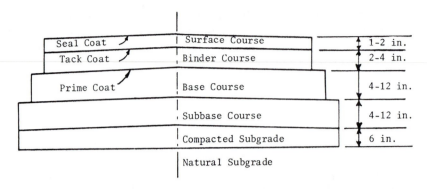

Figure 1.2 Typical cross section of a conventional flexible pavement (1 in. = 25.4 mm).

may or may not be covered with aggregate. Details about skid resistance are presented in Section 9.3.

Surface Course

The surface course is the top course of an asphalt pavement, sometimes called the wearing course. It is usually constructed by dense graded HMA. It must be tough to resist distortion under traffic and provide a smooth and skid-resistant riding surface. It must be waterproof to protect the entire pavement and subgrade from the weakening effect of water. If the above requirements cannot be met, the use of a seal coat is recommended.

Binder Course

The binder course, sometimes called the asphalt base course, is the asphalt layer below the surface course. There are two reasons that a binder course is used in addition to the surface course. First, the HMA is too thick to be compacted in one layer, so it must be placed in two layers. Second, the binder course generally consists of larger aggregates and less asphalt and does not require as high a quality as the surface course, so replacing a part of the surface course by the binder course results in a more economical design. If the binder course is more than 3 in. (76 mm), it is generally placed in two layers.

Tack Coat and Prime Coat

A tack coat is a very light application of asphalt, usually asphalt emulsion diluted with water, used to ensure a bond between the surface being paved and the overlying course. It is important that each layer in an asphalt pavement be bonded to the layer below. Tack coats are also used to bond the asphalt layer to a PCC base or an old asphalt pavement. The three essential requirements of a tack coat are that it must be very thin, it must uniformly cover the entire surface to be paved, and it must be allowed to break or cure before the HMA is laid.

A prime coat is an application of low-viscosity cutback asphalt to an absorbent surface, such as an untreated granular base on which an asphalt layer will be placed. Its purpose is to bind the granular base to the asphalt layer. The difference between a tack coat and a prime coat is that the tack cost does not require the penetration of asphalt into the underlying layer, while the prime coat penetrates into the underlying layer, plugs the voids, and forms a watertight surface. Although the type and quantity of asphalt used are quite different, both are spray applications.

Base Course and Subbase Course

The base course is the layer of material immediately beneath the surface or binder course. It may be composed of crushed stone, crushed slag, or other untreated or stabilized materials. The subbase course is the layer of material beneath the base course. The reason that two different granular materials are used is for economy. Instead of using the more expensive base course material for the entire layer, local and cheaper materials can be used as a subbase course on top of the subgrade. If the base course is open graded, the subbase course with more fines can serve as a filter between the subgrade and the base course.

Subgrade

The top 6 in. (152 mm) of subgrade should be scarified and compacted to the desirable density near the optimum moisture content. This compacted subgrade may be the in situ soil or a layer of selected material.

Full-Depth Asphalt Pavements

Full-depth asphalt pavements are constructed by placing one or more layers of HMA directly on the subgrade or improved subgrade. This concept was conceived by the Asphalt Institute in 1960 and is generally considered the most cost-effective and dependable type of asphalt pavement for heavy traffic. This type of construction is quite popular in areas where local materials are not available. It is more convenient to purchase only one material, i.e., HMA, rather than several materials from different sources, thus minimizing the administration and equipment costs.

Figure 1.3 shows the typical cross section for a full-depth asphalt pavement. The asphalt base course in the full-depth construction is the same as the binder course in conventional pavement. Similar to conventional pavement, a tack coat must be applied between two asphalt layers to bind them together.

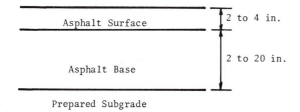

Asphalt Surface   2 to 4 in.

Asphalt Base   2 to 20 in.

Prepared Subgrade

Figure 1.3 Typical cross section of a full-depth asphalt pavement (1 in. = 25.4 mm).

According to the Asphalt Institute (AI, 1987), full-depth asphalt pavements have the following advantages:

1. They have no permeable granular layers to entrap water and impair performance.
2. Time required for construction is reduced. On widening projects, where adjacent traffic flow must usually be maintained, full-depth asphalt can be especially advantageous.
3. When placed in a thick lift of 4 in. (102 mm) or more, construction seasons may be extended.
4. They provide and retain uniformity in the pavement structure.
5. They are less affected by moisture or frost.
6. According to limited studies, moisture contents do not build up in subgrades under full-depth asphalt pavement structures as they do under pavements with granular bases. Thus, there is little or no reduction in subgrade strength.

Contained Rock Asphalt Mats

Another type of construction is the contained rock asphalt mat (CRAM), which is composed of four layers (Miller et al., 1986). Starting from the bottom, a modified dense-graded HMA layer is spread over a conventionally prepared subgrade, followed by a layer of open-graded aggregate, then a dense-graded aggregate, and finally a dense-graded HMA wearing surface, as shown in Figure 1.4.

A major advantage of CRAM construction is that the bottom asphalt layer significantly reduces the vertical compressive strain on the subgrade and the horizontal tensile stress in the overlying granular layer. The reduction of tensile stress in the granular material makes it stronger and thus reduces the tensile stress and strain in the asphalt surface layer. Benefits of the CRAM section include controlling surface water via the open-graded aggregate, preventing the contamination of aggregates by the infiltration of subgrade soils, improving fatigue resistance of bottom asphalt layer by the possible use of softer asphalts,

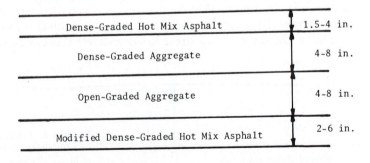

Dense-Graded Hot Mix Asphalt	1.5-4 in.
Dense-Graded Aggregate	4-8 in.
Open-Graded Aggregate	4-8 in.
Modified Dense-Graded Hot Mix Asphalt	2-6 in.

Figure 1.4 Typical CRAM section (1 in. = 25.4 mm).

and reducing crack propagation due to a more favorable distribution of tensile stress and strain in the surface layer. CRAM sections are further analyzed in Section 3.4.3.

1.2.2 Rigid Pavements

Rigid pavements are constructed of portland cement concrete and should be analyzed by the plate theory, instead of the layered theory. Plate theory is a simplified version of the layered theory that assumes the concrete slab to be a medium thick plate with a plane before bending and to remain a plane after bending. If the wheel load is applied in the interior of a slab, either plate or layered theory can be used and both should yield nearly the same flexural stress or strain, as discussed in Section 5.3.4. If the wheel load is applied near to the slab edge, say less than 2 ft (0.61 m) from the edge, only the plate theory can be used for rigid pavements. The reason that the layered theory is applicable to flexible pavements but not to rigid pavements is that PCC is much stiffer than HMA and distributes the load over a much wider area. Therefore, a distance of 2 ft (0.61 m) from the edge is considered quite far in a flexible pavement but not far enough in a rigid pavement. The existence of joints in rigid pavements also makes the layered theory inapplicable. Details of plate theory are presented in Chapters 4 and 5.

Figure 1.5 shows a typical cross section for rigid pavements. In contrast to flexible pavements, rigid pavements are placed either directly on the prepared subgrade or on a single layer of granular or stablized material. Since there is only one layer of material under the concrete and above the subgrade, some call it a base course while others call it a subbase.

Use of Base Course

Early concrete pavements were constructed directly on the subgrade without using a base course. As the weight and volume of traffic increased, pumping began to occur and the use of a granular base course became quite popular. When pavements are subject to a large number of very heavy wheel loads with free water on top of the base course, even granular materials may be eroded by the pulsative action of water. For heavily traveled pavements, the use of a cement-treated or asphalt-treated base course has now become a common practice.

Although the use of a base course can reduce the critical stress in the concrete, it is uneconomical to build a base course for the purpose of reducing the concrete stress. Because the strength of concrete is much greater than that of the base course, the same critical stress in the concrete slab can be obtained without a

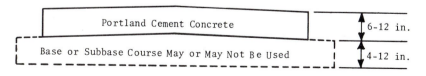

Figure 1.5 Typical cross section of a rigid pavement (1 in. = 25.4 mm).

base course by slightly increasing the concrete thickness. The following reasons have been frequently cited for using a base course.

Control of Pumping

Pumping is defined as the ejection of water and subgrade soil through joints, cracks, and along the edges of pavements caused by downward slab movements due to heavy axle loads. The sequence of events leading to pumping includes the creation of void space under the pavement caused by the temperature curling of the slab and the plastic deformation of the subgrade, the entrance of water, the ejection of muddy water, the enlargement of void space, and finally the faulting and cracking of the leading slab ahead of traffic. Pumping occurs under the leading slab when the trailing slab rebounds, which creates a vacuum and sucks the fine material from underneath the leading slab, as shown in Figure 1.6. The corrective measures of pumping include joint sealing, undersealing with asphalt cements, and muck jacking with soil cement.

Three factors must exist simultaneously to produce pumping:

1. The material under the concrete slab must be saturated with free water. If the material is well drained, no pumping will occur. Therefore, good drainage is one of the most efficient ways to prevent pumping.
2. There must be frequent passage of heavy wheel loads. Pumping will take place only under heavy wheel loads with large slab deflections. Even under very heavy loads, pumping will occur only after a large number of load repetitions.
3. The material under the concrete slab must be erodible. The erodibility of a material depends on the hydrodynamic forces created by the dynamic action of moving wheel loads. Any untreated granular materials, and even some weakly cemented materials, are erodible because the large hydrodynamic pressure will transport the fine particles in the subbase or subgrade to the surface. These fine particles will go into suspension and cause pumping.

Control of Frost Action

Frost action is detrimental to pavement performance. It results in frost heave, which causes concrete slabs to break and softens the subgrade during the frost melt period. In northern climates, frost heave can reach several inches or

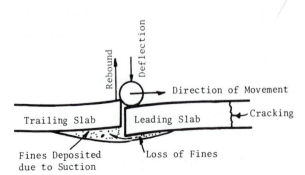

Figure 1.6 Pumping of rigid pavement.

more than one foot. The increase in volume of 9% when water becomes frozen is not the real cause of frost heave. For example, if a soil has a porosity of 0.5 and is subjected to a frost penetration of 3 ft (0.91 m), the amount of heave due to 9% increase in volume is $0.09 \times 3 \times 0.5 = 0.135$ ft or 1.62 in. (41 mm), which is much smaller than the 6 in. (152 mm) or more of heave experienced in such climate.

Frost heave is caused by the formation and continuing expansion of ice lenses. After a period of freezing weather, frost penetrates into the pavement and subgrade, as indicated by the depth of frost penetration in Figure 1.7. Above the frost line, the temperature is below the ordinary freezing point for water. The water will freeze in the larger voids but not in the smaller voids where the freezing point may be depressed as low as 23°F (-5°C).

When water freezes in the larger voids, the amount of liquid water at that point decreases. The moisture deficiency and the lower temperature in the freezing zone increase the capillary tension and induce flow toward the newly formed ice. The adjacent small voids are still unfrozen and act as conduits to deliver the water to the ice. If there is no water table or if the subgrade is above the capillary zone, only scattered and small ice lenses can be formed. If the subgrade is above the frost line and within the capillary fringe of the groundwater table, the capillary tension induced by freezing sucks up water from the water table below. The result is a great increase in the amount of water in the freezing zone and the segregation of water into ice lenses. The amount of heave is at least as much as the combined lens thicknesses.

Three factors must be present simultaneously to produce frost action:

1. The soil within the depth of frost penetration must be frost susceptible. It should be recognized that silt is more frost susceptible than clay because it has both high capillarity and high permeability. Although clay has a very high capillarity, its permeability is so low that very little water can be attracted from the water table to form ice lenses during the freezing period. Soils with more than 3% finer than 0.02 mm are generally frost susceptible, except that uniform fine sands with more than 10% finer than 0.02 mm are frost susceptible.

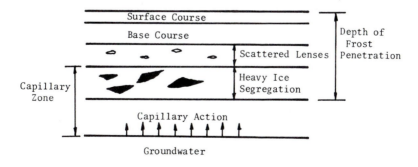

Figure 1.7 Formation of ice lenses due to frost action.

2. There must be a supply of water. A high water table can provide a continuous supply of water to the freezing zone by capillary action. Lowering the water table by subsurface drainage is an effective method to minimize frost action.

3. The temperature must remain freezing for a sufficient period of time. Due to the very low permeability of frost-susceptible soils, it takes time for the capillary water to flow from the water table to the location where the ice lenses are formed. A quick freeze does not have sufficient time to form ice lenses of any significant size.

Improvement of Drainage

When the water table is high and close to the ground surface, a base course can raise the pavement to a desirable elevation above the water table. When water seeps through pavement cracks and joints, an open-graded base course can carry it away to the road side. Cedergren (1988) recommends the use of an open-graded base course under every important pavement to provide an internal drainage system capable of rapidly removing all water that enters.

Control of Shrinkage and Swell

When moisture changes cause the subgrade to shrink and swell, the base course can serve as a surcharge load to reduce the amount of shrinkage and swell. A dense-graded or stabilized base course can serve as a waterproofing layer, and an open-graded base course can serve as a drainage layer. Thus, the reduction of water entering the subgrade further reduces the shrinkage and swell potentials.

Expedition of Construction

A base course can be used as a working platform for heavy construction equipment. Under inclement weather conditions, a base course can keep the surface clean and dry and facilitate the construction work.

As can be seen from the above reasoning, there is always a necessity to build a base course. Consequently, base courses have been widely used for rigid pavements.

Types of Concrete Pavement

Concrete pavements can be classified into four types: jointed plain concrete pavement (JPCP), jointed reinforced concrete pavement (JRCP), continuous reinforced concrete pavement (CRCP), and prestressed concrete pavement (PCP). Except for PCP with lateral prestressing, a longitudinal joint should be installed between two traffic lanes to prevent longitudinal cracking. Figure 1.8 shows the major characteristics of these four types of pavements.

Jointed Plain Concrete Pavements

All plain concrete pavements should be constructed with closely spaced contraction joints. Dowels or aggregate interlocks may be used for load transfer across the joints. The practice of using or not using dowels varies among the states. Dowels are used most frequently in the southeastern states, aggregate

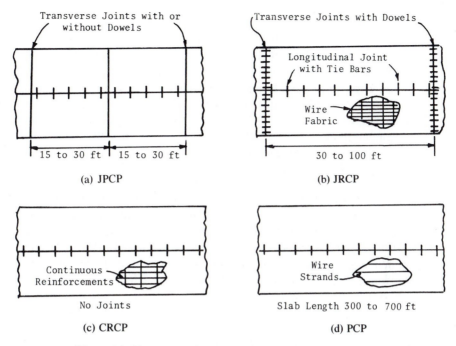

Figure 1.8 Four types of concrete pavements (1 ft = 0.305 m).

interlocks in the western and southwestern states, and both are used in other areas. Depending on the type of aggregate, climate, and prior experience, joint spacings between 15 and 30 ft (4.6 and 9.1 m) have been used. However, as the joint spacing increases, the aggregate interlock decreases, and there is also an increased risk of cracking. Based on the results of a performance survey, Nussbaum and Lokken (1978) recommended maximum joint spacings of 20 ft (6.1 m) for doweled joints and 15 ft (4.6 m) for undoweled joints.

Jointed Reinforced Concrete Pavements

Steel reinforcements in the form of wire mesh or deformed bars do not increase the structural capacity of pavements but allow the use of longer joint spacings. This type of pavement is used most frequently in the northeastern and north central part of the United States. Joint spacings vary from 30 to 100 ft (9.1 to 30 m). Because of the longer panel length, dowels are required for load transfer across the joints.

The amount of distributed steel in JRCP increases with the increase in joint spacing and is designed to hold the slab together after cracking. However, the number of joints and dowel costs decrease with the increase in joint spacing. Based on the unit costs of sawing, mesh, dowels, and joint sealants, Nussbaum and Lokken (1978) found that the most economical joint spacing was about 40 ft (12.2 m). Since maintenance costs generally increase with the increase in joint spacing, the selection of 40 ft (12.2 m) as the maximum joint spacing appears to be warranted.

Continuous Reinforced Concrete Pavements

It was the elimination of joints that prompted the first experimental use of CRCP in 1921 on Columbia Pike near Washington, D.C. The advantages of the joint-free design were widely accepted by many states, and more than two dozen states have used CRCP with a two-lane mileage totaling over 20,000 miles (32,000 km). It was originally reasoned that joints were the weak spots in rigid pavements and that the elimination of joints would decrease the thickness of pavement required. As a result, the thickness of CRCP has been empirically reduced by 1 to 2 in. (25 to 50 mm) or arbitrarily taken as 70 to 80% of the conventional pavement.

The formation of transverse cracks at relatively close intervals is a distinctive characteristic of CRCP. These cracks are held tightly by the reinforcements and should be of no concern as long as they are uniformly spaced. The distress that occurs most frequently in CRCP is punchout at the pavement edge. This type of distress takes place between two parallel random transverse cracks or at the intersection of Y cracks. If failures occur at the pavement edge instead of at the joint, there is no reason why a thinner CRCP should be used. The 1986 AASHTO design guide suggests using the same equation or nomograph for determining the thickness of JRCP and CRCP. However, the recommended load transfer coefficients for CRCP are slightly smaller than those of JPCP or JRCP, which may result in a slightly smaller thickness of CRCP. The amount of longitudinal reinforcing steel should be designed to control the spacing and width of cracks and the maximum stress in the steel. Details on the design of CRCP are presented in Section 12.4.

Prestressed Concrete Pavements

Concrete is weak in tension but strong in compression. The thickness of concrete pavement required is governed by its modulus of rupture which varies with the tensile strength of the concrete. The preapplication of a compressive stress to the concrete greatly reduces the tensile stress caused by the traffic loads and thus decreases the thickness of concrete required. The prestressed concrete pavements have less probability of cracking and fewer transverse joints and therefore result in less maintenance and longer pavement life.

The first known prestressed highway pavement in the United States was a 300-ft (91-m) pavement in Delaware built in 1971 (*Roads and Streets,* 1971). This was followed in the same year by a demonstration project on a 3200-ft (976-m) access road at Dulles International Airport (Pasko, 1972). In 1973, a 2.5-mile (4-km) demonstration project was constructed in Pennsylvania (Brunner, 1975). These projects were preceded by a construction and testing program on an experimental prestressed pavement constructed in 1956 at Pittsburgh, Pennsylvania (Moreell, 1958). These projects have the following features:

1. Slab length varied from 300 to 760 ft (91 to 232 m).
2. Slab thickness was 6 in. (152 mm) on all projects.
3. A post tension method with seven wire steel strands was used for all projects. In the post tension method, the compressive stress was imposed

after the concrete had gained sufficient strength to withstand the applied forces.

4. Longitudinal prestress varied from 200 to 331 psi (1.4 to 2.3 MPa) and no transverse or diagonal prestressing was used.

Prestressed concrete has been used more frequently for airport pavements than for highway pavements because the saving in thickness for airport pavements is much greater than that for highway pavements. The thickness of prestressed highway pavements has generally been selected as the minimum necessary to provide sufficient cover for the prestressing steel (Hanna et al., 1976). Since the prestressed concrete pavements are still at the experimental stage and their design is primarily the application of experience and engineering judgment, they will not be further discussed in this book.

1.2.3 Composite Pavements

A composite pavement is composed of both HMA and PCC. The use of PCC as a bottom layer and HMA as a top layer results in an ideal pavement with the most desirable characteristics. The PCC provides a strong base and the HMA provides a smooth and nonreflective surface. However, this type of pavement is very expensive and is rarely used as a new construction. As of 1990, there are about 100,000 miles (160,000 km) of composite pavements in the United States, practically all of which are the rehabilitation of concrete pavements using asphalt overlays. The design of overlay is discussed in Chapter 13.

Design Methods

When an asphalt overlay is placed over a concrete pavement, the major load-carrying component is the concrete, so the plate theory should be used. Assuming that the HMA is bonded to the concrete, an equivalent section can be used with the plate theory to determine the flexural stress in the concrete slab, as is described in Section 5.1.2. If the wheel load is applied near to the pavement edge or joint, only the plate theory can be used. If the wheel load is applied in the interior of the pavement far from the edges and joints, either layered or plate theory can be used. The concrete pavement can be either JPCP, JRCP, or CRCP.

Composite pavements also include asphalt pavements with stabilized bases. For flexible pavements with untreated bases, the most critical tensile stress or strain is located at the bottom of asphalt layer, while for asphalt pavements with stabilized bases, the most critical location is at the bottom of the stabilized bases.

Pavement Sections

The design of composite pavements varies a great deal. Figure 1.9 shows two different cross sections that have been used. Section (a) (Ryell and Corkill, 1973) shows the HMA placed directly on the PCC base, which is a more conventional type of construction. A disadvantage of this construction is the occurrence of reflection cracks on the asphalt surface due to the joints and cracks in the

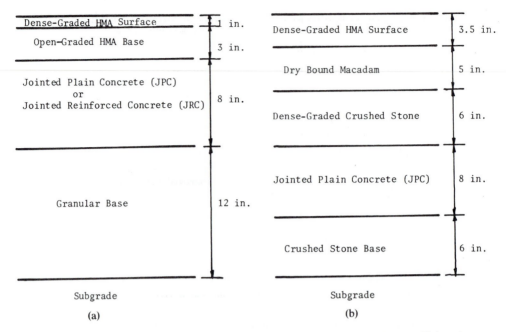

Dense-Graded HMA Surface	1 in.
Open-Graded HMA Base	3 in.
Jointed Plain Concrete (JPC) or Jointed Reinforced Concrete (JRC)	8 in.
Granular Base	12 in.
Subgrade	

(a)

Dense-Graded HMA Surface	3.5 in.
Dry Bound Macadam	5 in.
Dense-Graded Crushed Stone	6 in.
Jointed Plain Concrete (JPC)	8 in.
Crushed Stone Base	6 in.
Subgrade	

(b)

Figure 1.9 Two different cross sections for composite pavements (1 in. = 25.4 mm).

concrete base. The open-graded HMA serves as a buffer to reduce the amount of reflection cracking. Placing thick layers of granular materials between the concrete base and the asphalt layer, as in Section (*b*) (Baker, 1973), can eliminate reflection cracks, but the placement of a stronger concrete base under a weaker granular material may not be an effective design. The use of a 6-in. (152-mm) dense-graded crushed stone base beneath the more rigid macadam base, consisting of 2.5-in. (64-mm) stone choked with stone screenings, prevents reflection cracking. The 3.5-in. (89-mm) HMA is composed of a 1.5-in. (38-mm) surface course and a 2-in. (51-mm) binder course.

Figure 1.10 shows the composite structures recommended for premium pavements (Von Quintus et al., 1980). Section (*a*) shows the composite pavement with a jointed plain concrete base, and Section (*b*) shows the composite pavement with a continuous reinforced concrete base. Premium pavements are also called zero-maintenance pavements. They are designed to carry very heavy traffic with no maintenance required during the first 20 years and normal maintenance for the next 10 years before resurfacing. The ranges of thickness indicated in the figure depend on traffic, climate, and subgrade conditions.

A salient feature of the design is the 4-in. (102-mm) drainage layer on top of the natural or improved subgrade. The drainage layer is a blanket extending full width between shoulder edges. In addition, a perforated collector pipe should be placed longitudinally along the edge of the pavement. The pipe should be placed in a trench section cut in the subgrade. Filter materials should be placed between the drainage layer and the subgrade, if needed. Because the concrete base is protected by the HMA, a leaner concrete with a modulus of rupture from 450 to 550 psi (3.1 to 3.8 MPa) may be used. Note that a 3-in. (7.6-mm) asphalt crack relief layer

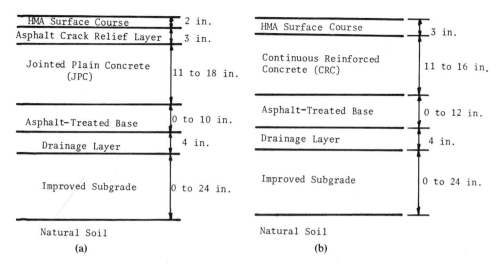

Figure 1.10 Typical cross sections for premium composite pavements (1 in. = 25.4 mm). (After Von Quintus et al. (1980).)

should be placed above the jointed plain concrete base. This is a layer of coarse open-graded HMA containing 25 to 35% interconnecting voids and made up of 100% crushed material. It is designed to reduce reflection cracking. Because of the large amount of interconnecting voids, this crack relief layer provides a medium that resists the transmission of differential movements of the underlying slab. The asphalt relief layer is not required for continuous reinforced concrete base. When the subgrade is poor, an asphalt-treated base may be required beneath the concrete slab.

1.3 ROAD TESTS

Because the observed performance under actual conditions is the final criterion to judge the adequacy of a design method, three major road tests under controlled conditions were conducted by the Highway Research Board from the mid 1940s to the early 1960s.

1.3.1 Maryland Road Test

The objective of this project was to determine the relative effects of four different axle loadings on a particular concrete pavement (HRB, 1952). The tests were conducted on a 1.1-mile (1.76-km) section of concrete pavement constructed in 1941 on US 301 approximately 9 miles (14.4 km) south of La Plata, Maryland.

General Layout

The pavement consisted of two 12-ft (3.66-m) lanes, having a 9–7–9-in. (229–178–229-mm) thickened edge cross section and reinforced with wire mesh.

The soils under the pavement ranged from A-1 to A-7-6 with the A-6 group predominant. There were four separate test sections. The west and east lanes of the southern 0.5 mile (0.8 km) were subjected to 18,000 and 22,400 lb (80 and 100 kN) single-axle loads, respectively, whereas the west and east lanes of the northern 0.6 mile (0.96 km) were subjected to 32,000 and 44,800 lb (142 and 200 kN) tandem-axle loads, respectively, as shown in Figure 1.11. Controlled traffic tests were conducted from June through December 1950. The total cost of the project was $245,000.

It should be noted that thickened edge pavements were very popular at the time of the road test but are rarely in use today. This type of pavement is more costly to construct because of the grading operations required at the thickened edge. Thickened edge pavements were popular in the old days when pavement widths were in the neighborhood of 18 to 20 ft (5.5 to 6.1 m) and most trucks

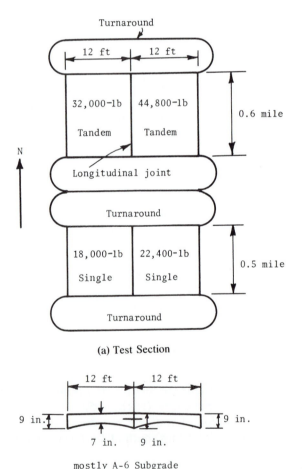

(a) Test Section

(b) Pavement Section

mostly A-6 Subgrade

Figure 1.11 Maryland Road Test (1 in. = 25.4 mm, 1 ft = 0.305 m, 1 mile = 1.6 km, 1 lb = 4.45 N).

Introduction   Chap. 1

traveled very close to the pavement edge. With current 24-ft (7.3-m) wide pavements, traffic concentration is between 3 and 4 ft (0.91 to 1.22 m) from the edge, thus significantly reducing the stress at the edge.

Major Findings

1. Both the average cracking and the average settlement of slab at the joint increased in the order of 18,000 lb (80 kN) single axle, 32,000 lb (142 kN) tandem axle, 22,400 lb (100 kN) single axle, and 44,800 lb (200 kN) tandem axle.

2. Pumping occurred on plastic clay soils but not on granular subgrades with low percentages of silt and clay. Prior to the development of pumping, the stresses for all cases of loading investigated were below generally accepted design limits, i.e., 50% of the modulus of rupture of the concrete. Pumping and the accompanying loss of subgrade support caused large increases in the stresses for the corner case of loading.

3. Under creep speed, the deflections for the corner loading averaged approximately 0.025 in. (0.64 mm), while those for the edge loading averaged approximately 0.014 in. (0.36 mm). After pumping has developed, the maximum deflections increased considerably. In some instances, the corner deflection approached 0.2 in. (5 mm) before cracking occurred. The failure of the slabs on the fine-grained soil can be explained by the fact that the deflections of these slabs under all of the test loads were sufficient to cause pumping when the other requisites for pumping were present. As pumping developed, the deflections increased with corresponding more rapid increase of stress to a magnitude sufficient to cause rupture of the slab.

4. With the exception of the corner case of loading for pumping soils, the stress and deflection resulting at vehicle speed of 40 mph (64 km/h) averaged approximately 20% less than those at creep speed. For pumping slabs under corner loading, the stresses averaged approximately the same for both vehicle speeds.

5. The stresses and deflections caused by loads acting at the corners and edges of slabs were influenced to a marked degree by temperature curling. For the corner loading, the stresses and deflections for a severe downward curled condition were observed to be only approximately one-third of those for the critical upward curled condition. For the case of edge loading, the effect of temperature curling, although appreciable, was not as pronounced as for the case of corner loading.

1.3.2 WASHO Road Test

After the successful completion of the Maryland Road Test sponsored by the eleven midwestern and eastern states, the Western Association of State Highway Officials (WASHO) conducted a similar test but on sections of flexible pavements in Malad, Idaho, with the same objective in mind (HRB, 1955).

General Layout

Two identical test loops were constructed, each having two 1900-ft (580-m) tangents made up of five 300-ft (92-m) test sections separated by four 100-ft (30-m) transition sections, as shown in Figure 1.12. One tangent in each loop was surfaced with 4 in. (102 mm) of HMA and 2 in. (51 mm) of crushed gravel base, while the other tangent was surfaced with 2 in. (51 mm) of HMA and 4 in. (102 mm) of crushed gravel base. The thicknesses of the gravel subbase were 0, 4, 8, 12, and 16 in. (0, 102, 203, 205, and 406 mm). Thus, the total thickness of pavement over the A-4 basement soil varied from 6 to 22 in. (152 to 559 mm) in 4-in. (102-mm) increments for the five test sections. In one loop, 18,000-lb (80-kN) single-axle loads were operated in the inner lane and 22,400 lb (100 kN) in the outer lane. In the other loop, 32,000-lb (142-kN) tandem-axle loads were in the inner lane and 40,000 lb (178 kN) in the outer lane.

Construction began in April 1952 and the completed pavement was accepted on September 30, 1952. On November 6, 1952, after completion of preliminary testing and installation of instruments, regular traffic operation was started. Except for one spring and two winter periods, it was continued until May 29, 1954. The total cost of the project was $840,000.

Major Findings

1. The amount of damage to the pavement increased in the order of 18,000 lb (80 kN) single axle, 32,000 lb (142 kN) tandem axle, 22,400 lb (100 kN) single axle, and 40,000 lb (178 kN) tandem axle. For example, the results of an analysis showed that the minimum thicknesses of pavement with 2-in. (51-mm) HMA that would have been adequate to carry the 238,000 applications of the above loads were 16, 17, 19, and 20 in. (406, 432, 483, and 508 mm), respectively.

2. The behavior of the pavement with 4-in. (102-mm) HMA was far superior to that of equal total thickness with 2-in. (51-mm) HMA.

3. Distress in the outer wheelpath was more than that in the inner wheelpath. Surfacing of the shoulders in three of the test sections in July 1953 proved to be highly effective in retarding distress in the outer wheelpath. Both facts suggest that the outer wheelpath with paved shoulders is the equivalent of the inner wheelpath and testify to the advantages of shoulder paving.

4. Development of structural distress was confined largely to two critical periods of traffic operation. One period was from June 11 to July 7, 1953, in which 27% of the total distress developed under 0.7% of the total load applications. The second period was from February 17 to April 7, 1954, in which 40% of the total distress developed under 13% of the total applications.

5. Based on pavement distress, a tandem axle with a total load about 1.5 times that of a single-axle load is equivalent to the single-axle load; whereas a tandem axle with a total load about 1.8 times a single-axle load produced equal maximum deflections.

6. Deflection of the pavement surface under traffic was influenced by vehicle speed, temperature of the surfacing, load, and moisture content of the top

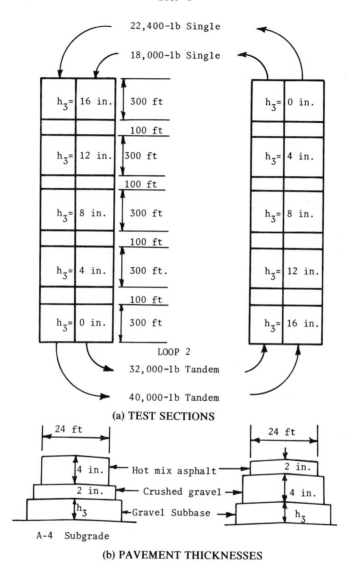

Figure 1.12 WASHO Road Test (1 in. = 25.4 mm, 1 ft = 0.305 m, 1 lb = 4.45 N).

layers of basement soil. The deflection was maximum under a static load. Deflection decreased as speed increased up to about 15 mph (24 km/h), after which deflections decreased but slightly as speed increased. Deflections were greater as temperature of surfacing increased. Deflections of the pavement surface under traffic were approximately proportional to the applied load. When the moisture content at the basement soil exceeded 22%, deflections increased with the increase in moisture contents.

7. For the type of the loading employed, there was no significant difference between the magnitudes of the wheel loads transmitted in the outer wheelpath and those in the inner wheelpath due to the crown of the pavement.

1.3.3 AASHO Road Test

The objective of this project was to determine the significant relationship between the number of repetitions of specified axle loads of different magnitudes and arrangements and the performance of different thicknesses of flexible and rigid pavements (HRB, 1962). The test facility was constructed along the alignment of Interstate 80 near Ottawa, Illinois, about 80 miles (128 km) southwest of Chicago.

General Layout

The test consisted of four large loops, numbered 3 through 6, and two smaller loops, 1 and 2. Each loop was a segment of a four-lane divided highway whose parallel roadways, or tangents, were connected by a turnaround at each end. Tangent lengths were 6800 ft (2070 m) in loops 3 through 6, 4400 ft (1340 m) in loop 2, and 2000 ft (610 m) in loop 1. In all loops, the north tangents were surfaced with HMA and south tangents with PCC. Centerlines divided the pavements into inner and outer lanes, called lane 1 and lane 2. Each tangent was constructed as a succession of pavement sections called structural sections. Pavement designs varied from section to section. The minimum length of a section was 100 ft (30.5 m) in loops 2 through 6 and 15 ft (4.6 m) in loop 1. The axle loads on each loop and lane are shown in Table 1.1.

Construction began in August 1956 and test traffic was inaugurated on October 15, 1958. Test traffic was operated until November 30, 1960, at which time 1,114,000 axle loads had been applied. The total cost of the project was $27,000,000.

TABLE 1.1 APPLICATIONS OF AXLE LOADS ON VARIOUS LANES
AT AASHO ROAD TEST

Loop no.	1		2		3	
Lane no.	1	2	1	2	1	2
Axle load (lb)	None	None	2000 single	6000 single	12,000 single	24,000 tandem
Loop no.	4		5		6	
Lane no.	1	2	1	2	1	2
Axle load (lb)	18,000 single	32,000 tandem	22,400 single	40,000 tandem	30,000 single	48,000 tandem

Note: 1 lb = 4.45 N.

Major Findings

One important contribution of the AASHO Road Test was the development of the pavement serviceability concept which is discussed in Section 9.2.1 together with the equations relating serviceability, load, and thickness design of both flexible and rigid pavements. Major findings for flexible and rigid pavements are summarized separately as follows.

Flexible Pavements

1. The superiority of the four types of base under study fell in the following order: bituminous treated, cement treated, crushed stone, and gravel bases. Most of the sections containing the gravel base failed very early in the test and their performance was definitely inferior to that of the sections with crushed stone base.

2. The pavement needed to maintain a certain serviceability at a given number of axle load applications would be considerably thinner in the inner than in the outer wheelpath.

3. Rutting of the pavement was due principally to decrease in thickness of the component layers. About 91% of the rutting occurred in the pavement itself with 32% in the surface, 14% in the base, and 45% in the subbase. Thus, only 9% of a surface rut could be accounted for by rutting of the embankment. Data also showed that changes in thickness of the component layers were not caused by the increase in density but due primarily to lateral movements of the materials.

4. More surface cracking occurred during periods when the pavement was in a relatively cold state than during periods of warm weather. Generally, cracking was more prevalent in sections having deeper ruts than in sections with shallower ruts.

5. The deflection occurring within the pavement structure (surface, base, and subbase), as well as that at the top of the embankment soil, was greater in the spring than during the succeeding summer months. This was due to the higher moisture contents of the base, subbase, and embankment soil that existed in the spring. A high degree of correlation was found between the deflection at the top of the embankment and the total surface deflection as well as between deflection and rutting. A pronounced reduction in deflection accompanied an increase in vehicle speed. Increasing the speed from 2 to 35 mph (3.2 to 56 km/h) reduced the total deflection 38% and the embankment deflection 35%.

Rigid Pavements

1. Of the three design variables, viz., reinforcement or panel length, subbase thickness, and slab thickness, only slab thickness has an appreciable effect on measured strains.

2. Inspections of the pavements were made weekly and after each rain. Faulting occasionally occurred at cracks, never at the transverse joints, because all joints were doweled. No part of the cracking in the traffic loops was attributed solely to environmental changes, since no cracks appeared in the nontraffic loop, or loop 1. (*Note:* although no cracks appeared during and immediately after the road test, cracks did occur many years later.)

3. Pumping of subbase material, including the coarser fractions, was the major factor causing failures of sections with subbase. The amount of materials pumped through joints and cracks was negligible when compared with the amount ejected along the edge.

4. Twenty-four-hour studies of the effect of fluctuating air temperature showed that the deflection of panel corners under vehicles traveling near the pavement edge might increase severalfold from afternoon to early morning. Edge strains and deflections were affected to a lesser extent.

5. Corner deflections of a 40-ft (12.2-m) reinforced panel usually exceeded those of a 15-ft (4.6-m) nonreinforced panel, if all other conditions were the same. Edge deflections and strains were not affected significantly by panel length or reinforcement. An increase in vehicle speed from 2 to 60 mph (3.2 to 96 km/h) resulted in a decrease in strain or deflection of about 29%.

1.4 DESIGN FACTORS

Design factors can be divided into four broad categories: traffic and loading, environment, materials, and failure criteria. The factors to be considered in each category will be described and how the design process fits into an overall pavement management system will be discussed.

1.4.1 Traffic and Loading

The traffic and loading to be considered include axle loads, the number of load repetitions, tire contact areas, and vehicle speeds.

Axle Loads

Figure 1.13 shows the wheel spacing for a typical semitrailer consisting of single axle with single tires, single axle with dual tires, and tandem axles with dual tires. For special heavy-duty haul trucks, tridem axles consisting of a set of three axles, each spaced at 48 to 54 in. (1.22 to 1.37 m) apart, also exist.

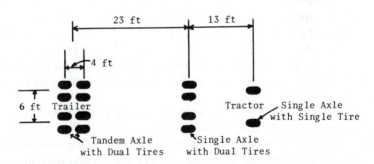

Figure 1.13 Wheel configuration for a typical semitrailer unit (1 ft = 0.305 m).

The spacings of 23 and 13 ft (7 and 4 m) shown in Figure 1.13 should have no effect on pavement design because the wheels are so far apart that their effect on stresses and strains should be considered independently. Unless an equivalent single-axle load is used, the consideration of multiple axles is not a simple matter. The design may be unsafe if the tandem and tridem axles are treated as a group and considered as one repetition. The design is too conservative if each axle is treated independently and considered as one repetition. A method for analyzing multiple-axle loads is presented in Section 3.1.3.

In the design of flexible pavements by layered theory, only the wheels on one side, say at the outer wheelpath, need be considered; whereas in the design of rigid pavements by plate theory, the wheels on both sides, even at a distance of more than 6 ft (1.8 m) apart, are usually considered.

Number of Repetitions

With the use of a high-speed computer, it is no problem to consider the number of load repetitions for each axle load and evaluate its damage. The method of dividing axle loads into a number of groups has been frequently used for the design of rigid pavements, as illustrated by the PCA method in Section 12.2. However, its application to flexible pavements is not widespread because of the empirical nature of the design and the large amount of computer time required.

Instead of analyzing the stresses and strains due to each axle load group, a simplified and widely accepted procedure is to develop equivalent factors and convert each load group into an equivalent 18-kip (80-kN) single axle load, as illustrated by the Asphalt Institute method in Section 11.2 and the AASHTO method in Sections 11.3 and 12.3. It should be noted that the equivalency between two different loads depends on the failure criterion employed. Equivalent factors based on fatigue cracking may be different from those based on permanent deformation. Therefore, the use of a single equivalent factor for analyzing different types of distress is empirical and should be considered as approximate only.

Contact Area

In the mechanistic method of design, it is necessary to know the contact area between tire and pavement, so the axle load can be assumed to be uniformly distributed over the contact area. The size of contact area depends on the contact pressure. As indicated by Figure 1.14, the contact pressure is greater than the tire pressure for low-pressure tires, because the wall of tires is in compression and the sum of vertical forces due to wall and tire pressure must be equal to the force due to contact pressure; the contact pressure is smaller than the tire pressure for high-pressure tires, because the wall of tires is in tension. However, in pavement design the contact pressure is generally assumed to be equal to the tire pressure. Because heavier axle loads have higher tire pressures and more destructive effects on pavements, the use of tire pressure as the contact pressure is therefore on the safe side.

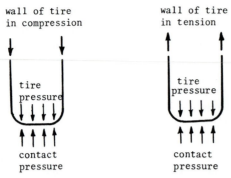

wall of tire
in compression

tire
pressure

contact
pressure

(a) Low Pressure Tire

wall of tire
in tension

tire
pressure

contact
pressure

(b) High Pressure Tire

Figure 1.14 Relationship between contact pressure and tire pressure.

Heavier axle loads are always applied on dual tires. Figure 1.15a shows the approximate shape of contact area for each tire, which is composed of a rectangle and two semicircles. By assuming a length of L and a width of $0.6L$, the area of contact $A_c = \pi(0.3L)^2 + (0.4L)(0.6L) = 0.5227L^2$, or

$$L = \sqrt{\frac{A_c}{0.5227}} \tag{1.1}$$

in which A_c = contact area, which can be obtained by dividing the load on each tire by the tire pressure.

The contact area shown in Figure 1.15a was used previously by PCA (1966) for the design of rigid pavements. The current PCA (1984) method is based on the finite element procedure and a rectangular area is assumed with a length of $0.8712L$ and a width of $0.6L$, which has the same area of $0.5227L^2$, as shown in Figure 1.15b. These contact areas are not axisymmetric and cannot be used with the layered theory. When the layered theory is used for flexible pavement design, it is assumed that each tire has a circular contact area. This assumption is not correct, but the error incurred is believed to be small. To simplify the analysis of flexible pavements, a single circle with the same contact area as the duals is frequently used to represent a set of dual tires, instead of using two circular areas. This practice usually results in a more conservative design, but may become unconservative for thin asphalt surface because the horizontal tensile strain at the

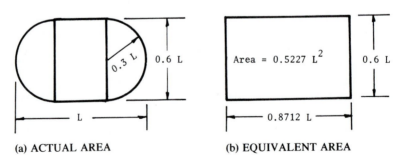

(a) ACTUAL AREA

(b) EQUIVALENT AREA

Figure 1.15 Dimension of tire contact area.

bottom of asphalt layer under the larger contact radius of single wheel is smaller than that under the smaller contact radius of dual wheels, as is illustrated in Section 3.4.1. For rigid pavements, it is more reasonable to use a larger circular area to represent a set of duals, as discussed in Section 4.2.1.

Example 1.1:

Draw the most realistic contact area for an 18-kip (80-kN) single-axle load with a tire pressure of 80 psi (552 kPa). What are the other configurations of contact area that have been used for pavement design?

Solution: The 18-kip (80-kN) single-axle load is applied over four tires, each having a load of 4500 lb (20 kN). The contact area of each tire is $A_c = 4500/80 = 56.25$ in.^{2} (3.6×10^4 mm^2). From Eq. 1.1, $L = \sqrt{56.25/0.5227} = 10.37$ in. (263 mm). The width of the tire is $0.6L = 0.6 \times 10.37 = 6.22$ in. (158 mm). The configuration of various contact areas is shown in Figure 1.16.

Figure 1.16a is the most realistic contact area consisting of a rectangle and two semicircles, as used previously by PCA (1966). Figure 1.16b is the rectangular contact area for use in the finite element analysis of rigid pavements with a length of $0.8712L$, or 9.03 in. (229 mm), and a width of 6.22 in. (158 mm). Figure 1.16c shows the contact area as two circles, each having a radius of $\sqrt{56.25/\pi}$, or 4.23 in. (107 mm). This assumption was also made by the Asphalt Institute (AI, 1981a), although they used a tire pressure of 70 psi (483 kPa) and a contact radius of 4.52 in. (115 mm). Figure 1.16d considers the contact area as a single circle with a contact radius of $\sqrt{2 \times 56.25/\pi} = 5.98$ in. (152 mm). This contact area was used in VESYS (FHWA, 1978).

Vehicle Speed

Another factor related to traffic is the speed of traveling vehicles. If the viscoelastic theory is used, such as in VESYS and KENLAYER, speed is directly related to the duration of loading. If the elastic theory is used, the resilient

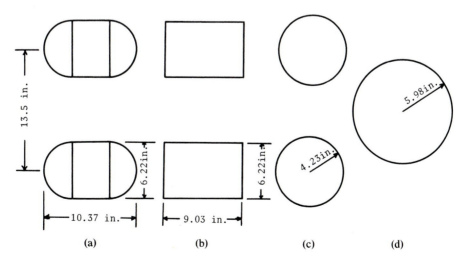

Figure 1.16   Example 1.1 (1 in. = 25.4 mm).

modulus of each paving material should be properly selected in commensurate with the vehicle speed. Generally, the greater the speed, the larger the modulus, and the smaller the strains in the pavement.

1.4.2 Environment

The environmental factors that influence pavement design include temperature and precipitation, both affecting the elastic moduli of the various layers. In the mechanistic–empirical method of design, each year can be divided into 24 periods, 12 months, or several seasons, each having a different set of layer moduli. The damage during each period, month, or season is evaluated and summed throughout the year to determine the design life.

Temperature

The effect of temperature on asphalt pavements is different from that on concrete pavements. Temperature affects the resilient modulus of asphalt layers and induces curling of concrete slabs. In cold climates, the resilient moduli of unstablilized materials also vary with the freeze–thaw cycles. The severity of cold climate is indicated by the freezing index, which can be correlated with the depth of frost penetration.

Effect on Asphalt Layer

The elastic and viscoelastic properties of HMA are affected significantly by pavement temperature. Any mechanistic methods of flexible pavement design must consider pavement temperature, which can be related to air temperature. During the winter when temperature is low, the HMA becomes rigid and reduces the strains in the pavement. However, stiffer HMA has less fatigue life, which may neutralize the beneficial effect of smaller strains. Low temperature can cause asphalt pavements to crack.

Effect on Concrete Slab

The temperature gradient in concrete pavements affects not only the curling stress but also the slab–subgrade contact. During the day when the temperature at top is higher than that at bottom, the slab curls down so that its interior may not be in contact with the subgrade. At night when the temperature at top is lower than that at bottom, the slab curls upward so that its edge and corner may be out of contact with the subgrade. The loss of subgrade contact will affect the stresses in concrete due to wheel loads. The change between maximum and minimum temperatures also determines the joint and crack openings and affects the efficiency of load transfer.

Frost Penetration

Another effect of temperature on pavement design in cold climate is the frost penetration, which results in a stronger subgrade in the winter but a much weaker subgrade in the spring. Figure 1.17 shows the maximum depth of frost penetration in the United States. Although frost heave causes differential settlements and

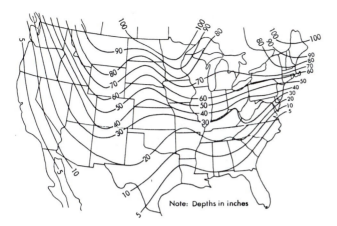

Figure 1.17 Maximum depth of frost penetration in the United States (1 in. = 25.4 mm).

pavement roughness, the most detrimental effect of frost penetration occurs during the spring breakup period when the ice melts and the subgrade is in a saturated condition. It is desirable to protect the subgrade by using non-frost-susceptible materials within the zone of frost penetration. If this cannot be done, the design method should take into consideration the weakening of subgrade during spring breakup.

Freezing Index

The severity of frost in a given region can be expressed as a freezing index in terms of degree days. A negative one-degree day represents one day with a mean air temperature one degree below freezing, while a positive one-degree day indicates one day with a mean air temperature one degree above freezing. The mean air temperature for a given day is the average of high and low temperatures during that day. If the mean air temperature is 25°F on the first day and 22°F on the second and third days, the total degree days for the three-day period are $(25 - 32) + 2 \times (22 - 32) = -27$ degree days. Given the mean air temperature of each day, the degree days for each month can be similarly calculated. A plot of cumulative degree days versus time results in a curve, as shown in Figure 1.18. The difference between the maximum and minimum points on the curve during one year is called the freezing index for that year. The freezing index has been correlated with the depth of frost penetration and can be used as a factor of pavement design and evaluation.

Example 1.2:

The monthly degree day data are September, 540; October, −130; November, −450; December, −770; January, −540; February, −450; March, −290; April, −70; and May, 170. Calculate the freezing index.

Solution: The monthly and cumulative degree day data are shown in Table 1.2. The cumulative degree days are plotted in Figure 1.18. Freezing index = 2160 + 540 = 2700 degree days, which can be calculated from the table or measured from Figure 1.18:

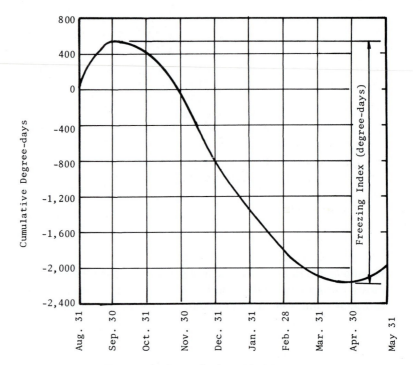

Figure 1.18 Determination of freezing index.

TABLE 1.2 MONTHLY AND
CUMULATIVE DEGREE DAYS

| Month | Degree days (°F) | |
	Monthly	Cumulative
September	540	540
October	−130	410
November	−450	−40
December	−770	−810
January	−540	−1350
February	−450	−1800
March	−290	−2090
April	−70	−2160
May	170	−1990

Precipitation

The precipitation from rain and snow affects the quantity of surface water infiltrating into the subgrade and the location of the groundwater table. Every effort should be made to improve drainage and alleviate the detrimental effect of water. If water from rainfalls can be drained out within a short time, its effect can be minimized, even in regions of high precipitation.

The location of the groundwater table is also important. The water table should be kept at least 3 ft (0.91 m) below the pavement surface. In seasonal frost areas, the depth from the pavement surface to the groundwater table should be much greater. For example, the recommended minimum depths are 7 ft (2.1 m) in Massachusetts, 5 ft (1.5 m) in Michigan and Minnesota, 8 to 12 ft (2.4 to 3.7 m) in Saskatchewan, and 3 to 7 ft (0.9 to 2.1 m) in Nebraska (Ridgeway, 1982).

If proper drainage cannot be provided, smaller elastic moduli must be selected for the component layers affected by poor drainage. However, this measure may not solve the problem because poor drainage may still incur damages other than the lack of shear strength, such as the pumping and the loss of support. Details about drainage are presented in Chapter 8.

1.4.3 Materials

In the mechanistic–empirical methods of design, the properties of materials must be specified, so that the responses of the pavement, such as stresses, strains, and displacements in the critical components, can be determined. These responses are then used with the failure criteria to predict whether failures will occur or the probability that failures will occur. Details of material characterization are presented in Chapter 7.

General Properties

The following general material properties should be specified for both flexible and rigid pavements:

1. When pavements are considered as linear elastic, the elastic moduli and Poisson ratios of the subgrade and each component layer must be specified. Since the Poisson ratios have relatively small effects on pavement responses, their values can be reasonably assumed.
2. If the elastic modulus of a material varies with the time of loading, the resilient modulus, which is the elastic modulus under repeated loads, must be selected in accordance with a load duration corresponding to the vehicle speed.
3. When a material is considered nonlinear elastic, the constitutive equation relating the resilient modulus to the state of stresses must be provided.

Flexible Pavements

The following properties may be specified for flexible pavements:

1. When the HMA is considered linear viscoelastic, the creep compliance, which is the reciprocal of the moduli at various loading times, must be specified. If the temperature at the creep test is not the same as the temperature used for pavement design, the time temperature shift factor, which indicates the sensitivity of asphalt mixtures to temperature as described in Section 2.3.2, must also be provided.

2. If the design is based on fatigue cracking, the fatigue properties of asphalt mixtures, as described in Section 7.3.1, must be specified.

3. If the design is based on rut depth by summing the permanent deformations over all layers, the permanent deformation parameters of each layer must be specified. These parameters can be obtained from permanent deformation tests, as described in Section 7.4.2.

4. If other distresses, such as low-temperature cracking, are used as a basis for design, appropriate properties, such as the asphalt stiffness at the winter design temperature, should be specified.

Rigid Pavements

The following properties may be specified for rigid pavements:

1. For rigid pavements on liquid foundations, the modulus of subgrade reaction, as described in Section 7.5.1, must be specified.

2. To consider the effect of temperature curling, the coefficient of thermal expansion of the concrete must be specified.

3. The most common distress in rigid pavements is fatigue cracking, so the modulus of rupture and the fatigue properties of concrete, as described in Section 7.3.2, must be specified.

4. If other distresses, such as faulting caused by excessive bearing stress on dowel bars, are used as a basis for design, appropriate properties, such as the diameter and spacing of dowels, must be specified.

1.4.4 Failure Criteria

In the mechanistic–empirical methods of pavement design, a number of failure criteria, each directed to a specific type of distress, must be established. This is in contrast to the AASHTO method where the present serviceability index (PSI), which indicates the general pavement conditions, is used. The failure criteria for mechanistic–empirical methods are described below.

Flexible Pavements

It is generally agreed that fatigue cracking, rutting, and low-temperature cracking are the three principal types of distress to be considered for flexible pavement design. These criteria are fully discussed in Section 11.1.4 and are briefly described below.

Fatigue Cracking

The fatigue cracking of flexible pavements is based on the horizontal tensile strain at the bottom of HMA. The failure criterion relates the allowable number of load repetitions to the tensile strain based on the laboratory fatigue test on small

HMA specimens. Due to the difference in geometric and loading conditions, the allowable number of repetitions for actual pavements is much greater than that obtained from laboratory tests. Therefore, the failure criterion must incorporate a shift factor to account for the difference.

Rutting

Rutting occurs only on flexible pavements, as indicated by the permanent deformation or rut depth along the wheelpaths. Two design methods have been used to control rutting: one to limit the vertical compressive strain on the top of subgrade and the other to limit the rutting to a tolerable amount, say 0.5 in. (13 mm).

The first method, which requires a failure criterion based on correlations with road tests or field performance, is much easier to apply and has been used by Shell Petroleum (Claussen et al., 1977) and the Asphalt Institute (Shook et al., 1982). This method is based on the contention that, if the quality of the surface and base courses is well controlled, rutting can be reduced to a tolerable amount by limiting the vertical compressive strain on the subgrade.

The second method, which computes directly the rut depth, can be based on empirical correlations with road tests, as used in PDMAP (Finn et al., 1986) and MICH-PAVE (Harichandran et al., 1989), or theoretical computations from the permanent-deformation parameters of each component layer, as incorporated in VESYS (FHWA, 1978). The Shell method also includes a procedure for estimating the rut depth in HMA (Shell, 1978). If rutting is due primarily to the decrease in thickness of the component layers above the subgrade, as was found in the AASHO Road Test, the use of this method should be more appropriate.

Thermal Cracking

This type of distress includes both low-temperature cracking and thermal fatigue cracking. Low-temperature cracking is usually associated with flexible pavements in northern regions of the United States and much of Canada, where winter temperatures can fall below $-10°F(-23°C)$. Thermal fatigue cracking can occur in much milder regions if an excessively hard asphalt is used or the asphalt becomes hardened due to aging.

The most comprehensive study on low-temperature cracking has been conducted in Canada, as reported by Christison et al. (1972). The potential of low-temperature cracking for a given pavement can be evaluated if the mix stiffness and fracture strength characteristics as a function of temperature and time of loading are known and temperature data on the site are available. The pavement will crack when the computed thermal stress is greater than the fracture strength.

Thermal fatigue cracking is similar to the fatigue cracking caused by repeated loads. It is caused by the tensile strain in the asphalt layer due to daily temperature cycle. The cumulative damage can be evaluated by Miner's hypothesis.

Rigid Pavements

Fatigue cracking has long been considered the major or only criterion for rigid pavement design. Only recently has pumping or erosion been considered. Other criteria in consideration include faulting and joint deterioration of JPCP and JRCP and edge punchout of CRCP. These criteria are fully discussed in Sections 12.1.2 through 12.1.6 and are briefly described below.

Fatigue Cracking

Fatigue cracking is most likely caused by the edge stress at the midslab. The allowable number of load repetitions to cause fatigue cracking depends on the stress ratio between flexural tensile stress and the concrete modulus of rupture. Because the design is based on the edge loading and only a small portion of the traffic loads is applied at the pavement edge, the total number of load repetitions must be reduced to an equivalent number of edge loads so that the same fatigue damage is obtained. This approach is different from the fatigue analysis of flexible pavements where a shift factor is used to adjust the allowable number of load repetitions.

Pumping or Erosion

Although permanent deformations are not considered in rigid pavement design, the resilient deformation under repeated wheel loads will cause pumping of the slabs. Consequently, corner deflections have been used in the latest version of the PCA method (PCA, 1984) as an erosion criterion in addition to the fatigue criterion. The applicability of the PCA method is quite limited because it is based on the results of the AASHO Road Test which employed a highly erodible subbase. Since pumping is caused by many other factors, such as types of subbase and subgrade, precipitation, and drainage, a more rational method for analyzing pumping is needed.

Other Criteria

Other major types of distress in rigid pavements include faulting, spalling, and joint deterioration. These distresses are difficult to analyze mechanistically and a great effort has been made recently in developing regression models to predict them. These empirical models are applicable only under the conditions by which the models were derived. Unless an extensive data base containing a sufficient number of pavement sections with widely different design characteristics is available, the usefulness of these models in practice may be limited due to the large amount of error involved.

1.4.5 Reliability

In view of the fact that the predicted distress at the end of a design period varies a great deal, depending on the variability of predicted traffic and the quality control on materials and construction, it is more reasonable to use a probabilistic approach based on the reliability concept. If PSI is used as a failure criterion, the reliability of the design, or the probability that the PSI is greater than the terminal serviceability index, can be determined by assuming the PSI at the end of a design period to be a normal distribution with a mean and a standard deviation. Conversely, given the required reliability and terminal serviceability index, the acceptable PSI at the end of the design period can be computed. More details about reliability are presented in Chapter 10.

1.4.6 Pavement Management Systems

It has long been recognized that pavement design is a part of total pavement management process, which includes planning, design, construction, maintenance, evaluation, and rehabilitation. With the use of a computer, a pavement management system (PMS) can be developed to assist decision makers in finding optimum strategies for providing, evaluating, and maintaining pavements in a serviceable condition over a given period of time. Pavement management can be divided into two generalized levels: network and project. At the network level, the pavement management system provides information on the development of an overall program of new construction, maintenance, or rehabilitation that will optimize the use of available resources. At the project level, consideration is given to alternative design, construction, maintenance, or rehabilitation activities for a particular project within the overall program.

Figure 1.19 is a flowchart for a project level pavement management system (AASHTO, 1986). The traffic and loading, environment, and materials are the design factors that have just been discussed. Models of pavement structure may be a mechanistic or an empirical model for flexible or rigid pavements. Behavior is characterized by stresses, strains, or deformations; distress is evaluated by failure criteria; and performance is based on PSI. For a given reliability, the life of the pavement before the serviceability index drops below the minimum acceptable value can be evaluated. Even if the life is less than the design period, say 20 years, the option is still open because an overlay at a later date will bring the serviceability index up and prolong the life of the pavement to more than 20 years. As long as the design meets the constraints, it will move on to the life-cycle costs block of the process.

Life-cycle costs refer to all costs, including construction, maintenance, and rehabilitation costs, as well as all benefits and indirect costs. An economic evaluation will be made on all possible options and an optimized design at the lowest overall cost will be selected. After the pavement has been constructed, information on performance, such as distress, roughness, traffic loading, skid characteristic, and deflection, should be monitored and put into a data bank. The feedback of these performance data into the PMS information system is crucial to the development of mechanistic–empirical design procedures. In this book only the structural design subsystem prior to the life-cycle cost is discussed. Readers interested in pavement management systems can refer to the textbook by Haas and Hudson (1978).

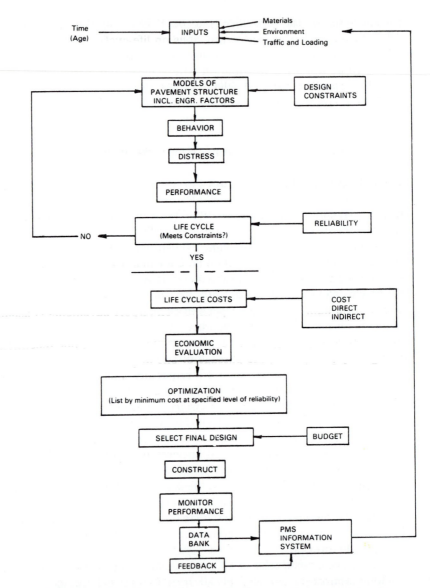

Figure 1.19 Flow diagram of a project level pavement management system. (From the *AASHTO Guide for Design of Pavement Structures*. Copyright 1986. American Association of State Highway and Transportation Officials, Washington, DC. Used by permission.)

It should be noted that pavement design is a critical part of pavement management. Poor design practice will result in higher pavement maintenance and rehabilitation costs throughout the years and has by far the greatest effect on life-cycle costs. Recent developments in pavement technology such as the improvement in laboratory and field testing equipment and the availability of high-speed microcomputers have provided pavement designers with more tools to evaluate the consequences of design alternatives on life-cycle costs.

1.5 HIGHWAY PAVEMENTS, AIRPORT PAVEMENTS, AND RAILROAD TRACKBEDS

The principles used for the design of highway pavements can also be applied to those of airport pavements and railroad trackbeds with some modifications.

1.5.1 Highway Versus Airport

Airport pavements are generally thicker than highway pavements and require better surfacing materials because the loading and tire pressure of aircraft are much greater than those of highway vehicles. The effect of loading and tire pressure can be taken care of automatically in any mechanistic method of design, whether the pavement is used for a highway or an airport. However, the following differences should be noted in applying the mechanistic methods:

1. The number of load repetitions on airport pavements is usually smaller that that on highway pavements. On airport pavements, due to the wander of aircraft, several passages of a set of gears are counted as one repetition, whereas on highway pavements, the passage of one axle is considered as one repetition. The fact that highway loadings are not really applied at the same location is considered in the failure criteria by increasing the allowable number of load repetitions, such as the incorporation of a shift factor for the fatigue of flexible pavements, as described in Section 7.3.1, and an equivalent damage ratio for the fatigue of rigid pavements, as described in Section 12.1.2.

2. The design of highway pavements is based on moving loads with the loading duration as an input for viscoelastic behaviors and the resilient modulus under repeated loads for elastic behaviors. The design of airport pavements is based on moving loads in the interior of runways but stationary loads at the end of runways. As a result, thicker pavements are used at the runway end than in the interior.

3. Although loads are applied near to the edge of highway pavements but far away from the outside edge of airport pavements, this fact is not considered in the design of flexible pavements. It is assumed that the edge effect is insignificant if a load is at a distance of 2 to 3 ft (0.6 to 0.9 m) from the edge, so the layered theory can still be applied. However, this fact should be considered in the design of rigid pavements. The Portland Cement Association employs edge loading for the design of highway pavements (PCA, 1984) but interior loading for the design of airport pavements (PCA, 1955). The Federal Aviation Administration considers edge loading but the edge stress is reduced by 25% to account for load transfer across the joint (FAA, 1988), so the loading is applied at the longitudinal joint not really at the outside pavement edge. Even if the loads can be applied near the outside edge of airport pavements in certain situations, the number of load repetitions is small and may be neglected. The above contention is based on the

assumption that the design is based on fatigue and the fact that the stresses at the edge and the interior are greater than those at the joints. This is not true if the design is based on the erosion caused by the corner deflection at the joints.

1.5.2 Highway Versus Railroad

Two methods have been used to incorporate HMA in railroad trackbeds, as shown in Figure 1.20. The first method, which is similar to the construction of flexible highway pavements, is called overlayment. The HMA is placed on top of the subgrade or above a layer of base course and the ties are placed directly on the asphalt mat. In the second method, called underlayment, the asphalt mat is placed under the ballast.

Based on the same design principles as used in highway pavements, Huang et al. (1984b) developed a computer model called KENTRACK for the design of HMA railroad trackbeds. The design considers traffic and loading, environment, and materials as the major factors and applies Burmister's layered theory and Miner's hypothesis of cumulative damage (Miner, 1945). However, there is a major difference between a highway pavement and a railroad trackbed, namely the distribution of wheel loads to the layered system. On highway pavements, wheel loads are applied over small areas and the magnitude of loads on each area is a constant independent of the stiffness of the layered system. On railroad trackbeds, wheel loads are distributed through rails and ties over a large area and the load on the most critical tie under the heaviest wheel load depends strongly on the stiffness of the layered system. Therefore, the use of thicker HMA for highway pavements is very effective in reducing both the tensile strain at the bottom of HMA and the compressive strain on the top of subgrade, but not very effective for railroad trackbeds. In fact, for an underlayment with a given combined thickness of ballast and HMA, the tensile strain increases as the HMA

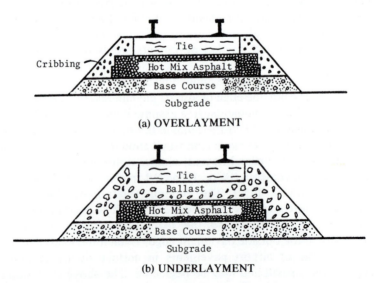

(a) OVERLAYMENT

(b) UNDERLAYMENT

Figure 1.20 Hot mix asphalt railroad trackbeds.

thickness increases, which indicates that the use of ballast is more effective than the use of HMA in reducing tensile strains (Huang et al., 1985). That the replacements of ballast by HMA increases the tensile strain is due to the load concentration, as indicated by the tremendous increase in the maximum contact pressure between tie and ballast caused by the stiffer trackbed.

It was also found that overlayment cannot be used for heavy haul trackbeds because the required thickness of asphalt mat is just too excessive, while an underlayment with a thick layer of ballast and a thin layer of HMA can easily satisfy the design requirements (Huang et al., 1987). For the same reason, the use of full-depth construction, which is popular for highway pavements, is not recommended for railroad trackbeds. Although the vertical compressive strain on the top of subgrade has been used most frequently for the design of highway pavements, it was found that the use of vertical compressive stress is more appropriate for railroad trackbeds (Huang et al., 1984a, 1986a).

Portland cement concrete can be used for the construction of slab tracks. The design of slab tracks is similar to that of rigid highway pavements except that loads are applied to the rails connected directly to the concrete slab or through rubber booted block ties. The KENTRACK computer program, originally developed for the design of HMA and conventional ballasted trackbeds, was later expanded for the design of slab tracks based on the finite element method and the fatigue principle (Huang et al., 1986b).

1.6 SUMMARY

This chapter provides an introduction to pavement analysis and design. It covers five major sections including historical development, pavement types, road tests, design factors, and the differences between highway pavements and airport pavements and between highway pavements and railroad trackbeds. The three major road tests constitute an important part of historical development but are presented in a separate section after pavement types, so that the conclusions drawn from the road tests can be more easily understood.

Important Points Discussed in Chapter 1

1. There has been a dramatic change in the design methods for flexible pavements, from the early purely empirical methods to the modern mechanistic–empirical methods. With the availability of high-speed microcomputers and sophisticated testing methods, the trend toward mechanistic methods is apparent.

2. The most practical and widely used mechanistic–empirical method for flexible pavement design is based on Burmister's elastic layered theory of limiting the horizontal tensile strain at the bottom of asphalt layer and the vertical compressive strain on the surface of subgrade. With modifications, the method can be applied to layered systems consisting of viscoelastic and nonlinear elastic materials.

3. Westergaard's plate theory has been used for the design of rigid pavements since the 1920s. The method can be applied only to a single slab on a liquid foundation with full slab–subgrade contact. Most design methods are based on the

flexural stress in the concrete. Earlier designs considered stress due to corner loading to be the most critical. With wider traffic lanes and more efficient load transfer, edge stress due to edge loadings near the midslab is now being considered more critical. The current PCA method considers both the edge stress to prevent fatigue cracking and the corner deflection to minimize pumping or erosion of the subgrade.

4. The most practical method to analyze multiple slabs on a layer foundation with partial slab–foundation contact is by the finite element computer programs. With the increase in speed and storage of modern computers, it is no longer necessary to consider the foundation as a liquid. More use of solid or layer foundation in design is expected in years to come.

5. There are two major types of pavements: flexible and rigid. The design of flexible pavements is based on the layered theory by assuming that the layers are infinite in areal extent with no discontinuities. Due to the rigidity of the slab and the existence of joints, the design of rigid pavements is based on the plate theory. Layered theory can also be applied to rigid pavements if the loads are applied in the interior of a slab. A third type of pavement is called the composite pavement. Composite pavements should be designed by the plate theory because concrete is the main load-carrying component.

6. Three types of flexible pavements are conventional, full-depth, and contained rock asphalt mat. Conventional pavements are layer systems with better materials on top and are most suited to regions where local materials are available. Full-depth asphalt pavements, though more expensive, have gained popularity because of their many advantages over conventional pavements. Contained rock asphalt mats, in which granular materials are sandwiched between two asphalt layers, are still in the experimental stage and their performance has yet to be evaluated.

7. Four types of rigid pavements are jointed plain concrete pavement (JPCP), jointed reinforced concrete pavement (JRCP), continuous reinforced concrete pavement (CRCP), and prestressed concrete pavement (PCP). The most economical construction is JPCP with closely spaced joints. Steel reinforcements in JRCP do not increase its structural capacity but allow the use of longer joint spacings; and if the pavement cracks, the steel can tie the concrete together. The advantage of CRCP is the complete elimination of transverse joints. Previous experience has indicated that the thickness required for CRCP should be the same as that for JPCP and JRCP. Although several experimental projects have been constructed in the United States, the use of PCP for highway pavements is limited because the savings in materials is not sufficient to compensate for the extensive labor required.

8. The three major road tests conducted from the mid 1940s to the early 1960s are the Maryland Road Test for rigid pavements, the WASHO Road Test for flexible pavements, and the AASHO Road Test for both flexible and rigid pavements. The first two tests were initiated for taxation purposes to determine the relative effects of four different axle loads on pavement distress, while the last test was more comprehensive and involved the development of equations for design purposes.

9. A significant outcome for each test is summarized as follows: The Maryland Road Test indicated that pumping was the major cause of distress for pavements on fine-grained soils and that the stresses for all cases of loading were less than 50% of the modulus of rupture of the concrete prior to the development of pumping, but increased tremendously after pumping and caused the rupture of the slab. The WASHO Road Test indicated that the distress in the outer wheelpath was more than that in the inner wheelpath and that the outer wheelpath with paved shoulders was the equivalent of the inner wheelpath, thus testifying to the advantage of shoulder paving. The AASHO Road Test developed the pavement serviceability concept and the equations relating the change in serviceability to the number of load repetitions, thus adding a new dimension to pavement design.

10. Most of the design methods in use today are based on the 18-kip equivalent single-axle load and a fixed set of material properties for the entire design period. However, with the use of modern computers and the mechanistic–empirical methods, these limitations and approximations can be overcome. Each axle load group can be considered separately and each year can be divided into a number of periods with widely different material properties for damage analysis.

11. Many of the mechanistic–empirical methods for flexible pavement design are based on the horizontal tensile strain at the bottom of the asphalt layer and the vertical compressive strain on the surface of subgrade. With heavier axle loads and higher tire pressures, much of the permanent deformation occurs in the upper layers above the subgrade. Consequently, the determination of permanent deformation in each individual layer, instead of the vertical compressive strain on the subgrade, is a better approach.

12. The fatigue principle has long been used throughout the world for the design of concrete pavements. In view of the fact that pumping is the main cause of distress, the inclusion of corner deflection as a failure criterion, in addition to fatigue cracking, is warranted.

13. Other than the AASHTO empirical methods, the concepts of serviceability and reliability have not been widely accepted in current methods of pavement design. As more research is directed to these concepts and more efficient computer programs are developed, it is expected that these concepts will become more popular and widely accepted in future mechanistic–empirical methods.

14. With the availability of high-speed computers and the improvement in laboratory and field testing equipment, pavement design should be integrated into the pavement management system, so that the consequences of design alternatives on life-cycle costs can be evaluated.

15. The design principles used for highway pavements can be equally applied to airport pavements with only a few exceptions, such as the consideration of aircraft wandering on the number of load repetitions and the use of stationary loads at the end of runways.

16. The use of full-depth asphalt, which is popular for highway pavements, is ineffective for railroad trackbeds. It is more economical to place the HMA under the ballast rather than over the ballast.

PROBLEMS AND QUESTIONS

1-1. Sketch and show the dimensions of the most realistic contact areas for a dual tandem axle load of 40,000 lb with a tire pressure of 100 psi. If the contact areas are assumed as rectangles, what should be the dimension of the rectangular area? [Answer: most realistic area consisting of a rectangle and two semicircles with a length of 9.78 in. and a width of 5.87 in.; rectangular area with a length of 8.52 in. and a width of 5.87 in.]

1-2. Mean monthly temperatures at a given paving project are September, 50°F; October, 32°F; November, 24°F; December, −3°F; January, 14°F; February, 16°F; March, 22°F; April, 25°F; and May, 40°F. Calculate the freezing index. Is this value likely to be different from the freezing index calculated using mean daily temperature? [Answer: 2851 degree days]

1-3. Why is the limiting shear failure method of flexible pavement design rarely in use today? Under what situations could the method be used?

1-4. What is meant by the mechanistic–empirical method of pavement design? Is it possible to develop a wholly mechanistic method of pavement design? Why?

1-5. What are the purposes of using a seal coat?

1-6. What is the difference between a tack coat and a prime coat? Which requires the use of a less viscous asphalt?

1-7. Why are flexible pavements generally built in layers with better materials on top?

1-8. What is the difference between a surface course and a binder course? Under what conditions should a binder course be used?

1-9. What are the advantages of placing a layer of HMA under a granular base or subbase?

1-10. Describe the mechanics of pumping.

1-11. Explain why rigid pavements pump whereas very few flexible pavements do so. Under what conditions would a flexible pavement result in pumping?

1-12. Which is more frost susceptible, silt or clay? Why?

1-13. Is frost heave due solely to the 9% volume expansion of soil water during freezing? Why?

1-14. Discuss the pros and cons of JPCP versus JRCP.

1-15. Discuss the pros and cons of JRCP versus CRCP.

1-16. What are the advantages and disadvantages of prestressed pavements?

1-17. What is the major distress in a composite pavement? How can it be prevented?

1-18. Why are modern concrete pavements no longer built with a thickened edge?

1-19. List any two conclusions from the Maryland Road Test that are similar to those found in the AASHO Road Test.

1-20. List any two conclusions from the WASHO Road Test that are similar to those found in the AASHO Road Test.

1-21. List any two conclusions from the Maryland Road Test on rigid pavements that are similar to the WASHO Road Test on flexible pavements.

1-22. Explain why, for a high-pressure tire, contact pressure is smaller than tire pressure.

1-23. Why is the shape of contact area used for the design of flexible pavements different from that of rigid pavements?

1-24. On which pavement does pumping occur more frequently, airport or highway?

1-25. For a given wheel load, which should be thicker, a highway or an airport pavement? Why?

1-26. Discuss the basic design differences between an airport and a highway pavement.

1-27. Discuss the basic design differences between a highway pavement and a railroad trackbed.

1-28. What is meant by full-depth construction? Why is full-depth construction not recommended for railroad trackbeds?

2

Stresses and Strains
in Flexible Pavements

2.1 HOMOGENEOUS MASS

The simplest way to characterize the behavior of a flexible pavement under wheel loads is to consider it as a homogeneous half-space. A half-space has an infinitely large area and an infinite depth with a top plane on which the loads are applied. The original Boussinesq (1885) theory was based on a concentrated load applied on an elastic half-space. The stresses, strains, and deflections due to a concentrated load can be integrated to obtain those due to a circular loaded area. Before the development of layered theory by Burmister (1943), much attention was paid to Boussinesq solutions because they were the only ones available. The theory can be used to determine the stresses, strains, and deflections in the subgrade if the modulus ratio between the pavement and the subgrade is close to unity, as exemplified by a thin asphalt surface and a thin granular base. If the modulus ratio is much greater than unity, the equation must be modified, as demonstrated by the earlier Kansas design method (Kansas State Highway Commission, 1947).

Figure 2.1 shows a homogeneous half-space subjected to a circular load with a radius a and a uniform pressure q. The half-space has an elastic modulus E and a Poisson ratio v. A small cylindrical element with center at a distance z below the surface and r from the axis of symmetry is shown. Due to axisymmetry, there are only three normal stresses, σ_z, σ_r, and σ_t, and one shear stress, τ_{rz}, which is equal to τ_{zr}. These stresses are functions of q, r/a, and z/a.

2.1.1 Solutions by Charts

Foster and Ahlvin (1954) presented charts for determining vertical stress σ_z, radial stress σ_r, tangential stress σ_t, shear stress τ_{rz}, and vertical deflection w, as shown in Figures 2.2 through 2.6. The load is applied over a circular area with a radius a

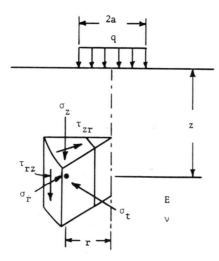

Figure 2.1 Component of stresses under axisymmetric loading.

and an intensity q. Because Poisson ratio has a relatively small effect on stresses and deflections, Foster and Ahlvin assumed the half-space to be incompressible with a Poisson ratio of 0.5, so only one set of charts is needed instead of one for each Poisson ratio. This work was later refined by Ahlvin and Ulery (1962) who presented a series of equations and tables so that the stresses, strains, and deflections for any given Poisson ratio can be computed. These equations and tables are not presented here because the solutions can be easily obtained from KENLAYER by assuming the homogeneous half-space to be a two-layer system

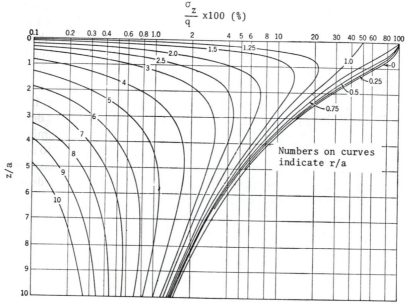

Figure 2.2 Vertical stresses due to circular loading. (After Foster and Ahlvin (1954).)

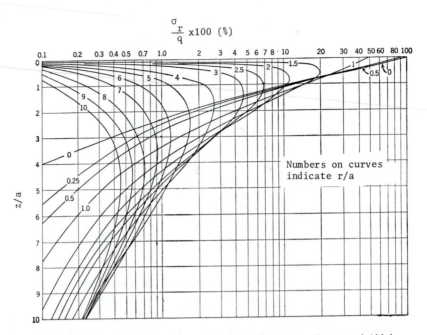

Figure 2.3 Radial stresses due to circular loading. (After Foster and Ahlvin, (1954).)

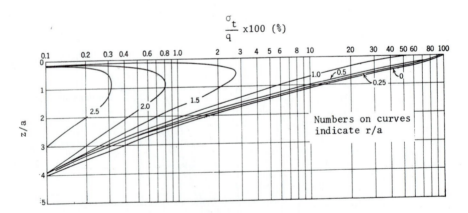

Figure 2.4 Tangential stresses due to circular loading. (After Foster and Ahlvin (1954).)

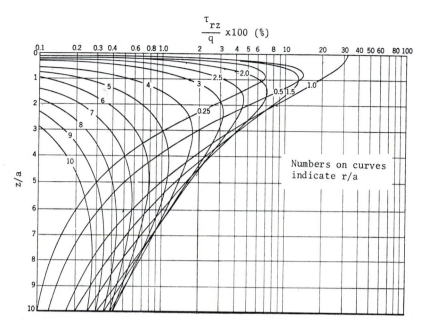

Figure 2.5 Shear stresses due to circular loading. (After Foster and Ahlvin (1954).)

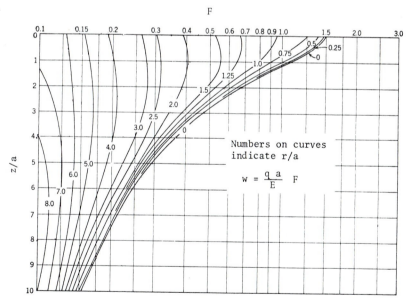

Figure 2.6 Vertical deflections due to circular loading. (After Foster and Ahlvin (1954).)

with any thickness but with the same elastic modulus and Poisson ratio for both layers.

After the stresses are obtained from the charts, the strains can be obtained by

$$\epsilon_z = \frac{1}{E}[\overset{1}{\sigma_z} - \nu(\overset{2}{\sigma_r} + \overset{3}{\sigma_t})] \tag{2.1a}$$

$$\epsilon_r = \frac{1}{E}[\overset{2}{\sigma_r} - \nu(\overset{3}{\sigma_t} + \overset{1}{\sigma_z})] \tag{2.1b}$$

$$\epsilon_t = \frac{1}{E}[\overset{3}{\sigma_t} - \nu(\overset{1}{\sigma_z} + \overset{2}{\sigma_r})] \tag{2.1c}$$

If the contact area consists of two circles, the stresses and strains can be computed by superposition.

Example 2.1:

Figure 2.7 shows a homogeneous half-space subjected to two circular loads, each 10 in. (254 mm) in diameter and spaced at 20 in. (508 mm) on centers. The pressure on the circular area is 50 psi (345 kPa). The half-space has an elastic modulus of 10,000 psi (69 MPa) and a Poisson ratio of 0.5. Determine the vertical stress, strain, and deflection at point A, which is located 10 in. (254 mm) below the center of one circle.

Solution: Given a = 5 in. (127 mm), q = 50 psi (345 kPa), and z = 10 in. (254 mm), from Figures 2.2, 2.3, and 2.4, the stresses at point A due to the left load with r/a = 0 and z/a = 10/5 = 2 are σ_z = 0.28 × 50 = 14.0 psi (96.6 kPa) and σ_r = σ_t = 0.016 × 50 = 0.8 psi (5.5 kPa); and those due to the right load with r/a = 20/5 = 4 and z/a = 2 are σ_z = 0.0076 × 50 = 0.38 psi (2.6 kPa), σ_r = 0.026 × 50 = 1.3 psi (9.0 kPa), and σ_t = 0. By superposition, σ_z = 14.0 + 0.38 = 14.38 psi (99.2 kPa), σ_r = 0.8 + 1.3 = 2.10 psi (14.5 kPa), and σ_t = 0.8 psi (5.5 kPa). From Eq. 2.1a, ϵ_z = [14.38 − 0.5(2.10 + 0.8)]/10,000 = 0.00129. From Figure 2.6, the deflection factor at point A due to the left load is 0.68 and that due to the right load is 0.21. The total deflection w = (0.68 + 0.21) × 50 × 5/10,000 = 0.022 in. (0.56 mm). The final answer is σ_z = 14.38 psi (99.2 kPa), ϵ_z = 0.00129, and w = 0.022 in. (0.56 mm). The results obtained from KENLAYER are σ_z = 14.6 psi (100.7 kPa), ϵ_z = 0.00132, and w = 0.0218 in. (0.554 mm), which check closely with those from the charts.

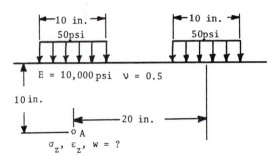

Figure 2.7   Example 2.1 (1 in. = 25.4 mm, 1 psi = 6.9 kPa).

In applying Boussinesq's solutions, it is usually assumed that the pavement above the subgrade has no deformation, so the deflection on the pavement surface is equal to that on the top of the subgrade. In the above example, if the pavement thickness is 10 in. (254 mm) and point A is located on the surface of the subgrade, the deflection on the pavement surface is 0.022 in. (0.56 mm).

2.1.2 Solutions at Axis of Symmetry

When the load is applied over a single circular loaded area, the most critical stress, strain, and deflection occur under the center of the circular area on the axis of symmetry, where $\tau_{rz} = 0$ and $\sigma_r = \sigma_t$, so σ_z and σ_r are the principal stresses.

Flexible Plate

The load applied from tire to pavement is similar to a flexible plate with a radius a and a uniform pressure q. The stresses beneath the center of the plate can be determined by

$$\sigma_z = q\left[1 - \frac{z^3}{(a^2 + z^2)^{1.5}}\right] \tag{2.2}$$

$$\sigma_r = \frac{q}{2}\left[1 + 2v - \frac{2(1 + v)z}{(a^2 + z^2)^{0.5}} + \frac{z^3}{(a^2 + z^2)^{1.5}}\right] \tag{2.3}$$

Note that σ_z is independent of E and v, and σ_r is independent of E. From Eq. 2.1

$$\epsilon_z = \frac{(1 + v)q}{E}\left[1 - 2v + \frac{2vz}{(a^2 + z^2)^{0.5}} - \frac{z^3}{(a^2 + z^2)^{1.5}}\right] \tag{2.4}$$

$$\epsilon_r = \frac{(1 + v)q}{2E}\left[1 - 2v - \frac{2(1 - v)z}{(a^2 + z^2)^{0.5}} + \frac{z^3}{(a^2 + z^2)^{1.5}}\right] \tag{2.5}$$

The vertical deflection w can be determined by

$$w = \frac{(1 + v)qa}{E}\left\{\frac{a}{(a^2 + z^2)^{0.5}} + \frac{1 - 2v}{a}\left[(a^2 + z^2)^{0.5} - z\right]\right\} \tag{2.6}$$

When $v = 0.5$, Eq. 2.6 can be simplified to

$$w = \frac{3qa^2}{2E(a^2 + z^2)^{0.5}} \tag{2.7}$$

On the surface of the half-space, $z = 0$, from Eq. 2.6

$$w_0 = \frac{2(1 - v^2)qa}{E} \tag{2.8}$$

Example 2.2:

Same as Example 2.1, except that only the left loaded area exists and the Poisson ratio is 0.3, as shown in Figure 2.8. Determine the stresses, strains, and deflection at point A.

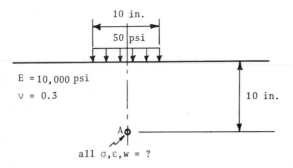

10 in.

50 psi

E = 10,000 psi
ν = 0.3

10 in.

A○
all σ,ε,w = ?

Figure 2.8 Example 2.2 (1 in. =
25.4 mm, 1 psi = 6.9 kPa).

Solution: Given $a = 5$ in. (127 mm), $q = 50$ psi (345 kPa), and $z = 10$ in. (254 mm), from Eq. 2.2, $\sigma_z = 50[1 - 1000/(25 + 100)^{1.5}] = 14.2$ psi (98.0 kPa). With $\nu = 0.3$, from Eq. 2.3, $\sigma_r = 25[1 + 0.6 - 2.6 \times 10/(125)^{0.5} + 1000/(125)^{1.5}] = -0.25$ psi (-1.7 kPa). The negative sign indicates tension, which is in contrast to a compressive stress of 0.8 psi (5.5 kPa) when $\nu = 0.5$. From Eq. 2.4, $\epsilon_z = 1.3 \times 50/10,000$ $[1 - 0.6 + 0.6 \times 10/(125)^{0.5} - 1000/(125)^{1.5}] = 0.00144$. From Eq. 2.5, $\epsilon_r = 1.3 \times$ $50/20,000 [1 - 0.6 - 1.4 \times 10/(125)^{0.5} + 1000/(125)^{1.5}] = -0.00044$. From Eq. 2.6, $w = 1.3 \times 50 \times 5/10,000 \{5/(125)^{0.5} + 0.4/5 [(125)^{0.5} - 10]\} = 0.0176$ in. (0.447 mm). The results obtained from KENLAYER are $\sigma_z = 14.2$ psi (98.0 kPa), $\sigma_r = -0.249$ psi (-1.72 kPa), $\epsilon_z = 0.00144$, $\epsilon_r = -0.000444$, and $w = 0.0176$ in. (0.447 mm), which are nearly the same as those derived by the formulas.

Rigid Plate

All the above analyses are based on the assumption that the load is applied on a flexible plate such as a rubber tire. If the load is applied on a rigid plate such as that used in a plate loading test, the deflection is the same at all points on the plate, but the pressure distribution under the plate is not uniform. The differences between a flexible and a rigid plate are shown in Figure 2.9.

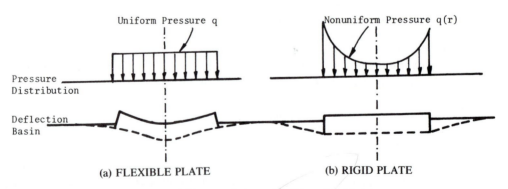

Uniform Pressure q

Nonuniform Pressure q(r)

Pressure
Distribution

Deflection
Basin

(a) FLEXIBLE PLATE

(b) RIGID PLATE

Figure 2.9 Differences between flexible and rigid plates.

The pressure distribution under a rigid plate can be expressed as (Ullidtz, 1987)

$$q(r) = \frac{qa}{2(a^2 - r^2)^{0.5}} \qquad (2.9)$$

in which r is the distance from center to the point where pressure is to be determined and q is the average pressure, which is equal to the total load divided by the area. The smallest pressure is at the center and equal to one-half of the average pressure. The pressure at the edge of the plate is infinity. By integrating the point load over the area, it can be shown that the deflection of the plate is

$$w_0 = \frac{\pi(1 - v^2)qa}{2E} \qquad (2.10)$$

A comparison of Eq. 2.10 with Eq. 2.8 indicates that the surface deflection under a rigid plate is only 79% of that under the center of a uniformly distributed load. This is reasonable because the pressure under the rigid plate is smaller near the center of the loaded area but greater near the edge. The pressure near the center has a greater effect on the surface deflection at the center. Although Eqs. 2.8 and 2.10 are based on a homogeneous half-space, the same factor of 0.79 can be applied if the plates are placed on a layer system, as indicated by Yoder and Witczak (1975).

Example 2.3:

A plate loading test using a plate of 12-in. (305-mm) diameter was performed on the surface of the subgrade, as shown in Figure 2.10. A total load of 8000 lb (35.6 kN) was applied to the plate and a deflection of 0.1 in. (2.54 mm) was measured. Assuming that the subgrade has a Poisson ratio of 0.4, determine the elastic modulus of the subgrade.

Solution: The average pressure on the plate is $q = 8000/(36\pi) = 70.74$ psi (488 kPa). From Eq. 2.10, $E = \pi(1 - 0.16) \times 70.74 \times 6/(2 \times 0.1) = 5600$ psi (38.6 MPa).

2.1.3 Nonlinear Mass

Boussinesq's solutions are based on the assumption that the material that constitutes the half-space is linear elastic. It is well known that subgrade soils are not elastic and result in permanent deformation under stationary loads. However, under the repeated application of moving traffic loads, most of the deformations are recoverable and can be considered elastic. It is therefore possible to select a

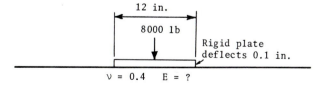

Figure 2.10 Example 2.3
(1 = 25.4 mm, 1 lb = 4.45 N).

reasonable elastic modulus commensurate with the speed of moving loads. Since linearity implies the applicability of the superposition principle, the elastic constant must not vary with the state of stresses. In other words, the axial deformation of a linear elastic material under an axial stress should be independent of the confining pressure. This is evidently not true for soils, because their axial deformation depends strongly on the magnitude of confining pressures. Consequently, the effect of nonlinearity on Boussinesq's solution is of practical interest.

Iterative Method

To show the effect of nonlinearity of granular materials on vertical stresses and deflections, Huang (1968a) divided the half-space into seven layers, as shown in Figure 2.11, and applied Burmister's layered theory to determine the stresses at the midheight of each layer. Note that the lowest layer is a rigid base with a very large elastic modulus.

After the stresses are obtained, the elastic modulus of each layer is determined by

$$E = E_0(1 + \beta\theta) \tag{2.11}$$

in which θ is the stress invariant, or the sum of three normal stresses; E is the elastic modulus under the given stress invariant; E_0 is the initial elastic modulus, or the modulus when the stress invariant is zero; and β is a soil constant indicating the increase in elastic modulus per unit increase in stress invariant. Note that the stress invariant should include the effects of the applied load as well as the geostatic stresses and can be expressed as

$$\theta = \sigma_z + \sigma_r + \sigma_t + \gamma z(1 + 2K_0) \tag{2.12}$$

in which σ_z, σ_r, and σ_t are the vertical, radial, and tangential stresses due to loading; γ is the unit weight of soil; z is the distance below ground surface at which the stress invariant is computed; and K_0 is the coefficient of earth pressure at rest. The problem can be solved by a method of successive approximations. First, an

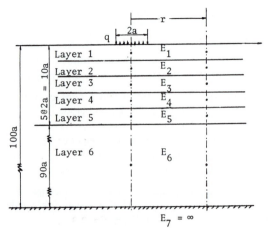

Figure 2.11 Division of half-space into a seven-layer system.

Stresses and Strains in Flexible Pavements   Chap. 2

elastic modulus is assumed for each layer and the stresses are obtained by the layered theory. Based on the stresses thus obtained, a new set of moduli is determined from Eq. 2.11 and a new set of stresses is then computed. The process is repeated until the moduli between two consecutive iterations converge to a specified tolerance.

In applying the layered theory for nonlinear analysis, a question immediately arises: Which radial distance r should be used to determine the stresses and the moduli? Huang (1968a) showed that the vertical stresses are not affected significantly by whether the stresses at $r = 0$ or $r = \infty$ are used to determine the elastic modulus, but the vertical displacements are tremendously affected. He later used the finite element method and found that the nonlinear behavior of soils has a large effect on vertical and radial displacements, an intermediate effect on radial and tangential stresses, and a very small effect on vertical and shear stresses (Huang, 1969a). Depending on the depth of the point in question, the vertical stresses based on nonlinear theory may be greater or smaller than those based on linear theory and, at a certain depth, both theories may yield the same stresses. This may explain why Boussinesq's solutions for vertical stress based on linear theory have been applied to soils with varying degrees of success, even though soils themselves are basically nonlinear.

Approximate Method

One approximate method to analyze a nonlinear half-space is to divide it into a number of layers and determine the stresses at the midheight of each layer by Boussinesq's equations based on linear theory. From the stresses thus obtained, the elastic modulus E for each layer is determined from Eq. 2.11. The deformation of each layer, which is the difference in deflection between the top and bottom of each layer based on the given E, can then be obtained. Starting from the rigid base, or a depth far from the surface where the vertical displacement can be assumed zero, the deformations are added to obtain the deflections at various depths. The assumption of Boussinesq's stress distribution was used by Vesic and Domaschuk (1964) to predict the shape of deflection basins on highway pavements and satisfactory agreements were reported.

It should be noted that Eq. 2.11 is one of the many constitutive equations for sands. Other constitutive relationships for sands or clays can also be used, as discussed in Section 3.1.4.

Example 2.4:

A circular load with a radius of 6 in. (152 mm) and a contact pressure of 80 psi (552 kPa) is applied on the surface of a subgrade. The subgrade soil is a sand with the relationship between the elastic modulus and the stress invariant, as shown in Figure 2.12a. The soil has a Poisson ratio of 0.3, a mass unit weight of 110 pcf (17.3 kN/m³), and a coefficient of earth pressure at rest of 0.5. The soil is divided into 6 layers, as shown in Figure 2.12b. Determine the vertical surface displacement at the axis of symmetry.

Solution: At the midheight of layer 1, $z = 6$ in. (152 mm). From Eq. 2.2, $\sigma_z = 80\,[1 - 216/(36 + 36)^{1.5}] = 51.7$ psi (357 kPa). From Eq. 2.3, $\sigma_r = \sigma_t = 40\,[1 + 2 \times$

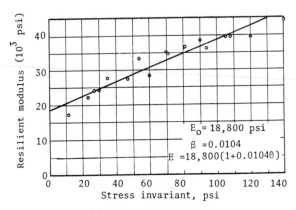

(a) RESILIENT MODULUS

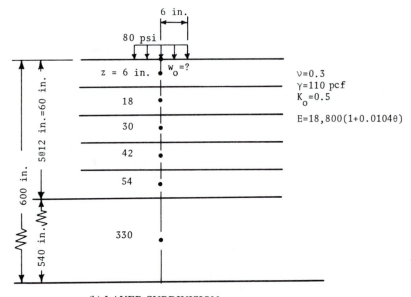

(b) LAYER SUBDIVISION

Figure 2.12 Example 2.4 (1 in. = 25.4 mm, 1 psi = 6.9 kPa, 1 pcf = 157.1 N/m³).

$0.3 - 2.6 \times 6/(72)^{0.5} + 216/(72)^{1.5}] = 4.60$ psi (31.7 kPa). From Eq. 2.12, $\theta = 51.7 + 4.6 + 4.6 + 110 \times 6 (1 + 2 \times 0.5)/(12)^3 = 61.7$ psi (426 kPa). From Eq. 2.11 with $E_0 = 18,800$ psi (130 MPa) and $\beta = 0.0104$, as shown in Figure 2.12a, $E = 18,800 (1 + 0.0104 \times 61.7) = 30,900$ psi (213 MPa). From Eq. 2.6, the deflection at top, when $z = 0$, $w = 1.3 \times 80 \times 6 (1 + 1 - 0.6)/30,900 = 0.0283$ in. (0.719 mm), and the deflection at bottom, when $z = 12$ in. (305 mm), $w = 1.3 \times 80 \times 6 \{6/(36 + 144)^{0.5} + 0.4 [(180)^{0.5} - 12]/6\}/30,900 = 0.0109$ in. (0.277 mm). The deformation for layer 1 is $0.0283 - 0.0109 = 0.0174$ in. (0.442 mm). The deformations for other layers can be determined similarly and the results are tabulated in Table 2.1.

TABLE 2.1 COMPUTATION OF DEFORMATION FOR EACH LAYER

Layer no.	Thickness (in.)	z at mid-height (in.)	σ_z (psi)	σ_r (psi)	θ (psi) Loading	θ (psi) Geostatic	E (psi)	wE (lb/in.)	Deformation (in.)
1	12	6	51.72	4.60	60.92	0.76	30,860	873.6	0.0174
2	12	18	11.69	−0.51	10.67	2.29	21,330	338.0	0.0073
3	12	30	4.57	−0.27	4.03	3.82	20,330	182.1	0.0029
4	12	42	2.39	−0.15	2.09	5.35	20,250	123.2	0.0015
5	12	54	1.46	−0.09	1.28	6.88	20,400	92.9	0.0009
6	540	330	0.04	0.00	0.04	42.01	27,020	74.5	0.0025
							Total	7.5	0.0325

Note. 1 in. = 25.4 mm, 1 psi = 6.9 kPa, 1 lb/in. = 175 N/m.

TABLE 2.2 DIFFERENCES IN STRESSES AND MODULI BETWEEN BOUSSINESQ AND BURMISTER SOLUTIONS

z at midheight (in.)	Boussinesq			Burmister		
	σ_z (psi)	σ_r (psi)	E (psi)	σ_z (psi)	σ_r (psi)	E (psi)
6	51.72	4.60	30,860	50.46	4.50	30,580
18	11.69	−0.51	21,330	10.61	−0.65	21,070
30	4.57	−0.27	20,330	4.26	−0.27	20,280
42	2.39	−0.15	20,250	2.31	−0.11	20,260
54	1.46	−0.09	20,400	1.47	0.01	20,440
330	0.04	0.00	27,020	0.04	0.00	27,020

Note. 1 in. = 25.4 mm, 1 psi = 6.9 kPa.

To compute the deformation of each layer, the product of w and E at each layer interface is determined first by Eq. 2.6. The difference in wE between the two interfaces divided by E gives the deformation of the layer. The surface deflection is the sum of all layer deformations and equals to 0.0325 in. (0.826 mm). It is interesting to note that the stress invariant θ due to the applied load decreases with depth, while that due to geostatic stresses increases with depth. As a result, the elastic moduli for all layers, except layers 1 and 6, become nearly the same. Note also that more than 50% of the surface deflections are contributed by the deformation at the top 12 in. (305 mm).

The same problem was solved by KENLAYER after incorporating Eq. 2.11 into the program. The differences in stress distribution between Boussinesq and Burmister theory and the resulting moduli are shown in Table 2.2. It can be seen that both solutions check quite well. The surface deflection based on layered theory is 0.0310 in. (0.787 mm), which also agrees with the 0.0325 in. (0.826 mm) from Boussinesq theory.

2.2 LAYERED SYSTEMS

Flexible pavements are layered systems with better materials on top and cannot be represented by a homogeneous mass, so the use of Burmister's layered theory is more appropriate. Burmister (1943) first developed solutions for a two-layer system and then extended them to a three-layer system (Burmister, 1945). With the advent of computers, the theory can be applied to a multilayer system with any number of layers (Huang, 1967, 1968a).

Figure 2.13 shows an n-layer system. The basic assumptions to be satisfied are:

1. Each layer is homogeneous, isotropic, and linearly elastic with an elastic modulus E and a Poisson ratio v.
2. The material is weightless and infinite in areal extent.

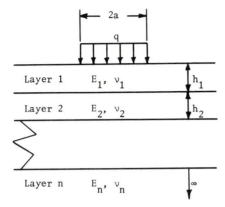

Figure 2.13 An *n*-layer system subjected to a circular load.

3. Each layer has a finite thickness h, but the lowest layer is infinite in thickness.

4. A uniform pressure q is applied on the surface over a circular area of radius a.

5. Continuity conditions are satisfied at the layer interfaces, as indicated by the same vertical stress, shear stress, vertical displacement, and radial displacement. For frictionless interface, the continuity of shear stress and radial displacement is replaced by zero shear stress at each side of the interface.

In this section, only some of the solutions on two- and three-layer systems with bonded interfaces are presented. The theoretical development of multilayer systems is discussed in Appendix B.

2.2.1 Two-Layer Systems

The exact case of a two-layer system is the full-depth construction in which a thick layer of HMA is placed directly on the subgrade. If a pavement is composed of three layers, e.g., an asphalt surface course, a granular base course, and a subgrade, it is necessary to combine the base course and the subgrade into a single layer for computing the stresses and strains in the asphalt layer or to combine the asphalt surface course and base course for computing the stresses and strains in the subgrade.

Vertical Stress

The vertical stress on the top of subgrade is an important factor in pavement design. The function of a pavement is to reduce the vertical stress on the subgrade so that detrimental pavement deformations will not occur. The allowable vertical stress on a given subgrade depends on the strength or modulus of the subgrade. To combine the effect of stress and strength, the vertical compressive strain has been used most frequently as a design criterion. This simplification is valid for highway and airport pavements because the vertical strain is caused primarily by the

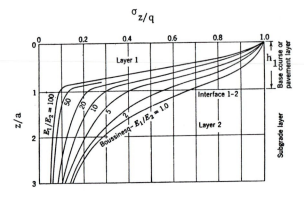

Figure 2.14 Vertical stress distribution in a two-layer system. (After Burmister (1958).)

vertical stress and the effect of horizontal stress is relatively small. As pointed out in Section 1.5.2, the design of railroad trackbeds should be based on vertical stress instead of vertical strain, because, due to the large horizontal stress caused by the distribution of wheel loads through rails and ties over a large area, the vertical strain is not a good indicator of the vertical stress.

The stresses in a two-layer system depend on the modulus ratio E_1/E_2 and the thickness-radius ratio h_1/a. Figure 2.14 shows the effect of pavement layer on the distribution of vertical stresses under the center of a circular loaded area. The chart is applicable to the case when the thickness h_1 of layer 1 is equal to the radius of contact area, or $h_1/a = 1$. As in all charts presented in this section, a Poisson ratio of 0.5 is assumed for all layers. It can be seen that the vertical stresses decrease significantly with the increase in modulus ratio. At the pavement–subgrade interface, the vertical stress is about 68% of the applied pressure if $E_1/E_2 = 1$, as indicated by Boussinesq's stress distribution, and reduces to about 8% of the applied pressure if $E_1/E_2 = 100$.

Figure 2.15 shows the effect of pavement thickness and modulus ratio on the vertical stress σ_c at the pavement–subgrade interface under the center of a circular loaded area. For a given applied pressure q the vertical stress increases with the

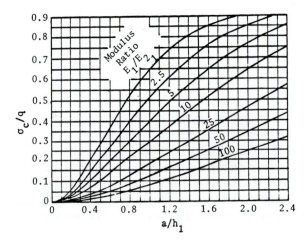

Figure 2.15 Vertical interface stresses for two-layer systems. (After Huang (1969b).)

Stresses and Strains in Flexible Pavements   Chap. 2

increase in contact radius and decreases with the increase in thickness. The reason that the ratio of a/h_1 instead of h_1/a was used is for the purpose of preparing influence charts (Huang, 1969b) for two-layer elastic foundations.

Example 2.5:

A circular load with a radius of 6 in. (152 mm) and a uniform pressure of 80 psi (552 kPa) is applied on a two-layer system, as shown in Figure 2.16. The subgrade has an elastic modulus of 5000 psi (35 MPa) and can support a maximum vertical stress of 8 psi (55 kPa). If the HMA has an elastic modulus of 500,000 psi (3.45 GPa), what is the required thickness of a full-depth pavement? If a thin surface treatment is applied on a granular base with an elastic modulus of 25,000 psi (173 MPa), what is the thickness of base course required?

Solution: Given $E_1/E_2 = 500,000/5000 = 100$ and $\sigma_c/q = 8/80 = 0.1$, from Figure 2.15, $a/h_1 = 1.15$, or $h_1 = 6/1.15 = 5.2$ in. (132 mm), which is the minimum thickness for full depth. Given $E_1/E_2 = 25,000/5000 = 5$ and $\sigma_c/q = 0.1$, from Figure 2.15, $a/h_1 = 0.4$, or $h_1 = 6/0.4 = 15$ in. (381 mm), which is the minimum thickness of granular base required.

In this example, an allowable vertical stress of 8 psi (55 kPa) is arbitrarily selected to show the effect of the modulus of reinforced layer on the thickness required. The allowable vertical stress should depend on the number of load repetitions. Based on the Shell design criterion and the AASHTO equation, Huang et al. (1984b) developed the following relationship:

$$N_d = 4.873 \times 10^{-5} \, \sigma_c^{-3.734} \, E_2^{3.583} \tag{2.13}$$

in which N_d is the allowable number of stress repetitions to limit permanent deformation, σ_c is the vertical compressive stress on the surface of subgrade in psi, and E_2 is the elastic modulus of subgrade in psi. For a stress of 8 psi (5 kPa) and an elastic modulus of 5000 psi (35 MPa), the allowable number of repetitions is 3.7×10^5.

Vertical Surface Deflection

Vertical surface deflections have been used as a criterion of pavement design. Figure 2.17 can be used to determine the surface deflections for two-layer systems. The deflection is expressed in terms of the deflection factor F_2 by

$$w_0 = \frac{1.5qa}{E_2} F_2 \qquad \text{tire} \tag{2.14}$$

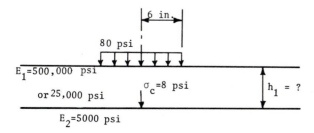

Figure 2.16 Example 2.5
(1 in. = 25.4 mm, 1 psi = 6.9 kPa).

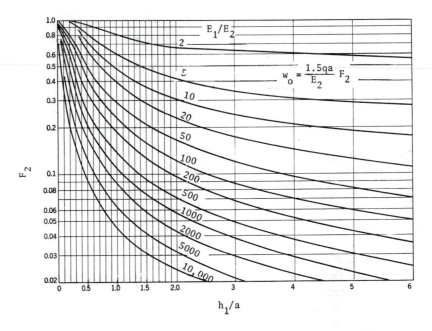

Figure 2.17 Vertical surface deflections for two-layer systems. (After Burmister (1943).)

The deflection factor is a function of E_1/E_2 and h_1/a. For a homogeneous half-space with $h_1/a = 0$, $F_2 = 1$, so Eq. 2.14 is identical to Eq. 2.8 when $v = 0.5$. If the load is applied by a rigid plate, then from Eq 2.10

$$w_0 = \frac{1.18qa}{E_2} F_2 \qquad (2.15)$$

Example 2.6:

A total load of 20,000 lb (89 kN) was applied on the surface of a two-layer system through a rigid plate 12 in. (305 mm) in diameter, as shown in Figure 2.18. Layer 1 has a thickness of 8 in. (203 mm) and layer 2 has an elastic modulus of 6400 psi

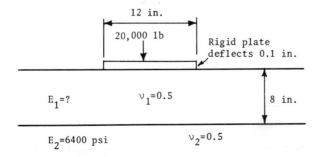

Figure 2.18 Example 2.6 (in. = 25.4 mm, 1 psi = 6.9 kPa, 1 lb = 4.45 N).

Stresses and Strains in Flexible Pavements   Chap. 2

(44.2 MPa). Both layers are incompressible with a Poisson ratio of 0.5. If the deflection of the plate is 0.1 in. (2.54 mm), determine the elastic modulus of layer 1.

Solution: The average pressure on the plate is $q = 20,000/(36\pi) = 176.8$ psi (1.22 MPa). From Eq. 2.15, $F_2 = 0.1 \times 6400/(1.18 \times 176.8 \times 6) = 0.511$. Given $h_1/a = 8/6 = 1.333$, from Figure 2.17, $E_1/E_2 = 5$, or $E_1 = 5 \times 6400 = 32,000$ psi (221 MPa).

Vertical Interface Deflection

The vertical interface deflection has also been used as a design criterion. Figure 2.19 can be used to determine the vertical interface deflection in a two-layer system (Huang, 1969c). The deflection is expressed in terms of the deflection factor F by

$$w = \frac{qa}{E_2} F \tag{2.16}$$

Note that F in Eq. 2.16 is different from F_2 in Eq. 2.14 by a factor of 1.5. The deflection factor is a function of E_1/E_2, h_1/a, and r/a, where r is the radial distance from the center of loaded area. Seven sets of charts with modulus ratios of 1, 2.5, 5, 10, 25, 50, and 100 are shown, so the deflection for any intermediate modulus ratio can be obtained by interpolation. The case of $E_1/E_2 = 1$ is Boussinesq's solution.

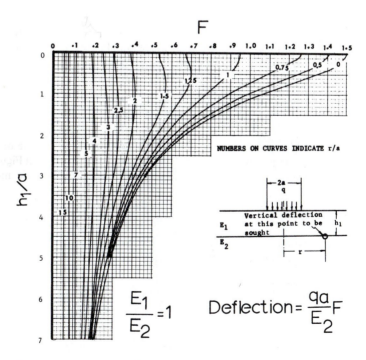

Figure 2.19 Vertical interface deflections for two-layer systems. (After Huang (1969c).)

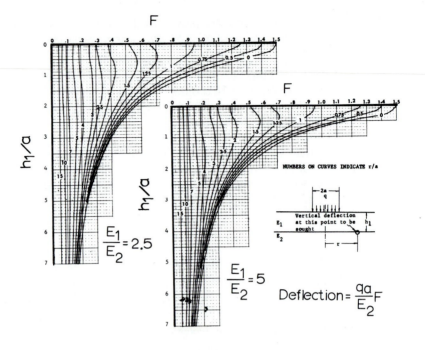

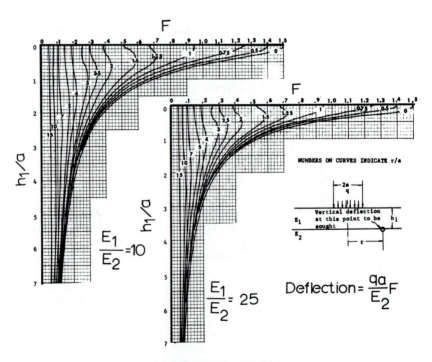

Figure 2.19 (Continued)

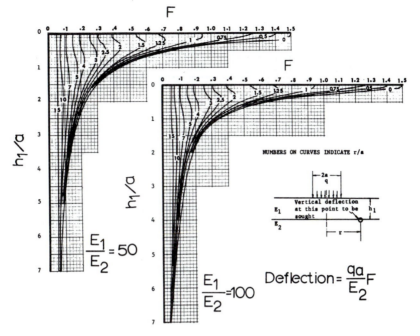

Figure 2.19 (Continued)

Example 2.7:

Figure 2.20 shows a set of dual tires, each having a contact radius of 4.52 in. (115 mm) and a contact pressure of 70 psi (483 kPa). The center to center spacing of the dual is 13.5 in. (343 mm). Layer 1 has a thickness of 6 in. (152 mm) and an elastic modulus of 100,000 psi (690 MPa) and layer 2 has an elastic modulus of 10,000 psi (69 MPa). Determine the vertical deflection at point A, which is on the interface beneath the center of one loaded area.

Solution: Given $E_1/E_2 = 100,000/10,000 = 10$ and $h_1/a = 6/4.52 = 1.33$, from Figure 2.19, the deflection factor at point A due to the left load with $r/a = 0$ is 0.56 and that due to the right load with $r/a = 13.5/4.52 = 2.99$ is 0.28. By superposition, $F = 0.56 + 0.28 = 0.84$. From Eq. 2.16, $w = 70 \times 4.52/10,000 \times 0.84 = 0.027$ in. (0.69 mm). The interface deflection obtained from KENLAYER is 0.0281 in. (0.714 mm), which checks well with the chart solution.

It should be pointed out that the maximum interface deflection under dual tires may not occur at point A. To determine the maximum interface deflection, it is necessary to compute the deflection at several points, say one under the center of one tire, one at the center between two tires, and the other under the edge of one tire, and find out which is maximum.

Critical Tensile Strain

The tensile strains at the bottom of asphalt layer have been used as a design criterion to prevent fatigue cracking. Two types of principal strains may be considered. One is the overall principal strain based on all six components of

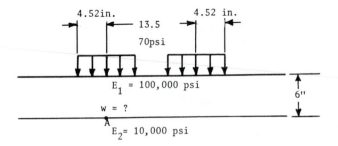

4.52in. 13.5 4.52 in.

70psi

$E_1 = 100,000$ psi

w = ?

A

$E_2 = 10,000$ psi

6"

Figure 2.20 Example 2.7
(1 in. = 25.4 mm, 1 psi = 6.9 kPa).

normal and shear stresses. The other, which is more popular and was used in KENLAYER, is the horizontal principal strain based on the horizontal normal and shear stresses only. The overall principal strain is slightly greater than the horizontal principal strain, so the use of overall principal strain is on the safe side.

Huang (1973a) developed charts for determining the critical tensile strain at the bottom of layer 1 for a two-layer system. The critical tensile strain is the overall strain and can be determined by

$$e = \frac{q}{E_1} F_e \tag{2.17}$$

in which e is the critical tensile strain and F_e is the strain factor, which can be determined from the charts.

Single Wheel

Figure 2.21 presents the strain factor for a two-layer system under a circular loaded area. In most cases, the critical tensile strain occurs under the center of the loaded area, where the shear stress is zero. However, when both h_1/a and E_1/E_2

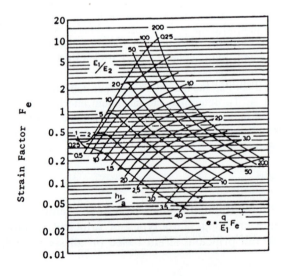

Figure 2.21 Strain factor for single wheel. (After Huang (1973a).)

are small, the critical tensile strain occurs at some distance from the center, due to the predominant effect of the shear stress. Under such situations, the principal tensile strains at radial distances of 0, 0.5a, a, and 1.5a from the center were computed and the critical value was obtained and plotted in Figure 2.21.

Example 2.8:

Figure 2.22 shows a full-depth asphalt pavement 8 in. (203 mm) thick subjected to a single-wheel load of 9000 lb (40 kN) having a contact pressure of 67.7 psi (467 kPa). If the elastic modulus of the asphalt layer is 150,000 psi (1.04 GPa) and that of the subgrade is 15,000 psi (104 MPa), determine the critical tensile strain in the asphalt layer.

Solution: Given $a = \sqrt{9000/(\pi \times 67.7)} = 6.5$ in. (165 mm), $h_1/a = 8/6.5 = 1.23$, and $E_1/E_2 = 150,000/15,000 = 10$, from Figure 2.21, $F_e = 0.72$. From Eq. 2.17, the critical tensile strain $e = 67.7 \times 0.72/150,000 = 3.25 \times 10^{-4}$, which checks well with the 3.36×10^{-4} obtained by KENLAYER.

It is interesting to note that, due to the bonded interface, the horizontal tensile strain at the bottom of layer 1 is equal to the horizontal tensile strain at the top of layer 2. If layer 2 is incompressible and the critical tensile strain occurs on the axis of symmetry, then the vertical compressive strain is equal to twice the horizontal strain, as shown by Eq. 2.21 (this concept is discussed later). Therefore, Figure 2.21 can be used to determine the vertical compressive strain on the surface of subgrade as well.

Dual Wheels

Because the strain factor for dual wheels with a contact radius a and a dual spacing S_d depends on S_d/a in addition to E_1/E_2 and h_1/a, the most direct method is to present charts similar to Figure 2.21, one for each value of S_d/a. However, this requires a series of charts and the interpolation may be quite time-consuming. To avoid these difficulties, a unique method was developed that requires only one chart, as shown in Figure 2.23.

In this method, the dual wheels are replaced by a single wheel with the same contact radius a, so that Figure 2.21 can still be used. Because the strain factor for dual wheels is generally greater than that for a single wheel, a conversion factor C, which is the ratio between dual- and single-wheel strain factors, must be determined. Multiplication of the conversion factor by the strain factor obtained from Figure 2.21 will yield the strain factor for dual wheels.

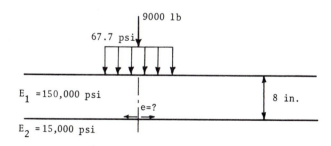

Figure 2.22 Example 2.8 (1 in. = 25.4 mm, 1 psi = 6.9 kPa, 1 lb = 4.45 N).

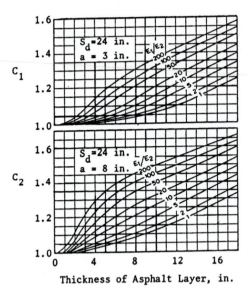

Figure 2.23 Conversion factor for dual wheel (1 in. = 25.4 mm). (After Huang (1973a).)

The two-layer theory indicates that the strain factor for dual wheels depends on h_1/a, S_d/a, and E_1/E_2. As long as the ratios h_1/a and S_d/a remain the same, the strain factor will be the same, no matter how large or small the contact radius a may be. Consider a set of dual wheels with $S_d = 24$ in. (610 mm) and $a = 3$ in. (76 mm). The strain factors for various values of h_1 and E_1/E_2 were calculated and the conversion factors were obtained and plotted as a set of curves on the upper part of Figure 2.23. Another set of curves based on the same S_d but with $a = 8$ in. (203 mm) is plotted at the bottom. It can be seen that for the same dual spacing, the larger the contact radius, the larger the conversion factor. However, the change in conversion factor due to the change in contact radius is not very large so that a straight-line interpolation should give a fairly accurate conversion factor for any other contact radii. Although Figure 2.23 is based on $S_d = 24$ in. (610 mm), it can be applied to any given S_d by simply changing a and h_1 in proportion to the change in S_d, so that the ratios h_1/a and S_d/a remain the same. The procedure can be summarized as follows:

1. From the given S_d, h_1, and a, determine the modified radius a' and the modified thickness h_1':

$$a' = \frac{24}{S_d} a \qquad (2.18a)$$

$$h_1' = \frac{24}{S_d} h_1 \qquad (2.18b)$$

2. Using h_1' as the pavement thickness, find conversion factors C_1 and C_2 from Figure 2.23.

3. Determine the conversion factor for a' by a straight-line interpolation between 3 and 8 in. (76 and 203 mm), or

$$C = C_1 + 0.2 \times (a' - 3) \times (C_2 - C_1) \qquad (2.19)$$

Stresses and Strains in Flexible Pavements   Chap. 2

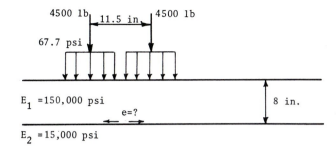

4500 lb |←11.5 in.→| 4500 lb

67.7 psi

E_1 =150,000 psi

e=?

E_2 =15,000 psi

8 in.

Figure 2.24 Example 2.9
(1 in. = 25.4 mm, 1 psi = 6.9 kPa,
1 lb = 4.45 N).

Example 2.9:

For the same pavement as in Example 2.8, if the 9000-lb (40-kN) load is applied over a set of dual tires with a center to center spacing of 11.5 in. (292 mm) and a contact pressure of 67.7 psi (467 kPa), as shown in Figure 2.24, determine the critical tensile strain in the asphalt layer.

Solution: Given $a = \sqrt{4500/(\pi \times 67.7)}$ = 4.6 in. (117 mm), S_d = 11.5 in. (292 mm), and h_1 = 8 in. (203 mm), from Eq. 2.18, a' = 24 × 4.6/11.5 = 9.6 in. (244 mm) and h_1' = 24 × 8/11.5 = 16.7 in. (424 mm). With E_1/E_2 = 10 and an asphalt layer thickness of 16.7 in. (424 mm), from Figure 2.23, C_1 = 1.35 and C_2 = 1.46. From Eq. 2.19, C = 1.35 + 0.2 (9.6 − 3)(1.46 − 1.35) = 1.50. From Figure 2.21, the strain factor for a single wheel = 0.47, and that for dual wheels = 1.50 × 0.47 = 0.705, so the critical tensile strain e = 67.7 × 0.705/150,000 = 3.18 × 10^{-4}, which checks closely with the 3.21 × 10^{-4} obtained by KENLAYER.

By comparing the results of Examples 2.8 and 2.9, it can be seen that, in this particular case when the asphalt layer is thick and the dual spacing is small, a load applied on a set of dual tires yields a critical strain that is not very different from that on a single wheel. However, this is not true when thin asphalt layers or large dual spacings are involved.

Huang (1972) also presented a simple chart for determining directly the maximum tensile strain in a two-layer system subjected to a set of dual tires spaced at a distance of $3a$ on center. A series of charts relating tensile strains to curvatures was also developed, so that the tensile strain under a design dual wheel load can be evaluated in the field by simply measuring the curvature on the surface (Huang, 1971).

Dual-Tandem Wheels

Charts similar to Figure 2.23 with dual spacing S_d of 24 in. (610 mm) and tandem spacings S_t of 24 in. (610 mm), 48 in. (1220 mm), and 72 in. (1830 mm) were developed for determining the conversion factor due to dual-tandem wheels, as shown in Figures 2.25, 2.26, and 2.27. The use of these charts is similar to the use of Figure 2.23. Because the conversion factor for dual-tandem wheels depends on h_1/a, S_d/a, and S_t/a, and the actual S_d may not be equal to 24 in. (610 mm), it is necessary to change S_d to 24 in. (610 mm) and then change the contact radius a proportionately according to Eq. 2.18a, thus keeping the ratio S_d/a unchanged.

Sec. 2.2   Layered Systems

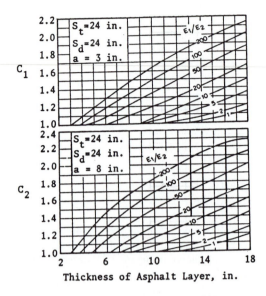

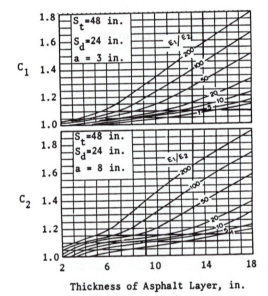

Figure 2.25 Conversion factor for dual-tandem wheels with 24-in. tandem spacing (1 in. = 25.4 mm). (After Huang (1973a).)

Figure 2.26 Conversion factor for dual-tandem wheels with 48-in. tandem spacing (1 in. = 25.4 mm). (After Huang (1973a).)

The values of h_1 and S_t must also be changed accordingly to keep h_1/a and S_t/a unchanged. Therefore, the original problem is changed to a new problem with $S_d = 24$ in. (610 mm) and a new S_t. Since the conversion factor for $S_t = 24$, 48, and 72 in. (0.61, 1.22, and 1.83 m) can be obtained from the charts, that for other values of S_t can be determined by interpolation.

If the new values of S_t are greater than 72 in. (1.83 m), Figure 2.23 based on dual wheels can be used for interpolation. In fact, Figure 2.23 is a special case of dual-tandem wheels when the tandem spacing approaches infinity. It was found that when $S_t = 120$ in. (3.05 m) the conversion factor due to dual-tandem wheels

Stresses and Strains in Flexible Pavements   Chap. 2

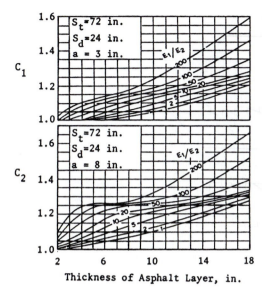

Figure 2.27 Conversion factor for dual-tandem wheels with 72-in. tandem spacing (1 in. = 25.4 mm). (After Huang (1973a).)

does not differ significantly from that due to dual wheels alone, so Figure 2.23 can be considered to have a tandem spacing of 120 in. (3.05 m).

A comparison of Figure 2.23 with Figures 2.25 through 2.27 clearly indicates that in many cases the addition of tandem wheels reduces the conversion factor, thus decreasing the critical tensile strain. This is due to the compensative effect caused by the additional wheels. The interaction among these wheels is quite unpredictable, as indicated by the irregular shape of the curves in the lower part of Figures 2.26 and 2.27.

Example 2.10:

Same as example 2.9 except that an identical set of duals is added to form dual-tandem wheels with a tamdem spacing of 49 in. (1.25 m), as shown in Figure 2.28.

Solution: Given $S_d = 11.5$ in. (292 mm) and $S_t = 49$ in. (1.25 m), modified tandem spacing = $49 \times 24/11.5 = 102.3$ in. (2.60 m). Values of a' and h' are the same as in Example 2.8. When $S_t = 72$ in. (1.83 m), $a' = 9.6$ in. (244 mm), and $h'_1 = 16.7$ in. (424 mm), from Figure 2.27, $C = 1.23 + 0.2 (9.6 - 3)(1.30 - 1.23) = 1.32$, which is smaller than the 1.5 for the dual wheels alone. With a conversion factor of 1.32 for $S_t = 72$ in. (1.83 m) and 1.50 for $S_t = 120$ in. (3.05 m), by straight-line interpolation, $C = 1.32 + (1.50 - 1.32)(102.3 - 72)/(120 - 72) = 1.43$. The strain factor due to dual-tandem wheels = $1.43 \times 0.47 = 0.672$. Critical tensile strain = $67.7 \times 0.672/150,000 = 3.03 \times 10^{-4}$, which checks closely with the 3.05×10^{-4} obtained from KENLAYER.

2.2.2 Three-Layer Systems

Figure 2.29 shows a three-layer system and the stresses at the interfaces on the axis of symmetry. These stresses include vertical stress at interface 1, σ_{z1}, vertical stress at interface 2, σ_{z2}, radial stress at bottom of layer 1, σ_{r1}, radial stress at top of layer 2, σ'_{r1}, radial stress at bottom of layer 2, σ_{r2}, and radial stress at top of layer

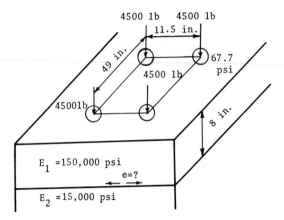

4500 lb   4500 lb

11.5 in.

49 in.

4500 lb

67.7 psi

4500 lb

8 in.

E_1 = 150,000 psi

e=?

E_2 = 15,000 psi

Figure 2.28 Example 2.10
(1 in. = 25.4 mm, 1 psi = 6.9 kPa,
1 lb = 4.45 N).

3, σ'_{r2}. Note that on the axis of symmetry, tangential and radial stresses are identical and the sheer stress is equal to 0.

When Poisson ratio is 0.5, from Eq. 2.1

$$\epsilon_z = \frac{1}{E}(\sigma_z - \sigma_r) \tag{2.20a}$$

$$\epsilon_r = \frac{1}{2E}(\sigma_r - \sigma_z) \tag{2.20b}$$

Equation 2.20 indicates that the radial strain equals one-half of the vertical strain and is opposite in sign, or

$$\epsilon_z = -2\epsilon_r \tag{2.21}$$

Equation 2.21 can be visualized physically from the fact that when a material is incompressible with a Poisson ratio of 0.5, the horizontal strain is equal to one-half of the vertical strain and the sum of ϵ_z, ϵ_r, and ϵ_t must be equal to 0.

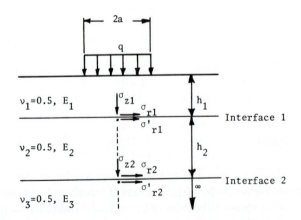

2a

q

ν_1=0.5, E_1

σ_{z1}

σ_{r1}

σ'_{r1}

h_1

Interface 1

ν_2=0.5, E_2

h_2

σ_{z2} σ_{r2}

σ'_{r2}

Interface 2

∞

ν_3=0.5, E_3

Figure 2.29 Stresses at interfaces of a three-layer system.

Jones' Tables

The stresses in a three-layer system depend on the ratios k_1, k_2, A, and H, defined as

$$k_1 = \frac{E_1}{E_2} \qquad k_2 = \frac{E_2}{E_3} \qquad (2.22a)$$

$$A = \frac{a}{h_2} \qquad H = \frac{h_1}{h_2} \qquad (2.22b)$$

Jones (1962) presented a series of tables for determining σ_{z1}, $\sigma_{z1} - \sigma_{r1}$, σ_{z2}, and $\sigma_{z2} - \sigma_{r2}$. His tables also include values of $\sigma_{z1} - \sigma_{r1}'$ at the top of layer 2 and $\sigma_{z2} - \sigma_{r2}'$ at the top of layer 3, but these tabulations are actually not necessary because they can be easily determined from those at the bottom of layers 1 and 2. The continuity of horizontal displacement at the interface implies that the radial strains at the bottom of one layer are equal to that at the top of the next layer, or from Eq. 2.20b

$$\sigma_{z1} - \sigma_{r1}' = \frac{\sigma_{z1} - \sigma_{r1}}{k_1} \qquad (2.23a)$$

$$\sigma_{z2} - \sigma_{r2}' = \frac{\sigma_{z2} - \sigma_{r2}}{k_2} \qquad (2.23b)$$

The tables presented by Jones consist of four values of k_1 and k_2, i.e., 0.2, 2, 20, and 200, so solutions for intermediate values of k_1 and k_2 can be obtained by interpolation. In view of the fact that solutions for three-layer systems can be easily obtained by KENLAYER and the interpolation from the tables is impractical and requires a large amount of time and effort, only the more realistic cases with k_1 of 2, 20, and 200 and k_2 of 2 and 20 are presented to conserve space.

Table 2.3 presents the stress factors for three-layer systems. The sign convention is positive in compression and negative in tension. Four sets of stress factors, i.e., ZZ1, ZZ2, ZZ1 − RR1, and ZZ2 − RR2, are shown. The product of the contact pressure and the stress factors gives the stresses.

$$\sigma_{z1} = q \quad (ZZ1) \qquad (2.24a)$$

$$\sigma_{z2} = q \quad (ZZ2) \qquad (2.24b)$$

$$\sigma_{z1} - \sigma_{r1} = q \quad (ZZ1 - RR1) \qquad (2.24c)$$

$$\sigma_{z2} - \sigma_{r2} = q \quad (ZZ2 - RR2) \qquad (2.24d)$$

Example 2.11:

Given the three-layer system shown in Figure 2.30 with a = 4.8 in. (122 mm), q = 120 psi (828 kPa), h_1 = 6 in. (152 mm), h_2 = 6 in. (203 mm), E_1 = 400,000 psi (2.8 GPa), E_2 = 20,000 psi (138 MPa), and E_3 = 10,000 psi (69 MPa), determine all the stresses and strains at the two interfaces on the axis of symmetry.

Solution: Given k_1 = 400,000/20,000 = 20, k_2 = 20,000/10,000 = 2, A = 4.8/6 = 0.8, and H = 6/6 = 1.0, from Table 2.3, ZZ1 = 0.12173, ZZ2 = 0.05938, ZZ1 −

TABLE 2.3 STRESS FACTORS FOR THREE-LAYER SYSTEMS

			$k_1 = 2$				$k_1 = 20$				$k_1 = 200$			
H	k_2	A	ZZ1	ZZ2	(ZZ1 − RR1)	(ZZ2 − RR2)	ZZ1	ZZ2	(ZZ1 − RR1)	(ZZ2 − RR2)	ZZ1	ZZ2	(ZZ1 − RR1)	(ZZ2 − RR2)
0.125	2	0.1	0.42950	0.00896	0.70622	0.01716	0.14529	0.00810	1.81178	0.01542	0.03481	0.00549	3.02259	0.00969
		0.2	0.78424	0.03493	0.97956	0.06647	0.38799	0.03170	3.76886	0.06003	0.11491	0.02167	8.02452	0.03812
		0.4	0.98044	0.12667	0.70970	0.23531	0.78651	0.11650	5.16717	0.21640	0.33218	0.08229	17.64175	0.14286
		0.8	0.99434	0.36932	0.22319	0.63003	1.02218	0.34941	3.43631	0.60493	0.72695	0.27307	27.27701	0.45208
		1.6	0.99364	0.72113	−0.19982	0.97707	0.99060	0.69014	1.15211	0.97146	1.00203	0.63916	23.38638	0.90861
		3.2	0.99922	0.96148	−0.28916	0.84030	0.99893	0.93487	−0.06894	0.88358	1.00828	0.92560	11.87014	0.91469
	20	0.1	0.43022	0.00228	0.69332	0.03467	0.14447	0.00182	1.80664	0.02985	0.03336	0.00128	3.17763	0.01980
		0.2	0.78414	0.00899	0.92086	0.13541	0.38469	0.00716	3.74573	0.11697	0.10928	0.00509	8.66097	0.07827
		0.4	0.97493	0.03392	0.46583	0.49523	0.77394	0.02710	5.05489	0.43263	0.31094	0.01972	20.12259	0.29887
		0.8	0.97806	0.11350	−0.66535	1.49612	0.98610	0.09061	2.92533	1.33736	0.65934	0.07045	36.29943	1.01694
		1.6	0.96921	0.31263	−2.82859	3.28512	0.93712	0.24528	−1.27093	2.99215	0.87931	0.20963	49.40857	2.64313
		3.2	0.98591	0.68433	−5.27906	5.05952	0.96330	0.55490	−7.35384	5.06489	0.93309	0.49938	57.84369	4.89895
0.25	2	0.1	0.15524	0.00710	0.28362	0.01353	0.04381	0.00530	0.63215	0.00962	0.00909	0.00259	0.96553	0.00407
		0.2	0.42809	0.02783	0.70225	0.05278	0.14282	0.02091	1.83766	0.03781	0.03269	0.01027	3.10763	0.01611
		0.4	0.77939	0.10306	0.96634	0.19178	0.37882	0.07933	3.86779	0.14159	0.10684	0.04000	8.37852	0.06221
		0.8	0.96703	0.31771	0.66885	0.55211	0.75904	0.26278	5.50796	0.44710	0.30477	0.14513	18.95534	0.21860
		1.6	0.98156	0.66753	0.17331	0.95080	0.98743	0.61673	4.24281	0.90115	0.66786	0.42940	31.18909	0.58553
		3.2	0.99840	0.93798	−0.05691	0.89390	1.00064	0.91258	1.97494	0.93254	0.98447	0.84545	28.98500	0.89191
	20	0.1	0.15436	0.00179	0.25780	0.02728	0.04236	0.00123	0.65003	0.01930	0.00776	0.00065	1.08738	0.00861
		0.2	0.42462	0.00706	0.67115	0.10710	0.13708	0.00488	1.90693	0.07623	0.02741	0.00257	3.59448	0.03421
		0.4	0.76647	0.02697	0.84462	0.39919	0.35716	0.01888	4.13976	0.29072	0.08634	0.01014	10.30923	0.13365
		0.8	0.92757	0.09285	0.21951	1.26565	0.68947	0.06741	6.48948	0.98565	0.23137	0.03844	26.41442	0.49135
		1.6	0.91393	0.26454	−1.22411	2.94860	0.85490	0.20115	6.95639	2.55231	0.46835	0.13148	57.46409	1.53833
		3.2	0.95243	0.60754	−3.04320	4.89878	0.90325	0.48647	6.05854	4.76234	0.71083	0.37342	99.29034	3.60964
0.5	2	0.1	0.04330	0.00465	0.08250	0.00878	0.01122	0.00259	0.17997	0.00440	0.00215	0.00094	0.26620	0.00128
		0.2	0.15325	0.01836	0.28318	0.03454	0.04172	0.01028	0.64779	0.01744	0.00826	0.00373	0.98772	0.00509
		0.4	0.42077	0.06974	0.70119	0.12954	0.13480	0.03998	1.89817	0.06722	0.02946	0.01474	3.19580	0.01996
		0.8	0.75683	0.23256	0.96681	0.41187	0.35175	0.14419	4.09592	0.23476	0.09508	0.05622	8.71973	0.07434
		1.6	0.93447	0.56298	0.70726	0.85930	0.70221	0.42106	6.22002	0.62046	0.27135	0.19358	20.15765	0.23838
		3.2	0.98801	0.88655	0.33878	0.96353	0.97420	0.82256	5.41828	0.93831	0.62399	0.52912	34.25229	0.54931
	20	0.1	0.04193	0.00117	0.08044	0.01778	0.00990	0.00063	0.19872	0.00911	0.00149	0.00023	0.31847	0.00257
		0.2	0.14808	0.00464	0.27574	0.07027	0.03648	0.00251	0.72264	0.03620	0.00564	0.00094	1.19598	0.01025
		0.4	0.40086	0.01799	0.67174	0.26817	0.11448	0.00988	2.19520	0.14116	0.01911	0.00372	1.02732	0.04047
		0.8	0.69098	0.06476	0.86191	0.91168	0.27934	0.03731	5.24726	0.51585	0.05574	0.01453	12.00885	0.15452
		1.6	0.79338	0.19803	0.39588	2.38377	0.50790	0.12654	10.30212	1.59341	0.13946	0.05399	32.77028	0.53836
		3.2	0.85940	0.49238	−0.41078	4.47022	0.70903	0.35807	16.38520	3.69109	0.30247	0.18091	77.62943	1.56409

1	2	0.1	0.01083	0.00241	0.02179	0.00453	0.00263	0.00100	0.04751	0.00160	0.00049	0.00029	0.06883	0.00035
		0.2	0.04176	0.00958	0.08337	0.01797	0.01029	0.00347	0.18481	0.00637	0.00195	0.00116	0.26966	0.00138
		0.4	0.14665	0.03724	0.28491	0.06934	0.03810	0.01565	0.66727	0.02498	0.00746	0.00460	1.00131	0.00545
		0.8	0.39942	0.13401	0.71341	0.24250	0.12173	0.05938	1.97428	0.09268	0.02647	0.01797	3.24971	0.02092
		1.6	0.71032	0.38690	1.02680	0.63631	0.31575	0.20098	4.37407	0.29253	0.08556	0.06671	8.92442	0.07335
		3.2	0.92112	0.75805	0.90482	0.97509	0.66041	0.53398	6.97695	0.65446	0.25186	0.22047	20.83387	0.21288
	20	0.1	0.00963	0.00061	0.02249	0.00920	0.00193	0.00024	0.05737	0.00322	0.00027	0.00007	0.08469	0.00062
		0.2	0.03697	0.00241	0.08618	0.03654	0.00751	0.00098	0.22418	0.01283	0.00104	0.00028	0.33312	0.00248
		0.4	0.12805	0.00950	0.29640	0.14241	0.02713	0.00387	0.82430	0.05063	0.00384	0.00110	1.25495	0.00985
		0.8	0.33263	0.03578	0.76292	0.51815	0.08027	0.01507	2.59672	0.19267	0.01236	0.00436	4.26100	0.03825
		1.6	0.52721	0.12007	1.25168	1.56503	0.17961	0.05549	6.77014	0.66326	0.03379	0.01683	12.91809	0.13989
		3.2	0.65530	0.33669	1.70723	3.51128	0.34355	0.18344	15.23252	1.88634	0.08859	0.06167	36.04291	0.45544
2	2	0.1	0.00250	0.00100	0.00555	0.00188	0.00059	0.00033	0.01219	0.00051	0.00011	0.00008	0.01737	0.00009
		0.2	0.00991	0.00397	0.02199	0.00750	0.00235	0.00130	0.04843	0.00203	0.00045	0.00033	0.06913	0.00036
		0.4	0.03832	0.01569	0.08465	0.02950	0.00922	0.00518	0.18857	0.00803	0.00179	0.00131	0.27103	0.00142
		0.8	0.13516	0.05974	0.29365	0.11080	0.03412	0.02023	0.68382	0.03093	0.00685	0.00520	1.00808	0.00553
		1.6	0.36644	0.20145	0.75087	0.35515	0.10918	0.07444	2.04134	0.10864	0.02441	0.02003	3.27590	0.02043
		3.2	0.67384	0.51156	1.17294	0.77434	0.29183	0.23852	4.60426	0.30709	0.08061	0.07248	9.02195	0.06638
	20	0.1	0.00181	0.00025	0.00652	0.00378	0.00033	0.00008	0.01568	0.00094	0.00005	0.00002	0.02160	0.00014
		0.2	0.00716	0.00099	0.02586	0.01507	0.00130	0.00031	0.06236	0.00374	0.00018	0.00007	0.08604	0.00058
		0.4	0.02746	0.00394	0.10017	0.05958	0.00503	0.00123	0.24425	0.01486	0.00071	0.00030	0.33866	0.00229
		0.8	0.09396	0.01535	0.35641	0.22795	0.01782	0.00485	0.90594	0.05789	0.00261	0.00119	1.27835	0.00901
		1.6	0.23065	0.05599	1.00785	0.78347	0.05012	0.01862	2.91994	0.21190	0.00819	0.00467	4.35311	0.03390
		3.2	0.37001	0.17843	2.16033	2.13215	0.11331	0.06728	7.95104	0.67732	0.02341	0.01784	13.26873	0.11666
4	2	0.1	0.00057	0.00034	0.00147	0.00065	0.00013	0.00010	0.00312	0.00015	0.00003	0.00002	0.00437	0.00002
		0.2	0.00228	0.00137	0.00587	0.00260	0.00054	0.00039	0.01245	0.00029	0.00011	0.00009	0.01746	0.00009
		0.4	0.00905	0.00544	0.02324	0.01032	0.00214	0.00154	0.04944	0.00235	0.00042	0.00036	0.06947	0.00036
		0.8	0.03500	0.02135	0.08957	0.04031	0.00837	0.00610	0.19247	0.00924	0.00168	0.00142	0.27221	0.00144
		1.6	0.12354	0.07972	0.31215	0.14735	0.03109	0.02358	0.69749	0.03488	0.00646	0.00560	1.01140	0.00553
		3.2	0.34121	0.25441	0.81908	0.43632	0.10140	0.08444	2.09049	0.11553	0.02332	0.02126	3.28913	0.01951
	20	0.1	0.00030	0.00008	0.00201	0.00128	0.00005	0.00002	0.00413	0.00025	0.00001	0.00000	0.00545	0.00003
		0.2	0.00119	0.00034	0.00803	0.00510	0.00021	0.00009	0.01651	0.00099	0.00003	0.00002	0.02178	0.00014
		0.4	0.00469	0.00134	0.03191	0.02032	0.00083	0.00035	0.06569	0.00396	0.00013	0.00008	0.08673	0.00054
		0.8	0.01790	0.00532	0.12427	0.07991	0.00321	0.00138	0.25739	0.01565	0.00050	0.00031	0.34131	0.00215
		1.6	0.06045	0.02049	0.45100	0.29991	0.01130	0.00542	0.95622	0.05993	0.00186	0.00124	1.28773	0.00833
		3.2	0.14979	0.07294	1.36427	0.97701	0.03258	0.02061	3.10980	0.20906	0.00612	0.00483	4.38974	0.03010

Source. After Jones (1962).

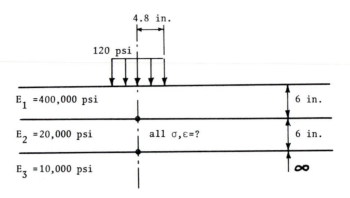

Figure 2.30 Example 2.11 (1 in. = 25.4 mm, 1 psi = 6.9 kPa).

RR1 = 1.97428, and ZZ2 − RR2 = 0.09268. From Eq. 2.24, σ_{z1} = 120 × 0.12173 = 14.61 psi (101 kPa), σ_{z2} = 120 × 0.05938 = 7.12 psi (49.1 kPa), $\sigma_{z1} − \sigma_{r1}$ = 120 × 1.97428 = 236.91 psi (1.63 MPa), and $\sigma_{z2} − \sigma_{r2}$ = 120 × 0.09268 = 11.12 psi (76.7 kPa). From Eq. 2.23, $\sigma_{z1} − \sigma'_{r1}$ = 236.91/20 = 11.85 psi (81.8 kPa) and $\sigma_{z2} − \sigma'_{r2}$ = 11.12/2 = 5.56 psi (38.4 kPa). At bottom of layer 1: σ_{r1} = 14.61 − 236.91 = −222.3 psi (−1.53 MPa), from Eq. 2.20, ϵ_z = 236.91/400,000 = 5.92 × 10^{-4} and ϵ_r = −2.96 × 10^{-4}. At top of layer 2: σ'_{r1} = 14.61 − 11.85 = 2.76 psi (19.0 kPa), ϵ_z = 11.85/20,000 = 5.92 × 10^{-4}, and σ_r = −2.96 × 10^{-4}. At bottom of layer 2: σ_{r2} = 7.12 − 11.12 = −4.0 psi (−28 kPa), ϵ_z = 11.12/20,000 = 5.56 × 10^{-4}, and ϵ_r = −2.78 × 10^{-4}. At top of layer 3: σ'_{r2} = 7.12 − 5.56 = 1.56 psi (10.8 kPa), ϵ_z = 5.56/10,000 = 5.56 × 10^{-4} and ϵ_r = − 2.78 × 10^{-4}.

In the above example, the parameters k_1, k_2, A, and H are exactly the same as those shown in the table, so no interpolation is needed. Because each interpolation requires three points, the interpolation of only one parameter requires at least three times the effort. If all four parameters are different from those in the table, the total effort required will be 3 × 3 × 3 × 3, or 81 times.

Peattie's Charts

Peattie (1962) plotted Jones' table in graphical forms. Figure 2.31 shows one set of charts for radial strain factors, (RR1 − ZZ1)/2, at the bottom of layer 1. As indicated by Eq. 2.20b, the radial strain can be determined by

$$\epsilon_r = \frac{q}{E}\left(\frac{RR1 − ZZ1}{2}\right) \tag{2.25}$$

The radial strains at the bottom of layer 1 should be in tension. Although the solutions obtained from the charts are not as accurate as those from the table, the chart has the advantage that interpolation for A and H can be easily done. However, interpolation for k_1 and k_2 is still cumbersome.

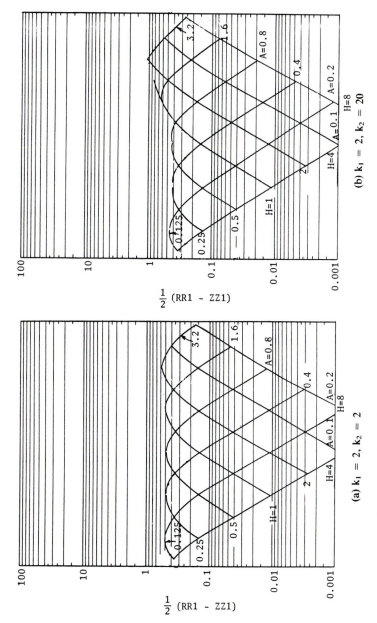

(a) $k_1 = 2$, $k_2 = 2$

(b) $k_1 = 2$, $k_2 = 20$

Figure 2.31 Charts for horizontal strain factors at bottom of layer 1. (After Peattie (1962).)

79

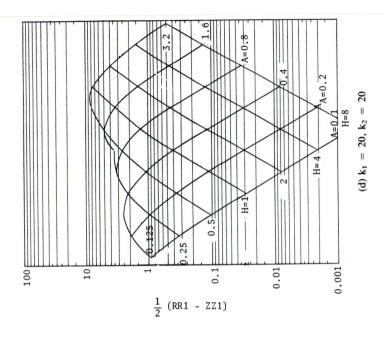

(d) $k_1 = 20$, $k_2 = 20$

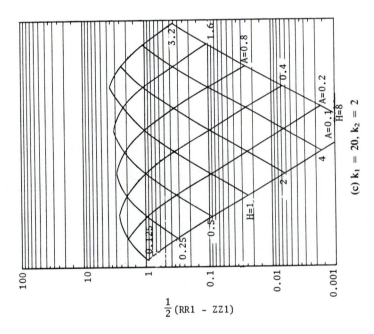

(c) $k_1 = 20$, $k_2 = 2$

Figure 2.31 (Continued)

80

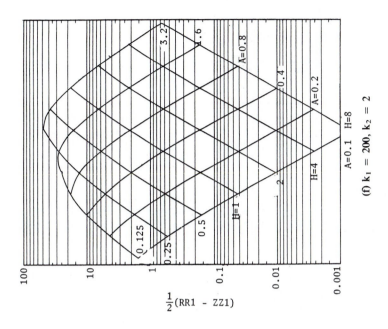

(f) $k_1 = 200$, $k_2 = 2$

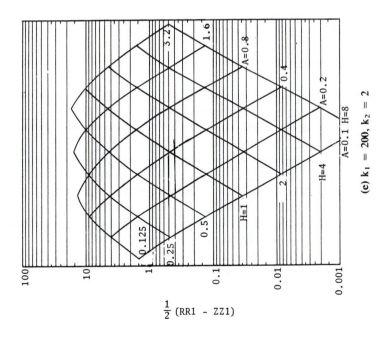

(e) $k_1 = 200$, $k_2 = 2$

Figure 2.31 (Continued)

81

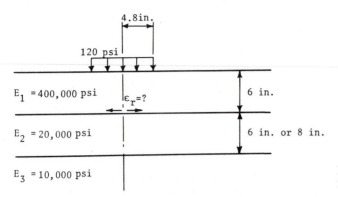

$E_1 = 400,000$ psi

$\epsilon_r = ?$

$E_2 = 20,000$ psi

$E_3 = 10,000$ psi

120 psi

4.8 in.

6 in.

6 in. or 8 in.

Figure 2.32 Example 2.12 (1 in. = 25.4 mm, 1 psi = 6.9 kPa).

Example 2.12:

For the same case as Example 2.11, determine the radial strain at the bottom of layer 1, as shown in Figure 2.32. If $h_2 = 8$ in. (203 mm), what is the radial strain at the bottom of layer 1?

Solution: Given $k_1 = 20$, $k_2 = 2$, $A = 0.8$, and $H = 1.0$, from Figure 2.31c, (RR1 − ZZ1)/2 = 1. From Eq. 2.25, $\epsilon_r = 120/400,000 = 3 \times 10^{-4}$ (tension), which checks closely with the 2.96×10^{-4} from the table. Given $h_2 = 8$ in. (203 mm), $A = 4.8/8 = 0.6$, and $H = 6/8 = 0.75$, from Figure 2.31c, the strain factor is still close to 1, indicating that the thickness of layer 2 has very little effect on the tensile strain due to the predominant effect of layer 1. The radial strain obtained from KENLAYER is 2.91×10^{-4}.

2.3 VISCOELASTIC SOLUTIONS

A viscoelastic material possesses both the elastic property of a solid and the viscous behavior of a liquid. Suppose that a material is formed into a ball. If the ball is thrown on the floor and rebounds, it is said to be elastic. If the ball is left on the table and begins to flow and flatten gradually under its own weight, it is said to be viscous. Due to the viscous component, the behavior of viscoelastic materials is time dependent and the longer the time the more the flow. As HMA is a viscoelastic material whose behavior depends on the time of loading, it is natural to apply the theory of viscoelasticity to the analysis of layered systems. The general procedure is based on the elastic–viscoelastic correspondence principle by applying the Laplace transform to remove the time variable t with a transformed variable p, thus changing a viscoelastic problem to an associated elastic problem. The Laplace inversion of the associated elastic problem from the transformed variable p to the time variable t results in the viscoelastic solutions. Details about the theory of viscoelasticity are presented in Appendix A. A simple collocation method to obtain the viscoelastic solutions from the elastic solutions is presented in this section.

2.3.1 Material Characterization

There are two general methods to characterize viscoelastic materials: one by a mechanical model and the other by a creep compliance curve. The latter is used in KENLAYER due to its simplicity. Because Poisson ratio v has a relatively small effect on pavement behavior, it is assumed to be elastic independent of time. Therefore, only modulus E is considered to be viscoelastic and time dependent.

Mechanical Models

Figure 2.33 shows various mechanical models for characterizing viscoelastic materials. The models are formed of two basic elements: a spring and a dashpot.

Basic Models

An elastic material is characterized by a spring, as indicated in Figure 2.33a, and obeys Hooke's law, whereby stress is proportional to strain:

$$\sigma = E\epsilon \tag{2.26}$$

in which σ is stress, ϵ is strain, and E is the elastic modulus.

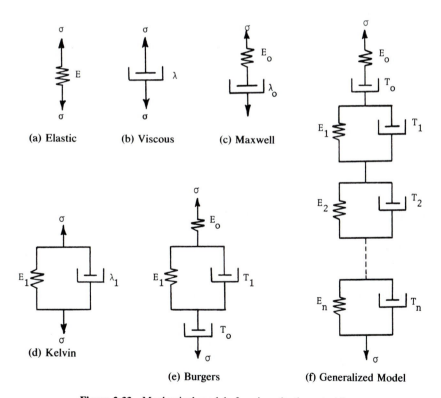

(a) Elastic   (b) Viscous   (c) Maxwell

(d) Kelvin   (e) Burgers   (f) Generalized Model

Figure 2.33 Mechanical models for viscoelastic materials.

A viscous material is characterized by a dashpot, as indicated in Figure 2.33b, and obeys Newton's law, whereby stress is proportional to the time rate of strain:

$$\sigma = \lambda \frac{\partial \epsilon}{\partial t} \tag{2.27}$$

in which λ is viscosity and t is time. Under a constant stress, Eq. 2.27 can be easily integrated and becomes

$$\epsilon = \frac{\sigma t}{\lambda} \tag{2.28}$$

Maxwell Model

A Maxwell model is a combination of spring and dashpot in series, as indicated in Figure 2.33c. Under a constant stress, the total strain is the sum of the strains of both spring and dashpot, or from Eqs. 2.26 and 2.28

$$\epsilon = \frac{\sigma}{E_0} + \frac{\sigma t}{\lambda_0} = \frac{\sigma}{E_0}\left(1 + \frac{t}{T_0}\right) \tag{2.29}$$

in which $T_0 = \lambda_0/E_0$ = relaxation time. A subscript 0 is used to indicate a Maxwell model. If a stress σ_0 is applied instantaneously to the model, the spring will experience an instantaneous strain, σ_0/E_0. If this strain is kept constant, the stress will gradually relax and, after a long period of time, will become zero. This can be shown by solving the differential equation

$$\frac{\partial \epsilon}{\partial t} = \frac{1}{E_0}\frac{\partial \sigma}{\partial t} + \frac{\sigma}{\lambda_0} \tag{2.30}$$

The first term on the right side of Eq. 2.30 is the rate of strain due to the spring and the second term due to the dashpot. If strain is kept constant, $\partial \epsilon/\partial t = 0$, or after integration

$$\sigma = \sigma_0 \exp\left(-\frac{t}{T_0}\right) \tag{2.31}$$

It can be seen from Eq. 2.31 that when $t = 0$, then $\sigma = \sigma_0$; when $t = \infty$, then $\sigma = 0$; and when $t = T_0$, then $\sigma = 0.368\,\sigma_0$. Consequently, the relaxation time T_0 of a Maxwell model is the time required for the stress to reduce to 36.8% of the original value. It is more convenient to specify relaxation time than viscosity, because of its physical meaning. A relaxation time of 10 min gives an idea that the stress will relax to 36.8% of the original value in 10 min.

Kelvin Model

A Kelvin model is a combination of spring and dashpot in parallel, as indicated in Figure 2.33d. Both the spring and the dashpot have the same strain but the total stress is the sum of the two stresses, or, using subscript 1 to indicate a Kelvin model,

$$\sigma = E_1\epsilon + \lambda_1 \frac{\partial \epsilon}{\partial t}$$

If a constant stress is applied, then

$$\int_0^\epsilon \frac{d\epsilon}{\sigma - E_1\epsilon} = \int_0^t \frac{dt}{\lambda_1}$$

or

$$\epsilon = \frac{\sigma}{E_1}\left[1 - \exp\left(-\frac{t}{T_1}\right)\right] \qquad (2.32)$$

in which $T_1 = \lambda_1/E_1$ = retardation time. It can be seen from Eq. 2.32 that when $t = 0$, then $\epsilon = 0$; when $t = \infty$, then $\epsilon = \sigma/E_1$, or the spring is fully stretched to its total retarded strain; and when $t = T_1$, then $\epsilon = 0.632\sigma/E_1$. Thus, the retardation time T_1 of a Kelvin model is the time to reach 63.2% of the total retarded strain.

Burgers Model

A Burgers model is a combination of Maxwell and Kelvin models in series, as indicated in Figure 2.33e. Under a constant stress, from Eqs. 2.29 and 2.32,

$$\epsilon = \frac{\sigma}{E_0}\left(1 + \frac{t}{T_0}\right) + \frac{\sigma}{E_1}\left[1 - \exp\left(-\frac{t}{T_1}\right)\right] \qquad (2.33)$$

The total strain is composed of three parts: an instantaneous elastic strain, a viscous strain, and a retarded elastic strain, as shown in Figure 2.34. Qualitatively, a Burgers model well represents the behavior of a viscoelastic material. Quantitatively, a single Kelvin model is usually not sufficient to cover the long period of time over which the retarded strain takes place and a number of Kelvin models may be needed.

Generalized Model

Figure 2.33f shows a generalized model which can be used to characterize any viscoelastic materials. Under a constant stress, the strain of a generalized model can be written as

$$\epsilon = \frac{\sigma}{E_0}\left(1 + \frac{t}{T_0}\right) + \sum_{i=1}^{n} \frac{\sigma}{E_i}\left[1 - \exp\left(-\frac{t}{T_i}\right)\right] \qquad (2.34)$$

in which n is the number of Kelvin models. This model explains the effect of load duration on pavement responses. Under a single load application, the in-

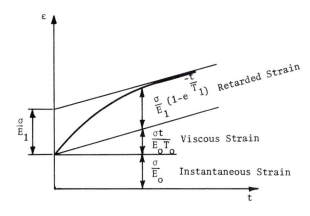

Figure 2.34 Three components of strain for a Burgers model.

stantaneous and the retarded elastic strains predominate, while the viscous strain is negligible. However, under a large number of load repetitions, the accumulation of viscous strains is the cause of permanent deformation.

Creep Compliance

Another method to characterize viscoelastic materials is the creep compliance at various times, $D(t)$, defined as

$$D(t) = \frac{\epsilon(t)}{\sigma} \tag{2.35}$$

in which $\epsilon(t)$ is the time-dependent strain under a constant stress.

Under a constant stress, the creep compliance is the reciprocal of Young's modulus. For the generalized model, the creep compliance can be expressed as

$$D(t) = \frac{1}{E_0}\left(1 + \frac{t}{T_0}\right) + \sum_{i=1}^{n}\frac{1}{E_i}\left[1 - \exp\left(-\frac{t}{T_i}\right)\right] \tag{2.36}$$

Given the various viscoelastic constants, E_0, T_0, E_i, and T_i, for a generalized model, the creep compliances at various times can be computed from Eq. 2.36.

Example 2.13:

A viscoelastic material is characterized by one Maxwell model and three Kelvin models connected in series with the viscoelastic constants shown in Figure 2.35a. Determine the creep compliance at various times and plot the creep compliance curve.

Solution: In Figure 2.35a, no units are given for the viscoelastic constants. If E is in lb/in.^{2}, then the creep compliance is in in.^{2}/lb. If E is in kN/m^2, then the creep compliance is in m^2/kN. If T is in seconds, then the actual time t is also in seconds.

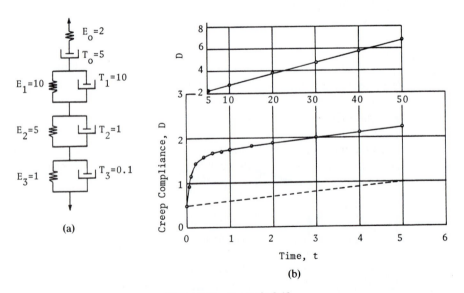

Figure 2.35 Example 2.13.

Stresses and Strains in Flexible Pavements   Chap. 2

TABLE 2.4 CREEP COMPLIANCE AT
VARIOUS TIMES

Time	Creep compliance	Time	Creep compliance
0	0.500	2	1.891
0.05	0.909	3	2.016
0.1	1.162	4	2.129
0.2	1.423	5	2.238
0.4	1.592	10	2.763
0.6	1.654	20	3.786
0.8	1.697	30	4.795
1.0	1.736	40	5.798
1.5	1.819	50	6.799

From Eq. 2.36, when $t = 0$, $D = 1/E_0 = \frac{1}{2} = 0.5$; and when $t = 0.1$, $D = 0.5$ $(1 + 0.1/5) + 0.1 (1 - e^{-0.01}) + 0.2 (1 - e^{-0.1}) + (1 - e^{-1}) = 1.162$. The creep compliances at various times are tabulated in Table 2.4 and plotted in Figure 2.35b. It can be seen that after $t = 5$ all the retarded strains have nearly completed and only the viscous strains exist, as indicated by a straight line. If the retarded strain lasts much longer, more Kelvin models with longer retardation times will be needed.

If a creep compliance curve is given, the viscoelastic constants of a generalized model can be determined by the method of successive residuals, as described in Appendix A. However, it is more convenient to use an approximate method of collocation, as described below.

2.3.2 Collocation Method

The collocation method is an approximate method to collocate the computed and actual responses at a predetermined number of time durations. Instead of determining both E_i and T_i by the tedious method of successive residuals, several values of T_i are arbitrarily assumed and the corresponding E_i values are determined by solving a system of simultaneous equations. The method can also be used to obtain the viscoelastic solutions from the elastic solutions.

Elastic Solutions

Given the creep compliance of each viscoelastic material at a given time, the viscoelastic solutions at that time can be easily obtained from the elastic solutions, as illustrated by the following example.

Example 2.14:

Figure 2.36 shows a viscoelastic two-layer system under a circular loaded area with a radius of 10 in. (254 mm) and a uniform pressure of 100 psi (690 kPa). The thickness of layer 1 is 10 in. (254 mm) and both layers are incompressible with a Poisson ratio of 0.5. The creep compliances of the two materials at five different times are tabulated in Table 2.5. Determine the surface deflection under the center of the loaded area at the given times.

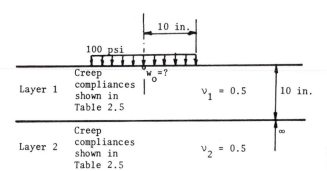

Figure 2.36 Example 2.14
(1 in. = 25.4 mm, 1 psi = 6.9 kPa).

TABLE 2.5 CREEP COMPLIANCES AND SURFACE DEFLECTIONS

Time (s)	0.01	0.1	1	10	100
Layer 1 $D(t)$ $(10^{-6}/\text{psi})$	1.021	1.205	2.683	9.273	18.320
Layer 2 $D(t)$ $(10^{-6}/\text{psi})$	1.052	7.316	19.520	73.210	110.000
Deflection w_0 (in.)	0.0016	0.0064	0.016	0.059	9.096

Note. 1 psi = 6.9 kPa, 1 in. = 25.4 mm.

Solution: If the modulus ratio is greater than 1, the surface deflection w_0 at any given time can be determined from Figure 2.17. Take $t = 1$ s, for example. The elastic modulus is the reciprocal of creep compliance. For layer 1, $E_1 = 1/(2.683 \times 10^{-6}) = 3.727 \times 10^5$ psi (2.57 GPa) and for layer 2, $E_2 = 1/(19.52 \times 10^{-6}) = 5.123 \times 10^4$ psi (353 MPa), so $E_1/E_2 = 3.727 \times 10^5/(5.123 \times 10^4) = 7.27$. From Figure 2.17, $F_2 = 0.54$, so $w_0 = 1.5 \times 100 \times 10 \times 0.54/(5.123 \times 10^4) = 0.016$ in. (4.1 mm). The same procedure can be applied to other time durations and the results are shown in Table 2.5.

It should be noted that the above procedure is not the exact viscoelastic solution. It is a quasi-elastic solution which provides a close approximation to the viscoelastic solution.

Dirichlet Series

Pavement design is based on moving loads with a short duration. The creep compliance $D(t)$ caused by the viscous strain is negligible, so Eq. 2.36 can be written as

$$D(t) = \frac{1}{E_0} + \sum_{i=1}^{n} \frac{1}{E_i}\left[1 - \exp\left(-\frac{t}{T_i}\right)\right] \tag{2.37}$$

It is therefore convenient to express the creep compliance as a Dirichlet series, or

$$D(t) = \sum_{i=1}^{n} G_i \exp\left(-\frac{t}{T_i}\right) \tag{2.38}$$

A comparison of Eqs. 2.37 and 2.38 with $T_n = \infty$ shows that

$$G_i = -\frac{1}{E_i} \qquad (2.39a)$$

$$G_n = \frac{1}{E_0} + \sum_{i=1}^{n} \frac{1}{E_i} \qquad (2.39b)$$

In KENLAYER, the collocation method is applied at two occasions. First, the creep compliances at a reference temperature are specified at a number of time durations and fitted with a Dirichlet series, so that the compliances at any other temperature can be obtained by the time–temperaure superposition principle. Second, the elastic solutions obtained at these durations are fitted with a Dirichlet series to be used later for analyzing moving loads.

Collocation of Creep Compliances

The creep compliances of viscoelastic materials are determined from creep tests. A 1000-s creep test with compliances measured at 11 different time durations of 0.001, 0.003, 0.01, 0.03, 0.1, 0.3, 1, 3, 10, 30, and 100 s is recommended (FHWA, 1978), although any number of time durations may be used.

Because moving loads usually have a very short duration, retardation times T_i of 0.01, 0.03, 0.1, 1, 10, 30, and ∞ seconds are specified in KENLAYER. If creep compliances are specified at seven durations, the coefficients G_1 through G_7 can be determined from Eq. 2.38 by solving 7 simultaneous equations. If the creep compliances are specified at 11 time durations, there are 11 equations but 7 unknowns, so the 11 equations must be reduced to 7 equations by multiplying both sides with a 7×11 matrix, which is the transpose of the 11×7 matrix, or

$$
\begin{bmatrix} \exp\left(-\frac{t_1}{T_1}\right) & \cdots & \exp\left(-\frac{t_{11}}{T_1}\right) \\ \cdot & & \cdot \\ \cdot & 7 \times 11 \text{ matrix} & \cdot \\ \cdot & & \cdot \\ \exp\left(-\frac{t_1}{T_7}\right) & \cdots & \exp\left(-\frac{t_{11}}{T_7}\right) \end{bmatrix}
\begin{bmatrix} \exp\left(-\frac{t_1}{T_1}\right) & \cdots & \exp\left(-\frac{t_1}{T_7}\right) \\ \cdot & & \cdot \\ \cdot & 11 \times 7 \text{ matrix} & \cdot \\ \cdot & & \cdot \\ \exp\left(-\frac{t_{11}}{T_1}\right) & \cdots & \exp\left(-\frac{t_{11}}{T_7}\right) \end{bmatrix}
\begin{Bmatrix} G_1 \\ \cdot \\ \cdot \\ \cdot \\ G_7 \end{Bmatrix}
$$

$$
= \begin{bmatrix} \exp\left(-\frac{t_1}{T_1}\right) & \cdots & \exp\left(-\frac{t_{11}}{T_1}\right) \\ \cdot & & \cdot \\ \cdot & 7 \times 11 \text{ matrix} & \cdot \\ \cdot & & \cdot \\ \exp\left(-\frac{t_1}{T_7}\right) & \cdots & \exp\left(-\frac{t_{11}}{T_7}\right) \end{bmatrix}
\begin{Bmatrix} D_1 \\ \cdot \\ \cdot \\ \cdot \\ D_{11} \end{Bmatrix} \qquad (2.40)
$$

After coefficients G_1 through G_7 are obtained, the creep compliance at any time t can be computed by Eq. 2.38.

Example 2.15:

It is assumed that the creep compliance of a viscoelastic material be represented by

$$D(t) = G_1 \exp(-10t) + G_2 \tag{2.41}$$

If the creep compliances at $t = 0.01$, 0.07, and 0.4 s are 9.516×10^{-5}, 5.034×10^{-4} and 9.817×10^{-4} in.^{2}/lb (13.8, 72.9, and 142.3 mm^2/kN), respectively, determine the coefficients G_1 and G_2.

Solution: With $t_1 = 0.01$, $t_2 = 0.07$, $t_3 = 0.4$, $T_1 = 0.1$, $T_2 = \infty$, $D_1 = 9.516 \times 10^{-5}$, $D_2 = 5.034 \times 10^{-4}$, and $D_3 = 9.817 \times 10^{-4}$. From Eq. 2.40

$$\begin{bmatrix} e^{-0.1} & e^{-0.7} & e^{-4} \\ 1 & 1 & 1 \end{bmatrix} \begin{bmatrix} e^{-0.1} & 1 \\ e^{-0.7} & 1 \\ e^{-4} & 1 \end{bmatrix} \begin{Bmatrix} G_1 \\ G_2 \end{Bmatrix} = \begin{bmatrix} e^{-0.1} & e^{-0.7} & e^{-4} \\ 1 & 1 & 1 \end{bmatrix} \begin{Bmatrix} 9.516 \times 10^{-5} \\ 5.034 \times 10^{-4} \\ 9.817 \times 10^{-4} \end{Bmatrix}$$

or

$$\begin{bmatrix} 1.066 & 1.420 \\ 1.420 & 3.000 \end{bmatrix} \begin{Bmatrix} G_1 \\ G_2 \end{Bmatrix} = \begin{Bmatrix} 3.541 \times 10^{-4} \\ 1.580 \times 10^{-3} \end{Bmatrix} \tag{2.42}$$

The solution of Eq. 2.42 is $G_1 = -0.001$ in.^{2}/lb (-145 mm^2/kN) and $G_2 = 0.001$ in.^{2}/lb (145 mm^2/kN), which is as expected because the given creep compliances are actually computed from a Kelvin model with

$$D(t) = 0.001 \ (1 - e^{-10t}) \tag{2.43}$$

Time–Temperature Superposition

It has been demonstrated that asphalt mixes subjected to a temperature increase experience an accelerated deformation as if the time scale were compressed. Figure 2.37 shows the plot of creep compliance D versus time t on log scales. At a given time, the creep compliance at a lower temperature is smaller

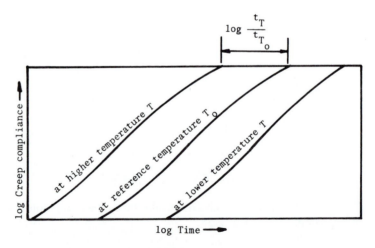

Figure 2.37 Creep compliance at different temperatures.

Stresses and Strains in Flexible Pavements   Chap. 2

than that at a higher temperature. There is a parallel shift between the curves at various temperatures.

If the creep compliances under a reference temperature T_0 are known, those under any given temperature T can be obtained by using a time–temperature shift factor a_T, defined as (Pagen, 1965)

$$a_T = \frac{t_T}{t_{T_0}} \tag{2.44}$$

in which t_T is the time to obtain a creep compliance at temperature T, and t_{T_0} is the time to obtain a creep compliance at reference temperature T_0.

Various laboratory tests on asphalt mixes have shown that a plot of log a_T versus temperature results in a straight line, as shown in Figure 2.38. The slope of the straight line β varies from 0.061 to 0.170, with an average about 0.113 (FHWA, 1978). From Figure 2.38

$$\beta = \frac{\log(t_T/t_{T_0})}{T - T_0} \tag{2.45}$$

or
$$t_T = t_{T_0} \exp[2.3026\beta(T - T_0)] \tag{2.46}$$

If the creep compliance based on the reference temperature T_0 is

$$D(t) = \sum_{i=1}^{n} G_i \exp\left(-\frac{t_{T_0}}{T_i}\right) \tag{2.47}$$

then the creep compliance based on temperature T is

$$D(t) = \sum_{i=1}^{n} G_i \exp\left(-\frac{t_T}{T_i}\right) \tag{2.48}$$

The relationship between t_T and t_{T_0} is indicated by Eq. 2.46.

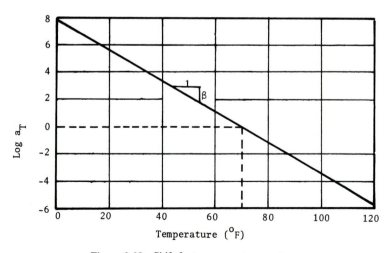

Figure 2.38 Shift factor versus temperature.

Example 2.16:

The expression for the creep compliance at a temperature of 70°F (21.1°C) is represented by Eq. 2.43. What is the expression for creep compliance at a temperature of 50°F (10°C) if the time–temperature shift factor β is 0.113?

Solution: From Eq. 2.46, $t_T = t_{T_0} \exp[2.3026 \times 0.113 \times (50 - 70)] = 0.0055 t_{T_0}$. From Eq. 2.43, $D(t) = 0.001[1 - \exp(-0.055t)]$. It can be seen that the creep compliance at 50°F (10°C) is much smaller than that at 70°F (21.1°C).

Collocation for Viscoelastic Solutions

Even though the exact viscoelastic solutions is not known, the viscoelastic response R can always be expressed approximately as a Dirichlet series:

$$R = \sum_{i=1}^{7} c_i \exp\left(-\frac{t}{T_i}\right) \tag{2.49}$$

If elastic solutions at 11 time durations are obtained, Eq. 2.40 can be applied to reduce the number of equations to seven, which is the number of unknowns to be solved. If the responses at seven time durations are obtained from the elastic solutions, the coefficients c_1 through c_7 can be solved directly by

$$
\begin{bmatrix}
e^{-\frac{0.01}{0.01}} & e^{-\frac{0.01}{0.03}} & e^{-\frac{0.01}{0.1}} & e^{-\frac{0.01}{1}} & e^{-\frac{0.01}{10}} & e^{-\frac{0.01}{30}} & 1 \\
e^{-\frac{0.03}{0.01}} & e^{-\frac{0.03}{0.03}} & e^{-\frac{0.03}{0.1}} & e^{-\frac{0.03}{1}} & e^{-\frac{0.03}{10}} & e^{-\frac{0.03}{30}} & 1 \\
e^{-\frac{0.1}{0.01}} & e^{-\frac{0.1}{0.03}} & e^{-\frac{0.1}{0.1}} & e^{-\frac{0.1}{1}} & e^{-\frac{0.1}{10}} & e^{-\frac{0.1}{30}} & 1 \\
e^{-\frac{1}{0.01}} & e^{-\frac{1}{0.03}} & e^{-\frac{1}{0.1}} & e^{-\frac{1}{1}} & e^{-\frac{1}{10}} & e^{-\frac{1}{30}} & 1 \\
e^{-\frac{10}{0.01}} & e^{-\frac{10}{0.03}} & e^{-\frac{10}{0.1}} & e^{-\frac{10}{1}} & e^{-\frac{10}{10}} & e^{-\frac{10}{30}} & 1 \\
e^{-\frac{30}{0.01}} & e^{-\frac{30}{0.03}} & e^{-\frac{30}{0.1}} & e^{-\frac{30}{1}} & e^{-\frac{30}{10}} & e^{-\frac{30}{30}} & 1 \\
1 & 1 & 1 & 1 & 1 & 1 & 1
\end{bmatrix}
\begin{Bmatrix}
c_1 \\ c_2 \\ c_3 \\ c_4 \\ c_5 \\ c_6 \\ c_7
\end{Bmatrix}
=
\begin{Bmatrix}
(R)_{0.01} \\ (R)_{0.03} \\ (R)_{0.1} \\ (R)_{1} \\ (R)_{10} \\ (R)_{30} \\ (R)_{\infty}
\end{Bmatrix}
\tag{2.50}
$$

After the coefficients c_i are obtained, the response under a stationary load can be determined by Eq. 2.49.

Example 2.17:

The creep compliance of a homogeneous half-space is expressed as a Dirichlet series shown by Eq. 2.41 with $G_1 = -0.001$ in.^{2}/lb (-145 mm^2/kN), $G_2 = 0.001$ in.^{2}/lb (145 mm^2/kN), $T_1 = 0.1$ s, and $T_2 = \infty$. Assuming that the half-space has a Poisson ratio of 0.5 and is subjected to a circular load with a contact radius of 6 in. (152 mm) and a contact pressure of 80 psi (552 kPa), as shown in Figure 2.39, determine the maximum surface deflection after a loading time of 0.1 s by the collocation method.

Solution: The maximum deflection occurs under the center of the loaded area. With $\nu = 0.5$, from Eq. 2.8

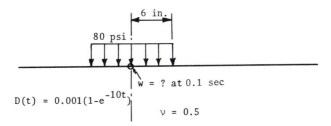

6 in.

80 psi

w = ? at 0.1 sec

$D(t) = 0.001(1-e^{-10t})$

$\nu = 0.5$

Figure 2.39   Example 2.15
(1 in. = 25.4 mm, 1 psi = 6.9 kPa).

$$w_0 = \frac{1.5qa}{E} = 1.5qaD(t) \qquad (2.51)$$

Substituting Eq. 2.43 and the values of q and a into Eq. 2.51 yields

$$w_0 = 0.72(1 - e^{-10t}) \qquad (2.52)$$

When $t = 0.1$, from Eq. 2.52, then $w_0 = 0.455$ in. (11.6 mm).

The above solution is simple and straightforward. However, to illustrate the collocation method, it is assumed that the surface deflections be expressed as a Dirichlet series as shown by Eq. 2.49. The elastic response on the right side of Eq. 2.53 is obtained from Eq. 2.52. From Eq. 2.50

$$\begin{bmatrix} 0.368 & 0.717 & 0.905 & 0.990 & 0.999 & 1.000 & 1 \\ 0.050 & 0.368 & 0.741 & 0.970 & 0.997 & 0.999 & 1 \\ 0.000 & 0.036 & 0.368 & 0.905 & 0.990 & 0.997 & 1 \\ 0.000 & 0.000 & 0.000 & 0.368 & 0.905 & 0.967 & 1 \\ 0.000 & 0.000 & 0.030 & 0.000 & 0.368 & 0.717 & 1 \\ 0.000 & 0.000 & 0.000 & 0.000 & 0.050 & 0.368 & 1 \\ 1 & 1 & 1 & 1 & 1 & 1 & 1 \end{bmatrix} \begin{Bmatrix} c_1 \\ c_2 \\ c_3 \\ c_4 \\ c_5 \\ c_6 \\ c_7 \end{Bmatrix} = \begin{Bmatrix} 0.069 \\ 0.187 \\ 0.455 \\ 0.720 \\ 0.720 \\ 0.720 \\ 0.720 \end{Bmatrix} \qquad (2.53)$$

The solution of Eq. 2.53 is $c_1 = 2.186$, $c_2 = -3.260$, $c_3 = 2.214$, $c_4 = -2.055$, $c_5 = 2.446$, $c_6 = -2.229$, and $c_7 = 1.418$. From Eq. 2.49, the surface deflection can be expressed as

$$w_0 = 2.186e^{-t/0.01} - 3.260e^{-t/0.03} + 2.214e^{-t/0.1} - 2.055e^{-t}$$
$$+ 2.446e^{-t/10} - 2.229e^{-t/30} + 1.418 \qquad (2.54)$$

When $t = 0.1$, then $w_0 = 0 - 0.116 + 0.814 - 1.859 + 2.422 - 2.222 + 1.418 = 0.457$ in. (11.6 mm), which checks with the exact solution of 0.455 in. (11.6 mm).

2.3.3 Analysis of Moving Loads

The elastic–viscoelastic correspondence principle can be applied directly to moving loads, as indicated by Perloff and Moavenzadeh (1967) for determining the surface deflection of a viscoelastic half-space, by Chou and Larew (1969) for the stresses and displacements in a viscoelastic two-layer system, by Elliott and Moavenzadeh (1971) in a three-layer system, and by Huang (1973b) in a multilayer

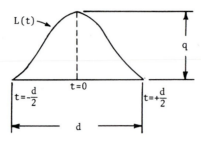

L(t)

q

t=0

$t = -\dfrac{d}{2}$

$t = +\dfrac{d}{2}$

d

Figure 2.40   Moving load as a function of time.

system. Due to the complexities of the analysis and the large amount of computer time required, these methods are not suited for practical use. Therefore, a simplified method has been used in both VESYS and KENLAYER.

In this method, it is assumed that the intensity of load varies with time according to a haversine function as shown in Figure 2.40. With $t = 0$ at the peak, the load function is expressed as

$$L(t) = q \sin^2\left(\frac{\pi}{2} + \frac{\pi t}{d}\right) \tag{2.55}$$

in which d is the duration of load. When the load is at a considerable distance from a given point, or $t = \pm d/2$, the load above the point is zero, or $L(t) = 0$. When the load is directly above the given point, or $t = 0$, the load intensity is q.

The duration of load depends on the vehicle speed s and the tire contact radius a. A reasonable assumption is that the load has practically no effect when it is at a distance of $6a$ from the point, or

$$d = \frac{12a}{s} \tag{2.56}$$

If $a = 6$ in. and $s = 40$ mph (64 km/h) $= 58.7$ ft/s (17.9 m/s), $d = 0.1$ s.

The response under static load can be expressed as a Dirichlet series:

$$R(t) = \sum_{i=1}^{7} c_i \exp\left(-\frac{t}{T_i}\right) \tag{2.49}$$

The response under moving load can be obtained by Boltzmann's superposition principle:

$$R = \int_{-d/2}^{0} R(t)\frac{dL}{dt}\,dt \tag{2.57}$$

From Eq. 2.55

$$\frac{dL}{dt} = -\frac{q\pi}{d} \sin\left(\frac{2\pi t}{d}\right) \tag{2.58}$$

Substituting Eqs. 2.49 and 2.58 into Eq. 2.57 and integrating yields

$$R = \frac{q\pi^2}{2} \sum_{i=1}^{n} c_i \frac{1 + \exp\left(-d/2T_i\right)}{\pi^2 + (d/2T_i)^2} \tag{2.59}$$

Example 2.18:

Same as the problem in Example 2.17, but the load is moving at a speed of 40 mph (64 km/h). Determine the maximum deflection.

Solution: According to Eq. 2.52 in Example 2.17, the surface deflection under a static load can be expressed as

$$w = 0.72 (1 - e^{-10r}) \qquad (2.52)$$

The first term is independent of time and therefore remains the same whether the load is moving or not. From Eq. 2.59, the second term with $T = 0.1$ and $d = 0.1$ s for 40 mph (64 km/h) should be changed to $0.5 \times \pi^2 \times 0.72 (1 + e^{-0.5})/(\pi^2 + 0.25) = 0.564$ in. (14.3 mm), so maximum deflection $= 0.72 - 0.564 = 0.156$ in. (3.96 mm).

2.4 SUMMARY

This chapter discusses the stresses and strains in flexible pavements and their determinations. An understanding of this subject is indispensable for any mechanistic methods of design.

Important Points Discussed in Chapter 2

1. Boussinesq theory can be applied only to an elastic homogeneous half-space, such as the analysis of plate bearing test on a subgrade or a wheel load on a thin pavement.

2. An approximate method to determine the deflection on the surface of a nonlinear elastic half-space, in which the elastic modulus varies with the state of stresses, is to assume the same stress distribution as in the linear theory but vary the moduli according to the state of stresses.

3. The most practical mechanistic method for analyzing flexible pavements is Burmister's layered theory. Based on two-layer elastic systems, various charts were developed for determining pavement responses. The vertical interface stress beneath the center of a circular loaded area can be determined from Figure 2.15 and the vertical interface deflection at various radial distances from Figure 2.19. The critical tensile strain at the bottom of layer 1 under a single wheel can be determined from Figure 2.21, under dual wheels from Figure 2.23, and under dual tandem wheels from Figures 2.25, 2.26, and 2.27. Based on three-layer elastic systems, the stresses and strains at the interfaces beneath the center of a circular loaded area can be determined from Table 2.3 and Figure 2.31.

4. Two methods can be used to characterize viscoelastic materials: a mechanical model and a creep compliance curve. Both are closely related and one can be converted to the other. The advantage of using a mechanical model is that the stress–strain relationship can be visualized physically to develop the governing differential equations, while the advantage of using a creep compliance curve is that it can be easily obtained by a laboratory creep test.

5. The elastic–viscoelastic correspondence principle based on Laplace transforms, as described in Appendix A, can be used to analyze layered systems consisting of viscoelastic materials. However, a more convenient method is to

obtain the elastic solutions at a number of time durations and fit them with a Dirichlet series as a function of time.

    6. Instead of using the tedious method of successive residuals, as described in Appendix A, a collocation method can be applied to convert the creep compliance to a mechanical model, as indicated by a Dirichlet series. The time–temperature superposition principle, as indicated by Eqs. 2.46 and 2.48, can then be applied to convert the creep compliance from a reference temperature to any given temperature.

    7. The responses of viscoelastic layered systems under moving loads can be obtained from those under static loads by applying Boltzmann's superposition principle, in which the moving load is expressed as a haversine function and the static response as a Dirichlet series.

PROBLEMS

2-1. A uniformly distributed load of intensity q is applied through a circular area of radius a on the surface of an incompressible ($\nu = 0.5$) homogeneous half-space with an elastic modulus E, as shown in Figure P2.1. In terms of q, a, and E, determine the vertical displacement, three principal stresses, and three principal strains at a point $2a$ below the surface under the edge ($r = a$) of the loaded area. [Answer: $w = 0.58\ qa/E$, $\sigma_1 = 0.221q$, $\sigma_2 = 0.011q$, $\sigma_3 = 0.004q$, $\epsilon_1 = 0.214q/E$, $\epsilon_2 = -0.102q/E$, $\epsilon_3 = -0.112q/E$]

2-2. A 100-psi pressure is applied through a circular area 12 in. in diameter on a granular half-space, as shown in Figure P2.2. The half-space has a mass unit weight of 110 pcf, a coefficient of earth pressure at rest of 0.6, a Poisson ratio of 0.35, and an elastic modulus varying with the sum of normal stresses according to the equation shown in the figure. Assuming that the Boussinesq stress distribution is valid and that the stresses at a point 12 in. below the center of the loaded area are used to compute the elastic modulus, determine the maximum surface displacement. [Answer: 0.054 in.]

2-3. A plate bearing test using a 12-in.-diameter rigid plate is made on a subgrade, as shown in Figure P2.3a. The total load required to cause a settlement of 0.2 in. is 10,600 lb. After placing 10 in. of gravel base course on the subgrade, a plate bearing test is made on the top of the base course, as shown in Figure P2.3b. The total load required to

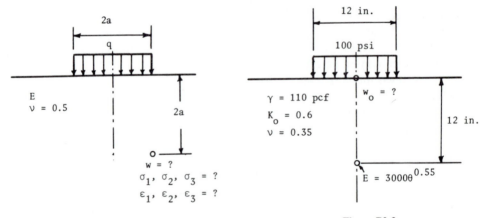

Figure P2.1   Figure P2.2

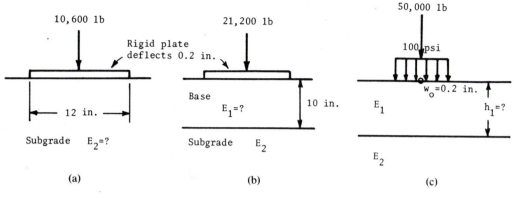

Figure P2.3

cause a settlement of 0.2 in. is 21,200 lb. Assuming a Poisson ratio of 0.5, determine the required thickness of base course to sustain a 50,000-lb tire with a contact pressure of 100 psi over a circular area, as shown in Figure P2.3c, and maintain a deflection of 0.2 in. [Answer: 70 in.]

2-4. A 10,000-lb wheel load with a contact pressure of 80 psi is applied on an elastic two-layer system, as shown in Figure P2.4. Layer 1 has an elastic modulus of 200,000 psi and a thickness of 8 in. Layer 2 has an elastic modulus of 10,000 psi. Both layers are incompressible with a Poisson ratio of 0.5. Assuming that the loaded area is a single circle, determine the maximum surface deflection, interface deflection, and interface stress. [Answer: 0.025 in., 0.024 in., 11 psi]

2-5. A full-depth asphalt pavement, consisting of an 8-in.-thick asphalt layer with an elastic modulus of 1,500,000 psi and a soil subgrade with an elastic modulus of 30,000 psi, is subjected to dual-tandem wheel loads, as shown in Figure P2.5. Each load weighs 50,000 lb with a tire pressure of 100 psi and center to center spacings of 28 in. between dual and 60 in. between tandem. Assuming a Poisson ratio of 0.5, determine the maximum tensile strain at the bottom of asphalt layer under the center of one wheel, and the vertical deflection on the surface of subgrade under the center of one wheel. [Answer: 2.05×10^{-4}, 0.057 in.]

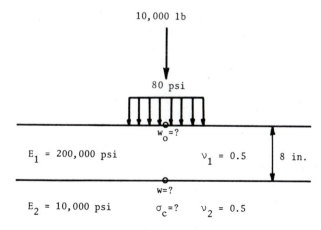

Figure P2.4

Sec. 2.4   Summary

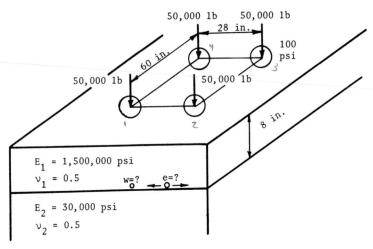

Figure P2.5

2-6. Figure P2.6 shows a pavement structure composed of the following three layers: 5.75 in. HMA with an elastic modulus of 400,000 psi, 23 in. granular base with an elastic modulus of 20,000 psi, and a subgrade with an elastic modulus of 10,000 psi. All layers are assumed to have a Poisson ratio of 0.5. Calculate the maximum horizontal tensile strain at the bottom of HMA and the maximum vertical compressive strain on the top of subgrade under a 40,000-lb wheel load and 150-psi contact pressure, assuming that the contact area is a circle. [Answer: -7.25×10^{-4}, 1.06×10^{-3}]

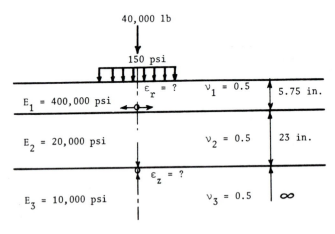

Figure P2.6

2-7. In Problem 2-6 if the base and subgrade are combined as one layer, as shown in Figure P2.7*a*, what should be the equivalent elastic modulus of this combined layer so that the same tensile strain at the bottom of HMA can be obtained? If the HMA and base are combined as one layer with the same total thickness of 28.75 in., as shown in Figure 2.7*b*, what should be the equivalent elastic modulus of this combined layer so that the same compressive strain on the top of subgrade can be obtained? [Answer: 20,000 psi, 35,000 psi]

Stresses and Strains in Flexible Pavements   Chap. 2

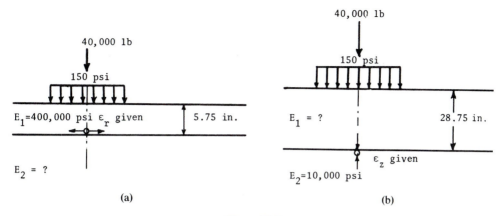

E_1=400,000 psi ε_r given
5.75 in.
E_2 = ?

(a)

E_1 = ?
28.75 in.
ε_z given
E_2=10,000 psi

(b)

Figure P2.7

2-8. A circular load with an intensity of 80 psi and a radius of 6 in. moves over point *A* on the surface of a homogeneous half-space. The half-space has a Poisson ratio of 0.3 and a modulus characterized by the Maxwell model shown in Figure P2.8. If the intensity of load varies with time according to the triangular function with a duration of 10 s, determine the surface deflection when the load arrives at point *A*. If the variation of load is a haversine function indicated by Eq. 2.55, determine the surface deflection. [Answer: 0.109 in., 0.109 in.]

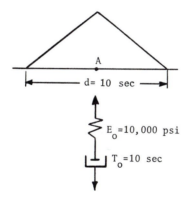

A

d= 10 sec

E_o=10,000 psi

T_o=10 sec

Figure P2.8

3

Kenlayer
Computer Program

3.1 THEORETICAL DEVELOPMENTS

The KENLAYER computer program can be applied only to flexible pavements with no joints or rigid layers. For pavements with rigid layers, such as PCC and composite pavements, the KENSLABS program described in Chapter 5 should be used. The backbone of KENLAYER is the solution for an elastic multilayer system under a circular loaded area. The solutions are superimposed for multiple wheels, applied iteratively for nonlinear layers, and collocated at various times for viscoelastic layers. As a result, KENLAYER can be applied to layered systems under single, dual, dual-tandem, or dual-tridem wheels with each layer behaving differently, either linear elastic, nonlinear elastic, or viscoelastic. Damage analysis can be made by dividing each year into a maximum of 24 periods, each with a different set of material properties. Each period can have a maximum of 24 load groups, either single or multiple. The damage caused by fatigue cracking and permanent deformation in each period over all load groups is summed up to evaluate the design life.

3.1.1 Elastic Multilayer System

Figure 3.1 shows an n-layer system in cylindrical coordinates, the nth layer being of infinite thickness. The modulus of elasticity and the Poisson ratio of the ith layer are E_i and v_i, respectively.

For axisymmetric problems in elasticity, a convenient method is to assume a stress function that satisfies the governing differential equation and the boundary and continuity conditions. After the stress function is found, the stresses and displacements can be determined (Timoshenko and Goodier, 1951).

100

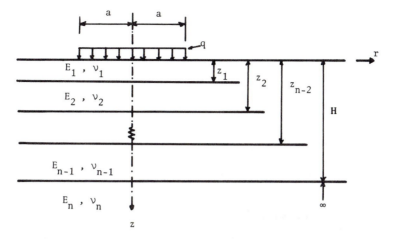

Figure 3.1 An *n*-layer system in cylindrical coordinates.

The governing differential equation to be satisfied is a fourth-order differential equation, as described in Appendix B. The stress function for each layer has four constants of integration, A_i, B_i, C_i, and D_i, where the subscript i is the layer number. Because the stress function must vanish at an infinite depth, the constants A_n and C_n should be equal to zero, i.e., the bottom most layer has only two constants. For an *n*-layer system, the total number of constants or unknowns is $4n - 2$, which must be evaluated by two boundary conditions and $4(n - 1)$ continuity conditions. The two boundary conditions are that the vertical stress under the circular loaded area is equal to q and that the surface is free of shear stress. The four conditions at each of the $n - 1$ interfaces are the continuity of vertical stress, vertical displacement, shear stress, and radial displacement. If the interface is frictionless, the continuity of shear stress and radial displacement is replaced by the vanish of shear stress both above and below the interface. The equations to be used in KENLAYER for computing the stresses and displacement in a multilayer system under a circular loaded area are presented in Appendix B.

3.1.2 Superposition of Wheel Loads

Solutions for elastic multilayer systems under a single load can be extended to cases involving multiple loads by applying the superposition principle. Figure 3.2a shows the plan view of a set of dual-tandem wheels. The vertical stress and vertical displacement under point A due to the four loads can be easily obtained by adding those due to each of the loads because they are all in the same vertical, or z, direction. However, the radial stress σ_r, the tangential stress σ_t, and the shear stress τ_{rz}, due to each load cannot be added directly because they are not in the same direction, as indicated by the four different radial directions at point A. Therefore, σ_r, σ_t, and τ_{rz} must be resolved into components in the x and y directions, as shown in Figure 3.2b for stresses at Point A due to load at point B.

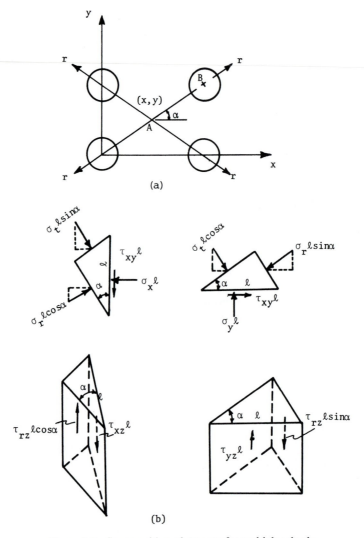

Figure 3.2 Superposition of stresses for multiple wheels.

The use of point A is for illustrative purposes and other points should also be tried in order to find the maximum stresses.

Resolution of Stresses into X and Y Components

By equating the forces in the x and y directions to zero, it can be easily proved from Figure 3.2b that

$$\sigma_x = \sigma_r \cos^2\alpha + \sigma_t \sin^2\alpha \qquad (3.1a)$$

$$\sigma_y = \sigma_r \sin^2\alpha + \sigma_t \cos^2\alpha \qquad (3.1b)$$

Kenlayer Computer Program   Chap. 3

$$\tau_{xy} = (\sigma_r - \sigma_t) \sin \alpha \cos \alpha \qquad (3.1c)$$

$$\tau_{yz} = \tau_{rz} \sin \alpha \qquad (3.1d)$$

$$\tau_{xz} = \tau_{rz} \cos \alpha \qquad (3.1e)$$

After resolving the stresses due to each load into σ_x, σ_y, τ_{xy}, τ_{yz}, and τ_{xz} components, those due to multiple loads can be obtained by superposition. During superposition, care should be taken in determining the proper sign of each stress.

Computation of Principal Stresses and Strains

After σ_x, σ_y, σ_z, τ_{xy}, τ_{yz}, and τ_{xz} are obtained by superposition, the three principal stresses, σ_1, σ_2, and σ_3, can be determined by solving the following cubic equation:

$$\sigma^3 - (\sigma_x + \sigma_y + \sigma_z)\sigma^2 + (\sigma_x\sigma_y + \sigma_y\sigma_z + \sigma_x\sigma_z - \tau_{yz}^2 - \tau_{xz}^2 - \tau_{xy}^2)\sigma$$
$$- (\sigma_x\sigma_y\sigma_z + 2\tau_{yz}\tau_{xz}\tau_{xy} - \sigma_x\tau_{yz}^2 - \sigma_y\tau_{xz}^2 - \sigma_z\tau_{xy}^2) = 0 \qquad (3.2)$$

These principal stresses can be used for nonlinear analysis. The principal strains, ϵ_1, ϵ_2, and ϵ_3, are then determined by

$$\epsilon_1 = \frac{1}{E}[\sigma_1 - v(\sigma_2 + \sigma_3)] \qquad (3.3a)$$

$$\epsilon_2 = \frac{1}{E}[\sigma_2 - v(\sigma_3 + \sigma_1)] \qquad (3.3b)$$

$$\epsilon_3 = \frac{1}{E}[\sigma_3 - v(\sigma_1 + \sigma_2)] \qquad (3.3c)$$

In the fatigue analysis, the horizontal minor principal strain, instead of the overall minor principal strain, is used. The strain is called minor because tensile strain is considered negative. Horizontal principal tensile strain is used because it is the strain that causes the crack to initiate at the bottom of asphalt layer. The horizontal principal tensile strain is determined from

$$\epsilon_t = \frac{\epsilon_x + \epsilon_y}{2} - \sqrt{\left(\frac{\epsilon_x - \epsilon_y}{2}\right)^2 + \gamma_{xy}^2} \qquad (3.4)$$

in which ϵ_t is the horizontal principal tensile strain at the bottom of asphalt layer, ϵ_x is the strain in the x direction, ϵ_y is the strain in the y direction, γ_{xy} is the shear strain on the x plane in the y direction, and

$$\epsilon_x = \frac{1}{E}[\sigma_x - v(\sigma_y + \sigma_z)] \qquad (3.5a)$$

$$\epsilon_y = \frac{1}{E}[(\sigma_y - v(\sigma_x + \sigma_z)] \qquad (3.5b)$$

$$\gamma_{xy} = \frac{2(1 + v)}{E}\tau_{xy} \qquad (3.5c)$$

3.1.3 Damage Analysis

Damage analysis is performed for both fatigue cracking and permanent deformation.

Failure Criteria

The failure criterion for fatigue cracking is expressed as

$$N_f = f_1(\epsilon_t)^{-f_2}(E_1)^{-f_3} \tag{3.6}$$

in which N_f is the allowable number of load repetitions to prevent fatigue cracking; ϵ_t is the tensile strain at the bottom of asphalt layer; E_1 is the elastic modulus of asphalt layer; and $f_1, f_2,$ and f_3 are constants determined from laboratory fatigue tests with f_1 modified to correlate with field performance observations. The Asphalt Institute used 0.0796, 3.291, and 0.854 for $f_1, f_2,$ and f_3, respectively, in their analytically based design procedure; the corresponding values used by Shell are 0.0685, 5.671, and 2.363 (Shook et al., 1982). In view of the fact that the number of load repetitions required to progress from the onset of cracking to limiting failure conditions is fewer for thin asphalt layers than for thicker layers, Craus et al. (1984) suggested that f_1 in the Asphalt Institute criterion be reduced to 0.0636 for HMA layers less than 4 in. (102 mm) in thickness.

The failure criterion for permanent deformation is expressed as

$$N_d = f_4(\epsilon_c)^{-f_5} \tag{3.7}$$

in which N_d is the allowable number of load repetitions to limit permanent deformation, ϵ_c is the compressive strain on the top of subgrade, and f_4 and f_5 are constants determined from road tests or field performance. Value of f_4 and f_5 are suggested as 1.365×10^{-9} and 4.477 by the Asphalt Institute (AI, 1982), 6.15×10^{-7} and 4.0 by Shell (Claussen et al., 1977) and 1.13×10^{-6} and 3.571 by the University of Nottingham (Brown et al., 1977).

Multiple Axles

Due to the large spacing between two axles, the critical tensile and compressive strains under multiple axles are only slightly different from those under a single axle. If the passage of each set of multiple axles is assumed to be one repetition, the damage caused by an 18-kip (80-kN) single axle is nearly the same as that caused by 36-kip (160-kN) tandem axles or 54-kip (240-kN) tridem axles. If one passage of tandem axles is assumed to be two repetitions and that of tridem axles to be three repetitions, the damage caused by 36-kip (160-kN) tandem and 54-kip (240-kN) tridem axles are two and three times greater than that by an 18-kip (80-kN) single axle. Both assumptions are apparently incorrect. The equivalent factors suggested by the Asphalt Institute are 1.38 for tandem axles and 1.66 for tridem axles, as indicated in Table 6.4.

The following procedure is used in KENLAYER to analyze damage due to tandem axle loads. First, determine the tensile and compressive strains at three points under dual-tandem wheels, as shown in Figure 3.3a, and find out which

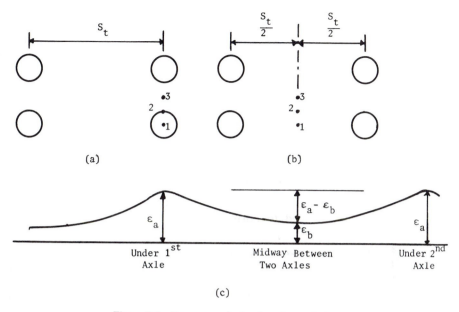

Figure 3.3 Damage analysis of tandem-axle loads.

point results in the maximum tensile strain and which point results in the maximum compressive strain. These maximum strains are then used with Eqs. 3.6 and 3.7 to determine the allowable number of load repetitions due to the first axle load.

Next, determine the tensile and compressive strains at the corresponding point that lies midway between the two axles, as shown in Figure 3.3b. The strain for damage analysis due to the second axle load is $\epsilon_a - \epsilon_b$, where ϵ_a is the strain due to the loading shown in Figure 3.3a and ϵ_b is the strain due to the loading shown in Figure 3.3b. This can be easily explained in Figure 3.3c, where the strain due to the second axle load is $\epsilon_a - \epsilon_b$. The same procedure was incorporated in VESYS (Jordahl and Rauhut, 1983), although VESYS can be applied only to a single tire and the point under the center of the load is used to determine the strains.

A similar but more approximate procedure is used for tridem axles. First, determine the maximum strain ϵ_a by comparing the strains at three points, as shown in Figure 3.4a. Then, determine the corresponding strain ϵ_b, as shown in Figure 3.4b. The strains to be used for the damage analysis of the three axle loads are ϵ_a, $\epsilon_a - \epsilon_b$, and $\epsilon_a - \epsilon_b$, respectively.

3.1.4 Nonlinear Layers

It is well known that granular materials and subgrade soils are nonlinear with an elastic modulus varying with the level of stresses. The elastic modulus to be used with the layered systems is the resilient modulus obtained from repeated unconfined or triaxial compression tests. Details about resilient modulus are presented in Section 7.1. The resilient modulus of granular materials increases with the

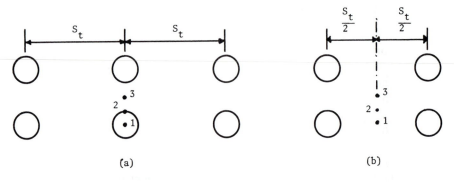

Figure 3.4 Damage analysis of tridem-axle loads.

increase in stress intensity, while that of fine-grained soils decreases with the increase in stress intensity. If the relationship between the resilient modulus and the state of stresses is given, a method of successive approximations can be used, as explained previously for the nonlinear homogeneous mass in Section 2.1.3. The nonlinear material properties, which have been incorporated in KENLAYER, are described below.

Granular Materials

The resilient modulus of granular materials increases with the increase in the first stress invariant, as indicated by Eq. 2.11. However, KENLAYER employs a more popular relationship which is described below.

Constitutive Relationship

A simple relationship between resilient modulus and the first stress invariant can be expressed as

$$E = K_1 \, \theta^{K_2} \tag{3.8}$$

in which K_1 and K_2 are experimentally derived constants and θ is the stress invariant, which can be either the sum of three normal stresses, σ_x, σ_y, and σ_z, or the sum of three principal stresses, σ_1, σ_2, and σ_3:

$$\theta = \sigma_1 + \sigma_2 + \sigma_3 = \sigma_x + \sigma_y + \sigma_z \tag{3.9}$$

Including the weight of a layered system gives

$$\theta = \sigma_x + \sigma_y + \sigma_z + \gamma z(1 + 2K_0) \tag{3.10}$$

in which γ is the average unit weight, z is the distance below surface at which the modulus is to be determined, and K_0 is the coefficient of earth pressure at rest. The reason σ_1, σ_2, and σ_3 are not used in Eq. 3.10 is that they may not be in the same direction as the geostatic stresses. In contrast to other computer programs, KENLAYER uses the soil mechanics sign convention for stresses and strains. Therefore, θ is positive when in compression and negative when in tension.

Two Methods of Analysis

It is well known that most granular materials cannot take any tension. Unfortunately, when they are used as a base or subbase on a weaker subgrade, the horizontal stresses at the bottom of these materials are most likely in tension. Two methods have been incorporated in KENLAYER for nonlinear analysis. In method 1, the nonlinear granular layer is subdivided into a number of layers and the stresses at the middepth of each layer are used to determine the modulus. If the horizontal stress, including the geostatic stress, is negative or in tension, it is set to 0. This stress modification is necessary to avoid negative θ. In method 2, all the granular materials are considered as a single layer and an appropriate point, usually between the upper quarter and the upper third of the layer, is selected to compute the modulus. Because the point is at the upper part of the layer, the chance of negative θ is rare, so no stress modification is needed. If θ turns out to be negative, an arbitrary or minimum modulus (EMIN) is assigned.

The use of method 1 by subdividing the nonlinear layer into several layers yields more accurate results but requires much more computer time. By selecting an appropriate point for computing the modulus, method 2 can produce comparable results with only a fraction of the computer time required. Whether method 1 with stress modification or method 2 without stress modification is to be used for a granular layer depends on the input parameter EMIN. If EMIN is specified as 0, method 1 is implied and no tension is allowed. If EMIN is nonzero, method 2 is used with no stress modification needed.

It should be noted that the use of layered system for nonlinear analysis is an approximate approach. It is desirable to have more exact solutions so that the results of KENLAYER can be compared. Theoretically, the finite element method should provide the best solutions for such nonlinear problems. Unfortunately, the finite element computer programs currently available have serious defects and cannot be used to check the accuracy of a solution, as is described in Section 3.3.2. If accurate solutions can be obtained in the future, it is possible to use method 2 by selecting appropriate points for computing the modulus so that the same strains can be obtained.

TABLE 3.1 NONLINEAR CONSTANTS K_1 AND K_2 FOR GRANULAR MATERIALS

Material type	No. of data points	K_1 (psi) Mean	K_1 (psi) Standard deviation	K_2 Mean	K_2 Standard deviation
Silty sand	8	1620	78	0.62	0.13
Sand–gravel	37	4480	4300	0.53	0.17
Sand–aggregate blend	78	4350	2630	0.59	0.13
Crushed stone	115	7210	7490	0.45	0.23

Note. 1 psi = 6.9 kPa.

Source. After Rada and Witczak (1981).

TABLE 3.2 RANGES OF K_1 AND K_2 FOR UNTREATED GRANULAR MATERIALS

Reference	Material	K_1 (psi)	K_2
Hicks (1970)	Partially crushed gravel, crushed rock	1600–5000	0.57–0.73
Hicks and Finn (1970)	Untreated base at San Diego Test Road	2100–5400	0.61
Allen (1973)	Gravel, crushed stone	1800–8000	0.32–0.70
Kalcheff and Hicks (1973)	Crushed stone	4000–9000	0.46–0.64
Boyce et al. (1976)	Well-graded crushed limestone	8000	0.67
Monismith and Witczak (1980)	In service base and subbase materials	2900–7750	0.46–0.65

Note. 1 psi = 6.9 kPa.

Source. After Shook et al. (1982).

Typical Values

Based on a statistical analysis of published data, Rada and Witczak (1981) presented the mean and standard deviation of resilient modulus for several granular materials, as shown in Table 3.1. It was reported (Finn et al., 1986) that the resilient moduli of base and subbase aggregates in the AASHO Road Test can be represented by Eq. 3.8 with K_2 equal to 0.6 and K_1 ranging from 3200 to 8000 psi, depending on the moisture contents. Other reported values of K_1 and K_2 are shown in Table 3.2.

Fine-Grained Soils

The resilient modulus of fine-grained soils decreases with the increase in deviator stress σ_d. In laboratory triaxial tests, $\sigma_2 = \sigma_3$, so the deviator stress is defined as

$$\sigma_d = \sigma_1 - \sigma_3 \tag{3.11}$$

In a layered system, σ_2 may not be equal to σ_3, so the average of σ_2 and σ_3 is considered as σ_3. Including the weight of layered system yields

$$\sigma_d = \sigma_1 - 0.5(\sigma_2 + \sigma_3) + \gamma z(1 - K_0) \tag{3.12}$$

Equation 3.12 is not theoretically correct because the principal loading stresses may not be in the same direction as the geostatic stresses. Since the loading stresses in the subgrade are usually small and do not have significant effect on the computed modulus, KENLAYER uses the three normal stresses, σ_x, σ_y, and σ_z, to replace the three principal stresses, σ_1, σ_2, and σ_3 in Eq. 3.12. If the point selected for computing the modulus is on the axis of symmetry for a single tire or on the plane of symmetry between two dual tires, the three normal stresses and the three principal stresses are identical.

Figure 3.5 shows the general relationship between resilient modulus and deviator stress of fine-grained soils obtained from laboratory repeated load tests.

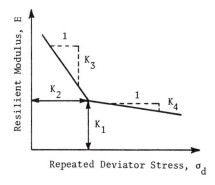

Figure 3.5 General relationship between resilient modulus and deviator stress for fine-grained soils.

The bilinear behavior can be expressed as

$$E = K_1 + K_3(K_2 - \sigma_d) \qquad \text{when } \sigma_d < K_2 \qquad (3.13a)$$

$$E = K_1 - K_4(\sigma_d - K_2) \qquad \text{when } \sigma_d > K_2 \qquad (3.13b)$$

in which K_1, K_2, K_3, and K_4 are material constants.

Thompson and Elliott (1985) indicated that the value of resilient modulus at the breakpoint in the bilinear curve, as indicated by K_1 in Figure 3.5 , is a good indicator of resilient behavior, while other constants, K_2, K_3, and K_4, display less variability and influence pavement response to a smaller degree than K_1. They classified fine-grained soils into four types, viz., very soft, soft, medium, and stiff, with the resilient modulus–deviator stress relationship shown in Figure 3.6. The

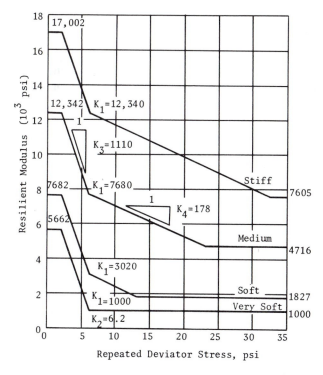

Figure 3.6 Resilient modulus–deviator stress relationship for four types of subgrade (1 psi = 6.9 kPa). (After Thompson and Elliott (1985).)

Sec. 3.1   Theoretical Developments

109

maximum resilient modulus is governed by a deviator stress of 2 psi (13.8 kPa). The minimum resilient modulus is limited by the unconfined compressive strengths, which are assumed to be 6.21 psi (42.8 kPa), 12.90 psi (89.0 kPa), 22.85 psi (157 kPa), and 32.8 psi (226 kPa) for the four soils. Equation 3.13 has also been incorporated in KENLAYER. Note that K_1 and K_2 defined herein are different from those by Thompson and Elliott in that they are interchanged in meaning.

Stress Point for Nonlinear Layer

The elastic modulus of each nonlinear layer is determined from the stresses at a designated point according to Eq. 3.8 or 3.13. This point is called a stress point and is defined by a point on the pavement surface, a slope of load distribution, SLD, and a z coordinate, ZCNOL, as shown in Figure 3.7.

The point on the surface may be located by the radial coordinate, RCNOL, for single wheel or the x and y coordinates, XCNOL and YCNOL, for multiple wheels. For a given ZCNOL, the r coordinate of the stress point is

$$r = \text{RCNOL} + (\text{SLD})(\text{ZCNOL}) \tag{3.14}$$

and the x and y coordinates of the stress point are

$$x = \text{XPTNOL} + (\text{SLD})(\text{ZCNOL}) \tag{3.15}$$

$$y = \text{YPTNOL} + (\text{SLD})(\text{ZCNOL}) \tag{3.16}$$

If only the maximum stresses, strains, or deflections are required, then the stress point should be located under the center of a single wheel with RCNOL = 0 and SLD = 0 or between the center of two duals with XPTNOL = 0, YPTNOL = YW/2, and SLD = 0. Note that YW is the center to center spacing between the duals, as shown in Figure 3.8. If the average responses such as the deflection basin under a circular area are required, then the use of RCNOL = 0.5a, where a is the radius of loaded area, and SLD = 0.25 is recommended. This is based on the general assumption that, starting from the edge of loaded area, the load is distributed downward at a slope of 0.5. Therefore, the stresses along a line with

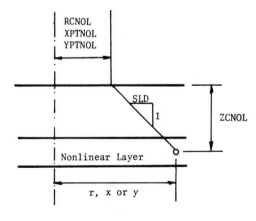

Figure 3.7 Loacation of stress point for nonlinear layer.

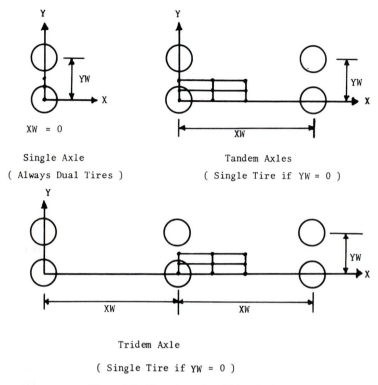

Figure 3.8 Plan view of multiple wheels.

RCNOL = 0.5*a* and SLD = 0.25 can be considered the average within the zone of influence.

For tandem- and tridem-axle loads, only the load on one axle is used to compute the modulus because the axles are so far apart and the zone of influence does not overlap significantly. After the moduli are determined, all the axle loads will be used for analysis.

3.2 PROGRAM DESCRIPTION

KENLAYER was written in FORTRAN 77 and requires a storage of 509K. In its present dimensions, it can be applied to a maximum of 19 layers with output at 10 different radial coordinates and 19 different vertical coordinates, or a total of 190 points. For multiple wheels, in addition to the 19 vertical coordinates, solutions can be obtained at a total of 25 points by specifying the *x* and *y* coordinates of each point. Creep compliances can be specified at a maximum of 15 time durations. Damage analysis can be made by dividing each year into a maximum of 24 periods, each with a maximum of 24 load groups.

Before running KENLAYER, the input program LAYERINP must be run first to set up a data file. The use of the input programs is described in Appendix

D. After a file has been established, KENLAYER can be run by simply typing KENLAYER. The program will ask for a file name and will start execution when the file name is entered. If the PC has a storage of 640K but the message "Program too big to fit in memory" appears on the screen, some of the memory must be cleared out first before KENLAYER can be run. A more convenient way is to start DOS over from the beginning using the original DOS diskette.

3.2.1 General Features

The capabilities of KENLAYER can be demonstrated by the following four input parameters which must be specified at the very outset:

MATL = 1 for linear elastic, 2 for nonlinear elastic, 3 for linear viscoelastic, and 4 for combination of nonlinear elastic and linear viscoelastic.
NDAMA = 1 with damage analysis and 0 without damage analysis.
NPY = number of periods per year.
NLG = number of load groups.

Materials

Unless indicated otherwise, all layers are assumed to be linear elastic with a constant elastic modulus. For the linear elastic case, solutions for multiple wheels are obtained by superposition of those due to single wheels.

For the nonlinear elastic case, the elastic moduli of some layers are stress dependent and these layers must be identified. An iterative procedure is used, in which the moduli of nonlinear layers are adjusted as the stresses vary, while the moduli of linear layers remain the same. During each iteration, a constant set of moduli is computed based on the stresses obtained from the previous iteration, so the problem is considered linear and the superposition principle can still be applied to multiple wheels. After the stresses due to single or multiple wheels are determined, the elastic moduli of nonlinear layers are recalculated and a new set of stresses is determined. The process is repeated until the moduli converge to a specified tolerance.

For the linear viscoelastic case, the viscoelastic layers must be identified and their creep compliances specified, while the other layers are linear elastic. Solutions due to either moving or stationary loads can be obtained. Moving loads based on a haversine function, as indicated by Eq. 2.55, are used for damage analysis. If stationary loads are specified, the elastic solutions corresponding to the specified creep times will be computed and the strains at the last time duration will be used for damage analysis. The use of stationary loads is equivalent to a square wave, instead of a haversine wave.

For a combination of nonlinear elastic and linear viscoelastic case, some layers are nonlinear elastic, some are viscoelastic, and the remaining, if any, are linear elastic. If only one layer is viscoelastic, the stresses in every layer will be time dependent. The stresses to be used for determining the modulus of nonlinear layer are the peak stresses at $t = 0$ under a moving load, as shown in Figure 2.40, or the stresses at the last time duration under a stationary load.

Damage Analysis

Damage analysis is based on the horizontal tensile strain at the bottom of specified layers, usually the HMA or layer 1, and the vertical compressive strain on the top of specified layers, usually the subgrade or the lowest layer. Instead of reading in the z coordinates, simply specifying the total number of layers for bottom tension (NLBT), the total number of layers for top compression (NLTC), the layer number for bottom tension (LNBT), and the layer number for top compression (LNTC), the program will determine the z coordinates of all necessary points and compute the required strains. If several radial coordinate points are specified under single wheel or several x and y coordinate points under multiple wheels, the program will compare the strains at these points and select the most critical ones for damage analysis.

Each year can be divided into a maximum of 24 periods. The elastic modulus, nonlinear coefficient K_1, and creep compliances may differ from period to period. The predicted number of load repetitions for each load group is specified for each period. The allowable number of load repetitions for fatigue cracking is determined by Eq. 3.6 and that for permanent deformation by Eq. 3.7. The elastic modulus of viscoelastic layer to be used in Eq. 3.6 is computed from the three principal stresses and the minor principal strain (minor because tension is negative) by

$$E = \frac{\sigma_3 - \nu(\sigma_1 + \sigma_2)}{\epsilon_3} \tag{3.17}$$

The same E is obtained if the major principal strain is used:

$$E = \frac{\sigma_1 - \nu(\sigma_2 + \sigma_3)}{\epsilon_1} \tag{3.18}$$

The damage ratio, which is the ratio between the predicted and allowable number of repetitions, is computed for each load group in each period and summed over the year by

$$D_r = \sum_{i=1}^{p} \sum_{j=1}^{m} \frac{n_{i,j}}{N_{i,j}} \tag{3.19}$$

in which D_r is the damage ratio at the end of a year, $n_{i,j}$ is the predicted number of load repetitions for load j in period i, $N_{i,j}$ is the allowable number of load repetitions based on Eqs. 3.6 and 3.7, p is the number of periods in each year, and m is the number of load groups. The design life, which is equal to $1/D_r$, is evaluated for both fatigue cracking and permanent deformation and the one with a shorter life controls the design.

Number of Periods Per Year

The number of periods per year (NPY) is used mainly for damage analysis. Each year can be considered as one period with the same material properties throughout the year or divided into four seasons, 12 months, or 24 periods, each with a different set of material properties. If no damage analysis is to be made, or NDAMA = 0, NPY is usually specified as 1. However, NPY can be more than 1 even if no damage analysis is required. In this case, NPY should be interpreted as the number of material properties, so several different sets of material properties can be run one after the other.

Some of the material properties that do not change over the periods are entered first. The properties that vary with each period will then be read in. To avoid the input of creep compliances for each period, the creep compliances are specified at a reference temperature and fitted with a Dirichlet series, as shown by Eq. 2.38. The creep compliances at any other temperature are obtained by the time–temperature superposition principle, as indicated by Eqs. 2.46 and 2.48. If creep compliances are used directly with no temperature conversions needed, an arbitrary reference temperature the same as the actual temperature may be specified, or a zero temperature shift coefficient, BETA, may be used.

Loads

Whether a load has a single or multiple wheels is identified by the parameter LOAD, with 0 for single wheel, 1 for single axle with dual tires, 2 for tandem axles, and 3 for tridem axles. The two dual tires must be oriented in the y direction with dual spacing YW. If YW is specified as 0, the tandem- or tridem-axle loads are applied on a single tire; otherwise, they are applied on dual tires.

For a single-wheel load, NR radial distances and NZ vertical distances must be specified, so solutions at NR × NZ points can be obtained. For example, if a point is located by the coordinates (r, z) and it is desired to obtain solutions at two points, (0, 0) and (10, 10), then two radial coordinates of 0 and 10 and two vertical coordinates of 0 and 10 must be specified and solutions at (0, 0), (0, 10), (10, 0), and (10, 10) will be obtained, even though the ones at (0, 10) and (10, 0) are not needed. The extra time required for these unnecessary points is minimal because KENLAYER computes all points at the same time and a large part of the program is the same no matter how many radial and vertical coordinates are used. To save the computer time for numerical integration, each point is checked for convergence. If the results at any given point have converged to the desired tolerance, no more integration will be performed at that point.

For multiple wheels, instead of radial distances, the x and y coordinates of each point must be specified. Figure 3.8 shows a plan view of multiple wheels. It is required to use the center of the left and lowest wheel as the origin for x and y coordinates. The dual wheels must be oriented along the y axis. The spacing YW must be specified for dual tires and XW for multiple axles. The (x, y) coordinates of all points at which solutions are to be sought are also needed. In Figure 3.8, three points are specified for dual wheels and nine points for tandem and tridem axles. Based on these coordinates, the radial distances from each point to each of the multiple wheels are determined and the solutions due to all single-wheel loads are computed at the same time. These solutions are then superimposed to obtain those caused by multiple wheels.

3.2.2 Input Parameters

For ease of reference, input parameters are listed in alphabetical order. Any unit can be used as long as it is consistent. In U.S. customary units, length is in inches (in.); pressure, stress, or modulus is in pounds per square inch (psi); and unit weight is in pounds per cubic inch (pci). In SI units, length is in meters (m), pressure is in kilonewtons per square meter (kN/m²), and unit weight is in kilonewtons per cubic meter (kN/m³). The values of some empirical constants, such as the nonlinear and fatigue coefficients, depend on the units used, and care should be taken in selecting the appropriate values. The maximum dimension of each array is shown and the parameter that determines the actual dimension is indicated immediately after the definition.

BETA(19)   temperature shift coefficient, or the slope of time–temperature shift factor versus temperature on a semilog plot, as shown in Figure 2.38 (NL). For hot mix asphalt, use 0.113.

CP(24)   contact pressure on circular loaded area (NLG).

CR(24)   contact radius of circular loaded area (NLG).

CREEP(19,15,1)   creep compliance of viscoelastic materials at reference temperature (NL, NTYME, 1).

DEL   tolerance for integration involving Bessel functions, 0.001 suggested.

DELNOL   tolerance for nonlinear analysis, 0.01 suggested.

DUR   duration of moving loads, assign 0 for stationary loads, 0.1 s for 40 mph, and increase proportionately with decreasing speed.

E(19,24)   elastic modulus of each layer (NL, NPY). Use as assumed modulus when the layer is nonlinear. Assign 0 or any value for viscoelastic layer.

EMIN(19,24)   minimum elastic modulus of nonlinear clayey or granular layer (NL, NPY). When a granular layer is subdivided into two or more layers, EMIN must be specified as 0, so no tension will be allowed.

FT(5,19)   damage coefficients (5, NL). Subscripts 1, 2, and 3 are the fatigue coefficients for bottom tension, as shown by Eq. 3.6, and subscripts 4 and 5 are the permanent deformation coefficients for top compression, as shown by Eq. 3.7. Values suggested by the Asphalt Institute are 0.0796, 3.291, 0.854, 1.365×10^{-9}, and 4.477.

GAM(19)   unit weight of each layer (NL). Suggested values are 0.084 pci (22.8 kN/m³) for HMA, 0.078 pci (21.2 kN/m³) for granular materials, and 0.072 pci (19.6 kN/m³) for soil.

ICL   maximum number of integration cycles involving Bessel functions, 80 suggested.

INT(18)   condition at each interface (NL − 1). Assign 1 for bonded interface and 0 for frictionless interface.

ITENOL   maximum number of iterations for nonlinear analysis, 10 suggested.

KO(19)   coefficient of earth pressure at rest (NL). Suggested values are 0.6 for granular materials and 0.8 for fine-grained soils.

K1(19,24)   nonlinear coefficient of granular layer or break point modulus for clayey layer (NL, NPY).

K2(19)   nonlinear exponent for granular layer or deviator stess at break point for clayey layer (NL).

K3(19)   slope of straight-line relationship between resilient modulus and deviator stress for clayey layer when the deviator stress is smaller than K2 (NL).

K4(19)   slope of straight-line relationship between resilient modulus and deviator stress for clayey layer when the deviator stress is greater than K2 (NL).

LAYNO(19) nonlinear layer number at which elastic modulus is stress dependent (NL).

LNBT(19) layer number for damage analysis of bottom tension (NL).

LNTC(19) layer number for damage analysis of top compression (NL).

LNV(19) layer number which is viscoelastic (NL).

LOAD(24) type of loadings, 0 for single axle with single tire, 1 for single axle with dual tires, 2 for tandem axles, and 3 for tridem axles (NLG).

MATL material property, 1 for linear elastic, 2 for nonlinear elastic, 3 for linear viscoelastic, and 4 for combination of nonlinear elastic and linear viscoelastic.

NCLAY(19) type of nonlinear layer (NL). Assign 1 for clayey layer and 0 for granular layer.

NDAMA index for damage analysis, 1 with damage analysis and 0 without damage analysis.

NBOND type of interface between two layers, 1 when all layers are bonded and 0 when some interfaces are unbonded or frictionless.

NL total number of layers, maximum 19.

NLBT number of layers for bottom tension.

NLG number of load groups, maximum 24.

NLTC number of layers for top compression.

NOLAY number of nonlinear layers.

NPROB number of problems to be solved.

NPT(24) number of points in x and y coordinates to be analyzed under multiple wheels, maximum 25 (NLG).

NPY number of periods per year, maximum 24.

NR(24) number of radial distances to be analyzed under a single wheel, maximum 10 (NLG).

NSTD computing code, 1 for vertical displacement only, 5 for vertical displacement and four stresses, and 9 for vertical displacement, four stresses, and four strains.

NTYME number of times at which creep compliances are to be inputted, maximum 15, suggested 11.

NVL number of viscoelastic layers.

NZ number of vertical distances to be analyzed, maximum 19. Assign 0 when NDAMA = 1.

PR(19) Poisson's ratio of each layer (NL). Suggested values are 0.35 for HMA and granular materials and 0.45 for fine-grained soils.

RC(10,25,24) radial distance, or r coordinate, of each point to be analyzed (NR, NPT, NLG).

RCNOL radial coordinate on pavement surface for computing elastic modulus of nonlinear layer under a single wheel. Assign 0 or any value if there is no single wheel. The same RCNOL is used for all load groups with a single wheel.

RELAX(24) relaxation factor for nonlinear analysis, 0.5 suggested (NPY). The use of relaxation factor is to ensure the convergence of elastic modulus. If the results diverge, smaller relaxation factors should be used.

SLD slope of load distribution.

TEMP(19) pavement temperature for each viscoelastic layer during each period (NL).

TEMREF(19) reference temperature of each viscoelastic layer at which creep compliances are determined (NL).

TH(18) thickness of each layer (NL-1). The last layer is infinite in thickness and need not be inputted.

TITLE any title or comments typed within columns 1 to 80.

TNLR(24,24) total predicted number of load repetitions for each load group in each period (NPY,NLG).

TYME(15)   times at which creep compliances are to be inputted (NTYME). Suggested values at 0.001, 0.003, 0.01, 0.03, 0.1, 0.3, 1, 3, 10, 30, 100.

XPT(25) x coordinates of points to be analyzed (NPT).

XPTNOL x coordinate of point on pavement surface for computing elastic modulus of nonlinear layer under multiple wheels. Assign 0 or any value if there are no multiple wheels. The same XPTNOL is used for all load groups with multiple wheels.

XW(24)   center to center spacing between two axles along the x axis (NLG). Assign 0 if only one axle exists.

YPT(25) y coordinates of points to be analyzed (NPT).

YPTNOL y coordinate of point on pavement surface for computing elastic modulus of nonlinear layer under multiple wheels. Assign 0 or any value if there are no multiple wheels. The same YPTNOL is used for all load groups with multiple wheels.

YW(24)   center to center spacing between two dual wheels along the y axis (NLG). Assign 0 if there is only one wheel.

ZC(19)   vertical distance, or z coordinate, of each point (NZ). When the point is located exactly at the interface between two layers, the results are at the bottom of the upper layer. If the results at the top of the lower layer are desired, a slightly larger z coordinate, say 0.0001 larger, should be used.

ZCNOL(19) z coordinate of points for computing elastic modulus of nonlinear layer (NL). If the base or subbase is subdivided into several layers, the points are located at the midheight of each layer with EMIN = 0. If the base or subbase is not subdivided, the upper quarter point is usually used with EMIN > 0.

3.2.3 Data File

Three types of format are used for input. All integers are in I5 format, each occupying five columns of an 80-column line, whereas all real numbers are in F10.5 or E10.3 format, each occupying ten columns. As long as a real number is typed within the ten designated columns with a decimal point, any F or E format can be used. The reason that F10.5 and E10.3 are specified is to arrange the data file in an orderly manner for easy checking.

To facilitate entering and editing data, a user-friendly program named LAYERINP can be used. The program uses menus and data entry forms to create and edit the data file. Details about LAYERINP are presented in Appendix D.

The arrangement of the data file is listed below. The file can be set up automatically by LAYERINP or entered manually according to the given format. By using LAYERINP, it is really not necessary to know the data arrangement and the required format. This listing is useful only to experienced users who may like to edit the data more expediently by an editor rather than by LAYERINP. When the data file is set up by LAYERINP, each line is identified at the end by a number in parentheses, which indicates the line number in the listing. By comparing these numbers, the data to be changed can be easily located. The listing can also be used to check the correctness of the data file without using LAYERINP.

The number of lines indicated in the listing is correct only when the given data can be accommodated in an 80-column line. If the data require more than 80 columns, they should continue on the next line until the given data are exhausted. If an asterisk appears before the step number, the step may be skipped under the condition specified.

(1) 1 line (I5) NPROB

(2) 1 line (80 columns) TITLE

(3) 1 line (4I5) MATL,NDAMA,NPY,NLG

(4) 1 line (F10.5) DEL

DEL = 0.001 suggested.

(5) 1 line (4I5) NL,NZ,ICL,NSTD

NZ = 0 when NDAMA = 1. ICL = 80 suggested.

(6) 1 line (8F10.5) (TH(I),I=1,NL−1)

(7) 1 line (8F10.5) (PR(I),I=1,NL)

*(8) 1 line (8F10.5) (ZC(I),I=1,NZ)

Skip if NDAMA = 1.

(9) 1 line (I5) NBOND

*(10) 1 line (16I5) (INT(I),I=1,NL−1)

Skip if NBOND = 1.

(11) NPY lines (8E10.3) (E(I,J),I=1,NL)

Each period, as indicated by the subscript J from 1 to NPY, should begin with a new line. E = 0 or any value if the layer is viscoelastic.

(12) There is a do loop for load from steps 13 to 21. The load number is indicated by the subscript J which starts from 1 to NLG. Note that XPT(I), and YPT(I) also vary with J. Immediately after their input, the radial distances RC(NR,NPT,J) and the direction cosines are computed so it is not necessary to save XPT and YPT by adding index J.

(13) 1 line (I5) LOAD(J)

(14) 1 line (2F10.5) CR(J),CP(J)

(15) If LOAD(J) > 0, go to (19)

(16) 1 line (I5) NR(J)

(17) 1 line (8F10.5) (RC(I,1,J),I=1,NR)

(18) go to (21)

(19) 1 line (I5) NPT(J)

(20) 1 line (8F10.5) XW(J),YW(J),(XPT(I),YPT(I),I=1,NPT)

(21) Go to (13) NLG times, one for each load

(22) If MATL = 1, go to (45)

(23) If MATL = 2 or 4, go to (25)

(24) If MATL = 3, go to (36)

(25) 1 line (2I5) NOLAY,ITENOL

ITENOL = 10 suggested.

(26) 1 line (16I5) (LAYNO(I),NCLAY(LAYNO(I)),I=1,NOLAY)

(27) 1 line (8F10.5) (ZCNOL(I),I=1,NOLAY)

(28) 1 line (5F10.5) RCNOL,XPTNOL,YPTNOL,SLD,DELNOL

DELNOL = 0.01 suggested

(29) 1 line (8F10.5) (RELAX(I),I=1,NPY)

RELAX = 0.5 suggested

(30) 1 line (8F10.5) (GAM(I),I=1,NL)

In the English system use pounds per cubic inch.

(31) NOLAY lines. Check whether each nonlinear layer is of clayey or granular material. Each nonlinear layer, as indicated by the subscript I from 1 to NOLAY, should begin with a new line.

If NCLAY(LAYNO(I)) = 0, or the layer is granular, then

1 line (2F10.5) K2(LAYNO(I)),K0(LAYNO(I))

If NCLAY(LAYNO(I)) = 1, or the layer is clay, then

1 line (4F10.5) K2(LAYNO(I)),K3(LAYNO(I)),K4(LAYNO(I)),K0(LAYNO(I))

(32) There is a do loop for period from steps 33 to 34. The period number is indicated by the subscript J from 1 to NPY

(33) NOLAY lines. Check whether each nonlinear layer is of clayey or granular material. Each nonlinear layer, as indicated by the subscript I from 1 to NOLAY, should begin with a new line.
If NCLAY(LAYNO(I)) = 0, or the layer is granular, then
1 line (2E10.3) EMIN(LAYNO(I),J),K1(LAYNO(I),J)
If NCLAY(LAYNO(I)) = 1, or the layer is clay
1 line (3E10.3) EMIN(LAYNO(I),J),EMAX(LAYNO(I),J),K1(LAYNO(I),J)

(34) Go to (33) NPY times, one for each period

(35) If MATL = 2, go to (45)

(36) 1 line (F10.5) DUR
DUR = 0 for stationary load.

(37) 1 line (16I5) NVL,(LNV(I),I = 1,NVL)

(38) 1 line (I5) NTYME

(39) 1 line (8F10.5) (TYME(J),J = 1,NTYME)

(40) There is a do loop for viscoelastic layer from steps 41 to 43. The viscoelastic layer number is indicated by the subscript I from 1 to NVL

(41) 1 line (2F10.5) BETA(LNV(I)),TEMPREF(LNV(I))

(42) 1 line (8E10.3) (CREEP(LNV(I),J,1),J = 1,NTYME)

(43) Go to (41) NVL times, one for each viscoelastic layer

(44) NPY lines (8F10.3) (TEMP(LNV(I)),I = 1,NVL)

(45) If NDAMA = 0, go to (52)

(46) 1 line (2I5) NLBT,NLTC

*(47) 1 line (16I5) (LNBT(I),I = 1,NLBT)
Skip if NLBT = 0.

*(48) 1 line (16I5) (LNTC(I),I = 1,NLTC)
Skip if NLTC = 0.

(49) NPY lines (8F10.5) (TNLR(I,J),J = 1,NLG)
Each period, as indicated by the subscript I from 1 to NPY, should begin with a new line. If values are too large and cannot be accommodated by F10.5 format, use other format with a decimal point at the end.

*(50) NLBT lines (3F10.5) (FT(1,I),FT(2,I),FT(3,I)
Each layer, as indicated by the subscript I from 1 to NLBT, should begin with a new line.
Skip if NLBT = 0.

*(51) NLTC lines (E10.3,F10.5) FT(4,I),FT(5,I)
Each layer, as indicated by the subscript I from 1 to NLTC, should begin with a new line
Skip if NLTC = 0.

(52) Go to (2) NPROB times

3.2.4 Printed Output

Every input parameter, after being read in, will be printed out for inspection. In most cases, the definition of the parameter, in addition to the variable name, is printed for easy understanding. The output is stored automatically in a file called LAYER.TXT. This file will be destroyed and replaced by a new file whenever KENLAYER is run again. If you want to keep the file, be sure to change its name.

Single Wheel Versus Multiple Wheels

The form of output depends on whether single or multiple loads are specified. The stresses, strains, and displacements are stored in STD(10,19,9) for single load and in STD(1,19,9) for multiple loads. The first subscript indicates the number of radial coordinates, the second indicates the number of vertical coordinates, and the third indicates the type of stress, strain, or displacement.

The points under multiple loads are not identified by radial coordinates, so the first subscript is always 1. These points are identified by point numbers defined by x and y coordinates. The computation is proceeded from point to point. As soon as the values of STD(1,19,9) are determined at a given point, they will be printed and compared with the previous results to determine the maximum for damage analysis. The program will then go to the next point until all points are completed. This is in contrast to the case of a single wheel in which all points at various radial coordinates are computed and printed at the same time.

The third subscript of STD is defined differently for single and multiple wheels. For a single wheel, a subscript of 1 stands for vertical displacement, 2 for vertical stress, 3 for radial stress, 4 for tangential stress, 5 for shear stress, 6 for vertical strain, 7 for radial strain, 8 for tangential strain, and 9 for shear strain. For multiple wheels, a subscript of 1 stands for vertical displacement, 2 for vertical stress, 3 for major principal stress, 4 for intermediate principal stress, 5 for minor principal stress, 6 for vertical strain, 7 for major principal strain, 8 for minor principal strain, and 9 for horizontal principal strain. If NSTD = 1, only vertical displacements will be computed; if NSTD = 5, only one displacement and four stresses will be computed at each point; and if NSTD = 9, all nine items will be computed. The soil mechanics sign convention is used with compression as positive and tension as negative.

Types of Analysis

During the execution of the program, some information will be printed on the screen to show that the program is running. Each time the subroutine LAYERS is executed, the actual number of cycles, IC, for numerical integration is printed on the screen. This number should be smaller than the maximum ICL specified, otherwise the results have not converged. If the probelm is nonlinear, the previous and the new elastic moduli of each nonlinear layer will be printed on the screen during each iteration. These moduli should be carefully checked for convergence. If the moduli diverge, the program should be stopped and a smaller relaxation factor be used. When a run is completed but the moduli have not converged to the desired accuracy, a rerun may be made using the moduli obtained from the last run as the assumed elastic moduli. Printouts for nonlinear elastic systems are the same as linear elastic systems except that the vertical geostatic stress at each stress point, the elastic moduli of all nonlinear layers during each iteration, and the three normal stresses to evaluate the modulus of each granular layer are printed.

In the case of viscoelastic systems under moving loads, the output is the same as that in the case of elastic systems. However, if the load is stationary, the solution at each time duration will be printed.

If a damage analysis is required, the critical strain in each specified layer together with the allowable number of repetitions and damage ratio is printed for each load group in each period. The damage ratios at the end of a year, the maximum damage ratio, and the design life in years are also printed.

3.3 COMPARISON WITH AVAILABLE SOLUTIONS

KENLAYER can be applied to a homogeneous half-space by assuming that all layers have the same elastic modulus and Poisson ratio. As indicated in Section 2.1, the solutions obtained by KENLAYER checked very closely with the Boussinesq solutions for a homogeneous half-space. In this section, the solutions obtained by KENLAYER are compared with ELSYM5 for multiple wheels, MICH-PAVE for nonlinear layers, VESYS for moving loads, and DAMA for damage analysis.

3.3.1 Multiple Wheels

The computation of principal stresses under multiple wheels is complex and requires the resolution of stresses into x and y components before superposition, as described in Section 3.1.2. To be sure that the solutions are correct, a comparison of solutions was made between KENLAYER and ELSYM5 (Kopperman et al., 1986). ELSYM5 is a linear elastic layer system composed of a maximum up to five layers. The pavement may be loaded with one or more identical uniform circular loads normal to the surface. The program superimposes the various loads and computes the stresses, strains, and displacements in three dimensions, along with the three principal stresses and strains, at locations specified by the user.

An elastic three-layer system, with the thicknesses, elastic moduli, and Poisson ratios shown in Figure 3.9a, is subjected to six-wheel loads, as shown in Figure 3.9b. Each loaded area has a radius of 4 in. (102 mm) and a contact pressure of 100 psi (690 kPa). The vertical displacements at each layer interface, the principal tensile strains at the bottom of asphalt layer, and the vertical compressive strains on the top of subgrade under the four points shown in Figure 3.9b were determined by both KENLAYER and ELSYM5 and are tabulated in Table 3.3. Also shown in Table 3.3 are the three principal stresses used to determine the strains. Values without parentheses were obtained by KEN-LAYER, while those within parentheses were obtained by ELSYM5. It can be seen that both solutions check quite closely with a maximum discrepancy not over 2%.

It should be noted that any computer solutions based on Burmister's layered theory are not exact and involve the numerical integration of an infinite series. Because the accuracy of the solutions depends on the interval and tolerance

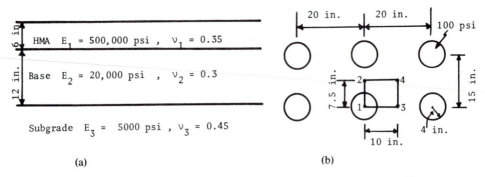

Figure 3.9 A layer system under six circular loaded areas (1 in. = 25.4 mm, 1 psi = 6.9 kPa).

specified for the integration, the solutions obtained from the two programs are not expected to be exactly the same.

3.3.2 Nonlinear Layers

The extension of linear elastic layered systems to include nonlinear, or stress sensitive, materials is certainly an improvement of practical significance. However, the use of the stresses at a single point in each nonlinear layer to compute the modulus of the layer is not theoretically correct. As the stresses vary with the radial distance from the load, the modulus should also change with the radial distance and is not uniform throughout the layer. Theoretically, this problem may be solved more accurately by the finite element method, such as the ILLI-PAVE program developed at the University of Illinois (Raad and Figueroa, 1980) and the MICH-PAVE program developed at the Michigan State University (Harichandran et al., 1989). If only the most critical tensile strain at the bottom of asphalt layer and the most critical compressive strain on the top of subgrade are required, it is possible to select a point in each nonlinear layer to compute the modulus, so that these critical strains obtained from KENLAYER can match reasonably well with those from the finite element programs. Unfortunately, this cannot be done at the present time due to the inaccuracy of the finite element solutions currently available.

ILLI-PAVE Model

The ILLI-PAVE computer program considers the pavement as an axisymmetric solid of revolution and divides it into a number of finite elements, each as a section of concentric rings. Incorporated in ILLI-PAVE are the stress-dependent resilient modulus and the failure criteria for granular materials and fine-grained soils. The principal stresses in the granular and subgrade layers are modified at the end of each iteration, so that they do not exceed the strength of the materials, as defined by the Mohr–Coulomb theory of failure (Raad and Figueroa, 1980).

TABLE 3.3 COMPARISON OF RESULTS BETWEEN KENLAYER AND ELSYM5

Point	Vertical coordinate	Vertical deflection (10^{-3} in.)	Strain (10^{-6})		Principal stress (psi)		
			Vertical compressive	Horizontal tensile	Major	Intermediate	Minor
1	6.00	70.50 (70.67)	244.3 (243.3)	−241.4 (−240.7)	13.9 (13.9)	−143.3 (−142.7)	−166.0 (−165.4)
	18.01	63.83 (64.06)	962.7 (949.3)	−502.7 (−496.4)	6.1 (6.1)	1.4 (1.5)	0.9 (0.9)
2	6.00	72.20 (72.08)	175.0 (175.2)	−157.3 (−157.5)	11.8 (11.8)	−104.9 (−105.0)	−111.2 (−111.4)
	18.01	65.45 (65.32)	1038.0 (1039.0)	−515.8 (−516.7)	6.3 (6.3)	1.6 (1.5)	1.0 (1.0)
3	6.00	67.81 (67.79)	119.5 (119.6)	−184.4 (−184.5)	9.2 (9.2)	−41.6 (−41.6)	−103.5 (−103.6)
	18.01	62.19 (62.17)	906.8 (907.8)	−482.7 (−483.2)	5.9 (5.9)	1.4 (1.4)	0.9 (0.9)
4	6.00	69.76 (69.77)	125.3 (125.3)	−173.7 (−173.6)	9.6 (9.6)	−50.5 (−50.6)	−101.2 (−101.2)
	18.01	63.75 (63.75)	979.3 (979.8)	−496.0 (−496.2)	6.1 (6.1)	1.5 (1.5)	0.9 (0.9)

Note. Values without parentheses are obtained by KENLAYER and those within parentheses by ELSYM5. Compression is positive and tension negative. 1 in. = 25.4 mm, 1 psi = 6.9 kPa.

Stress Modification

The procedure for stress modification as incorporated in ILLI-PAVE and MICH-PAVE can be best illustrated by a granular material with a cohesion of 0 and an angle of internal friction ϕ. The Mohr–Coulomb failure criterion can be written as

$$s = \sigma \tan \phi \qquad (3.20)$$

in which s is the shear strength, or the shear stress at the time of failure, σ is the normal stress, and ϕ is the angle of internal friction. Under a vertical loading stress σ_z and a vertical geostatic stress γz, the total vertical stress σ_v is

$$\sigma_v = \sigma_z + \gamma z \qquad (3.21)$$

The minimum and maximum allowable principal stresses $(\sigma_3)_{min}$ and $(\sigma_1)_{max}$ can be determined from the Mohr's circles, as shown by circles A and B in Figure 3.10 and expressed as

$$(\sigma_3)_{min} = \sigma_v \tan^2\left(45° - \frac{\phi}{2}\right) \qquad (3.22)$$

$$(\sigma_1)_{max} = \sigma_v \tan^2\left(45° + \frac{\phi}{2}\right) \qquad (3.23)$$

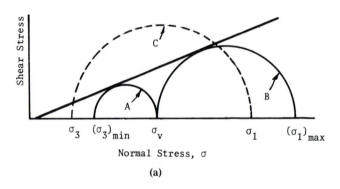

(a)

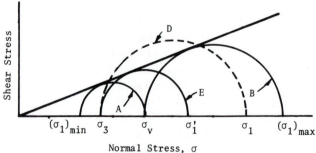

(b)

Figure 3.10 Stress modification to satisfy Mohr–Coulomb failure criterion.

Kenlayer Computer Program   Chap. 3

If the computed minor principal stresses σ_3, including the horizontal geostatic stress, is smaller than $(\sigma_3)_{min}$, as shown by circle C in Figure 3.10a, it must be increased to $(\sigma_3)_{min}$, so circle C is reduced to circle A with $\sigma_3 = (\sigma_3)_{min}$ and $\sigma_1 = \sigma_v$. If the minor principal stress σ_3 is greater than $(\sigma_3)_{min}$, as shown by circle D in Figure 3.10b, the maximum allowable major principal stress can be determined by circle E and expressed as

$$\sigma_1' = \sigma_3 \tan^2\left(45° + \frac{\phi}{2}\right) \tag{3.24}$$

If σ_1 is greater than σ_1', it must be reduced to σ_1', so circle D is reduced to circle E. If the computed σ_1 is smaller than σ_1', then the computed σ_1 should be used.

There are some concerns on the validity of the above method for modifying the principal stresses to satisfy the Mohr–Coulomb failure criterion. If the stress in one finite element is changed, it must be transferred to the adjoining elements, so that the equations of equilibrium are still satisfied. Arbitrarily changing the stresses in each element without considering overall equilibrium does not appear to be theoretically correct.

When the granular base or subbase is divided into several layers and the points used to evaluate the moduli are located at the midheight of each layer on the axis of symmetry for a single wheel and the plane of symmetry for dual wheels, the major principal stress is equal to σ_v and the minor principal stress may be very small in the upper layers but become tension in the lower layers. Therefore, the replacement of the small or negative σ_3 by a much larger positive $(\sigma_3)_{min}$ results in a much higher modulus. As indicated by Eq. 3.22, $(\sigma_3)_{min}$ decreases with the increase in ϕ. Therefore, after stress modifications, a granular material with a higher ϕ results in a lower modulus because of smaller σ_3. This is totally unreasonable and is the reason why KENLAYER replaces a negative σ_3 by 0, instead of $(\sigma_3)_{min}$. The assumption of $(\sigma_3)_{min} = 0$, which is equivalent to a ϕ angle of 90°, is actually on the safe side.

Based on experiments conducted with a layer of sand on a soft clay, Selig et al. (1986) indicated that the development of horizontal residual stresses under repeated loads is the key to the stability of the two-layer system. If this is true, then the stresses due to the applied load should not be changed, although the layer moduli may be adjusted due to the presence of residual stresses.

ILLI-PAVE Algorithms

In view of the fact that the computational techniques of ILLI-PAVE are too costly, complex, and cumbersome to be used for routine design, Thompson and Elliott (1985) developed simple regression equations or algorithms for predicting the responses of typical flexible pavements. The resilient modulus of the crushed stone base is represented by Eq. 3.8 with $K_1 = 9000$ psi (62 MPa) and $K_2 = 0.33$. The relationship between resilient modulus and deviator stress for four different subgrade soils is shown by Eq. 3.13 and Figure 3.6. Because parameters K_2, K_3, and K_4 are the same for all soils, the only soil property represented in the regression equation is K_1, which is the resilient modulus at the breakpoint. To apply the Mohr–Coulomb failure criteria, the crushed stone base is assumed

TABLE 3.4 SOME MATERIAL PROPERTIES FOR DEVELOPING ILLI-PAVE ALGORITHMS

Property	Hot mix asphalt			Crushed stone base	Subgrade soils			
	40°F	70°F	100°F		Stiff	Medium	Soft	Very soft
Unit weight (pcf)	145.0	145.0	145.0	135.0	125.0	120.0	115.0	110.0
Coefficient of earth pressure at rest	0.37	0.67	0.85	0.60	0.82	0.82	0.82	0.82
Poisson ratio	0.27	0.40	0.46	0.38	0.45	0.45	0.45	0.45
Modulus (10^3 psi)	1400	500	100	$9000\theta^{0.33}$		See Figure 3.6		

Note. 1 psi = 6.9 kPa, 1 pcf = 157.1 N/m³.
Source. After Thompson and Elliott (1985).

cohesionless with a friction angle of 40°, while the stiff, medium, soft, and very soft subgrade soils have a zero friction angle and a cohesion of 16.4, 11.425, 6.45, and 3.105 psi (113, 78.8, 44.5, and 21.4 kPa), respectively, which are equal to one-half of the unconfined compressive strength. Other material properties are shown in Table 3.4.

The regression equations were based on a 9000-lb (40.1-kN) circular load with a contact pressure of 80 psi (552 kPa), which represents one dual wheel of the standard 18,000-lb (81-kN) single-axle load. A total of 168 pavement configurations were solved by ILLI-PAVE, with thicknesses ranging from 1.5 to 8 in. (38 to 203 mm) for HMA and from 4 to 24 in. (102 to 610 mm) for the granular base. The factors included in the analysis as independent variables are the thickness and modulus of HMA, the thickness of granular base, and the subgrade K_1. The predicted responses include radial strain at the bottom of HMA, vertical strain on the top of subgrade, subgrade deviator stress, surface deflection, and subgrade deflection. Only algorithms for the HMA tensile strain ϵ_t and the subgrade compressive strain ϵ_c are presented below.

$$\log \epsilon_t = 2.9496 + 0.1289h_1 - \frac{0.5195}{h_1} \log h_2$$

$$- 0.0807h_1 \log E_1 - 0.0408 \log K_1 \qquad (3.25)$$

$$\log \epsilon_c = 4.5040 - 0.0738h_1 - 0.0334h_2$$

$$- 0.3267 \log E_1 - 0.0231K_1 \qquad (3.26)$$

in which ϵ_t and ϵ_c are in microinch per inch, or 10^{-6}, h_1 is the HMA thickness in inches, h_2 is the base thickness in inches, E_1 is the HMA modulus in ksi, and K_1 is the breakpoint resilient modulus of subgrade in ksi. The strains computed by Eqs. 3.25 and 3.26 will be used to compare with those obtained by MICH-PAVE and KENLAYER.

MICH-PAVE Model

The MICH-PAVE computer program is very similar to ILLI-PAVE and uses the same methods to characterize granular materials and fine-grained soils and the same Mohr–Coulomb failure criteria to adjust the state of stresses. A major improvement is the use of a flexible boundary at a limited depth beneath the surface of subgrade, instead of a rigid boundary at a large depth below the surface. The subgrade below the flexible boundary is considered as a homogeneous half-space, whose stiffness matrix can be determined and superimposed to the stiffness matrix of the pavement above the flexible boundary to form the overall stiffness matrix. This procedure is quite similar to the case of slabs on solid foundation, as described in Section 5.1.1. The use of a flexible boundary greatly reduces the number of finite elements required, especially those oblong elements at the bottom. Consequently, the storage requirement is significantly reduced and the program can be implemented on personal computers. The fewer number of simultaneous equations to be solved and the elimination of those oblong elements also yield more accurate results.

Comparison with ILLI-PAVE

Figure 3.11 shows the cross section of a three-layer system subjected to a circular load with a contact radius of 6 in. (152 mm) and a contact pressure of 80 psi (552 kPa). Layer 1 is HMA with $E_1 = 500,000$ psi (3.5 GPa), layer 2 is a 12-in. (305-mm) granular base with $K_1 = 9000$ psi (62 MPa) and $K_2 = 0.33$, and layer 3 is a clayey subgrade with $K_1 = 3020$ psi (20.8 MPa). Other properties are the same as those used in developing the ILLI-PAVE algorithms. This cross section was used to compare the results obtained from different models.

Figure 3.12 shows a comparison of tensile strain at the bottom of asphalt layer. The tensile strains for various HMA thicknesses h_1 were obtained by both ILLI-PAVE and MICH-PAVE and plotted as a series of curves. The curves for ILLI-PAVE model and ILLI-PAVE algorithm were originally presented by Thompson and Elliott (1985) to show the difference between the two. As can be seen from the figure, there are significant differences between ILLI-PAVE model and ILLI-PAVE algorithm. Of particular concern is the large difference between ILLI-PAVE and MICH-PAVE models.

In running MICH-PAVE, the HMA was divided into four layers, the base into six layers, and the 45-in. (1.14-m) subgrade above the flexible boundary by six layers. Figure 3.13 shows the finite element mesh for the pavement with 5-in. (127-mm) HMA, as plotted by MICH-PAVE.

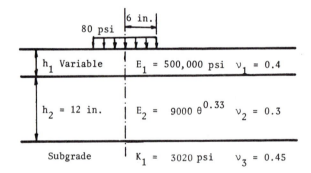

Figure 3.11 Cross section of a three-layer nonlinear system (1 in. = 25.4 mm, 1 psi = 6.9 kPa).

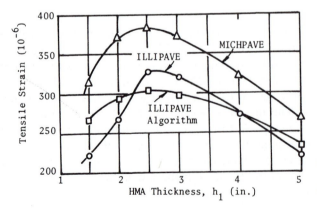

Figure 3.12 Comparison of tensile strains at bottom of asphalt layer (1 in. = 25.4 mm).

FINITE ELEMENT MESH

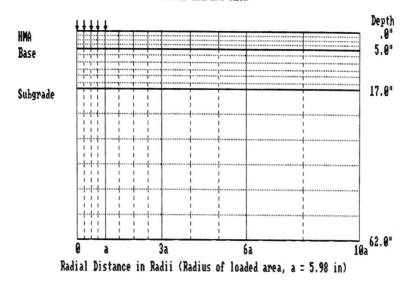

Figure 3.13 Finite element mesh for pavement with 5-in. HMA (1 in. = 25.4 mm).

Example 3.1:

For the pavement shown in Figure 3.11 with $h_1 = 2.5$ in. (64 mm), determine the tensile strains at the bottom of asphalt layer and the compressive strain on the top of subgrade by ILLI-PAVE algorithms.

Solution: From Eq. 3.25, $\log \epsilon_t = 2.9496 + 0.1289 \times 2.5 - (0.5195/2.5) \log 12 - 0.0807 \times 2.5 \times \log 500 - 0.0408 \log 3.02 = 2.483$, or $\epsilon_t = 304 \times 10^{-6}$, which checks with the value shown in Figure 3.12. From Eq. 3.26, $\log \epsilon_c = 4.5040 - 0.0738 \times 2.5 - 0.0334 \times 12 - 0.3267 \times \log 500 - 0.0231 \times 3.02 = 2.967$, or $\epsilon_c = 926 \times 10^{-6}$, which checks with the value shown in Figure 3.17.

Linear Analysis

The results of MICH-PAVE are more reasonable than those of ILLI-PAVE and check more favorably with KENLAYER. However, there are some concerns on the accuracy of MICH-PAVE. Take the pavement with 5-in. (127-mm) HMA for example. MICH-PAVE gives an equivalent resilient modulus of 19,720 psi (136 MPa) for the granular base and 8076 psi (55.7 MPa) for the subgrade. By considering the pavement as linear elastic with the above base and subgrade moduli, the responses obtained by MICH-PAVE, KENLAYER, and ELSYM5 are compared and presented in Table 3.5. The responses to be compared include the vertical deflection on the surface, the four components of strains at the bottom of asphalt layer, and the same strains on the top of subgrade. Comparisons were made at the axis of symmetry as well as at a distance of 0.7 in. (18 mm) from the axis, which is the center of the first column of finite elements. A distance of 0.7 in. (18 mm) is selected for comparison because it is located at the center of elements and the results are supposed to be more accurate.

TABLE 3.5 COMPARISON OF LINEAR SOLUTIONS BY MICH-PAVE, KENLAYER, AND ELSYM5

Model	Surface deflection (10⁻³ in.)	Strains at bottom of HMA (10⁻⁶ in./in.)				Strain on top of subgrade (10⁻⁶ in./in.)			
		ϵ_z	ϵ_r	ϵ_t	γ_{rz}	ϵ_z	ϵ_r	ϵ_t	γ_{rz}
		At Axis of Symmetry, or $r = 0$							
MICH-PAVE	23.33	369.9	−266.7	−266.7	129.2	457.0	−201.5	−201.5	37.4
KENLAYER	25.09	386.0	−276.9	−276.9	0.0	560.4	−245.2	−245.2	0.0
ELSYM5	25.19	386.1	−277.0	−277.0	0.0	560.8	−245.6	−245.6	0.0
		At $r = 0.7$ in.							
MICH-PAVE	23.24	285.8	−266.7	−266.7	52.6	454.7	−201.5	−201.5	43.9
KENLAYER	25.05	383.8	−274.5	−276.0	4.0	559.3	−244.5	−245.0	39.1
ELSYM5	25.16	384.0	−274.7	−276.2	4.0	559.7	−244.9	−245.4	39.1

Note. ϵ_z, ϵ_r, ϵ_t, and γ_{rz} are vertical, radial, tangential, and shear strains with compression as positive and tension as negative. 1 in. = 25.4 mm.

As can be seen from Table 3.5, the results of KENLAYER and ELSYM5 check very closely but are significantly different from those of MICH-PAVE, especially for the strains on the top of subgrade. Based on the equivalent moduli obtained by MICH-PAVE, a comparison of linear solutions between KEN-LAYER and MICH-PAVE for various HMA thicknesses is presented in Figure 3.14. When the HMA thickness is greater than 2.5 in. (64 mm), the tensile strains at the bottom of asphalt layer check reasonably well. However, when the HMA thickness is smaller than 2.5 in. (64 mm), the tensile strains obtained by MICH-PAVE are much greater than those obtained by KENLAYER. For all HMA thicknesses, the compressive strains on the top of subgrade obtained by MICH-PAVE are only 80% of those obtained by KENLAYER.

There are several possibilities that may cause inaccurate results in the finite element method:

1. The shape of finite elements has a significant effect on the accuracy obtained. The necessity of simulating layer systems by using oblong elements can yield inaccurate results.
2. The stresses and strains can be evaluated most accurately at the center of elements. Evaluations at element boundaries, particularly at element corners, are more prone to error.
3. The stresses at layer interfaces are obtained by linear extrapolation from those at the center of the two elements below or above the interface. This procedure can lead to error if the mesh is not fine enough.

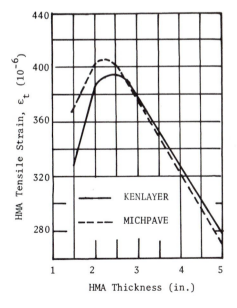 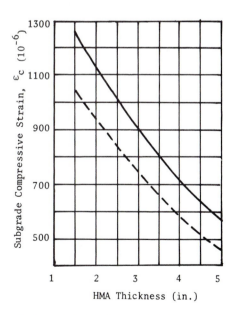

Figure 3.14 Comparison of linear solutions between KENLAYER and MICH-PAVE (1 in. = 25.4 mm).

KENLAYER Model

Two methods can be used in KENLAYER to compare with MICH-PAVE. The first method is similar to MICH-PAVE in that the granular base is subdivided into a number of layers and the stresses at the midheight are used to determine the modulus of each layer. In the second method the base course is considered as one layer and an appropriate point is selected for evaluating the base modulus.

Method 1 with Subdivision

In this method, the base was divided into six layers, each 2 in. (50.8 mm) thick, as shown in Figure 3.15. The vertical coordinates of the stress points are located at the midheight of each layer and at 6 in. (152 mm) below the top of subgrade. Because the maximum tensile strain at the bottom of asphalt layer and the maximum compressive strain on the top of subgrade are affected principally by the modulus of the materials lying directly beneath the loaded area, it is recommended that the stress points be located on the axis of symmetry, as shown in Figure 3.15.

Method 2 Without Subdivision

When the granular base is divided into a number of layers, the modulus of the uppermost layer, which lies directly below the HMA, has most effect on the HMA tensile strain; while the modulus of the lowest layer, which lies directly above the subgrade, has most effect on the subgrade compressive strain. To obtain comparable strains by method 2 without subdivision and thus save a large amount of computer time, the stress point in the granular base must be above the midheight close to the HMA, while the stress point in the subgrade must move up close to the top of subgrade. It was found that when the stress points for method 1 with subdivision were located at the midheight of each granular layer and at 6 in.

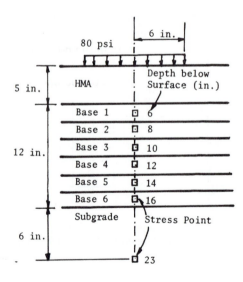

Figure 3.15 Location of stress points for method 1 with subdivision (1 in. = 25.4 mm, 1 psi = 6.9 kPa).

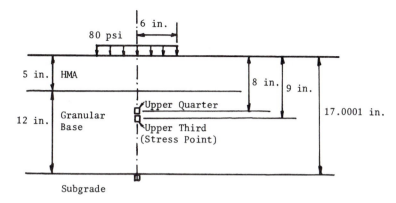

Figure 3.16 Location of stress point for method 2 without subdivision (1 in. = 25.4 mm, 1 psi = 6.9 kPa).

(152 mm) below the top of subgrade, those for method 2 without subdivision should be located between the upper quarter and the upper third points in the granular layer and at the top of subgrade, as shown in Figure 3.16. If the same stress point for the subgrade is used in both methods, method 2 without subdivision will result in a much smaller subgrade compressive strain because the modulus of the granular base computed at the upper quarter or third point is much greater than the moduli of the lowest granular layer, which lies directly above the subgrade. To increase the subgrade strain in method 2, the stress point in the subgrade must move up to increase the deviator stress.

Comparison with MICH-PAVE

Figure 3.17 shows a comparison of nonlinear solutions between method 1 of KENLAYER, as indicated by the solid curve, and MICH-PAVE, as indicated by the dashed curve. The differences between the two nonlinear solutions are similar to those of the two linear solutions shown in Figure 3.14. This is as expected because the nonlinear model is a simple extension of the linear model and any discrepancies in the linear solutions will show up in the nonlinear solutions. Also shown in Figure 3.17 are the KENLAYER solutions based on method 2 without subdivision. It can be seen that the results obtained by method 1 lie between those obtained by method 2, with one stress point at the upper quarter and the other at the upper third. Except for the case with 1.5 in. (38 mm) HMA, the use of upper quarter point appears to give results closer to method 1. The figure also shows a large difference in subgrade compressive strains between KENLAYER and ILLI-PAVE algorithm.

Analysis of Deflection Basins

One important usage of the KENLAYER program is to back-calculate the moduli of various layers, so that the deflection basin obtained by the program can match with field measurements.

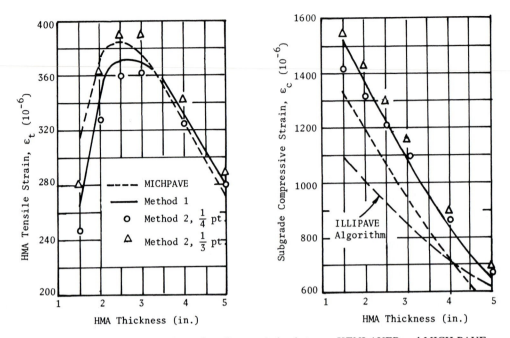

Figure 3.17 Comparison of nonlinear solution between KENLAYER and MICH-PAVE (1 in. = 25.4 mm).

Analysis by ILLI-PAVE

Chua and Lytton (1985) employed the ILLI-PAVE program to back-calculate the moduli of light pavement structures consisting of a thin asphalt surface on a granular base. They used a Dynatest FWD test system which delivered an impulse load to the pavement through a plate 12 in. (305 mm) in diameter resting on a thick rubber pad. The deflections were measured by velocity

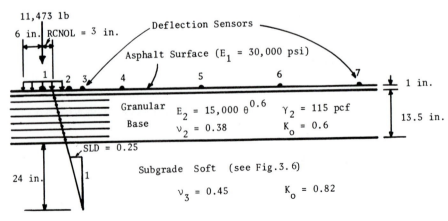

Figure 3.18 Cross section of pavement for deflection anaylsis (1 in. = 25.4 mm, 1 lb = 4.45 N, 1 psi = 6.9 kPa, 1 pcf = 157.1 N/m³).

Kenlayer Computer Program   Chap. 3

transducers at seven locations: 0, 7.9, 11.8, 23.6, 47.2, 70.9, and 94.5 in. (0, 0.2, 0.3, 0.6, 1.2, 1.8, and 2.4 m) from the center of load. Figure 3.18 is one of the pavement sections Chua and Lytton analyzed. Using the parameters shown in Figure 3.18 as input to ILLI-PAVE, the deflections at the seven locations were obtained and checked reasonably with the field measurements.

Table 3.6 is a comparison of ILLI-PAVE solutions and field measurements, as reported by Chua and Lytton (1985). Also shown in the table are the deflections obtained by MICH-PAVE and KENLAYER as well as those for a homogeneous half-space. In running KENLAYER, the granular base was divided into seven layers with the stress points shown in Figure 3.18. As indicated previously, the stress points for computing the maximum strains are located on the axis of symmetry, or RCNOL = 0 and SLD = 0. However, for analyzing deflection basins, average layer moduli based on RCNOL = 3 in. (76 mm) and SLD = 0.25 is recommended. The stress point for the subgrade is assumed at a depth of 24 in. (6.1 m) below the top of subgrade, instead of the 6 in. (152 mm) for computing maximum strains. It was found that the use of 24 in. (6.1 m) results in a subgrade modulus of 7668 psi (52.9 kPa), while the use of 6 in. (152 mm) results in 6608 psi (45.6 MPa). For a homogeneous half-space, an elastic modulus of 7682 psi (53 MPa) is assumed, which is the maximum allowable for a soft subgrade with a K_1 of 3020 psi (20.8 MPa), as assumed by Chua and Lytton and indicated in Figure 3.6.

As can be seen in Table 3.6, the deflections obtained by KENLAYER check very well with those obtained by MICH-PAVE but are quite different from those obtained by ILLI-PAVE. When a point is far away from the load, the surface deflection depends on the modulus of the subgrade and is practically independent of the modulus of the overlying layers. This is shown in Table 3.6 where the deflections based on layered system and homogeneous half-space are significantly different at the first few locations but become nearly the same at the last few locations. The deflections obtained by ILLI-PAVE at the last few locations do not appear reasonable because they are much smaller than those based on a homogeneous half-space, even if the half-space is assumed to have the largest possible modulus.

TABLE 3.6 COMPARISON OF DEFLECTION BASIN

Sensor number	Radial distance (in.)	Deflection (10^{-3} in.)				
		Field	ILLI-PAVE	MICH-PAVE	KENLAYER	Half-space
1	0	26.57	26.99	32.55	34.21	127.87
2	7.9	19.45	22.57	27.08	26.92	52.39
3	11.8	16.02	19.96	23.69	23.80	32.02
4	23.6	10.12	4.80	15.64	16.31	16.30
5	47.2	4.57	2.40	8.41	9.03	8.08
6	70.9	2.40	2.15	*	5.73	5.37
7	94.5	2.17	1.58	*	4.17	4.02

Note. * points outside the boundary. 1 in. = 25.4 mm.

Analysis by KENLAYER

Because the results obtained by ILLI-PAVE are doubtful, KENLAYER was used to match the field measurements. It was found that a close match could be obtained by reducing the K_1 of the granular base from 15,000 to 10,000 psi (104 to 69 MPa) and increasing the K_1 of the subgrade from 3020 to 12,340 (20.8 to 85.1 MPa), as shown in Figure 3.19. Using the same input data as in KEN-LAYER, the results by MICH-PAVE also check reasonably with the field measurements. Also shown in the figure are the solutions reported by Chua and Lytton using ILLI-PAVE. It can be seen that even the shape of deflection basin obtained by ILLI-PAVE does not match with the field data.

After values of K_1 for both the granular base and subgrade are back-calculated from deflection measurements, they can be used to determine the maximum tensile strain at the bottom of asphalt layer and the maximum compressive strain on the top of subgrade by selecting stress points directly under the load. This is the major advantage of nonlinear analysis over the linear analysis. In the linear analysis, the back-calculated elastic moduli are used directly for design, regardless of the fact that the layer modulus is not uniform and varies with the magnitude of the load and the distance from the load. In the nonlinear analysis, average values of K_1 based on the matching of deflection basin are back-calculated and later used to determine the moduli based on the magnitude of load and the location at which responses are to be evaluated.

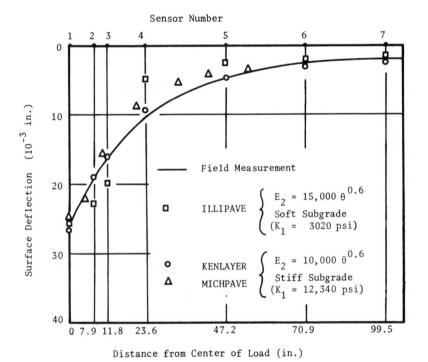

Figure 3.19 Comparison of deflection basin by different models (1 in. = 25.4 mm, 1 psi = 6.9 kPa).

3.3.3 Moving Loads

The use of KENLAYER for the analysis of viscoelastic layered systems under a moving load involves several steps. First, the creep compliances of HMA at several time durations are specified at a reference temperature. The compliances are then fitted with a Dirichlet series and the time–temperature superposition principle is applied to obtain the compliances at any given temperature and time. Next, the viscoelastic responses under a static load are determined and expressed again as a Dirichlet series and the Boltzmann's superposition principle is then applied to obtain the responses under a moving load. Because these procedures are complex and invoke repeatedly the approximate method of collocation, it is desirable to check the entire process from the input of creep compliances at a reference temperature to the output of responses under moving loads at various pavement temperatures. This can be accomplished by comparing the solutions between KENLAYER and VESYS. Details about the VESYS computer model are described in Section 10.5.1.

Figure 3.20 shows a viscoelastic three-layer system subjected to a moving load with a duration of 0.1 s. The load is applied over a circular area with a radius of 6 in. (152 mm) and a contact pressure of 74.43 psi (514 kPa). The HMA layer is considered viscoelastic with its creep compliances specified at 11 time durations under a reference temperature of 70°F (21°C). The base and the subgrade are considered linear elastic with their moduli shown in the figure.

The system shown in Figure 3.20 was employed by Kenis (1977) to illustrate the application of VESYS. Each year was divided into twelve months and the maximum deflections on the pavement surface and the radial tensile strains at the bottom of asphalt layer at twelve different temperatures were presented and are tabulated in Table 3.7. Also shown in Table 3.7 are the solutions obtained by KENLAYER. It can be seen that some solutions by VESYS are nearly the same as those by KENLAYER while others are somewhat different. However, the discrepancies are not more than 5%. In view of the approximate procedures

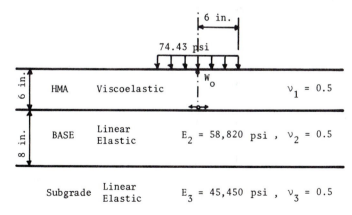

Figure 3.20 A viscoelastic three layer system under a moving load (1 in. = 25.4 mm, 1 psi = 6.9 kPa).

TABLE 3.7 COMPARISON OF VESYS AND KENLAYER

Pavement temperature (°F)	Deflection (10^{-3} in.)		Radial strain (10^{-6})	
	VESYS	KENLAYER	VESYS	KENLAYER
49.7	3.884	3.764	38.86	36.89
53.3	3.922	3.816	39.97	38.20
59.5	4.152	4.152	46.82	46.69
68.6	5.526	5.615	90.16	90.89
75.2	6.827	7.160	141.9	147.8
81.6	8.232	8.753	201.0	210.5
84.6	8.861	9.335	226.5	237.9
84.7	8.882	9.359	227.3	238.7
78.9	7.658	8.081	177.2	184.8
70.1	5.824	5.950	100.8	102.2
59.1	4.126	4.116	46.01	45.73
52.3	3.908	3.795	39.55	37.70

Note. Radial strains are in tension but negative signs are not shown.
1 in. = 25.4 mm.

employed by both programs, these discrepancies are expected and should be considered acceptable.

3.3.4 Damage Analysis

The DAMA computer program developed by the Asphalt Institute also employs the damage concept to determine the design life. It is interesting to compare the results obtained from DAMA with those from KENLAYER.

DAMA Model

The DAMA computer program can be used to analyze a multilayer elastic pavement structure by cumulative damage techniques for a single- or dual-wheel system. Any pavement structure comprised of hot mix asphalt, emulsified asphalt mixtures, untreated granular materials, and subgrade soils can be analyzed provided the maximum number of layers does not exceed five. Environmental effects are characterized as input variables by mean monthly temperatures and variable monthly material modulus.

Pavement Temperature

The air temperature data are used to account for the effect of temperature on moduli of asphalt mixtures. The relationship between mean pavement temperature M_p and mean monthly air temperature M_a is based on the depth below pavement surface by

$$M_p = M_a\left(1 + \frac{1}{z + 4}\right) - \frac{34}{z + 4} + 6 \tag{3.27}$$

in which z is the depth below surface in inches. The temperature at the upper third point of each layer is used as the weighted average pavement temperature.

The capability to account for monthly variations in the properties of base, subbase, and subgrade allows one to assess the effects of freeze–thaw or variable moisture conditions on pavement life. For pavements using emulsified asphalt stabilized materials, the effect of modulus change over any specified cure time is also considered in the damage analysis.

Nonlinear Analysis

In DAMA, the subgrade and all asphalt stabilized layers are considered to be linear elastic and the untreated granular base to be nonlinear elastic. The consideration of subgrade as linear elastic is a reasonable approximation because the variation of modulus due to the change of subgrade stresses is usually quite small and a reasonable subgrade modulus can be assumed. Instead of using the more accurate method of iterations and Eq. 3.8 to determine the modulus of granular layer, the following predictive equation based on multiple regression is used to account for stress dependency (Smith and Witczak, 1981):

$$E_2 = 10.447 h_1^{-0.471}\ h_2^{-0.041}\ E_1^{-0.139}\ E_3^{0.287}\ K_1^{0.868} \tag{3.28}$$

in which E_1, E_2, and E_3 are the modulus of asphalt layer, granular base, and subgrade, respectively; h_1 and h_2 are the thickness of the asphalt layer and granular base, respectively; and K_1 is the nonlinear constant in Eq. 3.8 with the exponent K_2 equal to 0.5.

Equation 3.28 was developed from pavement structures composed of only three layers under an 18,000-lb (80-kN) single-axle load. When two different asphalt stabilized layers with thicknesses of h_{1a} and h_{1b} and moduli of E_{1a} and E_{1b} are present above the aggregate layer, the equivalent modulus E_1 of the combined asphalt layer with thickness of $h_{1a} + h_{1b}$ is determined by

$$E_1 = \left[\frac{h_{1a}(E_{1a})^{1/3} + h_{1b}(E_{1b})^{1/3}}{h_{1a} + h_{1b}}\right]^3 \tag{3.29}$$

Example 3.2:

For the four-layer system shown in Figure 3.21 with moduli of 966,000 psi (6.7 GPa) for the surface course, 1,025,000 psi (7.1 GPa) for the binder course, and 12,000 psi (8.3 MPa) for the subgrade, determine the modulus of granular layer by Eq. 3.28 for a K_1 of 8000 psi (55 MPa).

Solution: In Eq. 3.28, E_1 is the equivalent modulus for the surface and binder courses combined and can be determined from Eq. 3.29, or $E_1 = \{[1.5 \times \sqrt[3]{966,000} + 4.0 \times \sqrt[3]{1,025,000}]/(1.5 + 4.0)\}^3 = 1,009,000$ psi (7.0 GPa). From Eq. 3.28 with $h_1 = 1.5 + 4.0 = 5.5$ in. (140 mm), the modulus of granular base $= 10.447 \times (5.5)^{-0.471} \times (6)^{-0.041} \times (1,009,000)^{-0.139} \times (12,000)^{0.287} \times (8000)^{0.868} = 23,040$ psi (159 MPa).

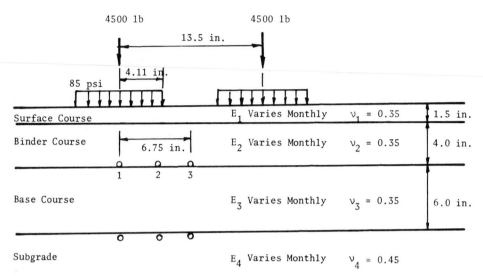

Figure 3.21 A four-layer linear system for damage analysis (1 in. = 25.4 mm, 1 psi = 6.9 kPa, 1 lb = 4.45 N).

Comparison of Strains by Linear Analysis

Figure 3.21 shows a four-layer linear system subjected to a 9000-lb (40-kN) dual-wheel load with a contact pressure of 85 psi (587 kPa). Using the Asphalt Institute's failure criteria, as shown in Eqs. 3.6 and 3.7 with $f_1 = 0.0796$, $f_2 = 3.291$, $f_3 = 0.854$, $f_4 = 1.365 \times 10^{-9}$, and $f_5 = 4.477$, a damage analysis was made by both DAMA and KENLAYER and the results were compared.

Table 3.8 shows a comparison of tensile strains at the bottom of the HMA binder course and the compressive strains on the top of subgrade between DAMA and KENLAYER. It can be seen that both solutions check very well.

Comments on Table 3.8

1. The monthly air temperature is used to determine the HMA modulus, which is described in Section 7.2.3. These are the temperatures at South Carolina and were applied by the Asphalt Institute to develop design charts.

2. The granular base is assumed nonlinear. The nonlinear constants K_1 vary throughout the year and are used to determine the modulus of the granular base. The value of K_1 is 8000 at normal months, but may increase to 400% in the winter and decrease to 25% during the spring breakup. After these values are entered, DAMA computes the modulus for each month by Eqs. 3.28 and 3.29. A modulus of 23,039 psi (159 MPa) for December checks with Example 3.2.

3. The moduli of HMA surface and binder courses are based on a standard mix and vary with the pavement temperature. Although the same mix properties

TABLE 3.8 COMPARISON OF STRAINS DETERMINED BY DAMA AND KENLAYER

Month	Air temperature (°F)	Granular base K_1	Modulus (psi)				Tensile strain (10^{-6})		Compressive strain (10^{-6})	
			HMA surface	HMA binder	Granular base	Sub-grade	DAMA	KENLAYER	DAMA	KENLAYER
Jan	45	8000	1,436,929	1,478,458	21,870	12,000	106.1	106.2	286.2	285.6
Feb	38	16,000	1,875,179	1,891,803	50,610	31,000	67.2	67.2	152.8	153.0
Mar	43	24,000	1,561,968	1,598,808	84,539	50,000	61.9	61.9	123.9	124.2
Apr	45	2000	1,436,929	1,478,458	5670	7200	128.9	129.7	308.8	307.2
May	56	3500	878,127	937,209	10,277	8400	171.1	171.1	410.1	409.4
Jun	70	5000	415,886	473,105	16,042	9600	250.7	250.6	553.0	553.8
Jul	78	6500	254,835	302,849	22,206	10,800	295.8	295.8	613.3	613.5
Aug	81	8000	209,688	253,517	28,111	12,000	292.9	293.2	599.4	599.5
Sep	78	8000	254,835	302,849	27,408	12,000	271.3	271.5	568.2	568.3
Oct	73	8000	347,698	402,234	26,322	12,000	236.5	236.6	516.2	516.9
Nov	58	8000	798,268	857,035	23,630	12,000	152.7	152.7	377.0	377.1
Dec	54	8000	965,976	1,024,883	23,039	12,000	136.2	136.2	346.0	345.9

Note. Negative signs for tensile strains not shown. 1 psi = 6.9 kPa.

are specified for surface and binder courses, the moduli are slightly different due to the difference in pavement temperatures at different depths.

4. The subgrade is considered linear elastic. The moduli of the subgrade is 12,000 psi (82.8 MPa) at normal months, but may range from 7200 to 50,000 psi (50 to 345 MPa).

5. The tensile strains at the bottom of the binder course and the compressive strains on the top of the subgrade were computed at three points: one under the center of one wheel, one at the edge of one wheel, and the other at the center between the two duals, as shown in Figure 3.21. Only the maximum among the three is presented in the table. It is not necessary to compute the strains at the bottom of the surface course because they are not critical and may be in compression.

6. The tensile strain for DAMA is the overall principal strain but that for KENLAYER is the horizontal principal strain. As can be seen, both strains check very closely. However, this is not true at the bottom of the HMA surface course, where a large difference exists between the two. It was found that the overall principal strain at the bottom of the surface course was in tension but the horizontal principal strain was in compression. If the surface and binder courses are combined as a unit, the bottom of the thinner surface course lies above the neutral axis and should be in compression. This is the reason the horizontal principal strain is used for fatigue analysis. For multiple wheels, KENLAYER prints out both the overall principal strain and the horizontal principal strain but only the horizontal strain is used for damage analysis.

7. KENLAYER solutions are based on linear elastic layers using the modulus values obtained from DAMA, as shown in the table. The close agreement between DAMA and KENLAYER indicates the correctness of the linear elastic solutions.

Comparison of Damage Ratios by Nonlinear Analysis

Table 3.9 is a comparison of damage ratios between DAMA and KENLAYER. The number of axle load repetitions during each month is assumed to be 5000. For each month, the damage ratios are computed at three points, as indicated in Figure 3.21, and the maximum among the three is shown in the table. In applying KENLAYER, both the HMA layers and the subgrade are assumed to be linear elastic, having the same modulus values as in DAMA, but the granular base is assumed to be nonlinear elastic with its modulus obtained by iterations based on the monthly K_1 values specified.

Two methods can be used with KENLAYER, as shown in Figure 3.22. In method 1, the granular base is divided into three layers, each 2 in. (51 mm) thick. In method 2, the granular base is considered as one layer with a stress point located at the upper quarter. The load is applied over dual wheels and another wheel in the transverse, or y, direction is not shown in the figure. In both methods, the stress points are placed laterally between the dual wheels with XPTNOL = 0, YPTNOL = 6.75 in. (171 mm), and SLD = 0.

TABLE 3.9 COMPARISON OF DAMAGE RATIOS DETERMINED BY DAMA AND KENLAYER

Month	Base Modulus (psi)			Damage ratio for ϵ_t (%)			Damage ratio for ϵ_c (%)		
	DAMA	Method 1	Method 2	DAMA	Method 1	Method 2	DAMA	Method 1	Method 2
Jan	21,870	22,080 / 18,920	21,850	0.0972	0.0998	0.0977	0.0502	0.0489	0.0497
Feb	50,610	50,940 / 43,390	50,490	0.0267	0.0275	0.0267	0.0030	0.0030	0.0030
Mar	84,539	86,030 / 71,510	83,960	0.0176	0.0183	0.0178	0.0012	0.0012	0.0012
Apr	5670	5494 / 5364	5472	0.1844	0.1901	0.1897	0.0705	0.0653	0.0660
May	10,277	10,500 / 9488	10,360	0.3172	0.3198	0.3165	0.2510	0.2452	0.2505
Jun	16,042	16,660 / 14,050	16,320	0.6225	0.6319	0.6147	0.9509	0.9662	0.9683
Jul	22,206	23,840 / 19,860	22,230	0.7313	0.7294	0.7318	1.5220	1.5619	1.5237
Aug	28,111	30,610 / 25,220	27,730	0.6098	0.5958	0.6223	1.3730	1.4173	1.3814
Sep	27,408	29,690 / 24,660	27,170	0.5515	0.5430	0.5581	1.0810	1.1131	1.0842
Oct	26,322	28,220 / 23,710	26,170	0.4473	0.4456	0.4502	0.7018	0.7260	0.7084
Nov	23,630	24,380 / 20,940	23,610	0.2023	0.2061	0.2022	0.1722	0.1733	0.1724
Dec	23,039	23,530 / 20,270	23,030	0.1617	0.1652	0.1617	0.1173	0.1169	0.1171
Sum of damage ratios (%)				3.9695	3.9725	3.9894	6.2941	6.4383	6.3259

Note. 1 psi = 6.9 kPa.

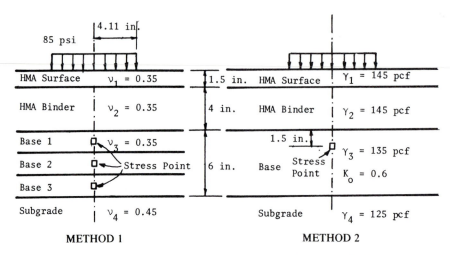

Figure 3.22 Two methods for characterizing nonlinear granular layers (1 in. = 25.4 mm, 1 pcf = 157.1 N/m³).

Comments on Table 3.9

1. Based on K_1 of the base course material, DAMA computes the base modulus by Eqs. 3.28 and 3.29, while KENLAYER determines the modulus by iterations. Two base moduli are shown for method 1: the top value for the top layer and the bottom value for the bottom layer. The modulus of the middle layer lies between the two and is not presented.

2. During each month, DAMA computes the damage ratios for fatigue cracking and permanent deformation at three points. These monthly ratios are summed separately over a year and the maximum ratio among the three at the end of one year is used to determine the design life. Because the maximum damage ratio for each month does not occur at the same point, the maximum damage ratios at the end of a year as obtained by DAMA are actually 3.943% for fatigue cracking and 6.244% for permanent deformation, which are slightly smaller than the sum of 3.970 and 6.294% shown in Table 3.9.

3. Since the damage analysis by KENLAYER is not limited to dual wheels with three fixed points but can also be applied to a combination of single-, dual-, and multiple-wheel loads, the maximum damage ratio during each month for each load group is determined and summed over the year to compute the design life. This procedure, although not theoretically correct, does give a clear picture of the damage during each month for each load group. The analysis is more conservative because the maximum damage ratio during each month for each load group may not occur at the same point.

4. Table 3.9 shows that the damage ratios obtained by KENLAYER check closely with those by DAMA, no matter which method is used. To save the computer time, it is suggested that method 2 be used for pavement design and that the stress point be placed laterally between the dual wheels and vertically at the upper quarter of the granular layer.

3.4 SENSITIVITY ANALYSIS

The information presented in this chapter is sufficient to run the KENLAYER computer program. Details of program development including layered theory, subroutines, flowcharts, and six illustrative examples are presented in Appendix B.

With the use of KENLAYER, sensitivity analyses were made on both three- and four-layer systems to illustrate the effect of various parameters on pavement responses. Due to the complex interactions among the large number of parameters, it is difficult to present a concise but accurate picture on the effect of a given parameter because the effect depends not only on the parameter itself but also on all other parameters. Conclusions based on a set of parameters may not be valid if some of the other parameters are changed. The best approach is to fix all other parameters at their most reasonable values while varying the parameter in question to show its effect.

3.4.1 Linear Analysis

This analysis is based on the assumption that all layers are linear elastic. Although HMA layers are viscoelastic and granular layers are nonlinear elastic, an approximate procedure is to assume them to be linear elastic by selecting appropriate moduli for HMA, based on vehicle speeds and pavement temperatures, and for granular materials, based on the level of loading.

Three-Layer Systems

To illustrate the effect of some design factors on pavement responses, an elastic three-layer system is used, as shown in Figure 3.23. The variables to be considered include layer thicknesses h_1 and h_2 and layer moduli E_1, E_2, and E_3.

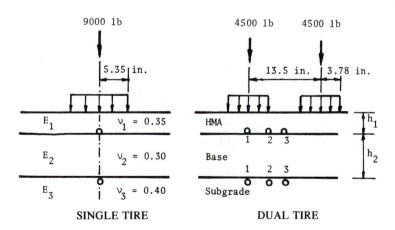

Figure 3.23 Three-layer systems subjected to single- and dual-wheel loads (1 in. = 25.4 mm, 1 lb = 4.45 N).

Two types of wheel loads are considered: one on a single tire and the other on a set of dual tires with a dual spacing of 13.5 in. (343 mm). A contact radius a of 5.35 in. (136 mm) is assumed for a single tire and 3.78 in. (96 mm) for dual tires. These radii are based on an 18-kip (80-kN) single-axle load with a contact pressure of 100 psi (690 kPa). The Poisson ratios for the three layers are 0.35, 0.3, and 0.4, respectively. For a single tire, the critical strains occur under the center of the loaded area. For a set of dual tires, the strains at points 1, 2, and 3, as shown in Figure 3.23, are computed and the largest among the three is selected as the most critical.

Effect of Layer Thickness

The effect of HMA thickness h_1 and base thickness h_2 on the tensile strain ϵ_t at the bottom of asphalt layer and the compressive strain ϵ_c at the top of subgrade was investigated.

Figure 3.24 shows the effect of h_1 on ϵ_t and ϵ_c when $E_1 = 500,000$ psi (34.5 GPa), $E_2 = 20,000$ psi (138 MPa), $E_3 = 7500$ psi (51.8 MPa), and $h_2 = 4$ or 16 in. (102 or 406 mm). The reason that two different thicknesses h_2 are used is to check whether the trend on a very thin base is also applicable to that on a thick base. The legend for various cases and a typical cross section are shown on the right side of the figure. A review of Figure 3.24 reveals the following trends:

1. For the same total load and contact pressure, single-wheel loads always result in greater ϵ_c, but not necessarily greater ϵ_t. When the asphalt surface is

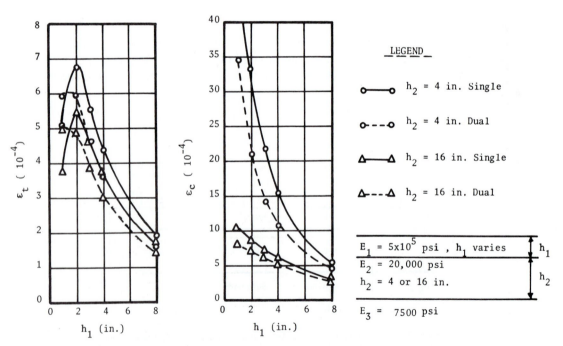

Figure 3.24 Effect of HMA thickness on pavement responses (1 in. = 25.4 mm, 1 psi = 6.9 kPa).

Kenlayer Computer Program   Chap. 3

very thin, ϵ_t under dual-wheel loads is greater than that under a single-wheel load. Therefore, the use of a single tire to replace a set of duals, as has been practiced in ILLI-PAVE and MICH-PAVE, is unsafe when analyzing the fatigue cracking of a thin asphalt surface.

2. Under a single-wheel load, there is a critical thickness at which ϵ_t is maximum. Above the critical thickness, the thicker the asphalt layer, the smaller is the tensile strain; whereas below this critical thickness, the thinner the asphalt layer, the smaller is the strain. The same trend is shown in Figures 2.21 and 3.12 and was also reported by Freeme and Marais (1973) who indicated that fatigue cracking could be minimized by keeping the asphalt surface as thin as possible. The critical thickness is not as pronounced under dual wheels as under single wheels.

3. Above the critical thickness, increasing h_1 is very effective in reducing ϵ_t, regardless of the base thickness. Unless the asphalt surface is less than 2 in. (51 mm) thick, the most effective way to prolong the fatigue life is to increase the HMA thickness.

4. Increasing h_1 is effective in reducing ϵ_c only when the base course is thin, not when the base course is thick.

Figure 3.25 shows the effect of h_2 on ϵ_t and ϵ_c when $E_1 = 500,000$ psi (34.5 GPa), $E_2 = 20,000$ (138 MPa), $E_3 = 7500$ psi (51.8 MPa), and $h_1 = 2$ or 8 in. (51 and 203 mm). The following trends can be found in Figure 3.25:

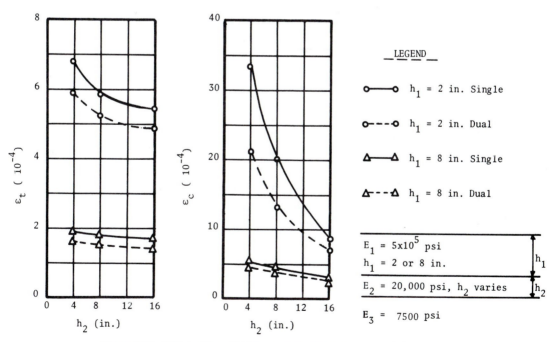

Figure 3.25 Effect of base thickness on pavement responses (1 in. = 25.4 mm, 1 psi = 6.9 kPa.)

1. When h_1 is 2 in. (51 mm) or more, the replacement of dual wheels by a single wheel increases both ϵ_t and ϵ_c.
2. An increase in h_2 does not cause a significant decrease in ϵ_t, especially when h_1 is large.
3. An increase in h_2 causes a significant decrease in ϵ_c only when h_1 is small. Unless a full-depth or thick layer of HMA is used, the most effective way to reduce ϵ_c is to increase h_2.

Effect of Layer Modulus

The effect of the base modulus E_2 and subgrade modulus E_3 on the tensile strain ϵ_t and the compressive strain ϵ_c is now discussed. The effect of HMA modulus E_1 is not presented, because it is well known that an increase in E_1 results in a decrease in ϵ_t and ϵ_c. However, an increase in E_1 also causes a decrease in the allowable number of repetitions for fatigue cracking, as indicated by Eq. 3.6. Whether a smaller ϵ_t due to a larger E_1 should increase or decrease the fatigue life depends on the material properties and failure criterion.

Figure 3.26 shows the effect of E_2 on ϵ_t and ϵ_c when $h_1 = 4$ in. (102 mm), $h_2 = 8$ in. (203 mm), $E_1 = 200,000$ or $1,000,000$ psi (1.4 or 6.9 GPa), and $E_3 = 7500$ psi (51.8 MPa). It can be seen that E_2 has more effect on ϵ_t than on ϵ_c and that the effect is greater when E_1 is smaller.

Figure 3.27 shows the effect of E_3 on ϵ_t and ϵ_c when $h_1 = 4$ in. (102 mm), $h_2 = 8$ in. (203 mm), $E_1 = 200,000$ or $1,000,000$ psi (1.4 or 6.9 GPa), and $E_2 = 20,000$ psi

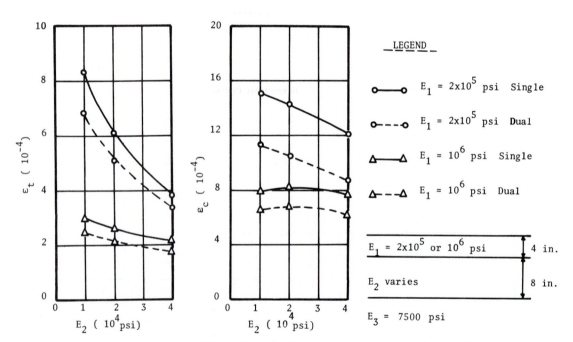

Figure 3.26 Effect of base modulus on pavement responses (1 in. = 25.4 mm, 1 psi = 6.9 kPa.).

Kenlayer Computer Program   Chap. 3

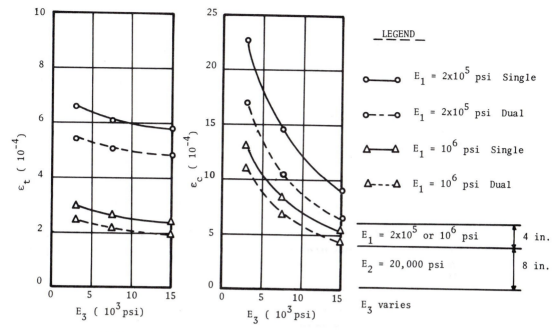

Figure 3.27 Effect of subgrade modulus on pavement responses (1 in. = 25.4 mm, 1 psi = 6.9 kPa).

(138 MPa). It can be seen that E_3 has a large effect on ϵ_c but a very small effect on ϵ_t. The effect of E_3 is nearly the same no matter how large or small E_1 may be.

Four-Layer Systems

Figure 3.28 shows a standard pavement consisting of 4 in. (102 mm) of hot mix asphalt surface course, 8 in. (203 mm) of crushed stone base course, and 8 in. (203 mm) of gravel subbase course which is subjected to a 9000-lb (40-kN) single-wheel load with a contact pressure of 70 psi (483 kPa). The elastic modulus and Poisson ratio of each layer are shown in the figure.

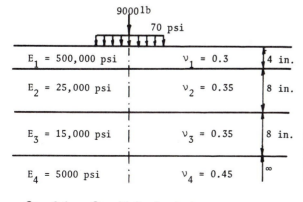

Figure 3.28 Four-layer elastic system for sensitivity analysis (1 in. = 25.4 mm, 1 psi = 6.9 kPa, 1 lb = 4.45 N).

TABLE 3.10 SENSITIVITY ANALYSIS OF ELASTIC MULTILAYER SYSTEMS

Location	Response	Standard case	Case 1 high pressure $q = 140$ psi	Case 2 strong subgrade $E_4 = 15{,}000$ psi	Case 3 $\nu = 0.5$	Case 4 asphalt base $E_2 = 300{,}000$ psi	Case 5 cement base $E_2 = 1{,}000{,}000$ psi	Case 6 6 in. lime $E_4 = 50{,}000$ psi $E_5 = 5000$ psi
Top of HMA	w_0 (in.)	0.0341 (0.034)	0.0360 (0.036)	0.0212 (0.021)	0.0336 (0.034)	0.0202 (0.021)	0.0161 (0.017)	0.0290 (0.029)
Bottom of HMA	σ_r (psi)	−199.9 (−200)	−291.4 (−291.0)	−190.8 (−191)	−269.6 (−270)	6.51 (6.3)	39.1 (38.4)	−188.1 (−188)
	$\epsilon_t(10^{-6})$	−295.7 (−296)	−428.7 (−429)	−283.5 (−284)	−293.8 (−294)	−21.16 (−21.5)	19.77 (18.8)	−279.8 (−280)
Top of base	σ_z (psi)	26.48 (26.5)	34.5 (34.4)	27.28 (27.3)	24.21 (24.2)	50.44 (50.4)	58.18 (58.2)	27.35 (27.4)
	σ_r (psi)	2.88	2.07	3.79	9.51	6.51	61.74	3.97
Bottom of base	σ_r (psi)	−6.95 (−7.0)	−7.56 (−7.4)	−4.29 (−4.3)	−7.67 (−7.7)	−61.08 (−60.6)	−102.1 (−100)	−3.99 (−3.9)
Top of subbase	σ_z (psi)	8.59 (8.6)	9.30 (9.3)	10.62 (10.6)	8.42 (8.4)	3.77 (3.7)	2.21 (2.2)	10.45 (10.5)
	σ_r (psi)	−2.32	−2.53	−0.29	−1.24	−1.13	−0.36	−0.15
Bottom of subbase	σ_r (psi)	−5.25 (−5.3)	−5.47 (−5.3)	−0.56 (−0.6)	−5.92 (5.9)	−2.22 (−2.2)	−1.20 (−1.1)	0.92 (1.0)
Top of subgrade	σ_z (psi)	3.48 (3.5)	3.58 (3.5)	5.66 (5.7)	3.49 (3.5)	1.74 (1.7)	1.22 (1.2)	2.13 (2.2)
	σ_r (psi)	0.39 (3.5)	0.02	3.65	0.35	0.18	0.26	0.40
	$\epsilon_c(10^{-6})$	688.8	713.8	355.2	627.4	314.9	196.0	354.8
36 in. below surface	σ_z (psi)	1.59 (1.6)	1.61 (1.6)	2.41 (2.4)	1.66 (1.6)	0.99 (0.9)	0.78 (0.7)	1.44 (1.4)
	σ_r (psi)	0.03	0.03	0.05	1.37	0.11	0.14	0.16

Note. Numbers in parentheses were reported by ERES Consultant, Inc. using ELSYM5. 1 in. = 25.4 mm, 1 psi = 6.9 kPa.

In addition to the standard case, six nonstandard cases, each with only one parameter different from the standard case, were also analyzed. The results are presented in Table 3.10. Case 1 has the same total load as the standard case but the contact pressure is doubled, thus resulting in a smaller contact radius. Case 2 has a strong subgrade with an elastic modulus three times greater than the standard case. In case 3, all layers are assumed incompressible with a Poisson ratio of 0.5. The granular base is replaced by an asphalt base in case 4 and by a cement-treated base in case 5. Theoretically, the Poisson ratio of these treated bases should be different from that of the granular base. However, since the effect of Poisson ratio is small, the same Poisson ratio of 0.35 is used. Case 6 is a five-layer system with the top 6 in. (152 mm) of subgrade replaced by a lime-stablilized layer with the same Poisson ratio of 0.45. The values in parentheses were obtained from the ELSYM5 (Kopperman et al., 1986) program as reported by ERES Consultant, Inc. (1987). It can be seen that the solutions obtained from KEN-LAYER check closely with those from ELSYM5.

The responses to be compared include the surface deflection w_0, the radial stress σ_r and the tensile strain ϵ_t at the bottom of the HMA, the vertical stress σ_z at the top of each layer, the radial stress at the top and bottom of each layer, the vertical compressive strain ϵ_c at the top of the subgrade, and the vertical and radial stresses in the subgrade 36 in. (914 mm) below the surface. The surface deflection is a good indication of the overall strength of a pavement. The tensile strain at the bottom of the asphalt layer and the compressive strain at the top of the subgrade have frequently been used as design criteria. The vertical stresses contribute to the consolidation of each layer and the rutting on the surface. The radial stresses are important because they cause the rupture of rigid layers and control the resilient modulus of unbounded granular layers.

Comments on Table 3.10

1. A comparison of case 1 with the standard case indicates that the effect of contact pressure or contact radius is significant only in the top layers. If the design is governed by the radial tensile strain at the bottom of the asphalt layer, case 1 results in a radial strain 45% greater than the standard case, thus requiring the use of thicker or better HMA to prevent fatigue cracking. If the design is governed by the vertical compressive strain at the top of subgrade, case 1 has a vertical strain only 3.6% greater than the standard case, so the effect of contact radius is insignificant.

2. A comparison of case 2 with the standard case indicates that a strong subgrade reduces the surface deflection by 38% but the tensile strain at the bottom of the asphalt layer is reduced by only 4.1%. Due to the stiffer foundation, the vertical and radial compressive stresses increase while the radial tensile stesses decrease. The effect is most significant in the subbase, which lies directly above the subgrade, and diminishes in the layers that are farther away from the subgrade. Because the granular base and subbase are actually nonlinear, the change in vertical and radial stresses will increase their resilient moduli, so the effect of stronger subgrade on the tensile strain at the bottom of the asphalt layer should be greater than the 4% obtained from the linear theory.

3. A comparison of case 3 with the standard case indicates that the Poisson ratios have very little effect on the surface deflection, vertical stresses, and the radial tensile strain at the bottom of the asphalt layer, but a significant effect on the radial stresses. The vertical compression strain at the top of the subgrade is reduced by 9.1%, due to the greater Poisson ratio.

4. The replacement of granular base by asphalt base, as shown in case 4, results in a significant decrease in the surface deflection and the stresses in the subbase and subgrade. The radial stress at the bottom of the HMA surface course changes from tension to compression, so the most critical tensile strain occurs at the bottom of the asphalt base course, instead of the surface course.

5. The replacement of granular base by cement base, as shown in case 5, further reduces the surface deflection and the stresses in the subbase and sub-grade, as compared to case 4. The radial tensile stress at the bottom of the base course increases with the rigidity of the base.

6. A comparison of case 6 with the standard case indicates that the placement of a 6-in. (152-mm) lime-stabilized layer above the subgrade reduces the tensile strain at the bottom of layer 1 by only 5.4%. The effect is similar to case 2 with a stronger subgrade.

7. The radial stresses in the subgrade are always in compression, while those in the subbase and at the bottom of the base are in tension. One exception is in case 6 where the radial stress at the bottom of the subbase is in compression, due to the existence of a very stiff lime-stabilized layer underneath.

3.4.2 Nonlinear Analysis

This analysis is based on the assumption that one or more layers are nonlinear elastic with a stress-dependent resilient modulus. The nonlinear granular layer can be considered as a signle layer or subdivided into a number of layers, each not over 2 in. (51 mm) thick.

Three-Layer Systems

Figure 3.29 shows a three-layer system subjected to a total load P, which is applied on a single tire and a set of dual tires. Under a given contact radius, the stresses, strains, and deflections in a linear system is proportional to the contact pressure or the magnitude of total load P. However, for a nonlinear system with stress-sensitive granular materials, the increase in responses is not as rapid as the increase in load due to the stiffening effect of granular materials under greater loads. The purpose herein is to find the effect of load magnitude on pavement responses. The information necessary for the analysis is shown in the figure.

Figure 3.30 illustrates the effect of P on ϵ_t and ϵ_c when $h_1 = 2$ or 8 in. (51 or 203 mm). The granular base is considered as a single layer with a stress point located at the upper quarter. The reason the subgrade is considered linear elastic instead of nonlinear elastic is that the nonlinear effect is usually quite small and can be neglected.

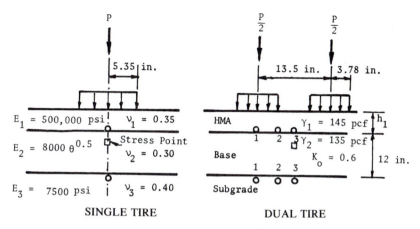

Figure 3.29 Pavement section for nonlinear analysis (1 in. = 25.4 mm, 1 psi = 6.9 kPa, 1 pcf = 157.1 N/m³).

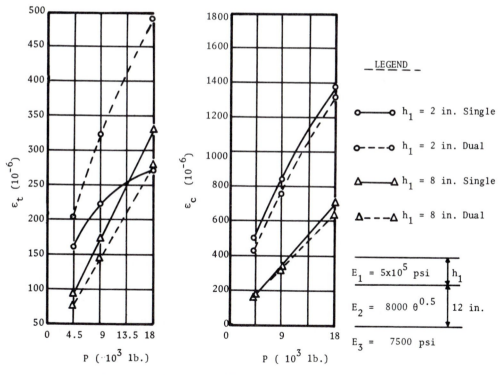

Figure 3.30 Effect of wheel loads on nonlinear responses (1 in. = 25.4 mm, 1 psi = 6.9 kPa, 1 lb = 4.45 N).

Comments on Figure 3.30

1. For a thin HMA layer with $h_1 = 2$ in. (51 mm), ϵ_t under dual-wheel loads is greater than that under a single-wheel load. This trend is noted in Figure 3.24 but is more pronounced when the base is nonlinear. Figure 3.30 further shows that the greater the load, the more difference there is in ϵ_t between single and dual wheels. This is because the single-wheel load results in greater stresses in the granular base and makes the base stronger, thus further decreasing ϵ_t.

2. The nonlinear effect, as indicated by the curvilinear relationship between ϵ_t and P, is more pronounced for thinner HMA than for thicker HMA. For thicker HMA with $h_1 = 8$ in. (203 mm), the relationship between ϵ_c and P is nearly linear.

3. The differences in responses between single and dual wheels are more significant when the HMA is thin and become less significant as the HMA thickness increases.

Four-Layer Systems

Figure 3.31 is the standard case for a nonlinear elastic system similar to the linear system shown in Figure 3.28. Even though layer 1 is actually viscoelastic, it is always possible to find a vehicle speed or load duration such that the modulus is equal to 500,000 psi (3.5 GPa). The elastic modulus of base, subbase, and subgrade are stress dependent, as indicated by the equations shown in the figure. For the fine-grained subgrade soil, only the equation for a deviator stress smaller than 6.2 psi (42.8 kPa) is shown because the actual stress is always smaller than this value. The constants in these nonlinear equations were selected so that the same

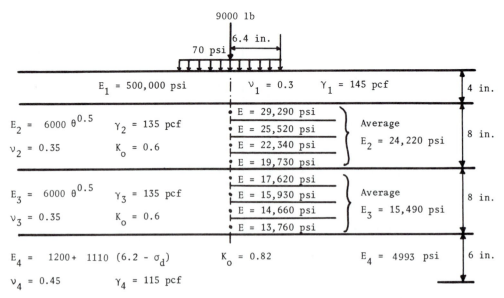

Figure 3.31 Four-layer nonlinear elastic system for sensitivity analysis (1 in. = 25.4 mm, 1 psi = 6.9 kPa, 1 lb = 4.45 N, 1 pcf = 151.7 N/m³).

Kenlayer Computer Program   Chap. 3

TABLE 3.11 SENSITIVITY ANALYSIS OF NONLINEAR ELASTIC MULTILAYER SYSTEMS

Response	Standard case	Case 1 high pressure $q = 140$ psi	Case 2 strong subgrade $K_1 = 12{,}340$ psi	Case 3 strong base $K_1 = 12{,}000$ psi	Case 4 strong base $K_2 = 0.7$	Case 5 strong surface $E_1 = 10^6$ psi	Case 6 subgrade $K_0 = 0.4$
Avg E_2 (psi)	24,220	26,360	25,870	53,000	46,110	20,740	24,020
Avg E_3 (psi)	15,490	15,720	17,930	30,090	21,600	14,510	15,160
E_4 (psi)	4993	4984	14,700	5514	5359	5211	4277
w_0 (in.)	0.0341	0.354	0.0205	0.0252	0.0277	0.0307	0.0371
	(0.0341)	(0.0360)	(0.0212)				
ϵ_t (10^{-6})	−290.0	−403.4	−267.5	−173.6	−193.2	−218.9	−293.4
	(−295.7)	(−428.7)	(−283.5)				
ϵ_c (10^{-6})	691.0	704.5	357.6	507.5	564.0	584.4	748.7
	(688.8)	(713.8)	(355.2)				

Note. Figures in parentheses are from Table 3.10 based on the elastic system. 1 psi = 6.9 kPa, 1 in. = 25.4 mm.

moduli as in the linear system could be obtained. To achieve a modulus of 25,000 psi (173 MPa) for the base and 15,000 psi (104 MPa) for the subbase, the same K_1 of 6000 should be used. The base and subbase courses are each subdivided into four layers. The modulus of each layer, as obtained by KENLAYER, is shown in the figure.

In addition to the standard case shown in Figure 3.31, six more cases, each with only one parameter different from the standard case, were also analyzed. The results are presented in Table 3.11. The responses include the average base and subbase moduli E_2 and E_3, the subgrade modulus E_4, the surface deflection w_0, the radial tensile strain ϵ_t at the bottom of layer 1, and the vertical compressive strain ϵ_c at the top of the subgrade. For comparison, the corresponding w_0, ϵ_t, and ϵ_c based on the linear theory, as presented in Table 3.10, are shown in parentheses.

Comments on Table 3.11

1. The standard case for nonlinear analysis is very similar to that for linear analysis with nearly the same layer moduli. A comparison between linear and nonlinear solutions shows that the nonlinear solution results in the same w_0, a slightly smaller ϵ_t, and a slightly greater ϵ_c. These results are reasonable because w_0 depends on the average moduli, ϵ_t depends to a large extent on the modulus of the material immediately under the asphalt layer, while ϵ_c depends on the modulus of the material immediately above the subgrade. Although the average moduli of the nonlinear system are the same as those of the linear system, the modulus of the granular layer immediately below the asphalt layer is 29,290 psi (201 MPa), which is greater than the average base modulus of 24,220 psi (167 MPa), and the modulus of the granular layer immediately above the subgrade is 13,760 psi (95 MPa), which is smaller than the average subbase modulus of 15,490 psi (107 MPa). The greater the base modulus immediately below the asphalt layer, the smaller is the ϵ_t, and the smaller the subbase modulus immediately above the subgrade, the greater is the ϵ_c.

2. With the same total load, an increase in tire pressure causes an increase in E_2 but has practically no effect on E_3 and E_4. This is reasonable because the subbase and subgrade are quite far away from the load and are not affected by contact pressure, as long as the total load is the same. Furthermore, the relatively large geostatic stress in the subbase and subgrade also reduces the effect of loading stress on the resilient modulus. Due to the increase in E_2, the nonlinear analysis results in a smaller w_0, ϵ_t, and ϵ_c compared to the linear analysis.

3. A strong subgrade causes an appreciable increase in E_2 and E_3. An increase of E_4 from 4993 psi (34.5 MPa) to 14,700 psi (101.4 MPa) results in a 16% increase in E_3 and 7% increase in E_2. Consequently, the nonlinear analysis results in a reduction of ϵ_t of 7.8% compared to the 4.1% in the linear analysis.

4. A stronger base or subbase can be obtained by increasing the nonlinear coefficients K_1 and K_2. The effect of K_1 and K_2 is more significant on ϵ_t than on ϵ_c.

5. A strong surface course causes a decrease in E_2, a slight decrease in E_3 and E_4, and a significant decrease in w_0, ϵ_t, and ϵ_c.

6. A decrease in K_0 of subgrade reduces E_4 and ϵ_c but has practically no effect on E_2, E_3, and ϵ_t. Because the horizontal stresses at all the stress points in

the granular layers are in tension, which must be set to 0 to compute the stress invariant, K_0 of the granular materials in the base and subbase has no effect on the analysis.

7. Due to the effect of geostatic stress, the modulus of the subgrade is not affected significantly by the loading stress or the moduli of the overlying layers. It appears reasonable to assume the subgrade to be linear with an elastic modulus independent of the state of stresses. This is not true for bases and subbases because their elastic moduli depend strongly on the stiffness of subgrade.

3.4.3 Contained Rock Asphalt Mats

The contained rock asphalt mat (CRAM) is a new concept of placing an additional asphalt layer below the untreated granular materials but above the subgrade. In this section, the differences in responses between conventional and CRAM sections are presented, followed by a discussion of the effect of HMA and aggregate thicknesses on the responses of CRAM. In the analysis that follows, it is assumed that the HMA and subgrade are linear elastic, while the granular materials are nonlinear elastic.

Figure 3.32 shows the cross section of a conventional pavement and a CRAM construction. Both pavements are essentially the same except that the bottom 2 in. (51 mm) of the open-graded aggregate in the conventional pavement is replaced by 2 in. (51 mm) of HMA in the CRAM section. In the non-linear analysis by KENLAYER, the granular materials are subdivided into 2-in. (51-mm) layers with the stress point located at the midheight of each layer. The material properties to be used for the analysis are shown in the figure.

Table 3.12 shows a comparison of responses between the conventional and CRAM sections. The surface deflection of the CRAM section is about 86% of the conventional section, the HMA tensile strain about 92%, and the subgrade com-

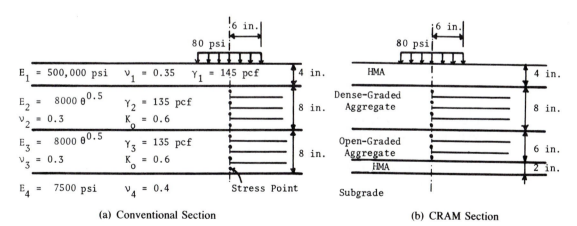

Figure 3.32 Cross sections of conventional and CRAM construction (1 in. = 25.4 mm, 1 psi = 6.9 kPa, 1 pcf = 157.1 N/m³).

TABLE 3.12 COMPARISON OF RESPONSES OF CONVENTIONAL AND CRAM SECTIONS

Pavement type	Surface deflection (in.)	HMA tensile strain (10^{-6})		Subgrade compressive strain (10^{-6})
		Top layer	Bottom layer	
Conventional	0.0247	244.9	—	530.6
CRAM	0.0212	225.1	68.4	272.4
	(0.86)	(0.92)		(0.51)

Note. Figures in parentheses are the ratios between CRAM and conventional.
1 in. = 25.4 mm.

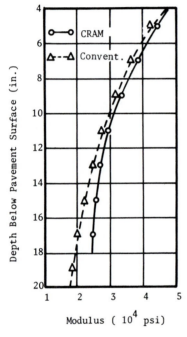

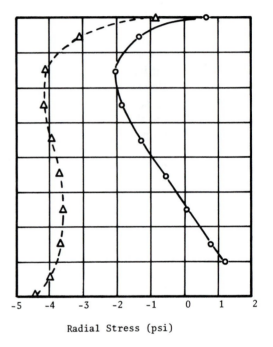

Figure 3.33 Moduli and radial stresses in conventional and CRAM sections (1 in. = 25.4 mm, 1 psi = 6.9 kPa).

pressive strain only 51%. Note that the tensile strain in the bottom HMA layer is very small compared to that in the top HMA layer.

Figure 3.33 compares the moduli and radial stresses in the granular layers of the conventional section with those of the CRAM section. The radial tensile stresses under the center of the loaded area in the CRAM section are much smaller than those in the conventional section. Consequently, the moduli of the granular layers in the CRAM section are greater than those in the conventional section, particularly for those layers near the bottom.

It can be seen that the major contribution of CRAM is the large reduction in the vertical compressive strain at the top of subgrade. This reduction is caused mainly by the very stiff HMA as well as the stiffer granular materials immediately above the subgrade. Even a 2-in. (51-mm) HMA layer can reduce the compressive strain by nearly 50%. It is believed that 2 in. (51 mm) of HMA is probably the minimum that can be used. Any thickness smaller than 2 in. (51 mm) is difficult to control and compact, especially when constructed on a poor subgrade.

Effect of HMA Thickness

The case to be analyzed is shown in Figure 3.34. It is assumed that the dense- and open-graded aggregates are each 4 in. (102 mm) thick and have the same properties. The HMA thicknesses h_1 at the top are 2, 4, 6, and 8 in. (51, 102, 152, and 203 mm) and the corresponding HMA thicknesses h_4 at the bottom are 6, 4, 2, and 0 in. (152, 102, 51, and 0 mm); thus the total thickness of HMA, or $h_1 + h_4$, is kept at 8 in. (203 mm). Because a single-wheel load yields an unrealistically small tensile strain in thin asphalt surfaces, the 18-kip (80-kN) single-axle load is applied on a set of dual tires. The maximum strains are determined by comparing the strains at three points, one under the center of one wheel, one at the edge of one wheel, and the other at the center between two wheels, as shown in Figure 3.34.

Figure 3.35 presents the effect of HMA thickness on the horizontal tensile strains at the bottom of layers 1 and 4 and the vertical compressive strain on the

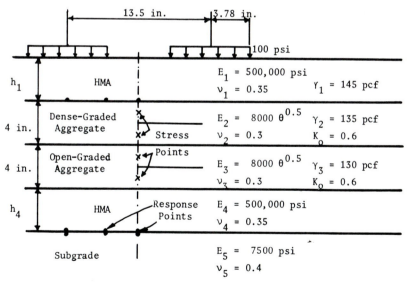

Figure 3.34 CRAM section for analyzing the effect of HMA thickness (1 in. = 25.4 mm, 1 psi = 6.9 kPa, 1 pcf = 157.1 N/m³).

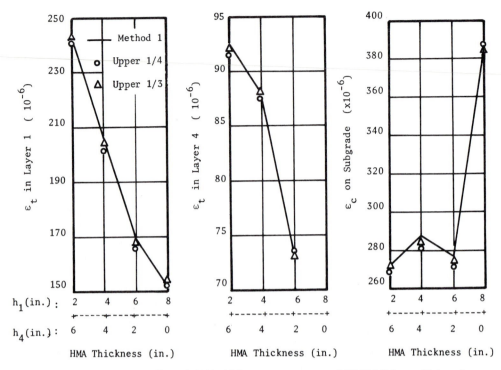

Figure 3.35 Effect of HMA thickness on responses of CRAM (1 in. = 25.4 mm).

top of subgrade. The solid lines are based on method 1 by subdividing each granular layer into two layers, while the individual points are based on method 2 by considering the granular materials as a single layer with the stress point located at the upper quarter or upper third of the layer.

Comments on Figure 3.35

1. The responses obtained by method 2, using either the upper quarter or upper third point to evaluate the modulus, check very well with those by method 1. The use of the upper third point appears to compare more favorably with method 1 and is recommended for CRAM, in contrast to the upper quarter point for conventional pavements.

2. For a given total thickness of HMA, it is more effective to place thicker HMA at the top (layer 1) than at the bottom (layer 4). The tensile strain at the bottom of both HMA layers decreases as the thickness of layer 1 increases. The most effective design is to place a thick layer of HMA at the top and a thin layer of HMA at the bottom.

3. For a given total HMA thickness, the vertical compressive strain on top of the subgrade is not affected significantly by how much thickness is placed at the

top or bottom. However, without the 2 in. (51 mm) of HMA at the bottom, the compressive strain increases drastically. This clearly indicates the advantage of placing a thin layer of HMA above the subgrade.

Effect of Aggregate Thickness

To illustrate the effect of aggregate thickness on the responses of CRAM, the sections shown in Figure 3.36 were analyzed. Two different thicknesses of 2 and 6 in. (51 and 152 mm) are assumed for layer 1, while a minimum HMA thickness of 2 in. (51 mm) is assumed for layer 3. The thickness of aggregate base h_2 varies from 4 to 16 in. (102 to 406 mm). The granular base is subdivided into 2-in. (51 mm) layers, i.e., from 2 layers for the 4 in. (102 mm) base to 8 layers for the 16 in. (406 mm) base. For comparison purposes, a conventional pavement without the 2-in. (51-mm) HMA at the bottom was also analyzed.

Figure 3.37 shows the effect of base thickness on the responses of conventional and CRAM sections. For conventional sections, both the HMA tensile strain and the subgrade compressive strain decrease with the increase in base thickness. For CRAM sections, both the tensile strain in the bottom HMA layer and the subgrade compressive strain decrease with the increase in base thickness, but the tensile strain in the top HMA layer increases with the increase in base thickness for the 2-in. (51-mm) HMA surface but is practically independent of the base thickness for the 6 in. (152 mm) HMA surface. Therefore, the base thickness is effective in reducing the HMA tensile strain for conventional pavements but not for CRAM sections. The figure also demonstrates the effectiveness of the bottom HMA layer in reducing the subgrade compressive strain. Note that the tensile

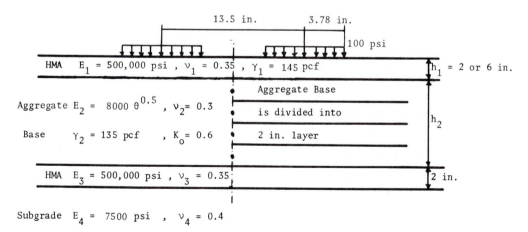

Figure 3.36 CRAM section for analyzing the effect of aggregate thickness (1 in. = 25.4 mm, 1 psi = 6.9 kPa, 1 pcf = 157.1 N/m³).

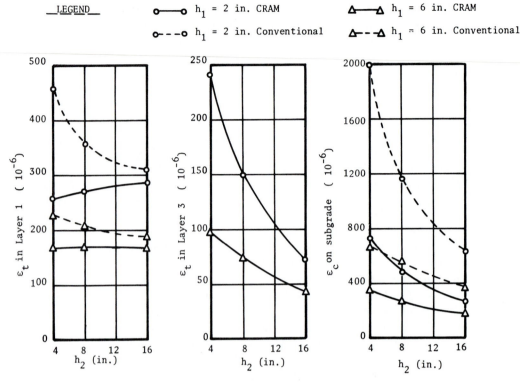

Figure 3.37 Effect of base thickness on pavement response (1 in. = 25.4 mm).

strains in the bottom HMA layer are much smaller than those in the top HMA layer, particularly when the granular base is thick.

3.5 SUMMARY

This chapter describes some features of the KENLAYER computer program. The theory presented in Chapter 2 and Section 3.1 was used in developing KEN-LAYER. More details about KENLAYER can be found in Appendix B.

Important Points Discussed in Chapter 3

1. The basic component of KENLAYER is the elastic multilayer system under a circular loaded area. Each layer is linear elastic, homogeneous, isotropic, and infinite in areal extent. The problem is axisymmetric and the solutions are in terms of cylindrical coordinates r and z.

2. For multiple wheels involving two to six circular loaded areas, the superposition principle can be applied because the system is linear. Since the stresses at a given point due to each of these loaded areas are not in the same direction, they must be resolved into x and y components and then superimposed.

3. The above superposition principle can also be applied to a nonlinear elastic system by a method of successive approximations. First, the system is considered to be linear and the stresses due to multiple-wheel loads are superimposed. Based on the stresses thus computed, a new set of moduli for each nonlinear layer is then determined. The system is considered linear again and the process is repeated until the moduli converge to a specified tolerance.

4. Because the most critical strains occur directly under or near the load, a point under the center of a single wheel or between the centers of dual wheels can be selected for computing the elastic modulus of each nonlinear layer. When the pavement is subjected to tandem- or tridem-axle loads, only one axle load is used to determine the modulus of nonlinear layers, because these axles are quite far apart and there is very little difference whether one or more axles are considered. After the moduli are determined, the responses due to tandem or tridem loads will then be computed.

5. Two methods can be used to determine the nonlinear modulus of granular layers. In method 1, the granular layer is subdivided into a number of layers, each with a maximum thickness of 2 in. (51 mm). The stresses at the midheight of each layer are used to evaluate the modulus. If the horizontal stress is in tension, it is set to zero to compute the stress invariant θ. In method 2, all granular materials, including the base and subbase, are considered as a single layer and a point at the upper quarter of conventional pavement or at the upper third of CRAM section is used to evaluate the modulus. When method 2 is used, a minimum modulus other than 0 should be specified, and no modification of negative horizontal stress will be made. To obtain more accurate results, especially for unconventional pavements, the use of method 1 is recommended.

6. If the layered system is viscoelastic, the responses under a static load can be expressed as a seven-term Dirichlet series, as indicated by Eq. 2.49, using the following seven T_i values: 0.01, 0.03, 0.1, 1, 10, 30, and ∞ seconds. The responses under a moving load are obtained by assuming the loading to be a haversine function and applying Boltzmann's superposition principle to the Dirichlet series, as indicated by Eq. 2.59.

7. A direct method for analyzing viscoelastic layer systems under static loads is to assume the viscoelastic layer to be elastic with a modulus varying with the loading time. For a given loading time, the elastic modulus is the reciprocal of the creep compliance at that loading time.

8. KENLAYER can be applied to layer systems with a maximum of 19 layers, each being either linear elastic, nonlinear elastic, or linear viscoelastic. If the layer is linear elastic, the modulus is a constant and no further work need be done to determine its value. If the layer is nonlinear elastic, the modulus varies with the state of stresses and a method of successive approximations is applied until it converges. If the layer is viscoelastic, elastic solutions under static loads are obtained at a specified number of time durations, usually 11, and then fitted with a Dirichlet series.

9. Damage analysis is based on the horizontal tensile strain at the bottom of a specified asphalt layer and the vertical compressive strain on the surface of a specified layer, usually the subgrade. To determine the allowable number of

repetitions to prevent fatigue cracking, it is necessary to know the elastic modulus of the hot mix asphalt. If the hot mix asphalt is specified as viscoelastic, its elastic modulus is not a constant but depends on the duration of loading and can be determined by Eq. 3.17.

10. In damage analysis, each year can be divided into several periods and each period can have a number of load groups. For tandem and tridem load groups, the allowable number of load repetitions for the first axle load is based on the total strain, while that for each additional axle is based on the difference between maximum and minimum strains. The damage ratios for fatigue cracking and permanent deformation in each season under each load group are evaluated and summed over a year and the one with a larger damage ratio controls the design. The reciprocal of the damage ratio is the design life of the pavement.

11. The results obtained by KENLAYER compare well with those by other layer system programs such as ELSYM5, VESYS, and DAMA but not with those by the finite element programs such as MICH-PAVE and ILLI-PAVE. For linear elastic systems, the finite element and layer system programs should yield the same results. The failure of MICH-PAVE and ILLI-PAVE to obtain linear elastic solutions needs to be resolved before they can be put into practical use.

12. KENLAYER can be used to evaluate the nonlinear coefficient K_1 of granular layers and fine-grained subgrade by a trial and error process to match the calculated and measured deflection basin. The deflection basin obtained by KEN-LAYER checks reasonably well with that by MICH-PAVE but deviates considerably from that by ILLI-PAVE. The use of a nonlinear layer system is a great improvement over the conventional linear system because it takes into consideration the load magnitude and the response location in determining the modulus to be used.

13. For thin asphalt surfaces, say less than 2 in. (51 mm) thick, the use of a single tire to replace the actual dual tires results in a smaller tensile strain and is unsafe for the prediction of fatigue cracking.

14. The most effective way to decrease the tensile strain at the bottom of the asphalt layer is to increase the HMA thickness or the modulus of the base course, while the most effective way to decrease the compressive strain on top of the subgrade is to increase the thickness of the granular base and subbase or the modulus of the subgrade.

15. A sensitivity analysis of elastic layered systems indicates that the incorporation of a stiff layer reduces appreciably the stresses and strains in all underlying layers, but among those overlying layers, only the one lying immediately above the stiff layer will be affected to a significant degree. For example, a strong subgrade significantly reduces the tensile stress in the subbase but not in the asphalt layer. However, a sensitivity analysis of nonlinear layered systems indicates that the above conclusion is only paritally true. A strong subgrade increases the moduli of base and subbase and has more effect in reducing the tensile stress and strain in the asphalt layer than that predicted by the linear theory.

16. The result of nonlinear analysis indicates that the modulus of subgrade is not affected significantly by the moduli of base and subbase but the moduli of base and subbase depend strongly on the modulus of subgrade. In pavement design, it

appears reasonable to assume the base and subbase to be nonlinear elastic and the subgrade to be linear elastic.

17. Other than waterproofing the subgrade and preventing the contamination of base and subbase, the major advantage of using the contained rock asphalt mat (CRAM) is to reduce drastically the compressive strain on the top of subgrade. The most effective design is to place a thin layer of HMA above the subgrade, then a minimum thickness of granular materials to provide drainage, and finally a thick layer of HMA on the surface, so that the tensile strains at the bottom of both asphalt layers and the compressive strain on top of the subgrade fall within the allowable limits.

PROBLEMS

3-1. Solve Problem 2-2 by KENLAYER. (*Hint:* Use a two-layer system with two nonlinear layers, the same nonlinear properties, and the same ZCNOL for each layer, ITENOL=2 and RELAX=1.0). [Answer: 0.0539 in.]

3-2. Solve Problem 2-4 by KENLAYER. [Answer: 0.0251 in., 0.0235 in., 11.4 psi]

3-3. Solve Problem 2-5 by KENLAYER. [Answer: 2.083×10^{-4}, 0.0571 in.]

3-4. Solve Problem 2-6 by KENLAYER. [Answer: -7.26×10^{-4}, 1.06×10^{-3}]

3-5. Figure P3.5 shows a three-layer system under a single-wheel load. Layer 1 is linear elastic, while layers 2 and 3 are nonlinear elastic. The loading, thicknesses, and material properties are shown in the figure. With the use of KENLAYER, determine the maximum tensile strain at the bottom of layer 1 and the maximum compressive strain at the top of layer 3 by (a) method 1 in which layer 2 is subdivided into two layers, each 2 in. thick, and (b) method 2 in which layer 2 is considered as one layer and the stress at the upper quarter point of layer 2 is used to determine E_2. [Answer: (a) 1.02×10^{-4}, 2.85×10^{-4}; (b) 1.04×10^{-4}, 2.85×10^{-4}]

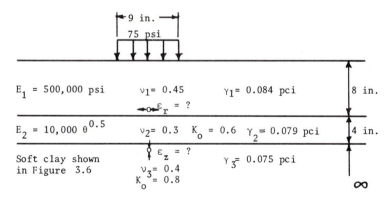

Figure P3.5

3-6. Figure P3.6 shows a three-layer system under a set of dual-wheel loads. Layers 1 and 3 are linear elastic, while layer 2 is nonlinear elastic. The loading, thicknesses, and material properties are shown in the figure. The stresses at the upper quarter of layer 2 between the two wheels are used to evaluate E_2. The maximum tensile strain at the bottom of layer 1 and the maximum compressive strain at the top of layer 3 are

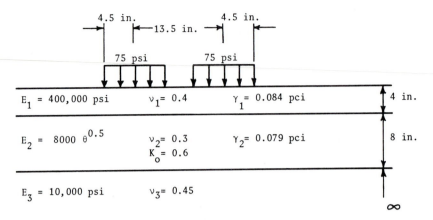

Figure P3.6

determined by comparing the results at three locations: one under the center of one wheel, one under the edge of one wheel, and the other at the center between two wheels. If the actual number of repetitions is 100 per day and the Asphalt Institute's failure criteria are used, determine the life of the pavement by KENLAYER. What should be the number of repetitions per day for a design life of 20 years? [Answer: 5.44 years, 368 per day]

3-7. Same as Problem 3-6 except that layer 2 is subdivided into four layers, each 2 in. thick, and the stresses at the midheight of each layer are used to determine the modulus of each layer. [5.06 years, 395 per day]

3-8. For the three-layer system shown in Figure P3.6, predict the modulus of layer 2 by Eq. 3.28. Compare the modulus with that obtained by KENLAYER, using the stress point at the upper quarter of layer 2 between the dual, and determine the three principal stresses at the stress point. [Answer: 28,550 versus 29,220 psi, 14.00 psi, 1.13 psi, −3.04 psi]

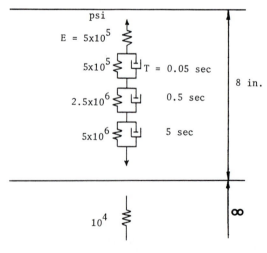

Figure P3.9

3-9. A two-layer system is subjected to a circular load with a radius of 6 in. and a contact pressure of 75 psi. The thickness of layer 1 is 8 in. Layer 1 is viscoelastic and layer 2 is elastic with their stress–strain relationship characterized by the models shown in Figure P3.9. Both layers are incompressible with a Poisson ratio of 0.5. If the load is stationary, determine the maximum vertical displacements at times of 0, 0.01, 0.1, 1, 10, and 100 s by KENLAYER. Check the displacement at 100 s by Figure 2.17. [Answer: 0.0156, 0.0166, 0.0196, 0.0206, 0.0209, 0.0209 in.]

3-10. If the load in Problem 3-9 is moving with a duration of 0.1 s and the number of load repetitions per year is 100,000, determine the life of the pavement based on the Asphalt Institute's failure criteria. [Answer: 21.0 years]

4

Stresses and Deflections in Rigid Pavements

4.1 STRESSES DUE TO CURLING

During the day when the temperature on the top of the slab is greater than that at the bottom, the top tends to expand with respect to the neutral axis while the bottom tends to contract. However, the weight of the slab restrains it from expansion and contraction; thus, compressive stresses are induced at the top while tensile stresses occur at the bottom. At night when the temperature on the top of the slab is lower than that at the bottom, the top tends to contract with respect to the bottom; thus, tensile stresses are induced at the top and compressive stresses at the bottom.

Another explanation of curling stress can be made by the theory of plate on a Winkler, or liquid, foundation. A Winkler foundation is characterized by a series of springs attached to the plate, as shown in Figure 4.1. When the temperature on the top is greater than that at the bottom, the top is longer than the bottom and the slab curls downward. The springs at the outside edge are in compression and push the slab up, while the springs in the interior are in tension and pull the slab down. As a result, the top of the slab is in compression and the bottom is in tension. When the temperature on the top is lower than that at the bottom, the slab curls upward. The exterior springs pull the slab down while the interior springs push the slab up, thus resulting in tension at the top and compression at the bottom. Westergaard (1926a) developed equations for determining the curling stress in concrete pavements based on the plate theory. The equations are very complex and are not presented here.

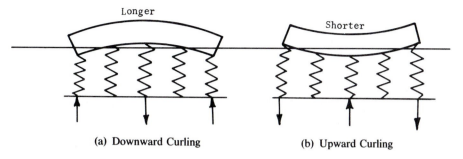

(a) Downward Curling   (b) Upward Curling

Figure 4.1 Curling of slab due to temperature gradient.

4.1.1 Bending of Infinite Plate

The difference between a beam and a plate is that the beam is stressed in only one direction, while the plate is stressed in two directions. For stresses in two directions, the strain ϵ_x in the x direction can be determined by the generalized Hooke's law

$$\epsilon_x = \frac{\sigma_x}{E} - \nu \frac{\sigma_y}{E} \tag{4.1}$$

in which E is the elastic modulus of concrete. In this chapter when only a Winkler foundation is considered, E is used to denote the elastic modulus of concrete. In later chapters involving plates on a solid foundation, the elastic modulus of concrete is denoted by E_c to distinguish it from the elastic modulus of foundation E_f.

The first term on the right side of Eq. 4.1 indicates the strain in the x direction caused by stress in the x direction, while the second term indicates the strain caused by stress in the y direction. Similarly,

$$\epsilon_y = \frac{\sigma_y}{E} - \nu \frac{\sigma_x}{E} \tag{4.2}$$

When the plate is bent in the x direction, ϵ_y should be equal to 0 because the plate is so wide and well restrained that no strain should ever occur unless near the very edge. Setting Eq. 4.2 to 0 yields

$$\sigma_y = \nu\sigma_x \tag{4.3}$$

Substituting Eq. 4.3 into Eq. 4.1 and solving for σ_x gives

$$\sigma_x = \frac{E\epsilon_x}{1 - \nu^2} \tag{4.4}$$

Eq. 4.4 indicates the stress in the bending direction, while Eq. 4.3 indicates the stress in the direction perpendicular to bending.

When bending occurs in both the x and y directions, as is the case for temperature curling, the stresses in both directions must be superimposed to obtain the total stress. The maximum stress in an infinite slab due to temperature curling can be obtained by assuming that the slab is completely restrained in both x and y directions.

Let Δt be the temperature differential between top and bottom of the slab and α_t be the coefficient of thermal expansion of concrete. If the slab is free to move and the temperature at top is greater than that at the bottom, the top will expand by a strain of $\alpha_t \Delta t/2$ and the bottom will contract by the same strain, as shown in Figure 4.2. If the slab is completely restrained and prevented from moving, a compressive strain will result at the top and a tensile strain at the bottom. The maximum strain is

$$\epsilon_x = \epsilon_y = \frac{\alpha_t\,\Delta t}{2} \tag{4.5}$$

From Eq. 4.4, the stress in the x direction due to bending in the x direction is

$$\sigma_x = \frac{E\alpha_t\Delta t}{2(1 - v^2)} \tag{4.6}$$

Because Eq. 4.6 is also the stress in the y direction due to bending in the y direction, from Eq. 4.3 the stress in the x direction due to bending in the y direction is

$$\sigma_x = \frac{vE\alpha_t\Delta t}{2(1 - v^2)} \tag{4.7}$$

The total stress is the sum of Eqs. 4.6 and 4.7:

$$\sigma_x = \frac{E\alpha_t\Delta t}{2(1 - v^2)}(1 + v) = \frac{E\alpha_t\Delta t}{2(1 - v)} \tag{4.8}$$

The above analysis is based on the assumption that the temperature distribution is linear throughout the depth of the slab. This is an approximation to reality because the actual temperature distribution is highly nonlinear.

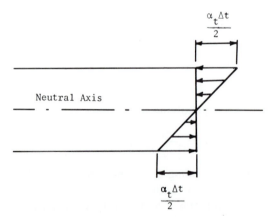

Neutral Axis

Figure 4.2 Temperature gradient in concrete slab.

4.1.2 Curling Stresses in Finite Slab

Figure 4.3 shows a finite slab with lengths L_x in the x direction and L_y in the y direction. The total stress in the x direction can be expressed as

$$\sigma_x = \frac{C_x E \alpha_t \Delta t}{2(1 - v^2)} + \frac{C_y v E \alpha_t \Delta t}{2(1 - v^2)} = \frac{E \alpha_t \Delta t}{2(1 - v^2)} (C_x + v C_y) \qquad (4.9a)$$

in which C_x and C_y are correction factors for a finite slab. The first term in Eq. 4.9a is the stress due to bending in the x direction and the second term is the stress due to bending in the y direction. Similarly, the stress in the y direction is

$$\sigma_y = \frac{E \alpha_t \Delta t}{2(1 - v^2)} (C_y + v C_x) \qquad (4.9b)$$

Based on Westergaard's analysis, Bradbury (1938) developed a simple chart for determining C_x and C_y, as shown in Figure 4.4. The correction factor C_x depends on L_x/ℓ and the correction factor C_y depends on L_y/ℓ, where ℓ is the radius of relative stiffness defined as

$$\ell = \left[\frac{E h^3}{12(1 - v^2)k} \right]^{0.25} \qquad (4.10)$$

in which E is the modulus of elasticity of concrete, h is the thickness of the slab, v is Poisson ratio of concrete, and k is the modulus of subgrade reaction. In all the examples presented in this chapter, a modulus of 4×10^6 psi (27.6 GPa) and a Poisson ratio of 0.15 are assumed for the concrete. Equation 4.9 gives the maximum interior stress at the center of a slab. The edge stress at the midspan of the slab can be determined by

$$\sigma = \frac{C E \alpha_t \Delta t}{2} \qquad (4.11)$$

in which σ may be σ_x or σ_y depending on whether C is C_x or C_y. Note that Eq. 4.11 is the same as Eq. 4.9 when Poisson ratio at the edge is considered as 0.

It can be seen from Figure 4.4 that the correction factor C increases as the ratio L/ℓ increases, having a value of $C = 1.0$ for $L = 6.7\ell$, reaching a maximum

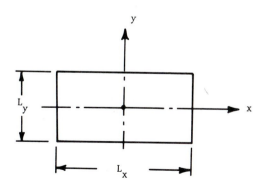

Figure 4.3 A finite slab.

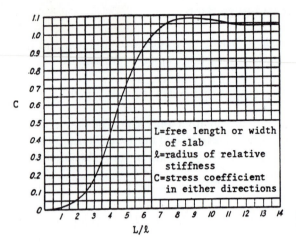

Figure 4.4 Stress correction factor for finite slab. (After Bradbury (1938).)

value of 1.084 for $L = 8.5\ell$, and then decreasing to 1 as L/ℓ approaches infinity. That the coefficient C can have a value greater than unity may be explained by the fact that in slabs longer than about 6.7ℓ the subgrade reaction actually reverses slightly the curvature that temperature curling tends to produce. However, this additional effect is relatively minor since the increase in stress is no more than 8.4% compared with that based on the assumption of zero curvature.

Example 4.1:

Figure 4.5 shows a concrete slab, 25 ft (7.62 m) long, 12 ft (3.66 m) wide, and 8 in. (203 mm) thick, subjected to a temperature differential of 20°F (11.1°C). Assuming that $k = 200$ pci (54.2 MN/m³) and $\alpha_t = 5 \times 10^{-6}$ in./in./°F (9 × 10⁻⁶ mm/mm/°C), determine the maximum curling stress in the interior and at the edge of the slab.

Solution: From Eq. 4.10, $\ell = [4 \times 10^6 \times 512/(12 \times 0.9775 \times 200)]^{0.25} = 30.57$ in. (776 mm). With $L_x/\ell = 25 \times 12/30.57 = 9.81$ and $L_y/\ell = 12 \times 12/30.57 = 4.71$, from Figure 4.4, $C_x = 1.07$ and $C_y = 0.63$. From Eq. 4.9a, the maximum stress in the interior is in the x direction, or $\sigma_x = 4 \times 10^6 \times 5 \times 10^{-6} \times 20 (1.07 + 0.15 \times 0.63)/1.955 = 238$ psi (1.64 MPa). From Eq. 4.11, the maximum stress at edge is also in the x direction, or $\sigma_x = 1.07 \times 4 \times 10^6 \times 5 \times 10^{-6} \times 20/2 = 214$ psi (1.48 MPa).

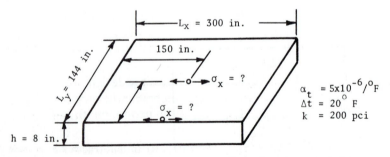

Figure 4.5 Example 4.1 (1 in. = 25.4 mm, 1 pci = 271.3 kN/m³).

4.1.3 Temperature Differentials

Curling stresses in concrete pavements vary with the temperature differentials between the top and bottom of a slab. Unless actual field measurements are made, it is reasonable to assume a maximum temperature gradient of 2.5 to 3.5°F per inch of slab (0.055 to 0.077°C/mm) during the day and about half the above values at night.

In the Arlington Road Test, the maximum temperature differentials between the top and bottom surfaces of slabs were measured during the months of April and May when there was probably as much curling as at any time during the year (Teller and Sutherland, 1935–1943). If the largest five measurements were averaged, the maximum temperature differential of a 6-in. (152-mm) slab was 22°F (12.2°C) and that of a 9-in. (229-mm) slab was 31°F (17.2°C); these values correspond to temperature gradients of 3.7°F/in. (0.080°C/mm) and 3.4°F/in. (0.074°C/mm), respectively.

In the AASHO Road Test (HRB, 1962), temperatures were measured in a 6.5-in. (165-mm) slab. The temperature at a point 0.25 in. (6.4 mm) below the top surface of the 6.5-in. (165-mm) slab minus the temperature at a point 0.5 in. (12.7 mm) above the bottom surface was referred to as the standard temperature differential. The maximum standard temperature differential for the months of June and July averaged about 18.5°F (10.2°C) when the slab curled down and −8.8°F (−4.9°C) when it curled up; these values correspond to temperature gradients of 3.2°F/in. (0.07°C/mm) and 1.5°F/in. (0.03°C/mm), respectively. Temperature measurements in slabs of other thicknesses at the AASHO test site also showed that the temperature differential was not proportional to the thickness of slab and that the increase in temperature differential was not as rapid as the increase in thickness. Therefore, greater temperature gradients should be used for thinner slabs.

4.1.4 Combined Stresses

Even though curling stresses may be quite large and cause concrete to crack when combined with loading stresses, they are usually not considered in the thickness design for the following reasons:

1. Joints and steel are used to relieve and take care of curling stresses. Curling stresses are relieved when the concrete cracks. Minute cracks will not affect the load-carrying capacity of pavements as long as the load transfer across cracks can be maintained.
2. When the fatigue principle is used for design, it is not practical to combine loading and curling stresses. A pavement may be subjected to millions of load repetitions during the design period, but the number of stress reversals due to curling is quite limited.
3. Curling stresses may be added to or subtracted from loading stresses to obtain the combined stresses. If the design is governed by the edge stress,

curling stresses should be added to loading stresses during the day but subtracted from the loading stresses at night. Due to this compensative effect and the fact that a large number of heavy trucks are driven at night, it may not be critical if curling stresses are ignored.

Whether the curling stress should be considered in pavement design is quite controversial. The Portland Cement Association does not consider curling stress in fatigue analysis, but many others indicate that it should be considered. Past experience has demonstrated that more cracks appear in longer slabs because longer slabs have much greater curling stress than shorter slabs. The nontraffic loop in the AASHO site did not have any cracks during the road test. However, when the site was surveyed after 16 years most of the 40-ft (12.2-m) long slabs had cracks, but not the 15-ft (4.6-m) slabs (Darter and Barenberg, 1977).

In designing zero-maintenance jointed plain concrete pavements, Darter and Barenberg (1977) suggested the inclusion of curling stress with loading stress for fatigue analysis. This is necessary because curling stresses are so large that when combined with the loading stresses they may cause the concrete to crack even under a few repetitions. The cracking of the slab will require proper maintenance, thus defeating the zero-maintenance concept. If curling stresses are really so important, it is more reasonable to consider the fatigue damage due to loading and curling separately and then combined, similar to the analysis of thermal cracking in flexible pavements described in Section 11.1.4.

The moisture gradient in concrete slabs also induces warping stresses. Determining the moisture gradient is difficult because it depends on a variety of factors, such as the ambient relative humidity at the surface, the free water in the concrete, and the moisture content of the subbase or subgrade. Since the moisture content at the top of a slab is generally lower than that at the bottom, the bottom of a slab is in compression, which compensates for the tensile stresses caused by edge loading. Furthermore, the moisture effect is seasonal and remains constant for a long time, thus resulting in very few stress reversals and very low fatigue damage. For this reason, warping stresses due to moisture gradient are not considered in the design of concrete pavements.

4.2 STRESSES AND DEFLECTIONS DUE TO LOADING

Three methods can be used to determine the stresses and deflections in concrete pavements: closed-form formulas, influence charts, and finite element computer programs. The formulas originally developed by Westergaard can be applied only to a single-wheel load with a circular, semicircular, elliptical, or semielliptical contact area. The influence charts developed by Pickett and Ray (1951) can be applied to multiple-wheel loads of any configuration. Both methods are applicable only to a large slab on a liquid foundation. If the loads are applied to multiple slabs on a liquid, solid, or layer foundation with load transfer across the joints, the finite element method should be used. The liquid foundation assumes the subgrade to be a set of independent springs. Deflection at any given point is proportional to the force at that point and independent of the forces at all other points. This assumption is unrealistic and does not represent soil behaviors. Due to its simplicity, it was used in Westergaard's analysis. However, with the ever increasing speed and storage of personal computers, it is no longer necessary to assume the foundation to be a liquid with a fictitious k value. The more realistic solid or layer foundation can be used. The KENSLABS computer program based on the finite element method and the various types of foundations are presented in Chapter 5.

4.2.1 Closed-Form Formulas

These formulas are applicable only to a very large slab with a single-wheel load applied near the corner, in the interior of a slab at a considerable distance from any edge, and near the edge far from any corner.

Corner Loading

The Goldbeck (1919) and Older (1924) formula is the earliest one for use in concrete pavement design. The formula is based on a concentrated load P applied at the slab corner, as shown in Figure 4.6a. When a load is applied at the corner, the stress in the slab is symmetrical with respect to the diagonal. For a cross section at a distance x from the corner, the bending moment is Px and the width of section is $2x$. When the subgrade support is neglected and the slab is considered as a cantilever beam, the tensile stress on top of the slab is

$$\sigma_c = \frac{Px}{\frac{1}{6}(2x)\,h^2} = \frac{3P}{h^2} \tag{4.12}$$

in which σ_c is the stress due to corner loading, P is the concentrated load, and h is the thickness of the slab. Note that σ_c is independent of x. In other words, every cross section, no matter how far from the corner, will have the same stress, as indicated by Eq. 4.12. If the load is really a concentrated load applied at the very

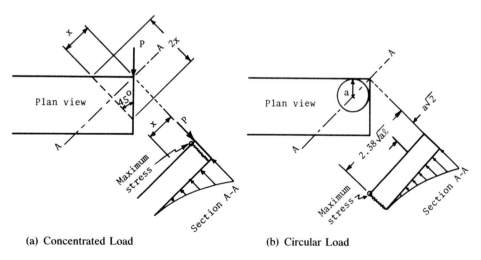

(a) Concentrated Load   (b) Circular Load

Figure 4.6 A slab subjected to corner load.

corner, Eq. 4.12 is an exact solution because at a cross section near to the load with x approaching 0, the subgrade reaction is very small and can be neglected.

Figure 4.6b shows a circular load applied near the corner of a slab. Because the section of maximum stress is not near the corner, the total subgrade reactive force is quite large and cannot be neglected. Westergaard (1926b) applied a method of successive approximations and obtained Eqs. 4.13 and 4.14:

$$\sigma_c = \frac{3P}{h^2}\left[1 - \left(\frac{a\sqrt{2}}{\ell}\right)^{0.6}\right] \tag{4.13}$$

$$\Delta_c = \frac{P}{k\ell^2}\left[1.1 - 0.88\left(\frac{a\sqrt{2}}{\ell}\right)\right] \tag{4.14}$$

in which Δ_c is the corner deflection, ℓ is the radius of relative stiffness, a is the contact radius, and k is the modulus of subgrade reaction. He also found that the maximum moment occurs at a distance of $2.38\sqrt{a\ell}$ from the corner. For a concentrated load with $a = 0$, Eqs. 4.13 and 4.12 are identical.

Ioannides et al. (1985) applied the finite element method to evaluate Westergaard's solutions. They suggested the use of Eqs. 4.15 and 4.16:

$$\sigma_c = \frac{3P}{h^2}\left[1 - \left(\frac{c}{\ell}\right)^{0.72}\right] \tag{4.15}$$

$$\Delta_c = \frac{P}{k\ell^2}\left[1.205 - 0.69\left(\frac{c}{\ell}\right)\right] \tag{4.16}$$

in which c is the side length of a square contact area. They found that the maximum moment occurs at a distance of $1.80c^{0.32}\ell^{0.59}$ from the corner. If a load is applied over a circular area, the value of c must be selected so that the square and the circle have the same contact area:

$$c = 1.772a \tag{4.17}$$

Example 4.2:

Figure 4.7 shows a concrete slab subjected to a corner loading. Given $k = 100$ pci (27.2 MN/m³), $h = 10$ in. (254 mm), $a = 6$ in. (152 mm), and $P = 10,000$ lb (44.5 kN), determine the maximum stress and deflection due to corner loading by Eqs. 4.13 through 4.16.

Solution: From Eq. 4.10, $\ell = [4 \times 10^6 \times 1000/(12 \times 0.9775 \times 100)]^{0.25} = 42.97$ in. (1.09 m). From Eq. 4.13, $\sigma_c = 3 \times 10,000/100 \times [1 - (6\sqrt{2}/42.97)^{0.6}] = 186.6$ psi

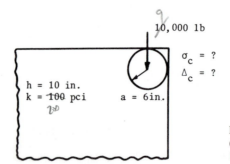

10,000 lb

$\sigma_c = ?$
$\Delta_c = ?$

h = 10 in.
k = 100 pci

a = 6 in.

Figure 4.7 Example 4.2
(1 in. = 25.4 mm, 1 lb = 4.45 N,
1 pci = 271.3 kN/m³).

Stresses and Deflections in Rigid Pavements   Chap. 4

(1.29 MPa). From Eq. 4.14, Δ_c = 10,000/(100 × 1846.4)[1.1 − 0.88(6√2/42.97)] = 0.0502 in. (1.27 mm). From Eqs. 4.15 and 4.17, σ_c = 3 × 10,000/100[1 − (1.772 × 6/42.97)$^{0.72}$] = 190.3 psi (1.31 MPa), which is 2% larger than the value obtained from Eq. 4.13. From Eqs. 4.16 and 4.17, Δ_c = 10,000/(100 × 1846.4)[1.205 − 0.69(1.772 × 6/42.97)] = 0.0560 in. (1.42 mm), which is 11% greater than the value obtained from Eq. 4.14.

Interior Loading

The earliest formula developed by Westergaard (1926b) for the stress in the interior of a slab under a circular loaded area of radius a is

$$\sigma_i = \frac{3(1 + v)P}{2\pi h^2}\left(\ln \frac{\ell}{b} + 0.6159\right) \tag{4.18}$$

in which ℓ is the radius of relative stiffness and

$$b = a \quad \text{when } a \geq 1.724h \tag{4.19a}$$

$$b = \sqrt{1.6a^2 + h^2} - 0.675h \quad \text{when } a < 1.724h \tag{4.19b}$$

For a Poisson ratio of 0.15 and in terms of base 10 log, Eq. 4.18 can be written as

$$\sigma_i = \frac{0.316P}{h^2}\left[4 \log\left(\frac{\ell}{b}\right) + 1.069\right] \tag{4.20}$$

The deflection equation due to interior loading (Westergaard, 1939) is

$$\Delta_i = \frac{P}{8k\ell^2}\left\{1 + \frac{1}{2\pi}\left[\ln\left(\frac{a}{2\ell}\right) - 0.673\right]\left(\frac{a}{\ell}\right)^2\right\} \tag{4.21}$$

Example 4.3:

Same as Example 4.2 except that the load is applied in the interior, as shown in Figure 4.8. Determine the maximum stress and deflection due to interior loading.

Solution: From Eq. 4.19b, $b = \sqrt{1.6 \times 36 + 100} - 0.675 \times 10 = 5.804$ in. (147 mm). From Eq. 4.20, $\sigma_i = 0.316 \times 10,000/100 \times [4 \log (42.97/5.804) + 1.069] = 143.7$ psi (992 kPa). The same result is obtained if Eq. 4.18 is used. From Eq. 4.21, $\Delta_i = 10,000/(8 \times 100 \times 1846.4)\{1 + (1/2\pi) \times [\ln(6/85.94) - 0.673] \times (6/42.97)^2\} = 0.0067$ in. (0.17 mm). Compared with the Westergaard solutions due to corner loading, the stress due to interior loading is 77% of that due to corner loading, while the deflection is only 13%. This is true only when there is no load transfer across the joint at the corner. If sufficient load transfer is provided, the stress due to corner

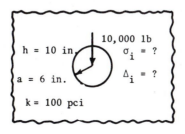

Figure 4.8 Example 4.3
(1 in. = 25.4 mm, 1 lb = 4.45 N, 1 pci = 271.3 kN/m³).

loading will be smaller than that due to interior loading but the deflection at the corner will still be greater.

Edge Loading

The stress due to edge loading was presented by Westergaard (1926b, 1933, 1948) in several different papers. In his 1948 paper, he presented generalized solutions for maximum stress and deflection produced by elliptical and semielliptical areas placed at the slab edge. Setting the length of both major and minor semiaxes of the ellipse to the contact radius a leads to the corresponding solutions for a circular or semicircular loaded area. In the case of a semicircle, its straight edge is in line with the edge of the slab. The results obtained from these new formulas differ significantly from those of the previous formulas. According to Ioannides et al. (1985), Eq. 4.22 through 4.25 are the correct ones to use:

$$\underset{\text{(circle)}}{\sigma_e} = \frac{3(1 + v)P}{\pi(3 + v)h^2}\left[\ln\left(\frac{Eh^3}{100ka^4}\right) + 1.84 - \frac{4v}{3} + \frac{1 - v}{2} + \frac{1.18(1 + 2v)a}{\ell}\right] \qquad (4.22)$$

$$\underset{\text{(semicircle)}}{\sigma_e} = \frac{3(1 + v)P}{\pi(3 + v)h^2}\left[\ln\left(\frac{Eh^3}{100ka^4}\right) + 3.84 - \frac{4v}{3} + \frac{(1 + 2v)a}{2\ell}\right] \qquad (4.23)$$

$$\underset{\text{(circle)}}{\Delta_e} = \frac{\sqrt{2 + 1.2v}P}{\sqrt{Eh^3k}}\left[1 - \frac{(0.76 + 0.4v)a}{\ell}\right] \qquad (4.24)$$

$$\underset{\text{(semicircle)}}{\Delta_e} = \frac{\sqrt{2 + 1.2v}P}{\sqrt{Eh^3k}}\left[1 - \frac{(0.323 + 0.17v)a}{\ell}\right] \qquad (4.25)$$

For $v = 0.15$, Eqs. 4.22 to 4.25 can be written as

$$\underset{\text{(circle)}}{\sigma_e} = \frac{0.803P}{h^2}\left[4 \log\left(\frac{\ell}{a}\right) + 0.666\left(\frac{a}{\ell}\right) - 0.034\right] \qquad (4.26)$$

$$\underset{\text{(semicircle)}}{\sigma_e} = \frac{0.803P}{h^2}\left[4 \log\left(\frac{\ell}{a}\right) + 0.282\left(\frac{a}{\ell}\right) + 0.650\right] \qquad (4.27)$$

$$\underset{\text{(circle)}}{\Delta_e} = \frac{0.431P}{k\ell^2}\left[1 - 0.82\left(\frac{a}{\ell}\right)\right] \qquad (4.28)$$

$$\underset{\text{(semicircle)}}{\Delta_e} = \frac{0.431P}{k\ell^2}\left[1 - 0.349\left(\frac{a}{\ell}\right)\right] \qquad (4.29)$$

Example 4.4:

Same as Example 4.2 except that the load is applied to the slab edge, as shown in Figure 4.9. Determine the maximum stress and deflection under both circular and semicircular loaded areas.

Solution: For a circular area, from Eq. 4.26, $\sigma_e = 0.803 \times 10,000/100 \times$ [4 log (42.97/6) + 0.666(6/42.97) − 0.034] = 279.4 psi (1.93 MPa); from Eq. 4.28, Δ_e = 0.431 × 10,000/(100 × 1846.4) × [1 − 0.82(6/42.97)] = 0.0207 in. (0.525 mm). For a semicircular area, from Eq. 4.27, σ_e = 0.803 × 10,000/100 × [4 log (42.97/6) + 0.282(6/42.97) + 0.650] = 330.0 psi (2.28 MPa); from Eq. 4.29, Δ_e = 0.431 × 10,000/(100 × 1846.4) × [1 − 0.349(6/42.97)] = 0.0222 in. (0.564 mm). It can be seen

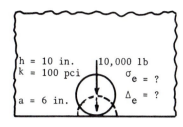

Figure 4.9 Example 4.4
(1 in. = 25.4 mm, 1 lb = 4.45 N,
1 pci = 271.3 kN/m³).

that the maximum stress due to edge loading is greater than that due to corner and interior loadings, while the maximum deflection due to edge loading is greater than that due to interior loading but much smaller than that due to corner loading. The fact that both the stress and deflection are greater under a semicircular loaded area than those under a circular area is reasonable because the centroid of a semicircle is closer to the pavement edge than that of a circle.

Dual Tires

With the exception of Eqs. 4.23, 4.25, 4.27, and 4.29 for a semicircular loaded area, all of the closed form formulas presented so far are based on a circular loaded area. When a load is applied over a set of dual tires, it is necessary to convert it into a circular area, so that the equations based on a circular loaded area can be applied. If the total load is the same but the contact area of the circle is equal to that of the duals, as has been frequently assumed for flexible pavements, the resulting stresses and deflection will be too large. Therefore, for a given total load, a much larger circular area should be used for rigid pavements.

Figure 4.10 shows a set of dual tires. It has been found that satisfactory results can be obtained if the circle has an area equal to the contact area of the duals plus the area between the duals, as indicated by the hatched area shown in the figure. If P_d is the load on one tire and q is the contact pressure, the area of each tire is

$$\frac{P_d}{q} = \pi(0.3L)^2 + (0.4L)(0.6L) = 0.5227L^2$$

or
$$L = \sqrt{\frac{P_d}{0.5227q}} \qquad (4.30)$$

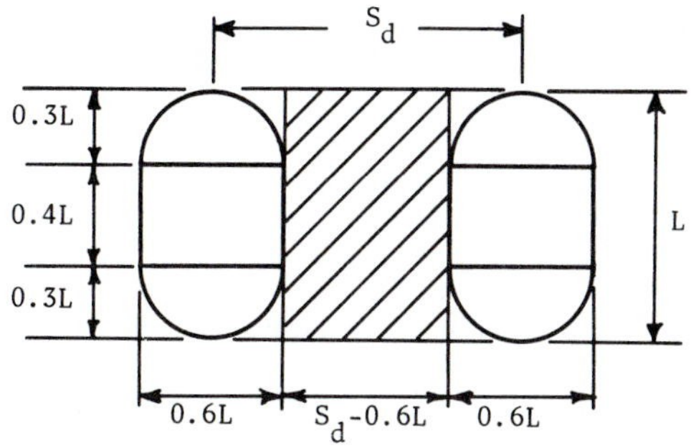

![Figure 4.10 diagram]

Figure 4.10 Method for converting duals into a circular area.

The area of an equivalent circle is

$$\pi a^2 = 2 \times 0.5227L^2 + (S_d - 0.6L)L = 0.4454L^2 + S_dL$$

Substituting L from Eq. 4.30 yields

$$\pi a^2 = \frac{0.8521P_d}{q} + S_d\sqrt{\frac{P_d}{0.5227q}}$$

So the radius of contact area is

$$a = \sqrt{\frac{0.8521P_d}{q\pi} + \frac{S_d}{\pi}\left(\frac{P_d}{0.5227q}\right)^{1/2}} \qquad (4.31)$$

Example 4.5:

Using Westergaard's formulas, determine the maximum stress in Examples 4.2, 4.3, and 4.4 if the 10,000-lb (44.5-kN) load is applied on a set of duals spaced at 14 in. (356 mm) on centers, as shown in Figure 4.11, instead of over a 6-in. (152-mm) circular area.

Solution: With S_d = 14 in. (356 mm), q = 10,000/(36π) = 88.42 psi (610 kPa), and P_d = 5000 lb (22.3 kN), from Eq. 4.31

$$a = \sqrt{\frac{0.8521 \times 5000}{88.42\pi} + \frac{14}{\pi}\left(\frac{5000}{0.5227 \times 88.42}\right)^{1/2}} = 7.85 \text{ in. (199 mm)}$$

which is greater than the original 6 in. (152 mm). From Eq. 4.13, σ_c = 3 × 10,000/100 × [1 − (7.85$\sqrt{2}$/42.97)$^{0.6}$] = 166.8 psi (1.15 MPa), which is about 89% of the stress in Example 4.2. From Eq. 4.19b, b = $\sqrt{1.6 \times 61.62 + 100}$ − 0.675 × 10 = 7.34 in. (186 mm). From Eq. 4.20, σ_i = 0.316 × 10,000/100 × [4 log (42.97/7.34) + 1.069] = 130.8 psi (902 kPa), which is about 91% of the stress in Example 4.3. From Eq. 4.26, σ_e = 0.803 × 10,000/100 × [4 log (42.97/7.85) + 0.666(7.85/42.97) − 0.034] = 244.2 psi (1.68 MPa), which is about 87% of the stress in Example 4.4.

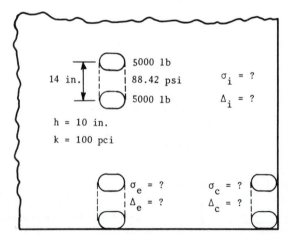

Figure 4.11 Example 4.5
(1 in. = 25.4 mm, 1 lb = 4.45 N, 1 pci = 271.3 kN/m³).

4.2.2 Influence Charts

Influence charts based on liquid foundations (Pickett and Ray, 1951) were used previously by the Portland Cement Association for rigid pavement design. The charts are based on Westergaard's theory with a Poisson ratio of 0.15 for the concrete slab. Only charts for interior and edge loadings are available, the interior loading being used for the design of airport pavements (PCA, 1955) and the edge loading for the design of highway pavements (PCA, 1966).

Interior Loading

Figure 4.12 shows the applications of influence charts for determining the moment at the interior of slab. The moment is at point O in the n direction. To use

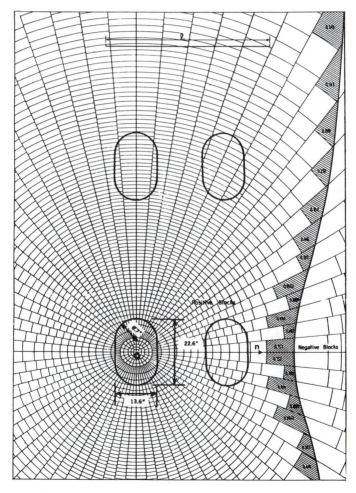

Figure 4.12 Application of influence chart for determining moment (1 in. = 25.4 mm). (After Pickett and Ray (1951).)

the chart, it is necessary to determine the radius of relative stiffness ℓ according to Eq. 4.10. For example, if ℓ is 57.1 in. (1.45 m), the scale on the top of Figure 4.12 is 57.1 in. (1.45 m). This scale should be used to draw the configuration of the contact area. If the actual length of tire imprint is 22.6 in. (574 mm), the length to be drawn on the influence chart is 22.6/57.1 or 39.6% of the length shown by the scale ℓ. The location of other tires is based on the same scale. By counting the number of blocks N covered by the tire imprints, the moment in the n direction M can be determined by

$$M = \frac{q\ell^2 N}{10,000} \tag{4.32a}$$

in which q is the contact pressure. The stress is determined by dividing the moment by the section modulus:

$$\sigma_i = \frac{6M}{h^2} \tag{4.32b}$$

For the tire imprints shown in Figure 4.12, the moment is under the center of the lower left tire in the lateral direction. If the moment in the longitudinal direction is desired, the tire assembly must rotate 90° clockwise so that two of the tires lie in the zone of negative blocks, and the moment becomes much smaller.

Figure 4.13 shows the influence chart for deflection due to interior loading.

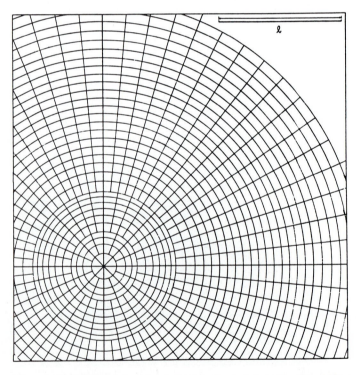

Figure 4.13 Influence chart for deflection due to interior loading. (After Pickett and Ray (1951).)

Stresses and Deflections in Rigid Pavements   Chap. 4

The chart is axisymmetric and the blocks are formed by concentric circles and radial lines. The deflection is at the center of the circles. The use of the chart is similar to that of Figure 4.12. After the number of blocks covered by the tire imprint are counted, the deflection can be determined by

$$\Delta_i = \frac{0.0005q\ell^4 N}{D} \tag{4.33}$$

in which D is the modulus of rigidity:

$$D = \frac{Eh^3}{12(1 - \nu^2)} \tag{4.34}$$

Example 4.6:

Same as Example 4.3. Determine the maximum stress and deflection due to interior loading by the influence charts shown in Figures 4.13 and 4.14.

Solution: With $\ell = 42.97$ in. (1.09 m), the scale at the top of Figure 4.14 is 42.97 in.

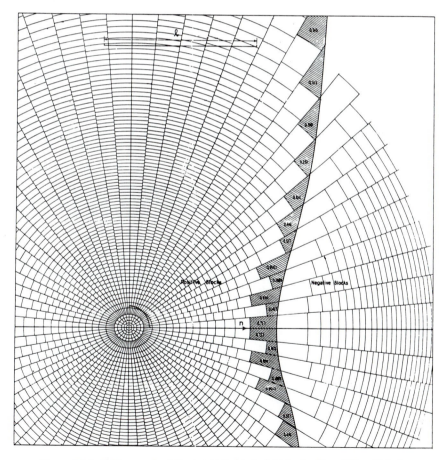

Figure 4.14 Influence chart for moment due to interior loading. (After Pickett and Ray (1951).)

(1.09 m). The radius of contact area $a = 6$ in. (152 mm), or $a/\ell = 6/42.97 = 0.14$. The number of blocks covered by a circle with center at 0 and a radius equal to 0.14ℓ is 148 and can be found from Figure 4.14. Due to symmetry, only the blocks within one-quarter of the circle need be counted and the result is then multiplied by 4. With $q = 10,000/(\pi \times 36) = 88.42$ psi (610 kPa), from Eq. 4.32a, $M = 88.42 \times (42.97)^2 \times 148/10,000 = 2416$ in.-lb/in. (10.8 m-kN/m), and from Eq. 4.32b, $\sigma_i = 6 \times 2416/100 = 145.0$ psi (1.0 MPa), which checks with the 143.7 psi (992 kPa) obtained in Example 4.3. The number of blocks for deflection is 16 and can be found from Figure 4.13. From Eq. 4.34, $D = 4 \times 10^6 \times 10^3/(12 \times 0.9775) = 3.41 \times 10^8$ in.-lb (38.5 m-MN). From Eq. 4.33, $\Delta_i = 0.0005 \times 88.42 \times (42.97)^4 \times 16/(3.41 \times 10^8) = 0.0071$ in. (0.18 mm), which checks with the 0.0067 in. (0.17 mm) obtained in Example 4.3.

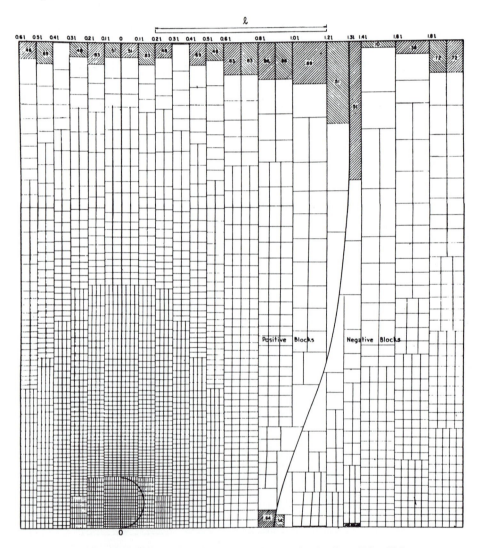

Figure 4.15 Influence chart for moment due to edge loading. (After Pickett and Ray (1951).)

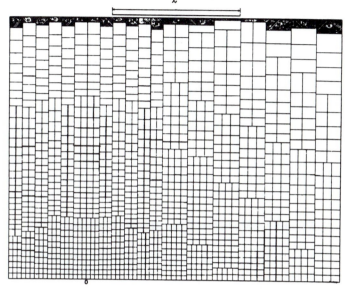

Figure 4.16 Influence chart for deflection due to edge loading. (After Pickett and Ray (1951).)

Edge Loading

Figures 4.15 and 4.16 show the influence charts for moment and deflection at point O on the edge of slab. The use of the charts is similar to the case of interior loading and the same formulas, Eqs. 4.32 to 4.34, apply.

Example 4.7:

Same as Example 4.4. Determine the maximum stress and deflection due to the circular loaded area tangent to the pavement edge.

Solution: By drawing a circle tangent to the edge at point O with a radius of 0.14ℓ, the number of blocks is 282 for moment and 46 for deflection. Because the stress and deflection are proportional to N, from Example 4.6, $\sigma_e = 145.0 \times 282/148 = 276.3$ psi (1.91 MPa), which checks with the 279.4 psi (1.93 MPa) obtained in Example 4.4, and $\Delta_e = 0.0071 \times 46/16 = 0.0204$ in. (0.518 mm), which checks with the 0.0207 in. (0.525 mm) in Example 4.4.

4.3 STRESSES DUE TO FRICTION

The friction between a concrete slab and its foundation causes tensile stresses in the concrete, in the steel reinforcements, if any, and in the tie bars. For plain concrete pavements, the spacing between contraction joints must be so chosen that the stresses due to friction will not cause the concrete to crack. For longer joint spacings, steel reinforcements must be provided to take care of the stresses

Sec. 4.3   Stresses Due to Friction

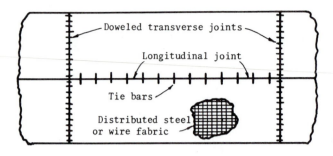

Figure 4.17 Steel and joints in concrete pavements.

caused by friction. The number of tie bars required is also controlled by the friction. Figure 4.17 shows the arrangement of joints and steel in concrete pavements.

4.3.1 Effect of Volume Change on Concrete

The volume change caused by the variation of temperature and moisture has two important effects on concrete. First, it induces tensile stresses and causes the concrete to crack. Second, it causes the joint to open and decreases the efficiency of load transfer.

Concrete Stress

Figure 4.18 shows a concrete pavement subject to a decrease in temperature. Due to symmetry, the slab tends to move from both ends toward the center, but the subgrade prevents it from moving; thus, frictional stresses are

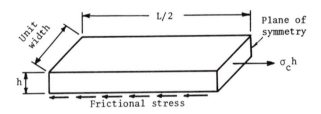

(a) Free Body Diagram

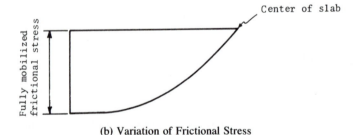

(b) Variation of Frictional Stress

Figure 4.18 Stresses due to friction.

developed between the slab and the subgrade. The amount of friction depends on the relative movement, being zero at the center where no movement occurs and maximum at some distance from the center where the movement is fully mobilized, as shown in Figure 4.18b. For practical purposes, an average coefficient of friction f_a may be assumed. The tensile stress in the concrete is greatest at the center and can be determined by equating the frictional force per unit width of slab, $\gamma_c h L f_a / 2$, to the tensile force $\sigma_c h$, as shown in Figure 4.18a:

$$\sigma_c = \frac{\gamma_c L f_a}{2} \tag{4.35}$$

in which σ_c is the stress in the concrete, γ_c is the unit weight of the concrete, L is the length of the slab, and f_a is the average coefficient of friction between slab and subgrade, usually taken as 1.5. Equation 4.35 implies that the stress in the concrete due to friction is independent of the slab thickness.

Example 4.8:

Given a concrete pavement with a joint spacing of 25 ft (7.6 m) and a coefficient of friction of 1.5, as shown in Figure 4.19, determine the stress in concrete due to friction.

Solution: With $\gamma_c = 150$ pcf $= 0.0868$ pci (23.6 kN/m³), $L = 25$ ft $= 300$ in. (7.6 m), and $f_a = 1.5$, from Eq. 4.35, $\sigma_c = 0.0868 \times 300 \times 1.5/2 = 19.5$ psi (135 kPa).

The tensile strength of concrete ranges from $3\sqrt{f_c'}$ to $5\sqrt{f_c'}$, where f_c' is the compressive strength of concrete (Winter and Nilson, 1979). If $f_c' = 3000$ psi (13.8 MPa), the tensile strength is from 164 to 274 psi (1.13 to 1.89 MPa), which is much greater than the tensile stress of 19.5 psi (135 kPa). Therefore, it does not appear that joint spacings in plain concrete pavements are dictated by the concrete stress due to friction.

Joint Opening

The spacing of joints in plain concrete pavements depends more on the shrinkage characteristics of the concrete rather than on the stress in the concrete. Longer joint spacings cause the joint to open wider and decrease the efficiency of load transfer. The opening of a joint can be computed approximately by (Darter and Barenberg, 1977)

$$\Delta L = CL(\alpha_t \Delta T + \epsilon) \tag{4.36}$$

in which ΔL is the joint opening caused by temperature change and drying shrinkage of concrete; α_t is the coefficient of thermal expansion of concrete, generally 5 to 6 $\times$ 10⁻⁶/°F (9 to 10.8 $\times$ 10⁻⁶/°C); ϵ is the drying shrinkage

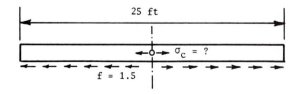

Figure 4.19 Example 4.8 (1 ft = 0.305 m).

coefficient of concrete, approximately 0.5 to 2.5×10^{-4}; L is the joint spacing or slab length; ΔT is the temperature range, which is the temperature at placement minus the lowest mean monthly temperature; and C is the adjustment factor due to slab–subbase friction, 0.65 for stabilized base and 0.8 for granular subbase.

Example 4.9:

Given $\Delta T = 60°F$ (33°C), $\alpha_t = 5.5 \times 10^{-6}/°F$ ($9.9 \times 10^{-6}/°C$), $\epsilon = 1.0 \times 10^{-4}$, $C = 0.65$, and the allowable joint openings for undoweled and doweled joints are 0.05 and 0.25 in. (1.3 and 6.4 mm), respectively, determine the maximum allowable joint spacing.

Solution: From Eq. 4.36, $L = \Delta L/[C(\alpha_t \Delta T + \epsilon)] = \Delta L/[0.65(5.5 \times 10^{-6} \times 60 + 0.0001)] = \Delta L/0.00028$. For the undoweled joint, $L = 0.05/0.00028 = 178.6$ in. $= 14.9$ ft (4.5 m). For the doweled joint, $L = 0.25/0.00028 = 892.9$ in. $= 74.4$ ft (22.7 m).

4.3.2 Steel Stress

Steel is used in concrete pavements as reinforcements, tie bars, and dowel bars. The design of longitudinal and transverse reinforcements and the tie bars across longitudinal joints is based on the stresses due to friction and is presented in this section. The design of dowel bars is not based on the stresses due to friction and is presented in Section 4.4.1.

Reinforcements

Wire fabric or bar mats may be used in concrete slabs for control of temperature cracking. These reinforcements do not increase the structural capacity of the slab but are used for two purposes: to increase the joint spacing and to tie the cracked concrete together and maintain load transfers through aggregate interlock. When steel reinforcements are used, it is assumed that all tensile stresses are taken by the steel alone, so $\sigma_c h$ in Figure 4.18a must be replaced by $A_s f_s$ and Eq. 4.35 becomes

$$A_s = \frac{\gamma_c h L f_a}{2 f_s} \tag{4.37}$$

in which A_s is the area of steel required per unit width and f_s is the allowable stress in steel. Equation 4.37 indicates that the amount of steel required is proportional to the length of slab.

The steel is usually placed at the middepth of the slab and discontinued at the joint. The amount of steel obtained from Eq. 4.37 is at the center of the slab and can be reduced toward the end. However, in actual practice the same amount of steel is used throughout the length of the slab.

Table 4.1 gives the allowable stress for different types and grades of steel. The allowable stress is generally taken as two-thirds of the yield strength. Table 4.2 shows the weight and dimensions of reinforcing bars and Table 4.3 shows those of welded wire fabric.

TABLE 4.1 YIELD STRENGTH AND ALLOWABLE STRESS FOR STEEL

Type and grade of steel	Yield strength (psi)	Allowable stress (psi)
Billet steel, intermediate grade	40,000	27,000
Rail steel or hard grade of billet steel	50,000	33,000
Rail steel, special grade	60,000	40,000
Billet steel, 60,000 psi minimum yield	60,000	40,000
Cold drawn wire (smooth)	65,000	43,000
Cold drawn wire (deformed)	70,000	46,000

Note. 1 psi = 6.9 kPa.

TABLE 4.2 WEIGHTS AND DIMENSIONS OF STANDARD REINFORCING BARS

Bar size designation	Weight (lb/ft)	Nominal dimensions, round sections		
		Diameter (in.)	Cross-sectional area (in.^{2})	Perimeter (in.)
No. 3	0.376	0.375	0.11	1.178
No. 4	0.668	0.500	0.20	1.571
No. 5	1.043	0.625	0.31	1.963
No. 6	1.502	0.750	0.44	2.356
No. 7	2.044	0.875	0.60	2.749
No. 8	2.670	1.000	0.79	3.142
No. 9	3.400	1.128	1.00	3.544
No. 10	4.303	1.270	1.27	3.990
No. 11	5.313	1.410	1.56	4.430

Note. 1 in. = 25.4 mm, 1 lb = 4.45 N, 1 ft = 0.305 m.

Welded wire fabric is prefabricated reinforcement consisting of parallel series of high-strength, cold-drawn wires welded together in square or rectangular grids. The spacings and sizes of wires are identified by "style." A typical style designation is 6 × 12 − W8 × W6, in which the spacing of longitudinal wires is 6 in. (152 mm), the spacing of transverse wires is 12 in. (305 mm), the size of longitudinal wire is W8 with a cross-sectional area of 0.08 in.^{2} (51.6 mm^2), and the size of transverse wires is W6 with a cross sectional area of 0.06 in.^{2} (38.7 mm^2). The typical style with deformed welded wire fabric is 6 × 12 − D8 × D6. The following standard practices on wire sizes, spacings, laps, and clearances are recommended by the Wire Reinforcement Institute (WRI, 1975):

1. Because the fabric is subjected to bending stresses as well as tensile stresses at cracks, neither the longitudinal nor the transverse wires should be less than W4 or D4.

2. To provide generous opening between wires to permit placement and vibra- tion of concrete, the minimum spacing between wires should not be less than

TABLE 4.3 WEIGHTS AND DIMENSIONS OF WELDED WIRE FABRIC

Wire size no.		Diameter (in.)	Weight lb/ft	Cross-sectional area (in.²/ft) center-to-center spacing (in.)						
Smooth	Deformed			2	3	4	6	8	10	12
W31	D31	0.628	1.054	1.86	1.24	.93	.62	.465	.372	.31
W30	D30	0.618	1.020	1.80	1.20	.90	.60	.45	.36	.30
W28	D28	0.597	.952	1.68	1.12	.84	.56	.42	.336	.28
W26	D26	0.575	.934	1.56	1.04	.78	.52	.39	.312	.26
W24	D24	0.553	.816	1.44	.96	.72	.48	.36	.288	.24
W22	D22	0.529	.748	1.32	.88	.66	.44	.33	.264	.22
W20	D20	0.504	.680	1.20	.80	.60	.40	.30	.24	.20
W18	D18	0.478	.612	1.08	.72	.54	.36	.27	.216	.18
W16	D16	0.451	.544	.96	.64	.48	.32	.24	.192	.16
W14	D14	0.422	.476	.84	.56	.42	.28	.21	.168	.14
W12	D12	0.390	.408	.72	.48	.36	.24	.18	.144	.12
W11	D11	0.374	.374	.66	.44	.33	.22	.165	.132	.11
W10.5		0.366	.357	.63	.42	.315	.21	.157	.126	.105
W10	D10	0.356	.340	.60	.40	.30	.20	.15	.12	.10
W9.5		0.348	.323	.57	.38	.285	.19	.142	.114	.095
W9	D9	0.338	.306	.54	.36	.27	.18	.135	.108	.09
W8.5		0.329	.289	.51	.34	.255	.17	.127	.102	.085
W8	D8	0.319	.272	.48	.32	.24	.16	.12	.096	.08
W7.5		0.309	.255	.45	.30	.225	.15	.112	.09	.075
W7	D7	0.298	.238	.42	.28	.21	.14	.105	.084	.07
W6.5		0.288	.221	.39	.26	.195	.13	.097	.078	.065
W6	D6	0.276	.204	.36	.24	.18	.12	.09	.072	.06
W5.5		0.264	.187	.33	.22	.165	.11	.082	.066	.055
W5	D5	0.252	.170	.30	.20	.15	.10	.075	.06	.05
W4.5		0.240	.153	.27	.18	.135	.09	.067	.054	.045
W4	D4	0.225	.136	.24	.16	.12	.08	.06	.048	.04

Note. Wire sizes other than those listed above may be produced provided the quantity required is sufficient to justify manufacture. 1 in. = 25.4 mm, 1 lb = 4.45 N, 1 ft = 0.305 m.

Source. After WRI (1975).

4 in. (102 mm). The maximum spacing should not be greater than 12 in. (305 mm) between longitudinal wires and 24 in. (610 mm) between transverse wires.

3. Because the dimensions of a concrete slab are usually greater than those of the welded wire fabric, the fabric should be installed with end and side laps. The end lap should be about 30 times the longitudinal wire diameter but not less than 12 in. (305 mm). The side laps should be about 20 times the transverse wire diameter but not less than 6 in. (152 mm).

4. The fabric should extend to about 2 in. (51 mm) but not more than 6 in. (152 mm) from the slab edges. The depth from the top of slab should not be less than 2.5 in. (64 mm) or more than middepth.

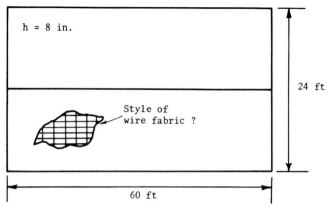

h = 8 in.

24 ft

Style of
wire fabric ?

60 ft

Figure 4.20 Example 4.10
(1 in. = 25.4 mm, 1 ft = 0.305 m).

Example 4.10:

Determine the wire fabric required for a two-lane concrete pavement, 8 in. (203 mm) thick, 60 ft (18.3 m) long, and 24 ft (7.3 m) wide, with a longitudinal joint at the center, as shown in Figure 4.20.

Solution: With γ_c = 0.0868 pci (23.6 kN/m^3), h_c = 8 in. (203 mm), L = 60 ft = 720 in. (18.3 m), f_a = 1.5, and f_s = 43,000 psi (297 MPa), from Eq. 4.37, the required longitudinal steel is A_s = 0.0868 × 8 × 720 × 1.5/(2 × 43,000) = 0.00872 in.^{2}/in. = 0.105 in.^{2}/ft (222 mm^2/m). The required transverse steel is A_s = 0.0868 × 8 × 24 × 12 × 1.5/(2 × 43,000) = 0.00349 in.^{2}/in. = 0.042 in.^{2}/ft (88.9 mm^2/m). From Table 4.3, use 6 × 12 − W5.5 × W4.5 with cross-sectional areas of 0.11 in.^{2} (71 mm^2) for longitudinal wires and 0.045 in.^{2} (29 mm^2) for transverse wires.

If the concrete pavement is used for a four-lane highway with all four slabs tied together at the three longitudinal joints, the transverse reinforcements in the two inside lanes should be doubled because the length L in Eq. 4.37 should be 48 ft (14.6 m) instead of 24 ft (7.3 m).

Tie Bars

Tie bars are placed along the longitudinal joint to tie the two slabs together so that the joint will be tightly closed and the load transfer across the joint can be ensured. The amount of steel required for tie bars can be determined in the same way as the longitudinal or transverse reinforcements by slightly modifying Eq. 4.37:

$$A_s = \frac{\gamma_c h L' f_a}{f_s} \qquad (4.38)$$

in which A_s is the area of steel required per unit length of slab and L' is the distance from the longitudinal joint to the free edge where no tie bars exist. For two- or three-lane highways, L' is the lane width. If tie bars are used in all three longitudinal joints of a four-lane highway, L' is equal to the lane width for the two outer joints and twice the lane width for the inner joint.

The length of tie bars is governed by the allowable bond stress. For deformed bars, an allowable bond stress of 350 psi (2.4 MPa) may be assumed. The length of bar should be based on the full strength of the bar:

$$t = 2\left(\frac{A_1 f_s}{\mu \, \Sigma o}\right) \qquad (4.39)$$

in which t is the length of the tie bar, μ is the allowable bond stress, A_1 is the area of one bar, and Σo is the bar perimeter. For a given bar diameter d, $A_1 = \pi d^2/4$ and $\Sigma o = \pi d$, so Eq. 4.39 can be simplified to

$$t = \frac{1}{2}\left(\frac{f_s d}{\mu}\right) \qquad (4.40)$$

The length t should be increased by 3 in. (76 mm) for misalignment.

It should be noted that many agencies use the standard tie bar design to simplify the construction. Tie bars 0.5 in. (13 mm) in diameter by 36 in. (914 mm) long spaced at intervals of 30 to 40 in. (762 to 1016 mm) are most commonly used.

Example 4.11:

Same pavement as Example 4.10. Determine the diameter, spacing, and length of tie bars required, as shown in Figure 4.21.

Solution: Assume $f_s = 27,000$ psi (186 MPa) for billet steel (see Table 4.1). With $L' = 12$ ft $= 144$ in. (3.66 m), from Eq. 4.38, $A_s = 0.0868 \times 8 \times 144 \times 1.5/27,000 = 0.00556$ in.²/in. If No. 4 (0.5 in. or 1.2 mm) bars are used, from Table 4.2, the cross-sectional area of one bar is 0.2 in.² (129 mm²). The spacing of the bar $= 0.2/0.00556 = 36$ in. (914 mm).

Assume that $\mu = 350$ psi (24 MPa), from Eq. 4.40, $t = 0.5 \times 27,000 \times 0.5/350 = 19.3$ in. (353 mm). After adding 3 in. (76 mm), $t = 19.3 + 3 = 22.3$ in. (use 24 in. or 610 mm). The design selected is No. 4 deformed bars, 24 in. (610 mm) long and 3 ft (0.9 m) on centers.

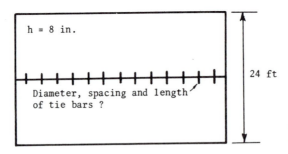

h = 8 in.

Diameter, spacing and length of tie bars ?

24 ft

Figure 4.21 Example 4.11 (1 in. = 25.4 mm, 1 ft = 0.305 m).

4.4 DESIGN OF DOWELS AND JOINTS

The design of dowels and joints is mostly based on experience, although some theoretical methods on the design of dowels are available. The size of dowels to be used depends on the thickness of slab. Table 4.4 shows the size and length of

TABLE 4.4 RECOMMENDED DOWEL SIZE AND LENGTH

Slab thickness (in.)	Dowel diameter (in.)	Dowel length (in.)
5	$\frac{5}{8}$	12
6	$\frac{3}{4}$	14
7	$\frac{7}{8}$	14
8	1	14
9	$1\frac{1}{8}$	16
10	$1\frac{1}{4}$	18
11	$1\frac{3}{8}$	18
12	$1\frac{1}{2}$	20

Note. All dowels spaced at 12 in. on centers,
1 in. = 25.4 mm.

Source. After PCA (1975).

dowels for different slab thicknesses as recommended by PCA (1975). It can be seen that the diameter of dowels is equal to one-eighth of the slab thickness. In a recent edition of joint design, PCA (1991) recommended the use of 1.25 in. (32 mm) diameter dowels for highway pavements less than 10 in. (254 mm) thick and 1.5 in. (38 mm) diameter dowels for pavements 10 in. (254 mm) thick or greater. A minimum dowel diameter of 1.25 to 1.5 in. (32 to 38 mm) is needed to control faulting by reducing the bearing stress in concrete.

4.4.1 Design of Dowels

Dowel bars are usually used across a transverse joint to transfer the loads to the adjoining slab. The stress and deflection at the joint are much smaller when the loads are carried by two slabs, instead of by one slab alone. The use of dowels can minimize faulting and pumping which has been considered by the Portland Cement Association (PCA, 1984) as a factor for thickness design.

Allowable Bearing Stress

Because concrete is much weaker than steel, the size and spacing of dowels required are governed by the bearing stress between dowel and concrete. The allowable bearing stress can be determined by

$$f_b = \left(\frac{4 - d}{3}\right) f'_c \tag{4.41}$$

in which f_b is the allowable bearing stress in psi, d is the dowel diameter in inches, and f'_c is the ultimate compressive strength of concrete.

Sec. 4.4   Design of Dowels and Joints

Bearing Stress on One Dowel

If the load applied to one dowel is known, the maximum bearing stress can be determined theoretically by assuming the dowel to be a beam and the concrete to be a Winkler foundation. Based on the original solution by Timoshenko, Friberg (1940) indicated that the maximum deformation of concrete under the dowel, as shown by y_0 in Figure 4.22, can be expressed by

$$y_0 = \frac{P_t(2 + \beta z)}{4\beta^3 \, E_d I_d} \tag{4.42}$$

in which y_0 is the deformation of the dowel at the face of the joint, P_t is the load on one dowel, z is the joint width, E_d is Young's modulus of the dowel, I_d is the moment of inertia of the dowel, and β is the relative stiffness of a dowel embedded in concrete. Note that

$$I_d = \frac{1}{64} \pi d^4 \tag{4.43}$$

and

$$\beta = \sqrt[4]{\frac{Kd}{4E_d I_d}} \tag{4.44}$$

in which K is the modulus of dowel support, which ranges from 300,000 to 1,500,000 pci (81.5 to 409 GN/m³), and d is the diameter of dowel. The bearing stress σ_b is proportional to the deformation:

$$\sigma_b = Ky_0 = \frac{KP_t(2 + \beta z)}{4\beta^3 E_d I_d} \tag{4.45}$$

The bearing stress obtained from Eq. 4.45 should compare with the allowable bearing stress computed by Eq. 4.41. If the actual bearing stress is greater than allowable, then larger dowel bars or smaller dowel spacing should be used. Recent studies have also shown that the bearing stress is related to the faulting of slabs, as described in Section 12.1.4. By limiting the bearing stress, the amount of faulting can be reduced to the allowable limit.

Dowel Group Action

When a load W is applied on one slab near the joint, as shown in Figure 4.23, part of the load will be transferred to the adjacent slab through the dowel group. If the dowels are 100% efficient, both slabs will deflect the same amount and the reactive forces under both slabs will be the same, each equal to $0.5W$, which is

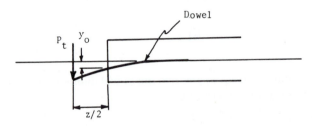

Figure 4.22 Dowel deformation under load.

Stresses and Deflections in Rigid Pavements   Chap. 4

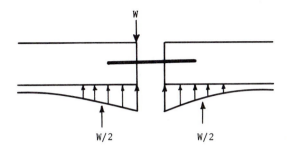

Figure 4.23 Load transfer through dowel group.

also the total shear force transferred by the dowel group. If the dowels are less than 100% efficient, as in the case of old pavements where some dowels become loose, the reactive forces under the loaded slab will be greater than 0.5W, while those under the unloaded slab will be smaller than 0.5W. As a result, the total shear force on the dowels is smaller than 0.5W. Therefore, the use of 0.5W for the design of dowels is more conservative.

Based on Westergaard's solutions, Friberg (1940) found that the maximum negative moment for both interior and edge loadings occurs at a distance of 1.8ℓ from the load, where ℓ is the radius of relative stiffness defined by Eq. 4.10. When the moment is maximum, the shear force is equal to zero. It is therefore reasonable to assume that the shear in each dowel decreases inversely with the distance of the dowel from the point of loading, being maximum for the dowel under or nearest to the point of loading and zero at a distance of 1.8ℓ. The application of the above principle for dowel design can be best illustrated by the following examples.

Example 4.12:

Figure 4.24 shows a concrete pavement 8 in. (203 mm) thick having a joint width of 0.2 in. (5.1 mm), a modulus of subgrade reaction of 100 pci (27 kN/m³), and a modulus of dowel support of 1.5×10^6 pci (407 GN/m³). A load of 9000 lb (40 kN) is applied over the outermost dowel at a distance of 6 in. (152 mm) from the edge. The dowels are $\frac{3}{4}$ in. (19 mm) in diameter and 12 in. (305 mm) on centers. Determine the maximum bearing stress between dowel and concrete.

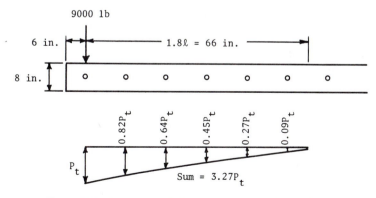

Figure 4.24 Example 4.12 (1 in. = 25.4 mm, 1 lb = 4.45 N).

Solution: From Eq. 4.10, $\ell = [4 \times 10^6 \times 512/(12 \times 0.9775 \times 100)]^{0.25} = 36.35$ in. (427 mm). If the dowel directly under the load is subjected to a shear force P_t, the forces on the dowels within a distance of 1.8ℓ, or 66 in. (1.68 m), can be determined by assuming a straight-line variation, as shown in Figure 4.24. The sum of the forces on all dowels is $3.27P_t$, which must be equal to one-half of the applied load based on 100% joint efficiency, or $P_t = 4500/3.27 = 1376$ lb (6.1 kN). From Eq. 4.43, $I_d = \pi(0.75)^4/64 = 0.0155$ in.^{4} (6450 mm^4). From Eq. 4.44, $\beta = [1.5 \times 10^6 \times 0.75/(4 \times 29 \times 10^6 \times 0.0155)]^{0.25} = 0.889$ in. (22.6 mm). From Eq. 4.45, $\sigma_b = 1.5 \times 10^6 \times 1376(2 + 0.889 \times 0.2)/(4 \times 0.703 \times 29 \times 10^6 \times 0.0155) = 3556$ psi (24.5 MPa). For a 3000-psi (20.7-MPa) concrete, the allowable bearing stress obtained from Eq. 4.41 is $f_b = (4 - 0.75) \times 3000/3 = 3250$ psi (22.4 MPa). Because the actual bearing stress is about 10% greater than the allowable, the design is not considered satisfactory.

In this example, only the left wheel load near the pavement edge is considered. The right wheel load is at least 6 ft (1.83 m) from the left wheel load, which is greater than 1.8ℓ, so the right wheel has no effect on the maximum force P_t on the dowel near the pavement edge. If the slab is thicker and stronger and the foundation is weaker, ℓ may become much larger and both wheels must be considered in determining the force P_t on the most critical dowel.

Example 4.13:

Figure 4.25a shows a 9.5-in. (241-mm) slab resting on a foundation with $k = 50$ pci (13.6 MN/m^3). Twelve dowels at 12 in. (305 mm) on centers are placed at the joint on the 12-ft (3.66-m) lane. Two 9000-lb (40-kN) wheel loads are applied at points A and B. Determine the maximum load on one dowel.

Solution: From Eq. 4.10, $\ell = [4 \times 10^6 \times (9.5)^3/(12 \times 0.9775 \times 50)]^{0.25} = 49.17$ in. (1.25 m), so $1.8\ell = 88$ in. (2.24 m). First, consider the 9000-lb (40-kN) load at A. If the dowel at A has a load factor of 1, the load factors at other dowels can be determined from similar triangles, as shown in Figure 4.25b. The sum of these factors results in 4.18 effective dowels, so the load carried by the dowel at A is 4500/4.18 or 1077 lb (4.8 kN). The loads carried by other dowels can be determined by proportion. Next, consider the 9000-lb (40-kN) load at B. If the dowel at B has a load factor of 1, the load factors at other dowels can be determined from the triangular distribution, as shown in Figure 4.25c. The sum of these factors results in 7.08 effective dowels. Note that the dowels on the other side of the longitudinal joint are not considered effective in carrying the load. The load carried by the dowel at B is 4500/7.08 or 636 lb (2.8 kN), and those carried by other dowels can be determined by proportion. Figure 4.25d shows the forces on each dowel due to the combined effect of both loads. It can be seen that the dowel nearest to the pavement edge is the most critical and should be used for design purposes. The load carried by this dowel can be determined directly by $P_t = 4500/4.18 + 0.18 \times 4500/7.08 = 1191$ lb (5.3 kN).

The above examples are based on the assumption that the maximum negative moment occurs at a distance of 1.8ℓ from the load. Recent studies by Heinrichs et al. (1989) have shown that the maximum negative moment occurs at 1.0ℓ, so the load carried by the most critical dowel should be larger than those shown in the examples. This has been proved by comparing the results with KENSLABS, as discussed in Section 5.4.2.

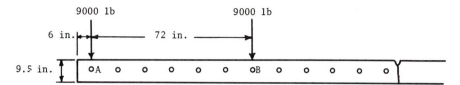

(a) Location of Loads and Dowels

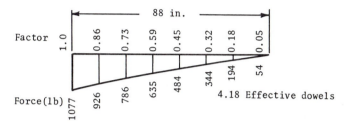

(b) Dowel Forces due to Load at A

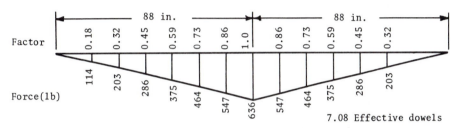

(c) Dowel Forces due to Load at B

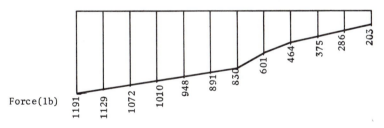

(d) Dowel Forces due to Both Loads

Figure 4.25 Example 4.13 (1 in. = 25.4 mm, 1 lb = 4.45 N).

4.4.2 Design of Joints

Joints should be provided in concrete pavements so premature cracks due to temperature or moisture changes will not occur. There are four types of joints in common use: contraction, expansion, construction, and longitudinal.

Contraction Joints

Contraction joints are transverse joints used to relieve tensile stresses. The spacing of joints should be based on local experience since a change in coarse aggregate types may have a significant effect on the concrete thermal coefficient and consequently the acceptable joint spacing. As a rough guide, the joint spacing in feet for plain concrete pavements should not greatly exceed twice the slab thickness in inches. For example, the maximum joint spacing for an 8-in. (203-mm) slab is 16 ft (4.9 m). Also, as a general guideline, the ratio of slab width to length should not exceed 1.25 (AASHTO, 1986).

Figure 4.26 shows typical contraction joints. In Figure 4.26a, a dummy groove is formed by placing a metal strip on the fresh concrete, which is later removed, or by sawing after the concrete is set. The groove is then sealed with a plastic material. If the joint spacing is small, the load transfer across the joint can be achieved by the aggregate interlock and no dowels may be needed. However, dowels are needed if the joint spacing is large or if the short panels are located near the end of the pavement. In such cases, the joint may open up and the load transfer through aggregate interlock may be lost. In lieu of a dummy groove, joints can be formed by placing a felt, asphalt ribbon, or asphalt board strip in the fresh concrete and leaving it there permanently, as shown in Figure 4.26b.

The sealant used in the joints must be capable of withstanding repeated extension and compression as the temperature and moisture in the slabs change. Sealants can be classified as field molded and preformed. Field-molded sealants are those applied in liquid or semiliquid form, and preformed sealants are shaped during manufacturing. Figure 4.27 shows the design of joint sealant reservoir for field-molded sealants. To maintain an effective field molded seal, the sealant reservoir must have the proper shape factor or depth to width ratio. The common

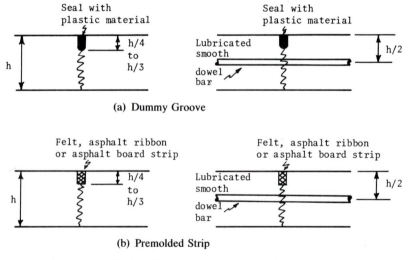

(a) Dummy Groove

(b) Premolded Strip

Figure 4.26 Typical contraction joints.

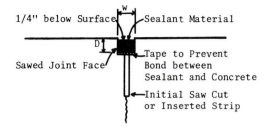

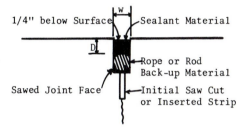

Figure 4.27 Design of joint sealant reservoir (1 in. = 25.4 mm). (After PCA (1975).)

practice is to have the ratio between 0.5 to 1. Table 4.5 shows the reservoir dimensions for field-molded sealants and Table 4.6 shows the joint and sealant widths for preformed seals as recommended by PCA (1975).

The preformed sealant is the type most recommended to achieve long-term performance. Preformed sealants can do an excellent job of keeping out incompressibles over a long period of time but may not be completely water tight compared to the field-molded sealants. The preformed sealants should be so designed that the seal will always be compressed at least 20% in the joint. The maximum allowable compression of the seal is 50%. Thus, the seal working range is 20 to 50% (Darter and Barenberg, 1977).

Example 4.14:

A concrete pavement 15 ft (4.6 m) long is placed on a gravel subbase. If the joint width is $\frac{1}{4}$ in. (6.4 mm), the design temperature range ΔT is 100°F (55.6°C), the

TABLE 4.5 RESERVOIR DIMENSIONS FOR FIELD-MOLDED SEALANTS

Joint spacing (ft)	Reservoir width (in.)	Reservoir depth (in.)
15 or less	$\frac{1}{4}$	$\frac{1}{2}$ minimum
20	$\frac{3}{8}$	$\frac{1}{2}$ minimum
30	$\frac{1}{2}$	$\frac{1}{2}$ minimum
40	$\frac{5}{8}$	$\frac{5}{8}$

Note. 1 ft = 0.305 m, 1 in. = 25.4 mm.
Source. After PCA (1975).

TABLE 4.6 JOINT AND SEALANT WIDTH FOR
PREFORMED SEALS

Joint spacing (ft)	Joint width (in.)	Sealant width (in.)
20 or less	$\frac{1}{4}$	$\frac{7}{16}$
30	$\frac{3}{8}$	$\frac{5}{8}$
40	$\frac{7}{16}$	$\frac{3}{4}$
50	$\frac{1}{2}$	$\frac{7}{8}$

Note. 1 ft = 0.305 m, 1 in. = 25.4 mm.

Source. After PCA (1975).

coefficient of thermal expansion α_t is $5 \times 10^{-6}/°F$ ($9 \times 10^{-6}/°C$), and the drying shrinkage coefficient ϵ is 1.0×10^{-4}, determine the width of preformed sealant required.

Solution: From Eq. 4.36, the joint opening due to temperature change is $\Delta L = 0.65 \times 15 \times 12(0.000005 \times 100 + 0.0001) = 0.07$ in. (1.8 mm). Try $\frac{7}{16}$ or 0.4375 in. (11.1 mm) sealant installed in summer, so the joint would not be further compressed. Check maximum compression of sealant: $(0.4375 - 0.25)/0.4375 = 0.43 < 50\%$, OK. Check minimum compression of sealant: $(0.4375 - 0.25 - 0.07)/0.4375 = 0.27 > 20\%$, OK. Therefore, the use of $\frac{7}{16}$ in. (11.1 mm) sealant for a $\frac{1}{4}$-in. (6.4-mm) joint is satisfactory, as shown in Table 4.6.

Contraction joints are usually placed at regular intervals perpendicular to the center line of pavements. However, skewed joints with randomized spacings, say 13–19–18–12 ft (4.0–5.8–5.5–3.7 m), have also been used. The abtuse angle at the outside pavement edge should be ahead of the joint in the direction of traffic since that corner receives the greatest impact from the sudden application of wheel loads. The advantage of skewed joints is that the right and left wheels do not arrive at the joint simultaneously, thus minimizing the annoyance of faulted joints. The use of randomized spacings can further reduce the resonance and improve the riding comfort.

Expansion Joints

Expansion joints are transverse joints for the relief of compressive stress. Because expansion joints are difficult to maintain and susceptible to pumping, they are no longer in use today except at the connection between pavement and structure. Experience has shown that the blowups of concrete pavements are related to a certain source and type of coarse aggregates. If proper precaution is exercised in selecting the aggregates, distress due to blowups can be minimized. Since the plastic flow of concrete can gradually relieve the compressive stress, if any, it is not necessary to install the expansion joint except at bridge ends.

Figure 4.28 shows a typical expansion joint. The minimum width of joint is $\frac{3}{4}$ in. (19 mm). Smooth dowel bars lubricated at least on one side must be used for load transfer. An expansion cap must be installed at the free end to provide space

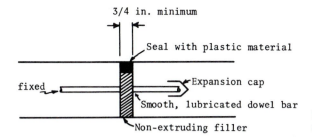

Figure 4.28 Expansion joint (1 in. = 25.4 mm).

for dowel movements. Nonextruding fillers, including fibrous and bituminous materials or cork, must be placed in the joint and the top sealed with a plastic material.

Construction Joints

If at all possible, the transverse construction joint should be placed at the location of contraction joint, as shown by the butt joint in Figure 4.29*a*. If the work must stop due to an emergency or machine breakdown, the key joint shown in Figure 4.29*b* may be used. This joint should be placed only in the middle third of normal joint interval. Key joints have not performed well and many failures have occurred.

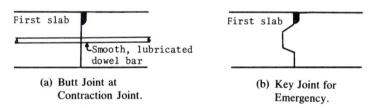

(a) Butt Joint at Contraction Joint.

(b) Key Joint for Emergency.

Figure 4.29 Construction joints.

Longitudinal Joints

Longitudinal joints are used in highway pavements to relieve curling and warping stresses. Different types of longitudinal joints may be used, depending on whether the construction is full width or lane-at-a-time.

In the full-width construction, as shown in Figure 4.30, the most convenient type is the dummy groove joint, in which tie bars are used to make certain that aggregate interlock is maintained, as shown in Figure 4.30*a*. These bars may be shoved into the wet concrete before the final finishing and placement of the dummy groove. The joint can also be formed by inserting a premolded strip into the fresh concrete and leaving it there permanently as an integral part of the warping joint, as shown in Figure 4.30*b*. Another method is to install deformed steel plates and tie bars at the center line before the pour of concrete, as shown in Figure 4.30*c*.

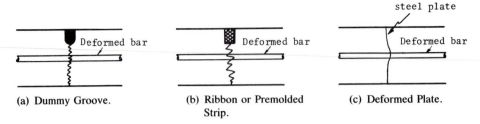

steel plate

(a) Dummy Groove. (b) Ribbon or Premolded Strip. (c) Deformed Plate.

Figure 4.30 Longitudinal joints for full-width construction.

Lane-at-a-time construction is used when it is necessary to maintain traffic on the other lane. To insure load transfer, key joints are usually used, as shown in Figure 4.31. In most cases, the keyed joints are tied together with tie bars. However, tie bars may be omitted if the longitudinal joint is at the interior of a multilane pavement and there is very little chance that the joint will be wide open.

Butt joints have also been used for lane-at-a-time construction. Current practice prefers the use of butt joints over keyed joints because keyed joints usually do not perform well due to the occurrence of cracks along the key and they are also difficult to construct with slipform paving.

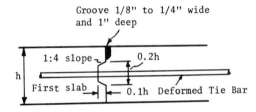

Figure 4.31 Longitudinal joints for lane-at-a-time construction (1 in. = 25.4 mm).

4.5 SUMMARY

This chapter discusses the stresses and deflections in rigid pavements based on Westergaard's theory. Westergaard viewed pavement as a plate on a liquid foundation with full subgrade contact. Analyses based on partial contact and other types of foundation are presented in Chapter 5.

Important Points Discussed in Chapter 4

1. Curling stresses in an infinite slab are caused by the restraining effect of the slab and can be determined easily from Hooke's law by assuming plane strain.

2. Curling stresses in a finite slab are caused by the curling of the slab and are difficult to compute. Based on Westergaard's theory, Bradbury developed a simple chart for determining the maximum warping stress in the interior and at the edge of a finite slab.

3. Whether curling stresses should be considered in rigid pavement design is controversial. The Portland Cement Association does not consider curling

stresses in fatigue analysis because the very few number of stress reversals does not contribute to fatigue cracking and also the curling stresses may be added to or subtracted from the loading stresses to neutralize the effect. Others think that curling stresses should be combined with loading stresses because past experience has shown that longer slabs with greater curling stresses always result in more cracking of the slab. A more reasonable approach is to consider the fatigue damage due to curling separately from that due to loading and then combine them.

4. Westergaard's closed-form formulas can be used to determine the maximum stresses and deflections in a concrete slab due to a circular loaded area applied at the corner, in the interior, or near to the edge. If the load is applied over a set of dual tires, the formulas can still be applied by using an equivalent circular area.

5. The stresses and deflections due to interior and edge loadings can also be determined by influence charts. When influence charts are employed, the actual tire imprints should be used, instead of assuming the imprints to be circular areas.

6. The design of steel reinforcements and tie bars is based on the stresses due to friction. These steel reinforcements, such as wire fabric and bar mats, do not increase the structural capacity of the slab but are used to increase the joint spacing and to tie the cracked concrete together to maintain load transfers through aggregate interlock.

7. The design of dowels is mostly based on experience. One rule of thumb is that the diameter of dowel be equal to $\frac{1}{8}$ of the slab thickness. However, a theoretical method is available to determine the bearing stress between dowel and concrete and check against the allowable bearing stress.

8. Joints should be provided in concrete pavements, so premature cracks due to temperature or moisture changes do not occur. As a rough guide, the joint spacing in feet for plain concrete pavements should not greatly exceed twice the slab thickness in inches and the ratio of slab width to length should not be greater than 1.25. Contraction joints are usually placed at regular intervals perpendicular to the center line of pavements. However, skewed joints with randomized spacings have also been used. Expansion joints are used only at the connection between pavement and structure. Longitudinal joints are used to relieve curling and warping stresses and different types may be used, depending on whether the construction is full width or lane-at-a-time.

PROBLEMS

In these problems the following data are assumed: modulus of elasticity of concrete = 4×10^6 psi, Poisson ratio of concrete = 0.15, modulus of elasticity of steel = 29×10^6 psi, Poisson ratio of steel = 0.3, modulus of dowel support = 1.5×10^6 pci, coefficient of thermal expansion of concrete = 5×10^{-6} in./in./°F, and coefficient of friction between slab and subgrade = 1.5.

4-1. Determine the curling stresses in an 8-in. slab during the day under a temperature gradient of 3°F per inch of slab for the following two cases: (a) at an interior point and at an edge point of an infinite slab, and (b) at points *A*, *B*, and *C* in a finite slab, as shown in Figure P4.1. The modulus of subgrade reaction is assumed to be 50 pci. [Answer: (a) 282.4 and 240 psi, (b) 211.4, 198.0, and 57.6 psi]

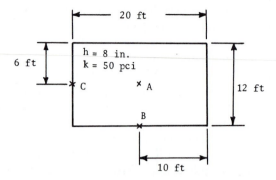

Figure P4.1

4-2. A concrete slab, 10 in. thick, is supported by a subgrade with a modulus of subgrade reaction of 200 pci. A 12,000-lb dual-wheel load (each wheel 6000 lb) spaced at 14 in. on centers is applied at the corner of the slab, as shown in Figure P4.2. The contact pressure is 80 psi. Determine the maximum stress in the concrete by Westergaard's equation with equivalent contact area. [Answer: 172.8 psi]

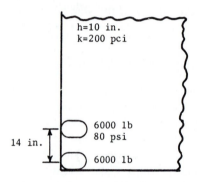

Figure P4.2

4-3. The pavement and loading are the same as in Problem 4-2 but the load is applied in the interior of an infinite slab, as shown in Figure P4.3. Determine the maximum stress in the concrete by (a) Westergaard's equation with equivalent contact area and (b) influence chart using dual tires. [Answer: 139.7 psi by equivalent area]

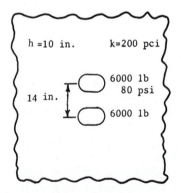

Figure P4.3

4-4. Same as Problem 4-3 except that the load is applied on the slab edge, as shown in Figure P4.4. [Answer: 252.5 psi by equivalent area]

Stresses and Deflections in Rigid Pavements   Chap. 4

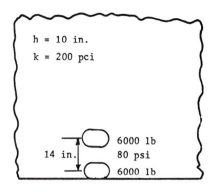

h = 10 in.

k = 200 pci

14 in. 6000 lb
       80 psi
       6000 lb

Figure P4.4

4-5. In Figure 4.12, what is the radius of relative stiffness of the pavement? If the thickness of slab is 11 in. and the contact pressure is 100 psi, determine the maximum stress due to the four tires under the center of one tire in the transverse, or n, direction. Estimate the maximum stress in the longitudinal direction under the center of one tire. [Answer: 57 in., 470 psi, 420 psi]

4-6. Figure P4.6 shows a set of dual tandem wheels with a total weight of 40,000 lb (10,000 lb per wheel), a tire pressure of 100 psi, a dual spacing of 20 in., and a tandem spacing of 40 in. The concrete slab is 8 in. thick and the modulus of subgrade reaction is 100 pci. Determine the interior stress in the y direction at point A under the center of the dual tandem wheels. [Answer: 279 psi]

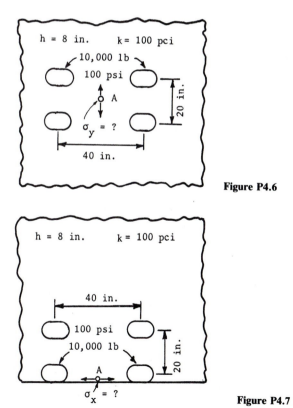

h = 8 in.   k = 100 pci

10,000 lb

100 psi

A

σ_y = ?

20 in.

40 in.

Figure P4.6

h = 8 in.   k = 100 pci

40 in.

100 psi

10,000 lb

A

σ_x = ?

20 in.

Figure P4.7

4-7. Same as Problem 4-6 except that the outside wheels are applied at the slab edge, as shown in Figure P4.7. Compute the edge stress at point A in an infinitely large slab by influence charts. [Answer: 317 psi]

4-8. A concrete slab, 40 ft long, 11 ft wide, and 9 in. thick, is placed on a subgrade having a modulus of subgrade reaction of 200 pci. A 9000-lb single-wheel load is applied on the edge of the slab over a circular area with a contact pressure of 100 psi, as shown in Figure P4.8. Compute (a) the curling stress at the edge during the night when the temperature differential is 1.5°F per inch of slab, (b) the loading stress due to the 9000-lb wheel load, and (c) the combined stress at the edge beneath the load due to (a) and (b). [Answer: 140.4, 290.3, 149.9 psi]

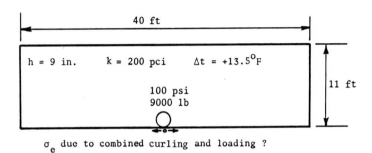

σ_e due to combined curling and loading ?

Figure P4.8

4-9. If the pavement in Problem 4-8 is one lane of a two-lane highway, as shown in Figure P4.9, design the welded wire fabric and tie bars. [Answer: area of fabric 0.0785 and 0.0432 in.^{2}/ft, $\frac{1}{2}$ in. tie bars, 2 ft long and 3 ft on centers]

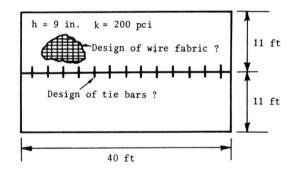

Figure P4.9

4-10. Develop a design chart for required area of temperature steel in terms of the thickness and length of slab, assuming an allowable tensile stress of 40,000 psi. [Answer: for 100-ft slab, steel areas are 0.1406, 0.1874, 0.2343, and 0.2812 in.^{2} for slab thicknesses of 6, 8, 10, and 12 in., respectively]

4-11. A concrete slab has a width of 12 ft, a thickness of 10 in., and a modulus of subgrade reaction of 300 pci. A 24,000-lb axle load with a wheel spacing of 6 ft is applied at the joint with one wheel 6 in. from the edge, as shown in Figure P4.11. Determine the maximum bearing stress between concrete and dowel, assuming 100% load transfer, 0.25-in. joint opening, and 1-in. dowel bars at 12 in. on centers. The maximum

Stresses and Deflections in Rigid Pavements   Chap. 4

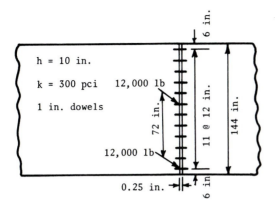

Figure P4.11

negative moment is assumed to occur at a distance of 1.8ℓ from the load, where ℓ is the radius of relative stiffness. [Answer: 3160 psi]

4-12. Repeat Problem 4-11 by assuming that the maximum negative moment occurs at a distance of 1.0ℓ from the load. [Answer: 4935 psi]

4-13. Same as Problem 4-11 except that each of the 12,000-lb wheels is placed at a distance of 2.5 ft from the edge, as shown in Figure P4.13. [Answer: 2147 psi]

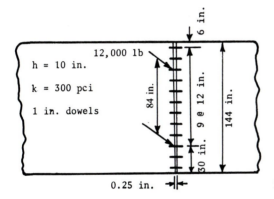

Figure P4.13

4-14. Repeat Problem 4-13 by assuming that the maximum negative moment occurs at a distance of 1.0ℓ from the load. [Answer: 3206 psi]

5

Kenslabs Computer Program

5.1 THEORETICAL DEVELOPMENTS

The KENSLABS computer program (Huang, 1985) is based on the finite element method, in which the slab is divided into rectangular finite elements with a large number of nodes. Both wheel loads and subgrade reactions are applied to the slab as vertical concentrated forces at the nodes.

5.1.1 Types of Foundation

Three different types of foundation can be assumed: liquid, solid, and layer. The Westergaard's theory and most of the finite element computer programs in use today are based on the liquid foundation. The use of liquid foundations results in a banded matrix for the simultaneous equations and requires very little computer time to solve. However, with the much faster speed and larger storage of personal computers, the more realistic solid and layer foundations should be used, if needed.

Liquid Foundation

The liquid foundation is also called a Winkler foundation, with the force–deflection relationship characterized by an elastic spring. The term "liquid" does not mean that the foundation is a liquid with no shear strength, but simply implies that the deformation of the foundation under a slab is similar to that of water under a boat. According to Archimedes' principle, the weight of the boat is equal to the weight of water displaced. This is similar to the case where a slab is placed on an infinite number of springs and the total volume of displacement is

208

proportional to the total load applied. The stiffness of a liquid foundation is defined by

$$k = \frac{p}{w} \qquad (5.1)$$

in which k is the modulus of subgrade reaction; p is the unit pressure, or force per unit area; and w is the vertical deflection. For water, $k = 62.4$ pcf $= 0.036$ pci (9.8 kN/m³); for the subgrade, k may range from 50 to 800 pci (13.6 to 217 MN/m³).

Figure 5.1 shows the replacement of the large number of springs under a rectangular plate element, with a length of 2a and a width of 2b, by four identical springs at the corners. The force on each spring is equal to the unit pressure p multiplied by the area, $a \times b$. From Eq. 5.1, $p = kw$, or the force at node i, F_{wi}, is related to the deflection at node i, w_i, by

$$F_{wi} = kabw_i \qquad (5.2)$$

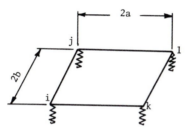

Figure 5.1 Liquid foundation under a plate element.

Equation 5.2 can be directly applied when the node is located at the corner of a slab. If the node is located at the edge or interior of a slab, superposition of two or four adjoining elements is required to obtain the force–displacement relationship. Equation 5.2 indicates that the vertical force at node i depends only on the vertical deflection at node i and is independent of the deflections at all other nodes, so the stiffness matrix of a liquid foundation is a diagonal matrix with zero entries everywhere except those relating the vertical force F_{wi} to the vertical deflection w_i itself.

Solid Foundation

A solid foundation is more realistic than a liquid foundation because the deflection at any nodal point depends not only on the force at the node itself but also on the forces at all other nodes. This type of foundation is also called a Boussinesq foundation because the Boussinesq equation for surface deflection, as indicated by Eq. 5.3, is used for determining the stiffness matrix:

$$w_{i,j} = \frac{P_j(1 - v_f^2)}{\pi E_f d_{i,j}} \qquad (5.3)$$

in which $w_{i,j}$ is the deflection at node i due to a force at node j, P_j is the force at node j, v_f is Poisson ratio of foundation, E_f is the elastic modulus of foundation, and $d_{i,j}$ is the distance between nodes i and j. The flexibility matrix of foundation is defined as the deflection at a given node due to the forces at all other nodes including the node itself. If $i \neq j$, the flexibility coefficient can be obtained directly from Eq. 5.3 by assigning $P_j = 1$. If $i = j$, then $d_{i,j} = 0$ and Eq. 5.3 cannot be applied. It is therefore necessary to distribute the unit concentrated load over one-quarter of the area as a uniform pressure of $1/(4ab)$ and then integrate numerically. A five-point Gaussian quadrature formula in both x and y directions with the responses computed at 25 points, as shown in Figure 5.2, was used by KENSLABS.

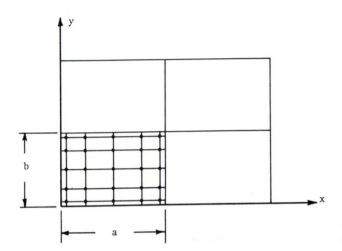

Figure 5.2 Solid foundation under a plate element.

For a total of n nodes on the foundation, the flexibility matrix of a solid foundation can be written as

$$
\begin{Bmatrix} w_1 \\ w_2 \\ \cdot \\ w_i \\ \cdot \\ w_n \end{Bmatrix} =
\begin{bmatrix}
g_{11} & g_{12} & \cdot & g_{1i} & \cdot & g_{1n} \\
g_{21} & g_{22} & \cdot & g_{2i} & \cdot & g_{2n} \\
\cdot & \cdot & \cdot & \cdot & \cdot & \cdot \\
g_{i1} & g_{i2} & \cdot & g_{ii} & \cdot & g_{in} \\
\cdot & \cdot & \cdot & \cdot & \cdot & \cdot \\
g_{n1} & g_{n2} & \cdot & g_{ni} & \cdot & g_{nn}
\end{bmatrix}
\begin{Bmatrix} F_{w1} \\ F_{w2} \\ \cdot \\ F_{wi} \\ \cdot \\ F_{wn} \end{Bmatrix}
\tag{5.4a}
$$

in which g_{ij} is the flexibility coefficient. If $i \neq j$, then $g_{i,j}$ can be determined by

$$
g_{i,j} = \frac{(1 - v_f^2)}{\pi E_f d_{i,j}}
\tag{5.4b}
$$

If $i = j$, then g_{ii} for each finite element must be integrated numerically and then superimposed over all the adjoining elements. It can be seen that the flexibility matrix of solid foundation is fully populated because the deflection at a given node is affected by the force at any other node. The inversion of the flexibility matrix results in the stiffness matrix of the foundation.

Layer Foundation

A layer foundation is also called a Burmister foundation because Burmister's layered theory is used to form the flexibility matrix. In Burmister's theory, the load is distributed over a circular area but can be easily converted to a concentrated load by letting the radius of contact approach zero, as illustrated in Appendix C. The procedure for layer foundation is the same as that for solid foundation except that the relationship between deflection and force of the layer foundation is computed from Eq. C.3 in Appendix C, instead of Eq. 5.3 for the solid foundation. To avoid the repeated evaluation of Eq. C.3 for several thousand times, the deflection at 21 different radial distances are computed by Eq. C.3 and those at any given distance can be obtained by a three-point interpolation formula.

Relationships Between Solid and Liquid Foundations

It is well known that the modulus of subgrade reaction k for a liquid foundation is a fictitious property not characteristic of soil behaviors. However, due to simplicity in application, k values have been used most frequently for the design of concrete pavements. It is useful if the k value can be related to the elastic modulus E_f and the Poisson ratio v_f of the solid foundation, so that liquid foundation can be used to replace solid foundation, thus resulting in a large saving of computer time and storage.

Vesic and Saxena (1974) indicated that the value of k depends on the relative flexibility of the slab with respect to the foundation and that there is no single value of k that can give stresses and deflections in a concrete pavement comparable to those obtained by considering the foundation as an elastic solid. For computing stresses, they suggested the use of

$$k = \left(\frac{E_f}{E}\right)^{1/3} \frac{E_f}{(1 - v_f^2)h} \tag{5.5}$$

in which E is the elastic modulus of concrete and h is the thickness of the slab. For computing deflections, they suggested that only 42% of the value obtained from Eq. 5.5 be used.

Huang and Sharpe (1989) found that Eq. 5.5 is applicable only to loads in the interior of a slab but not to loads near the slab edge. For edge loading, a factor of 1.75 must be applied to obtain the same edge stress:

$$k = 1.75\left(\frac{E_f}{E}\right)^{1/3} \frac{E_f}{(1 - v_f^2)h} \tag{5.6}$$

For corner loading at the transverse joint, a factor of 0.95 must be applied to Eq. 5.5 to obtain the same corner deflection:

$$k = 0.95\left(\frac{E_f}{E}\right)^{1/3} \frac{E_f}{(1 - v_f^2)h} \tag{5.7}$$

It should be noted that Eqs. 5.6 and 5.7 can only be used as a rough guide and are applicable only to slabs subjected to a single-axle load.

5.1.2 Two Layers of Slab

KENSLABS can have two layers of slab, either bonded or unbonded. The two layers can be an HMA on top of a PCC or a PCC over a cement-treated base. In the latter case, the cement-treated base can be considered as the second layer of slab or the first layer of the foundation. When considered as the foundation, it is assumed that there is no bond between the concrete slab and the foundation.

Bonded Slabs

Figure 5.3 shows a layer of hot mix asphalt with a thickness h_1, an elastic modulus E_1, and a Poisson ratio v_1 that has been placed on a concrete slab with a thickness h_2, an elastic modulus E_2, and a Poisson ratio v_2. The left figure is the original section with a unit width, and the right figure is the equivalent section in which the width of hot mix asphalt is reduced to E_1/E_2. When moment is taken at the bottom surface, the distance d from the neutral axis to the bottom of slab is

$$d = \frac{(E_1/E_2)h_1(0.5h_1 + h_2) + 0.5h_2^2}{(E_1/E_2)h_1 + h_2} \quad (5.8)$$

The composite moment of inertia I_c about the neutral axis is

$$I_c = \left(\frac{E_1}{E_2}\right)\left[\frac{1}{12}h_1^3 + h_1(0.5h_1 + h_2 - d)^2\right] + \frac{1}{12}h_2^3 + h_2(d - 0.5h_2)^2 \quad (5.9)$$

Given the moment M, the flexural stress f at the bottom of concrete slab is

$$f = \frac{Md}{I_c} \quad (5.10)$$

Equation 5.10 can also be used to determine the stress at any point in the composite section by considering d as the distance from that point to the neutral axis. When two layers are bonded, KENSLABS will print the stresses at four different locations, i. e., at both the top and bottom of each layer.

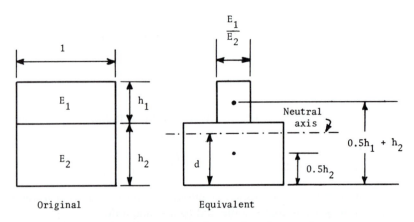

Figure 5.3 Original versus equivalent section of composite pavement.

Example 5.1:

A composite pavement with two bonded layers is subjected to a moment of 3000 in.-lb/in. of slab (13.4 kN-m/m). If $h_1 = 4$ in. (102 mm), $E_1 = 4 \times 10^5$ psi (2.8 GPa), $h_2 = 6$ in. (152 mm), and $E_2 = 4 \times 10^6$ psi (27.6 GPa), as shown in Figure 5.4, determine the maximum flexural stress in tension.

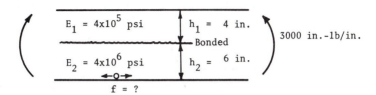

Figure 5.4 Example 5.1 (1 in. = 25.4 mm, 1 lb = 4.45 N, 1 psi = 6.9 kPa).

Solution: Given $E_1/E_2 = 0.1$, from Eq. 5.8, $d = [0.1 \times 4 \times (0.5 \times 4 + 6) + 0.5 \times 36]/(0.1 \times 4 + 6) = 3.31$ in. (84 mm). From Eq. 5.9, $I_c = 0.1 [(4)^3/12 + 4(0.5 \times 4 + 6 - 3.31)^2] + (6)^3/12 + 6(3.31 - 0.5 \times 6)^2 = 27.91$ in.^{3} (4.57×10^5 mm^3). From Eq. 5.10, $f = 3000 \times 3.31/27.91 = 355.8$ psi (2.46 MPa). If there is no asphalt overlay, $d = 3$ in. (76.2 mm) and $I_c = (6)^3/12 = 18$ in.^{3} (2.95×10^5 mm^3), so $f = 3000 \times 3/18 = 500$ psi (3.45 MPa). Note that the overlay reduces the flexural stress in concrete from 500 to 355.8 psi (3.45 to 2.46 MPa), or a reduction of about 30%. This is based on the assumption that the moment is the same in both cases. In fact, under the same load, the moment in thicker pavements should be greater than that in thinner pavements, so the actual reduction caused by thicker pavement should be smaller.

In the finite element method involving only one layer of slab, the stiffness matrix of a plate depends on the modulus of rigidity of the plate R, defined as

$$R = \frac{EI}{1 - v^2} = \frac{Eh^3}{12(1 - v^2)} \tag{5.11}$$

in which E, v, I, and h are the modulus of elasticity, Poisson ratio, moment of inertia, and thickness of the plate. When the slabs are composed of two bonded layers, the stiffness matrix of each layer is computed independently, based on their respective modulus of rigidity, and then added together to obtain the stiffness matrix of the slabs. The modulus of rigidity of each layer is computed by

$$R_1 = \frac{E_1[\frac{1}{12} h_1^3 + h_1(0.5h_1 + h_2 - d)^2]}{1 - v_1^2} \tag{5.12a}$$

$$R_2 = \frac{E_2[\frac{1}{12} h_2^3 + h_2(d - 0.5h_2)^2]}{1 - v_2^2} \tag{5.12b}$$

Unbonded Slabs

If there is no bond between the two layers, each layer is considered as an independent slab with the same displacements at the nodes. Therefore, the stiffness matrix of the slabs is the sum of the stiffness matrices of the two layers. After the displacements are determined, the moments at each node in each layer can be computed. After the moment M in each slab is found, the flexural stress f can be determined by

$$f = \frac{Md}{I} \qquad (5.13)$$

in which d is the distance from neutral axis to the top or bottom of each layer and I is the moment of inertia of each layer.

Example 5.2:

Same as Example 5.1 except that the two slabs are unbonded. The Poisson ratios are assumed 0.4 for HMA and 0.15 for PCC, as shown in Figure 5.5.

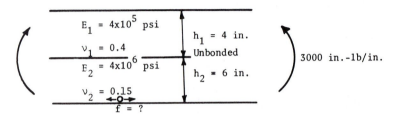

Figure 5.5 Example 5.2 (1 in. = 25.4 mm, 1 lb = 4.45 N, 1 psi = 6.9 kPa).

Solution: From Eq. 5.11, the modulus of rigidity of the asphalt layer is $4 \times 10^5 \times (4)^3/[12 \times (1 - 0.16)] = 2.54 \times 10^6$ in.^{2}-lb (7.3 m^2-kN) and that of the concrete is $4 \times 10^6 \times (6)^3/[12 \times (1 - 0.0225)] = 73.7 \times 10^6$ in.^{2}-lb (212 m^2-kN). Although the moments are not exactly proportional to the modulus of rigidity but are also affected slightly by the Poisson ratio, an approximate assumption can be made that the amount of moment carried by each layer be proportional to the modulus of rigidity. In the asphalt layer, the moment is $3000 \times 2.54/(2.54 + 73.7) = 100$ in.-lb/in. of slab (445 m-N/m) and, from Eq. 5.13, the flexural stress is $100 \times 2 \times 12/(4)^3 = 37.5$ psi (259 kPa). In the concrete slab, the moment is $3000 \times 73.7/(2.54 + 73.7) = 2900$ in.-lb/in. of slab (12.9 m-kN/m) and the flexural stress is $2900 \times 3 \times 12/(6)^3 = 483$ psi (3.33 MPa).

It can be seen that, if the hot mix asphalt is not bonded to the concrete, it has very little effect in reducing the flexural stress in the concrete. This is also based on the assumption that the total moment remains the same after the overlay. Since the total moment would increase after the overlay, the actual effect should be even smaller. By the use of KENSLABS computer program, it was found that a stress of 500 psi (3.45 MPa) could be obtained in the 6-in. (152-mm) slab with no overlay when the foundation had a modulus of subgrade reaction of 100 pci (27.2 MN/m³) and a concentrated load of 2715 lb (12.1 kN) was applied in the interior of the slab. Table 5.1 shows the effect of the 4-in. (102-mm) asphalt overlay on the maximum moments and tensile stresses in the concrete, as obtained by KENSLABS. It can be seen that the overlay increases the total moments by 5% in the bonded case and 1% in the unbonded case and that the decrease in concrete stress is 25 and 3%, respectively.

TABLE 5.1 EFFECT OF ASPHALT OVERLAY ON MOMENTS AND
STRESSES IN CONCRETE SLAB

	Total moment		Stress in concrete	
Type	in.-lb/in.	%	psi	%
No overlay	3000	100	500	100
Bonded	3162	105	375	75
Unbonded	3032	101	485	97

Note. 1 lb = 4.45 N, 1 psi = 6.9 kPa.

5.1.3 General Procedures

The stiffness matrix of the slab is combined with the stiffness matrices of the
foundation and the joint, if any, to form the overall stiffness matrix.

Stiffness Matrix of Slab

Figure 5.6 shows a rectangular finite element with nodes $i, j, k,$ and l. At
each node there are three fictitious forces and three corresponding displacements.
The three forces are a vertical force F_w, a moment about the x axis $F_{\theta x}$, and a
moment about the y axis $F_{\theta y}$. The three displacements are the vertical deflection in
the z direction w, a rotation about the x axis θ_x, and a rotation about the y axis θ_y.
The positive direction of the coordinates is shown in the figure and the positive
direction of moments and rotations can be determined by the right-hand rule. For
each element, the forces and displacements are related by

$$\begin{Bmatrix} F_i \\ F_j \\ F_k \\ F_l \end{Bmatrix} = [K_p]^e \begin{Bmatrix} \delta_i \\ \delta_j \\ \delta_k \\ \delta_l \end{Bmatrix} \tag{5.14a}$$

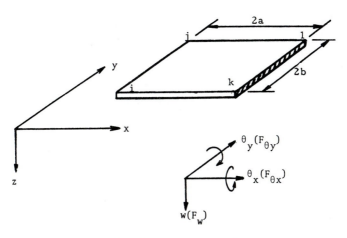

Figure 5.6 A rectangular plate element.

in which $[K_p]^e$ is the element stiffness matrix of a plate with a dimension of 12 × 12. At any given node

$$F_i = \left\{ \begin{matrix} F_{wi} \\ F_{\theta xi} \\ F_{\theta yi} \end{matrix} \right\} \qquad \delta_i = \left\{ \begin{matrix} w_i \\ \theta_{xi} \\ \theta_{yi} \end{matrix} \right\} \tag{5.14b}$$

The stiffness matrix of a slab is obtained by superimposing the element stiffness matrix over all elements. After combining the stiffness matrices of slab, foundation, and joint and replacing the fictitious nodal forces with the statical equivalent of the externally applied wheel loads, a set of simultaneous equations is obtained for solving the unknown nodal displacements:

$$[K] \{\delta\} = \{F\} \tag{5.15}$$

in which $[K]$ is the overall stiffness matrix, $\{\delta\}$ are the nodal displacements, and $\{F\}$ are the externally applied nodal forces. Since the overall stiffness matrix is symmetric, only the upper half of the matrix need be considered. The nodal moments and stresses can then be computed from the nodal displacements, using the stress matrix tabulated by Zienkiewicz and Cheung (1967). Because the stresses at a given node computed by means of one element are different from that by the neighboring elements, the stresses in all adjoining elements are computed and their average values obtained.

Stiffness of Joint

The stiffness of joint is represented by a shear spring constant C_w and a moment spring constant C_θ, defined as

$$C_w = \frac{\text{Shear force per unit length of joint}}{\text{Difference in deflections between two slabs}} \tag{5.16}$$

$$C_\theta = \frac{\text{Moment per unit length of joint}}{\text{Difference in rotations between two slabs}} \tag{5.17}$$

It is generally agreed that load is transferred across a joint principally by shear with $C_\theta = 0$. Ball and Childs (1975) reported that some moment may be transferred through the joints that remain closed, but moment transfer across joints with visible openings is negligible.

Figure 5.7 shows the shear transfer through a joint by aggregate interlock, as indicated by a spring having a spring constant C_w. After the load is applied, the left slab deflects an amount w_l, and the spring pushes the right slab down a distance w_r. The difference in deflection w_d is equal to $w_l - w_r$.

In the finite element method, the shear forces are concentrated at the nodes along the joint. From Eq. 5.16,

$$F_w = LC_w w_d \tag{5.18}$$

in which F_w is the nodal force applied to both slabs through the spring and L is the average nodal spacing at the joint. The forces F_w can then be substituted into Eq. 5.15 to solve the nodal displacements.

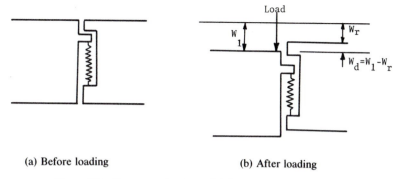

(a) Before loading   (b) After loading

Figure 5.7 Shear transfer through joint by aggregate interlock.

When dowel bars are used to transmit shear, it is assumed that they are concentrated at the nodes. If the dowel spacing is s_b, the number of dowels at each node is L/s_b. The force F_w is divided by the number of dowels needed to obtain the force P_t on each dowel:

$$P_t = \frac{S_b F_w}{L} \tag{5.19}$$

A simple procedure to include the effect of dowel bars in the finite element analysis was presented by Huang and Chou (1978). Figure 5.8 shows the shear transfer through a joint by a dowel bar. The difference in deflection w_d is caused by the shear deformation of the dowel ΔS and the deformation of concrete under the dowel y_0:

$$w_d = \Delta S + 2y_0 \tag{5.20}$$

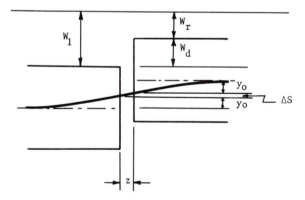

Figure 5.8 Shear transfer through joint by dowel bar.

The shear deformation of the dowel can be determined approximately by

$$\Delta S = \frac{P_t z}{GA} \tag{5.21}$$

in which P_t is the shear force on one dowel bar, z is the joint width, A is the area of the dowel, and G is the shear modulus of the dowel, which can be determined by

$$G = \frac{E_d}{2(1 + \nu_d)} \tag{5.22}$$

The deformation of concrete under the dowel can be determined from Eq. 4.42:

$$y_0 = \frac{P_t(2 + \beta z)}{4\beta^3 E_d I_d} \tag{4.42}$$

Substituting Eqs. 5.21 and 4.42 into Eq. 5.20 yields

$$w_d = \left(\frac{z}{GA} + \frac{2 + \beta z}{2\beta^3 E_d I_d}\right)P_t \tag{5.23}$$

Substituting Eq. 5.19 into Eq. 5.23 and comparing with Eq. 5.18 gives

$$C_w = \frac{1}{S_b\left(\dfrac{z}{GA} + \dfrac{2 + \beta z}{2\beta^3 E_d I_d}\right)} \tag{5.24}$$

Equation 5.24 indicates that, given the spacing and diameter of dowels and the joint width, the shear spring constant can be determined. The above analysis is based on the assumption that there is no gap between the spring (or dowel) and the concrete. If a gap w_g exists, then Eq. 5.18 should be written as

$$F_w = LC_w(w_d - w_g) \tag{5.25}$$

Therefore, a term, $LC_w w_g$, should be subtracted from the force vector $\{F\}$ in Eq. 5.15. Because $F_w = 0$ when $W_d < W_g$, Eq. 5.25 is valid only when $W_d > W_g$. Because w_d varies along the joint and it is not known whether W_d is greater than w_g, a trial and error method must be used if a gap exists.

Example 5.3:

A 10-in. (254-mm) concrete pavement has a joint width of 0.2 in. (5 mm), a dowel diameter of 1 in. (25.4 mm), and a dowel spacing of 12 in. (305 mm). Determine the spring constant of the joint.

Solution: Assume that $K = 1.0 \times 10^6$ pci (271 GN/m³) and $E_d = 29 \times 10^6$ psi (200 GPa). With $d = 1$ in. (25.4 mm), from Eq. 4.43, $I_d = \pi/64 = 0.0491$ in.⁴, and from Eq. 4.44, $\beta = [1.0 \times 10^6 \times 1/(4 \times 29 \times 10^6 \times 0.0491)]^{0.25} = 0.647$ in. (16.4 mm). Assuming that Poisson ratio of dowel $= 0.3$, from Eq. 5.22, $G = 0.5 \times 29 \times 10^6/(1 + 0.3) = 11.2 \times 10^6$ psi (76.9 GPa). With $s_b = 12$ in. (305 mm) and $z = 0.2$ in. (5 mm), from Eq. 5.24

$$C_w = \frac{1}{12\left(\dfrac{0.2}{11.2 \times 10^6 \times \pi} + \dfrac{2 + 0.647 \times 0.2}{2(0.647)^3 \times 29 \times 10^6 \times 0.0491}\right)}$$

$$= \frac{1}{12(5.68 \times 10^{-9} + 2.76 \times 10^{-6})} = 3.01 \times 10^4 \text{ psi (279 MPa)}$$

It can be seen that the term relating to the deformation of dowel is 5.68×10^{-9} while that relating to the deformation of concrete is 2.76×10^{-6}. Therefore, the deformation of concrete actually determines the spring constant of the joint.

Multiple Slabs

Figure 5.9 shows a four-slab system with load transfer across the joints. For simplicity, only a limited number of finite elements is used. Slab 1 is divided into four rectangular elements, slabs 2 and 3 are divided into two elements each, while slab 4 is considered as one element. The slabs are numbered consecutively from bottom to top and from left to right. Starting from slab 1, the nodes are also numbered from bottom to top and from left to right and continue to the next slab.

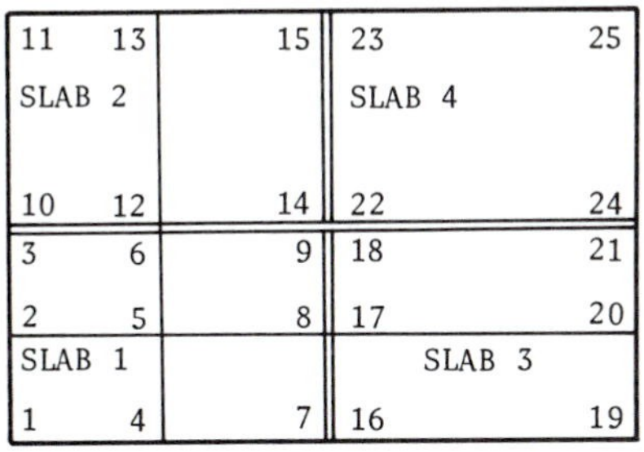

Figure 5.9 A four-slab system.

The stiffness matrix of the slab is a banded matrix because the forces at one node are affected by the displacements at another node only when the two nodes are located in the same element. For the system shown in Figure 5.9, the maximum difference between the two nodal numbers in the same element is 4 and each node has three degrees of freedom, so the minimum half-band width required is $(4 + 1) \times 3 = 15$. If the slabs are placed on liquid foundation with no load transfer across the joints, a half-band width of 15 is sufficient. The use of a small half-band width is highly desirable because it reduces both the computer time and storage in solving the simultaneous equations. If there is shear transfer across the joint, the maximum difference between the two nodes on the opposite side of a joint is 9, so that minimum half-band width required is $(9 + 1) \times 3 = 30$. If the slabs are placed on a solid or layer foundation, the stiffness matrix is fully populated with a half-band width of 75.

The procedure for computing the stiffness matrix of solid and layer foundations was described previously for a single slab. For multiple slabs, it is necessary to distribute the stiffness over those nodes meeting at the same point. Take Figure 5.9 for example. Although there are 25 nodes on the slab, only 16 nodes exist on the foundation, because many of the nodes, such as 9, 14, 18, and 22, are actually located at the same point and should be considered as one node. Therefore, the flexibility matrix of foundation has a dimension of 16×16. After inversion, the 16×16 stiffness matrix must be enlarged to a 25×25 matrix by distributing the stiffness to all nodes sharing the same point. For instance, the stiffness entry (3, 9) is evenly divided among (3, 9), (3, 14), (3, 18), (3, 22), (10, 9), (10, 14), (10, 18), and (10, 22); the sitffness entry (9, 9) is evenly divided among (9, 9), (14, 14), (18, 18) and (22, 22); and the stiffness entry (3, 21) is evenly divided among (3, 21) and (10, 24). When two or more nodes meet at a joint, the deflections of the slab at the nodes are different but the deflection of the foundation is the same.

The stiffness of a joint is represented by a shear spring constant and a moment spring constant as described previously. Taking nodes 7 and 16 in Figure 5.9 as an example, the shear and moment transfer across the joint can be expressed as

$$F_{w7} = (w_{16} - w_7)C_w \qquad (5.26a)$$

$$F_{\theta y7} = (\theta_{y16} - \theta_{y7})C_\theta \qquad (5.26b)$$

$$F_{w16} = (w_7 - w_{16})C_w \qquad (5.26c)$$

$$F_{\theta y16} = (\theta_{y7} - \theta_{y16})C_\theta \qquad (5.26d)$$

in which F_w is the shear force across joint, w is the vertical deflection, $F_{\theta y}$ is the moment about the y axis, and θ_y is the rotation about the y axis. These forces and moments can be placed in the force vector on the right side of Eq. 5.15 and the vertical deflections and rotations are then transferred to the left to form the stiffness matrix of the joint.

Simultaneous Equations

Equation 5.15 can be solved by the Gauss elimination method. For the system shown in Figure 5.9, the total number of equations is 75. If the foundation is a liquid, the half-band width is 30, instead of 15, due to the presence of joints, so the dimension for the stiffness matrix is $75 \times 30 = 2250$. If the foundation is solid or layer, the half-band width is 75, so the dimension for the stiffness matrix is $75 \times 75 = 5625$. In practical applications, the number of nodes is much greater than 25. Although the program in its present form can handle a maximum of 420 nodes, the limitation of total storage to less than 600K allows the use of this many nodes only in a liquid foundation. For solid and layer foundations, the dimension of stiffness matrix for 420 nodes is $420 \times 3 \times 420 \times 3 = 1,587,600$, which is equivalent to a computer storage of more than 6350K. It can be seen that a major problem for analyzing solid and layer foundations is computer storage. To solve this problem, an iteration method was developed (Huang, 1974b). In this method, an arbitrary half-band width can be assigned. Any entry in the stiffness matrix outside the half-band is stored and moved to the right side of Eq. 5.15:

$$[K_1]\{\delta\} = \{F\} - [K_2]\{\delta_a\} \qquad (5.27)$$

in which $[K_1]$ is the stiffness matrix within the assigned half-band, $[K_2]$ is the stiffness matrix outside the half-band, and $\{\delta_a\}$ are the assumed displacements outside the half-band. Because the stiffness matrix outside the half-band consists principally of the stiffness matrix of foundation with eight-ninths of the entries as zeros, which need not be stored, the storage required for $[K_2]$ is very small.

Equation 5.27 can be solved by an iteration method. First, a set of displacements $\{\delta_a\}$ is assumed and a new set of displacements $\{\delta\}$ is computed. Using $\{\delta\}$ as $\{\delta_a\}$, the process is repeated until the displacements converge to a specified tolerance. This iterative procedure can be applied to all three types of foundation. It was found that the solution converges quite rapidly if a large half-band width is assigned. If the half-band width is equal to the number of equations, no iteration will be required. However, if the half-band width is too small, the solution may diverge and a larger half-band width should be used. In KENSLABS, the largest possible half-band width within the allowable dimension will be selected automatically for the stiffness matrix.

5.1.4 Temperature Curling

In analyzing temperature curling it is assumed that each slab acts independently and is not restrained by the lubricated dowel bars. This assumption is reasonable if all the adjoining slabs are of the same size and thickness and curl the same amount at corresponding points along the joint.

General Formulation

The general formulation involving curling is similar to that for loading. After the stiffness matrix is superimposed over all elements and the nodal forces are replaced with the statical equivalent of externally applied loads, the following simultaneous equations can be obtained for solving the nodal displacements:

$$[K_p]\{\delta\} = \{F\} + [K_f]\{\delta'\} \tag{5.28a}$$

in which $[K_p]$ is the stiffness matrix of the slab including the joint, if any, $\{\delta\}$ are the nodal displacements of slab, $\{F\}$ are the nodal forces due to applied loads, $[K_f]$ is the stiffness matrix of foundation, and $\{\delta'\}$ are the nodal displacements of foundation. Note that the second term on the right side of Eq. 5.28a represents the nodal forces due to the foundation reaction. If the slab has a total of n nodes, then

$$\{\delta\} = \begin{Bmatrix} \delta_1 \\ \cdot \\ \delta_i \\ \cdot \\ \delta_n \end{Bmatrix} \qquad \{F\} = \begin{Bmatrix} F_1 \\ \cdot \\ F_i \\ \cdot \\ F_n \end{Bmatrix} \qquad \{\delta'\} = \begin{Bmatrix} \delta_1' \\ \cdot \\ \delta_i' \\ \cdot \\ \delta_n' \end{Bmatrix} \tag{5.28b}$$

and

$$\delta_i = \begin{Bmatrix} w_i \\ \theta_{xi} \\ \theta_{yi} \end{Bmatrix} \qquad F_i = \begin{Bmatrix} F_{wi} \\ 0 \\ 0 \end{Bmatrix} \qquad \delta_i' = \begin{Bmatrix} c_i - w_i \\ 0 \\ 0 \end{Bmatrix} \tag{5.28c}$$

in which the subscript i indicates the ith node; w is the vertical deflection, downward positive; θ_x is the rotation about the x axis; θ_y is the rotation about the y axis; F_w is the vertical force due to externally applied load, downward positive; and c is the initial curling of a weightless and unrestrained slab due to a temperature differential between the top and the bottom, upward positive. The subgrade displacement indicated by Eq. 5.28c is based on Westergaard's assumption of full contact. If the slab curls up an amount c_i, the subgrade will pull it down an amount w_i, so the vertical deflection of the subgrade is $c_i - w_i$. The reason that F_i and δ_i' contain only one nonzero element is that the nodal forces are determined by statics and only vertical loads and reactions are involved. If there is no curling, then $c_i = 0$, so w_i can be moved to the left side of Eq. 5.28a and Eq. 5.15 is obtained.

Initial Curling

The determination of initial curling was presented by Huang and Wang (1974). Figure 5.10 shows, in an exaggerated scale, a thin slab subjected to a temperature differential Δt between the top and the bottom. If the slab is weightless and unrestrained, it will form a spherical surface with a radius R. Because the slab is only slightly curved, the length of the arc on the upper surface is practically the same as that on the lower surface, so the length L of the upper surface is shown as the length of the lower surface. The length is actually greater at the bottom than at the top by $\alpha_t L \Delta t$, where α_t is the coefficient of thermal expansion. Since the radius R is much greater than the thickness h, and L is much greater than $\alpha_t L \Delta t$, it can be easily shown from geometry that

$$R = \frac{h}{\alpha_t \Delta t} \tag{5.29}$$

and

$$c = \frac{d^2}{2R} \tag{5.30}$$

in which d is the distance to the center of slab where curling is zero. Substituting Eq. 5.29 into Eq. 5.30 gives

$$c = \frac{\alpha_t \Delta t d^2}{2h} \tag{5.31}$$

Note that Δt is positive when the slab is curled up with a temperature at the bottom greater than that at the top and negative when it is curled down. The assumption that the slab remains in contact with the subgrade implies that the subgrade reaction always exists no matter how the slab is curled. If the slab is curled up, the subgrade will pull the slab down, and a deflection w is obtained, as shown in Figure 5.10. The displacement of the subgrade is thus $c - w$, as indicated by Eq. 5.28c. If w within $\{\delta'\}$ in the second term on the right side of Eq. 5.28a is moved to the left and combined with w on the left and c is combined with $\{F\}$, Eq. 5.15 is obtained to solve the nodal displacements. After the displacements are obtained, the stresses can also be computed.

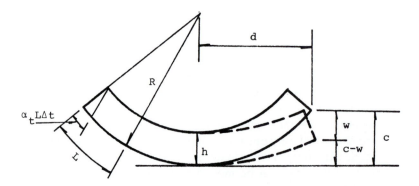

Figure 5.10 Curling of slab. (After Huang and Wang (1974).)

Kenslabs Computer Program   Chap. 5

The above derivation for upward curling also applies to downward curling. When the slab is curled down, the temperature differential is negative. If the temperature differential is the same for downward curling as for upward curling, the stresses and deflections will be the same in magnitude but opposite in sign.

5.1.5 Slab–Subgrade Contact

An important factor that affects the design of concrete pavements is the contact condition between slab and foundation. Both Westergaard's analysis for liquid foundations and Pickett's analysis for solid foundations (Pickett et al., 1951) are based on the assumption that the slab and foundation are in full contact. This assumption is valid if there are no gaps between slab and foundation, because the weight of the slab naturally imposes a large precompression on the foundation, which will keep the slab and foundation in full contact. However, this is not true when the slab is subjected to curling or pumping, which results in a separation between slab and foundation. The KENSLABS computer program is particularly useful for evaluating the effect of contact conditions on stresses and deflections. The analysis of partial contact for liquid foundations was presented by Huang and Wang (1974) and will be described here. Three cases of contact are discussed: full contact, partial contact without initial gaps, and partial contact with initial gaps. The modifications of the procedure for solid and layer foundations are also discussed.

Full Contact

Figure 5.11 shows a liquid or Winkler foundation consisting of a series of springs, each representing a nodal point in the finite element analysis. When a slab is placed on the foundation, the weight of the slab will cause a precompression of the springs, as shown in Figure 5.11b. Because the slab is uniform in thickness, each spring will deform the same amount, and no stresses will be induced in the slab. The amount of precompression can be determined directly by dividing the weight of slab per unit area by the modulus of subgrade reaction.

When the temperature on top of the slab is colder than that at the bottom, as is usually the case at night, part of the slab will deflect upward, as shown in Figure 5.11c. However, the slab and the springs remain in contact because the upward deflections are smaller than the precompression. The deflection of the slab due to curling can be determined by subtracting the precompression due to the weight of the slab from the deflection due to the weight and the curling combined, as indicated by the shaded area in Figure 5.11c. The result is exactly the same as when considering the curling alone. The same is true when a load is applied to a curled slab, as shown in Figure 5.11d. Therefore, when the slab and the subgrade are in full contact, the principle of superposition applies. The stresses and deflections due to curling and loading can be determined separately, one independent of the other, disregarding the weight of the slab. This principle forms the basis of Westergaard's analysis.

The major difference in procedure between full and partial contact is that in the case of full contact it is not necessary to consider the weight of the slab,

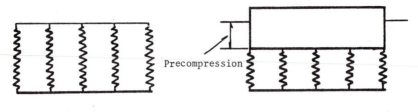

(a) Winkler Foundation before Paving   (b) Precompression due to Weight of Slab

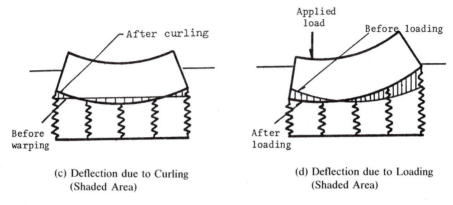

(c) Deflection due to Curling   (d) Deflection due to Loading
(Shaded Area)   (Shaded Area)

Figure 5.11   Spring analogy for full contact. (After Huang and Wang (1974).)

whereas in the case of partial contact, the weight of the slab must be considered. The latter case involves two steps. First, the gaps and precompressions of the subgrade due to the weight of the slab or due to the weight of the slab and the curling combined are determined. These gaps and precompressions are then used to determine the stresses and deflections due to applied loads.

It should be noted that full contact is a special case of partial contact. Every problem in partial contact is analyzed first by assuming the slab and the subgrade to be in full contact. If it turns out that they are actually in full contact, no iterations are needed. If some points are found out of contact, the reactive force at those points is set to zero. The process is repeated until the same contact conditions are obtained.

Partial Contact Without Initial Gaps

This case applies to new pavements not subjected to significant amounts of traffic and where there is no pumping or plastic deformation of the subgrade. Each spring in the Winkler foundation is in good condition and, if the slab is removed, will rebound to the same elevation with no initial gaps, as shown in Figure 5.12a. Under the weight of the slab, each spring is subjected to a precompression, as shown in Figure 5.12b. If the slab is curled up, gaps will form at the exterior springs, as indicated by a positive s in Figure 5.12c, and precompressions will

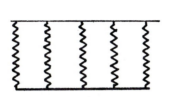

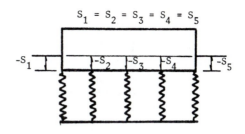

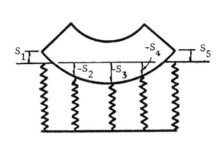

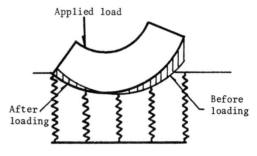

(a) Winkler Foundation without
    Initial Gap

(b) Precompression due to Weight
    of Slab

(c) Gap and Precompression due to
    Weight and Curling

(d) Deflection due to Loading
    (Shaded Area)

Figure 5.12 Partial contact without initial gap. (After Huang and Wang (1974).)

form at the interior springs, as indicated by a negative s. If the slab is curled down, all springs will be under precompression, similar to Figure 5.12b except that the precompressions are not equal. For very stiff springs, a gap may also form at the interior springs. The displacements due to the weight of the slab and curling combined can be determined from Eq. 5.28, except that the subgrade displacements be expressed as

$$\delta_i' = \left\{ \begin{array}{c} c_i - w_i \\ 0 \\ 0 \end{array} \right\} \quad \text{when } w_i > c_i \qquad (5.32a)$$

$$\delta_i' = \left\{ \begin{array}{c} 0 \\ 0 \\ 0 \end{array} \right\} \quad \text{when } w_i < c_i \qquad (5.32b)$$

Note that δ_i' in Eq. 5.32a is the same as that in Eq. 5.28c for full contact and will be used to start the iteration. After each iteration, a check is made on each node to find out whether any contact exists. If the deflection w is found to be smaller than the initial curling c, then the slab is not in contact with the subgrade and the subgrade displacement is set to zero, as indicated by Eq. 5.32b. Thus, after each iteration, a new set of simultaneous equations is established. The process is repeated until the same equations are obtained. In most cases, this can be achieved by five or six iterations. After the deflections due to the weight and curling are determined, the gaps and precompressions can be computed and used later for computing the stresses and deflections due to the load alone.

To determine the stresses and deflections due to the load alone, the gaps and precompressions shown in Figures 5.12b or 5.12c, depending on whether curling exists, must first be determined. Using these gaps and precompressions as s, the deflections due to the load alone, as shown in Figure 5.12d, can be determined from Eq. 5.28, except that the subgrade displacements are expressed as

$$\delta_i' = \left\{ \begin{array}{c} 0 \\ 0 \\ 0 \end{array} \right\} \quad \text{when } w_i < s_i \quad\quad (5.33a)$$

$$\delta_i' = \left\{ \begin{array}{c} s_i - w_i \\ 0 \\ 0 \end{array} \right\} \quad \text{when } w_i > s_i \quad \text{and} \quad s_i > 0 \quad\quad (5.33b)$$

$$\delta_i' = \left\{ \begin{array}{c} -w_i \\ 0 \\ 0 \end{array} \right\} \quad \text{when } w_i > s_i \quad \text{and} \quad s_i < 0 \quad\quad (5.33c)$$

In checking w with s, downward deflection is considered positive and upward negative, while the gap is considered positive and precompression negative. First, assume the slab and the subgrade to be in full contact, and determine the deflections of the slab due to the applied load. Then check the deflections with s, and form a new set of equations based on Eq. 5.33. Repeat the process until the same equations are obtained.

When the slab and the subgrade are in partial contact, the principle of superposition no longer applies. To determine the stresses and deflections due to an applied load, the deformed shape of the slab immediately before the application of the load must be computed first. Since the deformed shape depends strongly on the condition of curling, the stresses and deflections due to loading are affected appreciably by curling. This fact was borne out in both the Maryland (HRB, 1952) and the AASHO (HRB, 1962) road tests.

Partial Contact with Initial Gaps

This case applies to pavements that have been subjected to a high intensity of traffic. Because of pumping or plastic deformation of the subgrade, some springs in the Winkler foundation become defective and, if the slab is removed, will not return to the original elevation. Thus, initial gaps are formed, as indicated by the two exterior springs in Figure 5.13a. These gaps s must be assumed before an analysis can be made. The displacements due to the weight of the slab, as shown in Figure 5.13b, can be determined from Eq. 5.28, except that the subgrade displacements are expressed as

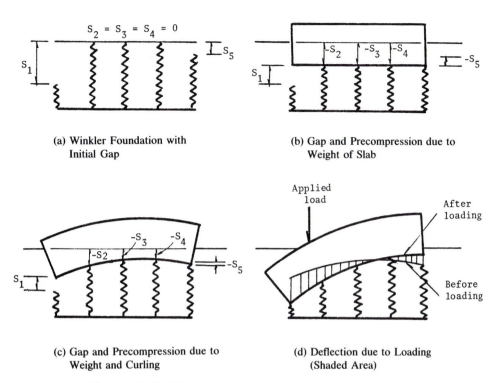

(a) Winkler Foundation with Initial Gap

(b) Gap and Precompression due to Weight of Slab

(c) Gap and Precompression due to Weight and Curling

(d) Deflection due to Loading (Shaded Area)

Figure 5.13 Partial contact with initial gap. (After Huang and Wang (1974).)

$$\delta_i' = \begin{Bmatrix} s_i - w_i \\ 0 \\ 0 \end{Bmatrix} \quad \text{when } w_i > s_i \qquad (5.34a)$$

$$\delta_i' = \begin{Bmatrix} 0 \\ 0 \\ 0 \end{Bmatrix} \quad \text{when } w_i < s_i \qquad (5.34b)$$

First, assume that the slab and the subgrade are in full contact, and determine the vertical deflections of the slab from Eq. 5.28a. Then check the deflection at each node against the gap s. If the deflection is smaller than the gap, as shown by the left-most spring in Figure 5.13b, then use Eq. 5.34b; if the deflection is greater than the gap, as shown by the other springs in Figure 5.13b, then use Eq. 5.34a. Repeat the process until the same equations are obtained. After the deflections are obtained, the gaps and precompressions can be computed and used later for computing the stresses and deflections due to loading, if no curling exists.

It can be seen that if the springs are of the same length, as shown in Figure 5.12, the weight of the slab will result in a uniform precompression, and no stresses will be set up in the slab. However, if the springs are of unequal lengths, the deflections will no longer be uniform, and stressing of the slab will occur.

Figure 5.13c shows the combined effect of weight and curling when the slab is curled down. Downward, rather than upward, curling is considered here because the case of upward curling is similar to Figure 5.12c except that the gaps are measured from the top of the defective springs. Because the method is applicable to both upward and downward curling, the case of downward curling is used for illustration. The procedure for determining the deflections is similar to that involving the weight of the slab alone except that the initial curling of the slab, as indicated by Eq. 5.31, is added to the gap shown in Figure 5.13a to form the total gap and precompression s for use in Eq. 5.34. Since the gap is either positive or zero and the initial curling may be positive or negative, depending on whether the slab is curled up or down, s may be positive or negative. After the deflections of the slab are obtained, the gaps and precompressions, as shown in Figure 5.13c, can be determined. These gaps and precompressions will be used for computing the stresses and deflections due to the load alone, as shown in Figure 5.13d.

Modifications for Solid and Layer Foundations

The above procedure for liquid foundations has to be modified when it is applied to solid or layer foundations. The actual criterion for deciding the contact condition is whether any tension exists between the slab and the foundation. In liquid foundations, tension is indicated if the slab moves up relative to the foundation, and compression is implied if the slab moves down. By merely comparing deflections, as indicated by Eqs. 5.32 and 5.33, the contact conditions can be ascertained. In solid or layer foundations, tension or compression at a given node is governed not only by the deflection of the node itself but also by the deflections of all other nodes. If the contact condition is purely based on the deflection at one node, there is the possibility that tension may develop due to the deflections at other nodes. Therefore, in addition to checking the deflection, the reactive force, either tension or compression, should also be checked.

First, the slab and foundation are assumed to be in full contact and the reactive force at each node is determined. If the reactive force is in tension and the tension is greater than the precompression (note that the term precompression used in solid or layer foundations indicates a force, while that in liquid foundation indicates a displacement), the node is not in contact and its reactive force is set to zero in the next iteration. When there is no reactive force at a given node, that node should be eliminated from the flexibility matrix and a new stiffness matrix of foundation is formed. If the slab and foundation are assumed not to be in contact and the deflection of the slab is smaller than the gap plus the deflection of foundation, the assumption is correct and the slab and foundation are not in contact. Otherwise, the slab and foundation should be assumed to be in contact and the iteration continues.

Two criteria are used for checking contact: If the slab and foundation are assumed not to be in contact, then the deflection of the slab must be smaller than the gap and the deflection of foundation combined; if the slab and foundation are assumed to be in contact, then the net effect of reactive force and precompression must not be in tension. If any one of the above criteria is violated, the contact condition should be changed and the iteration continued until both criteria are satisfied.

5.2 PROGRAM DESCRIPTION

KENSLABS was written in FORTRAN 77 and requires a storage of 522K. In its present dimensions, it can be applied to a maximum of 9 slabs, 12 joints, and 420 nodes. Each slab can have a maximum of 15 nodes in the x direction and 15 nodes in the y direction. The dimension of stiffness matrix is limited to 70,000. If the storage required for the stiffness matrix is more than 70,000, the iteration method shown in Eq. 5.27 will be activated automatically.

Before running KENSLABS, an input program SLABSINP may be run first to set up a data file. To run KENSLABS, simply type KENSLABS and the program will ask for a file name. If the PC has a storage of 640K but the message "Program too big to fit in memory" appears on the screen, some of the memory must be cleared out first before KENSLABS can be run. A more convenient way is to start DOS over from the beginning using the original DOS diskette.

5.2.1 General Features

The capabilities, limitations, and applications of the program together with the automatic checking of contact between slab and foundation are discussed in this section.

Capabilities

The capabilities of the program are as follows:

1. A maximum of nine slabs with shear and moment transfer across the joints can be analyzed. Shear transfer can be effected by specifying a shear spring constant or by providing information on the size, spacing, Young's modulus, and Poisson ratio of dowel bars together with the joint width and the modulus of the dowel support. Dowels can have nonuniform spacings by assigning zero, one, or more dowels at each node along a joint. Moment transfer is specified by a moment spring constant. The looseness of the dowel can be considered by specifying a gap between dowel and concrete. This gap can also be applied to the shear spring constant but not to the moment spring constant.

2. Each slab can have different thicknesses and sizes. In the same slab, thickness can vary from node to node. However, two adjoining slabs must have the same width, and all joints must be continuous throughout the slabs.

3. The slabs can have two rigid layers, either bonded or unbonded. Each layer has its own Young's modulus and Poisson ratio. The modulus of rigidity for determining the stiffness of the slabs is computed by Eq. 5.11 for unbonded layers and Eq. 5.12 for bonded layers. The stiffness matrix of the slabs is the sum of the stiffness matrices of both layers.

4. The load can be uniformly distributed over rectangular areas or concen-

trated at a given number of nodes. Each loaded area can have a different intensity of pressure.

5. If symmetry with respect to one or both axes exists, only one-half or one-quarter of the slab system need be considered (Huang, 1974b). This feature can save a great deal of computer time and storage. Figure 5.14 shows several cases of symmetry under single, dual, and dual-tandem wheel loads. Although the part to be used for analysis is indicated by hatched lines, any quadrant or half can be used.

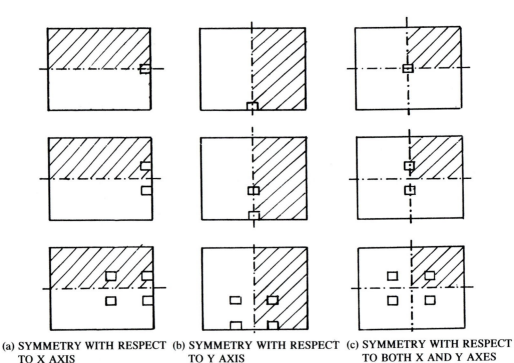

(a) SYMMETRY WITH RESPECT TO X AXIS

(b) SYMMETRY WITH RESPECT TO Y AXIS

(c) SYMMETRY WITH RESPECT TO BOTH X AND Y AXES

Figure 5.14 Use part of slab for cases of symmetry.

6. The effect of temperature curling and the gap between slab and foundation on the stresses and deflections can be analyzed.

7. The slab and foundation can be assumed to be in full contact at all nodes or not in contact at some designated nodes. The contact condition can also be evaluated automatically by iterations.

8. The program can analyze slabs on liquid, solid, or layer foundations.

Limitations

1. When considering temperature curling, each slab is assumed to curl into a spherical surface independent of the others. This occurs only when there is no moment transfer across the joints and each slab can move up and down freely, as in the case of lubricated dowel joints. It may not be applicable to hinged or

longitudinal joints, where the edges of adjoining slabs are held tightly together. Fortunately, the most critical area for a pavement is not in the vicinity of a longitudinal joint, so this inaccuracy should have very little effect on the final design. The analysis of temperature curling is based on the assumption that there is only one layer of slabs and that each slab is of uniform thickness. For slabs of nonuniform thickness, an average thickness must be assumed.

2. When slabs are composed of two layers, the joints through the two layers are exactly at the same locations. If the bottom layer has no joints and is much weaker than the top layer, this assumption should have very little effect on the stresses in the top layer.

3. The program does not permit the use of infinitely stiff joints, where the deflections or rotations on the two sides of the joint are equal. When a very large spring constant is applied to a joint, the equation for each node on the two opposite sides is identical and the system of simultaneous equations becomes singular. This situation will not occur in practice and can be easily detected if the printout shows that the sum of applied forces is significantly different from the sum of reactive forces. Methods based on the efficiency of load transfer (Chou, 1981; Huang and Deng, 1982) can be applied to infinitely stiff joints but are not used in KENSLABS because it is unreasonable to assume the same efficiency for all points along a joint.

4. The use of rectangular elements severely limits the size of elements to be employed. If small elements are used in the main slab, elements in the adjoining slabs will be of the same small width. The limitation that the length–width ratio of any element be not greater than 4 or 5 requires a lot of elements to be used throughout the slab system. If many slabs are involved, a relatively large grid should be used to save computer time and storage. Fortunately, a relatively large grid can still yield reasonable results.

5. The program can determine only the stresses in the concrete slabs and the deflections of the slab and the foundation. The stresses in the solid or layer foundation cannot be obtained.

Evaluation of Contact

To check the contact condition between slab and foundation under a given load, it is necessary to know the contact condition before the load is applied. Therefore, the gaps at those nodes not in contact and the precompressions at those nodes in contact must be specified. Because these gaps and precompressions prior to the application of the load are difficult to determine, the best way is to divide the analysis into two steps.

First, determine the gaps and precompressions between slab and foundation under weight, temperature curling, and initial gaps, if any. The initial gap is the gap before the slab is constructed. In fact, there are no gaps between slab and foundation when the concrete is poured. Gaps develop only after pumping or plastic deformation of the foundation has taken place. Initial gaps can be visualized as a change in grade should the slabs be removed. Because these gaps are difficult to predict, a rough estimate of their possible occurrences near the pavement edges or joints is all that is needed. Nondestructive deflection tests can be used to detect the voids under concrete pavements (AASHTO, 1986) and the information obtained from these tests on existing pavements can be used as a guide. Next, using the gaps and precompression obtained in the first step, determine the stresses and displacements under the applied load. These two steps should be executed in the same run, one immediately after the other, so the gaps and precompressions determined in step 1 can be used in step 2.

The reasons that curling and loading are analyzed separately in two steps are because these two cases do not occur at the same frequency, and because the modulus of the subgrade due to the slowly changing temperature gradient may be much smaller than that due to transient wheel loads. Although stresses due to temperature curling are significant in concrete pavements, many of the current design methods consider only the stresses due to loading, so a separation of loading and curling stresses is needed. However, the program can determine the combined effect of curling and loading, if so desired.

Applications

Before running KENSLABS, it is necessary to sketch a plan view of the slabs, divide them into rectangular finite elements of various sizes, and number the slabs, nodes, and joints. In dividing slabs into rectangular finite elements, it is not necessary to use very fine divisions. In most cases, elements with a width of 10 to 12 in. (254 to 305 mm) are sufficient for regions near the load or in the area of interest. Larger elements can be used when they are far away from the load. To obtain more accurate results, the length to width ratio of any element should not be greater than 5. To model an infinite slab, the boundary or edge of slab should be placed at least 10 ft (3.28 m) from the load.

Numbering of Slabs, Nodes, and Joints

To facilitate programming, it is necessary to number the slabs in a systematic manner. First, consider the slab at the lower left corner as number 1, and proceed column-wise from bottom to top. When one column of slabs has been numbered, move to the right column and continue numbering from bottom to top until all slabs are numbered. In each slab, the nodes are numbered consecutively from bottom to top along the y axis, starting from the lower left corner, and then moving to the right until all nodes in the slab are numbered. The numbering is then continued on the next slab. The joints can be numbered in any arbitrary way. Figure 5.15 shows the numbering of slabs, nodes, and joints. Also shown within the circles are the element numbers to be used internally and the uniform load applied at the corner of element 6 to be used later for illustration.

The above method for numbering nodal points is not very efficient for multiple slabs on liquid foundation because it usually results in a stiffness matrix with a larger half-band width. The half-band width is governed by the maximum difference between the nodal numbers on the opposite side of the joints. The maximum difference for the slabs shown in Figure 5.15 is 12 and occurs at joints 4 and 6. The half-band width is $(12 + 1) \times 3 = 39$. The half-band width can be reduced if the nodes are numbered vertically from bottom to top across the joints. For example, if nodes 19, 20, and 21 are changed to 13, 14, and 15 and nodes 13,

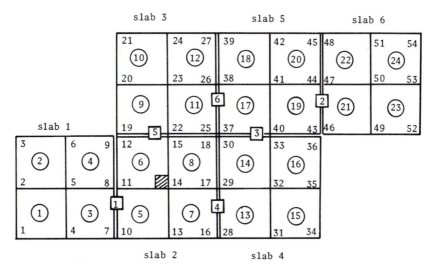

Figure 5.15 Numbering of slabs, nodes, and joints.

14, and 15 are changed to 16, 17, and 18, and the other nodes are changed accordingly, the maximum number of nodes in the vertical direction is 6, or the half-band width is reduced to $(6 + 2) \times 3 = 24$. To reduce the half-band width and save computer time and storage for liquid foundation, the program will renumber the nodes vertically from the bottom of one slab to the top of the other slab before forming the overall stiffness matrix. This change is done internally and the nodal numbers are switched back to the original without the knowledge of the program user. This additional procedure is applied only to liquid foundations with a dimension of stiffness matrix less than 70,000, so the equations can be solved without iterations. If after renumbering, the dimension of stiffness matrix is greater than 70,000, the original numbering will be maintained and the equations will be solved by iterations. For solid and layer foundations, it is not necessary to renumber the nodes because the stiffness matrix is always fully populated no matter how the nodes are numbered.

An input parameter JONO(J,I) is used to relate the location of joints to the slabs, where the first subscript, J, indicates the four sides (left, right, bottom, and top) of a slab, and the second subscript, I, indicates the slab number. For example, the left side of slab 1 has no joint, so JONO(1,1) = 0; the right side is joint 1, so JONO(2,1) = 1; the bottom and top sides have no joints, so JONO(3,1) = 0 and JONO(4,1) = 0. Similarly, for slab 2, JONO(1,2) = 1, JONO(2,2) = 4, JONO(3,2) = 0, and JONO(4,2) = 5. The same rule applies to all other slabs.

Types of Problems to Be Solved

The following input parameters are important for determining the type of problems to be solved:

INPUT = code indicating whether gaps and precompressions are obtained from a previous problem. Assign 0 for no and 1 for yes.

NCYCLE = maximum number of iteration cycles for checking subgrade contact. Assign 1 for full contact or when no iterations are required and 10 for partial contact or when iterations are required.

NOTCON = total number of nodes that are assumed not in contact or at which subgrade reactions are initially set to 0. If NCYCLE = 1, these nodes will never be in contact. If NCYCLE = 10, these nodes may or may not be in contact depending on the computed results.

NWT = code indicating whether slab weight is considered. Assign 0 when weight is not considered, as in the case of full contact with NCYCLE = 1, and 1 when weight is considered.

These parameters can be used in combinations to analyze the following cases:

Case 1: Slab and foundation are in full contact as originally assumed by Westergaard. Set INPUT to 0, NCYCLE to 1, NOTCON to 0, and NWT to 0. This case does not require iterations for checking contact conditions.

Case 2: Slab and foundation are in full contact at some nodes but completely out of contact at other nodes. Set INPUT to 0, NCYCLE to 1, NOTCON to the number of nodes not in contact, and NWT to 0. This case does not require iterations for checking contact conditions.

Case 3: Partial contact under wheel loads should be analyzed in two steps. First, determine the gaps and precompressions between slab and foundation under weight, temperature curling, and gaps, if any, by setting INPUT = 0, NCYCLE = 10, and NWT = 1. Then, using the gaps and precompressions thus computed, determine the stresses and deflections under wheel loads by setting INPUT = 1, NCYCLE = 10, and NWT = 0.

Case 4: If there is no gap between slab and foundation and no temperature curling, an approximate method with a single step can be used to analyze partial contact under wheel loads by specifying INPUT = 0, NCYCLE = 10, and NWT = 1. The stresses and deflections thus determined are due to the combined effect of the weight of slab and the wheel loads. Because the stress induced by the weight of slab is 0 for liquid foundations and very small for solid and layer foundations, when compared with the stresses due to wheel loads, the maximum stress obtained by this simplified procedure can be used directly for design purposes. However, this method cannot be used to determine the deflections under wheel loads because the deflections induced by the weight of slab are quite significant.

NOTCON can also be used to speed up the convergence of contact conditions. If it is definitely known that one or more nodes will not be in contact with the subgrade due to large temperature gradients, it is more efficient to specify these nodes as not in contact (NOTCON ≠ 0), so the reactive forces at these nodes will be set to 0 during the first iteration. If these nodes are assumed to be in contact, the program still works but the reactive forces will be set to 0 in the second iteration instead of the first iteration. If a gap is specified at a certain node, KENSLABS will automatically set the reactive force at that node to 0 in the first iteration. This is to avoid the unrealistic situation where full contact is assumed for a large gap and the slab may be pulled down by an amount greater than the gap specified.

Damage Analysis

Similar to KENLAYER, damage analysis can be performed by dividing each year into a maximum of 24 periods, each with a maximum of 24 load groups. Because only the properties of foundation vary with the season, a foundation seasonal adjustment factor (FSAF) is assigned to each period. The modulus of subgrade reaction of liquid foundation or the stiffness matrix of solid and layer foundations is multiplied by this factor to simulate the seasonal change in the stiffness of foundation.

Damage is based on fatigue cracking only and is defined by the cracking index (CI), which is the same as the damage ratio shown by Eq. 3.19. The allowable number of repetitions can be expressed as

$$\log N_f = f_1 - f_2 \left(\frac{\sigma}{S_c}\right) \tag{5.35}$$

in which N_f is the allowable number of repetitions, σ is the flexural stress in slab, and S_c is the modulus of rupture of concrete. In the design of zero-maintenance jointed plain concrete pavements, Darter and Barenberg (1977) recommended the use of $f_1 = 16.61$ and $f_2 = 17.61$. The following fatigue equations are recommended by the Portland Cement Association (Packard and Tayabji, 1985):

$$\text{For } \frac{\sigma}{S_c} \geq 0.55: \quad \log N_f = 11.737 - 12.077 \left(\frac{\sigma}{S_c}\right) \tag{5.36a}$$

$$\text{For } 0.45 < \frac{\sigma}{S_c} < 0.55: \quad N_f = \left(\frac{4.2577}{\sigma/S_c - 0.4325}\right)^{3.268} \tag{5.36b}$$

$$\text{For } \frac{\sigma}{S_c} \leq 0.45: \quad N_f = \text{unlimited} \tag{5.36c}$$

More about fatigue characteristics of concrete is presented in Section 7.3.2.

How damage analysis is to be performed is controlled by the input parameter NDAMA. If NDAMA = 0, no damage analysis will be made. If NDAMA = 1, Eq. 5.36 recommended by PCA will be used. If NDAMA = 2, Eq. 5.35 will be used and the coefficients f_1 and f_2 must be specified by the users. To simplify programming, NCYCLE must be equal to one when the number of periods (NPY) or the number of load groups (NLG) is greater than one.

One question on fatigue cracking that needs to be resolved is whether the passage of one set of tandem-axle loads should be considered as one or two repetitions. The PCA method (PCA, 1984) assumes one tandem-axle load as one repetition, while the zero-maintenance method (Darter and Barenberg, 1977) considers it as two repetitions. Theoretically, the damage analysis of tandem axles is similar to flexible pavements, as explained in Figure 3.3, except that the tensile stress caused by the passage of the first axle load is σ_a and that of the second axle load is $\sigma_a - \sigma_b$. If $\sigma_a - \sigma_b$ is much smaller than σ_a, the stress ratio (the ratio between the flexural stress and the modulus of rupture) due to the second load is most probably smaller than 0.45 and, according to the PCA failure criteria, should have no effect on fatigue damage. Therefore, the assumption of one tandem-axle load as one repetition is more reasonable. This is in line with the erosion analysis based on corner deflection in which one tandem-axle load is considered as one repetition. By the same reasoning, the passage of a set of tridem-axle loads should still be counted as one repetition.

If the stress ratio due to $\sigma_a - \sigma_b$ is greater than 0.45, or Eq. 5.35 without such a limit is used, then the second and third axle loads will have some effect on fatigue damage. This additional damage can be analyzed by KENSLABS and is discussed in Section 6.4.5.

5.2.2 Input Parameters

Input parameters are listed in alphabetical order. The maximum dimension of each array is shown and the actual dimension is indicated immediately after the definition.

BARNO(1,15,12) number of dowel bars at nodes along joint I (1,NNAJ(I),NJOINT). In most cases, BARNO is computed automatically based on the spacing of dowel bars. However, if NNAJ(I) for joint I is not 0, BARNO for joint I must be entered.

BD(12) bar diameter at joint (NJOINT). Assign 0 or any value if shear spring constant is used as indicated by a nonzero SPCON (1, NJOINT).

BS(12) bar spacing at joint (NJOINT). Assign 0 or any value if shear spring constant is used.

CT coefficient of thermal expansion of concrete, 0.000005 per °F suggested.

CURL(420) amount of gap between slab and subgrade (NGAP).

DEL tolerance for solving simultaneous equations by iteration, 0.001 suggested. Although DEL is not needed when the direct method of Gauss elimination is used, it is still suggested that 0.001 be specified in case the iteration method is activated due to storage limitations.

E(6) elastic modulus of each Burmister layer (NL).

FF(420) applied vertical nodal forces (NCNF(NLG)).

FMAX maximum allowable vertical displacement at node NNCK, 1.0 suggested for pavements and larger values for raft foundations. If the vertical displacement is greater than FMAX, something is apparently wrong and the program will stop.

FSAF(24) foundation seasonal adjustment factor (NPY). Use 1 if no adjustment is needed.

F1(2) fatigue coefficient, or log of the allowable number of repetitions when the stress ratio is 0 (NLAYER).

F2(2) fatigue coefficient, or the slope of log number of repetitions versus stress ratio curve (NLAYER).

GAMA(2) unit weight of each slab layer (NLAYER). If only one layer exists, GAMA(2) should be assumed to be 0. The unit weight is used to compute the nodal forces when NWT = 1.

GAP(420) gap between slab and foundation (NGAP).

GDC(12) gap between dowel and concrete (NJOINT). This parameter also applies to the shear spring constant but not to the moment spring constant.

INPUT code indicating whether the input of gaps or precompressions from a previous problem is needed, 1 for yes and 0 for no.

JONO(4,9) joint number on four sides of each slab (4,NSLAB). An index of 1 for the first subscript indicates the joint number on the left side of the slab, 2 on the right side, 3 on the bottom side, and 4 on the top side. If there is no joint on a given side, JONO for that side should be specified as 0.

LS(60) slab number on which load is applied (NUDL(NLG)).

MAXIC maximum number of cycles for numerical integration of layer foundation, 30 suggested.

MDPO more detailed printout, 1 for yes and 0 for no.

NAS number of additional subgrade moduli to be read in. Assign 0 if modulus is uniform throughout.

NAT(2) number of additional thicknesses to be read in for each layer (NLAYER). Assign 0 if thickness is uniform throughout.

NBOND bond between two layers. Use 1 when two layers are bonded, and 0 when two layers are not bonded or there is only one layer.

NCNF(24) number of concentrated nodal forces (NLG).

NCYCLE maximum number of cycles for checking subgrade contact. Use 1 for full contact or when no iterations are required, and 10 for partial contact or when iterations are required.

NDAMA 0 for no damage analysis, 1 for damage analysis based on PCA fatigue criteria, and 2 for damage analysis based on user specified fatigue coefficients.

NFOUND type of foundation; 0 for liquid, 1 for solid, and 2 for layer.

NG(420) nodal number at which a gap between slab and foundation is specified (NGAP).

NGAP total number of nodes at which a gap exists between slab and foundation. Assign 0 if no gap exists or when NCYCLE = 1.

NJOINT total number of joints, maximum 12. Assign 0 for single slab.

NL number of layers in Burmister's foundation, maximum 6.

NLAYER number of slab layers, either 1 or 2.

NLG number of load groups

NN(420) nodal number at which concentrated force is specified (NCNF(NLG)).

NNAJ(12) number of nodes at each joint (NJOINT). If NNAJ(I) for joint I is not 0, BARNO(1,NNAJ(I),I) must be specified.

NNCK nodal number for checking convergence when the iterative method is used for solving simultaneous equations. Use nodal number under the heaviest load, if possible. This node is also used to check FMAX and its number must be specified even if no iterations are needed.

NODNC(420) nodal number at which reactive force is initially assumed to be zero (NOTCON).

NODSX(50) nodal number on the x axis of symmetry (NSX). If the x axis of symmetry is along a joint, assign any one nodal number on the x axis.

NODSY(50) nodal number on the y axis of symmetry (NSY). If the y axis of symmetry is along a joint, assign any one nodal number on the y axis.

NOTCON total number of nodes assumed initially not in contact, maximum 420. If NCYCLE = 1, these nodes will never be in contact. If NCYCLE > 1, these nodes may or may not be in contact depending on the calculated results.

NP(420) nodal number at which stresses are computed and printed (NPRINT).

NPRINT number of nodes at which stresses are printed. If NPRINT = 0, the stresses at every node will be computed and printed.

NPROB number of problems to be solved. A maximum of two problems can be run at the same time. When analyzing the case of partial contact under wheel loads, it is convenient to set NPROB to 2. The first problem is run by specifying NCYCLE = 10, INPUT = 0, NWT = 1, and NTEMP = 1, so the gaps and precompressions due to temperature curling and weight are determined. The second problem is run by specifying NCYCLE = 10, INPUT = 1, NWT = 0, and NTEMP = 0, so the gaps and precompressions obtained from the first problem are used to determine the stresses and deflections due to wheel loads.

NPY number of periods per year.

NSLAB   total number of slabs, maximum 9.

NS(420)   nodal number at which subgrade modulus is different from the one specified (NAS).

NSX   number of points on the x axis of symmetry, maximum 50. Assign 0 if the x axis is not an axis of symmetry. Assign 1 if the x axis of symmetry is along a joint.

NSY   number of points on the y axis of symmetry, maximum 50. Assign 0 if the y axis is not an axis of symmetry. Assign 1 if y axis of symmetry is along a joint.

NT(420)   nodal number at which the thickness of the slab is different from the one specified (NAT(NLAYER)).

NTEMP   condition of curling. Assign 0 if there is no temperature curling and 1 if there is curling.

NUDL(24)   number of loaded areas (NLG), maximum 60. A load distributed over several elements is still considered as having one loaded area.

NWT   code indicating whether slab weight is considered. Assign 0 if weight is not considered, as in the cases of full contact. Assign 1 if weight is considered.

NX(9)   number of nodes in the x direction for each slab (NSLAB).

NY(9)   number of nodes in the y direction for each slab (NSLAB).

PMR(2)   modulus of rupture of pavement slab (NLAYER).

PR(2)   Poisson ratio of each slab layer (NLAYER).

PRBF(6)   Poisson ratio of each Burmister layer (NL).

PRS   Poisson ratio of solid foundation.

PRSB   Poisson ratio of steel dowel bar.

QQ(60)   tire contact pressure of each loaded area (NUDL(NLG)).

SCKV(12)   modulus of dowel support, or steel-concrete K value (NJOINT). Assign 0 or any value if dowel bars are not used as indicated by a nonzero spring constant SPCON (1,NJOINT).

SPCON(3,12)   spring constant for each joint (3,NJOINT). An index of 1 for the first subscript indicates shear transfer and a 2 indicates moment transfer. Assign 0 to SPCON (1,NJOINT) if dowel bars are used. If moment is with respect to the y axis, the program will change the first subscript from 2 to 3.

SUBMOD(420)   subgrade modulus of liquid foundation (LNP). Note that LNP is the total number of nodes. If the modulus is not uniform, SUBMOD(1) is the modulus of the uniform part, while the modulus SUBMOD(I) at node I, which is different from SUB-MOD(1), will be read in later.

T(2,420)   thickness of each slab layer (NLAYER,LNP). If the thickness is not uniform, T(J,1) is the thickness of layer J for the uniform part, while the thickness T(J,I) at node I, which is different from T(J,1), will be read in later. If there is temperature curling, T(J,1) is the thickness for determining the initial curling.

TEMP   temperature at bottom of slab minus temperature at top of slab, positive if curled upward and negative if curled downward.

TH(5)   thickness of each Burmister layer with finite thickness (NL-1). The last layer is infinite in thickness and need not be read in.

TITLE   any title or comments within column 1 to 80 of a line.

TNLR(24,24)   total or predicted number of load repetitions for each load group in each period (NPY,NLG).

WJ(12)   width of joint (NJOINT). Assign 0 or any value if a spring constant other than 0 is specified.

X(15,9) x local coordinate of each slab starting from 0 and increasing from left to right (NX(NSLAB),NSLAB).

XL(1,60) and XL(2,60)   left and right limits of loaded area in x local coordinates (2,NLOAD).

Y(15,9) y local coordinate of each slab, starting from 0 and increasing from bottom to top (NY(NSLAB),NSLAB).

YL(1,60) and YL(2,60)   lower and upper limits of loaded area in y local coordinates (2,NLOAD).

YM(2)   Young's modulus of each slab layer (NLAYER).

YMS   Young's modulus of solid foundation.

YMSB   Young's modulus of steel dowel bar.

5.2.3 Data File

The program was written in FORTRAN 77. Similar to KENLAYER, a data file can be set up automatically by SLABSINP or entered manually according to the given formats. Details about SLABSINP are presented in Appendix D.

The input data file is listed below. The number of lines indicated is correct only when the given data can be accommodated in an 80-column line. If the data require more than 80 columns, they should continue on the next line. If an asterisk appears before the step number, the step may be skipped under the condition specified.

(1)   1 line (I5) NPROB

(2)   1 line (80 columns) TITLE

(3)   1 line (4I5) NFOUND,NDAMA,NPY,NLG

(4)   1 line (2I5) NSLAB, NJOINT

(5)   NSLAB lines (6I5) NX(I),NY(I),(JONO(J,I),J=1,4).
Each slab, as indicated by the subscript I from 1 to NSLAB, should begin with a new line.

(6)   1 line (15I5) NLAYER,NNCK,NOTCON,NGAP,NPRINT,INPUT,NBOND,NTEMP,NWT, NCYCLE,NAT(1),NAT(2),NSX,NSY,MDPO

(7)   1 line (8F10.5) TEMP,GAMA(1),GAMA(2),PMR(1),PMR(2),CT,DEL,FMAX
GAMA(1) = 0.0868 pci, CT = 5×10^{-6} per °F, DEL = 0.001, FMAX = 1.0 in. for pavement and 10.0 in. for raft foundation.

(8)   1 line (4F10.5) F1(1),F1(2),F2(1),F2(2)

(9)   NSLAB lines (8F10.5) (X(J,I),J=1,NX(I)),(Y(J,I),J=1,NY(I))
Each slab, as indicated by the subscript I from 1 to NSLAB, should begin with a new line.

(10)   1 line (2(2F10.5,E10.3)) (T(I,1),PR(I),YM(I),I=1,NLAYER)

*(11)   NLAYER lines (5(I5,F10.5)) (NT(J),T(I,NT(J)),J=1,NAT(I))
Each layer, as indicated by the subscript I from 1 to NLAYER, should begin with a new line.
Skip layer I if NAT(I) = 0.

(12)   1 line (16I5) (NUDL(I),I=1,NLG)

(13)   1 line (16I5) (NCNF(I),I=1,NLG)

(14)   There is a do loop for load group from step 15 to 16. The load group number is indicated by the subscript J which starts from 1 to NLG.

*(15)   NUDL(J) lines (I5,5F10.5) LS(I,J),XL(1,I,J),XL(2,I,J),YL(1,I,J),YL(2,I,J), QQ(I,J)
Each loaded area, as indicated by the subscript I from 1 to NUDL(J), should begin with a new line.
Skip if NUDL(J) = 0.

*(16)   1 line (5(I5,F11.5)) (NN(I),FF(NN(I),J),I=1,NCNF(J))
Skip if NCNF(J) = 0.

*(17) 1 line (16I5) (NODNC(I),I = 1,NOTCON)
       Skip if NOTCON = 0.
*(18) 1 line (16I5) (NP(I),I = 1,NPRINT)
       Skip if NPRINT = 0.
*(19) 1 line (16I5) (NODSX(I),I = 1,NSX)
       Skip if NSX = 0.
*(20) 1 line (16I5) (NODSY(I),I = 1,NSY)
       Skip if NSY = 0.
*(21) 1 line (5(I5,F10.5)) (NG(I),CURL(NG(I)),I = 1,NGAP)
       Skip if NGAP = 0.
 (22) 1 line (8F10.5) (FSAF(I),I = 1,NPY)
 (23) If NFOUND = 1 or 2, go to (27)
 (24) 1 line (I5,E10.3) NAS,SUBMOD(1)
*(25) 1 line (5(I5,E10.3)) (NS(I),SUBMOD(NS(I)),I = 1,NAS)
       Skip if NAS = 0.
 (26) Go to (34)
 (27) If NFOUND = 2, go to (30)
 (28) 1 line (E10.3,F10.5) YMS,PRS
 (29) Go to (34)
 (30) 1 line (I5) NL, MAXIC
       MAXIC = 30
 (31) 1 line (5F10.5) (TH(I),I = 1,NL-1)
 (32) 1 line (6E10.3) (E(I),I = 1,NL)
 (33) 1 line (6F10.5) (PRBF(I),I = 1,NL)
*(34) 1 line (E10.3,F10.5) YMSB,PRSB
       Skip if NSLAB = 1.
*(35) NJOINT lines (3E10.3,4F10.5,I5) SPCON(1,I),SPCON(2,I),SCKV(I),BD(I),BS(I),WJ(I),
       GDC(I),NNAJ(I)
       Each joint, as indicated by the subscript I from 1 to NJOINT, should begin with a new
       line.
       Skip if NSLAB = 1.
*(36) NJOINT lines (8F10.5) (BARNO(1,J,I), J = 1, NNAJ(I))
       Each joint, as indicated by the subscript I from 1 to NJOINT, should begin with a new
       line. The subscript J from 1 to NNAJ(I) indicates the node along each joint.
       Skip if NSLAB = 1 or NNAJ(I) = 0.
*(37) NPY lines (8E10.3) (TNLR(I,J),J = 1,NLG)
       Each period, as indicated by the subscript I from 1 to NPY, should begin wtih a new
       line.
       Skip if NDAMA = 0.
 (38) Go to (2) NPROB times

5.2.4 Printed Output

Every input parameter, after being read in, will be printed out for inspection. The output is stored automatically in a file called SLABS.TXT. This file will be destroyed and replaced by new output data when KENSLABS is run again. If you want to keep the file, be sure to change its name.

Two types of output can be generated. When MDPO = 0, only the regular output will be printed. In almost all cases, the specification of regular output is sufficient. More detailed output may be specified for debugging purposes or by users who are not familiar with KENSLABS and would like to check all intermediate output to be sure that the program has been used properly. The names of output parameters, as they appear in the computer printout, are listed and explained below.

Regular Output

BEARS bearing stress between dowel and concrete.

CI cracking index or damage ratio

CURL amount of initial curling at each node due to temperature curling.

DF difference in deflection between iterations at a designated node when the iterative method is used for solving simultaneous equations.

F vertical deflection of the slab at each node.

FAJ1 shear force at each node along the joint.

FAJ2 moment at each node along the joint, which should equal to 0 if there is no moment transfer.

FAJPD shear force on one dowel bar.

FOSUM sum of applied forces.

GAP gap or precompression at each node. For liquid foundation, gap is positive and precompression is negative. For solid or layer foundation, gap is positive and precompression is zero.

IC iteration number for checking the convergence of displacement.

MAJOR major principal stress in slab.

MAX.SHEAR maximum shear stress in slab.

MINOR minor principal stress in slab.

NB half-band width.

NCY cycle number for checking slab–subgrade contact.

PRESSURE reactive pressure at each node.

SDIF stress differential due to multiple axle loads.

SHEARS shear stress in dowel.

SMAX maximum stress.

SPCON equivalent spring constant for doweled joints.

STRESS X stresses in the x direction, positive when bottom of slab is in tension.

STRESS Y stresses in the y direction, positive when bottom of slab is in tension.

STRESS XY shear stress on the xy plane.

SUBR subgrade reactive force at each node.

SUBSUM sum of subgrade reactions, which should be nearly equal to the sum of applied forces, FOSUM.

SUMCI sum of crackling index over all load groups in all periods.

More Detailed Output

Many of the output parameters need explanations. Figure 5.15 is used as an example to show the values of these parameters.

INITEN initial or smallest element number in each slab, e.g., 1 for slab 1 and 21 for slab 6.

INITNP initial or smallest nodal number in each slab, e.g., 1 for slab 1 and 46 for slab 6.

ISEN initial starting element number on both sides of each joint, e.g., 3 and 5 for joint 1, and 11 and 17 for joint 6.

ISNN initial starting nodal number on both sides of each joint, e.g., 7 and 10 for joint 1, and 25 and 37 for joint 6.

LASTEN last or largest element number in each slab, e.g., 4 for slab 1 and 24 for slab 6.

LASTNP last or largest nodal number in each slab, e.g., 9 for slab 1 and 54 for slab 6.

LFEN last or final element number on both sides of each joint, e.g., 4 and 6 for joint 1, and 12 and 18 for joint 6.

LFNN last or final nodal number on both sides of each joint, e.g., 9 and 12 for joint 1, and 27 and 39 for joint 6.

NDTY type of node. Assign 0 to the secondary node at the joint and 2, 3, or 4 to the primary node, depending on the number of slabs at the joints. A secondary node has a higher nodal number than the primary node on the opposite side of the joint. For example, node 18 is a primary node with NDTY = 4, while nodes 25, 30, and 37 are secondary nodes with NDTY = 0. Similarly, the values of NDTY are 2 for node 17 and 3 for node 36, while those for nodes 29, 43, and 46 are 0. This parameter is used for solid and layer foundations to reduce the number of nodes for forming the flexibility matrix and redistributing the stiffness matrix of foundation to the adjoining slabs. If a node is not at the joint, NDTY is assigned as 1.

NE element number over which uniformly distributed loads are applied, e.g., element 6.

NPOP nodal number on opposite sides of a joint. Assign 0 when the node is not at a joint or assign the primary nodal number when three or four slabs meet at the joint. For example, values of NPOP are 29 for node 17 and 17 for node 29; 18 for nodes 18, 25, 30, and 37; 36 for nodes 36, 43, and 46; and 0 for node 14. This parameter is used to superimpose the flexibility matrix of foundation and then redistribute the stiffness matrix of foundation to the slabs.

NPSM nodal numbers at which four slabs meet, e.g., nodes 18, 25, 30, and 37.

NPSM3 nodal numbers at which three slabs meet, e.g., nodes 9, 12, and 19 and nodes 36, 43, and 46.

NTSM number of times four slabs meet, e.g., 1.

NTSM3 number of times three slabs meet, e.g., 2.

QQ uniformly applied contact pressure.

ROTAT.X rotation of slab about the *x* axis.

ROTAT.Y rotation of slab about the *y* axis.

SUBD displacement of subgrade for solid or layer foundation. The subgrade displacement is the same for all nodes meeting at the joint because there is actually only one point on the subgrade.

SUMP sum of reactive forces and precompression for solid or layer foundation. The slab and subgrade are not in contact if SUMP is negative.

XDA left and right limits of loaded area in dimensionless *x* coordinates. If the loaded area covers the entire length of an element, XDA is from -1 to 1. If the loaded area covers one-fourth of the length, as shown in Figure 5.15, XDA is from 0.5 to 1.

XN *x* local coordinates using the lower left corner of each slab as origin.

XO *x* global coordinates using the lower left corner of slab 1 as origin, or revised global *x* coordinates in case of symmetry.

YDA lower and upper limits of loaded area in dimensionless *y* coordinates. If the loaded area covers the entire width of an element, YDA is from -1 to 1. if the loaded area covers one-fourth of the width, as shown in Figure 5.15, YDA is from -1 to -0.5.

YN *y* local coordinates using the lower left corner of each slab as origin.

YO *y* global coordinates using the lower left corner of slab 1 as origin, or revised global *y* coordinates in case of symmetry.

5.3 COMPARISON WITH AVAILABLE SOLUTIONS

It is generally agreed that the finite element method is one of the most powerful tools for analyzing rigid pavements, particularly those with partial contact. Many studies (Huang and Wang, 1973, 1974; Huang, 1974a; Darter and Barenberg, 1977; Chou and Huang, 1981) were made comparing finite element solutions with field measurements and close agreements were reported. In this section, the solutions obtained by KENSLABS are compared with other theoretical solutions available.

5.3.1 Analytical Solutions

The analytical solutions to be compared include Westergaard's solutions for edge loading and the solutions for temperature curling by Westergaard and Bradbury.

Edge Loading

The analytical solutions for a concentrated load applied at the edge of an infinite slab far from any corner were presented by Westergaard (1926b). In the finite element method, the infinite slab can be approximated by a large slab, 20ℓ long by 10ℓ wide, where ℓ is the radius of relative stiffness. Because the problem is symmetrical with respect to the y axis, only one-half of the slab need be considered. The slab is divided into rectangular finite elements, as shown by the insert in Figure 5.16. The x coordinates are 0, $\pi\ell/8$, $\pi\ell/4$, $\pi\ell/2$, $\pi\ell$, 5ℓ, 7ℓ, and 10ℓ, and the y coordinates are 0, $\pi\ell/4$, $\pi\ell/2$, $\pi\ell$, $1.5\pi\ell$, $2\pi\ell$, $2.5\pi\ell$, and 10ℓ. The Poisson ratio of the concrete is 0.25, as was assumed by Westergaard.

Figure 5.16 shows a comparison between Westergaard's exact solutions, as indicated by the curves, and the finite element solutions obtained by KEN-

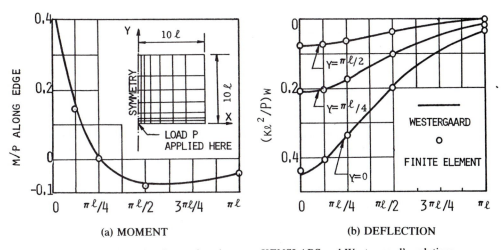

Figure 5.16 Comparison between KENSLABS and Westergaard's solutions. (After Huang and Wang (1974).)

SLABS, as indicated by the small circles. The moment M and the deflection w along the edge of slab and the deflection at a distance of $\pi\ell/4$ and $\pi\ell/2$ from the edge are presented. It can be seen that the solutions by KENSLABS check very closely with Westergaard's results.

Temperature Curling

Westergaard's solutions for a slab of infinite length and Bradbury's solutions for a finite slab can be used to compare with KENSLABS.

Slab of Infinite Length

Westergaard (1926a) presented exact solutions for the stress and deflection due to temperature curling in a concrete slab 4.2ℓ wide and infinitely long, where ℓ is the radius of relative thickness. His solutions for the stress and deflection along the y axis, which is perpendicular to the pavement edge, are shown by the two curves in Figure 5.17. The stress and deflection are expressed as dimensionless ratios in terms of σ_0 and w_0, respectively, where

$$\sigma_0 = \frac{E\alpha_t\Delta t}{2(1 - v)} \tag{5.37a}$$

$$w_0 = \frac{(1 + v)\alpha_t\,\ell^2}{h} \tag{5.37b}$$

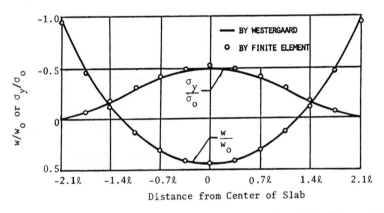

Figure 5.17 Stress and deflection along transverse direction. (After Huang and Wang (1974).)

In the finite element analysis, a long slab, 4.2ℓ wide and 16.8ℓ long, is employed. The slab is divided into finite elements, as shown in Figure 5.18. Due to symmetry, only the upper right quadrant of the slab is considered. It is assumed that $E = 3 \times 10^6$ psi (20.7 GPa), $k = 100$ pci (27.1 MN/m³), $h = 9$ in. (229 mm), $\alpha_t = 5 \times 10^{-6}/°F$ ($9 \times 10^{-6}/°C$), and $\Delta t = 10°F$ (5.6°C). The radius of relative stiffness ℓ is 36.95 in. (938.5 mm), which is used for determining the actual size of grid for computer input.

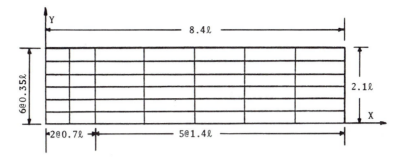

Figure 5.18 Division of slab into finite elements.

After the stress and deflection are computed, they are expressed as dimensionless ratios, as shown by the circles in Figure 5.17. It can be seen that the KENSLABS solutions check closely with Westergaard's exact solutions.

Westergaard's solutions are based on the assumption that the slab and the subgrade are in full contact. If the slab moves down, the subgrade will push it up; if the slab moves up, the subgrade will pull it down. This case is analyzed physically in KENSLABS by assuming that the slab is curled into a spherical surface and that the spring at each node will push or pull the slab up or down, depending on the movement of the slab. The close agreement in stress and deflection between the two solutions clearly indicates that the concept employed by KENSLABS is fundamentally correct.

Finite Slab

To compare Bradbury's solutions with those obtained by KENSLABS, it is assumed that the slab is 20 ft (6.1 m) long and 12 ft (3.7 m) wide with a Young's modulus of 5×10^6 psi (34.5 GPa), a Poisson ratio of 0.2, a coefficient of thermal expansion of 5×10^{-6}/°F (9×10^{-6}/°C), and a temperature differential of 3°F per in. (0.066°C per mm) of slab. Three different subgrade moduli of 50, 200, and 500 pci (7.9, 31.4, and 78.6 kN/m³) and three different thicknesses of 8, 10, and 14 in. (203, 254, and 356 mm) are assumed. The results are shown in Table 5.2.

TABLE 5.2 COMPARISON OF CURLING STRESS

Slab thickness (in.)	Method	Maximum curling stress (psi)		
		$k = 50$ pci	$k = 200$ pci	$k = 500$ pci
8	KENSLABS	248.8	361.3	391.8
	Bradbury	253.1	363.1	390.0
10	KENSLABS	224.5	394.4	466.4
	Bradbury	224.2	398.2	467.9
14	KENSLABS	158.3	382.4	538.5
	Bradbury	154.2	386.6	541.4

Note. 1 in. = 25.4 mm, 1 psi = 6.9 kPa, 1 pci = 271.3 kN/m³.

The stress shown in Table 5.2 is the maximum stress at the center of the slab. Bradbury's solutions are obtained by following the same procedure as illustrated in Example 4.1. The solutions from KENSLABS are obtained by using one-quarter of the slab and dividing it into 1 ft $\times$ 1 ft (0.3 $\times$ 0.3 m) square elements. The agreement between the two solutions for a variety of thicknesses and subgrade moduli further validates KENSLABS for handling temperature warping with full subgrade contact.

The assumption of full contact for temperature curling is unreasonable because, when the slab curls up, there is no way that the subgrade will pull it down. Experience has shown that the curling stresses obtained by Westergaard or Bradbury's theory based on full contact are much larger than those from field measurements (Moore, 1956). As is shown in Section 5.3.3, the finite element solutions based on partial contact result in much smaller stresses and check more closely with field measurements.

5.3.2 Influence Charts

The influence charts developed by Pickett and Ray (1951) for liquid foundations and by Pickett and Badaruddin (1956) for solid foundations can be used to compare with the solutions by KENSLABS.

Liquid Foundations

Figure 5.19 shows an influence chart used by PCA (1969) for determining the moment at point O under a 36-kip (160-kN) tandem-axle load having an 11.5-in. (292-mm) dual spacing, a 49-in. (1.25-m) tandem spacing, and a 71-in. (1.80-m) spacing between the center of the two sets of dual tires. The radius of relative stiffness ℓ is 50 in. (1.77 m) and the number of blocks N covered by each tire is shown in the figure. When ℓ, N, and a tire contact pressure q of 67.2 psi (433 kPa) are known, the moment can be computed by Eq. 4.32a and compared with KENSLABS.

In the finite element analysis by KENSLABS, each load was considered independently. The problem can be made symmetric by placing an image tire on the left corresponding to each tire on the right; therefore, only one-half of the slab is required. The moment thus determined must be divided by 2 to give the moment resulting from one tire only. A relatively large square grid was used near the loads, as shown in Figure 5.20.

Table 5.3 gives a comparison of the moments at point O due to six different tire positions. It can be seen that the solutions by KENSLABS check quite well with those by the influence chart, especially when the tire is close to point O. The large discrepancy for tire 7 is due to the fact that this tire straddles between positive and negative blocks and covers only two blocks. The percentage of discrepancy would become zero if the block enclosed was counted as 1.2 instead of the two blocks counted by PCA.

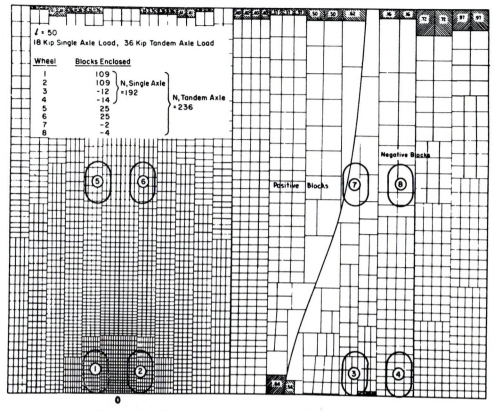

$\ell = 50$
18 Kip Single Axle Load, 36 Kip Tandem Axle Load

Wheel	Blocks Enclosed		
1	109		
2	109	N, Single Axle	
3	-12	=192	N, Tandem Axle
4	-14		=236
5	25		
6	25		
7	-2		
8	-4		

Figure 5.19 Contact imprints on influence chart based on liquid foundation. (After PCA (1969).)

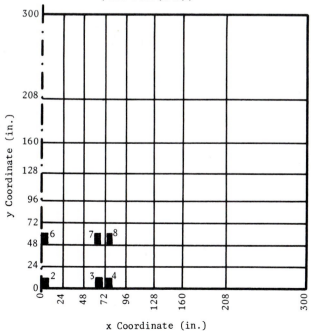

Figure 5.20 Finite element mesh for liquid foundation (1 in. = 25.4 mm).

TABLE 5.3 COMPARISON OF MOMENTS BETWEEN INFLUENCE CHART AND KENSLABS

Tire number	Moment (in.-lb)		Percentage of discrepancy
	Influence chart	KENSLABS	
2	1831	1861	+1.6
3	−202	−204	+1.0
4	−235	−260	+10.6
6	420	393	−6.4
7	−33	−20	−39.4
8	−67	−73	+9.0

Note. 1 in.-lb = 0.113 m-N.

Solid Foundation

Pickett and Badaruddin (1956) developed analytical solutions for stresses at the edge of an infinite slab on a solid foundation due to a load near the edge. Figure 5.21 is an influence chart they developed. To use the chart, it is necessary to determine the modulus of relative stiffness for the solid foundation defined as

$$\ell = h\left[\frac{1}{6}\frac{E_c}{E_f}\frac{1 - v_f^2}{1 - v_c^2}\right]^{1/3} \tag{5.38}$$

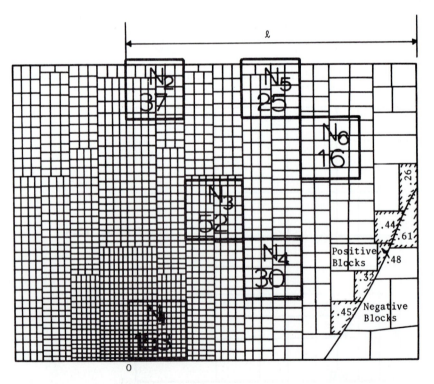

Figure 5.21 Contact areas on influence chart based on solid foundation.

The subscripts c and f denote the concrete slab and the solid foundation, respectively. After ℓ is determined, the loaded area can be plotted on the chart using the scale on the top. The number of blocks N covered by the loaded area is counted, and the stress at point O can be obtained by Eq. 4.32. The number of blocks covered by the six contact areas, each having a size of $0.2\ell \times 0.2\ell$, was tabulated by Pickett and Badaruddin (1956) and is shown in Figure 5.21.

In the finite element analysis by KENSLABS, a slab 8 in. (203 mm) thick, 24 ft (7.32 m) long, and 12 ft (3.66 m) wide was assumed. The problem can be solved by considering one-half of the slab, similar to the case of liquid foundation described previously. The slab was divided into finite elements as shown in Figure 5.22. Assuming that $E_c = 4 \times 10^6$ psi (27.6 GPa), $E_f = 1 \times 10^4$ psi (69 MPa), $\nu_c = 0.15$, and $\nu_f = 0.475$, then from Eq. 5.38, $\ell = 30$ in. (762 mm).

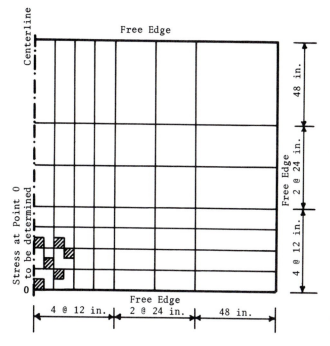

Figure 5.22 Finite element mesh for solid foundation (1 in. = 25.4 mm).

TABLE 5.4 COMPARISON OF STRESSES BETWEEN INFLUENCE CHART AND KENSLABS

Loading position	Edge stress/contact pressure, σ_e/q		Percentage of discrepancy
	Influence chart	KENSLABS	
1	2.580	2.653	+2.8
2	0.617	0.645	+4.5
3	0.870	0.909	+4.5
4	0.506	0.545	+7.7
5	0.417	0.443	+6.2
6	0.268	0.283	+5.6

Table 5.4 gives a comparison of edge stresses between the solutions by Pickett and Badaruddin's influence chart and those by KENSLABS. Both solutions check closely with an average discrepancy of about 5%.

5.3.3 ILLI-SLAB Computer Program

ILLI-SLAB was originally developed in 1977 for the structural analysis of one or two layers of slabs with or without mechanical load transfer systems at joints and cracks (Tabatabaie, 1977). It has since been continuously revised and expanded through several research studies to improve its accuracy and ease of application. Its capabilities are similar to KENSLABS, which was originally developed in 1973 and has been continuously updated and improved for classroom use.

Heinrichs et al. (1989) compared several available computer models for rigid pavements and concluded that both ILLI-SLAB or JSLAB, which is a similar finite element program developed by the PCA, were efficient to use and could structurally model many key design factors of importance. They also indicated that the ILLI-SLAB had extensive checking, revisions, and verification by many researchers and was more free of errors than any other available program. They presented the results of several cases by different models, with which KENSLABS can be compared.

Edge Loading

Figure 5.23 shows the cross section of a two-layer pavement with tied shoulders, which was analyzed by Heinrichs et al. In addition to the dimensions and material properties shown in the figure, other information includes joint width, 0.25 in. (6.4 mm); modulus of dowel support, 1.5×10^6 psi (10.4 GPa);

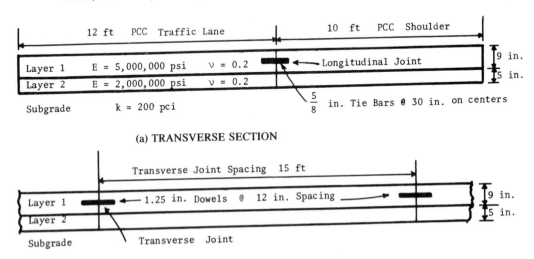

(a) TRANSVERSE SECTION

(b) LONGITUDINAL SECTION

Figure 5.23 Cross section of concrete pavement (1 ft = 0.305 m, 1 in. = 25.4 mm, 1 psi = 6.9 kPa, 1 pci = 271.3 kN/m³).

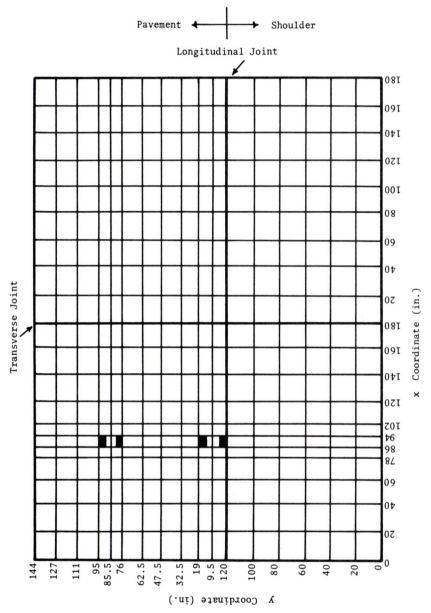

Figure 5.24 Finite element grid for analyzing longitudinal edge loading (1 in. = 25.4 mm).

TABLE 5.5 COMPARISON OF RESULTS BASED ON LONGITUDINAL EDGE LOADING

Load transfer	Type of interface	Model	Minimum longitudinal joint efficiency (%)	Maximum tensile stress at bottom of		Maximum deflection $(10^{-3}$ in.$)$
				Slab (psi)	Base (psi)	
Aggregate interlock	Unbonded	KENSLABS	74	169.6	37.7	7.7
		ILLI-SLAB	74	172.3	38.3	7.9
		JSLAB	74	166.6	37.0	7.9
	Bonded	KENSLABS	72	66.7	67.4	6.2
		ILLI-SLAB	72	66.8	68.1	6.3
		JSLAB	72	66.9	68.3	6.3
Dowel and tie bars	Unbonded	KENSLABS	43	195.7	43.5	9.4
		ILLI-SLAB	60	185.7	41.3	8.6
		JSLAB	44	193.2	42.9	9.6
	Bonded	KENSLABS	41	77.0	78.5	7.8
		ILLI-SLAB	57	71.9	73.4	6.8
		JSLAB	42	77.1	78.7	7.7

Note. 1 in. = 25.4 mm, 1 psi = 6.9 kPa.

Young's modulus of the dowel, 29×10^6 psi (200 GPa); Poisson ratio of the dowel, 0.3; single-axle load, 18 kip (80 kN); and contact pressure, 100 psi. The finite element grid and the loading position are shown in Figure 5.24.

Table 5.5 gives a comparison of results based on longitudinal edge loading. The slabs are assumed to be in full contact with the subgrade. This assumption is valid because there is no temperature curling and no initial gaps between the slab and the subgrade. The load transfer system can be either aggregate interlock, with spring constants of 100,000 psi (690 MPa) at transverse joints and 31,570 psi (218 MPa) at longitudinal joints, or dowel and tie bars, with 1.25-in. (32-mm) dowel bars at 12-in. (305-mm) spacing and $\frac{5}{8}$-in. (16-mm) tie bars at 30-in. (76-mm) spacing. The interface between the two layers can be either unbonded or bonded. The joint efficiency is defined as the ratio of deflections between the unloaded and loaded slabs on the opposite side of the joint. The minimum longitudinal joint efficiency, as well as the maximum stress and deflection, occurs under the wheel load at the longitudinal edge. It can be seen from Table 5.5 that the results obtained by KENSLABS, ILLI-SLAB, and JSLAB all check very well in the case of aggregate interlock. However, when dowel and tie bars are used, the ILLI-SLAB results in higher efficiencies of load transfer and smaller stresses and deflections compared to KENSLABS and JSLAB.

Corner Loading

The same pavement was analyzed for corner loading with the grid shown in Figure 5.25. Table 5.6 is a comparison of results between KENSLABS and ILLI-SLAB. The data without parentheses were obtained by KENSLABS, while those

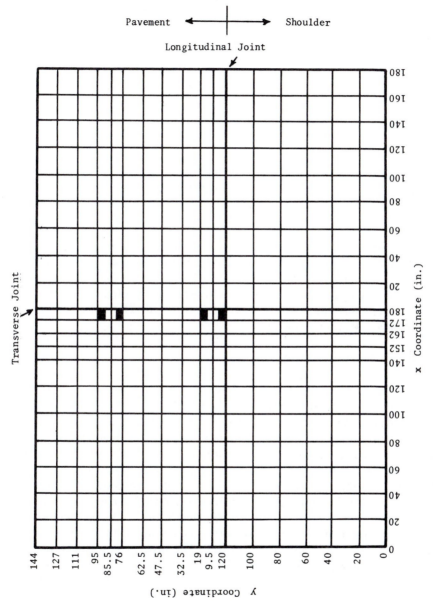

Figure 5.25 Finite element grid for analyzing corner loading (1 in. = 25.4 mm).

253

TABLE 5.6 COMPARISON OF RESULTS BASED ON CORNER LOADING

| Load transfer | Type of interface | Minimum joint efficiency | | Maximum principal tensile stress (psi) | | Maximum subgrade stress (psi) | Maximum deflection (10^{-3} in.) |
		Transverse (%)	Longitudinal (%)	Slab	Base		
Aggregate interlock	Unbonded	85	64	96.3	21.4	2.4	11.8
		(86)	(65)	(97.9)	(21.8)	(2.4)	(12.1)
	Bonded	84	60	52.9*	38.5	1.9	9.6
		(85)	(61)	(51.9*)	(37.1)	(1.9)	(9.7)
Dowel and tie bars	Unbonded	80	34	101.1	22.5	3.0	14.9
		(88)	(50)	(93.6)	(20.8)	(2.6)	(13.1)
	Bonded	79	31	67.0*	38.1	2.4	12.1
		(87)	(47)	(52.2*)	(35.2)	(2.1)	(10.5)

Note. Numbers without parentheses were obtained by KENSLABS and those within parentheses were obtained by ILLI-SLAB. Stresses are at the bottom of slab or base unless marked by *, when the critical tensile stress is at top. 1 in. = 25.4 mm, 1 psi = 6.9 kPa.

within parentheses were obtained by ILLI-SLAB. The results of JSLAB are not shown because they were not reported by Heinrichs et al. (1989). Similar to edge loading, ILLI-SLAB checks well with KENSLABS in the case of aggregate interlock but results in higher joint efficiencies and slightly lower stresses and deflections when the dowel and tie bars are used. Therefore, the use of KEN-SLABS is on the safe side.

Temperature Curling

In analyzing temperature curling, it is more reasonable to assume that the slab and subgrade are in partial contact. If the slab is weightless and the temperature at the top is colder than that at the bottom, the slab will curl up and form a spherical surface. There is no contact between the slab and the subgrade except at the very center. However, the weight of the slab will cause it to settle and part of the slab in the interior will finally come in contact. If the temperature gradient is small, the entire slab may be in contact with the subgrade. In the finite element method, the initial curling of a weightless slab at each node is first determined and the deflection at each node due to the weight of slab is then computed. If the deflection of the slab is smaller than the initial curling, the node is not in contact and its reactive force should be set to 0. The process is repeated until the same contact condition is obtained. Similar procedures can be applied when the slab curls down except that the interior of the slab may not be in contact.

Table 5.7 is a comparison of maximum curling stress based on partial contact. The slab is 20 ft (6.1 m) long and 12 ft (3.7 m) wide, which is similar to the case presented in Table 5.2 for full contact with the same material properties and the same mesh size of 1 ft × 1 ft (0.3 × 0.3 m). Two different temperature

TABLE 5.7 COMPARISON OF CURLING STRESS BASED ON PARTIAL CONTACT

Slab thickness (in.)	Computer model	Downward curling ($-3°$F/in.) Subgrade modulus (pci)			Upward curling (1.5°F/in.) Subgrade modulus (pci)		
		50	200	500	50	200	500
8	KENSLABS	219.6	270.2	284.6	120.4	165.3	182.1
	ILLI-SLAB	222.7	283.2	301.5	120.5	165.3	182.1
	JSLAB	180.2	225.0	233.8	94.0	127.4	133.5
10	KENSLABS	193.3	258.2	288.7	107.3	164.2	191.6
	ILLI-SLAB	198.7	276.6	315.2	107.2	164.5	191.9
	JSLAB	160.4	221.3	249.4	84.7	133.8	151.2
14	KENSLABS	139.1	203.3	223.6	76.1	133.0	163.6
	ILLI-SLAB	143.2	216.2	240.1	76.0	133.2	163.9
	JSLAB	116.0	184.2	204.9	60.1	114.6	141.7

Note. Curling stress is in psi. 1 in. = 25.4 mm, 1 psi = 6.9 kPa, 1 pci = 271.3 kN/m^3.

gradients of $-3°$F/in. ($-0.066°$C/mm) and 1.5°F/in. (0.033°C/mm) of slab were employed. The negative gradient indicates that the slab curls down and the positive gradient indicates that the slab curls up. The maximum curling stress occurs at the center of the slab.

As indicated in Table 5.7, the curling stresses obtained by KENSLABS are nearly the same as those by ILLI-SLAB when the slab curls up, but are slightly smaller when the slab curls down. However, the curling stresses obtained by JSLAB are much smaller than those by KENSLABS and ILLI-SLAB.

5.3.4 Layered Theory

Plate theory is used for the structural analysis of concrete pavements, and the layered theory is used to analyze asphalt pavements. When the modulus of elasticity of the pavement is much greater than that of the foundation and the load is applied in the interior of pavement far away from edges and corners, both theories should yield about the same results. Therefore, the results obtained by KENLAYER can be used to check those by KENSLABS.

In the following examples, a large slab, 24 ft (7.3 m) long, 24 ft (7.3 m) wide, and 8 in. (203 mm) thick, is assumed. The concrete has an elastic modulus of 4 × 10^6 psi (28 GPa) and a Poisson ratio of 0.15. The foundation is the solid type with an elastic modulus of 10,000 psi (69 MPa) and a Poisson ratio of 0.4. A 9000-lb (40-kN) load with a contact pressure of 75 psi (518 kPa) is applied at the center of the slab. In applying KENLAYER, the contact area is a circle with a radius of 6.16 in. (157 mm); while in KENSLABS, a square area with each side of 10.95 in. (278 mm) is assumed. Figure 5.26 shows the division of the slab into rectangular finite elements. Due to symmetry, only one-quarter of the slab need be considered.

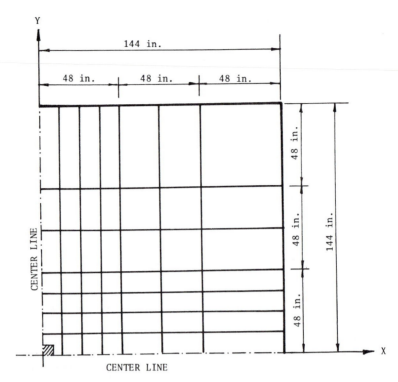

Figure 5.26 Finite element grid of a large slab with an interior load (1 in. = 25.4 mm).

One Layer of Slab

Table 5.8 gives a comparison between KENSLABS and KENLAYER when there is only one layer of slab on a solid foundation. In applying KENLAYER, the pavement is considered as a two-layer system with a frictionless interface between the concrete and the foundation, which is similar to a slab on a solid foundation. The radial and tangential stresses and the surface deflections are determined at distances of 0, 12, 24, 36, and 48 in. (0, 0.30, 0.61, 0.91, and 1.22 m) from the load, which correspond to the first five nodes in the finite element grid. The stresses shown in the table are at the bottom of the slab and are positive when in tension. The difference is expressed as a percentage using KENLAYER as a base. A negative difference implies that the solution by KENSLABS is smaller than that by KENLAYER. It can be seen that the solutions by KENSLABS check very well with those by KENLAYER.

Two Layers of Slabs

Figure 5.27 shows the cross sections and material properties for two layers of slab on a solid foundation. Two cases are considered: a cement-treated base (CTB) under a PCC slab and an HMA overlay on a PCC slab. The interface

TABLE 5.8 COMPARISON OF RESULTS FOR ONE LAYER OF SLAB

Parameter	Distance (in.)				
	0	12	24	36	48
Radial stress (psi)					
KENSLABS	171.9	58.6	13.8	−3.6	−11.3
KENLAYER	176.0	60.2	14.5	−3.6	−11.2
Difference (%)	−2.3	−2.7	−5.1	0	1.0
Tangential stress (psi)					
KENSLABS	171.9	101.7	61.5	37.9	24.1
KENLAYER	176.0	109.9	63.0	38.9	24.8
Difference (%)	−2.3	−8.1	−2.4	−2.6	−2.9
Surface deflection (0.001 in.)					
KENSLABS	9.16	8.66	7.68	6.62	5.63
KENLAYER	9.53	8.82	7.78	6.65	5.64
Difference (%)	−4.0	−1.8	−1.3	−0.5	−0.2

Note. 1 in. = 25.4 mm, 1 psi = 6.9 kPa.

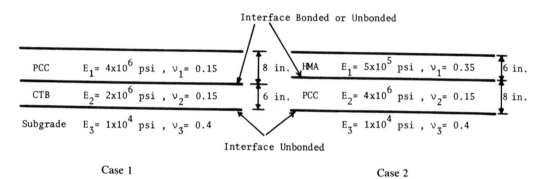

Figure 5.27 Two different cases for two layers of slabs on a solid foundation (1 in. = 25.4 mm, 1 psi = 6.9 kPa).

between the two layers can be either unbonded or bonded. The case of bonded slabs is of particular interest because it indicates the accuracy of the equivalent section method, as described in Section 5.1.2.

Cement-Treated Base

Figure 5.28 compares the results between KENLAYER and KENSLABS for case 1 with a cement-treated base. Similar to one layer of slab, the same finite element mesh shown in Figure 5.26 was used. The vertical deflection at the surface, the vertical stress on the top of subgrade, the tangential stress at the bottom of PCC, and the tangential stress at the bottom of CTB are compared at radial distances of 0, 1, 2, 3, and 4 ft from the load. The reason that the tangential stresses are presented is because they are larger than the radial stresses and control the design. In the figure, the results obtained from KENLAYER are indicated by the solid curves for the unbonded layers and the dashed curves for

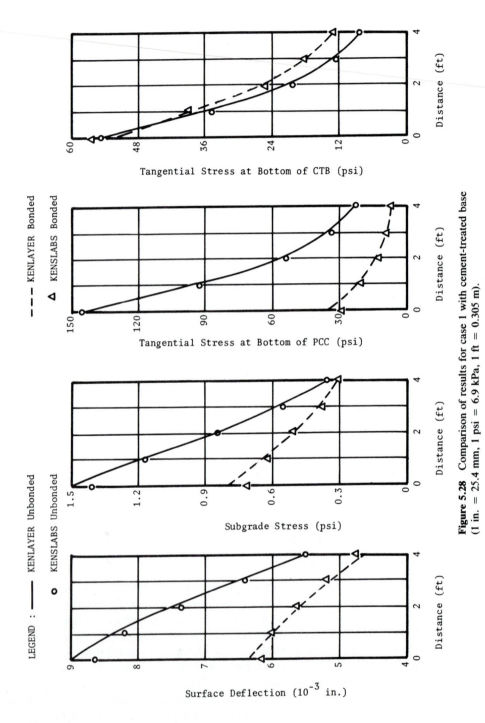

Figure 5.28 Comparison of results for case 1 with cement-treated base (1 in. = 25.4 mm, 1 psi = 6.9 kPa, 1 ft = 0.305 m).

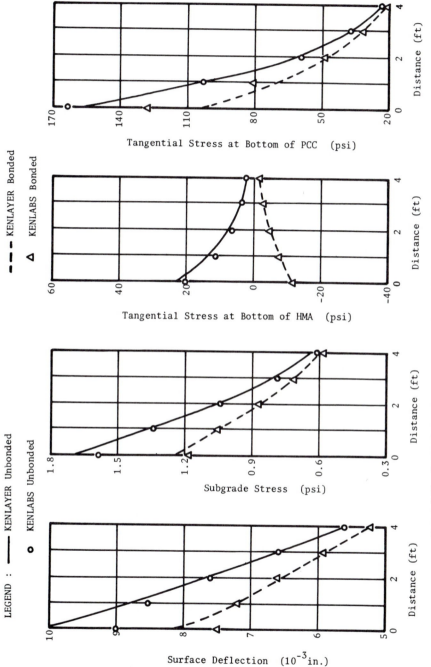

LEGEND : ——— KENLAYER Unbonded

o   KENLABS Unbonded

– – – KENLAYER Bonded

△   KENLABS Bonded

Tangential Stress at Bottom of PCC   (psi)

Tangential Stress at Bottom of HMA   (psi)

Subgrade Stress   (psi)

Surface Deflection   (10⁻³in.)

Figure 5.29 Comparison of results for case 2 with HMA overlay (1 in. = 25.4 mm, 1 psi = 6.9 kPa, 1 ft = 0.305 m).

the bonded layer, while those from KENSLABS are indicated by circles for the unbonded layers and triangles for the bonded layers. When the two layers are bonded, the tangential stresses at the bottom of both layers are in tension because the neutral axis lies in the upper layer.

It can be seen from Figure 5.28 that the results obtained from KENSLABS check very well with those from KENLAYER. Because the layered theory considers the vertical compression of the layers, it is not surprising that KEN-LAYER results in greater surface deflections and subgrade stress directly under the load. The agreement on tangential stresses is better for unbonded layers than for bonded layers. For unbonded layers, the tangential stresses in PCC are much greater than those in CTB, but the reverse is true for the bonded layers.

HMA Overlay

Figure 5.29 compares the results between KENLAYER and KENSLABS for case 2 with HMA overlay. Because HMA has a much lower modulus and cannot really be considered a plate, the discrepancy between KENLAYER and KENSLABS is much greater in case 2 with HMA overlay than in case 1 with cement-treated base. Unfortunately, the greatest discrepancy occurs directly under the load, which is also the most critical location for use in design. When the two layers are bonded, the tangential stress at the bottom of HMA is in compression because the neutral axis lies below the HMA layer. The use of KENSLABS to determine the tensile stress at the bottom of PCC in a composite pavement is conservative and results in a greater stress.

5.4 SENSITIVITY ANALYSIS

With the use of KENSLABS, sensitivity analyses were made to determine the effect of finite element mesh size, dowel distribution, and various design parameters on pavement responses.

5.4.1 Mesh Size

The effect of mesh size is discussed separately for both edge and corner loadings. The two layers of unbonded slabs with aggregate interlock joints, as discussed in Section 5.3.3, were used for analysis. The cross section of the slabs is the same as shown in Figure 5.23, except that a spring constant of 100,000 psi (690 MPa) was used for the transverse joint and 31,570 psi (218 MPa) for the longitudinal joint.

Edge Loading

Because transverse joints have practically no effect on the critical responses under edge loading, a two-slab system consisting of a main slab and a tied shoulder slab can be used to replace the four-slab system shown in Figure 5.24. Three different mesh sizes, designated as fine, medium, and coarse, were used, as shown in Figure 5.30. Because of symmetry, only one-half of the slabs need be considered. The fine mesh is similar to that shown in Figure 5.24, except that some adjustment was made near to the load. This is necessary because the x axis must pass through the centers of loaded areas in order to keep symmetry, whereas no such line was used in Figure 5.24. It is well known that the maximum edge stress occurs under the center, rather than at the edge, of the loaded areas, so the division shown in Figure 5.30 is more reasonable than that in Figure 5.24 and results in a greater edge stress.

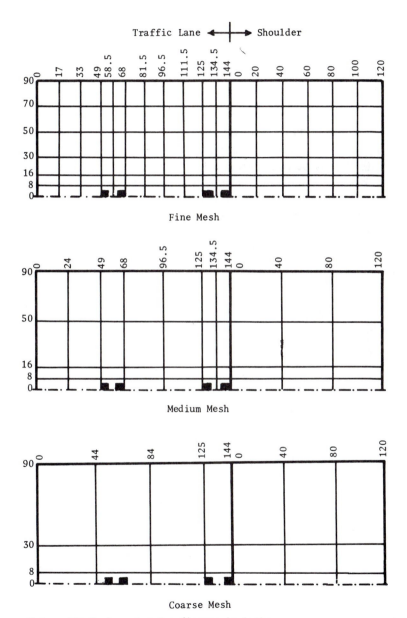

Figure 5.30 Fine, medium, and coarse mesh for analysis of edge loading (1 in. = 25.4 mm).

TABLE 5.9 EFFECT OF MESH SIZE ON RESPONSES UNDER EDGE LOADING

Mesh size	No. of nodes	Minimum longitudinal joint efficiency	Maximum tensile stress (psi)		Maximum subgrade stress (psi)	Maximum deflection (10^{-3} in.)
			Slab	Base		
Fine	133	73	204.1	45.4	1.57	7.83
			(1.00)	(1.00)	(1.00)	(1.00)
Medium	60	73	191.9	42.6	1.43	7.15
			(0.94)	(0.94)	(0.91)	(0.91)
Coarse	36	73	178.5	39.7	1.35	6.75
			(0.87)	(0.87)	(0.86)	(0.86)

Note. 1 in. = 25.4 mm, 1 psi = 6.9 kPa.

Table 5.9 shows the effect of mesh size on the responses under edge loading. Comparisons were made on the minimum longitudinal joint efficiency, the maximum tensile stress at bottom of slab and base, the maximum subgrade stress, and the maximum deflection. The values in parentheses are the ratios with respect to the fine mesh. It can be seen that the coarser the mesh, the smaller the stresses and deflections, so the use of coarse mesh is on the unsafe side. Note that the maximum difference between the coarse mesh with 36 nodes and the fine mesh with 133 nodes is only 14%.

Corner Loading

The fine mesh for corner loading is the same as that shown in Figure 5.25, while the medium and coarse meshes are shown in Figures 5.31 and 5.32. The

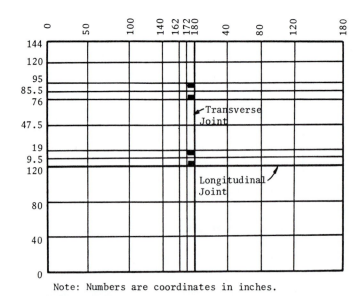

Note: Numbers are coordinates in inches.

Figure 5.31 Medium finite element mesh for corner loading (1 in. = 25.4 mm).

Kenslabs Computer Program   Chap. 5

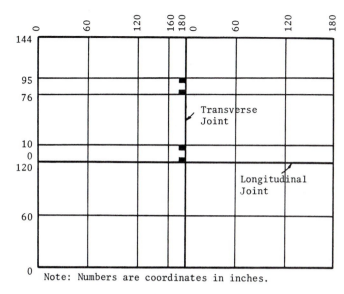

Note: Numbers are coordinates in inches.

Figure 5.32 Coarse finite element mesh for corner loading (1 in. = 25.4 mm).

TABLE 5.10 EFFECT OF MESH SIZE ON RESPONSES UNDER CORNER LOADING

| Mesh size | Number of nodes | Minimum joint efficiency | | Maximum principal tensile stress (psi) | | Maximum subgrade stress (psi) | Maximum deflection (10⁻³ in.) |
		Transverse (%)	Longitudinal (%)	Slab	Base		
Fine	418	85	64	96.3	21.4	2.4	11.8
				(1.00)	(1.00)	(1.00)	(1.00)
Medium	156	83	64	91.2	20.3	2.1	10.6
				(0.95)	(0.95)	(0.89)	(0.89)
Coarse	72	82	64	43.2	9.6	1.8	8.9
				(0.45)	(0.45)	(0.75)	(0.75)

Note. 1 in. = 25.4 mm, 1 psi = 6.9 kPa.

comparisons are presented in Table 5.10. It can be seen that the results obtained by the medium and fine meshes check quite well but those obtained by the coarse mesh are unreasonably small. This is as expected because one of the slabs is divided into six finite elements, which is certainly not enough.

5.4.2 Dowel Distribution

A basic rule for finite element analysis is to use small mesh near the load, or in the area of interest, and large mesh far away from the load. Because it is not economical, and sometimes impossible, to have a node at each dowel, it is assumed that dowels are concentrated at the nodes. For uniformly spaced dowels, the larger the distance between the nodes, the more the dowels will be placed thereon. How this assumption affects the accuracy of the results needs to be investigated. The two cases presented in Examples 4.12 and 4.13 are used for the analysis.

Figure 5.33 shows a two-slab system with a fine mesh so that each dowel location has a node. The dowels are spaced at 12 in. (305 mm) on centers with the first and last at 6 in. (153 mm) from the edge. One or two concentrated loads are applied at the joint, as indicated by the black dots in the figure. There is no load transfer at the two outside nodes because of the absence of dowels. With a total of 168 nodes and a half-band width of 48, the dimension of stiffness matrix required is 24,192.

Figure 5.34 shows a two-slab system with a coarse mesh, which is considered adequate for determining the corner deflection. It is assumed that dowels are distributed to every node, even though there are no dowels at the outside nodes. Because the load near the edge is not applied directly to a node, it must be evenly

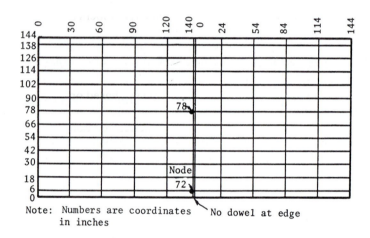

Figure 5.33 Fine mesh with a node at each dowel (1 in. = 25.4 mm).

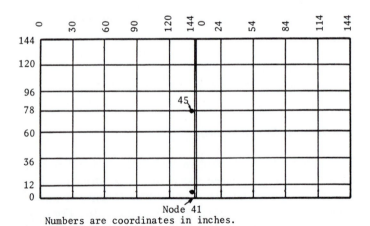

Figure 5.34 Coarse mesh for finite element analysis (1 in. = 25.4 mm).

distributed to the outside nodes. With a total of 96 nodes and a half-band width of 30, the required storage for the stiffness matrix is 8640, which is about one-third of that for the fine mesh.

One Load

In Example 4.12, the bearing stress on the dowel under one concentrated load of 9000 lb (40 kN) was determined by the manual method. Assuming that the load is distributed over a width of 1.8ℓ, the shear force on the most critical dowel is 1376 lb (6.1 kN) and the bearing stress is 3556 psi (24.5 MPa). If the width of load distribution is 1.0ℓ, as suggested by Heinrichs et al. (1989), then the shear force and bearing stress are 2250 lb (10 kN) and 5815 psi (40.1 MPa), respectively. It will be interesting to find out how these results check with the computer solutions.

Table 5.11 shows the shear force and bearing stress on one dowel under a single 9000-lb (40-kN) load. The two nodal numbers for the fine mesh are shown in Figure 5.33 and those for the coarse mesh are shown in Figure 5.34. The results based on the coarse mesh by assuming that the dowels be concentrated at the nodes check very well with those based on the fine mesh with one dowel at each node and no dowel at the edge node. The table also shows that the manually calculated shear forces and bearing stresses based on 1.0ℓ checks reasonably well with the computer solutions, while those based on 1.8ℓ are far too small. The fact that the shear forces on the interior nodes, or nodes 78 and 45, are negligible compared to those on the exterior nodes, or nodes 72 and 41, indicates that the load on the exterior nodes has very little effect on the interior node, which is as expected.

Two Loads

In Example 4.13 the procedures for computing the shear forces on dowels under two 9000-lb (40-kN) loads were illustrated. Using the same joint information as in Example 4.12, except that the 0.75-in. (19-mm) dowels are replaced by 1.25-in. (32-mm) dowels, the bearing stresses on the dowels can also be determined.

TABLE 5.11 COMPARISON OF DOWEL RESPONSES UNDER ONE LOAD

| Procedure | Method | Shear force | | Bearing stress (psi) |
		Nodal no.	lb	
KENSLABS	Fine mesh	72	2417	6242
		78	76	
	Coarse mesh	41	2502	6460
		45	58	
Manual	1.8ℓ		1376	3556
	1.0ℓ		2250	5815

Note. 1 lb = 4.45 N, 1 psi = 6.9 kPa.

TABLE 5.12 COMPARISON OF DOWEL RESPONSES UNDER TWO LOADS

| Procedure | Method | Shear force | | Bearing stress (psi) |
		Nodal no.	lb	
KENSLABS	Fine mesh	72	2685	2763
		78	1246	
	Coarse mesh	41	2756	2835
		45	1207	
Manual	1.8ℓ	Exterior	1190	1225
		Interior	636	
	1.0ℓ	Exterior	1800	1852
		Interior	1125	

Note. 1 lb = 4.45 N, 1 psi = 6.9 kPa.

Table 5.12 shows the shear force and bearing stress on one dowel under two 9000-lb (40-kN) loads. Similar to the case of one load, the results obtained by the coarse mesh also check well with the fine mesh. However, the shear force based on a width of 1.0ℓ checks reasonably with the computer solutions at the interior node, but is much smaller than the computer solutions at the exterior node. For pavements with a larger ℓ, as symbolized by a strong slab on a weak foundation, it appears that the width of load distribution for edge loading should be even smaller than 1.0ℓ.

Joint Deflections

The function of dowels is to transfer loads from a loaded slab to an unloaded slab. The efficiency of load transfer is indicated by the difference in deflections at the two opposite sides of the joint. Figure 5.35 shows a comparison of deflections on both sides of the joint for the above two cases. The deflections obtained from the coarse mesh also check very well with those obtained from the fine mesh. It can therefore be concluded that the common practice of distributing the dowels to the nodes is accurate enough for engineering applications.

5.4.3 Effect of Some Design Parameters

KENSLABS was used to determine the effect of various design parameters on pavement responses. The responses compared are the edge stress and the corner deflection, both of which have been used by PCA (1984) for the design of concrete pavements. The most critical edge stress occurs when the axle load is applied at the midspan of the slab far away from the joints so that, without tied concrete shoulders, only one slab is needed for use in the analysis. The most critical corner deflection occurs when the axle load is applied at the joint, so that a two-slab system with shear transfer across the joint is needed. The slab is assumed to be 12 ft (3.66 m) wide. To show the effect of panel length on stress and deflection, two different lengths, one 15 ft (4.57 m) and the other 23 ft (7.0 m), were used.

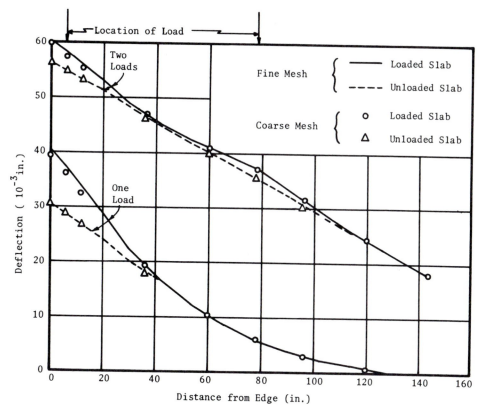

Figure 5.35 Deflections on two opposite sides of joint under one and two loads (1 in. = 25.4 mm).

Edge Stress

Figure 5.36 shows the division of slab into rectangular finite elements. Due to symmetry, only one-half of the slab need be considered. The standard case consists of an 18-kip (80-kN) single-axle load with a dual spacing of 13.5 in. (343 mm), a wheel spacing of 77 in. (1.96 m), and a contact pressure of 75 psi (517 kPa). The tire imprint is 9.3 × 6.5 in. (236 × 165 mm) and, because of symmetry, only one-half of the length is shown, as indicated by the shaded rectangles in the figure. The slab is 8 in. (203 mm) thick with a Young's modulus of 4 × 10⁶ psi (27.6 GPa) and a Poisson ratio of 0.15. The foundation has a modulus of subgrade reaction of 100 pci (27.2 MN/m³) and there is no temperature curling.

In addition to the standard case, eight more cases, each with only one parameter different from the standard case, were also analyzed. The results are presented in Table 5.13. In the table, the parameter ratio is the value of the parameter in the given case divided by that in the standard case. The stress is the edge stress directly under the axle load and the stress ratio is the ratio of the

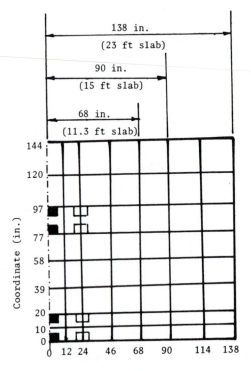

138 in.
(23 ft slab)

90 in.
(15 ft slab)

68 in.
(11.3 ft slab)

Coordinate (in.)

144
120
97
77
58
39
20
10
0

0 12 24   46   68   90   114   138

Figure 5.36 Analysis of edge stress by KENSLABS (1 in. = 25.4 mm, 1 ft = 0.305 m).

TABLE 5.13 SENSITIVITY ANALYSIS OF EDGE STRESS

Case	Parameter ratio	15-ft slab		23-ft slab	
		Stress (psi)	Stress ratio	Stress (psi)	Stress ratio
Standard case	1.00	326.10	1.00	320.79	1.00
1. Slab thickness, 10 in.	1.25	233.13	0.72	231.26	0.72
2. Concrete modulus, 8×10^6 psi	2.00	365.53	1.12	363.69	1.13
3. Subgrade k, 300 pci	3.00	265.21	0.82	264.65	0.83
4. Warpup $\Delta t = 12°F$		321.89	0.99	322.02	1.00
5. Warpdown $\Delta t = -24°F$		354.56	1.09	330.63	1.03
6. Tandem		284.98	0.87	282.34	0.88
7. Solid foundation, 8-in. slab $E_f = 4000$ psi, $v_f = 0.35$		329.46	1.01	325.31	1.01
8. Solid foundation, 10-in. slab $E_f = 4720$ psi, $v_f = 0.35$		237.51	0.73	234.67	0.73

Note. 1 psi = 6.9 kPa, 1 in. = 25.4 mm, 1 pci = 271.3 kN/m³, 1 ft = 0.305 m.

stresses between the case in question and the standard case. It can be seen that in all cases but case 4, the edge stress in the 23-ft (7.0-m) slab is smaller than that in the 15-ft (4.57-m) slab. A separate run of the standard case for a slab 11.3 ft (3.45 m) long results in an edge stress of 316.77 psi (2.19 MPa), indicating that the 15-ft (4.57-m) slab is the most critical with the largest edge stress.

Comments on Table 5.13

1. Increasing the slab thickness from 8 in. (203 mm) to 10 in. (254 mm) decreases the edge stress by 28%, indicating that thickness is very effective in reducing the edge stress.

2. Increasing the Young's modulus of concrete from 4×10^6 psi (27.6 GPa) to 8×10^6 psi (55.2 GPa) increases the edge stress by only 12 to 13%. Because the increase in stress is much smaller than the increase in strength resulting from the higher modulus, the use of high-quality concrete for the slab has a beneficial effect.

3. Increasing the modulus of subgrade reaction from 100 pci (27.2 MN/m³) to 300 pci (81.5 MN/m³) decreases the edge stress by only 17 to 18%. It is uneconomical to improve the subgrade modulus just for the purpose of reducing the edge stress. The stress can be reduced more effectively by simply increasing the slab thickness.

4. When the temperature at the bottom of slab is 12°F (6.7°C) higher than that at the top, the slab curls up, causing the loss of subgrade contact at the slab edge. However, this loss of subgrade contact has a significant effect on the stresses due to loading in the transverse direction, but little effect on the edge stress in the longitudinal direction. In fact, the stress in the 15-ft (4.57-m) slab actually decreases due to the transfer of subgrade reaction from the edge to the interior of the slab.

5. When the temperature on the top of the slab is 24°F (13.3°F) higher than that at the bottom, the slab curls down, causing the loss of subgrade contact at the interior of the slab. The edge stress due to loading increases about 9% in the 15-ft (4.57-m) slab and 3% in the 23-ft (7.0-m) slab. This increase is caused by the transfer of subgrade reaction from the interior to the edge of slab, thus resulting in greater edge stress in the longitudinal direction.

6. When the slab is subjected to a 36-kip (160-kN) tandem-axle load, as indicated by the unshaded rectangles in Figure 5.36, the maximum edge stress occurs under the tandem axles. Due to the negative moment caused by the other axle, the edge stress due to tandem axles is smaller than that due to a single axle.

7. The standard case of liquid foundation is equivalent to a solid foundation with an elastic modulus of 4000 psi (27.6 MPa) and a Poisson ratio of 0.35, as indicated by Eq. 5.6. The edge stress obtained by the liquid foundation checks well with that by the solid foundation.

8. When the thickness of slab is increased by 10 in. (254 mm), the elastic modulus of solid foundation should be 4720 psi (32.6 MPa) according to Eq. 5.6. The edge stress based on the solid foundation also agrees reasonably well with case 1 based on the liquid foundation.

Corner Deflection

Figure 5.37 shows the finite element mesh for a two-slab system under an 18-kip (80-kN) single-axle load applied at the joint, as indicated by the shaded rectangles. The standard case is the same as that for the analysis of edge stress with the following information on the doweled joint: dowel spacing, 12 in.

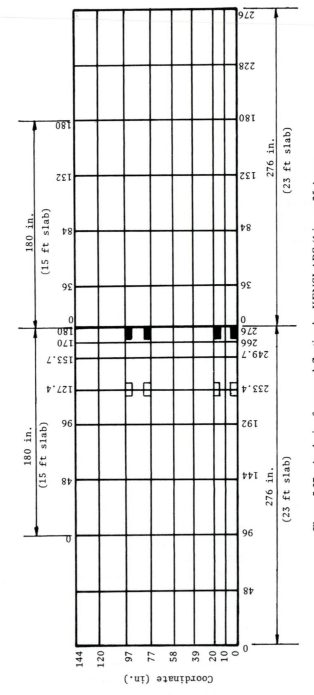

Figure 5.37 Analysis of corner deflection by KENSLABS (1 in. = 25.4 mm, 1 ft = 0.305 m).

TABLE 5.14 SENSITIVITY ANALYSIS OF CORNER DEFLECTION

Case	Parameter ratio	Deflection (in.)	Deflection ratio
Standard case	1.00	0.0353	1.00
1. Slab thickness, 10 in.	1.25	0.0291	0.82
2. Concrete modulus, 8×10^6 psi	2.00	0.0289	0.82
3. Subgrade $k = 300$ pci	3.00	0.0177	0.50
4. Dowel diameter, 1.5 in.	1.50	0.0343	0.97
5. Modulus of dowel support, 1.5×10^5 pci	0.10	0.0413	1.17
6. Gap between dowel and concrete, 0.01 in.		0.0404	1.14
7. Tandem		0.0458	1.30
8. Warpup $\Delta t = 12°F$ (15-ft slab)		0.0391	1.11
9. Warpup $\Delta t = 12°F$ (23-ft slab)		0.0391	1.11
10. Warpdown $\Delta t = -24°F$ (15-ft slab)		0.0366	1.04
11. Warpdown $\Delta t = -24°F$ (23-ft slab)		0.0361	1.02
12. Solid foundation, 8-in. slab, $E_f = 6400$ psi, $\nu_f = 0.35$		0.0362	1.03
13. Solid foundation, 10-in. slab, $E_f = 7560$ psi, $\nu_f = 0.35$		0.0288	0.82

Note. 1 in. = 25.4 mm, 1 psi = 6.9 kPa, 1 pci = 271.3 kN/m³, 1 ft = 0.305 m.

(305 mm); dowel diameter, 1.0 in. (25.4 mm); modulus of dowel support, 1.5×10^6 pci (407 GN/m³); and joint width, 0.125 in. (3.18 mm). In addition to the standard case, thirteen more cases were analyzed, with the results tabulated in Table 5.14. Unless noted otherwise, the results are based on the 15-ft (4.6-m) slabs. It was found that the same results were obtained for the 23-ft (7.0-m) slabs, indicating that the length of slab has practically no effect on corner deflections.

Comments on Table 5.14

1. Increasing the slab thickness by 25% decreases the corner deflection by only 18%, indicating that thickness is not as effective in reducing the corner deflection as in reducing the edge stress.

2. Increasing the Young's modulus of concrete decreases the corner deflection, which is in contrast to the increase in edge stress.

3. A threefold increase in the modulus of subgrade reaction reduces the corner deflection by 50%. Compared to the 18% reduction in edge stress, the improvement of subgrade is more effective in reducing the corner deflection.

4. Increasing the dowel diameter by 50% decreases the corner deflection by only 3%. However, large dowels reduce the bearing stress between dowel and concrete and prevent the faulting of joints.

5. Reducing the modulus of dowel support by one order of magnitude from 1.5×10^6 to 1.5×10^5 pci (407 to 40.7 GN/m³) increases the corner deflection by 17%.

6. The looseness of dowels with a gap of 0.01 in. (0.25 mm) between dowel and concrete increases the corner deflection by 14%.

7. The application of an additional 18-kip (80-kN) single-axle load, as indicated by the unshaded rectangles in Figure 5.37, to form a 36-kip (160-kN) tandem-axle load increases the corner deflection by 30%. The fact that the tandem-axle load decreases the edge stress by 13% but increases the corner deflection by 30% can well explain why single-axle loads are usually more critical in fatigue analysis and tandem-axle loads are more critical in erosion analysis (PCA, 1984).

8. The upward curling of the 15-ft (4.6-mm) slab by a temperature differential of 12°F (6.7°C) increases the corner deflection by 11%. The deflection is measured from the curled position. The actual deflection of the subgrade is only 0.0241 in. (0.61 mm) due to the existence of a gap of 0.0150 in. (0.38 mm) prior to loading.

9. It was originally suspected that longer slabs would have more curling at the corner and thus result in greater corner deflections due to loading. However, a comparison between cases 8 and 9 indicates that this is not true. Although the initial curling of a weightless slab is greater for longer slabs, the final curling is nearly the same due to the greater weight of longer slabs. Consequently, when the slabs curl up, the corner deflection is not affected by the length of slab.

10. When the slabs curl down under a temperature differential of 24°F (13.3°C), the corner deflection of the 15-ft (4.6-m) slab increases by 4%. This increase is caused by the loss of subgrade contact in the interior of the slab. Without the subgrade reaction to pull the slab down and counteract the corner loading, the deflection at the corner will increase.

11. Under the same temperature gradient as in case 10 for the 15-ft (4.6-m) slab, the corner deflection of the 23-ft (7.0-m) slab increases by only 2%. The smaller increase in longer slabs is reasonable because the interior of longer slabs is farther from the corner and the loss of subgrade contact in the interior should have only a small effect on the corner deflection.

12. The standard case of liquid foundation is equivalent to a solid foundation with an elastic modulus of 6400 psi (44.2 MPa) and a Poisson ratio of 0.35, as indicated by Eq. 5.7. The corner deflection obtained by the liquid foundation checks reasonably with that by the solid foundation.

13. Based on Eq. 5.7, the elastic modulus of solid foundation should be 7560 psi (52.2 MPa) for a 10-in. (254-mm) slab. The corner deflection based on the solid foundation also checks really well with that in case 1 based on the liquid foundation.

5.5 SUMMARY

This chapter presents the finite element method for analyzing concrete slabs on liquid, solid, or layer foundations and describes the KENSLABS computer program. More details about KENSLABS can be found in Appendix C.

Important Points Discussed in Chapter 5

1. KENSLABS can be applied to three different types of foundation: liquid, solid, or layer. The liquid foundation, even though very unrealistic, has been used most frequently because of its simplicity and because it requires less computer time and storage. With the availability of personal computers, it is no longer necessary to use liquid foundations. The more realistic solid foundations should be used if the slabs are placed directly on the subgrade. If the slabs are placed on one or more layers of granular materials, the use of layer foundations is recommended.

2. The modulus of subgrade reaction k used in liquid foundations is a fictitious property not characteristic of soil behaviors. It is not possible to find a single k value that will match both the stresses and deflections in slabs on liquid foundations with those on solid foundations. However, if the design is based on the maximum stress or deflection under a single-axle load, it is possible to replace the solid foundation with a fictitious k using Eq. 5.5 for interior stress, Eq. 5.6 for edge stress, and Eq. 5.7 for corner deflection, thus resulting in a large saving on computer time and storage.

3. The overall stiffness matrix is obtained by adding the stiffness matrices of slabs, foundation, and joints. The resulting simultaneous equations are solved by the Gauss elimination method. If the dimension of the overall stiffness matrix exceeds 70,000, an iterative method is activated automatically to solve the simultaneous equations.

4. The stiffness of joint is computed from a shear spring constant and a moment spring constant. The moment spring constant should be specified as 0 if there is visible opening at the joint. The shear spring constant can be entered directly or computed from the size and spacing of dowel bars and other joint information. A gap may be specified between the spring or dowel and the concrete.

5. The stresses and deflections due to temperature curling can be obtained by assuming that the slab is initially curled into a spherical surface and then subjected to the subgrade reactions at the nodes.

6. The finite element method is particularly suited for analyzing slabs not in full contact with the subgrade by simply setting the reactive forces at those nodes not in contact to zero. Whether a node is in contact with the subgrade or not is based on the deflection of the slab compared with the gap or precompression of the subgrade. However, for solid or layer foundations if the slab and subgrade are in contact, one more requirement is that the reactive force should not be in tension.

7. The analysis of partial contact under wheel loads by KENSLABS involves two steps. First, determine the gaps and precompressions between slab and foundation due to weight, temperature curling, and gaps, if any. Then, using the gaps and precompressions thus computed, determine the stresses and deflections under the wheel loads.

8. KENSLABS can be applied to a maximum of 9 slabs, 12 joints, and 420 nodes. If the slabs and loadings exhibit symmetry, only one-half or one-quarter of the slab need be considered. Thus, a large amount of computer time can be saved.

9. Results obtained by KENSLABS compare favorably with those from analytical solutions, influence charts, and the ILLI-SLAB model. When the load is applied in the interior of a slab, good agreement also exists between KENSLABS and KENLAYER solutions.

10. The size of finite element mesh has a significant effect on the results obtained. Therefore, the selection of an appropriate mesh requires careful consideration. It is really not necessary to use a very fine mesh. Satisfactory results can be obtained if a fine mesh is used near the load and a rather coarse mesh away from the load.

11. For uniformly spaced dowels, it is reasonable to assume that they be distributed to the nodes in proportion to the nodal spacing. To compute the shear force in the most critical dowel, the assumption that the load be distributed over a width of 1.0ℓ checks more favorably with KENSLABS than the previous assumption of 1.8ℓ.

12. Sensitivity analyses were made to determine the effect of various parameters on the edge stress and corner deflections for slabs on liquid foundations. An interesting finding is that the upward curling of slab has only a small effect on the edge stress due to loading, but has a significant effect on the corner deflection, while downward curling causes a slight increase in both edge stress and corner deflection.

PROBLEMS

Use KENSLABS to solve problems 5-3 to 5-16. The following material properties are assumed: modulus of elasticity of concrete $= 4 \times 10^6$ psi, Poisson ratio of concrete $= 0.15$, modulus of elasticity of steel $= 29 \times 10^6$ psi, Poisson ratio of steel $= 0.3$, modulus of dowel support $= 1.5 \times 10^6$ pci, and coefficient of thermal expansion of concrete $= 5 \times 10^{-6}$ in./in./°F.

5-1. A 7-in. concrete pavement with an elastic modulus of 4×10^6 psi is subjected to a wheel load with a bending stress of 400 psi, as shown in Figure P5.1a. What is the bending moment per in. width of the slab? Assuming that the moment remains the same after placing a 3-in. hot mix asphalt with an elastic modulus of 7×10^5 psi as an overlay, as shown in Figure P5.1b, what will be the maximum tensile stress in the

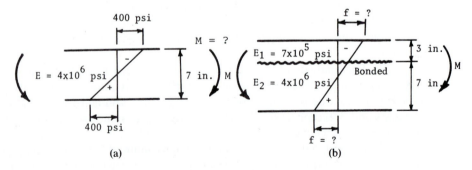

Figure P5.1

concrete and the maximum compressive stress in the hot mix asphalt when both layers are bonded? [Answer: 3267 in.-lb/in., 305 psi, and 85.4 psi].

5-2. Same as Problem 5-1 except that both layers are unbonded, as shown in Figure P5.2, and the Poisson ratios are 0.4 for the hot mix asphalt and 0.15 for the concrete. What will be the maximum stress in the concrete and in the hot mix asphalt? [Answer: 394 and 34 psi].

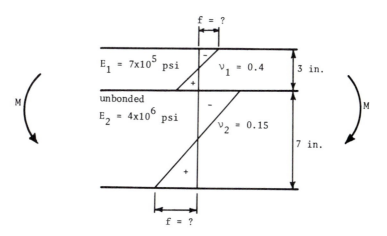

Figure P5.2

5-3. Determine the curling stresses during the day at points *A*, *B*, and *C* of a finite slab shown in Figure P5.3*a* using one-quarter of the slab. The slab has a thickness of 8 in. and is subjected to a temperature gradient of 3°F per inch of slab. The modulus of subgrade reaction is 50 pci. [Answer: 214.2, 207.4, and 57.8 psi based on the mesh shown in Figure P5.3*b*, which checks with Problem 4-1b]

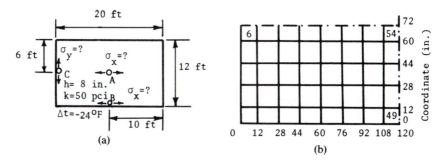

Figure P5.3

5-4. Figure P5.4*a* shows a concrete slab, 10 in. thick, supported by a subgrade with a modulus of subgrade reaction of 200 pci. A 12,000-lb dual-wheel load (each wheel 6000 lb) spaced at 14 in. on centers is applied at the corner of the slab. The contact pressure is 80 psi. Determine the maximum stress in the concrete. [Answer: 174.6 psi at node 20 based on the mesh shown in Figure P5.4*b*, which checks with Problem 4-2]

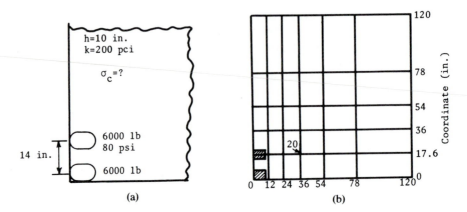

(a)   (b)

Figure P5.4

5-5. The pavement and loading are the same as in Problem 5-4, but the load is applied in the interior of an infinite slab, as shown in Figure P5.5a. Determine the maximum stress in the slab using one-quarter of the slab. [Answer: 128.3 psi at node 2 based on the mesh shown in Figure P5.5b, which is slightly smaller than the 139.7 psi in Problem 4-3 based on equivalent area]

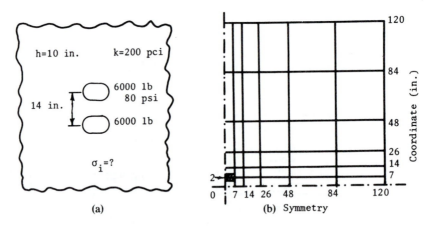

(a)   (b) Symmetry

Figure P5.5

5-6. Same as Problem 5-4 except that the load is applied at the slab edge, as shown in Figure P5.6a. Determine the maximum stress using one-half of the slab. [Answer: 251.5 psi at node 1 based on the mesh shown in Figure P5.6b, which checks with problem 4-4]

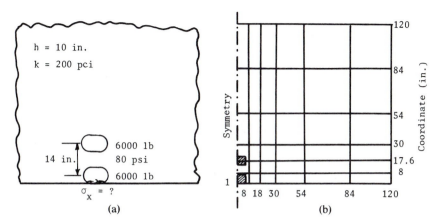

Figure P5.6

5-7. Figure P5.7*a* shows a set of dual-tandem wheels with a total weight of 40,000 lb (10,000 per wheel), a tire pressure of 100 psi, a dual spacing of 20 in., and a tandem spacing of 40 in. The concrete slab is 8 in. thick with a modulus of subgrade reaction of 100 pci. Determine the interior stress at point *A* in the *y* direction under the center of the dual-tandem wheels using one-quarter of the slab. [Answer: 299.3 psi in the *y* direction based on the mesh shown in Figure P5.7*b*, which checks with Problem 4-6]

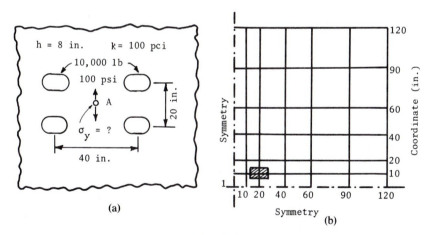

Figure P5.7

5-8. Same as Problem 5-7 except that the outside wheels are applied at the slab edge, as shown in Figure P5.8*a*. Determine the edge stress at point *A* in an infinite slab using one-half of the slab. [Answer: 314.7 psi based on the mesh shown in Figure P5.8*b*, which checks with Problem 4-7]

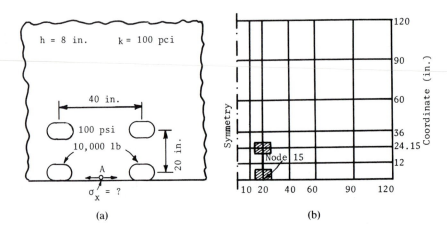

Figure P5.8

5-9. A concrete slab, 40 ft long, 11 ft wide, and 9 in. thick, is placed on a subgrade having a modulus of subgrade reaction of 200 pci. The pavement is subject to a temperature differential of 1.5°F per in. at night when a 9000-lb single-axle load is applied on the edge of the slab over a circular area with a contact pressure of 100 psi, as shown in Figure P5.9a. Determine the combined stress due to curling and loading at the edge beneath the load using one-half of the slab. [Answer: 158.3 psi based on the mesh shown in Figure P5.9b, which checks with Problem 4-8c]

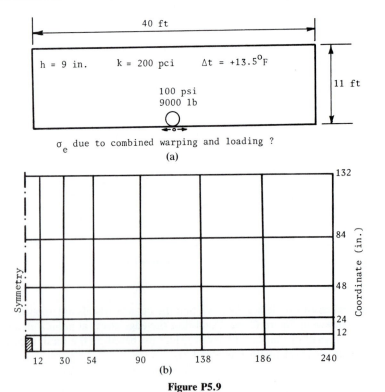

Figure P5.9

5-10. A concrete slab has a width of 12 ft, a thickness of 10 in., and a modulus of subgrade reaction of 300 pci. Two 1-ft × 1-ft loaded areas, each weighing 12,000 lb and spaced at 6 ft apart, are applied at the joint with the outside loaded area adjacent to the pavement edge, as shown in Figure P5.10a. Determine the maximum bearing stress between concrete and dowel. The joint has an opening of 0.25 in. and the dowels are 1 in. in diameter and 12 in. on centers. [Answer: 4530 psi based on the mesh shown in Figure P5.10b]

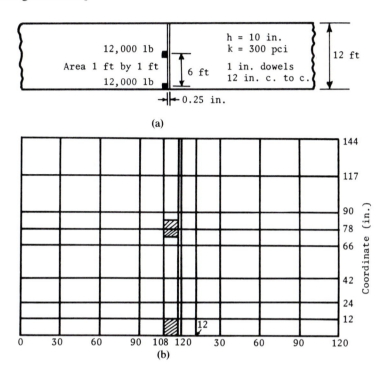

(a)

(b)

Figure P5.10

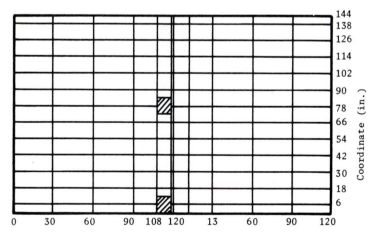

Figure P5.11

5-11. Same as Problem 5-10, but using a mesh such that the location of the nodes are in line with the dowels, as shown in Figure P5.11. [Answer: 4043 psi]

5-12. Same as Problem 5-10 except that a gap of 0.01 in. exists between dowel and concrete. [Answer: 3828 psi]

5-13. Same as Problem 5-10 except that each of the 12,000-lb wheels is considered as a concentrated load and placed at a distance of 2.5 ft from edge, as shown in Figure P5.13a. Determine the maximum bearing stress between concrete and dowel using one-half of the slabs. [Answer: 2267 psi based on the mesh shown in Figure P5.13b]

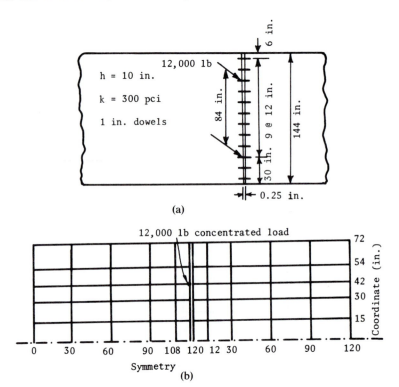

(a)

(b)

Figure P5.13

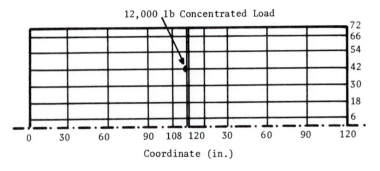

Coordinate (in.)

Figure P5.14

Kenslabs Computer Program   Chap. 5

5-14. Same as Problem 5-13, but using a mesh such that the location of the nodes is in line with the dowels, as shown in Figure P5.14. [Answer: 2284 psi]

5-15. Same as Problem 5-13 except that a gap of 0.01 in. exists between dowel and concrete. [Answer: 1656 psi]

5-16. A slab track with an 8-in. continuous reinforced concrete is placed on a subgrade with a k value of 150 pci. The rail is supported on 7-in. × 12-in. pads. The design load consists of four wheels, each weighing 12,600 lb. By considering one-quarter of the slab, as shown in Figure P5.16, determine the maximum stress in the concrete. [Answer: 283.8 psi]

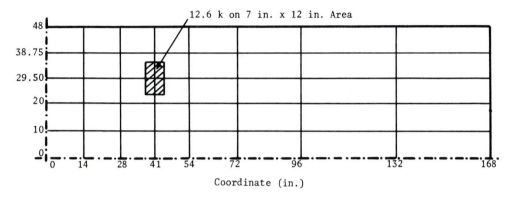

Figure P5.16

6

Traffic Loading and Volume

6.1 DESIGN PROCEDURES

Traffic is the most important factor in pavement design. The consideration of traffic should include both the loading magnitude and configuration and the number of load repetitions. There are three different procedures for considering vehicular and traffic effects in pavement design: fixed traffic, fixed vehicle, and variable traffic and vehicle.

6.1.1 Fixed Traffic

In fixed traffic, the thickness of pavement is governed by a single-wheel load and the number of load repetitions is not considered as a variable. If the pavement is subjected to multiple wheels, they must be converted to an equivalent single-wheel load (ESWL) so the design method based on a single wheel can be applied. This method has been used most frequently for airport pavements or for highway pavements with heavy wheel loads but light traffic volume. Usually the heaviest wheel load anticipated is used for design purposes. Although this method is rarely in use today for pavement design, the concept of converting multiple-wheel loads to a single-wheel load is important and is discussed in Section 6.2.

6.1.2 Fixed Vehicle

In the fixed vehicle procedure, the thickness of pavement is governed by the number of repetitions of a standard vehicle or axle load, usually the 18-kip (80-kN) single-axle load. If the axle load is not 18 kip (80 kN) or consists of tandem or tridem axles, it must be converted to an 18-kip single-axle load by an equivalent axle load factor (EALF). The number of repetitions under each single- or multiple-axle load must be multiplied by its EALF to obtain the equivalent effect based on an 18-kip (80-kN) single-axle load. A summation of the equivalent effects of all axle loads during the design period results in an equivalent single-axle load (ESAL), which is the single traffic parameter for design purposes. Due to the great varieties of axle loads and traffic volumes and their intractable effects on pavement performance, most of the design methods in use today are based on the fixed vehicle concept. Details about the equivalent axle load factor are discussed in Section 6.3.

6.1.3 Variable Traffic and Vehicle

In the variable traffic and vehicle procedure, both traffic and vehicle are considered individually, so there is no need to assign an equivalent factor for each axle load. The loads can be divided into a number of groups and the stresses, strains, and deflections under each load group can be determined separately and used for design purposes. This procedure is most suited for the mechanistic methods of design, wherein the responses of pavement under different loads can be evaluated by using a computer. The method has long been employed by the PCA with the use of design charts, and is discussed in Section 12.2.

6.2 EQUIVALENT SINGLE-WHEEL LOAD

The study of ESWL for dual wheels was first initiated during World War II when the B-29 bombers were introduced into combat missions. Because the design criteria for flexible airport pavements then available were based on single-wheel loads, the advent of these dual-wheel planes required the development of new criteria for this type of loading. Neither time nor economic considerations permitted the direct development of such criteria. It was necessary, therefore, to relate theoretically the new loading to an equivalent single-wheel load, so that the established criteria based on single-wheel load could be applied.

The ESWL obtained from any theory depends on the criterion selected to compare the single-wheel load with multiple-wheel loads. Based on Burmister's two-layer theory, Huang (1969c) conducted a theoretical study on the effect of various factors on ESWL by assuming that single and dual wheels have the same contact pressure. Similar studies were made by Gerrard and Harrison (1970) on single, dual, and dual-tandem wheels by assuming that all wheels have equal contact radii. It was found that the use of different criteria, based on stress, strain, or deflection, plays an important role in determining ESWL and, regardless of the criterion selected, the ESWL increases as the pavement thickness and modulus ratio increase or the multiple-wheel spacing decreases. In this section, some simple theoretical methods for determining ESWL of flexible pavements are presented. The determination of ESWL for rigid pavements can be made by comparing the critical flexural stress in the concrete and is not presented here.

The ESWL can be determined from theoretically calculated or experimentally measured stress, strain, or deflection. It can also be determined from pavement distress and performance such as the large-scale WASHO and AASHO road tests (HRB, 1955, 1962). Any theoretical method can be used only as a guide and should be verified by performance. This is particularly true for the empirical methods of design in which the ESWL analysis is an integral part of the overall design procedure. Erroneous results may be obtained if different ESWL methods are transposed for a given set of design curves.

6.2.1 Equal Vertical Stress Criterion

Based on a theoretical consideration of the vertical stress in an elastic half-space, Boyd and Foster (1950) presented a semirational method for determining ESWL, which had been used by the Corps of Engineers to produce dual-wheel design

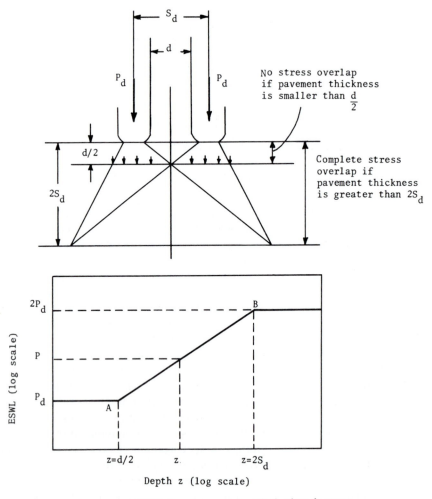

Figure 6.1 ESWL based on equal vertical subgrade stress.

criteria from single-wheel criteria. The method assumes that the ESWL varies with the pavement thickness, as shown in Figure 6.1. For thicknesses smaller than half the clearance between dual tires, the ESWL is equal to one-half the total load, indicating that the subgrade vertical stresses caused by the two wheels do not overlap. For thicknesses greater than twice the center to center spacing of tires, the ESWL is equal to the total load, indicating that the subgrade stresses due to the two wheels overlap completely. By assuming a straight-line relationship between pavement thickness and wheel load on logarithmic scales, the ESWL for any intermediate thicknesses can be easily determined. After the ESWL for dual wheels is found, the procedure can be applied to tandem wheels.

Instead of plotting, it is more convenient to compute the ESWL by

$$\log(\text{ESWL}) = \log P_d + \frac{0.301 \, \log(2z/d)}{\log(4S_d/d)} \tag{6.1}$$

in which P_d is the load on one of the dual tires, z is the pavement thickness, d is the clearance between dual tires, and S_d is the center to center spacing between dual tires.

Example 6.1:

A set of dual tires has a total load $2P_d$ of 9000 lb (40 kN), a contact radius a of 4.5 in. (114 mm), and a center to center tire spacing S_d of 13.5 in. (343 mm), as shown in Figure 6.2a. Determine the ESWL by Boyd and Foster's method for a 13.5-in. (343-mm) pavement.

Solution: Given $S_d = 13.5$ in. (343 mm) and $a = 4.5$ in. (114 mm), the clearance between the duals $d = 13.5 - 9.0 = 4.5$ in. (114 mm). When the thickness of pavement is equal to $d/2$, or 2.25 in. (57 mm), ESWL $= P_d = 9000/2 = 4500$ lb (20 kN). When the thickness is equal to $2S_d$ or 27 in. (689 mm), ESWL $= 2P_d = 9000$ lb (40 kN). After plotting thickness versus ESWL in Figure 6.2b, the ESWL for a 13.5 in. (343 mm) pavement is 7400 lb (32.9 kN). The ESWL can also be determined from Eq. 6.1 or $\log(\text{ESWL}) = \log 4500 + 0.301 \log(2 \times 13.5/4.5)/\log (4 \times 13.5/4.5) = 3.87$, or ESWL $= 7410$ lb (33.0 kN).

The vertical stress factor σ_z/q presented in Figure 2.2 can be used to determine the theoretical ESWL based on Boussinesq's theory. Figure 6.3 shows a pavement of thickness z under single and dual wheels that have the same contact radius a. The maximum subgrade stress under a single wheel occurs at point A with a stress factor of σ_z/q_s, where q_s is the contact pressure under a single wheel. The location of the maximum stress under dual wheels is not known and can be determined by comparing the stresses at three points: point 1 under the center of one tire, point 3 at the center between two tires, and point 2 midway between points 1 and 3. The stress factor at each point is obtained by superposition of the two wheels and the maximum stress factor σ_z/q_d is found, where q_d is the contact pressure under dual wheels. To obtain the same stress

$$q_s\left(\frac{\sigma_z}{q_s}\right) = q_d\left(\frac{\sigma_z}{q_d}\right) \tag{6.2}$$

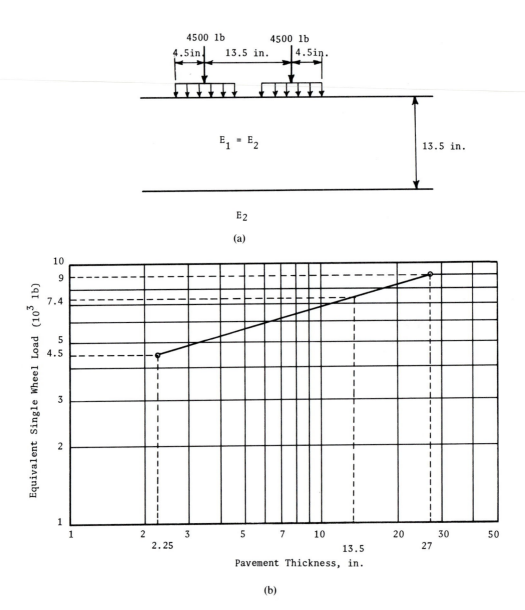

Figure 6.2 Example 6.1 (1 in. = 25.4 mm, 1 lb = 4.45 N).

For the same contact radius, contact pressure is proportioned to wheel load, or

$$\frac{P_s}{P_d} = \frac{\sigma_z/q_d}{\sigma_z/q_s} \tag{6.3}$$

in which P_s is the single-wheel load, which is the ESWL to be determined, and P_d is the load on each of the duals.

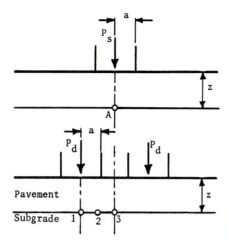

Figure 6.3 Location of maximum vertical stress or deflection on subgrade.

Example 6.2:

Same as Example 6.1 and Figure 6.2a. Determine the ESWL by Boussinesq's theory.

Solution: With $z/a = 13.5/4.5 = 3$, the stress factors can be obtained from Figure 2.2 and the results for dual wheels are presented in Table 6.1. It can be seen that the stresses at the three points are nearly the same with a maximum stress factor of 0.179 at point 2, which is slightly greater than the minimum of 0.173 at point 1. For a single wheel, the stress factor at point A is 0.143. From Eq. 6.3, ESWL $= P_s = 0.179/0.143 \times 4500 = 5630$ lb (25.1 kN), which differs significantly from the 7410 lb (32.9 kN) by Boyd and Foster's method. Although the original concept conceived by Boyd and Foster was based on the vertical subgrade stress, the method is more or less empirical in nature and, depending on the pavement thickness and the loading configuration, may not check well with the theoretical method based on Boussinesq stress distribution. However, the method gives an ESWL greater than the theoretical value and is therefore on the safe side.

TABLE 6.1 STRESS FACTORS UNDER DUAL WHEELS

Point no.	Left wheel		Right wheel		Sum
	r/a	σ_z/q_s	r/a	σ_z/q_s	σ_z/q_d
1	0	0.143	3	0.030	0.173
2	0.75	0.126	2.25	0.053	0.179
3	1.5	0.088	1.5	0.088	0.176

6.2.2 Equal Vertical Deflection Criterion

After the application of Boyd and Foster's method and the subsequent completion of accelerated traffic tests, it was found that the design method was not very safe and an improved method was developed by Foster and Ahlvin (1958). In this method, the pavement system is considered as a homogeneous half-space and the vertical deflections at a depth equal to the thickness of the pavement can be obtained from Boussinesq solutions. A single-wheel load that has the same contact radius as one of the dual wheels and results in a maximum deflection equal to that caused by the dual wheels is the ESWL.

The vertical deflection factor F presented in Figure 2.6 can be used to determine ESWL. Similar to the case of vertical stress, the deflection factors F_s at point A under the single wheel and F_d at points 1, 2, and 3 under the duals, as shown in Figure 6.3, are determined. The deflection can then be expressed as

$$w_s = \frac{q_s a}{E} F_s \qquad (6.4a)$$

$$w_d = \frac{q_d a}{E} F_d \qquad (6.4b)$$

in which the subscript s indicates single wheel and d indicates dual wheels. The deflection factor F_d is obtained by superposition of the duals. To obtain the same deflection, $w_s = w_d$, or

$$q_s F_s = q_d F_d \qquad (6.5)$$

For the same contact radius, contact pressure is proportional to wheel load:

$$\text{ESWL} = P_s = \frac{F_d}{F_s} P_d \qquad (6.6)$$

Example 6.3:

Same as Example 6.1 and Figure 6.2a. Determine the ESWL by Foster and Ahlvin's method.

Solution: The chart shown in Figure 2.6 can be used to determine vertical deflections. The deflection factors F at the three points shown in Figure 6.3 are calculated and presented in Table 6.2. The maximum deflection due to dual wheels occurs at point 3 with a deflection factor of 0.78. The maximum deflection due to a single wheel occurs under the center of the tire with a deflection factor of 0.478. From Eq. 6.6, ESWL = 0.78/0.478 × 4500 = 7340 lb (32.7 kN), which checks with the 7410 lb (32.9 kN) obtained in Example 6.1 by Boyd and Foster's method. The close agreement between the two methods is a coincidence. Depending on the pavement thickness and the geometry of wheel loads, Foster and Ahlvin's method may yield a much greater ESWL than Boyd and Foster's method.

TABLE 6.2 DEFLECTION FACTORS UNDER DUAL WHEELS FOR A HOMOGENEOUS HALF-SPACE

Point no.	Left wheel r/a	Left wheel F_s	Right wheel r/a	Right wheel F_s	Sum F_d
1	0	0.478	3	0.263	0.741
2	0.75	0.443	2.25	0.318	0.761
3	1.5	0.390	1.5	0.390	0.780

Although the improved method by Foster and Ahlvin results in a larger pavement thickness, which is more in line with traffic data than the earlier method by Boyd and Foster, the assumption of a homogeneous half-space instead of a layered system is not logical from a theoretical viewpoint. From the data presented by Foster and Ahlvin (1958), Huang (1968b) indicated that the improved method was still not safe, as evidenced by the fact that some of the pavements with thicknesses greater than those obtained by the method were considered inadequate or on the borderline. Since the ESWL for layered systems is greater than that for a homogeneous half-space, Huang (1968b) suggested the use of layered theory and presented a simple chart for determining ESWL based on the interface deflection of two layered systems, as shown in Figure 6.4. Given contact radius a, pavement thickness h_1, modulus ratio E_1/E_2, and dual spacing S_d, the chart gives a load factor L defined as

$$L = \frac{\text{Total load}}{\text{ESWL}} = \frac{2P_d}{P_s} \tag{6.7a}$$

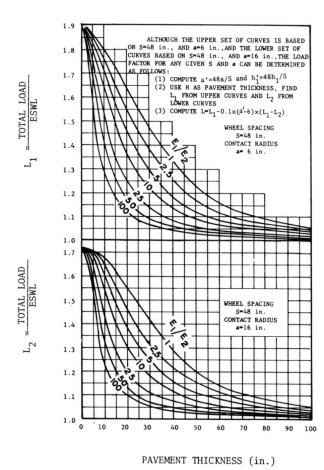

PAVEMENT THICKNESS (in.)

Figure 6.4 Chart for determining equivalent single-wheel load (1 in. = 25.4 mm). (After Huang (1968b).)

or
$$\text{ESWL} = \frac{2P_d}{L} \tag{6.7b}$$

Comparing Eq. 6.6 with Eq. 6.7b yields

$$L = \frac{2F_s}{F_d} \tag{6.8}$$

Equation 6.8 was used to develop the chart shown in Figure 6.4.

The ESWL can also be determined from the deflection factors presented in Figure 2.19 by following the same procedure illustrated in Example 6.3. However, the use of the chart shown in Figure 6.4 is much quicker. The chart is based on a dual spacing S_d of 48 in. (1.22 m). If the actual spacing is different, it must be changed to 48 in. (1.22 m), and the values of a and h_1 changed proportionally. As long as S_d/a and h_1/a remain the same, the load factor will be the same. The upper chart is for a contact radius of 6 in. (152 mm) and the lower chart is for a contact radius of 16 in. (406 mm). The load factor for any other contact radius can be obtained by a straight-line interpolation. The procedure can be summarized as follows:

1. From the given S_d, h_1, and a, determine the modified radius a' and the modified thickness h_1' by

$$a' = \frac{48}{S_d}a \tag{6.9a}$$

$$h_1' = \frac{48}{S_d}h_1 \tag{6.9b}$$

2. Using h_1' as the pavement thickness, find load factors L_1 and L_2 from the chart.
3. Determine the load factor L by

$$L = L_1 - (L_1 - L_2)\frac{a' - 6}{10} \tag{6.10}$$

and ESWL by Eq. 6.7b.

Example 6.4:

Same as Example 6.1 but the pavement is considered as a two-layer system, as shown in Figure 6.5. Determine the ESWL by equal interface deflection criterion for E_1/E_2 of 1 and 25, respectively.

Solution: Given $S_d = 13.5$ in. (343 mm), $a = 4.5$ in. (114 mm), and $h_1 = 13.5$ in. (343 mm), from Eq. 6.9, $a' = 48/13.5 \times 4.5 = 16$ in. (406 mm) and $h_1' = 48/13.5 \times 13.5 = 48$ in. (1.22 m). Because the modified contact radius is exactly 16 in. (406 mm), no interpolation is needed. From the lower chart of Figure 6.4 $L = 1.22$ when $E_1/E_2 = 1$ and $L = 1.06$ when $E_1/E_2 = 25$. With $2P_d = 9000$ lb (40 kN), from Eq. 6.7b, ESWL $= 9000/1.22 = 7380$ lb (32.8 kN) when $E_1/E_2 = 1$ and ESWL $= 9000/1.06 = 8490$ lb (37.8 kN) when $E_1/E_2 = 25$. It can be seen that ESWL increases as the modulus ratio increases. An ESWL of 7380 lb (32.8 kN) for $E_1/E_2 = 1$ checks well with the 7340 lb (32.7 kN) obtained in Example 6.3.

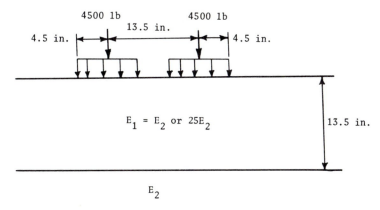

$$E_1 = E_2 \text{ or } 25E_2$$

$$E_2$$

Figure 6.5 Example 6.4 (1 in. = 25.4 mm, 1 lb = 4.45).

6.2.3 Equal Tensile Strain Criterion

The conversion factors presented in Figures 2.23, 2.25, 2.26, and 2.27 can be used to determine ESWL. According to Eq. 2.17, the tensile strain e at the bottom of layer 1 under a single-wheel load is

$$e = \frac{q_s}{E_1} F_e \qquad (6.11)$$

in which q_s is the contact pressure of a single wheel. The tensile strain under dual or dual-tandem wheels is

$$e = \frac{Cq_d}{E_1} F_e \qquad (6.12)$$

in which C is the conversion factor and q_d is the contact pressure of dual or dual-tandem wheels. Equate Eqs. 6.11 and 6.12 to obtain the same tensile strain:

$$q_s = Cq_d \qquad (6.13)$$

For equal contact radius, contact pressure is proportional to wheel load:

$$\text{ESWL} = P_s = CP_d \qquad (6.14)$$

Example 6.5:

A full-depth asphalt pavement, 8 in. (203 mm) thick, is loaded by a set of dual wheels with a total load $2P_d$ of 9000 lb (40 kN), a contact radius a of 4.5 in. (114 mm), and a center to center wheel spacing S_d of 13.5 in. (343 mm), as shown in Figure 6.6. If $E_1/E_2 = 50$, determine ESWL by equal tensile strain criterion.

Solution: Given $S_d = 13.5$ in. (343 mm), $a = 4.5$ in. (114 mm), and $h_1 = 8$ in. (203 mm), from Eq. 2.18, $a' = 24/13.5 \times 4.5 = 8$ in. (203 mm) and $h_1' = 24/13.5 \times 8 = 14.2$ in. (361 mm). Because the modified contact radius is exactly 8 in. (203 mm), no interpolation is needed. From the lower chart in Figure 2.23, $C = 1.50$. From Eq. 6.14, ESWL $= 1.50 \times 4500 = 6750$ lb (30.0 kN).

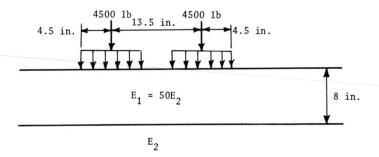

Figure 6.6 Example 6.5 (1 in. = 25.4 mm, 1 lb = 4.45 N).

6.2.4 Criterion Based on Equal Contact Pressure

All of the above analyses of ESWL are based on the assumption that the single wheel has the same contact radius as each of the dual wheels. Another assumption, which has been frequently made, is that the single wheel has a different contact radius but the same contact pressure as the dual wheels. Although this assumption is more reasonable, its solution is much more complicated and cannot be presented by a simple chart.

According to Eq. 2.16, the interface deflections for single and dual wheels with the same contact pressure can be written as

$$w_s = \frac{q a_s}{E_2} F_s \tag{6.15a}$$

$$w_d = \frac{q a_d}{E_2} F_d \tag{6.15b}$$

in which the subscript s indicates single wheel and d indicates dual wheels. The deflection factor F_d is obtained by superposition of the duals. To obtain equal deflection, $w_s = w_d$, or

$$\frac{q a_s}{E_2} F_s = \frac{q a_d}{E_2} F_d \tag{6.16}$$

Because

$$a_s = \sqrt{\frac{P_s}{\pi q}} \tag{6.17a}$$

and

$$a_d = \sqrt{\frac{P_d}{\pi q}} \tag{6.17b}$$

Eq. 6.16 becomes

$$\text{ESWL} = P_s = \left(\frac{F_d}{F_s}\right)^2 P_d \tag{6.18}$$

If P_d and q are given, a_d can be computed from Eq. 6.17b, and the maximum deflection factor F_d can be determined. However, the application of Eq. 6.18

presents a difficulty because F_s depends on the contact radius a, which varies with P_s according to Eq. 6.17a. Consequently, Eq. 6.18 can be solved only by a trial and error method.

Example 6.6:

A two-layer system with a thickness h_1 of 13.5 in. (343 mm) and a modulus ratio E_1/E_2 of 25 is loaded under a set of duals with a total load $2P_d$ of 9000 lb (40 kN), a contact pressure q of 70 psi (483 kPa), and a center to center tire spacing S_d of 13.5 in. (343 mm), as shown in Figure 6.7. Determine the ESWL based on the equal interface deflection criterion with equal contact pressure.

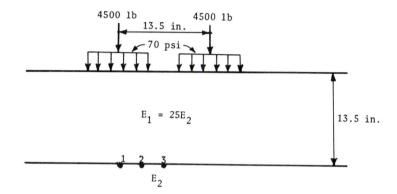

Figure 6.7 Example 6.6 (1 in. = 25.4 mm, 1 psi = 6.9 kPa, 1 lb = 4.45 N).

Solution: Given $P_d = 4500$ lb (20 kN) and $q = 70$ psi (483 kPa), from Eq. 6.17b, $a_d = \sqrt{4500/(\pi \times 70)} = 4.5$ in. (519 mm). With $h_1/a_d = 13.5/4.5 = 3.0$ and $E_1/E_2 = 25$, the deflection factors at points 1, 2, and 3, as shown in Figure 6.7, can be obtained from Figure 2.19 and the results are presented in Table 6.3. It can be seen that the same deflection factor of 0.36 is obtained at all three points, or $F_d = 0.36$.

Assume that $P_s = 8000$ lb (35.6 kN). From Eq. 6.17a, $a_s = \sqrt{8000/(\pi \times 70)} = 6.03$ in. (153 mm). With $h_1/a_s = 13.5/6.03 = 2.24$ and $E_1/E_2 = 25$, from Figure 2.19, $F_s = 0.26$. From Eq. 6.18, $P_s = (0.36/0.26)^2 \times 4500 = 8630$ lb (38.4 kN). Because the deflection factor cannot be read accurately from Figure 2.19, a P_s of 8300 lb (36.9 kN), which is midway between 8000 lb (35.6 kN) and 8630 lb (38.4 kN), is taken as the final solution. The ESWL based on equal contact radius, as illustrated in Example 6.4, is 8490 (37.8 kN), which is not much different from the 8300 lb (36.9 kN) based on equal contact pressure.

TABLE 6.3 DEFLECTION FACTORS UNDER DUAL WHEELS FOR A LAYERED SYSTEM

| Point no. | Left wheel | | Right wheel | | Sum |
	r/a	F_s	r/a	F_s	F_d
1	0	0.19	3	0.17	0.36
2	0.75	0.19	2.25	0.17	0.36
3	1.5	0.18	1.5	0.18	0.36

Huang (1968b) compared the ESWL based on equal contact radius with that based on equal contact pressure for a variety of cases. He found that unless the pavement is extremely thin and the modulus ratio close to unity, the differences between the two methods are not very significant.

Two-layer interface deflections based on equal contact pressure were also used by the Asphalt Institute to compute the ESWL for full-depth asphalt pavements. This procedure is applicable to aircraft having less than 60,000 lb (267 kN) gross weight. By the use of Figure 2.19, simplified charts were developed for determining the ESWL for dual wheels based on the CBR of the subgrade (AI, 1973).

6.3 EQUIVALENT AXLE LOAD FACTOR

An equivalent axle load factor (EALF) defines the damage per pass to a pavement by the axle in question relative to the damage per pass of a standard axle load, usually the 18-kip (80-kN) single-axle load. The design is based on the total number of passes of the standard axle load during the design period, defined as the equivalent single-axle load (ESAL) and computed by

$$\text{ESAL} = \sum_{i=1}^{m} F_i \, n_i \tag{6.19}$$

in which m is the number of axle load groups, F_i is the EALF for the ith-axle load group, and n_i is the number of passes of the ith-axle load group during the design period.

The EALF depends on the type of pavements, thickness or structural capacity, and the terminal conditions at which the pavement is considered failed. Most of the EALFs in use today are based on experience. One of the most widely used methods is based on the empirical equations developed from the AASHO Road Test (AASHTO, 1972). The EALF can also be determined theoretically based on the critical stresses and strains in the pavement and the failure criteria. In this section, the equivalent factors for flexible and rigid pavements are discussed separately.

6.3.1 Flexible Pavements

The AASHTO equations for computing EALF are described first, followed by a discussion of equivalent factor based on the results obtained from KENLAYER.

AASHTO Equivalent Factors

The following regression equations based on the results of road tests can be used for determining EALF:

$$\log\left(\frac{W_{tx}}{W_{t18}}\right) = 4.79 \log(18 + 1) - 4.79 \log(L_x + L_2)$$

$$+ 4.33 \log L_2 + \frac{G_t}{\beta_x} - \frac{G_t}{\beta_{18}} \tag{6.20a}$$

$$G_t = \log\left(\frac{4.2 - p_t}{4.2 - 1.5}\right) \tag{6.20b}$$

$$\beta_x = 0.40 + \frac{0.081(L_x + L_2)^{3.23}}{(SN + 1)^{5.19}L_2^{3.23}} \tag{6.20c}$$

in which W_{tx} is the number of x-axle load applications at the end of time t; W_{t18} is the number of 18-kip (80-kN) single-axle load applications to time t; L_x is the load in kip on one single axle, one set of tandem axles, or one set of tridem axles; L_2 is the axle code, 1 for single axle, 2 for tandem axles, and 3 for tridem axles; SN is the structural number, which is a function of the thickness and modulus of each layer and the drainage conditions of base and subbase; p_t is the terminal serviceability, which indicates the pavement conditions to be considered as failures; G_t is a function of P_t; and β_{18} is the value of β_x when L_x is equal to 18 and L_2 is equal to one. The method for determining SN is presented in Section 11.3.4. Note that

$$EALF = \frac{W_{t18}}{W_{tx}} \tag{6.21}$$

Equation 6.20 can be used to solve EALF. The effect of p_t and SN on EALF is erratic and is not completely consistent with theory. However, under heavy axle loads with an equivalent factor much greater than unity, the EALF increases as p_t or SN decreases. This is as expected because heavy axle loads are more destructive to poor and weaker pavements than to good and stronger ones. A disadvantage of using the above equations is that the EALF varies with the structural number, which is a function of layer thicknesses. Theoretically, a method of successive approximations should be used because the EALF depends on the structural number and the structual number depends on the EALF. Practically, EALF is not very sensitive to pavement thickness and a SN of 5 may be used for most cases. Unless the design thickness is significantly different, no iterations will be needed. The AASHTO equivalent factors with $p_t = 2.5$ and SN = 5 are used by the Asphalt Institute, as shown in Table 6.4. The original table has single and tandem axles only but the tridem axles are added based on the AASHTO design guide (AASHTO, 1986). Tables of equivalent factors for SN values of 1, 2, 3, 4, 5, and 6 and p_t values of 2, 2.5, and 3 can be found in the AASHTO design guide.

Example 6.7:

Given $p_t = 2.5$ and SN = 5, determine the EALF for a 32-kip (151-kN) tandem-axle load and a 48-kip (214-kN) tridem-axle load.

Solution: For the tandem axles, $L_x = 32$ and $L_2 = 2$, from Eq. 6.20, $G_t = \log(1.7/2.7) = -0.201$, $\beta_x = 0.4 + 0.081 (32 + 2)^{3.23}/[(5 + 1)^{5.19}(2)^{3.23}] = 0.470$, $\beta_{18} = 0.4 + 0.081 (18 + 1)^{3.23}/(5 + 1)^{5.19} = 0.5$, and $\log(W_{tx}/W_{t18}) = 4.79 \log 19 - 4.79 \log (32 + 2) + 4.33 \log 2 - 0.201/0.47 + 0.201/0.5 = 0.067$, or $W_{tx}/W_{t18} = 1.167$. From Eq. 6.21, EALF = 0.857, which is exactly the same as that shown in Table 6.4.

For the tridem axles, $L_x = 48$, $L_2 = 3$, from Eq. 6.20, $\beta_x = 0.4 + 0.081(48 + 3)^{3.23}/[(5 + 1)^{5.19}(3)^{3.23}] = 0.470$, and $\log (W_{tx}/W_{t18}) = 4.79 \log 19 - 4.79 \log (48 + 3) + 4.33 \log 3 - 0.201/0.47 + 0.201/0.5 = -0.0139$, or $W_{tx}/W_{t18} = 0.968$. From Eq. 6.21, EALF = 1.033, as shown in Table 6.4.

TABLE 6.4 ASPHALT INSTITUTE'S EQUIVALENT AXLE LOAD FACTORS

Axle load (lb)	Equivalent axle load factor			Axle load (lb)	Equivalent axle load factor		
	Single axles	Tandem axles	Tridem axles		Single axles	Tandem axles	Tridem axles
1000	0.00002			41,000	23.27	2.29	0.540
2000	0.00018			42,000	25.64	2.51	0.597
3000	0.00072			43,000	28.22	2.76	0.658
4000	0.00209			44,000	31.00	3.00	0.723
5000	0.00500			45,000	34.00	3.27	0.793
6000	0.01043			46,000	37.24	3.55	0.868
7000	0.0196			47,000	40.74	3.85	0.948
8000	0.0343			48,000	44.50	4.17	1.033
9000	0.0562			49,000	48.54	4.51	1.12
10,000	0.0877	0.00688	0.002	50,000	52.88	4.86	1.22
11,000	0.1311	0.01008	0.002	51,000		5.23	1.32
12,000	0.189	0.0144	0.003	52,000		5.63	1.43
13,000	0.264	0.0199	0.005	53,000		6.04	1.54
14,000	0.360	0.0270	0.006	54,000		6.47	1.66
15,000	0.478	0.0360	0.008	55,000		6.93	1.78
16,000	0.623	0.0472	0.011	56,000		7.41	1.91
17,000	0.796	0.0608	0.014	57,000		7.92	2.05
18,000	1.000	0.0773	0.017	58,000		8.45	2.20
19,000	1.24	0.0971	0.022	59,000		9.01	2.35
20,000	1.51	0.1206	0.027	60,000		9.59	2.51
21,000	1.83	0.148	0.033	61,000		10.20	2.07
22,000	2.18	0.180	0.040	62,000		10.84	2.85
23,000	2.58	0.217	0.048	63,000		11.52	3.03
24,000	3.03	0.260	0.057	64,000		12.22	3.22
25,000	3.53	0.308	0.067	65,000		12.96	3.41
26,000	4.09	0.364	0.080	66,000		13.73	3.62
27,000	4.71	0.426	0.093	67,000		14.54	3.83
28,000	5.39	0.495	0.109	68,000		15.38	4.05
29,000	6.14	0.572	0.126	69,000		16.26	4.28
30,000	6.97	0.658	0.145	70,000		17.19	4.52
31,000	7.88	0.753	0.167	71,000		18.15	4.77
32,000	8.88	0.857	0.191	72,000		19.16	5.03
33,000	9.98	0.971	0.217	73,000		20.22	5.29
34,000	11.18	1.095	0.246	74,000		21.32	5.57
35,000	12.50	1.23	0.278	75,000		22.47	5.86
36,000	13.93	1.38	0.313	76,000		23.66	6.15
37,000	15.50	1.53	0.352	77,000		24.91	6.46
38,000	17.20	1.70	0.393	78,000		26.22	6.78
39,000	19.06	1.89	0.438	79,000		27.58	7.11
40,000	21.08	2.08	0.487	80,000		28.99	7.45

Note. 1 lb = 4.45 N.

Theoretical Analysis

In the mechanistic method of design, the EALF can be determined from the failure critera. The failure criterion for fatigue cracking was shown in Eq. 3.6 with f_2 of 3.291 by the Asphalt Institute and 5.671 by Shell:

$$N_f = f_1 (\epsilon_t)^{-f_2} (E_1)^{-f_3} \tag{3.6}$$

Deacon (1969) conducted a theoretical analysis of EALF by layered theory based on an assumed f_2 of 4, or from Eq. 3.6

$$\text{EALF} = \frac{W_{t18}}{W_{tx}} = \left(\frac{\epsilon_x}{\epsilon_{18}}\right)^4 \tag{6.22}$$

in which ϵ_x is the tensile strain at the bottom of asphalt layer due to an x-axle load and ϵ_{18} is the tensile strain at the bottom of asphalt layer due to an 18-kip (80-kN) axle load. If W_{tx} is also a single axle, it is reasonable to assume that tensile strains are directly proportional to axle loads:

$$\text{EALF} = \left(\frac{L_x}{18}\right)^4 \tag{6.23}$$

in which L_x is the load in kip on a single axle. Equation 6.23 is valid only when L_x is on a single axle. For tandem or tridem axles, a more general equation is

$$\text{EALF} = \left(\frac{L_x}{L_s}\right)^4 \tag{6.24}$$

in which L_s is the load in kip on standard axles which have the same number of axles as L_x. If the EALF for one set of tandem or tridem axles is known, that for other axles can be determined by Eq. 6.24.

Example 6.8:

Given $p_t = 2.5$ and SN = 5, determine the EALF for 5000-lb (22.3-kN) and 50,000-lb (223-kN) single axles by Eq. 6.23. If the EALF of a 32-kip (142-kN) tandem axle is 0.857, determine the EALF for 15,000-lb (66.8-kN) and 80,000-lb (356-kN) tandem axles by Eq. 6.24.

Solution: For a single axle with $L_x = 5$, from Eq. 6.23, EALF = $(5/18)^4 = 0.006$, which compares with 0.005 from Table 6.4. For a single axle with $L_x = 50$, EALF = $(50/18)^4 = 59.54$ compared with 52.88 from Table 6.4. For tandem axles with $L_x = 15$ and $L_s = 32$, from Eq. 6.24, the equivalent factor between 15-kip (66.8-kN) and 32-kip (142-kN) tandem axles is $(15/32)^4 = 0.0483$, which must be multiplied by 0.857 to obtain the equivalent factor between 15-kip (66.8-kN) tandem axles and an 18-kip (80-kN) single axle, so EALF = $0.0483 \times 0.857 = 0.041$ compared with 0.036 from Table 6.4. For tandem axles with $L_x = 80$, EALF = $0.857 \times (80/32)^4 = 33.48$ compared with 28.99 from Table 6.4. It can be seen that the equivalent factors based on tensile strains at the bottom of asphalt layer check favorably with the AASHTO and Asphalt Institute's equivalent factors.

The other failure criterion is to control permanent deformation by limiting the vertical compressive strain on top of the subgrade, which can be expressed as

$$N_d = f_4 \, (\epsilon_c)^{-f_5} \tag{3.7}$$

Suggested values of f_5 are 4.477 by the Asphalt Institute, 4.0 by Shell, and 3.571 by the University of Nottingham. It can be seen that the use of 4 for f_5 is also reasonable. Therefore, when both L_x and L_s in Eq. 6.24 are of the same configuration, the EALF based on fatigue cracking may not be much different from that based on permanent deformation. However, this is not true when L_s is for a single axle but L_x is for multiple axles because the effect of additional axles on the tensile strains at the bottom of asphalt layer is quite different from that on the compressive strain at the top of the subgrade.

Table 6.5 shows the computation of EALF based on fatigue cracking, while Table 6.6 shows that based on permanent deformation. The damage ratio D_r is caused by the passage of one single axle or a set of tandem or tridem axles and can be obtained directly by KENLAYER. A full-depth asphalt pavement with two extreme HMA thicknesses of 2 and 12 in. (51 and 305 mm) and two extreme subgrade moduli of 5000 and 30,000 psi (35 and 207 MPa) is assumed. The single axle is a standard 18-kip (80-kN) axle with dual tires having a contact pressure of 100 psi (690 kPa) and a dual spacing of 13.5 in. (343 mm). Each of the tandem and tridem axles is the same as the single axle, so the total load on tandem axles is 36 kip (160 kN) and that on tridem axles is 54 kip (240 kN). The spacing between the two axles is 48 in. (1.2 m). The HMA has a resilient modulus of 450,000 psi (3.1 GPa) and a Poisson ratio of 0.35; the subgrade has a Poisson ratio of 0.45. The Asphalt Institute failure criteria were used for the analysis. The 12-in. (305-mm) pavement is shown in Figure B.10 as an example in Section B.4.5.

For tandem-axle loads, the primary D_r is the damage ratio caused by the strain ϵ_a due to the passage of the first axle, while the secondary D_r is the damage ratio caused by the strain $\epsilon_a - \epsilon_b$, due to the passage of the second axle, as

TABLE 6.5 ANALYSIS OF DAMAGE RATIO FOR FATIGUE CRACKING BY KENLAYER

HMA thickness (in.)	2		12	
Subgrade modulus (psi)	5000	30,000	5000	30,000
18k Single damage ratio	1.311×10^{-4}	7.030×10^{-6}	8.450×10^{-8}	1.974×10^{-8}
36k Tandem				
Primary D_r	1.282×10^{-4}	6.984×10^{-6}	6.740×10^{-8}	1.334×10^{-8}
Secondary D_r	1.138×10^{-4}	6.900×10^{-6}	6.199×10^{-10}	1.018×10^{-9}
Total D_r	2.420×10^{-4}	1.388×10^{-5}	6.802×10^{-8}	1.436×10^{-8}
EALF	1.85	1.97	0.80	0.73
54k Tridem				
Primary D_r	1.253×10^{-4}	6.939×10^{-6}	9.493×10^{-8}	8.555×10^{-9}
Secondary D_r	2.222×10^{-4}	1.371×10^{-5}	2.477×10^{-9}	7.034×10^{-10}
Total D_r	3.475×10^{-4}	2.065×10^{-5}	9.741×10^{-8}	9.258×10^{-9}
EALF	2.65	2.94	1.15	0.46

Note. 1 in. = 25.4 mm, 1 kip = 4.45 kN, 1 psi = 6.9 kPa.

Traffic Loading and Volume   Chap. 6

TABLE 6.6 ANALYSIS OF DAMAGE RATIO FOR PERMANENT DEFORMATION BY KENLAYER

HMA thickness (in.)	2		12	
Subgrade modulus (psi)	5000	30,000	5000	30,000
18k Single damage ratio	3.202×10^{-3}	4.843×10^{-5}	6.358×10^{-8}	4.408×10^{-9}
36k Tandem				
Primary D_r	3.178×10^{-3}	4.820×10^{-5}	9.634×10^{-8}	4.261×10^{-9}
Secondary D_r	3.178×10^{-3}	4.820×10^{-5}	1.456×10^{-9}	1.178×10^{-9}
Total D_r	6.356×10^{-3}	9.640×10^{-5}	9.780×10^{-8}	5.439×10^{-9}
EALF	1.99	1.99	1.54	1.23
54k Tridem				
Primary D_r	3.154×10^{-3}	4.797×10^{-5}	1.409×10^{-7}	4.118×10^{-9}
Secondary D_r	6.308×10^{-3}	9.594×10^{-5}	7.253×10^{-9}	2.250×10^{-9}
Total D_r	9.462×10^{-3}	1.439×10^{-4}	1.482×10^{-7}	6.368×10^{-9}
EALF	2.96	2.97	2.33	1.44

Note. 1 in. = 25.4 mm, 1 kip = 4.45 kN, 1 psi = 6.9 kPa.

illustrated in Figure 3.3. For tridem-axle loads, the primary damage ratio is based on the loading position shown in Figure 3.4a, while the secondary damage ratio is based on two passages of the loading shown in Figure 3.4b. The total damage ratio is the sum of primary and secondary damage ratios. Dividing the total damage ratio due to tandem or tridem axles by the damage ratio due to a single axle gives the values of EALF shown in the tables.

Review of Tables 6.5 and 6.6

1. For the 2-in. (51-mm) pavement with a subgrade modulus of 5000 psi (35 MPa), the damage ratios due to permanent deformation are greater than those due to fatigue cracking, whereas the reverse is true for the 12-in. (305-mm) pavement with a subgrade modulus of 30,000 psi (207 MPa). Therefore, the design is controlled by fatigue cracking for stronger pavements but by permanent deformation for weaker pavements.

2. For the 2-in. (51-mm) pavement with a subgrade modulus of 30,000 psi (207 MPa), the strains due to one axle load is not affected by the other axle loads, so the damage ratios for tandem axles are nearly twice as large as those for a single axle, with an EALF close to 2, while those for tridem axles are three times as large, with an EALF close to 3. When the pavement becomes thicker or the subgrade modulus becomes smaller, the two or three axle loads begin to interact and the EALF based on fatigue cracking may be quite different from that based on permanent deformation, as indicated by an EALF much smaller than 1 for fatigue cracking but much greater than 1 for permanent deformation.

3. The AASHTO equivalent factors for 36-kip (160-kN) tandem axles and 54-kip (240-kN) tridem axles, as obtained from Eq. 6.20, is practically independent of structural number. For a terminal serviceability of 2.5, the equivalent factor is 1.33 for 36-kip (160-kN) tandem axles and 1.66 for 54-kip (240-kN) tridem axles. However, the EALF obtained from KENLAYER may range from 0.73 to

1.99 for tandem axles and from 0.46 to 2.97 for tridem axles depending on pavement geometry, material properties, and failure criteria.

4. Due to the many factors involved, it is not possible to select an appropriate EALF that can be applied to all situations. For a truly mechanistic design method, each load group should be considered individually, instead of using an equivalent single axle load.

6.3.2 Rigid Pavements

The AASHTO equations for computing EALF are described first, followed by a discussion of equivalent factor based on the results obtained from KENSLABS.

AASHTO Equivalent Factors

The AASHTO equations for determining the EALF of rigid pavements are as follows:

$$\log\left(\frac{W_{tx}}{W_{t18}}\right) = 4.62 \log(18 + 1) - 4.62 \log(L_x + L_2)$$

$$+ 3.28 \log L_2 + \frac{G_t}{\beta_x} - \frac{G_t}{\beta_{18}} \quad (6.25a)$$

$$G_t = \log\left(\frac{4.5 - p_t}{4.5 - 1.5}\right) \quad (6.25b)$$

$$\beta_x = 1.00 + \frac{3.63(L_x + L_2)^{5.20}}{(D + 1)^{8.46}L_2^{3.52}} \quad (6.25c)$$

in which W_{tx}, W_{t18}, L_x, L_2, p_t, and β_{18} are as defined for Eq. 6.20 for flexible pavements and D is the slab thickness in inches. Note that Eqs. 6.20 and 6.25 are in the same form but with different values of regression constants and that the structural number SN in Eq. 6.20 is replaced by the slab thickness D in Eq. 6.25. The EALF can also be computed by Eq. 6.21. Table 6.7 shows the AASHTO equivalent factors for $p_t = 2.5$ and $D = 9$ in. (229 mm). If the thickness is not known in the design stage, a value of D equal to 9 can be used. A comparison of Tables 6.4 and 6.7 shows that for single-axle loads less than 18 kip the equivalent factor is smaller for a rigid pavement than for a flexible pavement, while for heavier loads the reverse is true.

Example 6.9:

A flexible pavement with SN = 5 is subjected to a single-axle load of 12 kip (53.4 kN) and a tandem-axle load of 40 kip (178 kN). Based on a p_t of 2.5, what are the single- and tandem-axle loads on a 9-in. (229-mm) rigid pavement that are equivalent to those on the flexible pavements?

Solution: From Table 6.4 for flexible pavements, the equivalent factor for a single-axle load of 12 kip (53.4 kN) is 0.189 and that for a 40-kip (178-kN) tandem-axle load is 2.08. From an interpolation of Table 6.7, the single-axle load for a rigid pavement with an EALF of 0.189 is 12.2 kip (54.3 kN), which is slightly greater than the 12 kip

TABLE 6.7 EQUIVALENT AXLE LOAD FACTORS FOR RIGID PAVEMENTS WITH D = 9 IN. AND p_t = 2.5

Axle load (kips)	Equivalent axle load factor			Axle load (kips)	Equivalent axle load factor		
	Single axles	Tandem axles	Tridem axles		Single axles	Tandem axles	Tridem axles
2	0.0002	0.0001	0.0001	48	56.8	7.73	2.49
4	0.002	0.0005	0.0003	50	67.8	9.07	2.94
6	0.01	0.002	0.001	52		10.6	3.44
8	0.032	0.005	0.002	54		12.3	4.00
10	0.082	0.013	0.005	56		14.2	4.63
12	0.176	0.026	0.009	58		16.3	5.32
14	0.341	0.048	0.017	60		18.7	6.08
16	0.604	0.082	0.028	62		21.4	6.91
18	1.00	0.133	0.044	64		24.4	7.82
20	1.57	0.206	0.067	66		27.6	8.83
22	2.34	0.308	0.099	68		31.3	9.9
24	3.36	0.444	0.141	70		35.3	11.1
26	4.67	0.622	0.195	72		39.8	12.4
28	6.29	0.850	0.265	74		44.7	13.8
30	8.28	1.14	0.354	76		50.1	15.4
32	10.7	1.49	0.463	78		56.1	17.1
34	13.6	1.92	0.596	80		62.5	18.9
36	17.1	2.43	0.757	82		69.6	20.9
38	21.3	3.03	0.948	84		77.3	23.1
40	26.3	3.74	1.17	86		86.0	25.4
42	32.2	4.55	1.44	88		95.0	27.9
44	39.2	5.48	1.74	90		105.0	30.7
46	47.3	6.53	2.09				

Note. 1 kip = 4.45 kN, 1 in. = 25.4 mm.
Source. After AASHTO (1986).

(53.4 kN) for the flexible pavement. The tandem-axle load for an equivalent factor of 2.08 is 34.7 kip (154 kN), which is much smaller than the 40 kip (178 kN) for the flexible pavement. This may indicate that heavier axle loads are more destructive to rigid pavements than to flexible pavements, probably due to the effect of pumping under heavier loads.

Theoretical Analysis

For a terminal serviceability index of 2.5, the AASHTO equivalent factors obtained from Eq. 6.25 for a 6-in. (152-mm) slab are 2.29 for 36-kip (160-kN) tandem axles and 3.79 for 54-kip (240-kN) tridem axles, while those for a 14-in. (356-mm) slab are 2.53 and 4.34, respectively. These equivalent factors are quite different from those obtained from KENSLABS because the AASHTO factors are based on the results of a road test, where the predominant type of failure is by pumping or the erosion of subbase, whereas the damage analysis by KENSLABS is based on fatigue cracking.

Slab thickness (in.)	6		14	
Subgrade k value (pci)	50	500	50	500
18k Single cracking index	3.085	7.090×10^{-6}	1.741×10^{-12}	4.526×10^{-14}
36k Tandem				
Primary CI	3.478×10^{-2}	6.323×10^{-8}	7.969×10^{-12}	2.192×10^{-14}
Secondary CI	4.054×10^{-7}	6.323×10^{-8}	1.049×10^{-15}	1.733×10^{-15}
Total CI	3.478×10^{-2}	1.265×10^{-7}	7.970×10^{-12}	2.365×10^{-14}
EALF	0.011	0.018	4.58	0.52
54k Tridem				
Primary CI	5.119×10^{-4}	5.249×10^{-9}	2.702×10^{-11}	1.102×10^{-14}
Secondary CI	8.360×10^{-6}	1.050×10^{-8}	9.647×10^{-15}	5.949×10^{-15}
Total CI	5.202×10^{-4}	1.575×10^{-8}	2.703×10^{-11}	1.697×10^{-14}
EALF	0.00017	0.0022	15.5	0.37

Note. 1 in. = 25.4 mm, 1 kip = 4.45 kN, 1 pci = 271.3 kN/m³.

Table 6.8 shows the computation of EALF from the cracking index obtained by KENSLABS. Two extreme thicknesses of 6 and 14 in. (152 and 356 mm) and k values of 50 and 500 pci (13.6 to 136 MN/m³) are assumed. The 18-kip (80-kN) single axle, 36-kip (160-kN) tandem axles, and 54-kip (240-kN) tridem axles are the same as those used for the analysis of flexible pavements presented in Tables 6.5 and 6.6. The center to center spacing between the left and right dual wheels is 77 in. (1.96 m). The concrete has a modulus of rupture of 600 psi (4.1 MPa), an elastic modulus of 4×10^6 psi (28 GPa), and a Poisson ratio of 0.15. The allowable number of repetitions for fatigue cracking is based on Eq. 5.35 with $f_1 = 16.61$ and $f_2 = 17.61$. The division of slab into finite elements is shown in Figure C-5 as an example in Section C.4.4.

The cracking index shown in Table 6.8 is the same as the damage ratio for fatigue cracking shown in Table 6.5 and is caused by the passage of a single axle or a set of tandem or tridem axles. The primary cracking index is based on the maximum stress, which occurs at the slab edge under either of the tandem axles or the central axle of the tridem. The secondary cracking index is caused by the stress differential due to the passage of one or two additional axles. Dividing the total cracking index of the tandem or tridem axles by the cracking index of a single axle gives the EALF.

It can be seen from Table 6.8 that the values of EALF vary a great deal and are very difficult to predict. For the 6-in. (152-mm) slab, the single axle is more destructive than the tandem axles, and the tandem axles are more destructive than the tridem axles. The same is true for the 14-in. (356-mm) slab with a subgrade k value of 500 pci (136 MN/m³) because the flexural stress in the concrete is largest under a single axle and smallest under a set of tridem axles. However, for the 14-in. (356-mm) slab with a k value of 50 pci (13.6 MN/m³), the reverse is true. The large EALF of 4.58 for tandem axles and 15.52 for tridem axles is because the maximum stresses are 187.8 psi (1.30 MPa) under tandem axles and 205.9 psi (1.42 MPa) under tridem axles, in contrast to the 165.3 psi (1.14

MPa) under a single axle. This analysis clearly indicates the sensitivity of EALF to the change in slab thickness and subgrade support and the difficulty of establishing an EALF based on empirical methods. The equivalent factor is valid only under the given conditions and is no longer applicable if any of the conditions are changed.

6.4 TRAFFIC ANALYSIS

To design a highway pavement, it is necessary to predict the number of repetitions of each axle load group during the design period. Information on initial traffic can be obtained from field measurements or from the W-4 form of a loadometer station that has traffic characteristics similar to those of the project in question. The initial daily traffic is in two directions over all traffic lanes and must be multiplied by the directional and lane distribution factors to obtain the initial traffic on the design lane. The traffic to be used for design is the average traffic during the design period, so the initial traffic must be multiplied by a growth factor. If n_i is the total number of load repetitions to be used in design for the ith load group, then

$$n_i = (n_0)_i \, (G)(D)(L)(365)(Y) \tag{6.26}$$

in which $(n_0)_i$ is the initial number of repetitions per day for the ith load group, G is the growth factor, D is the directional distribution factor, which is usually assumed to be 0.5 unless the traffic in two directions is different, L is the lane distribution factor which varies with the volume of traffic and the number of lanes, and Y is the design period in years. If the design is based on the equivalent 18-kip (80-kN) single-axle load, then the initial number of repetitions per day for the ith load group can be computed by

$$(n_0)_i = (p_i F_i)(\text{ADT})_0(T)(A) \tag{6.27}$$

in which p_i is the percentage of total repetitions for the ith load group, F_i is the equivalent axle load factor (EALF) for the ith load group, $(\text{ADT})_0$ is the average daily traffic at the start of the design period, T is the percentage of trucks in the ADT, and A is the average number of axles per truck. Substituting Eq. 6.27 into 6.26 and summing over all load groups, the equivalent axle load for the design lane is

$$\text{ESAL} = \left(\sum_{i=1}^{m} p_i \, F_i \right) (\text{ADT})_0(T)(A)(G)(D)(L)(365)(Y) \tag{6.28}$$

In computing ESAL, it is convenient to combine the first and fourth terms in Eq. 6.28 to form a new term called the truck factor:

$$T_f = \left(\sum_{i=1}^{m} p_i \, F_i \right) (A) \tag{6.29}$$

in which T_f is the truck factor, or the number of 18-kip (80-kN) single-axle load applications per truck. Thus, Eq. 6.28 becomes

$$\text{ESAL} = (\text{ADT})_0(T)(T_f)(G)(D)(L)(365)(Y) \tag{6.30}$$

6.4.1 Average Daily Truck Traffic

The minimum traffic information required for a pavement design is the average daily truck traffic (ADTT) at the start of the design period. The ADTT may be expressed as a percentage of ADT or as an actual value. This information can be obtained from the actual traffic counts on the existing roadway where the pavement is to be constructed or on nearby highways with similar travel patterns. Traffic volume maps showing the ADT, sometimes with the percentage of trucks, on various roadways within a given area may also be used, although they are far less accurate than the actual counts. The traffic counts must be adjusted for daily (weekday versus weekend) and seasonal (summer versus winter) variations to obtain the annual average daily traffic (AADT).

Traffic is the most important factor in pavement design. Every effort should be made to collect actual data on the project. This requires a portable device that can be taken to the project site. This device can be a static scale or the newer weigh in motion (WIM) scale. The WIM scale does not interrupt the traffic stream and can enhance the credibility of the data due to its concealment.

If actual traffic data are not available, Table 6.9 can be used as a guide to determine the distribution of ADTT on different classes of highways in the United States.

6.4.2 Truck Factor

A single truck factor can be applied to all trucks, or separate truck factors can be used for different classes of trucks. The latter case should be considered if the growth factors for different types of trucks are not the same. Table 6.10 shows the distribution of truck factors for different classes of highways and vehicles in the United States. Table 6.11 shows the computation of truck factors for trucks with five or more axles on a flexible pavement. The equivalent factors are based on an SN of 5 and a p_t of 2.5 and can be obtained from Table 6.4. The number of axles for each load group as well as the number of trucks weighed can be found from the W-4 form. The sum of ESALs for all trucks weighed divided by the number of trucks weighed gives the truck factor.

6.4.3 Growth Factor

One simple way to project the growth factor is to assume a yearly rate of traffic growth and use the average traffic at the start and end of the design period as the design traffic:

$$G = \tfrac{1}{2}[1 + (1 + r)^Y] \tag{6.31}$$

in which r is the yearly rate of traffic growth. The Portland Cement Association (1984) applies the traffic at the middle of the design period as the design traffic:

$$G = (1 + r)^{0.5Y} \tag{6.32}$$

Table 6.12 shows the growth factors for 20- and 40-year design periods based on Eq. 6.32.

TABLE 6.9 DISTRIBUTION OF TRUCKS ON DIFFERENT CLASSES OF HIGHWAYS IN THE UNITED STATES[a]

	Percent trucks											
	Rural systems						Urban systems					
Truck class	Interstate	Other Principal	Minor Arterial	Collectors		Range	Interstate	Other Freeways	Other Principal	Minor Arterial	Collectors	Range
				Major	Minor							
Single-unit trucks												
2-axle, 4 tire	43	60	71	73	80	43–80	52	66	67	84	86	52–86
2-axle, 6-tire	8	10	11	10	10	8–11	12	12	15	9	11	9–15
3-axle or more	2	3	4	4	2	2–4	2	4	3	2	<1	<1–4
All single units	53	73	86	87	92	53–92	66	82	85	95	97	66–97
Multiple-unit trucks												
4-axle or less	5	3	3	2	2	2–5	5	5	3	2	1	1–5
5-axle[b]	41	23	11	10	6	6–41	28	13	12	3	2	2–28
6-axle or more[b]	1	1	<1	1	<1	<1–1	1	<1	<1	<1	<1	<1–1
All multiple units	47	27	14	13	8	8–47	34	18	15	5	3	3–34
All trucks	100	100	100	100	100	100	100	100	100	100	100	100

[a] Compiled from data supplied by the Highway Statistics Division, U.S. Federal Highway Administration.
[b] Including full-trailer combinations in some states.
Source. After AI (1991).

TABLE 6.10 DISTRIBUTION OF TRUCK FACTORS FOR DIFFERENT CLASSES OF HIGHWAYS AND VEHICLES IN THE UNITED STATES[a]

	Truck factors											
	Rural systems						Urban systems					
				Collectors								
Vehicle type	Interstate	Other Principal	Minor Arterial	Major	Minor	Range	Interstate	Other Freeways	Other Principal	Minor Arterial	Collectors	Range
Single-unit trucks												
2-axle, 4-tire	0.003	0.003	0.003	0.017	0.003	0.003–0.017	0.002	0.015	0.002	0.006	—	0.006–0.015
2-axle, 6-tire	0.21	0.25	0.28	0.41	0.19	0.19–0.41	0.17	0.13	0.24	0.23	0.13	0.13–0.24
3-axle or more	0.61	0.86	1.06	1.26	0.45	0.45–1.26	0.61	0.74	1.02	0.76	0.72	0.61–1.02
All single units	0.06	0.08	0.08	0.12	0.03	0.03–0.12	0.05	0.06	0.09	0.04	0.16	0.04–0.16
Tractor semitrailers												
4-axle or less	0.62	0.92	0.62	0.37	0.91	0.37–0.91	0.98	0.48	0.71	0.46	0.40	0.40–0.98
5-axle[b]	1.09	1.25	1.05	1.67	1.11	1.05–1.67	1.07	1.17	0.97	0.77	0.63	0.63–1.17
6-axle or more[b]	1.23	1.54	1.04	2.21	1.35	1.04–2.21	1.05	1.19	0.90	0.64	—	0.64–1.19
All multiple units	1.04	1.21	0.97	1.52	1.08	0.97–1.52	1.05	0.96	0.91	0.67	0.53	0.53–1.05
All trucks	0.52	0.38	0.21	0.30	0.12	0.12–0.52	0.39	0.23	0.21	0.07	0.24	0.07–0.39

[a] Compiled from data supplied by the Highway Statistics Division, U.S. Federal Highway Administration.

[b] Including full-trailer combinations in some states.

Source. After AI (1991).

TABLE 6.11 COMPUTATION OF TRUCK FACTOR FOR TRUCKS WITH FIVE OR MORE AXLES

Axle load (lb)	EALF	Number of axles	ESAL
Single Axles			
Under 3000	0.0002	0	0.000
3000–6999	0.0050	1	0.005
7000–7999	0.0320	6	0.192
8000–11,999	0.0870	144	12.528
12,000–15,999	0.3600	16	5.760
16,000–29,999	5.3890	1	5.389
Tandem Axles			
Under 6000	0.0100	0	0.000
6000–11,999	0.0100	14	0.140
12,000–17,999	0.0440	21	0.924
18,000–23,999	0.1480	44	6.512
24,000–29,999	0.4260	42	17.892
30,000–32,000	0.7530	44	33.132
32,001–32,500	0.8850	21	18.585
32,501–33,999	1.0020	101	101.202
34,000–35,999	1.2300	43	52.890
ESALs for all trucks weighed			255.151

$$\text{Truck factor} = \frac{\text{18-kip ESALs for all trucks weighed}}{\text{Number of trucks weighed}} = \frac{255.151}{165} = 1.5464$$

Note. 1 lb = 4.45 N.

TABLE 6.12 TRAFFIC GROWTH FACTORS

Annual growth rate (%)	20-Year design period	40-Year design period
1.0	1.1	1.2
1.5	1.2	1.3
2.0	1.2	1.5
2.5	1.3	1.6
3.0	1.3	1.8
3.5	1.4	2.0
4.0	1.5	2.2
4.5	1.6	2.4
5.0	1.6	2.7
5.5	1.7	2.9
6.0	1.8	3.2

Source. After PCA (1984).

The Asphalt Institute (AI, 1981a) and the AASHTO design guide (AASHTO, 1986) recommend the use of traffic over the entire design period to determine the total growth factor, as indicated by

$$\text{Total growth factor} = (G)(Y) = \frac{(1 + r)^Y - 1}{r} \tag{6.33}$$

TABLE 6.13 TOTAL GROWTH FACTOR

Design period (years)	Annual growth rate (%)							
	No growth	2	4	5	6	7	8	10
1	1.0	1.0	1.0	1.0	1.0	1.0	1.0	1.0
2	2.0	2.02	2.04	2.05	2.06	2.07	2.08	2.10
3	3.0	3.06	3.12	3.15	3.18	3.21	3.25	3.31
4	4.0	4.12	4.25	4.31	4.37	4.44	4.51	4.64
5	5.0	5.20	5.42	5.53	5.64	5.75	5.87	6.11
6	6.0	6.31	6.63	6.80	6.98	7.15	7.34	7.72
7	7.0	7.43	7.90	8.14	8.39	8.65	8.92	9.49
8	8.0	8.58	9.21	9.55	9.90	10.26	10.64	11.44
9	9.0	9.75	10.58	11.03	11.49	11.98	12.49	13.58
10	10.0	10.95	12.01	12.58	13.18	13.82	14.49	15.94
11	11.0	12.17	13.49	14.21	14.97	15.78	16.65	18.53
12	12.0	13.41	15.03	15.92	16.87	17.89	18.98	21.38
13	13.0	14.68	16.63	17.71	18.88	20.14	21.50	24.52
14	14.0	15.97	18.29	19.16	21.01	22.55	24.21	27.97
15	15.0	17.29	20.02	21.58	23.28	25.13	27.15	31.77
16	16.0	18.64	21.82	23.66	25.67	27.89	30.32	35.95
17	17.0	20.01	23.70	25.84	28.31	30.84	33.75	40.55
18	18.0	21.41	25.65	28.13	30.91	34.00	37.45	45.60
19	19.0	22.84	27.67	30.54	33.76	37.38	41.45	51.16
20	20.0	24.30	29.78	33.06	36.79	41.00	45.76	57.28
25	25.0	32.03	41.65	47.73	54.86	63.25	73.11	98.35
30	30.0	40.57	56.08	66.44	79.06	94.46	113.28	164.49
35	35.0	49.99	73.65	90.32	111.43	138.24	172.32	271.02

Source. After AI (1981a).

Table 6.13 shows the total growth factor, which is the growth factor multiplied by the design period, as recommended by the Asphalt Institute. The same factor is used in the AASHTO design guide. If the growth rate is not uniform, different growth rates should be used for different load groups or types of vehicles.

To determine the annual growth rate, the following factors should be considered:

1. Attracted or diverted traffic due to the improvement of existing pavement
2. Normal traffic growth due to the increased number and usage of motor vehicles
3. Generated traffic due to motor vehicle trips that would not have been made if the new facility had not been constructed
4. Development traffic due to changes in land use as a result of the new facility

Example 6.10:

For an annual growth rate of 3.5% and a design period of 30 years, compute the growth factors by Eqs. 6.31, 6.32, and 6.33.

Solution: From Eq. 6.31, $G = 0.5[1 + (1 + 0.035)^{30}] = 1.90$; from Eq. 6.32, $G = (1 + 0.035)^{0.5 \times 30} = 1.68$; and from Eq. 6.33, $G = [(1 + 0.035)^{30} - 1]/(0.035 \times 30) = 1.72$. It can be seen that the use of Eq. 6.31 by averaging the traffic at the start and the end of a design period is most conservative and results in the greatest growth factor, while the growth factor based on Eq. 6.33 is only slightly greater than that based on Eq. 6.32.

6.4.4 Lane Distribution Factor

For two-lane highways, the lane in each direction is the design lane, so the lane distribution factor is 100%. For multilane highways, the design lane is the outside lane. Table 6.14 shows the truck distribution on a multiple-lane highway based on 129 counts from 1982 to 1983 in six states (Darter et al., 1985). For four-lane highways with two lanes in each direction, the lane distribution factors range from 66 to 94%. For multiple-lane highways with three or more lanes in each direction, the lane distribution factors range from 49 to 82%.

Based on the data in Table 6.14, the Portland Cement Association (PCA, 1984) developed a chart for determining the proportion of trucks in the design lane, as shown in Figure 6.8. The Asphalt Institute (AI, 1981a) combines the directional and lane distribution factors ($D \times L$) and determines the percentage of total truck traffic in the design lane by Table 6.15. A comparison of Table 6.15 and Figure 6.8 indicates that the lane distribution factors by the Asphalt Institute are

TABLE 6.14 TRUCK DISTRIBUTION FOR MULTIPLE-LANE HIGHWAYS

One-way ADT	Two lanes in each direction		Three or more lanes in each direction		
	Inner	Outer	Inner[a]	Center	Outer
2000	6	94	6	12	82
4000	12	88	6	18	76
6000	15	85	7	21	72
8000	18	82	7	23	70
10,000	19	81	7	25	68
15,000	23	77	7	28	65
20,000	25	75	7	30	63
25,000	27	73	7	32	61
30,000	28	72	8	33	59
35,000	30	70	8	34	58
40,000	31	69	8	35	57
50,000	33	67	8	37	55
60,000	34	66	8	39	53
70,000	—	—	8	40	52
80,000	—	—	8	41	51
100,000	—	—	9	42	49

[a] Combined inner one or more lanes.

Source. After Darter et al. (1985).

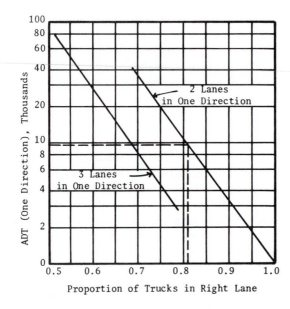

Figure 6.8 Proportion of trucks in design lane of multiple-lane highways. (After PCA (1984).)

TABLE 6.15 PERCENTAGE OF TOTAL TRUCK TRAFFIC IN DESIGN LANE

Number of traffic lanes in two directions	Percentage of trucks in design lane
2	50
4	45 (35–48)[a]
6 or more	40 (25–48)[a]

[a] Probable range.
Source. After AI (1981a).

TABLE 6.16 LANE DISTRIBUTION FACTOR

No. of lanes in each direction	Percentage of 18-kip ESAL in design lane
1	100
2	80–100
3	60–80
4	50–75

Source. After AASHTO (1986).

about the same as those by PCA for an ADT of 3000 in one direction. The lane distribution factors recommended by the AASHTO design guide are shown in Table 6.16. Note that the percentage in Table 6.15 is based on total traffic but the percentage in Table 6.16 is based on the traffic in one direction.

Example 6.11:

A two-lane major rural highway has an AADT of 4000 during the first year of traffic, 25% trucks, 4% annual growth rate, and 50% on the design lane. If the distribution of trucks is the same as shown in Table 6.9 and the distribution of truck factors is as shown in Table 6.10, compute the ESAL for a design period of 20 years.

Solution: For a growth rate of 4% and a design period of 20 years, from Table 6.13, or Eq. 6.33, $(G)(Y) = 29.78$. The truck factor for all trucks, as shown in Table 6.10, is the weighted average of the distribution of trucks shown in Table 6.9. If the distribution of trucks is given in Table 6.9, it is not necessary to breakdown the trucks into different classes and the use of $T_f = 0.38$ for all trucks in the "other principal" column of the rural systems, as shown in Table 6.10, is sufficient. From Eq. 6.30, ESAL = $4000 \times 0.25 \times 0.38 \times 29.78 \times 0.5 \times 365 = 2,065,200$.

6.4.5 Other Design Considerations

In considering the number of load repetitions for pavement design, two important questions must be answered. The first question is whether the passage of a set of tandem or tridem axles should be counted as one or more than one repetition. The second question is whether the effect of traffic wandering or lateral load placement should be considered in the design of highway pavements.

On the first question, if the design is based on the equivalent 18-kip (80-kN) single-axle load, the passage of a set of tandem or tridem axles should be counted as one repetition because the equivalent factor is based on one passage of the tandem- or tridem-axle loads. If the design is based on different load groups and the stresses or strains under single-, tandem-, or tridem-axle loads are used for damage analysis, the damage caused by the passage of a set of tandem or tridem axles should be analyzed by the method shown in Figure 3.3 or 3.4. The data presented in Tables 6.5 and 6.6 for flexible pavements and in Table 6.8 for rigid pavement can be used to determine the number of repetitions required. The number of repetitions can be computed from these tables by dividing the total damage ratio (or cracking index) by the primary damage ratio and the results are presented in Table 6.17.

As can be seen from Table 6.17, the required number of repetitions may range from 1 to 2 for tandem axles and from 1 to 3 for tridem axles, depending on the combined effect of pavement thickness and subgrade support. For flexible pavements, the passage of a set of tandem or tridem axles may be considered as one repetition for thicker pavements on a week subgrade, but as two or three repetitions for thin pavements, regardless of the subgrade support. For rigid pavements with poor subgrade support, the passage of a set of tandem or tridem axles may be considered as one repetition, but the number of repetitions is increased as the subgrade support improved. However, it should be pointed out that these general trends are based on the input data assumed. The interactions among all the design variables are so complex that the best way is to analyze the damage directly by KENLAYER or KENSLABS, instead of applying a fixed number to tandem or tridem axles.

TABLE 6.17 NUMBER OF REPETITIONS FOR ONE PASSAGE OF MULTIPLE AXLES

Flexible Pavements				
HMA thickness (in.)	2		12	
Subgrade modulus (pci)	5000	30,000	5000	30,000
Tandem				
Fatigue cracking	1.89	1.99	1.01	1.08
Permanent deformation	2.00	2.00	1.02	1.28
Tridem				
Fatigue cracking	2.77	2.98	1.03	1.08
Permanent deformation	3.00	3.00	1.05	1.55

Rigid Pavements				
Slab thickness (in.)	6		14	
Subgrade k value (pci)	50	500	50	500
Tandem	1.00	2.00	1.00	1.08
Tridem	1.02	3.00	1.00	1.54

Note. 1 in. = 25.4 mm, 1 psi = 6.9 kPa, 1 pci = 271.3 kN/m³.

As to the second question, traffic wandering certainly has a great effect on pavement damage. However, instead of reducing the number of load repetitions, the effect is incorporated in the failure criteria. For example, the damage coefficients for flexible pavement, such as f_1 in Eq. 3.6 and f_4 in Eq. 3.7, should include the wandering effect of traffic. In the PCA method of rigid pavement design, the wandering effect is considered in the fatigue analysis by reducing the edge stress and in the erosion analysis by increasing the allowable number of load repetitions, as described in Section 12.2.

6.5 SUMMARY

This chapter presents the load equivalent concept and the methods for traffic analysis.

Important Points Discussed in Chapter 6

1. The concept of load equivalency has been used most frequently in the empirical methods of pavement design. In the mechanistic methods, it is not necessary to apply the load equivalency concept because different loads can be considered separately in the design process. This is why the concept has been used more frequently on flexible pavements than on rigid pavements.

2. There are two types of load equivalency, one called the equivalent single-wheel load (ESWL) and the other called the equivalent axle load factor (EALF).

The ESWL is based on the fixed traffic procedure of converting the most critical wheel loads, usually in multiple wheels, into an equivalent single-wheel load and using it for design purposes. The EALF is based on the fixed vehicle procedure of converting the number of repetitions of a given axle load, either single, tandem, or tridem, into an equivalent number of repetitions of an 18-kip (80-kN) single-axle load.

3. Various criteria have been used for determining the ESWL for a two-layer system, e.g., equal vertical interface stress, equal interface deflection, and equal tensile strain. The equivalency based on one criterion may be quite different from that based on other criteria. Also, the equivalency based on equal contact radius is different from that based on equal contact pressure.

4. The values of EALF depend on the failure criterion employed. The EALF based on fatigue cracking is different from that based on permanent deformation. The use of a single value for both modes of failure is approximate at best. The most widely used method for determining the EALF is that which uses the empirical equations developed from the AASHO Road Test. However, damage analyses by KENLAYER and KENSLABS indicate that the values of EALF vary a great deal and are difficult to predict. Due to the complex interactions among a large number of variables, it is not possible to select an appropriate EALF that can be applied to all situations. For a truly mechanistic design method, each load group should be considered individually, instead of using equivalent single-axle loads.

5. The traffic to be used in design is not the average traffic during the first year, but the average during the design period. Three different methods based on the compound rate are presented to compute the traffic growth factor. The use of average traffic at the start and the end of the design period gives the highest growth factor, while the use of the traffic at the middle of the design period results in the lowest growth factor.

6. Traffic is the most important factor in pavement design. Every effort should be made to collect actual traffic data. If actual data are not available, Tables 6.9 and 6.10, which give average values for different classes of highways in the United States, can be used as a guide.

7. In addition to the initial traffic and the growth factor, the directional and lane distribution factors must also be considered. Unless the traffic loading or volume is heavier in one direction than in the other due to some special reason, a directional distribution of 0.5 is assumed. The lane distribution factor varies with the number of lanes and the ADT and can be obtained from Figure 6.8 or Tables 6.15 and 6.16, although these tables do not consider ADT as a factor.

8. When the design is based on the equivalent 18-kip (80-kN) single-axle load, the use of a truck factor is very convenient. The method for computing truck factors based on the number of axles and trucks weighed, number of trucks counted, and the AASHTO equivalent factors is illustrated.

9. Damage analyses by KENLAYER and KENSLABS indicate that the passage of a set of tandem or tridem axles may be considered as one repetition in some cases, but as more than one repetition in other cases, depending on pave-

ment thickness and subgrade support. The problem is so complex that it is best to analyze each load group, either tandem or tridem, directly, instead of applying a fixed number of repetitions.

PROBLEMS

6-1. A set of dual tires is spaced at 34 in. center to center and carries a total load of 45,000 lb with a tire pressure of 100 psi. Assuming the pavement to be a homogeneous half-space, determine the ESWL for a pavement of 25 in. using (a) the Boyd and Foster method, (b) the Foster and Ahlvin method, and (c) Huang's chart based on equal contact radius. [Answer: 32,200 lb, 32,800 lb, 32,800 lb]

6-2. A full-depth asphalt pavement is loaded by a set of dual wheels, each weighing 8000 lb and spaced at 20 in. on centers. The hot mix asphalt has a thickness of 10 in. and an elastic modulus of 250,000 psi; the subgrade has an elastic modulus of 10,000 psi. Both layers are incompressible with a Poisson ratio of 0.5. If the dual wheels and the equivalent single wheel have the same contact radius of 6 in., determine the ESWL based on (a) equal interface deflection and (b) equal tensile strain at the bottom of asphalt layer. [Answer: 13,800 lb, 10,900 lb]

6-3. A pavement is subjected to the single-axle loads shown in Table P6.3. Determine the ESAL for a design period of 20 years using (a) AI's equivalent axle load factors and (b) the equivalent axle load factors from Eq. 6.23. [Answer: 2.99×10^6, 3.07×10^6]

TABLE P6.3

Axle load (kip)	Number per day	Axle load (kip)	Number per day	Axle load (kip)	Number per day
12	200.0	20	47.2	28	2.9
14	117.4	22	21.4	30	1.2
16	84.5	24	12.9	32	0.7
18	61.4	26	6.1	34	0.3

6-4. Based on Eq. 6.25, discuss the effect of p_t on EALF.

6-5. Derive Eq. 6.33 and indicate the assumptions on which the equation is obtained. [*Hint:* $(G)(Y) = \int_0^n (1 + r)^n dn$]

6-6. Estimate the equivalent 18-kip single-axle load applications (ESAL) for a four-lane pavement (two lanes in each direction) of a rural interstate highway with a truck count of 1000 per day (including 2-axle, 4-tire panel, and pickup trucks), an annual growth rate of 5%, and a design life of 20 years. [Answer: 2.82×10^6]

6-7. Table P6.7 is abstracted from a W-4 table on tractor semitrailer combinations for a loadometer station from July 16 to August 8. It is assumed that the traffic during the recorded period represents the average over the entire year. If the pavement to be constructed has a structural number SN of 5, estimate the ESAL of the tractor semitrailer combinations during the first year in two directions over all lanes based on a p_t of 2.5. [Answer: 192,000]

6-8. Same as Problem 6-7 but for a rigid pavement with a concrete thickness of 9 in. [Answer: 257,000]

Axle loads (lb)	3-axle	4-axle	5-axle	Tractor semitrailer combinations probable no.
Single axle under 3000	—	—	—	—
3000–6999	256	227	60	3188
7000–7999	148	243	85	2843
8000–11,999	345	939	363	9942
12,000–15,999	174	288	54	3111
16,000–17,999	67	225	12	1899
18,000–19,000	22	141	7	1078
19,001–19,999	8	54	5	423
20,000–21,999	5	80	9	598
22,000–23,999	1	29		144
24,000–25,999	—	1	—	6
26,000–29,999	—	1	—	6
Total single axles weighed	1026	2228	595	
Total single axles counted	5541	14,526	3171	23,238
Tandem axle under 6000	—	1	—	7
6000–11,999	—	237	173	2631
12,000–17,999	—	189	209	2541
18,000–23,999	—	218	150	2362
24,000–29,999	—	270	214	3103
30,000–31,500	—	62	52	703
31,501–31,999	—	11	11	141
32,000–33,999	—	36	43	503
34,000–35,999	—	26	35	388
36,000–37,999	—	16	28	280
38,000–39,999	—	12	27	247
40,000–41,999	—	10	19	183
42,000–43,999	—	14	11	160
44,000–45,999	—	2	6	51
46,000–49,999	—	5	5	64
50,000–53,999	—	1	—	7
Total tandem axles weighed		1110	983	—
Total tandem axles counted		7263	6135	13,398
Total vehicles counted	1847	7263	3087	12,197

6-9. Based on the axles weighed and the axles and vehicles counted, as shown in Table P6.7, determine the truck factor of all tractor semitrailer combinations for a flexible pavement with SN = 5 and p_t = 2.5. [Answer: 1.04]

6-10. Same as Problem 6-9 but for a rigid pavement with D = 9. [Answer: 1.38]

7

Material Characterization

7.1 RESILIENT MODULUS

The resilient modulus is the elastic modulus to be used with the elastic theory. It is well known that most paving materials are not elastic but experience some permanent deformation after each load application. However, if the load is small compared to the strength of the material and is repeated for a large number of times, the deformation under each load repetition is nearly completely recoverable and proportional to the load and can be considered as elastic.

Figure 7.1 shows the straining of a specimen under a repeated load test. At the initial stage of load applications, there is considerable permanent deformation, as indicated by the plastic strain in the figure. As the number of repetitions

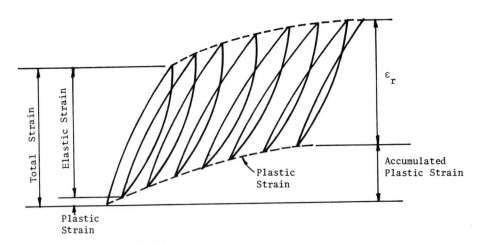

Figure 7.1 Strains under repeated loads.

increases, the plastic strain due to each load repetition decreases. After 100 to 200 repetitions, the strain is practically all recoverable, as indicated by ϵ_r in the figure.

The elastic modulus based on the recoverable strain under repeated loads is called the resilient modulus M_R, defined as

$$M_R = \frac{\sigma_d}{\epsilon_r} \tag{7.1}$$

in which σ_d is the deviator stress, which is the axial stress in an unconfined compression test or the axial stress in excess of the confining pressure in a triaxial compression test. Because the applied load is usually small, the resilient modulus test is a nondestructive test and the same sample can be used for many tests under different loading and environmental conditions.

7.1.1 Loading Waveform

The type and duration of loading used in the repeated load test should simulate that actually occurring in the field. When a wheel load is at a considerable distance from a given point in the pavement, the stress at that point is zero. When the load is directly above the given point, the stress at the point is maximum. It is therefore reasonable to assume the stress pulse to be a haversine or triangular loading, the duration of which depends on the vehicle speed and the depth of the point below the pavement surface.

Barksdale (1971) investigated the vertical stress pulses at different points in flexible pavements. The stress pulse can be approximated by a haversine or a triangular function, as shown in Figure 7.2. After considering the inertial and viscous effects based on the vertical stress pulses measured in the AASHO Road Test, the stress pulse time can be related to the vehicle speed and depth, as shown in Figure 7.3. Because of these effects, the loading time is not inversely proportional to the vehicle speed.

Brown (1973) derived the loading time for a bituminous layer as a function of vehicle speed and layer thickness. The loading time is based on the average pulse

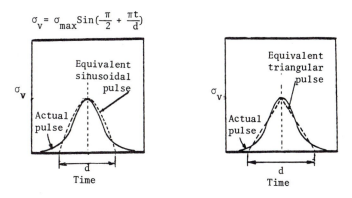

Figure 7.2 Equivalent haversine and triangular pulse.

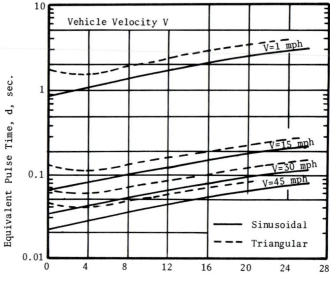

Figure 7.3 Vertical stress pulse time under haversine or triangular loading (1 in. = 25.4 mm. 1 mph = 1.6 km/h). (After Barksdale (1971).)

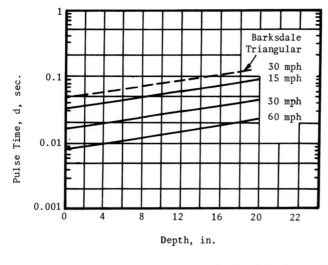

Figure 7.4 Vertical stress pulse time under square wave form (1 in. = 25.4 mm, 1 mph = 1.6 km/h). (After McLean (1974).)

time for stresses in the vertical and horizontal directions at various depths in the bituminous layer. For thicker layers, his loading times are slightly smaller than those obtained by Barksdale.

McLean (1974) determined the loading time for an equivalent square wave vertical pulse, as shown in Figure 7.4 on which the Barksdale's results for 30 mph (48 km/h) triangular loading are superimposed for comparison. It can be seen that the pulse time based on the square wave is shorter than that based on the triangular wave, which is as expected.

Example 7.1:

Repeated load compression tests are employed to determine the resilient moduli of the surface, base, and subbase materials in a flexible pavement, as shown in Figure 7.5. The points at the midheight of each layer are used to determine the stress pulse times. If the vehicle speed is 40 mph (64 km/h), what should be the load durations of haversine and square wave loadings for each material?

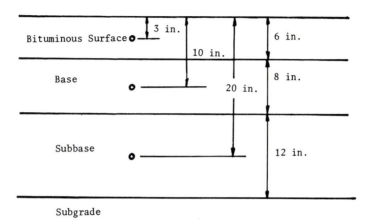

Figure 7.5   Example 7.1 (1 in. = 25.4 mm).

Solution: With a depth of 3 in. (76 mm) below the surface and a vehicle speed of 40 mph (64 km/h), the vertical stress pulse time is 0.028 s for a haversine load, as obtained from Figure 7.3, and 0.014 s for a square wave, as obtained from Figure 7.4. The results for all three materials are shown in Table 7.1.

It can be seen that the stress pulse time based on McLean's square wave loading is about one-half of that based on Barksdale's haversine loading. Note that a haversine pulse time of 0.028 to 0.064 s for a vehicle speed of 40 mph (64 km/h) is much smaller than the 0.1 s based on Eq. 2.55. The haversine pulse times indicated in Table 7.1 are based on $\sin(\pi/2 + \pi t/d)$, as shown in Figure 7.2, while Eq. 2.55 is based on $\sin^2(\pi/2 + \pi t/d)$, as shown in Figure 2.40. The use of $\sin^2(\pi/2 + \pi t/d)$ in KENLAYER results in a reverse curve with a longer pulse time, which checks more closely with the actual stress pulse in pavements.

When the elastic theory is employed to analyze pavements, the duration of loading for determining the resilient modulus under repeated loading can be selected by using Figure 7.3 or 7.4 as a guide, depending on whether the loading is haversine or square wave. In view of the fact that the vehicle speed varies a great

TABLE 7.1 VERTICAL STRESS PULSE TIMES FOR MATERIALS AT VARIOUS DEPTHS

Material	Bituminous surface	Base course	Subbase course
Depth (in.)	3	10	20
Haversine wave	0.028 s	0.041 s	0.064 s
Square wave	0.014 s	0.020 s	0.031 s

Note. 1 in. = 25.4 mm.

deal and the depth of the material may not be known during the design stage, it is recommended that a haversine load with a duration of 0.1 s and a rest period of 0.9 s be used. It should be noted that load duration has very little effect on the resilient modulus of granular materials, some effect on fine-grained soils depending on moisture contents, and a considerable effect on bituminous materials. The effect of rest period is not known, but is probably insignificant.

7.1.2 Equipment

The resilient modulus of granular materials and fine-grained soils can be determined by the repeated load triaxial test. Figure 7.6 shows the test setup recommended by FHWA (1978). The sample is 4 in. (102 mm) in diameter and 8 in. (203 mm) in height. The triaxial cell is similar to most standard cells except that it is larger to accommodate the internally mounted load and deformation measuring equipment and has additional outlets for the electrical leads from the measuring devices. With the internally mounted measuring devices, air, instead of water, should be used as the confining fluid. Other cells with suitable externally mounted load and deformation measuring equipment also may be used. However, the use of internal measuring equipment has the advantage that the effects of equipment deformation, end restraint, and piston friction can be eliminated.

The repetitive loading device can be an air-actuated piston assembly with electronic solenoid control or a sophisticated electrohydraulic testing machine with precise control on the shape of load pulse. The deformation measuring equipment consists of two linear variable differential transformers (LVDT) attached to the specimen by a pair of clamps at the upper and lower quarter points. In addition to the measuring devices, suitable signal excitation, conditioning, and recording equipment are needed.

The setup shown in Figure 7.6 can be used to determine the resilient modulus of asphalt mixtures. Unless the temperature or the level of stress is high, the confining pressure has very little effect on the resilient modulus, so the repeated load unconfined compression test without confining pressure, rubber membrane, and porous stones can be used. A temperature control system should be used to maintain a constant temperature of the specimen during the test.

The resilient modulus of asphalt mixtures can also be determined by the repeated load indirect tension test. A compressive load with a haversine or other suitable waveform is applied in the vertical diametric plane of a cylindrical specimen through a loading strip and the resulting horizontal recoverable deformation is measured. The repetitive loading device is the same as that used in the compression test. The transducer arrangement to measure the resilient modulus is shown in Figure 7.7. The resilient modulus is computed by

$$M_R = \frac{P(\nu + 0.2734)}{\delta t} \tag{7.2}$$

in which P is the magnitude of the dynamic load in pounds, ν is Poisson ratio, δ is the total recoverable deformation in inches, and t is the specimen thickness in inches. Poisson ratio is generally taken as 0.35.

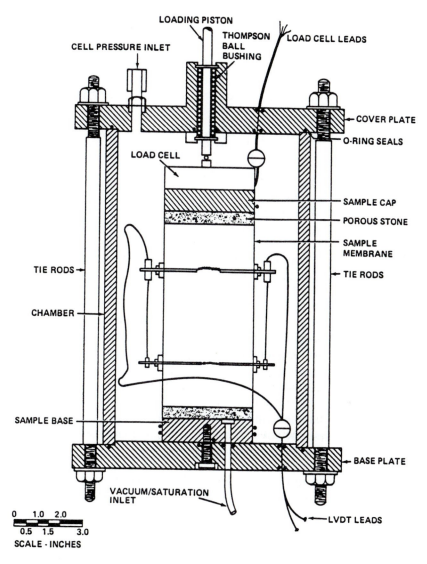

LOADING PISTON

CELL PRESSURE INLET

THOMPSON BALL BUSHING

LOAD CELL LEADS

COVER PLATE

O-RING SEALS

LOAD CELL

SAMPLE CAP

POROUS STONE

SAMPLE MEMBRANE

TIE RODS

TIE RODS

CHAMBER

SAMPLE BASE

BASE PLATE

VACUUM/SATURATION INLET

LVDT LEADS

0 1.0 2.0
0.5 1.5 3.0
SCALE - INCHES

Figure 7.6 Triaxial cell for testing cylindrical specimens. (After FHWA (1978).)

7.1.3 Granular Materials

The resilient modulus test for granular materials and fine-grained soils is specified by AASHTO (1989) in "T274-82 Resilient Modulus of Subgrade Soils." Sample conditioning can be accomplished by applying various combinations of confining pressures and deviator stresses as follows:

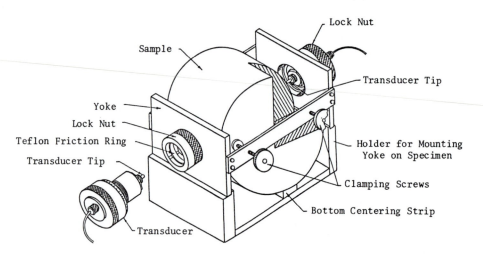

Figure 7.7 Transducer arrangement for indirect tensile test.

1. Set the confining pressure to 5 psi (35 kPa) and apply a deviator stress of 5 psi (35 kPa) and then 10 psi (69 kPa), each for 200 repetitions.
2. Set the confining pressure to 10 psi (69 kPa) and apply a deviator stress of 10 psi (69 kPa) and then 15 psi (104 kPa), each for 200 repetitions.
3. Set the confining pressure to 15 psi (104 kPa) and apply a deviator stress of 15 psi (104 kPa) and then 20 psi (138 kPa), each for 200 repetitions.

After sample conditioning, the following constant confining pressure-increasing deviator stress sequence is applied and the results are recorded at the 200th repetition of each deviator stress:

1. Set the confining pressure to 20 psi (138 kPa) and apply deviator stresses of 1, 2, 5, 10, 15, and 20 psi (6.9, 14, 35, 69, 104, and 138 kPa).
2. Reduce the confining pressure to 15 psi (104 kPa) and apply deviator stresses of 1, 2, 5, 10, 15, and 20 psi (6.9, 14, 35, 69, 104, and 138 kPa).
3. Reduce the confining pressure to 10 psi (69 kPa) and apply deviator stresses of 1, 2, 5, 10, and 15 psi (6.9, 14, 35, 69, and 104 kPa).
4. Reduce the confining pressure to 5 psi (35 kPa) and apply deviator stresses of 1, 2, 5, 10, and 15 psi (6.9, 14, 35, 69, and 104 kPa).
5. Reduce the confining pressure to 1 psi (6.9 kPa) and apply deviator stresses of 1, 2, 5, 7.5, and 10 psi (6.9, 14, 35, 52, and 69 kPa). Stop the test after 200 repetitions of the last deviator stress level or when the specimen fails.

Example 7.2:

Table 7.2 shows the results of resilient modulus tests on a granular material. The distance between the LVDT clamps is 4 in. (102 mm). The average recoverable deformations measured by the two LVDTs after 200 repetitions of each deviator

TABLE 7.2 COMPUTATION OF RESILIENT MODULUS FOR GRANULAR MATERIALS

Confining pressure σ_3 (psi)	Deviator stress σ_d (psi)	Recoverable deformation (0.001 in.)	Recoverable strain ϵ_r ($\times 10^{-3}$)	Resilient modulus M_R ($\times 10^3$ psi)	Stress invariant θ (psi)
20	1 _2_	0.264	0.066	15.2	61
	2 _4_	0.496	0.124	16.1	62
	5 _10_	1.184	0.296	16.9	65
	10 _15_	2.284	0.571	17.5	70
	15 _20_	3.428	0.857	17.5	75
	20 _30_	4.420	1.105	18.1	80
15	1	0.260	0.065	15.4	46
	2	0.512	0.128	15.6	47
	5	1.300	0.325	15.4	50
	10	2.500	0.625	16.0	55
	15	3.636	0.909	16.5	60
	20	4.572	1.143	17.5	65
10	1	0.324	0.081	12.3	31
	2	0.672	0.168	11.9	32
	5	1.740	0.435	11.5	35
	10	3.636	0.909	11.0	40
	15	3.872	0.968	15.5	45
5	1	0.508	0.127	7.9	16
	2	0.988	0.247	8.1	17
	5	2.224	0.556	9.0	20
	10	3.884	0.971	10.3	25
	15	5.768	1.442	10.4	30
1	1	0.636	0.159	6.3	4
	2	0.880	0.220	9.1	5
	5	2.704	0.676	7.4	8
	7.5	3.260	0.815	9.2	10.5
	10	4.444	1.111	9.0	13

Note. 1 psi = 6.9 kPa, 1 in. = 25.4 mm.

stress are shown in Table 7.2. Determine the nonlinear coefficient K_1 and exponent K_2 in Eq. 3.8.

Solution: The first three columns of data are given. The recoverable strains are obtained by dividing the recoverable deformation by an initial length of 4 in. (102 mm). The resilient modulus is obtained from Eq. 7.1. The stress invariant is obtained by

$$\theta = \sigma_1 + 2\sigma_3 = \sigma_d + 3\sigma_3 \tag{7.3}$$

The resilient modulus is plotted against the stress invariant on logarithmic scales, as shown in Figure 7.8. The least-square lines has an intercept of 3690 psi (25.5 MPa) at $\theta = 1$ psi (6.9 kPa), which is K_1, and a slope of $\log(18.58/3.69)/\log(100/1) = 0.351$, which is K_2, or

$$M_R = 3690\theta^{0.351} \tag{7.4}$$

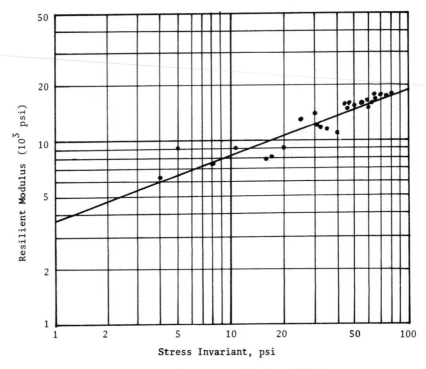

Figure 7.8 Example 7.2 (1 psi = 6.9 kPa).

7.1.4 Fine-Grained Soils

Sample conditioning for fine-grained soils is not as extensive as that for granular materials. AASHTO recommends the use of a confining pressure of 6 psi (41 kPa) followed by 200 repetitions each of deviator stresses of 1, 2, 4, 8, and 10 psi (6.9, 14, 28, 55, and 69 kPa). After sample conditioning, the following sequences of stresses are applied and the results are recorded at the 200th repetition of each deviator stress:

1. Apply a deviator stress of 1 psi (6.9 kPa) under confining pressures of 6, 3, and 0 psi (41, 21, and 0 kPa).
2. Apply a deviator stress of 2 psi (14 kPa) under the same decreasing confining pressures of 6, 3, and 0 psi (41, 21, and 0 kPa).
3. Continue the constant deviator stress–decreasing confining pressure sequence for deviator stresses of 4, 8, and 10 psi (28, 48, and 69 kPa).

It is believed that a deviator stress of 10 psi is probably the largest that may occur in a subgrade. If the resilient moduli at higher deviator stresses are required, larger deviator stresses should be used until the unconfined compressive strength is reached.

Example 7.3:

Resilient modulus tests were made on a fine-grained soil. The distance between the LVDT clamps is 4 in. (102 mm). The average recoverable deformations measured by the two LVDTs after 200 repetitions of each deviator stress are shown in Table 7.3. The unconfined compressive strength of the soil is 15.5 psi (107 kPa). Determine the coefficients K_1, K_2, K_3, and K_4 in Eq. 3.13 and the maximum and minimum resilient modulus.

TABLE 7.3 COMPUTATION OF RESILIENT MODULUS FOR FINE-GRAINED SOILS

Deviator stress σ_d (psi)	Confining pressure σ_3 (psi)	Recoverable deformation (0.001 in.)	Recoverable strain ϵ_r (10^{-3})	Resilient modulus M_R (10^3 psi)
2	6	0.392	0.098	10.2
	3	0.416	0.104	9.6
	0	0.456	0.114	8.8
4	6	0.816	0.204	9.8
	3	0.868	0.271	9.2
	0	1.040	0.260	7.7
4	6	2.052	0.513	7.8
	3	2.224	0.556	7.2
	0	3.020	0.755	5.3
8	6	5.712	1.428	5.6
	3	7.112	1.778	4.5
	0	9.412	2.353	3.4
10	6	7.692	1.923	5.2
	3	11.112	2.778	3.6
	0	16.000	4.000	2.5

Note. 1 psi = 6.9 kPa, 1 in. = 25.4 mm.

Solution: The first three columns of data are given. The recoverable strains are obtained by dividing the recoverable deformation by an initial length of 4 in. (102 mm). The resilient modulus is computed by Eq. 7.1. The resilient modulus is plotted against the deviator stress on arithmetic scales, as shown in Figure 7.9. Two straight lines with the best fit are drawn. The intersection of the two lines gives K_1 = 5600 psi (38.6 MPa) and K_2 = 5.2 psi (36 kPa). The slopes of the lines are K_3 = (10,800 − 5600)/5.2 = 1000 and K_4 = (5600 − 1600)/(15.5−5.2) = 388. The maximum resilient modulus is 8800 psi (61 MPa), which corresponds to a deviator stress of 2 psi (14 kPa). The minimum resilient modulus is 1600 psi (11 MPa), which corresponds to an unconfined compressive strength of 15.5 psi (107 kPa).

It should be noted that the procedures specified in AASHTO T-274 for the testing of granular and fine-grained soils have many shortcomings and were modified by the Strategic Highway Research Program to produce more repeatable and less complicated test procedures (Claros et al., 1990). The main changes are the use of external LVDTs for deformation measurements of all soil types and the complete modification of loading sequences, eliminating low deviator stresses, which produce high variability, and high deviator stresses, which cause sample failures.

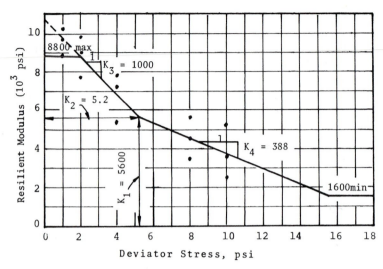

Figure 7.9 Example 7.3 (1 psi = 6.9 kPa).

7.1.5 Asphalt Mixtures

The specimens used for compression tests are usually 4 in. (102 mm) in diameter and 8 in. (203 mm) high, while those for indirect tensile tests are 4 in. (102 mm) in diameter and 2.5 in. (64 mm) thick. The advantage of indirect tensile tests is to use specimens of Marshall size, which can be easily fabricated in the laboratory or cored from the pavements.

Sample conditioning is required before the recoverable deformation is recorded. The conditioning can be effected by applying a repeated load to the specimen without impact for a minimum period sufficient to obtain uniform deformation readout. Depending upon the loading frequency and temperature, a minimum of 50 to 200 load repetitions is typical; however, the minimum for a given situation must be determined so that the resilient deformations are stable. Tests on the same specimen are usually made at three temperatures of 41, 77, and 104°F (5, 25, and 40°C) to generate design values over the range of temperatures normally encountered in pavements.

The resilient modulus of compression specimens is determined from Eq. 7.1. A 20-psi (138-kPa) haversine loading with a duration of 0.1 s and a rest period of 0.9 s has been most frequently used. The resilient modulus of indirect tension specimens is computed by Eq. 7.2. Dynamic load amplitudes of 40 to 60 lb (180 to 270 kN) with a load duration of 0.1 s applied every 3 s are typical. The test is specified by ASTM (1989b) in "D4123-82 Standard Test Method for Indirect Tension Test for Resilient Modulus of Bituminous Mixtures."

7.1.6 Correlations with Other Tests

Various empirical tests have been used to determine the material properties for pavement design. Most of these tests measure the strength of the material and are not a true representation of the resilient modulus. An extensive study was made by Van Til et al. (1972) to relate the resilient modulus as well as other test parameters to the soil support value or the layer coefficient employed in the AASHO design equation. These correlations can be used as a guide if other more reliable information is not available. It should be noted that any empirical correlation is based on a set of local conditions. The correlation is not valid if the actual conditions are different from those under which the correlation is established. Therefore, great care must be exercised in the judicious selection of the resilient modulus from these correlations.

Subgrade Soils

Figure 7.10 shows a correlation chart that can be used to estimate the resilient modulus of subgrade soils from the R value, CBR, Texas triaxial classification, and group index.

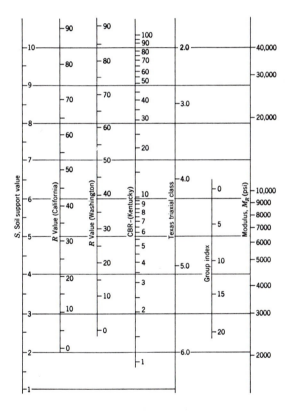

Figure 7.10 Correlation chart for estimating resilient modulus of subgrade soils (1 psi = 6.9 kPa). (After Van Til et al. (1972).)

R Value

The R value is the resistance value of a soil determined by a stabilometer. The stabilometer test was developed by the California Division of Highways and measures basically the internal friction of the material, while the cohesion for bounded materials is measured by the cohesiometer test. Figure 7.11 is a sche-

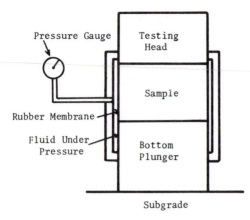

Figure 7.11 Schematic diagram of stabilometer

Pressure Gauge

Testing Head

Sample

Rubber Membrane

Fluid Under Pressure

Bottom Plunger

Subgrade

matic diagram of stabilometer, which is a closed-system triaxial test. A vertical pressure of 160 psi (1.1 MPa) is applied to a sample, 4 in. (102 mm) in diameter and about 4.5 in. (114 mm) in height, and the resulting horizontal pressures induced in the fluid within the rubber membrane are measured.

The resistance value is computed by

$$R = 100 - \frac{100}{(2.5/D_2)(p_v/p_h - 1) + 1} \tag{7.5}$$

in which R is the resistance value; p_v is the applied vertical pressure of 160 psi (1.1 MPa); p_h is the transmitted horizontal pressure at p_v of 160 psi (1.1 MPa); and D_2 is the displacement of stabilometer fluid necessary to increase horizontal pressure from 5 to 100 psi (35 to 690 kPa), measured in revolutions of a calibrated pump handle. The value of D_2 is determined after the maximum vertical pressure of 160 psi (1.1 MPa) is applied. If the sample is a liquid with no shear resistance, then $p_h = p_v$, or from Eq. 7.5, $R = 0$. If the sample is rigid with no deformation at all, then $p_h = 0$, or $R = 100$. Therefore, the R value ranges from 0 to 100. To ensure that the sample is saturated, California used an exudation pressure of 240 psi (1.7 MPa) while Washington used 300 psi (2.1 MPa).

CBR

The California Bearing Ratio test (CBR) is a penetration test wherein a standard piston, having an area of 3 in.^{2} (1935 mm^2), is used to penetrate the soil at a standard rate of 0.05 in. (1.3 mm) per minute. The pressure at each 0.1-in. (2.5-mm) penetration up to 0.5 in. (12.7 mm) is recorded and its ratio to the bearing value of a standard crushed rock is termed as the CBR. The standard values of a high-quality crushed rock are as follows:

Penetration	Pressure
0.1 in. (2.5 mm)	1000 psi (6.9 MPa)
0.2 in. (5.0 mm)	1500 psi (10.4 MPa)
0.3 in. (7.6 mm)	1900 psi (13.1 MPa)
0.4 in. (10.2 mm)	2300 psi (15.9 MPa)
0.5 in. (12.7 mm)	2600 psi (17.9 MPa)

In most cases, CBR decreases as the penetration increasees, so the ratio at the

0.1-in. (2.5-mm) penetration is used as the CBR. In some cases, the ratio at 0.2 in. (5.0 mm) may be greater than that at 0.1 in. (2.5 mm). If this occurs, the ratio at 0.2 in. (5.0 mm) should be used. In the Kentucky method, the specimen is molded at or near to the optimum moisture as determined by the standard proctor method. However, the sample is placed in a mold, 6 in. (152 mm) in diameter and 4.6 in. (117 mm) in height, and compacted in five equal layers, each subjected to 10 blows of a 10-lb (4.5-kg) hammer at 18 in. (457 mm) drop. The specimen is soaked for 4 days before testing.

Texas Triaxial Classification

The Texas triaxial test is used to classify soils based on the location of Mohr's envelope. The apparatus consists of a stainless cylinder with an inside diameter of $6\frac{3}{4}$ in. (171 mm) fitted with a tubular rubber membrane 6 in. (152 mm) in diameter. The lateral pressure σ_3 is applied by compressed air between the cylinder and the rubber membrane. The major principal stress σ_1 is the applied stress because the confining pressure is not applied to the top of the specimen. From the principal stresses at the time of failure, Mohr's circles for several tests with different confining pressures are constructed. Mohr's failure envelope is transferred to a classification chart, as shown in Figure 7.12, and the strength class of the material is determined to the nearest tenth.

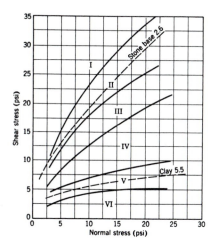

Figure 7.12 Chart for classification of subgrade and base by Texas triaxial test (1 psi = 6.9 kPa).

Group Index

The group index, which ranges from 0 to 20, is used in the AASHTO soil classification system. The values vary with the percentage passing through a No. 200 sieve, the plasticity index, and the liquid limit and can be found from charts or formulas.

Other Correlations

In addition to Figure 7.10, other correlations between M_R, CBR, and R values are also available. These correlations may be quite different from those shown in Figure 7.10.

Heukelom and Klomp (1962) show that

$$M_R = 1500 \ (CBR) \tag{7.6}$$

in which M_R is the resilient modulus in psi. The coefficient of 1500 may vary from 750 to 3000 with a factor of 2. Available data indicate that Eq. 7.6 provides better results at values of CBR less than about 20. In other words, the correlation appears to be more reasonable for fine-grained soils and fine sands rather than granular materials.

The Asphalt Institute (1982) proposed the following correlation between M_R and the R value:

$$M_R = 1155 + 555R \tag{7.7}$$

Laboratory data obtained from six different soil samples were used by the Asphalt Institute (1982) to illustrate the relationships, as shown in Table 7.4. The R values were obtained at an exudation pressure of 240 psi (1.7 MPa). The CBR samples were compacted at optimum moisture content to maximum density and soaked before testing. The repeated load triaxial tests were performed at optimum conditions using a deviator stress of 6 psi (41 kPa) and a confining pressure of 2 psi (14 kPa).

It can be seen from Table 7.4 that the equations for estimating M_R from CBR and R values have a very limited range. The resilient moduli estimated from CBR values of 5.2 and 7.6 and R values of 18 and 21 generally conform to the guidelines for accuracy within a factor of 2. Estimates from CBR values of 25 or higher and R values above 60 would appear to overestimate M_R by Eqs. 7.6 and 7.7.

It should be noted that the M_R of granular materials increases with the increase in confining pressure, while that of fine-grained soils decreases with the increase in deviator stress. Therefore, a large variety of correlations may be obtained depending on the confining pressure or the deviator stress to be used in the resilient modulus test.

TABLE 7.4 COMPARISON OF CBR, R VALUE, AND RESILIENT MODULUS

Soil description	CBR test		R value test		Triaxial test
	CBR	M_R (psi) by eq. 7.6	R	M_R (psi) by eq. 7.7	M_R (psi)
Sand	31	46,500	60	34,500	16,900
Silt	20	30,000	59	33,900	11,200
Sandy loam	25	37,500	21	12,800	11,600
Silt–clay loam	25	37,500	21	12,800	17,600
Silty clay	7.6	11,400	18	11,000	8200
Heavy clay	5.2	7800	<5	<3900	14,700

Note. 1 psi = 6.9 kPa.

Source: After AI (1982).

Hot Mix Asphalt

Figure 7.13 shows the relationships of the layer coefficient, Marshall stability, cohesiometer values, and resilient modulus.

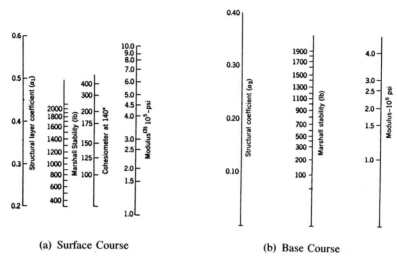

(a) Surface Course   (b) Base Course

Figure 7.13   Correlation charts for estimating resilient modulus of HMA (1 lb = 4.45 N, 1 psi = 6.9 kPa). (After Van Til et al. (1972).)

Structural Layer Coefficient

In the AASHTO design method, the quality of the HMA, base, and subbase is indicated by their structural layer coefficients. These corelation charts were originally developed to determine the layer coefficients, but can also be used to determine the resilient modulus. More about layer coefficients is presented in Section 11.3.4.

Marshall Test

The Marshall test is performed on cylindrical specimens, 4 in. (102 mm) in diameter and 2.5 in. (64 mm) in height, at a temperature of 140°F (60°C) and a rate of loading of 2 in. (51 mm) per minute. Two values are measured: the stability, which is the required load to fail the specimen, and the flow index, which is the vertical distortion at the time of failure. Due to the very fast rate of loading, the stability is a measure of the cohesion, while the flow index is a measure of the internal friction.

Cohesiometer Test

The cohesiometer test is used to measure the cohesion of HMA or rigidly cemented materials. Figure 7.14 is a schematic diagram of the cohesiometer setup. The load is applied at a control rate by the weight of shot until the sample breaks.

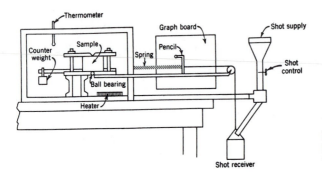

Figure 7.14 Schematic diagram of cohesiometer setup (After Yoder and Witczak (1975).)

The cohesiometer value is computed by

$$C = \frac{L}{W(0.2t + 0.044t^2)} \qquad (7.8)$$

in which C is the cohesiometer value in grams per inch width corrected to 3 in. (76 mm) height, L is the weight of lead shot in grams, W is the diameter or width of specimen in inches, and t is the thickness of the specimen in inches. Note that when $t = 3$ in. (76 mm), $C = L/W$.

Bases

Figure 7.15 shows the correlation charts for untreated granular base, bituminous-treated base, and cement-treated base. The resilient modulus of untreated bases is correlated with CBR, R value, and Texas triaxial classification, as shown in Figure 7.15a. The resilient modulus of bituminous-treated bases is correlated with the Marshall stability, as shown in Figure 7.15b, while that of cement-treated bases is correlated with the unconfined compressive strength, as shown in Figure 7.15c.

Subbases

Figure 7.16 shows the correlation chart for estimating the resilient modulus of granular subbases from CBR, R values, and Texas triaxial classification. For the same untreated granular materials, the correlations for base, subbase, and subgrade are different, as illustrated by the following example.

Example 7.4:

Given CBR values of 30 and 80, determine the corresponding R value, Texas classification, and resilient modulus when the materials are used as a base course, a subbase course, and a subgrade.

Solution: The correlations for base, subbase, and subgrade can be obtained from Figures 7.15a, 7.16, and 7.10. The results are tabulated in Table 7.5.

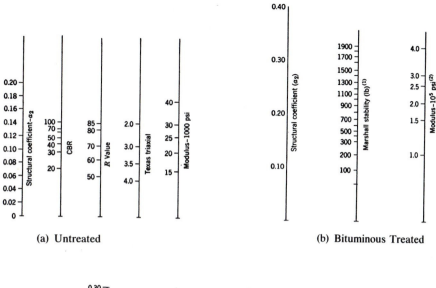

(a) Untreated

(b) Bituminous Treated

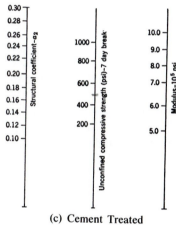

(c) Cement Treated

Figure 7.15 Correlation charts for estimating resilient modulus of bases (1 lb = 4.45 KN, 1 psi = 6.9 kPa). (After Van Til et al. (1972).)

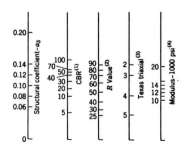

Figure 7.16 Correlation chart for estimating resilient modulus of subbases (1 psi = 6.9 kPa). (After Van Til et al. (1972).)

TABLE 7.5 CORRELATION BETWEEN CBR AND RESILIENT MODULUS

	CBR = 30			CBR = 80		
Location	R value	Texas classification	M_R (psi)	R value	Texas classification	M_R (psi)
Base	65	3.2	20,000	83	2.1	29,000
Subbase	61	3.4	14,700	85	2.3	20,000
Subgrade	64	3.2	19,000	83	2.1	39,000

Note. 1 psi = 6.9 kPa.

It can be seen from Table 7.5 that the correlations among CBR, *R* value, and Texas classification are practically the same no matter whether the material is used as a base, a subbase, or a subgrade. This is not true for the resilient modulus, where a large variation exists. This is reasonable because the resilient modulus depends on the state of stresses, which varies with the location where the material is to be placed.

7.2 DYNAMIC MODULUS OF BITUMINOUS MIXTURES

In addition to the resilient modulus, the dynamic complex modulus and the dynamic stiffness modulus have also been used for pavement design.

7.2.1 Dynamic Complex Modulus

The difference between a resilient modulus test and a complex modulus test for bituminous mixtures is that the former uses loadings of any waveform with a given rest period, while the latter applies a sinusoidal or haversine loading with no rest period. The complex modulus is one of the many methods for describing the stress–strain relationship of viscoelastic materials. The modulus is a complex quantity, of which the real part represents the elastic stiffness and the imaginary part characterizes the internal damping of the materials. The absolute value of the complex modulus is commonly referred to as the dynamic modulus. The theory of complex modulus and Fourier transforms is presented in Appendix A.

The complex modulus test is usually conducted on cylindrical specimens subjected to a compressive haversine loading (Papazian, 1962). The same equipment described previously for the resilient modulus can be used for the complex modulus test. The dynamic modulus varies with the loading frequency. A frequency that most closely simulates the actual traffic load should be selected for the test, so the dynamic modulus thus determined is equivalent to the resilient modulus for design purposes. The dynamic modulus test is specified by ASTM (1989b) in ''D3497-79 Standard Test Method for Dynamic Modulus of Asphalt Mixtures.'' In the ASTM method, a haversine compressive stress is applied to the specimen for a minimum of 30 s and not exceeding 45 s at temperatures of 41, 77,

Material Characterization   Chap. 7

and 104°F (5, 25, and 40°C) and at load frequencies of 1, 4, and 16 Hz for each temperature. The axial strains are measured by bonding two wire strain gauges at midheight of the specimen opposite each other. The ratio between the axial stress and the recoverable axial strain is the dynamic modulus.

Most of the complex modulus tests are made by applying a compressive haversine loading to the specimens. If the specimens are truly viscoelastic, any other testing modes should yield the same results. Kallas (1970) investigated the complex modulus of HMA under a tension haversine loading and a tension–compression full sine loading. The dynamic modulus and phase angle in tension and tension–compression were compared to those in compression. The following conclusions were drawn for dense-graded HMA with asphalt and air void contents within the normal ranges:

1. Differences in dynamic modulus are generally insignificant or relatively small among tension, tension–compression, and compression tests for temperatures ranging from 40 to 70°F (4 to 21°C) and loading frequencies from 1 to 16 Hz.

2. Differences in dynamic modulus are significant between the tension or tension–compression test and the compression test at a frequency of 1 Hz and temperatures ranging from 70 to 100°F (21 to 38°C). Under these conditions, the dynamic modulus in tension or tension–compression averages about one-half to two-thirds of that in compression.

3. Differences in phase angle are pronounced between tension and compression and are less pronounced between tension–compression and compression.

4. The phase angle is greatest in tension, least in compression, and intermediate in tension–compression. On the average, the phase angle in tension exceeds that in compression by about 50% and the phase angle in tension–compression exceeds that in compression by about 25%.

The above conclusions may indicate that if a design is based on the elastic theory with a given dynamic modulus for the HMA, either of the above three testing modes may be used. However, if the design is based on the viscoelastic theory with both the dynamic modulus and phase angle as design variables, a testing mode consistent with the actual loading conditions should be used. This will probably be a tension–compression test, as suggested by Witczak and Root (1974).

The dynamic modulus can also be determined from a bending test. A two-point bending apparatus was developed by Shell for determining the modulus of asphalt mixtures (Bonnaure et al., 1977). In this test a trapezoidal specimen fixed at the bottom is subjected to a sinusoidal load at the free end. A continuous plot of load and deformation at the free end is obtained and the stiffness modulus of the sample can be calculated. Another means is provided to calculate the stiffness modulus by measuring the strain on the surface at midlength of the beam with a strain gage. Due to the use of sinusoidal loads, the stiffness modulus obtained from Shell nomographs, described in Section 7.2.3, is actually the dynamic modulus. The test can also be used to determine the phase angle. From the stiffness modulus and the phase angle, the complex modulus can be obtained.

7.2.2 Dynamic Stiffness Modulus

In the fatigue testing of asphalt beam specimens, the elastic modulus at the initial stage must be determined, so that the initial strain can be computed. The elastic modulus based on the resilient deformation of the beam at the 200th repetition is called the dynamic stiffness modulus.

Test Method

The fatigue testing equipment and procedures are described in Section 7.3.1. Repeated haversine loadings with a load duration of 0.1 s and a rest period of 0.4 s are applied at the third points, as shown in Figure 7.17. The size of specimens used by the University of California at Berkeley (Deacon, 1965) was 1.5 in. (38 mm) in width and depth and 15 in. (381 mm) long. To reduce test variability, the width and depth of beams were increased to 3 in. (76 mm) by the Asphalt Institute (Kallas and Puzinauskas, 1972). The following formulas based on the elastic theory have frequently been used to compute the stress, stiffness modulus, and strain:

$$\sigma = \frac{3aP}{bh^2} \tag{7.9}$$

$$E_s = \frac{Pa(3L^2 - 4a^2)}{4bh^3\Delta} \tag{7.10}$$

$$\epsilon_t = \frac{\sigma}{E_s} = \frac{12h\Delta}{3L^2 - 4a^2} \tag{7.11}$$

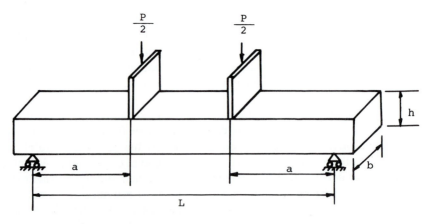

Figure 7.17 Third-point beam test for dynamic modulus.

in which σ is the extreme fiber stress, a is the distance between the load and the nearest support, P is the total dynamic load with $P/2$ applied at each third point, b is the specimen width, h is the specimen depth, E_s is the stiffness modulus based on center deflection, L is the span length between supports, Δ is the dynamic deflection at beam center, and ϵ_t is the extreme fiber tensile strain. When $a = L/3$, Eq. 7.10 becomes

$$E_s = \frac{23PL^3}{108bh^3\Delta} \tag{7.12}$$

After considering the shear deformation, Irwin and Gallaway (1974) suggested the use of the following equation for E_s:

$$E_s = \frac{23PL^3}{108bh^3\Delta} \left[1 + \frac{216h^2(1 + \nu)}{115L^2} \right] \tag{7.13}$$

in which ν is Poisson ratio of the beam. Note that the expression in the bracket is the correction factor for shear deformation. The correction factor varies with h/L. For a beam with $L = 12$ in. (305 mm) and $\nu = 0.35$, the correction factor is 1.04 for $h = 1.5$ in. (38 mm) but increases to 1.16 for $h = 3.0$ in. (76 mm).

Example 7.5:

A beam with a span length of 12 in. (305 mm) and a width and depth of 3 in. (76 mm) is subjected to a dynamic load of 300 lb (1.34 kN) at third points. The dynamic deflection measured at the center of the beam is 6.25×10^{-4} in. (0.016 mm). Determine the extreme fiber stress and the stiffness modulus by Eqs. 7.12 and 7.13. In applying Eq. 7.13, a Poisson ratio of 0.35 may be assumed.

Solution: Given $P = 300$ lb (1.34 kN), $a = 4$ in. (102 mm), and $b = h = 3$ in. (76 mm), from Eq. 7.9, $\sigma = 3 \times 4 \times 300/(3 \times 9) = 133.3$ psi (920 kPa). Given $L = 12$ in. (305 mm) and $\Delta = 0.000625$ in. (0.016 mm), from Eq. 7.12, $E_s = 23 \times 300 \times (12)^3/(108 \times 3 \times 27 \times 0.000625) = 2.18 \times 10^6$ psi (15 GPa), and from Eq. 7.13, $E_s = 2.18 \times 10^6 \times 1.16 = 2.53 \times 10^6$ psi (17.5 GPa).

Relation to Dynamic Modulus

All constant stress fatigue tests have shown that the dynamic stiffness modulus decreases with the increase in dynamic load, due to the relatively large strain in the flexural test. Witczak and Root (1974) indicated that a plot of log E_s versus σ results in a straight line:

$$E_s = E_0 A_1^\sigma \tag{7.14}$$

in which E_0 is the stiffness modulus when $\sigma = 0$, and A_1 is a regression constant depending upon the particular mix and test temperature. Figure 7.18 shows the typical plot of one type of asphalt mix at a temperature of 70°F (21°C). The value of E_0 for this case is 5.5×10^5 psi (38 GPa), which is the intercept of the regression line at zero flexural stress.

Because the stiffness modulus changes with the level of stresses, it cannot be used as a dynamic modulus in the linear elastic layer system. To define the

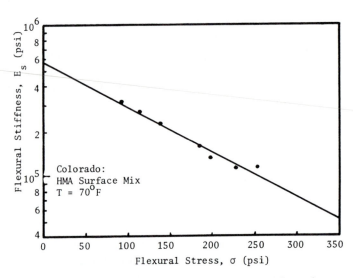

Figure 7.18 Relationship between stiffness modulus and flexural stress (1 psi = 6.9 kPa). (After Witczak and Root (1974).)

relationship between E_0 and $|E^*|$, a regression analysis was conducted by Witczak and Root (1974) on the laboratory results of 17 different asphalt mix and temperature combinations that had been tested by both procedures. The loading used for determining the dynamic modulus was haversine in compression and that for the stiffness modulus was 0.1 s haversine and 0.5 s rest period. The regression equation can be represented by

$$|E^*| = 0.18089 f^{2.1456} E_0^{\,(14.6918/f^{0.01} - 13.5739)} \qquad (7.15)$$

in which f is the frequency at which $|E^*|$ is desired.

Example 7.6:

For the results of dynamic stiffness shown in Figure 7.18, determine the dynamic modulus at a frequency of 8 Hz.

Solution: With $E_0 = 5.5 \times 10^6$ psi (38 GPa) and $f = 8$ Hz, from Eq. 7.15, $|E^*| = 0.18089(8)^{2.1456}(5.5 \times 10^6)^{[14.6918/(8)^{0.01} - 13.5739]} = 4.932 \times 10^6$ psi (34 GPa).

7.2.3 Nomographs and Formulas

Determination of the dynamic modulus of bituminous mixtures by laboratory tests not only is time-consuming, but also requires sophisticated equipment. It is highly desirable that the modulus can be predicted by nomographs or formulas based on the property of the asphalt and the volume concentration of the aggregate. If actual test data are not available, the Shell nomographs (Bonnaure et al., 1977) or the Asphalt Institute formulas (AI, 1982) can be used to determine the modulus without performing the modulus tests. Typical values of the dynamic modulus at different temperatures and frequencies are also presented.

Shell Nomographs

The term "stiffness modulus" is used by Shell in lieu of the dynamic modulus. Two nomographs are used. The first nomograph is applied to determine the stiffness modulus of bitumen based on the temperature, the time of loading, and the characteristics of bitumen actually present in the mix. The second nomograph is then applied to determine the stiffness modulus of the bituminous mix based on the stiffness modulus of the bitumen, the percent volume of the bitumen, and the percent volume of the mineral aggregate.

The characteristics of bitumen are expressed as a penetration index, PI, defined by

$$PI = \frac{20 - 500A}{1 + 50A} \tag{7.16}$$

in which A is the temperature susceptibility, which is the slope of the straight line plot between the logarithm of penetration and temperature, or

$$A = \frac{\log(\text{pen at } T_1) - \log(\text{pen at } T_2)}{T_1 - T_2} \tag{7.17}$$

in which T_1 and T_2 are two temperatures at which penetrations are measured. When the penetration of the recovered bitumen at two different temperatures is known, the penetration index can be determined by Eqs. 7.16 and 7.17.

A convenient temperature to use is the temperature at the ring and ball softening point as specified by AASHTO (1989) in "T53-84 Softening Point of Asphalt (Bitumen) and Tar in Ethylene Glycol (Ring and Ball)." This is a reference temperature at which all bitumens have the same viscosity or penetration of about 800. Replacing T_2 in Eq. 7.17 by $T_{R\&B}$ and pen at T_2 by 800 yields

$$A = \frac{\log(\text{pen at } T) - \log 800}{T - T_{R\&B}} \tag{7.18}$$

Figure 7.19 shows the nomograph for determining the stiffness modulus of bitumens (Van der Poel, 1954). The three factors to be considered are the time of loading, temperature, and penetration index. The temperature to be used is the normalized temperature, which is the difference between the testing temperature and the temperature when the penetration is 800, or $T_{R\&B}$.

The stiffness modulus of bitumen can be determined by either a creep test with a loading time t or a dynamic test under a sinusoidal loading with a frequency f. It was found by Van Der Poel (1954) that the same stiffness modulus is obtained when t is related to f by

$$t = \frac{1}{2\pi f} \tag{7.19}$$

It has been suggested by Shell (1978) that a loading time of 0.02 s, which corresponds to a frequency of 8 Hz according to Eq. 7.19, is representative of the range of loading times occurring in practice and equivalent to a vehicle speed of 30 to 40 mph.

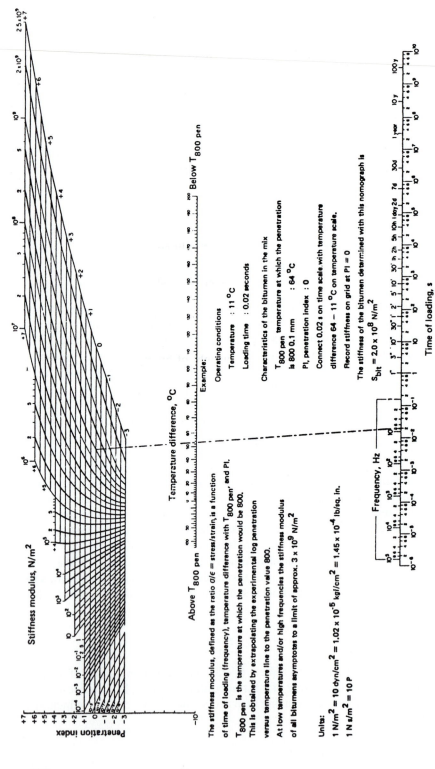

Figure 7.19 Nomograph for stiffness modulus of bitumens. (After Van der Poel (1954).)

Example 7.7:

An asphalt cement recovered from a mix has a penetration of 22 at 25°C (77°F) and a ring and ball softening point of 64°C (147°F). Determine the stiffness modulus of the asphalt under a temperature of 11°C (52°F) and a loading time of 0.02 s.

Solution: Since SI units are used in the Shell nomographs, all computations will be based on SI units. From Eq. 7.18, $A = (\log 22 - \log 800)/(25 - 64) = 0.04$. From Eq. 7.16, $PI = (20 - 500 \times 0.04)/(1 + 50 \times 0.04) = 0$. The temperature below $T_{800pen} = 64 - 11 = 53$°C. As shown in Figure 7.19, a straight line is drawn from 0.02 s on the time of loading scale to 53°C below T_{800pen} on the temperature difference scale and extended to intersect the horizontal line with a penetration index of 0. A stiffness modulus of 2×10^8 N/m^2 (2.9×10^4 psi) can be read from the curve.

Figure 7.20 shows the nomograph for determining the stiffness modulus of bituminous mixtures (Bonnaure et al., 1977). The three factors to be considered are the stiffness modulus of bitumen, the percent volume of bitumen, and the percent volume of aggregate. The percent volumes of aggregate, bitumen, and air can be computed from the percentage by weight of bitumen, the specific gravities of bitumen and aggregate, and the bulk specific gravity of the mixture. The latter can be determined by the water displacement method, as specified by AASHTO (1989) in "T166-83 Bulk Specific Gravity of Compacted Bituminous Mixtures."

Figure 7.21 shows the phase diagram of a bituminous mixture. Let W be the total weight of the mixture and P_b be the bitumen content expressed as a ratio of bitumen weight to total weight. The weight of bitumen is P_bW and the weight of aggregate is $(1 - P_b)W$, as shown on the right side of the diagram. For simplicity, it is assumed that no asphalt is absorbed by the aggregate. If the aggregate is porous and has a significant amount of absorption, the weight of asphalt absorbed should be deducted from W and not used for the computation.

If the specific gravity of the aggregate is G_g and that of the bitumen is G_b, the volumes of the aggregate and bitumen are $(1 - P_b)W/G_g$ and P_bW/G_b, respectively. If the bulk specific gravity of the mixture is G_m, the volume of the mixture is W/G_m. These volumes are shown on the left side of the diagram. By definition, the percent volume of aggregate V_g is

$$V_g = \frac{(1 - P_b)W/G_g}{W/G_m} \times 100 = \frac{100(1 - P_b)G_m}{G_g} \qquad (7.20)$$

The percent volume of bitumen V_b is

$$V_b = \frac{P_bW/G_b}{W/G_m} \times 100 = \frac{100P_bG_m}{G_b} \qquad (7.21)$$

The percent volume of air void V_a is

$$V_a = 100 - V_g - V_b \qquad (7.22)$$

If the aggregate is a combination of several fractions, each with a different specific gravity, the average specific gravity of the aggregate is

$$G_g = \frac{100}{P_1/G_1 + P_2/G_2 + \cdots + P_n/G_n} \qquad (7.23)$$

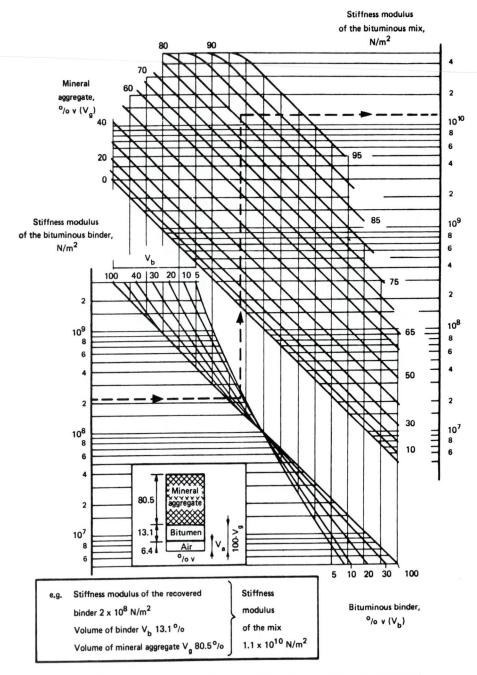

Figure 7.20 Nomograph for stiffness modulus of mixes. (After Shell (1978).)

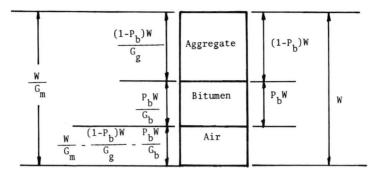

Figure 7.21 Phase diagram of bituminous mix.

in which $P_1, P_2, \ldots, P_n$ is the percentage by weight of each fraction, and $G_1, G_2,$ G_n is the specific gravity of each fraction. Note that the specific gravities to be used in the above equations are the bulk specific gravity for the aggregates and the mixtures and the apparent specific gravity for the bitumen and the mineral filler. The apparent specific gravity, which is based on the solid volume only, is larger than the bulk specific gravity, which is based on the total volume including the air void.

Example 7.8:

An asphalt mixture has a bitumen content of 5.5% by weight and a bulk specific gravity of 2.43. The specific gravity of bitumen is 1.02 and that of aggregate is 2.85. If the stiffness modulus of bitumen is 2×10^8 N/m² (2.9×10^4 psi), estimate the stiffness modulus of the mixture.

Solution: From Eq. 7.20, $V_g = 100 \times (1 - 0.055) \times 2.43/2.85 = 80.6\%$. From Eq. 7.21, $V_b = 100 \times 0.055 \times 2.43/1.02 = 13.1\%$. The stiffness modulus of the mixture is 1.1×10^{10} N/m² (1.6×10^6 psi) and can be read from Figure 7.20, as indicated by the dashed lines.

Bonnaure et al. (1977) also developed the following equations for predicting the stiffness modulus of the mix S_m, based on V_g, V_b, and the stiffness modulus of the bitumen S_b:

$$\beta_1 = 10.82 - \frac{1.342(100 - V_g)}{V_g + V_b} \tag{7.24a}$$

$$\beta_2 = 8.0 + 0.00568V_g + 0.0002135V_g^2 \tag{7.24b}$$

$$\beta_3 = 0.6 \log \left(\frac{1.37V_b^2 - 1}{1.33V_b - 1}\right) \tag{7.24c}$$

$$\beta_4 = 0.7582 (\beta_1 - \beta_2) \tag{7.24d}$$

For 5×10^6 N/m² $< S_b < 10^9$ N/m²,

$$\log S_m = \frac{\beta_4 + \beta_3}{2}(\log S_b - 8) + \frac{\beta_4 - \beta_3}{2} |\log S_b - 8| + \beta_2 \tag{7.25a}$$

For 10^9 N/m^2 < S_b < 3×10^9 N/m^2,

$$\log S_m = \beta_2 + \beta_4 + 2.0959(\beta_1 - \beta_2 - \beta_4)(\log S_b - 9) \qquad (7.25b)$$

Equation 7.25 is based on SI units with S_m and S_b in N/m^2. If S_m and S_b are in psi, Eq. 7.26 should be used. For 725 psi < S_b < 145,000 psi,

$$\log S_m = \frac{\beta_4 + \beta_3}{2} (\log S_b - 4.1612)$$

$$+ \frac{\beta_4 - \beta_3}{2} |\log S_b - 4.1612| + \beta_2 - 3.8383 \qquad (7.26a)$$

For 145,000 psi < S_b < 435,000 psi,

$$\log S_m = \beta_2 + \beta_4 + 2.0959(\beta_1 - \beta_2 - \beta_4)(\log S_b - 5.1612) - 3.8388 \qquad (7.26b)$$

Example 7.9:

For the nine cases with values of S_b, V_b, and V_g shown in Table 7.6, determine the stiffness modulus of the mixtures by Eqs. 7.24 and 7.25 and compare with the nomograph solutions.

Solution: For case 1 with an extremely small S_b of 6×10^6 N/m^2 (870 psi), the detailed calculations by the formulas are shown below. From Eq. 7.24, $\beta_1 = 10.82 - 1.342 (100 - 80)/(80 + 5) = 10.504$, $\beta_2 = 8.0 + 0.00568 \times 80 + 0.0002135(80)^2 = 9.821$, $\beta_3 = 0.6 \log[(1.37 \times 25 - 1)/(1.33 \times 5 - 1)] = 0.462$, $\beta_4 = 0.7582(10.504 - 9.821) = 0.518$. From Eq. 7.25a, $\log S_m = 0.5(0.518 + 0.462)[\log(6 \times 10^6) - 8] + 0.5(0.518 - 0.462) |\log(6 \times 10^6) - 8| + 9.821 = 9.188$, or $S_m = 1.5 \times 10^9$ N/m^2 (2.6×10^5 psi), which compares with 1.3×10^9 N/m^2 (1.9×10^5 psi) from the nomograph. The comparison of S_m obtained from the equations and the nomograph is shown at the bottom of Table 7.6. It can be seen that both solutions check quite well in most cases. However, for some extreme cases, a large discrepancy exists between the two solutions. Note that Eq. 7.25a should be used for cases 1 to 6 and Eq. 7.25b for cases 7 to 9.

According to Shell International Petroleum (1978), the accuracy of the nomograph, as checked by extensive measurements on a large number of different asphalt mixes, is a factor of 1.5 to 2, which is sufficient for practical design. Therefore, the use of Eqs. 7.24 and 7.25 to estimate the modulus in a computerized method of design should be considered acceptable.

Asphalt Institute Formulas

In developing the DAMA computer program for the Asphalt Institute, Hwang and Witczak (1979) applied the following regression formulas to determine the dynamic modulus of HMA, $|E^*|$:

$$|E^*| = 100,000 \times 10^{\beta_1} \qquad (7.27a)$$

$$\beta_1 = \beta_3 + 0.000005 \beta_2 - 0.00189 \beta_2 f^{-1.1} \qquad (7.27b)$$

$$\beta_2 = \beta_4^{0.5} T^{\beta_5} \qquad (7.27c)$$

TABLE 7.6 STIFFNESS MODULUS OF MIXTURES WITH VARIOUS COMPOSITIONS

	Case no.								
	1	2	3	4	5	6	7	8	9
S_b (N/m²)	6×10^6	6×10^6	6×10^6	1×10^8	1×10^8	1×10^8	2×10^9	2×10^9	2×10^9
V_b (%)	5	10	40	5	10	40	5	10	40
V_g (%)	80	85	60	80	85	60	80	85	60
S_m (N/m²)									
Equation	1.5×10^9	1.8×10^9	8.3×10^7	6.6×10^9	1.1×10^{10}	1.3×10^9	2.8×10^{10}	3.6×10^{10}	1.5×10^{10}
Nomograph	1.3×10^9	1.7×10^9	9.0×10^7	6.9×10^9	1.1×10^{10}	1.3×10^9	2.7×10^{10}	4.8×10^{10}	1.7×10^{10}

Note. 1 psi = 6900 N/m².

$$\beta_3 = 0.553833 + 0.028829 \ (P_{200}f^{-0.1703}) - 0.03476V_a$$
$$+ \ 0.070377\lambda + 0.931757f^{-0.02774} \tag{7.27d}$$

$$\beta_4 = 0.483V_b \tag{7.27e}$$

$$\beta_5 = 1.3 + 0.49825 \log f \tag{7.27f}$$

in which β_1 to β_5 are temporary constants, f is the load frequency in Hz, T is the temperature in °F, P_{200} is the percentage by weight of aggregate passing through a No. 200 sieve, V_v is the volume of air void in %, λ is the asphalt viscosity at 70°F in 10^6 poise, and V_b is the volume of bitumen in %. If sufficient viscosity data are not available to estimate λ at 70°F, the following equation may be used:

$$\lambda = 29,508.2 \ (P_{77°F})^{-2.1939} \tag{7.28}$$

in which $P_{77°F}$ is the penetration at 77°F (25°C). It can be seen that the factors considered by Asphalt Institute are mostly the same as those by Shell with the following exceptions:

1. The percentage of fines passing through a No. 200 sieve is considered by AI but not by Shell.
2. The viscosity or penetration of asphalt considered by Shell is determined from the recovered asphalt, or the asphalt actually present in the mix, while that by AI is the original asphalt.
3. The temperature and the viscosity of asphalt are considered by AI, whereas the normalized temperature, which is the temperature above or below $T_{R\&B}$, and the penetration index, which indicates the temperature sensitivity of the asphalt, are used by Shell.

Example 7.10:

An asphalt cement has an original penetration of 70 at 77°F (25°C), a reduced penetration of 50 after being recovered from the mix, and a ring and ball softening point of 140°F (60°C). The mix containing the asphalt has an asphalt volume of 11%, an air void volume of 5%, and 6% of aggregates passing through a No. 200 sieve. Determine the dynamic modulus of the mix at a temperature of 77°F (25°C) and a frequency of 8 Hz by the AI equations and compare with the stiffness modulus of the mix by the Shell nomographs. What are the values of $|E^*|$ when the percentages passing through a No. 200 sieve are 1 and 11%, respectively.

Solution: First consider the AI method. With $P_{77°F} = 70$, from Eq. 7.28, $\lambda = 29,508.2 \times 70^{-2.1939} = 2.64 \times 10^6$ poise. Given $f = 8$ Hz, $T = 77°F$ (25°C), $P_{200} = 6\%$, $V_v = 5\%$, $\lambda = 2.64 \times 10^6$ poise and $V_b = 11\%$, from Eq. 7.27, $\beta_5 = 1.3 + 0.49825 \log 8 = 1.750$, $\beta_4 = 0.483 \times 11 = 5.313$, $\beta_3 = 0.553833 + 0.028829 \times 6 \times (8)^{-0.1703} - 0.03476 \times 5 + 0.070377 \times 2.64 + 0.931757 \times (8)^{-0.02774} = 1.567$, $\beta_2 = (5.313)^{0.5} \times (77)^{1.750} = 4613.5$, $\beta_1 = 1.567 + 0.000005 \times 4613.5 - 0.00189 \times 4613.5 \times (8)^{-1.1} = 0.705$, and $|E^*| = 100,000 \times (10)^{0.705} = 5.07 \times 10^5$ psi (3.5 GPa).

Next consider the Shell method. With $P_{77°F} = 50$ for recovered asphalt and $T_{R\&B} = 60°C$ (140°F), from Eq. 7.18, $A = (\log 50 - \log 800)/(25 - 60) = 0.0344$; from Eq. 7.16, PI = $(20 - 500 \times 0.0344)/(1 + 50 \times 0.0344) = 1.029$. With a frequency of 8 Hz, a temperature difference of $60 - 25 = 35°C$ (95°F) and a penetration index of 1.029, from Figure 7.19, $S_b = 10^7$ N/m^2 (1450 psi). With $S_b = 10^7$ N/m^2 (1450 psi), $V_b = 11\%$, and $V_g = 100 - 11 - 5 = 84\%$, from Figure 7.20, $S_m = 2.1 \times 10^9$ N/m^2 (3.1×10^5 psi), which compares with 3.5×10^9 N/m^2 (5.1×10^5 psi) from the AI equations.

If $P_{200} = 1\%$, from Eq. 7.27, $\beta_3 = 0.553833 + 0.028829 \times 1 \times (8)^{-0.1703} - 0.03476 \times 5 + 0.070377 \times 2.64 + 0.931757 \times (8)^{-0.02774} = 1.466$, $\beta_1 = 1.466 + 0.000005 \times 4613.5 - 0.00189 \times 4613.5 \times (8)^{-1.1} = 0.604$, and $|E^*| = 100,000 \times (10)^{0.604} = 4.0 \times 10^5$ psi (2.7 GPa). If $P_{200} = 11\%$, from Eq. 7.27, $\beta_3 = 1.668$, $\beta_1 = 0.806$, and $|E^*| = 6.4 \times 10^5$ psi (4.4 GPa).

It can be seen that the dynamic modulus increases with the increase in fine contents. An increase of the percentage passing through a No. 200 sieve from 1 to 11% increases the dynamic modulus from 4.0×10^5 to 6.4×10^5 psi (2.7 to 4.4 GPa). This change is quite small compared to other factors such as temperature, viscosity, frequency, and the volume concentration of bitumen and aggregate.

Typical Ranges

Table 7.7 shows the typical ranges of an HMA modulus at different temperatures and load frequencies. As indicated previously, a frequency of 8 Hz is equivalent to a speed about 35 mph.

TABLE 7.7 TYPICAL VALUES OF DYNAMIC MODULUS FOR HMA

Temperature (°F)	Load frequency (Hz)					
	1		4		16	
	Range	Mean	Range	Mean	Range	Mean
40	6.0–18.0	12.0	9.0–27.0	16.0	10.0–30.0	18.0
70	2.0–6.0	3.0	4.0–9.0	5.0	5.0–11.0	7.0
100	0.5–1.5	0.7	0.7–2.2	1.0	1.0–3.2	1.6

Note. Modulus in terms of 10^5 psi, 1 psi = 6.9 kPa.

7.3 FATIGUE CHARACTERISTICS

Fatigue of bituminous mixtures and Portland cement concrete under repeated flexure is an important factor of pavement design. In this section, the use of laboratory fatigue tests, nomographs, and equations to predict the fatigue life is discussed.

7.3.1 Bituminous Mixtures

A variety of methods have been developed for the fatigue testing of bituminous mixtures. Most of the methods employ the bending of beams, although the bending of plates has also been used (Jimenez and Gallaway, 1962; Jimenez,

1972). In the beam tests, a simple beam with third-point (Deacon, 1965) or center-point loading (Franchen and Verstraeten, 1974) or a cantilever beam with rotating bending (Pell, 1962) has been used. The repeated load indirect tensile test has also been employed (Adedimila and Kennedy, 1976). In this section, only the beam test with third-point loading is described. The advantage of third-point loading over the center-point loading is the existence of a constant bending moment over the middle third of the specimen, so any weak spot due to nonuniform material properties will show up in the test results. In view of the fact that fatigue tests are expensive and require a large number of specimens, nomographs and equations for predicting fatigue life are also presented.

Test Procedures

Two types of controlled loading can be applied: constant stress and constant strain, as shown in Figure 7.22. In the constant stress test, the stress remains constant but the strain increases with the number of repetitions. In the constant strain test, the strain is kept constant, and the load or stress is decreased with the number of repetitions. The constant stress type of loading is applicable to thicker pavements, wherein the HMA is more than 6 in. (152 mm) thick and is the main load-carrying component. As the HMA becomes weaker under repeated loads, the strain should increase with the number of repetitions. The constant strain type of loading is applicable to thin pavements with HMA less than 2 in. (51 mm) thick because the strain in the asphalt layer is governed by the underlying layers and is

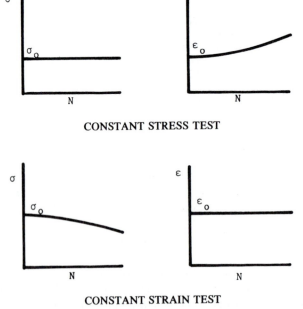

CONSTANT STRESS TEST

CONSTANT STRAIN TEST

Figure 7.22 Two types of controlled loading for fatigue testing.

Material Characterization   Chap. 7

not affected by the decrease in stiffness of HMA. For intermediate thicknesses, a combination of constant stress and constant strain exists.

As can be seen in Figure 7.22, both stress and strain are larger in the constant stress test, so the use of constant stress is more conservative. The use of constant stress has the further advantage that failure occurs more quickly and can be more easily defined, while an arbitrary failure criterion, such as a stress equal to 50% of the initial stress, is frequently used for the constant strain test.

Figure 7.23 is a schematic diagram of the fatigue testing equipment. The load is applied upward through the piston rod to a beam specimen, 15 in. (310 mm) long with a width and depth not exceeding 3 in. An electrohydraulic or pneumatic testing machine capable of applying repeated tension–compression loads in the form of haversine waves for 0.1-s duration with 0.4-s rest periods can be used. A sufficient load, approximately 10% of the upward load, is applied in the downward direction, forcing the beam to return to its original horizontal position and holding it at that position during the rest period. Adjustable stop nuts installed on the loading rods prevent the beam from bending below the initial horizontal position during the rest period. The dynamic deflection of the beam at the midspan is measured with a linear variable differential transformer (LVDT). The LVDT core is attached to a nut bonded with epoxy cement to the center of the specimen. The repeated flexure apparatus is enclosed in a controlled temperature cabinet.

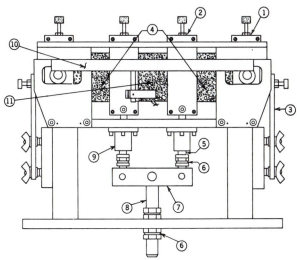

1. Reaction Clamp 5. Loading Rod 9. Ball Bushing
2. Load Clamp   6. Stop Nuts   10. LVDT Holder
3. Restrainer   7. Load Bar   11. LVDT
4. Specimen   8. Piston Rod

Figure 7.23 Fatigue testing equipment.

A range of stresses should be selected so that the specimens will fail within a range from 1000 to 1,000,000 repetitions. Normally, 8 to 12 specimens are required to establish the fatigue relationship for a given temperature. Tests at several different temperatures are required so the effect of stiffness or temperature on the fatigue life can be evaluated.

Analysis

The stiffness modulus and the initial strain of each test are determined at the 200th repetition by using Eqs. 7.10 and 7.11, respectively. The initial strains are plotted versus the number of repetitions to failure on log scales. The plot can be approximated by a straight line as shown in Figure 7.24 and expressed by

$$N_f = c_2(\epsilon_t)^{-f_2} \tag{7.29}$$

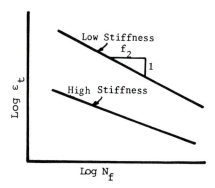

Figure 7.24 Relationship between strain and number of repetitions to failure.

in which N_f is the number of repetitions to failure, c_2 is a fatigue constant that is the value of N_f when $\epsilon_t = 1$, and f_2 is the inverse slope of the straight line. Under the same initial strain, laboratory tests show that the number of repetitions to failure decreases with the increase in stiffness modulus, so Eq. 7.29 can also be written as

$$N_f = c_1(\epsilon_t)^{-f_2}(E_s)^{-f_3} \tag{7.30}$$

Note that Eq. 7.30 is similar to Eq. 3.6. However, Eq. 7.30 is based purely on laboratory fatigue tests, whereas Eq. 3.6 is the extension from laboratory specimens to actual prototype pavements. The factor f_1 for prototype pavements should be much greater than c_1 for laboratory specimens due to the fact that wheel loads on actual pavements do not apply at the same location and have longer rest periods, both of which increase the fatigue life. Also, for thicker pavements, it takes more repetitions for cracks to appear on the surface to be considered as failure. The Asphalt Institute's fatigue criterion is based on the assumption that f_1 is 18.4 times greater than c_1.

Example 7.11:

Fatigue tests were performed on 3-in. × 3-in. (75-mm × 75-mm) beams by the third-point loading with a span of 12 in. (305 mm). The deflection at the center of beam Δ and the number of repetitions to failure N_f are shown in Table 7.8. Develop an equation relating the number of repetitions to failure and the tensile strain.

TABLE 7.8 RESULTS OF FATIGUE TESTS ON HMA SPECIMENS

	Test no.							
	1	2	3	4	5	6	7	8
Δ (0.001 in.)	12.9	8.03	8.24	6.03	4.42	2.81	1.97	1.48
N_f	1110	3140	4115	6010	30,625	89,970	289,110	915,060

Note. 1 in. = 25.4 mm.

Solution: The tensile strain for each test can be determined from Eq. 7.11, or $\epsilon_t = 12 \times 3 \times \Delta/(3 \times 144 - 4 \times 16) = 0.0978\Delta$. The strains for the eight tests are 12.6, 7.85, 8.06, 5.90, 4.32, 2.75, 1.93, and 1.45 $\times 10^{-4}$. Figure 7.25 is a plot of ϵ_t versus log N_f.

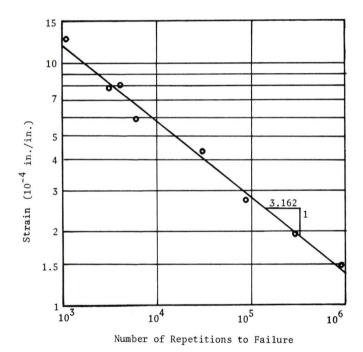

Figure 7.25 Example 7.11.

The inverse slope of the straight line $f_2 = 3/\log(11.9/1.34) = 3.162$. The intercept c_1 when $\epsilon_t = 1$ can be obtained from a given point (log ϵ_t, log N_f) on the straight line by

$$f_2 = \frac{\log N_f - \log c_1}{\log 1 - \log \epsilon_t}$$

or
$$\log c_1 = \log N_f + f_2 \log \epsilon_t \qquad (7.31)$$

Given $f_2 = 3.162$, $N_f = 1000$, and $\epsilon_t = 11.9 \times 10^{-4}$, from Eq. 7.31, log $c_1 =$

Sec. 7.3   Fatigue Characteristics

351

$\log(1000) + 3.162 \log(0.00119) = -6.247$ or $c_1 = 5.66 \times 10^{-7}$. Therefore, the fatigue equation is

$$N_f = 5.66 \times 10^{-7}(\epsilon_t)^{-3.162}$$

Nomographs and Equations

The fatigue test is a destructive test and is much more time-consuming than the nondestructive resilient modulus test. To obtain Eq. 7.30, more than two dozen specimens need to be tested. There is a need to estimate the fatigue properties by simple nomographs and equations without actually performing the fatigue tests.

Based on 146 fatigue lines covering a wide range of mixes, bitumens, and testing conditions, Shell (Bonnaure et al., 1980) developed separate equations for constant stress and constant strain tests. For constant stress tests

$$\epsilon_t = [36.43\text{PI} - 1.82\text{PI}(V_b) + 9.71V_b - 24.04]$$

$$\times 10^{-6}\left(\frac{S_m}{5 \times 10^9}\right)^{-0.28}\left(\frac{N_f}{10^6}\right)^{-0.2} \tag{7.32}$$

in which ϵ_t is the tensile strain, PI is the penetration index defined by Eq. 7.16, V_b is the percentage of bitumen volume in the mix, S_m is the stiffness modulus of the mix in N/m^2 as determined from Figure 7.20, and N_f is the number of repetitions to failure. If S_m is in psi, Eq. 7.32 can be written as

$$N_f = [0.0252\text{PI} - 0.00126\text{PI}(V_b) + 0.00673V_b - 0.0167]^5 \; \epsilon_t^{-5}S_m^{-1.4} \tag{7.33}$$

For constant strain tests

$$\epsilon_t = [36.43\text{PI} - 1.82\text{PI}(V_b) + 9.71V_b - 24.04]$$

$$\times 10^{-6}\left(\frac{S_m}{5 \times 10^{10}}\right)^{-0.36}\left(\frac{N_f}{10^6}\right)^{-0.2} \tag{7.34}$$

If S_m is in psi, Eq. 7.34 can be written as

$$N_f = [0.17\text{PI} - 0.0085\text{PI}(V_b) + 0.0454V_b - 0.112]^5 \; \epsilon_t^{-5}S_m^{-1.8} \tag{7.35}$$

The solutions of the above equations can be presented in a nomograph, as shown in Figure 7.26. The accuracy of the equations for constant stress tests was reported within $\pm40\%$ for 90% of the results and that for constant strain tests within $\pm50\%$.

Example 7.12:

Given $V_b = 13\%$, PI $= -0.7$, and $S_m = 3.3 \times 10^9 \; N/m^2$, determine the initial strains when failure occurs at 10^6 cycles of constant stress and 10^5 cycles of constant strain by using both the equations and the nomograph.

Solution: For constant stress tests, from Eq. 7.32, $\epsilon_t = [36.43(-0.7) - 1.82(-0.7) \times 13 + 9.71 \times 13 - 24.04] \times 10^{-6} \times [(3.3 \times 10^9)/(5 \times 10^9)]^{-0.28} \times (10^6/10^6)^{-0.2} = 1.1 \times 10^{-4}$. For constant strain tests, from Eq. 7.34, $\epsilon_t = [36.43(-0.7) - 1.82(-0.7) \times 13 + 9.71 \times 13 - 24.04] \times 10^{-6} \times [(3.3 \times 10^9)/$

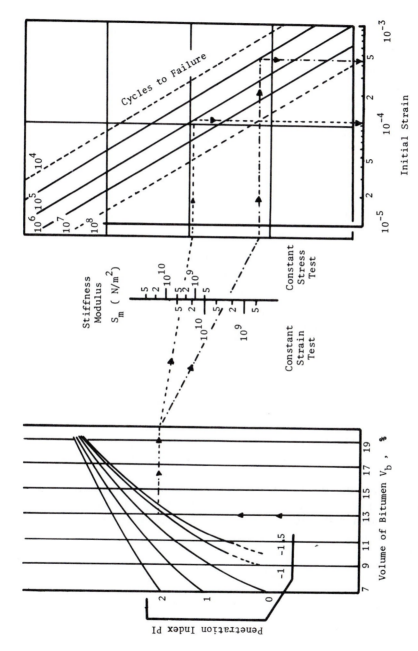

Figure 7.26 Nomograph for predicting fatigue life of bituminous mixes. (After Bonnaure et al. (1980).)

$(5 \times 10^{10})]^{-0.36} \times (10^5/10^6)^{-0.2} = 4.0 \times 10^{-4}$. The initial strains obtained from the nomograph shown in Figure 7.26 are 1.1×10^{-4} for the constant stress test and 4.2×10^{-4} for the constant strain test.

The laboratory fatigue equations developed by the Asphalt Institute (AI, 1982) are based on the constant stress criterion and can be expressed as

$$N_f = 0.00432C\epsilon_t^{-3.291}|E^*|^{-0.854} \qquad (7.36)$$

in which C is the correction factor expressed as

$$C = 10^M \qquad (7.37a)$$

and

$$M = 4.84 \left(\frac{V_b}{V_a + V_b} - 0.69 \right) \qquad (7.37b)$$

Note that for a standard mix with an asphalt volume V_b of 11% and an air void volume of 5%, $M = 0$ or $C = 1$. This standard mix is used by the Asphalt Institute in the ninth edition of the design manual (AI, 1981a). After multiplying by a factor of 18.4 to account for the differences between laboratory and field conditions, the fatigue failure criterion becomes

$$N_f = 0.0796\epsilon_t^{-3.291}|E^*|^{-0.854} \qquad (7.38)$$

in which $|E^*|$ is the dynamic modulus, which is equivalent to the elastic modulus of the asphalt layer in a layer system or the stiffness modulus in the Shell method. A comparison of Eq. 7.38 with Eq. 7.33 shows that the exponents in Eq. 7.38 are somewhat smaller than those in Eq. 7.33. This is not surprising in view of the large variability inherent in the fatigue tests.

7.3.2 Portland Cement Concrete

A fatigue test of portland cement concrete usually applies a repeated flexural loading on beam specimens 15 in. (381 mm) long, 3 in.(76 mm) wide, and 3 in. (76 mm) deep. Loading is generally applied at the third points with a rate of 1 to 2 repetitions per second and a duration of 0.1 s. The extreme fiber stress in the beam is computed by Eq. 7.9. The number of repetitions to failure in log scale, log N_f, is plotted against the stress ratio, which is a quotient of the stress σ and the modulus of rupture of the concrete S_c. The modulus of rupture is determined by the same beam test but with a steadily increasing static load, as specified by ASTM (1989a) in "C78-84 Standard Test Method for Flexural Strength of Concrete Using Simple Beam with Third Point Loading."

Figure 7.27 shows the fatigue data obtained by several investigators. It is speculated that concrete will not fail by fatigue when the stress ratio is smaller than 0.5, although no real limit has been found up to 10–20 million repetitions. The average line for 50% probability of failure is shown by the solid line in Figure 7.27 and expressed as

$$\log N_f = 17.61 - 17.61 \left(\frac{\sigma}{S_c} \right) \qquad (7.39)$$

Material Characterization   Chap. 7

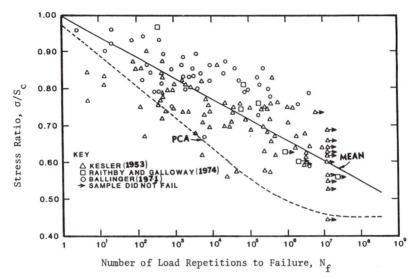

Figure 7.27 Results of fatigue tests on concrete from different sources.

Note that Eq. 7.39 is the same as Eq. 5.35 with $f_1 = 17.61$ and $f_2 = 17.61$. In the design of zero-maintenance pavements, Darter and Barenberg (1977) recommended the reduction of f_1 by one order of magnitude to 16.61. The design curve used by the Portland Cement Association, as indicated by Eq. 5.36, is shown by the broken line. It can be seen that the PCA fatigue curve lies below most of the failure points and is very conservative.

7.4 PERMANENT DEFORMATION PARAMETERS

Permanent deformation is an important factor in flexible pavement design. With the increase in traffic load and tire pressure, most of the permanent deformation occurs in the upper layers rather than in the subgrade. To estimate the rut depth, it is necessary to determine the permanent deformation parameters of the material for each layer.

Various methods have been used to determine the permanent deformation of paving materials. Most methods employ the repeated load test, which is similar to the resilient modulus test except that loads up to 100,000 repetitions have been applied and the permanent deformations at a number of designated cycles have been recorded.

7.4.1 Basic Principles

The permanent deformation of paving materials depends strongly on the testing methods and the procedures used for fabricating and testing the specimens. These variations together with the uncertainty in traffic and environmental conditions, makes the prediction of rut depth extremely difficult. Therefore, the use of simplified methods is warranted.

Direct Method

When the load is applied over a circular area, the maximum deformation occurs directly under the center of the area. The rut depth on the pavement surface can be determined by the following direct method.

1. Divide the pavement and subgrade into a number of layers and estimate the vertical and radial stresses at the midheight of each layer.
2. Using the vertical stress due to the applied load as the repeated deviator stress and the radial stress due to the applied load and the overburden combined as the confining pressure, conduct repeated load tests on laboratory specimens that represent the actual materials in the field and determine the permanent strains under a large number of load repetitions.
3. Compute the vertical deformation of each layer under any given number of repetitions by multiplying the permanent strains obtained from the laboratory tests with the thickness of the layer.
4. Sum the permanent deformation over all layers to obtain the rut depth on the surface.

Unless an equivalent single-axle load and an average set of material properties are used throughout the design period, it is not possible to apply the above method because the stresses in each layer can be any values depending on the axle loads and the environmental conditions throughout the year. A more practical approach is to perform a large number of permanent deformation tests under various stress and environment conditions and develop regression equations relating permanent strains to these conditions. These regression equations can then be incorporated into a multilayer computer program to compute the rut depth, as illustrated by Allen and Deen (1986).

The division of a pavement system into a large number of layers is cumbersome and the development of regression equations for each paving material is expensive and time-consuming. The original VESYS program (FHWA, 1978) used the following simple method to characterize the permanent deformation properties, although a direct method was later added as another option.

VESYS Method

The method incorporated in the VESYS computer program for the prediction of rut depth is based on the assumption that the permanent strain is proportional to the resilient strain by

$$\epsilon_p(N) = \mu \epsilon N^{-\alpha} \tag{7.40}$$

in which $\epsilon_p(N)$ is the permanent or plastic strain due to a single load application, i.e., at the Nth application; ϵ is the elastic or resilient strain at the 200th repetition;

N is the load application number; μ is a permanent deformation parameter representing the constant of proportionality between permanent and elastic strains; and α is a permenant deformation parameter indicating the rate of decrease in permanent deformation as the number of load applications increases. The total permanent deformation can be obtained by integrating Eq. 7.40:

$$\epsilon_p = \int_0^N \epsilon_p(N)\, dN = \epsilon\mu\frac{N^{1-\alpha}}{1-\alpha} \tag{7.41}$$

Equation 7.41 indicates that a plot of log ϵ_p versus log N results in a straight line, as shown in Figure 7.28. From Eq. 7.41

$$\log \epsilon_p = \log \left(\frac{\epsilon\mu}{1-\alpha}\right) + (1-\alpha)\log N \tag{7.42}$$

So the slope of the straight line $S = 1 - \alpha$, or

$$\alpha = 1 - S \tag{7.43}$$

The intercept at $N = 1$, $I = \epsilon\, \mu/(1 - \alpha)$, or

$$\mu = \frac{IS}{\epsilon} \tag{7.44}$$

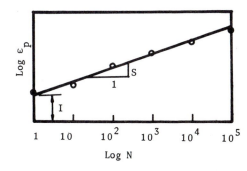

Figure 7.28 Log–log plot of permanent strain versus number of load repetitions.

To determine the permanent deformation parameters of the layer system, α_{sys} and μ_{sys}, from those of the individual layers, it is further assumed that the sum of permanent and recoverable strains due to each load application is a constant and equals to the elastic strain at the 200th repetition. This means that after the 200th repetition

$$\epsilon = \epsilon_p(N) + \epsilon_r(N) \tag{7.45}$$

in which $\epsilon_r(N)$ is the recoverable strain due to each load application. Substituting Eq. 7.40 into Eq. 7.45 yields

$$\epsilon_r(N) = \epsilon(1 - \mu N^{-\alpha}) \tag{7.46}$$

Under the same stresses, strains are inversely proportional to the moduli, so Eq. 7.46 can be written as

$$E_r(N) = \frac{E}{1 - \mu N^{-\alpha}} = \frac{EN^\alpha}{N^\alpha - \mu} \tag{7.47}$$

in which $E_r(N)$ is the elastic modulus due to unloading and E is the elastic modulus due to loading. Note that $E_r(N)$, which is the unloading modulus for each individual layer, is not a constant but increases with the increase of load repetitions. These unloading moduli are used to determine the recoverable deformation $w_r(N)$ at different values of N. The permanent deformation $w_p(N)$ can then be computed by

$$w_p(N) = w - w_r(N) \tag{7.48}$$

in which w is the elastic deformation due to loading. Similar to Eq. 7.40, $w_p(N)$ can be expressed by

$$w_p(N) = \mu_{sys} w N^{-\alpha_{sys}} \tag{7.49}$$

Combining Eqs. 7.48 and 7.49 gives

$$1 - \frac{w_r(N)}{w} = \mu_{sys} N^{-\alpha_{sys}} \tag{7.50}$$

Equation 7.50 shows that a plot of log $[1 - w_r(N)/w]$ versus log N results in a straight line, as shown in Figure 7.29. The slope of the straight line is α_{sys} and the intercept at $N = 1$ is log μ_{sys}.

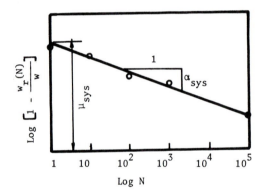

Figure 7.29 Log–log plot of $[1 - w_r(N)/w]$ versus number of load repetitions.

The determination of α_{sys} and μ_{sys} from the α and μ of each layer can be summarized as follows:

1. Assume several values of N, say 1, 10, 10^2, 10^3, 10^4, and 10^5, and determine the unloading modulus $E_r(N)$ of each layer by Eq. 7.47.
2. Using the unloading moduli as the elastic moduli, determine the recoverable deformation $w_r(N)$ at the surface by the layered theory.
3. For each N, compute $1 - w_r(N)/w$ and plot it against N on log scales.
4. Fit the plotted points by a least-squares line. The slope of the line is α_{sys} and the intercept at $N = 1$ is μ_{sys}.
5. Determine the permanent deformation by Eq. 7.49.

Because α and μ may change with the temperature of bituminous mixtures and the state of stresses, the VESYS IV-B program can input different values of α and μ for different temperatures and loadings to account for nonlinear effects.

Material Characterization   Chap. 7

Example 7.13:

A circular load of radius a and intensity q is applied on the surface of a homogeneous half-space with elastic modulus E, Poisson ratio v, and permanent deformation parameters α and μ. Derive an equation for determining the permanent deformation at the center of the loaded area as a function of the number of load applications.

Solution: The elastic deformation can be obtained from Eq. 2.8:

$$w = \frac{2(1 - v^2)qa}{E}$$

From Eq. 7.47, the unloading modulus is

$$E_r(N) = \frac{EN^\alpha}{N^\alpha - \mu}$$

and the elastic deformation due to unloading is

$$w_r = \frac{2(1 - v^2)qa(N^\alpha - \mu)}{EN^\alpha}$$

From Eq. 7.48, the permanent deformation is

$$w_p(N) = \frac{2(1 - v^2)qa}{E}\left(1 - \frac{N^\alpha - \mu}{N^\alpha}\right) = \frac{2(1 - v^2)qa}{E}\mu N^{-\alpha}$$

Note that for a homogeneous half-space α_{sys} and μ_{sys} are the same as α and μ because there is only one material.

7.4.2 Types of Tests

Three different types of compression tests, viz., incremental static, dynamic, and creep tests, are described. The test equipment and setup are the same as those used for the resilient modulus tests described previously. The specimens are usually 4 in. (101 mm) in diameter and 8 in. (203 mm) in height. The incremental static test is applicable only to bituminous mixtures and fine-grained soils that exhibit a predominant amount of viscous flow, while the dynamic and creep tests can be applied to all kinds of materials. In performing the tests, care should be taken to assure that test results reflect the effects of stress state, temperature, moisture content, and other conditions corresponding to those of the pavement in situ. The VESYS manual recommends the use of an unconfined test with an axial stress of 20 psi (138 kPa) for bituminous mixtures, a confined test with a confining pressure of 10 psi (69 kPa) and a deviator stress of 20 psi (138 kPa) for untreated granular materials, and a confined test with a confining pressure of 3 psi (21 kPa) and a deviator stress of 6 psi (41 kPa) for fine-grained soils. A haversine load with a duration of 0.1 s and a rest period of 0.9 s has been frequently used.

Incremental Static Test

This is a simplified test and requires much less time to perform than the dynamic test. With a dynamic loading time of 0.1-s duration, a static load of t

seconds is equivalent to $10t$ load repetitions. Therefore, the permanent deformation after a 1000-s creep test is equivalent to that of a dynamic test after 10,000 repetitions. The procedures for the testing of bituminous mixtures, as suggested in the VESYS manual (FHWA, 1978), are summarized below:

1. For preconditioning, apply two ramp loads of 20 psi (138 kPa) and hold each peak load for 10-min durations, with a minimum of unload time between them. A third load is then applied for 10 min, followed by a 10-min rest period.
2. Five different ramp loads with durations of 0.1, 1, 10, 100, and 1000 s and rest periods of 2, 2, 2, 4, and 8 min are applied successively, as shown in Figure 7.30. The total permanent strains at the end of each rest period are measured.

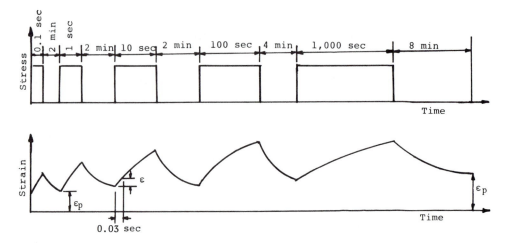

Figure 7.30 Stress and strain of incremental static test.

3. In the fifth ramp load or 1000-s creep test, measure the magnitude of creep deformations after 0.03, 0.1, 0.3, 1, 3, 10, 30, 100, and 1000 s. The 0.03-s creep strain is equivalent to the resilient strain ϵ under a dynamic haversine load of 0.1-s duration. The creep data should be extrapolated to obtain the strains at 0.001, 0.003, and 0.01 s so that the creep compliances at 11 time increments can be entered into VESYS or KENLAYER.
4. Plot the total permanent strain ϵ_p versus the incremental loading time and fit with a straight line, as shown in Figure 7.31.
5. Determine the slope S and the intercept I of the straight line and compute α from Eq. 7.43 and μ from Eq. 7.44, in which ϵ is the creep strain at 0.03 s.

The same procedures can be applied to fine-grained soils except that a confining pressure of 3 psi (21 kPa) and a deviator stress of 6 psi (41 kPa) is used. Note that temperature control is needed for testing bituminous mixtures but not for soils.

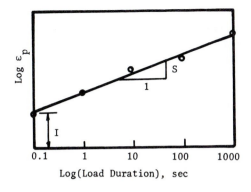

Figure 7.31 Log–log plot of incremental static test.

Example 7.14:

A load of 20 psi (138 kPa) is applied to a HMA specimen. The average permanent deformations after the five ramp loads of 0.1-, 1-, 10-, 100-, and 1000-s duration, as measured by two LVDTs attached to the specimen by a pair of clamps, are 0.0012, 0.0035, 0.0117, 0.0317, and 0.0635 mm, respectively. The gauge length, or the distance between the LVDT clamps, is 4 in. (102 mm). The creep deformation at 0.03 s is 0.0031 mm. Determine the permanent deformation parameters α and μ.

Solution: Dividing permanent deformation by a gauge length of 4 in. (102 mm) gives permanent strains of 1.2, 3.5, 11.5, 31.2, and 62.5 $\times$ 10^{-5} for the five ramp loads, respectively. The permanent strains are plotted versus the incremental loading times in log scales, as shown in Figure 7.32. The plot can be approximated by a straight line

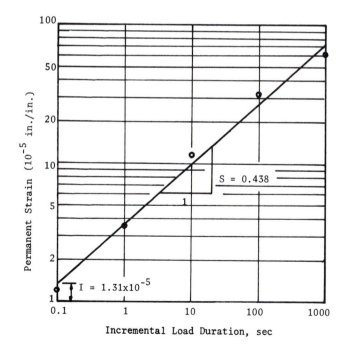

Figure 7.32 Example 7.14.

with a slope $S = (\log 74.3 - \log 1.31)/4 = 0.438$ and an intercept $I = 1.31 \times 10^{-5}$. From Eq. 7.43, $\alpha = 1 - 0.438 = 0.562$. The creep strain at 0.03 s $\epsilon = 0.0031/(4 \times 25.4) = 3.05 \times 10^{-5}$. From Eq. 7.44, $\mu = 1.31 \times 0.438/3.05 = 0.188$. It is interesting to note that both α and μ are dimensionless parameters not affected by the magnitude of applied load.

Dynamic Test

The dynamic test can be performed immediately after the incremental static test without further sample conditioning. If the test is run independently, the sample must be preconditioned by applying a sufficient number of repeated loads until the resilient deformation becomes more stable.

The level of the repeated dynamic load and the confining pressure, if any, should simulate what actually occurs in the field. A minimum of 100,000 load applications is usually required and the accumulated permanent deformation are measured at 1, 10, 100, 200, 1000, 10,000, and 100,000 repetitions. At the 200th repetition, both the permanent and the resilient deformations must be measured.

Example 7.15:

The resilient strain of a HMA specimen at the 200th repetition is 5.7×10^{-5} and the permanent strains at 1, 10, 100, 200, 1000, 10,000, and 100,000 repetitions are 4.5, 10.3, 18.4, 21.5, 41.5, 55.4, and 96.5×10^{-5}. Determine the permanent deformation parameters α and μ.

Solution: The permanent strains are plotted versus the number of load applications and fitted with a straight line, as shown in Figure 7.33. The slope of the straight line

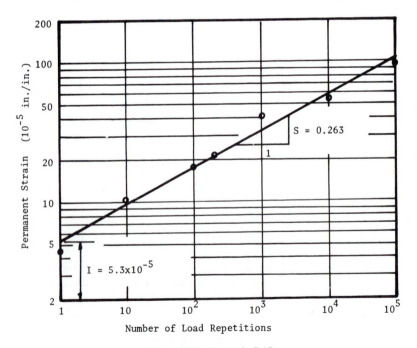

Figure 7.33 Example 7.15.

$S = (\log 109 - \log 5.3)/5 = 0.263$ and the intercept at 1 repetition $I = 5.3 \times 10^{-5}$. From Eq. 7.43, $\alpha = 1 - 0.263 = 0.737$. The resilient strain at the 200th repetition is 5.7×10^{-5}, so from Eq. 7.44, $\mu = 5.3 \times 0.263/5.7 = 0.245$.

Creep Test

Creep test can be used not only to obtain the creep compliances at various times as an input for the analysis of viscoelastic systems, but also to estimate the rut depth due to the permanent deformation of bituminous layer. Van de Loo (1974, 1978) developed a method for estimating rut depth which was incorporated in the *Shell Pavement Design Manual*. The rut depth is computed by

$$\text{Rut depth} = C_m h_1 \left(\frac{\sigma_{av}}{S_{mix}} \right) \tag{7.51}$$

in which C_m is a correction factor for dynamic effect with values ranging from 1 to 2 depending on the type of mix, h_1 is the thickness of the asphalt layer, σ_{av} is the average vertical stress in the asphalt layer, and S_{mix} is the stiffness modulus of the mix. Because permanent deformation is caused by the viscous component of the mix while S_{mix} obtained from the creep test includes both the elastic and viscous components, a modification of S_{mix} based on the Van der Poel's nomograph and the viscosity of the bitumen is needed. Finn et al. (1983) suggested a simplified procedure in which S_{mix} is determined directly from the creep test and C_m is assumed to be 1. If the number of load applications is N, then the creep time in seconds corresponding to N is $0.1N$ because the duration of each load application is 0.1 s.

Example 7.16:

Figure 7.34 shows a full-depth asphalt pavement 6 in. (305 mm) thick. A circular load with a radius of 6 in. (152 mm) and an intensity of 100 psi (690 kPa) is applied on the surface. The pavement has a modulus ratio E_1/E_2 of 100 and a Poisson ratio of 0.5 for both layers. If the HMA has a creep compliance of 2.1×10^{-5} in.^{2}/lb (3.0 mm^2/kN) at a loading time of 1000 s, estimate the rut depth after 10,000 load applications.

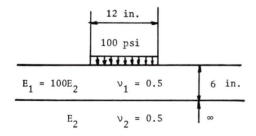

Figure 7.34 Example 7.16
(1 in. = 25.4 mm, 1 psi = 6.9 kPa).

Solution: The distribution of vertical stress in the asphalt layer can be found in Figure 2.14. For $E_1/E_2 = 100$, the average stress can be obtained quite accurately by averaging the stresses on the surface and at the interface, or $\sigma_{av} = (100 + 8)/2 = 54$ psi (373 kPa). With $S_{mix} = 1/(2.1 \times 10^{-5}) = 4.76 \times 10^4$ psi (328 MPa), $C_m = 1$, and $h_1 = 12$ in. (305 mm), from Eq. 7.51, rut depth $= 6 \times 54/(4.76 \times 10^4) = 0.0068$ in. (0.17 mm). Note that the rut depth is caused by the permanent deformation of the asphalt layer only, excluding that caused by the subgrade.

7.5 OTHER PROPERTIES

Some material properties that are related to pavement design but have not yet been discussed are presented here.

7.5.1 Modulus of Subgrade Reaction

The modulus of subgrade reaction k is determined from the loading test on a circular plate, 30 in. (762 mm) in diameter. To minimize bending, a series of stacked plates should be used. The load is applied to the plates by a hydraulic jack. A steel beam tied to heavy mobile equipment can be used as the reaction for the load. Deflections of the plate are measured by three dial gauges located at the outside edge about 120° apart. The support for the deflection dials must be located as far from the loaded area as possible, usually not less than 15 ft (4.5 m). Figure 7.35 is a schematic diagram of the plate loading test.

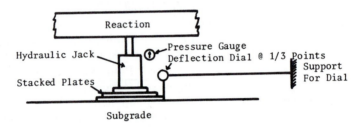

Figure 7.35 Schematic diagram of plate loading test.

The load is applied at a predetermined rate until a pressure of 10 psi (69 kPa) is reached. The pressure is held constant until the deflection increases not more than 0.001 in. (0.025 mm) per minute for three consecutive minutes. The average of the three dial readings is used to determine the deflection. The modulus of subgrade reaction is given by

$$k = \frac{p}{\Delta} \tag{7.52}$$

in which p is the pressure on the plate, or 10 psi, and Δ is the deflection of plate in in.

Since the k value is determined from a field test, it cannot be conducted at various moisture contents and densities to simulate the different service conditions or the worst possible condition during the design life. To modify the k value for conditions other than those during the field test, laboratory specimens can be fabricated, one having the same moisture content and density as those in the field and the other having a different moisture content and density to simulate the service conditions. The specimens are subjected to a creep or consolidation test under a pressure of 10 psi (69 kPa), and the deformations d at various times are measured until the increase in deformation becomes negligibly small. The modified k value can be computed by

$$k_s = \frac{d_u}{d_s} k_u \qquad (7.53)$$

in which subscript s indicates the service or saturated condition and u indicates the unsaturated or field condition.

Since the plate loading test is time-consuming and expensive, the k value is usually estimated by correlation to simpler tests such as CBR and R value tests. Figure 7.36 shows the approximate relationship between the k value and other soil properties.

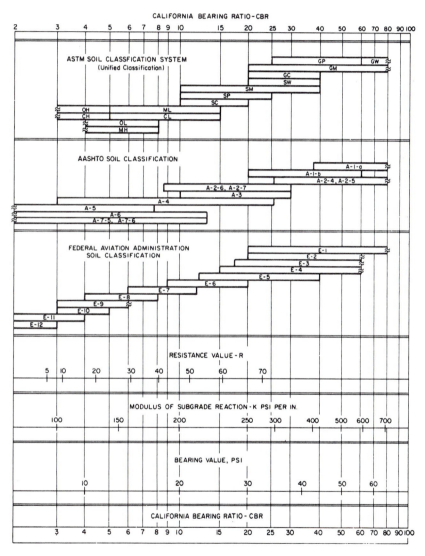

Figure 7.36 Approximate relationship between k values and other soil properties (1 psi = 6.9 kPa, 1 pci = 271.3 kN/m³). (After PCA (1966).)

7.5.2 Elastic Modulus

The ranges of K_1 and K_2 for untreated granular material are shown in Table 3.2 and those of the HMA dynamic modulus in Table 7.7. The modulus of fine-grained soil depends on the deviator stress. The maximum and minimum moduli shown in Figure 3.6 can be used as a guide. Table 7.9 shows the ranges of elastic modulus for different materials.

TABLE 7.9 ELASTIC MODULI FOR DIFFERENT MATERIALS

Material	Range	Typical
Portland cement concrete	3×10^6 to 6×10^6	4×10^6
Cement-treated bases	1×10^6 to 3×10^6	2×10^6
Soil cement materials	5×10^4 to 2×10^6	1×10^6
Lime-flyash materials	5×10^5 to 2.5×10^6	1×10^6
Stiff clay	7600 to 17,000	12,000
Medium clay	4700 to 12,300	8000
Soft clay	1800 to 7700	5000
Very soft clay	1000 to 5700	3000

Note. Modulus in psi, 1 psi = 6.9 kPa.

7.5.3 Poisson Ratio

The mechanistic method of pavement design requires the information on the Poisson ratio. The Poisson ratio v is defined as the ratio of the lateral strain to the axial strain. By measuring the axial and lateral strains during a resilient modulus test, the Poisson ratio can be determined. Because Poisson ratio has a relatively small effect on pavement responses, it is customary to assume a reasonable value for use in design, rather than to determine it from actual tests. Table 7.10 shows typical Poisson ratios for paving materials.

TABLE 7.10 POISSON RATIOS FOR DIFFERENT MATERIALS

Material	Range	Typical
Hot mix asphalt	0.30–0.40	0.35
Portland cement concrete	0.15–0.20	0.15
Untreated granular materials	0.30–0.40	0.35
Cement-treated granular materials	0.10–0.20	0.15
Cement-treated fine-grained soils	0.15–0.35	0.25
Lime-stablilized materials	0.10–0.25	0.20
Lime–flyash mixtures	0.10–0.15	0.15
Loose sand or silty sand	0.20–0.40	0.30
Dense sand	0.30–0.45	0.35
Fine-grained soils	0.30–0.50	0.40
Saturated soft clays	0.40–0.50	0.45

7.5.4 Portland Cement Concrete

The fatigue and modulus of rupture for portland cement concrete were discussed in Section 7.3.2. Some other concrete properties directly related to rigid pavement design are presented in this section.

The compressive strength is a universal measure of concrete quality and durability. As the information on compressive strength is readily available, studies have been made to correlate compressive strength with other properties. A general relationship between the modulus of rupture and the compressive strength is

$$S_c = 8\sqrt{f'_c} \quad \text{to} \quad 10\sqrt{f'_c} \tag{7.54}$$

in which S_c is the modulus of rupture in psi and f'_c is the compressive strength in psi. The relationship between the indirect tensile strength and the compressive strength, as suggested by the American Concrete Institute, is

$$f_t = 6.5\sqrt{f'_c} \tag{7.55}$$

in which f_t is the indirect tensile strength in psi. The relationship between the modulus of elasticity and the modulus of rupture is (ERES, 1987)

$$S_c = \frac{43.5E_c}{10^6} + 488.5 \tag{7.56a}$$

or $$E_c = (S_c - 488.5) \times 2.3 \times 10^4 \tag{7.56b}$$

For normal weight concrete, the American Concrete Institute suggested

$$E_c = 57,000 \sqrt{f'_c} \tag{7.57}$$

The coefficient of thermal expansion ranges from 3 to $8 \times 10^{-6}/°F$ (5.4 to 14.4 $\times$ $10^{-6}/°C$), with $5 \times 10^{-6}/°F$($9 \times 10^{-6}/°C$) as typical. The coefficient of drying shrinkage ranges from 0.5 to 2.5×10^{-4}, with 1×10^{-4} as typical.

7.6 SUMMARY

This chapter describes the various methods for characterizing paving materials, including resilient and dynamic moduli, fatigue characteristics, and permanent deformation parameters.

Important Points Discussed in Chapter 7

1. In the analysis of layer systems under moving loads, the elastic modulus to be used for each layer is the resilient modulus obtained from the repeated load tests.

2. Theoretically, the load pulse for repeated load tests should vary with the vehicle speed and the location of the material below the pavement surface. Practically, a haversine load with a duration of 0.1 s and a rest period of 0.9 s has been frequently used.

3. Both the resilient and dynamic moduli have been used as the elastic modulus for bituminous mixtures. The difference between a resilient modulus and a dynamic modulus is that the resilient modulus is measured under a haversine pulse with a rest period, while the dynamic modulus is measured under a sinusoidal or haversine pulse with no rest period. The effect of rest period is insignificant as evidenced by the fact that the same dynamic modulus is obtained no matter whether a sinusoidal or a haversine pulse is used.

4. The resilient modulus of untreated granular materials and fine-grained soils can be determined from the repeated load triaxial compression tests. Procedures for testing these materials are described and the results are analyzed to determine the various parameters.

5. Correlations between the resilient modulus and other properties, such as CBR, R value, Texas triaxial classification, group index, Marshall stability, and cohesiometer value, are presented. These correlations are only approximate and should be selected with great care because the properties are related more to the shear strength than the resilient modulus. The dynamic modulus of bituminous mixtures can be predicted by Shell nomographs and AI equations.

6. Due to the high tensile strains frequently used in the fatigue testing of bituminous mixtures, the dynamic stiffness modulus decreases with the increase in dynamic load and cannot be used as an elastic modulus. However, a correlation between the dynamic stiffness modulus and the dynamic modulus has been established for design purposes.

7. The fatigue testing of bituminous mixtures is time-consuming and expensive. To shorten the testing time, a haversine loading with a duration of 0.1 s and a rest period of 0.4 s, instead of the 0.9 s for the resilient modulus test, is usually used. Two types of fatigue test can be performed. The constant stress test, in which the strain increases with the increase in number of load repetitions, is applicable to thick pavements, and the constant strain test, in which the strain is kept constant by reducing the stress as the number of load repetitions increases, is applicable to thin pavements. The use of the constant stress test for thin pavements is more conservative and results in a shorter design life. Nomographs and equations developed by Shell and AI can be used to predict the number of repetitions to failure.

8. The fatigue life of portland cement concrete depends on the stress ratio, which is a quotient between the flexural stress and the modulus of rupture. It is impractical to perform fatigue tests for a specific project because the variability of the test results requires a large number of tests and a great expense involved. Furthermore, a sufficient amount of data, as shown in Figure 7.27, is available and can be used for design purpose. The PCA design criterion is very conservative and results in a fatigue life much lower than the average life.

9. The permanent deformation properties of bituminous mixtures and fine-grained soils with sufficient viscous flow can be evaluated by the incremental static test, the dynamic test, and the creep test, whereas those of granular materials and fine-grained soils with little viscous flow can be evaluated only by the dynamic test. In conducting the permanent deformation test, care should be taken that the stresses used in the test be representative of those actually occurring in the in situ pavement, so the effect of nonlinearity can be ignored.

10. The modulus of subgrade reaction k can be determined from a plate loading test using a circular plate 30 in. (762 mm) in diameter. Because the test is time-consuming and does not simulate the service conditions, it is customary to correlate the k value with other simpler tests such as CBR and R value, instead of performing the plate loading test.

PROBLEMS

7-1. The results of repeated load tests on a granular material are tabulated in Table P7.1. Develop an equation relating the resilient modulus to the first stress invariant or the sum of three principal stresses. [Answer: $M_R = 17500^{0.711}$]

TABLE P7.1

Confining pressure (psi)	2	5	10	20	2
Deviator stress (psi)	6	15	30	60	6
Recoverable strain (10^{-4})	5.8	7.4	9.5	11.4	6.0

7-2. A fine-grained soil has an unconfined compressive strength of 2.3 tsf. The results of repeated load tests are tabulated in Table P7.2. Develop an equation relating the resilient modulus to the deviator stress. What should be the maximum and minimum resilient moduli for this soil? [Answer: $K_1 = 8300$, $K_2 = 7.9$, $K_3 = 1010$, $K_4 = 145$, $E_{max} = 14,200$ psi and $E_{min} = 4850$ psi]

TABLE P7.2

Deviator stress (psi)	2	4	7	14	20	28
Recoverable strain (10^{-4})	1.41	3.20	7.61	18.90	30.32	51.81

7-3. The flexural stresses and the initial moduli for a series of fatigue tests on HMA specimens are shown in Table P7.3. Determine the dynamic modulus for a frequency of 8 Hz by Eq. 7.15. [Answer: 8×10^5 psi]

TABLE P7.3

Test No.	1	2	3	4	5	6	7	8	9
Stress (psi)	278	254	228	197	185	165	137	115	91
Initial stiffness modulus (10^5 psi)	1.00	1.14	1.15	1.34	1.60	2.87	2.22	2.78	3.15

7-4. The results of fatigue test on an asphalt treated base are shown in Table P7.4. Develop an equation relating the number of repetitions to failure and the initial tensile strain. [Answer: $N_f = 6.66 \times 10^{-8}(\epsilon_t)^{-3.5}$]

TABLE P7.4

Test No.	1	2	3	4	5	6	7	8
Initial strain (10^{-3})	3.15	2.61	2.49	1.92	1.59	1.27	1.09	0.873
Fracture life (N_f)	35	75	100	294	340	970	1630	3573

7-5. An asphalt mixture has an asphalt content of 7% and a bulk specific gravity 2.24. The recovered asphalt has a specific gravity of 1.02, a ring and ball softening point of 120°F, and a penetration of 50 at 77°F. The specific gravity of the aggregate is 2.61. Determine the stiffness modulus of the mixture at a temperature of 74°F and a loading time of 0.02 s by the Shell nomographs shown in Figures 7.19 and 7.20. Check the result by Eq. 7.25. [Answer: 2×10^9 N/m^2]

7-6. The asphalt mixture is the same as in Problem 7-5. The mineral aggregate has 5% passing through a No. 200 sieve and the original asphalt has a penetration of 75 at 77°F. Determine the dynamic modulus at a temperature of 74°F and a loading frequency of 8 Hz by the AI equations. [Answer: 3.8×10^5 psi]

7-7. The asphalt mixture in Problem 7-5 is subjected to an initial tensile strain of 0.00015. Determine the number of repetitions to failure by the nomograph shown in Figure 7.26 based on constant stress and constant strain tests, respectively, and check the results using Eqs. 7.33 and 7.35. [Answer: 8.8×10^5 and 8.1×10^7]

7-8. The asphalt mixture in Problem 7-6 is subjected to an initial strain of 0.00015. Determine the number of repetitions to failure by Eqs. 7.36 and 7.37. [Answer: 6.4×10^5]

7-9. The results of incremental static test on a HMA specimen are shown in Table P7.9. If the creep strain at a time duration of 0.03 s is 3.045×10^{-5}, determine the permanent deformation parameters α and μ. [Answer: $\alpha = 0.558$ and $\mu = 0.186$]

TABLE P7.9

Load duration (s)	0.1	1	10	100	1000
Permanent strain (10^{-5})	1.158	3.474	11.533	31.172	62.483

7-10. The results of dynamic test on a HMA specimen are shown in Table P7.10. If the elastic strain at the 200th repetition is 4.128×10^{-5}, determine the permanent deformation parameters α and μ. [Answer: $\alpha = 0.705$ and $\mu = 0.187$]

TABLE P7.10

No. of repetitions	1	10	100	200	1000	10,000	100,000
Accumulated permanent strain (10^{-5})	2.360	5.831	12.773	13.930	19.715	36.281	73.506

8

Drainage Design

8.1 GENERAL CONSIDERATION

Drainage is one of the most important factors in pavement design. Until recently, this factor has not received the attention it merits. One misconception is that good drainage is not required if the thickness design is based on saturated conditions. This concept may have been true during the old days when the traffic loading and volume were small. As the weight and number of axle loads increase, water may cause more damage to pavements, such as pumping and degradation of paving materials, other than the loss of shear strength. Theoretically, an internal drainage system is not required if the infiltration into the pavement is smaller than the drainage capacity of the base, subbase, and subgrade. Because the infiltration and drainage capacity vary a great deal and are difficult to estimate, it is suggested that drainage layers be used for all important pavements, such as the premium or zero-maintenance pavements described in Section 1.2.3.

8.1.1 Detrimental Effects of Water

Water enters the pavement structure either as infiltrations through cracks, joints, pavement surfaces, and shoulders, or as groundwater from high water table, interrupted aquifers, and localized springs. The detrimental effects of water, when entrapped in the pavement structure, can be summarized as follows:

1. It reduces the strength of unbounded granular materials and subgrade soils.
2. It causes pumping of concrete pavements with subsequent faulting, cracking, and general shoulder deterioration.

3. With the high hydrodynamic pressure generated by moving traffic, pumping of fines in the base course of flexible pavements may also occur with resulting loss of support.

4. In northern climates with a depth of frost penetration greater than the pavement thickness, high water table causes frost heave and the reduction of load-carrying capacity during the frost melting period.

5. Water causes differential heaving over swelling soils.

6. Continuous contact with water causes stripping of asphalt mixture and durability or "D" cracking of concrete.

8.1.2 Movement of Water

The movement of water in pavement structures can be caused by gravity, capillary action, vapor pressures, or a combination of the above. The flow in granular materials is basically by gravity, while that in fine-grained soil is mostly by capillary action. In the absence of gravity and capillary flows, water moves primarily in the vapor phase due to the difference in vapor pressures.

The movements of water by gravity obeys Darcy's law of saturated flow:

$$v = ki \tag{8.1}$$

in which v is the discharge velocity, k is the coefficient of permeability (simply called permeability), and i is the hydraulic gradient, which is the head loss between two points divided by the distance between them. The discharge velocity can be used to determine the discharge by

$$Q = vA \tag{8.2}$$

in which Q is the discharge, or the volume of flow per unit time, and A is a cross-sectional area normal to the direction of flow. It can be seen that the discharge velocity is not the actual seepage velocity through the pores, but is an imaginary velocity for computing the discharge.

Darcy's law can be used in conjunction with the continuity equation to form the governing differential equation of groundwater flow. A convenient and practical way to solve the equation is by drawing the flownets, as illustrated by Cedergren (1977). In many simple cases encountered in drainage design, Darcy's law can be applied directly to determine the amount of seepage.

Example 8.1:

Figure 8.1 shows a pavement with a thin granular drainage layer. The infiltration of surface water into the drainage layer is 0.5 ft³/day/ft² (0.15 m³/day/m²). The drainage layer is placed on a 5-ft (1.5-m) sand layer with a permeability of 0.15 ft/day (5.3 × 10⁻⁵ cm/s). The sand layer is underlain by a very pervious gravel deposit and there is no water table in the sand layer. Estimate the required discharge capacity of the drainage layer.

Solution: Because the thickness of the sand layer is small compared to the width of the pavement, it is reasonable to assume that the flow of water through the sand is in the vertical direction, so Darcy's law can be directly applied. Due to the small thickness of the drainage layer, the pressure head on the top of sand layer can be

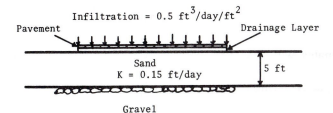

Figure 8.1 Example 8.1 (1 ft = 0.305 m).

assumed zero. Since both the top and the bottom of sand layer are subjected to the same atmospheric pressure, the head loss between top and bottom is equal to the difference in elevations, or 5 ft (1.5 m). The distance between the top and the bottom is also 5 ft (1.5 m), so the hydraulic gradient is 5/5 or 1. Note that when the atmospheric pressure exists at both top and bottom, the hydraulic gradient is always equal to 1, irrespective of the thickness of sand layer. From Eq. 8.1, $v = 0.15 \times 1 = 0.15$ ft/day (5.3×10^{-5} cm/s), which is the same as a discharge rate in ft^3/day/ft^2. Therefore, the required discharge capacity of the drainage layer is $0.5 - 0.15$, or 0.35 ft^3/day/ft^2 (0.11 m^3/day/m^2).

If the sand layer in this example is very thick, the flow channel becomes wider at greater depths. As more areas are open to the flow, the amount of seepage increases. Therefore, the use of Darcy's law for drainage design by assuming the flow lines to be vertical is on the conservative side. The analysis of unsaturated flow by capillary action is much more complicated. Both the permeability and the hydraulic gradient due to the change in capillary heads are functions of moisture content and are sometimes influenced by the past moisture conditions.

It is well known that moisture moves in vapor form from warmer to cooler regions. The vapor condenses under the pavement at night when the pavement becomes cool. In northern temperate climates, there is a downward migration of moisture vapor from the warmer surface to the cooler subsurface in the summer and a corresponding upward movement of moisture in the winter. This upward movement of water vapor is partially responsible for the high moisture content in granular bases.

8.1.3 Methods for Controlling Water in Pavements

The detrimental effects of water on pavements can be minimized by preventing it from entering the pavement, providing drainage to remove it quickly, or building the pavement strong enough to resist the combined effect of load and water.

Prevention

The prevention of water entering the pavement structure requires intercepting groundwater and sealing the pavement surface. The detrimental effects of groundwater have been well documented, and highway engineers have paid considerable attention to intercepting groundwater. Less attention has been given to sealing the surface to exclude infiltration from rain and snow melt. As a result, a

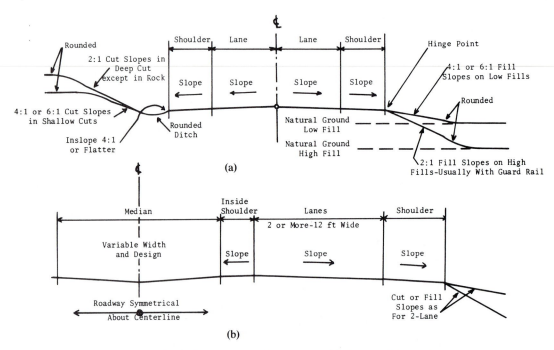

Figure 8.2 Cross sections of typical highways (1 ft = 0.305 m).

considerable amount of water often enters the pavement structure. Because complete prevention is not possible during the useful life of a pavement, the installation of internal drainage systems to remove water may be needed.

To minimize the infiltration of surface water into the pavement, a good surface drainage is always required. Figure 8.2 shows typical cross sections for a two-lane and a multilane highway. To facilitate surface drainage, transverse slopes are introduced in all tangent sections of roadway. Except where super-elevation of curves directs all water toward the inside, two-lane highways usually have each lane sloping in a different direction. To make driving easier, each half of a divided highway is sloped individually, usually with the outside edges lower than the inside edges. Table 8.1 shows the transverse slopes of pavements, shoulders, and ditches.

TABLE 8.1 TRANSVERSE SLOPES OF
PAVEMENTS, SHOULDERS, AND DITCHES

Feature	Pavement	Shoulder	Ditch
Slope (in./ft)	$\frac{3}{16}-\frac{3}{8}$	$\frac{3}{8}-\frac{3}{4}$	3–6
Slope (%)	1.5–3.0	3.0–6.0	25.0–50.0

Note. 1 in./ft = 82 mm/m.

Removal

If water enters the pavement structure either through infiltration or groundwater, it must be removed quickly before any damage can be initiated. Three different drainage installations, either individually or in combination, have been used most frequently for pavement design: drainage layer or blanket, longitudinal drain, and transverse drain.

Figure 8.3 shows a drainage blanket for intercepting surface infiltration. The drainage blanket can be terminated with longitudinal drains and pipe outlets, as shown in (a), or daylighted by extending to the side slope, as shown in (b). To minimize the intrusion of fines, all materials surrounding the drainage blanket and the longitudinal drain must meet the filter criteria (discussed later). The use of longitudinal drains is more reliable and may be even more economical than the daylighted construction. The disadvantages of the daylighted construction are the propensity to contamination and clogging of the outlet at the side slope during construction and maintenance operations, the smaller hydraulic gradient caused by the wider blanket, and the possibility of carrying water from the side ditch to the pavement structure, instead of from the pavement structure to the side ditch. The longitudinal drains can be French drains without the perforated or slotted pipes. However, these pipes are usually used because they can accommodate larger flows and thus reduce the dimensions of French drains required.

Figure 8.4 shows the location of drainage layer in a pavement structure. In (a), the base course is used as a drainage layer and satisfies both the requirements of strength for a base course and the permeability for a drainage layer. In (b), the drainage layer is placed on the top of subgrade, either as an extra layer not considered in thickness design or as a part of the subbase. The placement of drainage layer directly under the PCC or HMA is preferable because the water can be drained out more quickly, thus eliminating any chance for pumping to occur. However, it has the disadvantages that the deficiency of fines in the drainage layer may cause stability problems and that the water in the subbase cannot drain readily into the drainage layer. If the drainage layer is placed on the top of subgrade, the permeability of the base and subbase must be greater than the infiltration rate, so that water can flow freely to the drainage layer.

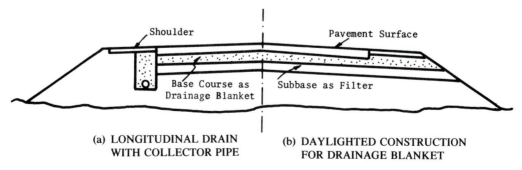

(a) LONGITUDINAL DRAIN
WITH COLLECTOR PIPE

(b) DAYLIGHTED CONSTRUCTION
FOR DRAINAGE BLANKET

Figure 8.3 Drainage blanket with alternate longitudinal drain.

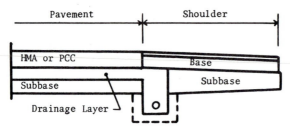

(a) Base as Drainage Layer

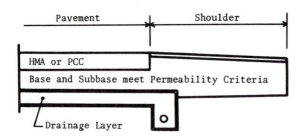

(b) Drainage Layer below Subbase or as Part of Subbase

NOTE: All Materials Surrounding Drainage Layer
Must Meet Filter Criteria.

Figure 8.4 Location of drainage layer and filter. (After Ridgeway (1982).)

Figure 8.5 shows the use of a longitudinal drain parallel to the roadway to lower the groundwater table. If the general direction of groundwater flow is parallel to the roadway, such as when the roadway is cut more or less perpendicular to the existing contours, then transverse drains can be more effective in intercepting the flow.

Longitudinal trench drains were used by Arkansas, Florida, Louisiana, and New Mexico for improving the drainage of existing rigid pavements (FHWA, 1986). The trench was placed adjacent to the concrete pavement and is therefore called an edge drain. Using similar designs with trench drains and collector pipes, these states reported that edge drains can remove water satisfactorily and are considered cost effective. A performance survey of five-year-old edge drains representing 120 miles (192 km) in Louisana indicated that 75% of the system continued to provide moderate to rapid drainage and the remaining 25% experienced clogging of the ends of lateral outlet pipes. Most of the clogged laterals began to drain upon cleaning, indicating no major problems in draining water from the trench. These pavements all contain cement-treated bases.

Stronger Pavement Sections

The use of thicker HMA or PCC can greatly reduce the hydrodynamic pressure and its detrimental effects. The Asphalt Institute strongly advocates the use of full-depth asphalt pavements. According to AI (1984), vapor movement is

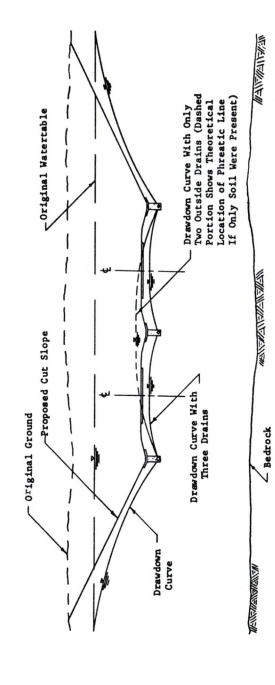

Figure 8.5 Longitudinal drains for lowering groundwater table. (After Moulton (1980).)

probably a primary cause of moisture entering and saturating granular bases, and this problem can be eliminated by the construction of a full-depth asphalt pavement placed directly on the native soil. The same is true for PCC pavements with stabilized bases. Because water can still enter into the subgrade through cracks and joints and along the pavement edges, the thickness design should be based on a saturated subgrade, if no drainage layer is installed.

8.2 DRAINAGE MATERIALS

Drainage materials include aggregates, geotextiles, and pipes. Aggregates can be used as drainage layers and French drains or as filter materials for their protection. Geotextiles are used mainly to replace aggregates as filters. Pipes can be perforated, slotted, or open-jointed types to be placed inside the French drain to collect water, or the conventional type to convey water laterally to the outlet.

8.2.1 Aggregates

Aggregates to be used for drainage layers and French drains should consist of sound, clean, and open-graded materials. They must have a high permeability to accommodate the free passage of water and be protected from clogging by means of a filter.

Permeability

Figure 8.6 shows the grain size curve and permeability of some typical open-graded bases and filter materials. The grain size is always plotted in log scale and the percent passing in arithmetic scale. The sizes in inches corresponding to the sieve numbers are also shown to facilitate the determination of various sizes for filter design.

Effect of Fines

It has long been recognized that proper gradation and density are vital to the stability of granular materials. The gradation required for stability usually ranges uniformly from coarse to fine. To obtain the desired permeability, the fine portions need be deleted; thus, the stability of the drainage layer may be adversely affected. This can be compensated for by stabilizing the drainage layer with a small amount of asphalt or portland cement. Table 8.2 shows the permeability of untreated and asphalt-treated open-graded aggregates. It can be seen that the use of 2% asphalt reduces the permeability only slightly.

Table 8.3 shows the permeability of graded aggregates with no fines. Sample 1 has a gradation similar to AASHTO grading C of soil aggregate mixtures for base course (AASHTO, 1989), except that grading C requires 5 to 15% passing No. 200. The other samples were obtained by successively eliminating small particles. As can be seen, the elimination of small particles significantly increases the permeability.

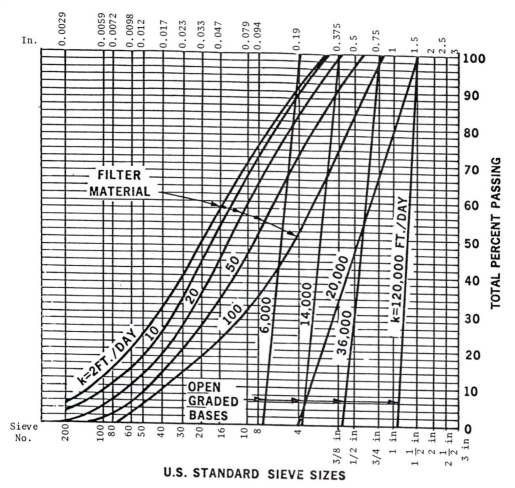

Figure 8.6 Typical gradations and permeabilities of open-graded bases and filter materials (1 in. = 25.4 mm, 1 ft/day = 3.5 × 10⁻⁴ cm/s. (After Cedergren et al. (1972).)

TABLE 8.2 PERMEABILITIES OF UNTREATED AND ASPHALT-TREATED OPEN-GRADED AGGREGATES

	Permeability (ft/day)	
Aggregate size	Untreated	Treated with 2% asphalt
1½ to 1 in.	140,000	120,000
¾ to ⅜ in.	38,000	35,000
No. 4 to No. 8	8000	6000

Note. 1 in. = 25.4 mm, 1 ft/day = 3.5 × 10⁻⁴ cm/s.

Source. After Lovering and Cedergren (1962).

TABLE 8.3 GRAIN SIZE, DENSITY, AND PERMEABILITY OF GRADED AGGREGATES

% Passing	Sample 1	Sample 2	Sample 3	Sample 4	Sample 5	Sample 6
¾ in.	100	100	100	100	100	100
½ in.	85	84	83	81.5	79.5	75
⅜ in.	77.5	76	74	72.5	69.5	63
No. 4	58.5	56	52.5	49	43.5	32
No. 8	42.5	39	34	29.5	22	5.8
No. 10	39	35	30	25	17	0
No. 20	26.5	22	15.5	9.8	0	0
No. 40	18.5	13.3	6.3	0	0	0
No. 60	13	7.5	0	0	0	0
No. 140	6	0	0	0	0	0
No. 200	0	0	0	0	0	0
Dry density (pcf)	121	117	115	111	104	101
k (ft/day)	10	110	320	1000	2600	3000

Note. 1 in. = 25.4 mm, 1 pcf = 157.1 N/m³, 1 ft/day = 3.5 × 10⁻⁴ cm/s.
Source. After Barber and Sawyer (1952).

The effect of fines passing through a No. 200 sieve on the permeability of sample 1 is shown in Figure 8.7. The effect depends on the type of fines. The permeability is about 10 ft/day (3.5×10^{-3} cm/s) with no fines, and decreases to 2×10^{-2} to 5×10^{-5} ft/day (7×10^{-6} to 1.7×10^{-8} cm/s) with 25% of fines.

Effect of Grain Size

A number of approximate relationships have been suggested between permeability and grain size. The most frequently used approximation is the one

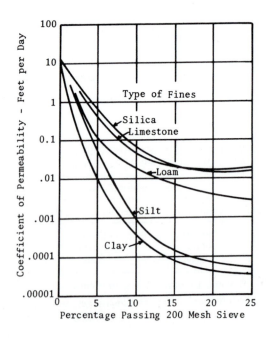

Figure 8.7 Effect of fines on permeability of sample 1 (1 ft/day = 3.5 × 10⁻⁴ cm/s). (After Barber and Sawyer (1952).)

TABLE 8.4 VALUES OF HAZEN'S COEFFICIENT

Soil type	D_{10} range (mm)	C_k (1/mm-s)
Uniform sand	0.06–3.0	8–12
Well-graded sands and silty sands	0.003–0.6	5–8

Source. After Whitlow (1990).

suggested by Hazen for filter sands:

$$k = C_k D_{10}^2 \qquad (8.3)$$

in which k is the permeability in mm/s, D_{10} is the effective size, or the grain size corresponding to 10% passing, and C_k is an experimental coefficient dependent on the nature of soil. Experimental evidence suggests that the acceptable approximate values for k can be obtained when Hazen's formula is applied over a wide range of soils. Table 8.4 gives the range of suggested values for the coefficient C_k.

In addition to the effective size D_{10}, other sizes, such as D_{15}, D_{50}, and D_{85}, have been used for the design of filters. If the values of percent passing, a and b, for two grain sizes, D_a and D_b, are given, then the size D_x for any percent passing x between a and b can be computed by linear interpolation, as shown in Figure 8.8:

$$\log D_x = \log D_a + \frac{x - a}{b - a} \log \frac{D_b}{D_a} \qquad (8.4)$$

Example 8.2:

If 3% of minus 200 fines is added to sample 1 in Table 8.3, determine D_{10} of the sample.

Solution: The two successive sieves for determining D_{10} are No. 140, or $D_a = 0.105$ mm, and No. 60, or $D_b = 0.25$ mm. In the original sample, 6% pass No. 140 and 13% pass No. 60. After adding 3% fines, $a = (6 + 3)/1.03 = 8.7\%$ and $b = (13 + 3)/1.03 = 15.5\%$. From Eq. 8.4, $\log D_{10} = \log 0.105 + (10 - 8.7)/(15.5 - 8.7)$ $\log(0.25/0.105) = -0.907$, or $D_{10} = 0.124$ mm.

Moulton (1980) developed the following empirical equation for determining the permeability of granular drainage and filter materials:

$$k = \frac{6.214 \times 10^5 (D_{10})^{1.478} (n)^{6.654}}{(P_{200})^{0.597}} \qquad (8.5)$$

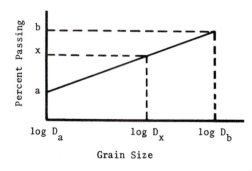

Figure 8.8 Interpolation between two grain sizes.

in which k is the permeability in ft/day, D_{10} is the effective size in mm, which is the size at 10% passing, and n is the porosity, which can be computed from the dry unit weight by

$$n = 1 - \frac{\gamma_d}{62.4G_s} \qquad (8.6)$$

in which γ_d is the dry unit weight in pcf and G_s is the specific gravity of solids. A major limitation of the formula is that it cannot be applied to materials with no fines passing the No. 200 sieve.

Example 8.3:

If 3% of minus 200 fines is added to sample 1 in Table 8.3, determine its permeability by Eq. 8.5. It is assumed that the specific gravity of solids is 2.7 and that the dry density increases by 3% due to the addition of fines.

Solution: With $\gamma_d = 1.03 \times 121 = 124.6$ pcf (19.6 kN/m²) and $G_s = 2.7$, from Eq. 8.6, $n = 1 - 124.6/(62.4 \times 2.7) = 0.260$. From Example 8.2, $D_{10} = 0.124$ mm and $P_{200} = 3/1.03 = 2.91$. From Eq. 8.5, $k = 6.214 \times 10^5 \, (0.124)^{1.478} \, (0.26)^{6.654}/(2.91)^{0.597} = 1.92$ ft/day.

Filter Criteria

Any aggregate used for drainage must satisfy the following filter criteria.

Clogging Criterion

The filter material must be fine enough to prevent the adjacent finer material from piping or migrating into the filter material, as indicated by

$$\frac{D_{15} \text{ filter}}{D_{85} \text{ soil}} \leq 5 \qquad (8.7)$$

in which D_{15} and D_{85} are the grain size corresponding to 15 and 85% passing, respectively, and can be obtained from the grain size curves of each material. This criterion should be applied not only to the filter material but also to the drainage layer. For example, if the subbase is designed as a filter, as shown in Figure 8.3, Eq. 8.7 should be applied first by considering the subbase as the filter and the subgrade as the soil and then by considering the drainage layer as the filter and the subbase as the soil. This will prevent migration of the subgrade material into the subbase and the subbase material into the drainage layer.

Permeability Criterion

The filter material must be coarse enough to carry water without any significant resistance, as indicated by

$$\frac{D_{15} \text{ filter}}{D_{15} \text{ soil}} \geq 5 \qquad (8.8)$$

This criterion need be applied only to the filter or the subbase. The drainage layer is so permeable and this criterion can certainly be satisfied.

Additional Criteria

Equations 8.7 and 8.8 were originally developed by Betram (1940), with the advice of Terzaghi and Casagrande. The work of Betram was later expanded by the U.S. Army Corps of Engineers (1955) and, to make the grain size curves of filters and protected soil materials somewhat parallel, an additional requirement, as indicated by Eq. 8.9, was added:

$$\frac{D_{50} \text{ filter}}{D_{50} \text{ soil}} \le 25 \tag{8.9}$$

To minimize segregation, the Corps of Engineers further specified that filter materials must have a coefficient of uniformity, which is a ratio between D_{60} and D_{10}, not greater than 25. To prevent the fines in the filter from infiltrating into the drainage layer, Moulton (1980) recommended that the amount of fines passing through a No. 200 sieve be not greater than 5%, or D_5 of filter ≥ 0.0029 in. (0.074 mm). If the protected soil contains a large percentage of gravels, Sherard et al. (1963) indicated that the filter should be designed on the basis of the material finer than 1 in. (25.4 mm).

Example 8.4:

Can the open-graded base with a permeability of 20,000 ft/day (70 mm/s), as shown in Figure 8.6, be placed directly on a subgrade soil with $D_{15} = 0.0013$ in. (0.033 mm), $D_{50} = 0.0055$ in. (0.038 mm), and $D_{85} = 0.021$ in. (0.53 mm)? If a subbase is to be placed between the subgrade and the open-graded base, which of the filter materials shown in Figure 8.6 can be used as a subbase?

Solution: When the base is placed directly on the subgrade, the base is considered as filter and the subgrade as soil. The D_{15} and D_{50} of base can be determined by Eq. 8.4. From the grain size curve in Figure 8.6, D_{15} is between $D_a = 0.19$ in. (4.8 mm) and $D_b = 0.375$ in. (9.5 mm) with $a = 1\%$ and $b = 32\%$. From Eq. 8.4, $\log D_{15} = \log 0.19 + [(15 - 1)/(32 - 1)] \log(0.375/0.19) = -0.588$, or $D_{15} = 0.26$ in. (6.6 mm). By following the same procedure, $\log D_{50} = \log 0.5 + [(50 - 47)/(66 - 47)] \log(0.75/0.5) = -0.273$, or $D_{50} = 0.53$ in. (13.5 mm). From Eq. 8.7, (D_{15} of base)/(D_{85} of subgrade) $= 0.26/0.021 = 12.4$, which is greater than the maximum of 5 required, so the criterion is not satisfied. From Eq. 8.8, (D_{15} of base)/(D_{15} of subgrade) $= 0.26/0.0013 = 200$, which is greater than 5 and is considered satisfactory. From Eq. 8.9, (D_{50} of base)/(D_{50} of subgrade) $= 0.53/0.0055 = 96.4$, which is greater than the maximum of 25 required, so the criterion is not satisfied. Therefore, the base should not be placed directly on the subgrade because Eqs. 8.7 and 8.9 are not satisfied.

When a subbase is placed between the subgrade and the base, the analysis should be divided into two steps. First consider the subbase as filter and the subgrade as soil. From Eqs. 8.7, 8.8, and 8.9, the requirements of the subbase are as follows: $D_{15} \le 5 \times 0.021 = 0.105$ in. (2.67 mm), $D_{15} \ge 5 \times 0.0013 = 0.0065$ in. (0.17 mm), and $D_{50} \le 25 \times 0.0055 = 0.138$ in. (3.5 mm). Next consider the base as filter and the subbase as soil. Based on Eqs. 8.7, 8.8, and 8.9, the requirements of the subbase are $D_{85} \ge 0.26/5 = 0.052$ in. (1.32 mm), $D_{15} \le 0.26/5 = 0.052$ in. (1.32

mm), and $D_{50} \geq 0.53/25 = 0.0212$ in. (0.54 mm). The above six requirements can be reduced into the following three:

$$0.0065 \text{ in. } (0.17 \text{ mm}) \leq D_{15} \leq 0.052 \text{ in. } (1.32 \text{ mm})$$

$$0.0212 \text{ in. } (0.54 \text{ mm}) \leq D_{50} \leq 0.138 \text{ in. } (3.5 \text{ mm})$$

$$D_{85} \geq 0.052 \text{ in. } (1.32 \text{ mm})$$

A review of the five filter materials in Figure 8.6 indicates that, except for the coarsest material with a D_{50} of 0.18 in. (4.6 mm), all the above requirements are satisfied. Because the finest filter material has slightly more than 5% passing No. 200 and is not considered acceptable, only the three intermediate materials with permeabilities of 10, 20, and 50 ft/day (3.5×10^{-3}, 7.0×10^{-3}, and 1.8×10^{-2} cm/s) can be used.

8.2.2 Geotextiles

Geotextiles are filter fabrics which can be used to protect the drainage layer from clogging. In addition to the two general requirements to retain soil and to allow water to flow, geotextiles must have sufficient opening areas to prevent them from clogging. They must be made from strong, tough, polyester, polypropylene, or other rot-proof polymeric fibers formed into a fabric of the woven or nonwoven type. They must be free of any treatment or coating that might significantly alter their physical properties. They must be dimensionally stable to keep the fibers at their relative position with respect to each other and provide adequate inservice performance.

Filter Criteria

A variety of filter criteria have been developed by a number of organizations and researchers. Based on a review of these criteria, the Geotextile Engineering Manual (FHWA, 1989) suggested the use of more stringent criteria when the hydraulic loadings are severe or the performance of the drainage system is critical to the protection of roadways and pertinent structures. For less severe and less critical applications, cost savings can be realized by using the less stringent criteria. In view of the fact that the fabrics used for drainage are neither severe nor critical, only the less stringent criteria suggested by the manual is presented here.

The most important dimension of geotextiles is the apparent opening size, AOS, defined as the size of glass beads when 5% pass through the geotextile. The method for determining AOS is specified by the ASTM (1989a). There are two conflicting requirements for AOS. To achieve the maximum retention of soils, the smallest opening size, or the largest AOS in terms of sieve number, should be used; while to minimize the clogging of fabrics, the maximum opening size, or the lowest AOS, should be used.

Retention or Pumping Resistance Criteria

1. For fine-grained soils with more than 50% passing through a No. 200 sieve:

   Woven: $AOS \leq D_{85}$ (8.10a)

   Nonwoven: $AOS \leq 1.8 D_{85}$ (8.10b)

   $AOS \geq$ No. 50 sieve or opening smaller than 0.297 mm   (8.10c)

2. For granular materials with 50% or less passing through a No. 200 sieve:

$$AOS \leq B \times D_{85} \qquad (8.11)$$

   in which $B = 1$ when $C_u \leq 2$ or ≥ 8, $B = 0.5C_u$ when $2 \leq C_u \leq 4$, $B = 8/C_u$ when $4 < C_u < 8$, and C_u is the coefficient of uniformity $= D_{60}/D_{10}$.

3. When the protected soil contains particles from 1 in. (25.4 mm) to those passing through a No. 200 sieve, only the portion passing through a No. 4 sieve should be used to determine the grain size.

Permeability Criteria

Methods for determining the permeability of fabric are specified by the ASTM. Similar to soils, both the constant and the falling head methods can be used. The result gives the permittivity of the fabric, which must be multiplied by the thickness of the fabric to obtain its permeability. To reduce hydraulic head loss in the filter and increase drainage efficiency, the fabric must be more permeable than the adjacent soil:

$$k(\text{fabric}) \geq k(\text{soil}) \qquad (8.12)$$

Equation 8.12 can usually be satisfied unless the soil is extremely permeable.

Clogging Criteria

Woven: Percent open area $\geq 4\%$ (8.13a)

Nonwoven: Porosity $\geq 30\%$ (8.13b)

Example 8.5:

Figure 8.9 shows the grain size curves of two different soils, as indicated by the solid curves. Determine the required AOS of woven fabrics to retain the soils.

Solution: For the fine-grained soil, from Figure 8.9, $D_{85} = 0.24$ mm. From Eq. 8.10a, $AOS \leq 0.24$ mm, so an AOS of No. 70, which corresponds to an apparent opening size of 0.21 mm, can be used. The selection of No. 70 also satisfies Eq. 8.10c.

For the granular material, only the fraction passing No. 4 is used for determining the grain size, as shown by the dashed curve in the figure. For any given grain size, the percent finer on the dashed curve can be obtained by dividing that on the solid curve with 0.57, which is the fraction passing through the No. 4 sieve. From the dashed curve, $D_{85} = 2.4$ mm, $D_{60} = 0.78$ mm, and $D_{10} = 0.042$ mm. From Eq. 8.11, with $C_u = 0.78/0.042 = 18.6$, $B = 1$, or $AOS \leq 1 \times 2.4 = 2.4$ mm. Therefore, an AOS greater than No. 16, which corresponds to an apparent opening size of 1.19 mm, can be used.

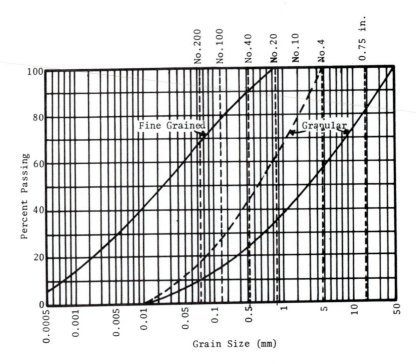

Figure 8.9 Example 8.5 (1 in. = 25.4 mm).

Applications

Geotextiles for subsurface drainage may be used as an envelope of trench drains, a wrapping of pipe drains, or a filter of drainage layers. Figure 8.10 shows the various uses of geotextiles for pavement subdrainage.

In the design of subdrainage, geotextiles should be given consideration as an alternative design. Due to the relative ease of installation as compared to the difficulty of placing a filter aggregate and a coarse aggregate in separate layers without contamination, the use of geotextiles may be more cost effective.

Healey and Long (1972) described the use of prefabricated "fin" drains for longitudinal drains. These drains basically consist of a fin with vertical channels covered with drainage fabric. The channels are connected to the pipe, as shown in Figure 8.11. Water enters through the drainage fabric, runs down the channels into the pipe, and is carried away from the site. This system can be installed in very narrow trenches and does not require any special fill, thus saving excavation and aggregate costs.

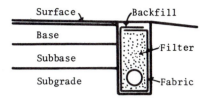

(a) Edge Drain

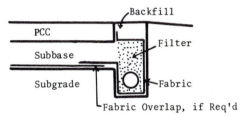

Fabric Overlap, if Req'd

(b) Subbase Drain

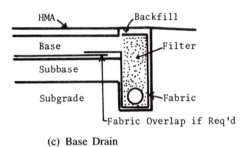

Fabric Overlap if Req'd

(c) Base Drain

Figure 8.10 Use of geotextiles for subsurface drainage. (After Haliburton and Lawmaster (1981).)

8.2.3 Pipes

The pipes to be used for subdrainage may be made of concrete, clay, bituminized fiber, metal, or various plastics with smooth or corrugated surfaces. When used as underdrains, most are perforated or slotted and a few have open joints, so the water can flow freely into the pipes. They must be surrounded by suitable aggregates or fabrics as filter materials to prevent the openings from clogging. When used as outlet pipes, they do not need perforations, slots, or open joints and can be placed in a trench and backfilled with native soils.

Filter Criteria

When perforated or slotted pipes are used for the collection and removal of water, the material in contact with the pipes must be coarse enough that no appreciable amount of this material can enter into the pipes. The criteria used by

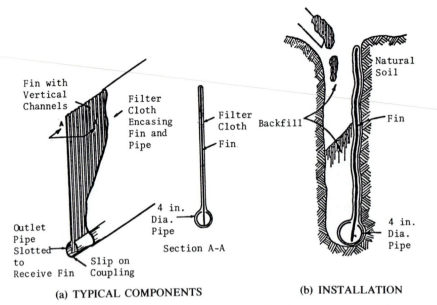

Figure 8.11 Use of prefabricated fin drain in trench. (1 in. = 25.4 mm). (After Healey and Long (1972).)

the U.S. Army Corps of Engineers (1955) for pipes with slots and circular holes are as follows:

For slots:
$$\frac{D_{85} \text{ of filter material}}{\text{Slot width}} > 1.2 \tag{8.14}$$

For circular holes:
$$\frac{D_{85} \text{ of filter material}}{\text{Hole diameter}} > 1.0 \tag{8.15}$$

For open-jointed pipes, the U.S. Bureau of Reclamation (1973) specifies that

$$\frac{D_{85} \text{ of filter material}}{\text{Maximum opening of pipe}} \geq 2 \tag{8.16}$$

Applications

Various types and sizes of pipes are available and can be used for the collection system. Physical strength is a major concern because the pipes may be subjected to heavy construction loading and rough handling. They must be durable under the physical and chemical environments to which they will be exposed. For example, metal pipes should not be used in areas of mine wastes and coal seams to avoid the corrosion by acid water. Some plastic pipes should not be placed in areas with gnawing rodents. Existing ASTM and AASHTO specifications and manufacturer's design recommendations should be consulted in selecting the proper types. Precedents based on past experience and history of performance together with economic considerations generally play an important role in the selection process.

Collector pipes are generally placed on compacted bedding material with perforations or slots down to reduce the possibility of sedimentation in the pipe and to lower the static water level in the trench. However, in extremely wet and muddy situations, where maintaining the trench and bedding materials in a free draining condition is difficult, it may be advisable to place collector pipes with the perforations and slots up or oriented somewhat laterally toward the direction of flow.

Outlet pipes should be installed at convenient intervals to convey the collected water to a suitable and safe exit point. The exit must be protected from natural and man-made hazards. This protection generally consists of a headwall and a combination of screens or valves and markers. Screens are usually adequate to prevent small animals or birds from nesting or depositing debris in the pipes. If high flows with a level above the outlet pipe may be expected to occur in the outfall ditches, flap valves can be used to prevent backflow or deposition of debris. Suitable markers should be installed at each outlet site to facilitate inspection and maintenance.

8.3 DESIGN PROCEDURES

Figure 8.12 shows the cross section of a two-lane highway with asphalt shoulders. The concrete pavement is 9 in. (229 mm) thick with an open-graded subbase that also serves as a drainage layer. The drainage layer is 6 in. (152 mm) thick and is connected to the perforated or slotted collector pipes for lowering the water table and removing the inflow. This figure is used in the examples that follow to illustrate the design procedures.

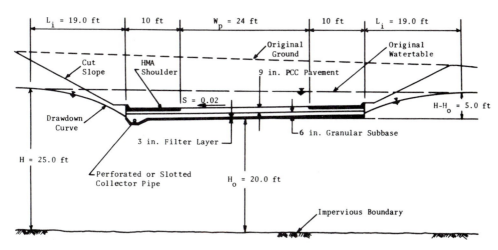

Figure 8.12 Subdrainage of rigid pavement in cut section (1 ft = 0.305 m, 1 in. = 25.4 mm).

8.3.1 Estimation of Inflow

The major sources of inflow are surface infiltration, groundwater seepage, and meltwater from ice lenses. Surface infiltration is the most important source of water and should always be considered in subdrainage design. Whenever possible, groundwater should be lowered by deep longitudinal drains and not be allowed to seep into the pavement structure, as shown in Figure 8.5. If this is not feasible, the amount of seepage entering the drainage layer should be estimated. The meltwater from ice lenses needs to be considered only in northern climates with frost heave. Because frozen fine-grained soils are very impermeable, it is unlikely that flow from both groundwater and meltwater would occur at the same time. Therefore, only one of the larger inflows need be considered.

Surface Infiltration

When water table is at a considerable distance below the pavement surface, as in the case of a high fill, surface infiltration is probably the only water to be considered for subdrainage design. Cedergren et al. (1973) recommended that the design infiltration rate be found by multiplying the 1-h duration/1-yr frequency rain rate by a coefficient varying from 0.33 to 0.50 for asphalt pavements and 0.50 to 0.67 for concrete pavements. Figure 8.13 shows the 1-h duration/1-yr frequency precipitation rates in the United States.

Based on the results of infiltration tests in Connecticut, Ridgeway (1976) indicated that the duration of rainfall is a more critical factor than the intensity. He found that the amount of infiltration can be related directly to cracking and

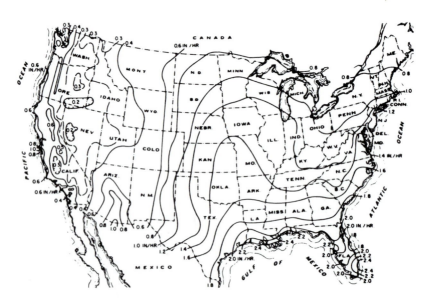

Figure 8.13 Maximum 1-h duration/1-yr frequency precipitation in the United States (1 in. = 25.4 mm). (After Cedergren et al. (1973).)

suggested that an infiltration rate I_c of 2.4 ft³/day/ft of crack (0.22 m³/day/m) be used for design. The infiltration rate per unit area q_i can be expressed as

$$q_i = I_c \left(\frac{N_c}{W_p} + \frac{W_c}{W_p C_s} \right) + k_p \tag{8.17}$$

in which I_c is the crack infiltration rate, N_c is the number of longitudinal cracks, W_p is the width of pavement subjected to infiltration, W_c is the length of transverse cracks or joints, C_s is the spacing of transverse cracks or joints, and k_p is the rate of infiltration through uncracked pavement surface, which is numerically equal to the coefficient of permeability of HMA or PCC and is exceedingly small. By assuming that $N_c = N + 1$, where N is the number of traffic lanes, $W_c = W_p$, $k_p = 0$, and an infiltration rate of 0.1 ft³/h/ft of crack, the inflow rate can be written as (Ridgeway, 1982)

$$q = q_i W_p = 0.1 \left(N + 1 + \frac{W_p}{C_s} \right) \tag{8.18}$$

in which q is the inflow rate in ft³/h/linear ft of pavement, and C_s is the joint spacing for concrete pavements and is 40 ft (12.2 mm) for asphalt pavements.

Example 8.5:

The two-lane pavement shown in Figure 8.12 has a width of 24 ft (7.3 m) and a joint spacing of 15 ft (4.6 m). Compute the amount of surface infiltration q_i using Eq. 8.18. Assume that the pavement is located in Connecticut. Determine the surface infiltration by Cedergren's method.

Solution: Given $N = 2$, $W_p = 24$ ft (7.3 m), and $C_s = 15$ ft (4.6 m), from Eq. 8.18 $q = 0.1 \times (2 + 1 + 24/15) = 0.46$ ft³/h/ft of pavement (0.042 m³/h/m), or $q_i = 0.46/24 = 0.0192$ ft³/h/ft² (0.00585 m³/h/m²), which is equivalent to 0.23 in./h (5.8 mm/h), or 0.46 ft/day (0.14 m/day). In determining q_i, the effect of paved shoulders is not considered. It is assumed that the infiltration through the shoulders is the same as that through the pavement. When the drainage layer extends over the entire roadway, the infiltration through the pavement per unit area is applied over the entire roadway, including the shoulders.

If the pavement is located in Connecticut, from Figure 8.13, the 1-h duration/1-yr frequency precipitation rate is 1.1 in. (28 mm). Based on Cedergren's recommendation by applying a coefficient of 0.50 to 0.67, the surface infiltration ranges from 0.55 to 0.74 in./h (14.0 to 18.8 mm/h).

In this example, the infiltration obtained by Cedergren's method is much greater than that by Eq. 8.18. The results will check better in the western part of the United States where there is less precipitation. It is recommended that Eq. 8.18 be used in the eastern part of the United States because it is more rational and is based on field measurements. However, Cedergren's method can be used as a check and, if necessary, the larger of the two may be used.

Groundwater Seepage

If the drainage layer is used to lower the water table, in addition to providing drainage for surface infiltration, the chart shown in Figure 8.14 can be applied to

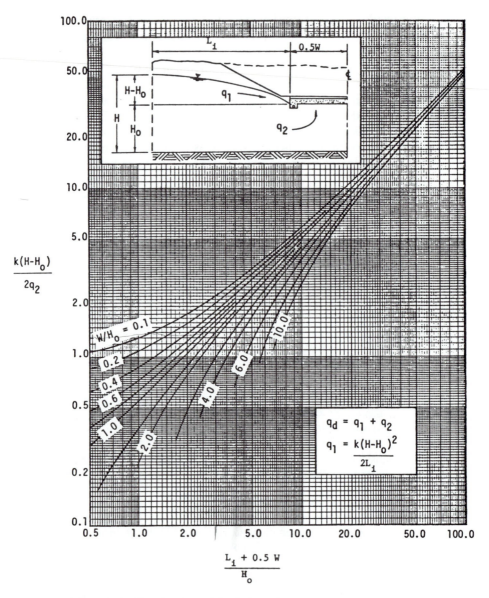

Figure 8.14 Chart for determining inflow of groundwater. (After Moulton (1980).)

determine the inflow from groundwater. The chart is applicable to the general case where an impervious boundary lies at a certain distance below the drainage layer. The inflow is divided into two parts. The inflow above the bottom of the drainage layer is q_1 and can be determined by

$$q_1 = \frac{k(H - H_o)^2}{2L_i} \tag{8.19}$$

in which k is the permeability of the soil in the cut slope or in the subgrade, H is the initial height of groundwater table above the impervious layer, H_o is the vertical distance between the bottom of drainage layer and the impervious layer, and L_i is the distance of influence, which can be determined by

$$L_i = 3.8(H - H_o) \qquad (8.20)$$

The inflow below the drainage layer is q_2 and can be determined from Figure 8.14. Note that q_1 and q_2 are the volume of flow per unit time per unit length of the longitudinal drain. The lateral flow q_L is

$$q_L = q_1 + q_2 \qquad (8.21)$$

The groundwater inflow q_g to the drainage layer per unit area is

$$q_g = \frac{2q_2}{W} \qquad (8.22)$$

in which W is the width of the roadway. However, if the pavement is sloped to one side and the collector pipes are installed only on one side, as shown in Figure 8.12, the lateral inflow per unit length of pipe is

$$q_L = 2(q_1 + q_2) \qquad (8.23)$$

$$q_g = \frac{q_1 + 2q_2}{W} \qquad (8.24)$$

Example 8.6:

For the situations shown in Figure 8.12, if the native soil is a silty sand with a permeability of 0.34 ft/day (1.2×10^{-4} cm/s), determine the inflow q_L to the longitudinal drain and q_g to the drainage layer.

Solution: Given $H = 25$ ft (7.6 m) and $H_o = 20$ ft (6.1 m), from Eq. 8.20, $L_i = 3.8 \times (25 - 20) = 19$ ft (5.8 m). From Eq. 8.19, $q_1 = 0.34 \times (25 - 20)^2/(2 \times 19) = 0.22$ ft^3/day/ft of longitudinal drain (0.021 m^3/day/m).

With $W/H_o = 44/20 = 2.2$ and $(L_i + 0.5W)/H_o = (19 + 0.5 \times 44)/20 = 2.05$, from Figure 8.14, $k(H - H_o)/(2q_2) = 0.74$, or $q_2 = 0.34 \times (25 - 20)/(2 \times 0.74) = 1.15$ ft^3/day/ft of drainage layer (0.107 m^3/day/m). From Eq. 8.23, $q_L = 2(0.22 + 1.15) = 2.74$ ft^3/day/ft (0.25 m^3/day/m). From Eq. 8.24, $q_g = (0.22 + 2 \times 1.15)/44 = 0.057$ ft^3/day/ft^2 (0.017 m^3/day/m^2). Compared with a q_i of 0.46 ft^3/day/ft^2 (0.14 m^3/day/ m^2), the contribution by q_g is very small.

Meltwater from Ice Lenses

Based on the theory of consolidation, Moulton (1980) developed a simple chart for estimating the inflow of meltwater from ice lenses, as shown in Figure 8.15. The rate of seepage from the consolidating soil is at maximum immediately following thawing and decreases rapidly as the time increases. Since the maximum rate of drainage exists for only a short period of time, the design inflow rate q_m is taken as the average during the first day after thawing. The inflow q_m depends on the average rate of heave and the permeability k of subgrade soil as well as the consolidation pressure σ_p on the subgrade. The average rate of heave

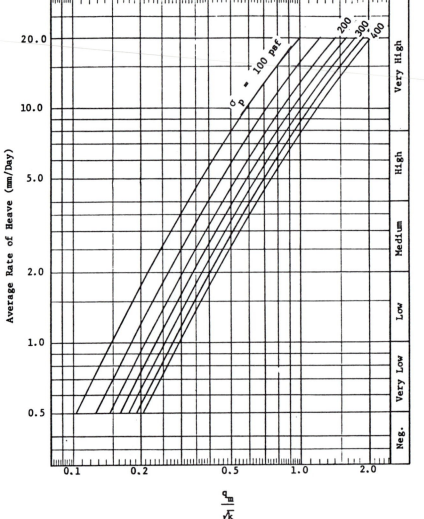
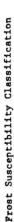

Figure 8.15 Chart for estimating inflow of meltwater from ice lenses (1 psf = 0.048 kPa). (After Moulton (1980).)

can be determined from laboratory tests or estimated using Table 8.5. The value of σ_p can be determined simply by calculating the weight per unit area of pavement structure above the subgrade. In using the chart, the unit of q_m is ft³/day/ft² and that of k is ft/day.

Example 8.7:

If the silty sand subgrade of the previous example has 9% of particles finer than 0.02 mm and is subjected to frost action, determine the inflow q_m to the drainage layer due to the melting of ice lenses.

TABLE 8.5 GUIDELINES FOR SELECTING HEAVE RATE FOR USE WITH FIGURE 8.15

Unified classification		Percent passing 0.02 mm	Heave rate (mm/day)	Frost susceptibility classification
Soil type	Symbol			
Gravel and	GP	0.4	3.0	Medium
sandy gravel	GW	0.7–1.0	0.3–1.0	Negligible to low
		1.0–1.5	1.0–3.5	Low to medium
		1.5–4.0	3.5–2.0	Medium
Silty and	GP-GM	2.0–3.0	1.0–3.0	Low to medium
sandy gravel	GW-GM	3.0–7.0	3.0–4.5	Medium to high
	GM	7.0–10.0	4.5–3.0	High to medium
Clayey and	GW-GC	4.2	2.5	Medium
Silty gravel	GM-GC	15.0	5.0	High
	GC	15.0–30.0	2.5–5.0	Medium to high
Sand and	SP	1.0–2.0	0.8	Very low
gravelly sand	SW	2.0	3.0	Medium
Silty and	SP-SM	1.5–2.0	0.2–1.5	Negligible to low
gravelly sand	SW-SM	2.0–5.0	1.5–6.0	Low to high
	SM	5.0–9.0	6.0–9.0	High to very high
	SM	9.0–22.0	9.0–5.5	Very high to high
Clayey and	SM-SC	9.5–35.0	5.0–7.0	High
silty sand	SC	9.5–35.0	5.0–7.0	High
Silt and	ML-OL	23.0–33.0	1.1–14.0	Low to very high
organic silt	ML	33.0–45.0	14.0–25.0	Very high
	ML	45.0–65.0	25.0	Very high
Clayey silt	ML-CL	60.0–75.0	13.0	Very high
Gravelly and sandy clay	CL	38.0–65.0	7.0–10.0	High to very high
Lean clay	CL	65.0	5.0	High
	CL-OL	30.0–70.0	4.0	High
Fat clay	CH	60.0	0.8	Very low

Solution: The silty sand can be classified as SM. With 9% finer than 0.02 mm, from Table 8.5, the heave rate is 9 mm/day. If unit weights are 150 pcf (23.6 kN/m^3) for the concrete and 115 pcf (18.1 kN/m^3) for the drainage layer, then $\sigma_p = 150(9/12) + 115(6/12) = 170$ psf. From Figure 8.15, $q_m/\sqrt{k} = 0.74$, or $q_m = 0.74 \times \sqrt{0.34} = 0.43$ ft^3/day/ft^2 (0.13 m^3/day/m^2). It can be seen that the value of q_m is as large as q_i in Example 8.5. In areas of seasonal frost, the meltwater from ice lenses constitutes an important part of the total inflow.

Design Inflow

The design inflow is the sum of the inflows from all sources minus the outflow through the subgrade soil. The outflow through subgrade depends on the permeability of the soil and the water table at the boundary and can be determined by the use of flownets or other simplified design charts, as presented by Moulton (1980). When the subgrade is not affected by any water table, a simple and conservative method to estimate the outflow capacity is to assume a hydraulic gradient of 1, as illustrated in Example 8.1, so the outflow rate is equal to the permeability of the soil. In Example 8.6, the soil has a permeability of 0.34 ft/day $(1.2 \times 10^{-4}$ cm/s), so the outflow rate will be 0.34 ft^3/day/ft^2 (0.10 m^3/day/m^2), which is quite large compared to the surface infiltration of 0.46 ft^3/day/ft^2 (0.14 m^2/day/m^2).

If the outflow through the subgrade is neglected, the design inflow can be determined by one of the following combinations:

1. If there in no frost action, the design inflow q_d is the sum of surface infiltration q_i and groundwater flow q_g:

$$q_d = q_i + q_g \qquad (8.25)$$

If there is no groundwater flow, then $q_g = 0$, so $q_d = q_i$.

2. If there is frost action, q_d is the sum of surface infiltration q_i and inflow from meltwater q_m:

$$q_d = q_i + q_m \qquad (8.26)$$

If q_d obtained from Eq. 8.25 is greater than that from Eq. 8.26, then Eq. 8.25 should be used.

Example 8.8:

Based on the results of the three previous examples, determine the design inflow q_d to the drainage layer and q_L to the longitudinal drain.

Solution: The inflows to the drainage layer are $q_i = 0.46$ ft^3/day/ft^2 (0.14 m^3/day/m^2), $q_g = 0.057$ ft^3/day/ft^2 (0.017 m^3/day/m^2), and $q_m = 0.43$ ft^3/day/ft^2 (0.13 m^3/day/m^2), and the inflow to the longitudinal drain q_l is 0.22 ft^3/day/ft (0.021 m^3/day/m). If there is no frost action, from Eq. 8.25, $q_d = 0.46 + 0.057 = 0.517$ ft^3/day/ft^2 (0.158 m^3/day/m^2) and $q_L = 0.517 \times 44 + 0.22 = 23.0$ ft^3/day/ft (2.14 m^3/day/m). If there is frost action, from Eq. 8.26, $q_d = 0.46 + 0.43 = 0.89$ ft^3/day/ft^2 (0.27 m^3/day/m^2) and $q_L = 0.89 \times 44 = 39.2$ ft^3/day/ft (3.64 m^3/day/m).

8.3.2 Determination of Drainage Capacity

The capacity of both the drainage layer and the collector pipe must be designed so that the outflow rate is greater than the inflow rate and the water can be carried out safely from the sources to the outlet sites.

Drainage Layer

There are two design requirements for a drainage layer. First, the steady-state capacity must be greater than the inflow rate. Second, the unsteady-state capacity must be such that the water can be drained quickly after each precipitation event.

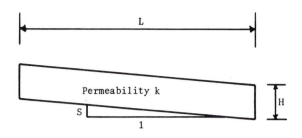

Figure 8.16 Dimension of drainage layer.

Steady-State Flow

Figure 8.16 shows the dimensions of the drainage layer. The steady-steady capacity of the drainage layer can be computed by (Baber and Sawyer, 1952)

$$q = kH\left(S + \frac{H}{2L}\right) \tag{8.27}$$

in which q is the discharge capacity of the drainage layer, k is the permeability of the drainage layer, S is the slope of the drainage layer, H is the thickness of the drainage layer, and L is the length of the drainage layer. Equation 8.27 indicates that the discharge is composed of two terms. The first term is the discharge through area H caused by the hydraulic gradient S, and the second term is that through area $H/2$ caused by a hydraulic gradient H/L. When $S = 0$, $q = 0.5kH^2/L$, which is a direct application of Darcy's law by assuming that the phreatic surface is at the top of the drainage layer on one end and at the bottom of the layer on the other end with an average flow area of $H/2$. The drainage capacity q must be greater than q_dL. If the subgrade is permeable with significant outflow, then the capacity of the drainage layer must be greater than the design inflow minus the subgrade outflow.

Unsteady-State Flow

The unsteady-state flow capacity is defined by the degree of drainage, which is a ratio between the volume of water drained since the rain stops and the total storage capacity of the drainage layer. Casagrande and Shannon (1952) showed that the time for 50% degree of drainage can be computed by

$$t_{50} = \frac{n_e L^2}{2k(H + SL)} \tag{8.28}$$

in which t_{50} is the time for 50% drainage and n_e is the effective porosity, which is the porosity occupied by drainable water. For open-graded materials, all water is drainable, so the effective porosity is the same as the total porosity. Equation 8.28 was applied by Casagrande and Shannon to the design of the base course. They recommended that the time for 50% drainage of a saturated base be not greater than 10 days. This criterion is certainly inadequate for modern highways under heavy traffic. For example, the AASHTO (1986) design guide divides the quality of drainage into five categories: excellent, good, fair, poor, and very poor. If 50% of the water is removed within 10 days, several months will be required to remove most of the water and the drainage is classified as very poor. For excellent drainage, AASHTO requires that water be removed within 2 h. For the design of drainage layer, the requirement that the time for complete or 95% drainage be less than 1 h, as used by Ridgeway (1982) to compare several different criteria, appears to be more appropriate.

In discussing the paper by Casagrande and Shannon (1952), Barber presented a simple chart to determine the time required for any degree of drainage, as shown in Figure 8.17. The degree of drainage U depends on a time factor T_f and a slope factor S_f, defined as

$$T_f = \frac{kHt}{n_e L^2} \qquad (8.29)$$

$$S_f = \frac{LS}{H} \qquad (8.30)$$

in which t is the time since the rain stopped and drainage began. Based on Figure 8.17, the relationship between the time factor for 95% drainage and the slope factor can be plotted in Figure 8.18 and used directly for design purposes.

Example 8.9:

For a drainage layer with $S = 0.02$, $H = 0.5$ ft (152 mm), $L = 22$ ft (6.7 m), a permeability k of 6000 ft/day (2.1 cm/s), and a porosity n_e of 20%, determine the steady-state capacity of the drainage layer and the time for 50 and 95% drainage, respectively.

Solution: The steady-state capacity can be determined by Eq. 8.27, or $q = 6000 \times 0.5 \times (0.02 + 0.5/44) = 94.1$ ft³/day/ft of drainage layer (8.7 m³/day/m), which is 2.4 times greater than the maximum inflow of 39.2 ft³/day/ft (3.64 m³/day/m) obtained in Example 8.8.

The time for 50% drainage can be obtained either directly from Eq. 8.28 or from Figure 8.17. From Eq. 8.28, $t_{50} = 0.2 \times (22)^2/[2 \times 6000 \times (0.5 + 0.02 \times 22)] = 0.0086$ day, or 0.21 h. If Figure 8.17 is used, it can be found that the time factor T_f corresponding to $U = 0.5$ and $S_f = 22 \times 0.02/0.5 = 0.88$ is 0.265. From Eq. 8.29, $t_{50} = 0.265 \times 0.2 \times (22)^2/(6000 \times 0.5) = 0.0086$ day, or 0.21 h, which is exactly the same as the result obtained from Eq. 8.28.

The time for 95% drainage can be obtained from Figure 8.18. With $S_f = 0.88$, from Figure 8.18, $T_f = 1.8$, or $t_{95} = 1.8 \times 0.2 \times (22)^2/(6000 \times 0.5) = 0.058$ day, or 1.39 h. If the design is based on a maximum t_{95} of 1 h, the permeability of 6000 ft/day (2.1 cm/s) is not sufficient and should be increased to 1.39×6000, or 8340 ft/day (2.9 cm/s). If open-graded aggregates are used as drainage layers, the permeability will be much greater than the above value and there is no difficulty in satisfying the criterion that the time required for 95% drainage be less than 1 h.

The above example clearly illustrates that if open-graded aggregates are used as drainage layers, the following two design criteria can be easily satisfied:

1. The lateral outflow rate, as computed by Eq. 8.27, must be greater than the inflow rate.
2. The time for 95% drainage, as obtained from Figure 8.18, must be smaller than 1 h.

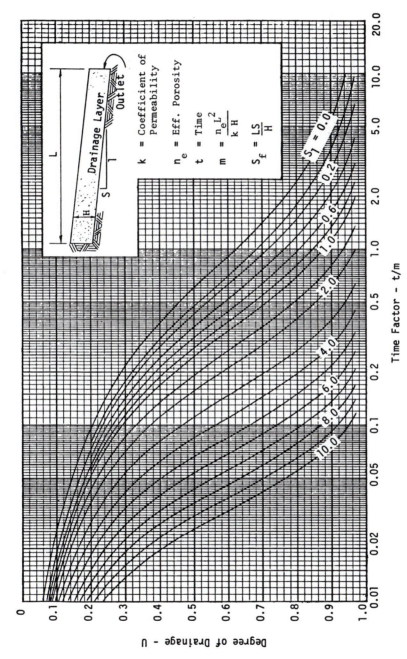

Figure 8.17 Time-dependent drainage of saturated layer. (After Barber and Sawyer (1952).)

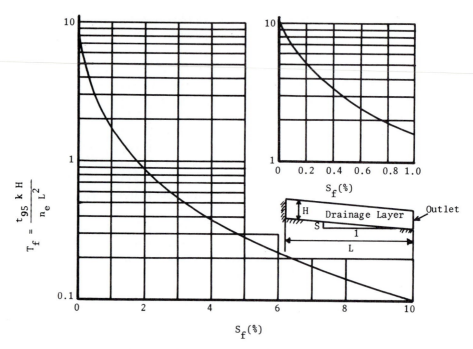

Figure 8.18 Relationship between time and slope factors for 95% drainage.

Collector Pipes

A system of longitudinal collectors with some transverse collectors at critical points is generally used to remove the free water from the drainage layer as well as from groundwater, if any. Transverse collectors may be used when the combination of transverse and longitudinal grades is such that flow tends to take place more in the longitudinal direction than in the transverse direction. The collection system consists of a set of perforated, slotted, or open-jointed pipes that are used to remove water from the pavement structure and to convey it to suitable outlets outside the roadway limits. The design of such systems should consider the type of pipe used, the location and depth of transverse and longitudinal collectors and their outlets, the slope and size of the collector pipes, and the provision of adequate filter protection for the pipes.

The longitudinal roadway grades or the cross slopes usually govern the slopes of the collector pipes. The pipes are simply set at a constant depth below the roadway surface. However, practical construction and operational factors dictate that slopes of collector pipes be not less than 1% for smooth bore pipes and 2% for corrugated pipes. Thus, in areas where the longitudinal grades or cross slopes are very flat, it may be necessary to steepen the grades of collector pipes to meet these minimum requirements. Since the size and flow capacity of the collector depend in part on the pipe gradient, it may be necessary to steepen the pipe gradient to achieve a reduction in pipe size. Minimum recommended diameters are 3 in. (76 mm) for PVC pipes and 4 in. (102 mm) for all other pipes (Cedergren et al., 1972).

Figure 8.19 shows the location and depth of longitudinal collectors. If there is no significant frost penetration and no groundwater, the longitudinal collector pipes may be placed in shallow trenches, as shown in (a). If there is significant frost penetration or it is desirable to drawdown the groundwater table, deeper trenches should be used, as shown in (b). In either case, the longitudinal collector drain can be placed just outside the pavement edge, which is the most critical location for pumping. It can also be placed at the outer edge of the shoulder to provide drainage for the entire shoulder area. However, this will significantly increase the length of flow, or decrease the hydraulic gradient, in the drainage layer, which may result in an increase in the thickness required.

When deeper trenches are used, the trench backfill must have an adequate permeability so that the water from the drainage layer can be freely transmitted to the collector pipe. The required permeability k can be determined directly from Darcy's law by assuming a hydraulic gradient of unity, or

$$k = \frac{\text{Discharge per unit length of trench}}{\text{Width of trench}} \tag{8.31}$$

Equation 8.31 can be used to determine the width of the trench if the permeability of the aggregate in the trench is given.

The discharge capacity of collector pipes can be calculated by Manning's formula for channel flow, or

$$Q = 86,400\left(\frac{1.486}{n}A\ R^{2/3}\ S^{1/2}\right) \tag{8.32}$$

in which Q is the discharge in ft^3/day, n is the roughness coefficient, A is the area of pipe in ft^2, R is the hydraulic radius of pipe in ft, S is the slope of pipe in ft/ft. Table 8.6 shows typical roughness coefficients for underdrains.

The hydraulic radius is a ratio between flow area and wetted perimeter. When the flow in the pipe is full, $A = \pi D^2/4$ and the wetted perimeter $= \pi D$, so $R = D/4$, where D is the pipe diameter. Substituting the above A and R into Eq. 8.32 and noting that $Q = q_L L_o$, where L_o is the distance between two outlet points, yields

$$nq_L L_o = 53\ S^{0.5}\ D^{2.667} \tag{8.33}$$

in which n is the roughness coefficient, q_L is the lateral flow in ft^3/day/ft, L_o is the distance between outlets in ft, S is the slope in ft/ft, and D is the diameter of the pipe in inches. Equation 8.33 shows the relationships among n, q_L, L_o, S, and D. Given any four of the above values, the other one can be computed. To determine the diameter of the pipe, Eq. 8.33 can be written as

$$D = \left(\frac{nq_L L_o}{53S^{0.5}}\right)^{0.375} \tag{8.34}$$

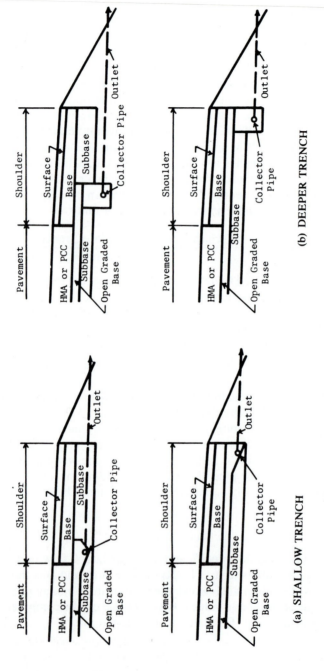

Figure 8.19 Location and depth of collector pipes. (After Cedergren et al. (1972).)

(a) SHALLOW TRENCH

(b) DEEPER TRENCH

TABLE 8.6 TYPICAL ROUGHNESS COEFFICIENTS
FOR UNDERDRAIN

Clay drain tile	0.014–0.018
Concrete drain tile	0.011–0.015
Asbestos cement perforated drain tile	0.013
Bituminous fiber tile	0.012
Corrugated metal pipe	0.017–0.024
Corrugated polyethylene tubing	0.020
Smooth plastic pipe	0.008–0.012

Example 8.10:

Corrugated metal pipe with a roughness coefficient of 0.024 is placed at 2% slope in a trench drain. Outlets are provided at every 500 ft (152 m) and the lateral flow is 39.2 ft^3/day/ft (3.64 m^3/day/m), as obtained in Example 8.8. Determine the diameter of the pipes required.

Solution: Given $n = 0.024$, $q_L = 39.2$ ft^3/day/ft (3.64 m^3/day/m), $L_o = 500$ ft (152 m), and $S = 0.02$, from Eq. 8.34, $D = [0.024 \times 39.2 \times 500/(53 \times \sqrt{0.02})]^{0.375} = 4.72$ in. (120 mm). Use 5-in. (127-mm) pipes.

8.4 SUMMARY

Drainage is one of the most important factors in pavement design. Although highway engineers have long paid much attention to surface and groundwater drainage, very little has been done to provide drainage for surface infiltrations through joints and cracks and, possibly, for the meltwater from ice lenses. The use of drainage layers and longitudinal drains to take care of surface infiltration and meltwater from ice lenses is the main theme of this chapter.

Important Points Discussed in Chapter 8

1. Water entrapped in the pavement structure not only weakens pavements and subgrades, but also generates high hydrodynamic pressures which pump out the fine materials under the pavement and result in loss of support.

2. The movement of water in pavements is governed by Darcy's law. In most instances, Darcy's law can be applied directly to the design of drainage systems.

3. The detrimental effects of water can be minimized by preventing it from entering the pavement, providing drainage to remove it quickly, or building the pavement strong enough to resist the combined effect of load and water. Of the above three methods, the first is difficult to achieve and the last is expensive, so the provision of drainage is the most practical and economical approach. All important highways should be provided with a drainage layer under the pavement.

4. The aggregates used for drainage purposes must satisfy the filter requirements. They must be fine enough to prevent the adjacent soil from piping or migrating into them but also coarse enough to carry water through them with no significant resistance. To meet the filter criteria, it may be necessary to use several different aggregates, one placed adjacent to the other. This procedure is difficult to construct without contamination and can be replaced by using geotextiles.

5. Geotextiles for subsurface drainage may be used as an envelope for trench drains, a wrapping for pipe drains, or a filter for drainage layers. A geotextile can easily be selected to satisfy the filter criteria without using one or more aggregates.

6. The most effective location for the drainage layer is directly under the HMA or PCC, but drainage layers may also be placed on the top of subgrade. To increase shear strength and facilitate compaction, the open-graded drainage material may be treated with a small amount of asphalt. The drainage layer is usually connected to longitudinal drains with collector pipes, although a daylighted construction extending over the entire roadway to the side slope may also be used.

7. The design inflow q_d for drainage layers should be based on either surface infiltration plus groundwater inflow, $q_i + q_g$, or surface infiltration plus inflow from melting ice lenses, $q_i + q_m$, whichever is greater.

8. The lateral outflow from the drainage layer must satisfy two requirements. First, the steady-state outflow per unit width, as indicated by Eq. 8.27, should be greater than $q_d L$, where L is the length of drainage layer. Second, the unsteady-state flow after the cease of rain should remove 95% of the initially saturated water within a very short time, say, 1 h.

9. Manning's formula, as indicated by Eq. 8.32, can be used to determine the discharge capacity of collector pipes. The diameter of the pipe required can be computed directly by Eq. 8.34, given the lateral inflow q_L, distance between outlets L_o, roughness coefficient n, and slope of the pipe S.

PROBLEMS

8-1. Water is sprayed uniformly over a 400-ft^2 soil surface. If the soil has a permeability of 0.001 ft/min, what is the maximum spray rate in gal/min so that all water will be absorbed by the soil and there is no overland flow? [Answer: 2.99 gpm]

8-2. Determine D_{15} and D_{85} of sample 6 in Table 8.3. [Answer: 3.7 mm, 15 mm]

8-3. Figure P8.3 shows the grain size curves for three different soils. Can soil A be used as a filter for soil C? If soil B is placed between soils A and C, can all the filter criteria be satisfied? [Answer: A and C no; A, B, and C yes]

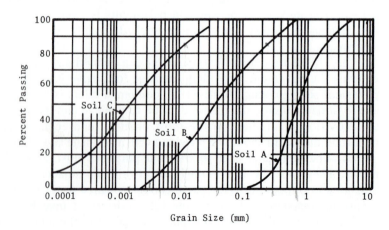

Figure P8.3

8-4. Can geotextiles be used as a filter for soil C in Figure P8.3? Why? If geotextiles are used for soils A and B, what AOS do you recommend? [Answer: no for C, greater than No. 60 for B, greater than No. 8 for A]

8-5. Estimate the permeability in ft/day of soil B in Figure P8.3 by Eq. 8.3. If the dry unit weight is 110 pcf, estimate its permeability in ft/day by Eq. 8.5. [Answer: 0.037 to 0.059 ft/day by Eq. 8.3, 0.019 ft/day by Eq. 8.5]

8-6. Figure P8.6 show a two-lane highway paved with HMA on the traffic lanes and the shoulders. Estimate the amount of surface infiltration in $ft^3/h/ft^2$ by Eq. 8.18. Assume that the pavement is located in Kentucky and estimate surface infiltration by Cedergren's method. [Answer: 0.016 $ft^3/h/ft^2$ by Eq. 8.18, 0.033 to 0.05 ft^3h/ft^2 by Cedergren]

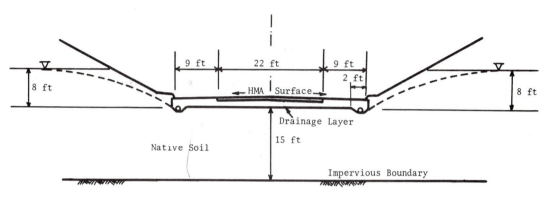

Figure P8.6

8-7. If the native soil above the impervious boundary, as shown in Figure P8.6, has a permeability of 0.5 ft/day, determine the groundwater inflow q_g into the drainage layer in $ft^3/day/ft^2$ and the lateral inflow q_L into the longitudinal drains in $ft^3/day/ft$. [Answer: 0.084 $ft^3/day/ft^2$, 1.67 $ft^3/day/ft$]

8-8. The highway in Figure P8.6 is located in a cold region with severe frost action and the subgrade consists of a well-graded clayey gravel, GW-GC, with 4% finer than 0.02 mm. Determine the inflow q_m to the drainage layer due to the melting of ice lenses. It is assumed that the HMA is 4 in. thick with a unit weight of 145 pcf, the drainage layer, including the filter, is 10 in. thick with an average unit weight of 120 pcf, and the subgrade has a permeability of 0.05 ft/day. [Answer: 0.067 $ft^3/day/ft^2$]

8-9. The drainage layer shown in Figure P8.6 has a thickness H of 8 in., a cross slope S of 4%, a porosity n_e of 25%, and a permeability k of 10,000 ft/day. Determine the steady-state capacity of the drainage layer and the time for 50 and 95% drainage based on a drainage length L of 18 ft. [Answer: 390 $ft^3/day/ft$, 0.07 h, 0.44 h]

8-10. Smooth plastic pipe, 4 in. in diameter with a roughness coefficient of 0.01, is placed at 2.5% slope in a trench drain. If outlets are provided every 300 ft, what is the maximum allowable lateral inflow into the plastic pipe in $ft^3/day/ft$? [Answer: 112.7 $ft^3/day/ft$]

Pavement Performance

9.1 DISTRESS

Distress is an important factor of pavement design. In the mechanistic–empirical methods, each failure criterion should be developed separately to take care of each specific distress. Unfortunately, many of the distresses are caused by the deficiencies in construction, materials, and maintenance and are not related directly to design. However, a knowledge of the various types of distress is important to pavement designers because it can help them to identify the causes of the distress. If distress is due to improper design, improvements in the design method can be made. Furthermore, the evaluation of pavement distress is an important part of the pavement management system by which a most effective strategy for maintenance and rehabilitation can be developed.

An excellent reference for identifying pavement distress is the *Highway Pavement Distress Identification Manual* published by the Federal Highway Administration (Smith et al., 1979), in which each distress is described by its general mechanism, levels of severity (low, medium, and high), and measurement criteria. In this section, only the most significant types of distress directly related to design are described in detail and illustrated by photographs, while other types of distress are only briefly discussed. The description and photographs presented herein are reproduced from the *Highway Pavement Distress Identification Manual*. To provide a data base for all regions of North America, the manual was revised by the Strategic Highway Research Program for use in the long-term pavement performance study (Smith et al., 1987).

9.1.1 Asphalt Pavements

Because many distresses in composite pavements, or asphalt overlays on concrete pavements, are similar to those in asphalt pavements, both types of pavements are discussed in this section. A typical pattern of deterioration in asphalt pavements is rutting, which develops somewhat rapidly during the first few years and then levels off to a much slower rate. Fatigue or alligator cracking does not normally occur until after considerable loadings and then increases rapidly as the pavement weakens. In climates with either large variations in temperature or very cold temperatures, asphalt pavements develop transverse and longitudinal cracks from temperature stresses. These cracks usually break down and spall under traffic.

The most common problem with composite pavements is reflection cracking from joints and cracks in the underlying concrete slab. Infiltration of water into the cracks, along with freezing, thawing, and repeated loadings, usually results in breakup and spalling of the asphalt surface. Typical load-associated fatigue cracking is not common on composite pavements. If fatigue cracking does occur, it usually initiates at the bottom of the concrete slab and may eventually deteriorate and spoil the asphalt overlay.

Alligator or Fatigue Cracking

Alligator or fatigue cracking is a series of interconnecting cracks caused by the fatigue failure of asphalt surface or stabilized base under repeated traffic loading. The cracking initiates at the bottom of the asphalt surface or stabilized base where the tensile stress or strain is highest under a wheel load. The cracks propagate to the surface initially as one or more longitudinal parallel cracks. After repeated traffic loading, the cracks connect and form many-sided, sharp-angled pieces that develop a pattern resembling chicken wire or the skin of an alligator, as shown in Figure 9.1. The pieces are usually less than 1 ft on the longest side. Alligator cracking occurs only in areas that are subjected to repeated traffic

Figure 9.1 Alligator cracking in wheelpaths.

loadings. It would not occur over an entire area unless the entire area was subjected to traffic loading. Alligator cracking does not occur in asphalt overlays over concrete slabs. Pattern-type cracking, which occurs over an entire area that is not subjected to loading, is rated as block cracking, which is not a load-associated distress. Alligator cracking is considered a major structural distress. Alligator cracking is measured in square feet or square meters of surface area.

Block Cracking

Block cracks divide the asphalt surface into approximately rectangular pieces, as shown in Figure 9.2. The blocks range in size from approximately 1 to 100 ft² (0.1 to 9.3 m²). Cracking into larger blocks is generally rated as longitudinal and transverse cracking. Block cracking is caused mainly by the shrinkage of hot mix asphalt and daily temperature cycling, which results in cyclic stress and strain. It is not load associated, although loads can increase the severity of the cracks. The occurrence of block cracking usually indicates that the asphalt has hardened significantly. Block cracking normally occurs over a large portion of pavement area, but sometimes will occur only in nontraffic areas. Block cracking is measured in square feet or square meters of surface area.

Joint Reflection Cracking from Concrete Slab

This distress occurs only on pavements that have an asphalt surface over a jointed concrete slab. Cracks occur at transverse joints as well as at the longitudinal joints where the old concrete pavement has been widened before overlay. This distress does not include the reflection cracking away from a joint or from any other types of stabilized bases, which is identified as longitudinal and transverse

Figure 9.2 Block cracking.

Figure 9.3 Joint reflection cracking from longitudinal widening joint.

cracking. Joint reflection cracking is shown in Figure 9.3. Joint reflection cracking is caused mainly by the movement of concrete slab beneath the asphalt surface because of thermal or moisture changes and is generally not load initiated. However, traffic loading may cause a breakdown of the hot mix asphalt near the initial crack, resulting in spalling. A knowledge of slab dimensions beneath the asphalt surface will help to identify these cracks. Joint reflection cracking is measured in linear feet or linear meters.

Lane/Shoulder Dropoff or Heave

This distress occurs where there is a difference in elevation between traffic lane and shoulder, as seen in Figure 9.4. Typically, the outside shoulder settles due to the consolidation, settlement, or pumping of the underlying granular or subgrade material. Heave of the shoulder may be caused by frost action or swelling soils. Dropoff of granular or soil shoulder is generally caused from blowing away of shoulder material from passing trucks. Lane/shoulder dropoff or heave is measured every 100 ft (30 m) in inches (or mm) along the pavement edge. The mean difference in elevation is computed and used to determine severity level.

Longitudinal and Transverse Cracking

Longitudinal cracks are parallel to the pavement centerline, while transverse cracks extend across the centerline, as shown in Figure 9.5. They may be caused by the shrinkage of asphalt surface due to low temperatures or asphalt hardening or result from reflective cracks caused by cracks beneath the asphalt surface,

Figure 9.4 Lane/shoulder dropoff.

Figure 9.5 Transverse cracking.

including cracks in concrete slabs but not at the joints. Longitudinal cracks may also be caused by a poorly constructed paving lane joint. These types of cracks are not usually load associated. Longitudinal and transverse cracks are measured in linear feet or linear meters.

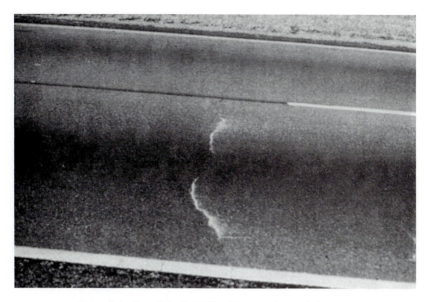

Figure 9.6 Pumping of stabilized base under asphalt pavement.

Pumping and Water Bleeding

Pumping is the ejection of water and fine materials under pressure through cracks under moving loads, as shown in Figure 9.6. As the water is ejected, it carries fine materials, thus resulting in progressive material deterioration and loss of support. Surface staining or accumulation of material on the surface close to cracks is evidence of pumping. Water bleeding occurs where water seeps slowly out of cracks on the pavement surface. Pumping and water bleeding are measured by counting the number that exist.

Rutting

A rut is a surface depression in the wheel paths; see Figure 9.7. Pavement uplift may occur along the sides of the rut. However, in many instances ruts are noticeable only after a rainfall, when the wheel paths are filled with water. Rutting stems from the permanent deformation in any of the pavement layers or the subgrade, usually caused by the consolidation or lateral movement of the materials due to traffic loads. Rutting may be caused by plastic movement of the asphalt mix in hot weather or inadequate compaction during construction. Significant rutting can lead to major structural failures and hydroplaning potentials. Rutting is measured in square feet or square meters of surface area for a given severity level based on rut depth.

Swell

Swell is characterized by an upward bulge on the pavement surface. A swell may occur sharply over a small area or as a long, gradual wave. Either type of

Figure 9.7 Rutting at the wheelpath.

swell can be accompanied by surface cracking. A swell is usually caused by frost action in the subgrade or by swelling soils, but a swell can also occur on the surface of an asphalt overlay on concrete pavement as a result of blowup in the concrete slab, as shown in Figure 9.8. Swell can often be identified by oil droppings due to the bumpy surface. Swells are measured in square feet or square meters of surface area.

Other Types of Distress

The following types of distress may be caused by deficiencies in construction, materials, or maintenance:

1. *Bleeding*. Bleeding is a film of bituminous material on the pavement surface, which creates a shiny, glass-like, reflecting surface that usually becomes sticky. It is caused by high asphalt content or low air void content. Since the bleeding process is not reversible during cold months, asphalt will accumulate on the surface and lower the skid resistance.

2. *Corrugation*. Corrugation is a form of plastic movement typified by ripples across the asphalt surface. It occurs usually at bus stops where vehicles accelerate or decelerate and is the result of shear action in the pavement surface or between the pavement surface and the base material.

3. *Depression*. Depressions are localized pavement surface areas having elevations slightly lower than those of the surrounding pavement. They can be caused by the settlement of foundation soil or can be "built in" during construction. Depressions cause roughness and, when filled with water of sufficient depth, could cause hydroplaning of vehicles.

Figure 9.8 Swell at a patch due to buckling of concrete slab.

4. *Lane/shoulder joint separation.* The widening of the joint between the traffic lane and the shoulder is not considered a distress if the joint is well sealed and water cannot enter through it.

5. *Patch deterioration.* Deteriorations occur in a patch, which is an area where the original pavement has been removed and replaced with either similar or different material. Traffic load, material, or poor construction practices can all cause patch deterioration.

6. *Polished aggregate.* A portion of the aggregates extending above the asphalt surface is either very small or without rough or angular particles to provide good skid resistance. This type of distress occurs mainly in the wheel path due to repeated traffic loads.

7. *Potholes.* Potholes are bowl-shaped holes of various sizes on the pavement surface. They are caused by the broken pavement surface due to alligator cracking, localized disintegration, or freeze–thaw cycles.

8. *Raveling and weathering.* Raveling and weathering are the wearing away of the pavement surface caused by the dislodging of aggregate particles due to stripping and the loss of asphalt binder due to hardening.

9. *Slippage cracking.* Slippage cracks are crescent- or half-moon-shaped with both ends pointed into the direction of traffic. They are caused by the low strength of HMA or a weaker bond between the surface course and the layer below. Part of the asphalt surface moves laterally away from the rest of the surface due to induced lateral and shear stresses caused by traffic loads.

Table 9.1 lists all possible types of distress or failure in asphalt pavements and indicates whether they are structural or functional failures and load-associ-

TABLE 9.1 DISTRESSES IN ASPHALT PAVEMENTS

Types of Distress	Structural	Functional	Load-associated	Non-load-associated
Alligator or fatigue cracking	×		×	
Bleeding		×		×
Block cracking	×			×
Corrugation		×		×
Depression		×		×
Joint reflection cracking	×			×
Lane/shoulder dropoff or heave		×		×
Lane/shoulder separation		×		×
Longitudinal and transverse cracking	×			×
Patch deterioration	×	×	×	
Polished aggregate		×	×[a]	
Potholes	×	×	×	
Pumping and water bleeding	×	×	×	×
Raveling and weathering		×		×
Rutting		×	×	
Slippage cracking	×		×	
Swell	×	×		×

[a] Tire abrasion.

ated or non-load-associated distresses. Structural failure is associated with the ability of the pavement to carry the design load, whereas functional failure is associated with ride quality and safety. When structural failure increases in severity, it always results in functional failure as well. The non-load-associated distress is caused by climates, materials, or construction. A non-load-associated distress may be increased in severity by traffic loads.

9.1.2 Concrete Pavements

Concrete pavements include jointed plain concrete pavements (JPCP), jointed reinforced concrete pavements (JRCP), and continuous reinforced concrete pavements (CRCP). Common types of distress in these pavements are pumping, faulting, and cracking. Joint deterioration is a major problem in JRCP due to longer joint spacing, while edge punchout is a major problem in CRCP due to the closely spaced cracks.

Blowup

Blowups occur in hot weather at a transverse joint or crack that does not permit the expansion of concrete slabs. The insufficient width of joints for expansion is usually caused by the infiltration of incompressible materials into the joint space. When the compressive expansive pressure cannot be relieved, a localized upward movement of the slab edges or shattering occurs in the vicinity of the joint; see Figure 9.9. Blowup can also occur at utility cut patches and drainage inlets. Blowup are accelerated due to the spalling away of the slab at the bottom

Figure 9.9 Blowup.

causing the reduction of contact areas between the two slabs. The presence of "D" cracking also weakens the concrete near the joint, thus resulting in increased spalling and blowup potential. Blowups are measured by counting the number that exist.

Corner Break

A corner break is a crack that intersects the joint at a distance less than 6 ft (1.8 m) on either side measured from the corner of the slab, as shown in Figure 9.10. A corner break differs from a corner spall in that the crack extends vertically through the entire slab thickness, while a corner spall intersects the joint at an angle. Load repetitions combined with the loss of support, poor load transfer across the joint, and thermal curling and moisture warping stresses usually cause corner breaks. Corner breaks are measured by counting the number that exist.

Faulting of Transverse Joints and Cracks

Faulting is the difference of elevation across a joint or crack; see Figure 9.11. Faulting is caused in part by a buildup of loose materials under the trailing slab near the joint or crack as well as the depression of the leading slab. The buildup of eroded or infiltrated materials is caused by pumping due to heavy loadings. The upward warp and curl of the slab near the joint or crack due to moisture and temperature gradients contribute to the pumping condition. Lack of load transfer contributes greatly to faulting. Faulting is determined by measuring the difference in elevation of slabs 1 ft from the pavement edge.

Figure 9.10 Corner break.

Figure 9.11 Joint faulting.

Joint Load Transfer System Associated Deterioration

This distress develops as a transverse crack a short distance from a transverse joint, usually at the end of joint load transfer dowels, as shown in Figure 9.12. This usually occurs when the dowel system fails to function properly due to extensive corrosion or misalignment. It may also be caused by a combination of small-diameter dowels and heavy traffic loadings. The deterioration is measured by counting the number of deteriorated joints.

Lane/Shoulder Dropoff or Heave

The description of this distress is the same as that for asphalt pavements. Lane/shoulder dropoff is shown in Figure 9.13.

Longitudinal Cracks

Longitudinal cracks generally occur parallel to the centerline of the pavement; see Figure 9.14. They are often caused by a combination of heavy load repetition, loss of foundation support, and curling and warping stresses. Improper construction of longitudinal joints can also cause longitudinal cracks. Cracks are measured in linear feet or linear meters.

Figure 9.12 Joint load transfer system associated deterioration.

Figure 9.13 Lane/shoulder dropoff.

Figure 9.14 Longitudinal crack.

Pavement Performance   Chap. 9

Longitudinal Joint Faulting

Longitudinal joint faulting is a difference in elevation at the longitudinal joint between two traffic lanes, as shown in Figure 9.15. Where the longitudinal joint has faulted, the length of the affected area and the maximum joint faulting should be recorded.

Pumping and Water Bleeding

Pumping is the ejection of material by water through joints or cracks, caused by the deflection of slab under moving loads. As the water is ejected, it carries particles of gravel, sand, clay, or silt, resulting in a progressive loss of pavement support. Surface staining or accumulation of base or subgrade material on the pavement surface close to joints and cracks is evidence of pumping; see Figure 9.16. Pumping can occur without such evidence, particularly when stabilized bases are used. The observation of water being ejected by heavy traffic loads after a rainstorm can also be used to identify pumping. Water bleeding occurs when water seeps out of joints or cracks. Pumping and water bleeding are measured by counting the number that exist.

Spalling (Transverse and Longitudinal Joint or Crack)

Spalling of cracks and joints is the cracking, breaking, or chipping of the slab edges within 2 ft (0.6 m) of the joint or crack. A joint spall, shown in Figure 9.17,

Figure 9.15 Longitudinal joint faulting.

Figure 9.16 Pumping of fines at pavement edge and joint.

Figure 9.17 Joint spall.

usually does not extend vertically through the whole slab thickness, but extends to intersect the joint at an angle. Spalling usually results from excessive stresses at the joint or crack caused by the infiltration of incompressible materials and subsequent expansion or traffic loading. It can also be caused by the disintegration of concrete, weak concrete at the joint caused by overworking, or poorly designed or constructed load transfer devices. Spalling is measured by counting the number of spalls that exist.

Spalling (Corner)

Corner spalling is the raveling or breakdown of the slab within approximately 2 ft (0.6 m) of the corner; see Figure 9.18. A corner spall differs from a corner break in that the spall usually angles downward at about 45° to intersect the joint, while a break extends vertically through the slab. Corner spalling can be caused by freeze–thaw conditions, "D" cracking, and other factors. Corner spalling is measured by counting the number that exist.

Swell

Swell is characterized by an upward bulge on the pavement surface. A swell may occur sharply over a small area or as a long, gradual wave. Either type of swell is usually accompanied by slab cracking. A swell is usually caused by frost action in the subgrade, as shown in Figure 9.19, or by swelling soils. Swells can often be identified by oil droppings due to the bumpy surface. Swells are counted by the number that exist.

Transverse and Diagonal Cracks

These cracks are usually caused by a combination of heavy load repetitions and stresses due to temperature gradient, moisture gradient, and drying shrinkage. A transverse crack is shown in Figure 9.20. Transverse and diagonal

Figure 9.18 Corner spall.

Figure 9.19 Swell due to frost heave.

Figure 9.20 Transverse crack.

cracks are measured by counting the number that exist. Hairline cracks that are less than 6 ft (1.8 m) long are ignored.

Edge Punchout

An edge punchout, a major structural distress of CRCP, is first characterized by a loss of aggregate interlock at one or two closely spaced cracks, usually less than 2 ft (1.2 m) apart, at the edge. The crack or cracks begin to fault and spall slightly, which causes the portion of the slab between the closely spaced cracks to act essentially as a cantilever beam. As the applications of heavy truck load continue, a short longitudinal crack forms between the two transverse cracks about 2 to 5 ft (0.6 to 1.5 m) from the pavement edge. Eventually the transverse cracks breakdown further, the steel ruptures, and pieces of concrete punch downward under load into the subbase and subgrade; see Figure 9.21. There is generally evidence of pumping near the edge punchouts, which is sometimes extensive. The distressed area will expand in size to adjoining cracks and develop into a very large area if not repaired. Edge punchouts are measured by counting the number that exist.

Localized Distress

Localized distress occurs in CRCP where the concrete breaks up into pieces or spalls in a localized area, as shown in Figure 9.22. The localized distress takes

Figure 9.21 Edge punchout.

Figure 9.22 Localized distress.

many shapes and forms. It usually occurs within an area between intersecting, Y-shaped, or closely spaced cracks. Localized distress can occur anywhere on the slab surface, but is frequently located in the wheelpaths. Inadequate consolidation of concrete is often a primary cause of localized distress. Localized distress is measured by counting the number that exist.

Other Types of Distress

Other types of distress caused possibly by the deficiencies in construction, materials, and maintenance are described below:

1. *Depression.* Similar to asphalt pavements, depressions can be caused by the settlement of foundation soil or can be "built in" during construction. The exception is that the depression in concrete pavements is generally associated with significant cracking.
2. *Durability or "D" cracking.* "D" cracking is a series of closely spaced, crescent-shaped, hairline cracks that appear at the concrete surface adjacent and roughly parallel to joints and cracks, as well as along the slab edge. The fine surface cracks contain calcium hydroxide residue, which causes a dark coloring of the crack in the immediate surrounding area. "D" cracking is caused by freeze–thaw expansive pressures of certain types of coarse aggregates.
3. *Joint seal damage of transverse joints.* Typical evidence of joint seal damage includes stripping and extrusion of joint sealant, weed growth, hardening of

the filler, loss of bond to the slab edge, and lack or absence of sealant in the joint.

4. *Lane/shoulder joint separation.* The widening of the joint between the traffic lane and shoulder is not considered a distress if the joint is well sealed and water cannot enter through it.

5. *Patch deterioration.* A concrete patch is an area where a portion of the original concrete slab has been removed and replaced by concrete or other epoxy materials. Poor construction of the patch, loss of support, heavy load repetitions, lack of load transfer devices, improper or lack of joints, and moisture or thermal gradients can all cause the distress.

6. *Patch adjacent slab deterioration.* This is similar to patch deterioration except that the deterioration occurs in the original concrete slab adjacent to a permanent patch.

7. *Popouts.* A popout is a small piece of concrete that breaks loose from the surface. It can be caused by expansive, nondurable or unsound aggregates or by freeze and thaw action.

8. *Reactive aggregate distress.* Reactive aggregates either expand in alkaline environments or develop prominent siliceous reaction rims in concrete. The reaction may be alkali–silica or alkali–carbonate. As the expansion occurs, the cement matrix is disrupted and cracks. It appears as a map-cracked area; however, the cracks may go deeper into the concrete than the normal map cracking.

9. *Scaling, map cracking, or crazing.* Map cracking or crazing refers to a network of shallow, fine, or hairline cracks that extend only through the upper surface of the concrete. Map cracking or crazing is usually caused by overfinishing the concrete and may lead to scaling of the surface, which is the breakdown of the slab surface to a depth of approximately $\frac{1}{4}$ to $\frac{1}{2}$ in. (6 to 13 mm). Scaling may be caused by deicing salts, traffic, improper construction, free–thaw cycles, and steel reinforcements too close to the surface.

10. *Construction joint deterioration.* Construction joint distress is a breakdown of the concrete or steel at a CRCP construction joint. It often results in a series of closely spaced transverse cracks near the construction joint or a large number of interconnecting cracks. The primary causes of construction joint distress are poorly consolidated concrete and inadequate steel content and placement.

Table 9.2 lists all possible types of distress for rigid pavements and indicates whether they are structural or functional, load associated or non-load associated.

9.2 SERVICEABILITY

Serviceability is the ability of a specific section of pavement to serve traffic in its exiting conditions. There are two ways to determine the serviceability. One method is to use the present serviceability index (PSI) developed at the AASHO

TABLE 9.2 DISTRESSES IN CONCRETE PAVEMENTS

Types of distress	Structural	Functional	Load-associated	Non-load-associated
Blowup	×	×		×
Corner break	×	×	×	
Depression		×		×
Durability "D" cracking	×	×		×
Faulting of transverse joints and cracks		×	×	
Joint load transfer associated deterioration	×	×	×	×
Joint seal damage of transverse joints				×
Lane/shoulder dropoff and heave		×		×
Lane/shoulder joint separation		×		×
Longitudinal cracks		×		×
Longitudinal joint faulting		×		×
Patch deterioration	×	×	×	×
Patch adjacent slab deterioration	×	×	×	×
Popouts		×		×
Pumping and water bleeding	×	×	×	×
Reactive aggregate distress	×	×		×
Scaling, map cracking, and crazing	×			×
Spalling (transverse and longitudinal joints or cracks)	×	×		×
Spalling at corner	×	×		×
Swell	×	×		×
Transverse and diagonal cracks	×		×	×
Construction joint deterioration	×			×
Edge punchout	×		×	
Localized distress	×	×		×

Road Test, which is based on pavement roughness as well as distress conditions, such as rutting, cracking, and patching. The other method is to use a roughness index based on the roughness only.

9.2.1 Present Serviceability Index

The pavement serviceability–performance concept was developed during the AASHO Road Test (Carey and Irick, 1960). Prior to that time, there were no widely accepted definitions of performance that could be considered in the design of pavements. The inclusion of serviceability as a factor of pavement design was an outstanding feature of the AASHO design methods.

Definitions

To correlate the subjective rating of pavement performance with objective measurements, the following terms need be defined:

1. *Present serviceability*. The ability of a specific section of pavement to serve high-speed, high-volume, mixed traffic in its existing conditions.

2. *Individual present serviceability rating*. An independent rating by an individual of the present serviceability of a specific section of roadway. The ratings range from 0 to 5, as shown in Figure 9.23, which is the rating form used during the AASHO Road Test. The form also includes provision for the rater to indicate whether the pavement is acceptable as a primary highway. It was found that the 50th percentile for acceptibility occurred when the PSR was in the neighborhood of 2.9, whereas the 50th percentile for unacceptibility corresponded roughly to a PSR of 2.5.

3. *Present serviceability rating (PSR)*. The mean of the individual ratings made by the members of a specific panel.

4. *Present serviceability index (PSI)*. A mathematical combination of values obtained from certain physical measurements so formulated as to predict the PSR for those pavements within prescribed limits.

5. *Performance index (PI)*. A summary of PSI over a period of time, which can be represented by the area under the PSI versus time curve, as shown in Figure 9.24. There are many possible ways in which the summary value can be computed. Perhaps the simplest summary consists the mean ordinate of the curve of PSI against time.

Steps in Formulating PSI

The following represents the minimum requirements for the establishment, derivation, and validation of PSI. The procedures are general and can be applied to obtain similar indices for other purposes.

1. *Establishment of definitions*. The precise meaning of terms and exactly what is to be rated and what should be included and excluded from consideration must be clearly understood by those involved in rating and in formulating and using the index.

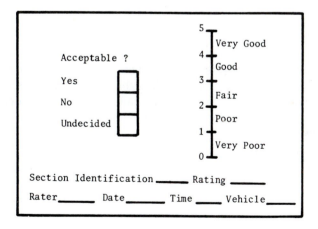

Figure 9.23 Individual present serviceability rating form. (After Carey and Irick (1960).)

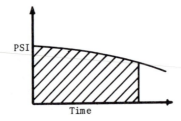

Figure 9.24 PSI as a function of time.

2. *Establishment of rating group or panel.* Since the raters represent highway users, they should be selected from various segments of the users with divergent views and attitudes.

3. *Orientation and training of rating panel.* Members of the panel must understand the pertinent definitions and the rules of the game. Practice rating sessions may be conducted in which the raters can discuss their ratings.

4. *Selection of pavement for rating.* The pavements selected for rating should range from very poor to very good and contain all the various types and degrees of distress that are likely to influence serviceability. The minimum desirable length of a pavement to be rated shall be 1200 ft (366 m), so the raters can ride over the section at high speed and not be influenced by the condition of the pavement at both ends. A total of 74 sections of flexible pavements and 49 sections of rigid pavements in Illinois, Minnesota, and Indiana, as well as some flexible pavements at the AASHO Road Test site, were used for developing the PSI equations.

5. *Field rating.* Members of the panel are taken in small groups to the sections to be rated. They are allowed to ride over each section in a vehicle of their choice and to walk and examine the pavement as they wish. Each rater works independently and no discussion between raters is permitted.

6. *Replication.* To determine the ability of the Road Test panel to rate the pavements consistently, many sections were rated twice within a short period of time so that the section did not change physically. It was found that the difference between two ratings ranged from 0 to 0.5 with an average of 0.2.

7. *Validity of rating panel.* The sections of flexible pavements in Illinois were also rated by two professional drivers based on the rides they obtained when driving their own fully loaded tractor semitrailers and by 20 Canadians, who were ordinary automobile drivers, not professionally associated with highways. It was found that the ratings given by the panel were similar to those given by the other user groups.

8. *Physical measurements.* The purpose of making physical measurements is to relate them to PSR, so the tedious task of panel rating can be replaced by mechanical measurements. It was found that present serviceability was a function primarily of longitudinal and transverse profiles, with some likelihood that cracking, patching, and faulting would also contribute. There-

fore, measurements can be divided into two categories: those that describe surface deformation and those that describe surface deterioration. Details about objective measurements are discussed in the next section.

9. *Summaries of measurements.* There are many different ways to summarize surface deformation and surface deterioration. For example, longitudinal profile may be expressed as mean slope variance, total deviation of the record from some baseline in in./mile (m/km), number of bumps greater than some minimum, a combination of several different measurements, or any other summary statistics involving variance of the record. Transverse profile may be summarized by mean rut depth, variance of transverse profile, or the variance of rut depth along the wheelpath. Cracking and deterioration occur in different levels of severity, and their measurements can be expressed in one unit or another.

10. *Derivation of PSI.* After PSRs and measurement summaries have been obtained for all the selected pavements, the final step is to combine the measurement summaries into a PSI formula that predicts the PSRs to a satisfactory approximation. The technique of multiple linear regression analysis may be used to arrive at the formula as well as to decide which measurements may be neglected. For example, if the longitudinal profile summary is found sensitive to faulting, then faulting measurements need not appear in the formula whenever the profile measure is included.

Objective Measurements

Many measurement summaries were tried in the Road Test, but those finally selected were mean slope variance $\overline{SV}$ for the longitudinal profile, mean rut depth $\overline{RD}$ for the transverse profile, and cracking C and patching P for surface deterioration.

Mean Slope Variance

The symbol $\overline{SV}$ is used for the summary statistics of wheelpath roughness as measured by the Road Test longitudinal profilometer. For each wheelpath, the profilometer produces a continuous record of the pavement slope between two points 9 in. apart, as shown by angle A in Figure 9.25. The slopes are generally sampled at 1-ft (0.3-m) intervals over the length of the record. A variance SV is computed for the sample slopes in each wheelpath by

$$SV = \frac{\Sigma(S - \overline{S})^2}{n - 1} \tag{9.1}$$

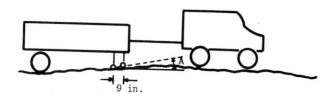

Figure 9.25 Measurement of slope variance by profilometer (1 in. = 25.4 mm).

in which S is the sample slopes, N is the number of samples, and $\overline{S}$ is the mean of all slopes $= \Sigma S/n$. The SVs of the two wheelpaths are averaged to give the mean slope variance $\overline{SV}$.

Mean Rut Depth

The transverse profile of the flexible pavement sections was measured by a rut depth gage. The gage is used to determine the differential elevation between the wheelpath and a line connecting two points each 2 ft (0.6 m) away from the center of the wheelpath in the transverse direction, as shown in Figure 9.26. Rut depth measurements were obtained at 20-ft (6.1-m) intervals in both wheelpaths, which were averaged to give the mean rut depth $\overline{RD}$.

Cracking and Patching

Cracking and patching were combined as a single variable. Cracking is expressed as linear feet and patching as square feet, both per 1000 ft² of pavement area.

Development of PSI Equation

The basic equation is in the following linear form:

$$PSI = A_0 + (A_1R_1 + A_2R_2 + \cdots) + (B_1D_1 + B_2D_2 + \cdots) \qquad (9.2)$$

in which R_1, R_2, . . . are functions of profile roughness and D_1, D_2, . . . are functions of surface deterioration. Note that PSI is an approximation of PSR within prescribed limits.

Choice of Functions

Because the measurement summaries are not linearly related to PSR, a transformation must be applied to each of the summaries. This can be achieved by plotting the measurements versus PSR in various forms and finding out which plot results in a straight line. The following linearizing transformations can be used:

$$R_1 = \log(1 + \overline{SV}) \qquad (9.3)$$

$$R_2 = \overline{RD}^2 \qquad (9.4)$$

$$D_1 = \sqrt{C} + P \qquad (9.5)$$

Figures 9.27, 9.28, and 9.29 show the plot of PSR versus R_1, R_2, and D_1 for the 74 flexible sections and Figures 9.30 and 9.31 show the plot of PSR versus R_1 and D_1 for the 49 rigid sections. It can be seen that the data scatter a great deal, especially in Figures 9.28 and 9.29. However, these plots were found to result in

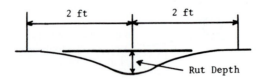

Figure 9.26 Measurement of rut depth by a gage (1 ft = 0.305 m).

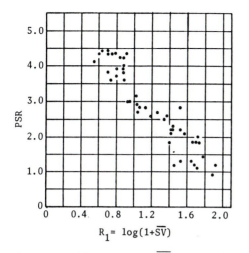

Figure 9.27 PSR vs. log(1 + $\overline{SV}$) for flexible pavements. (After Carey and Irick (1960).)

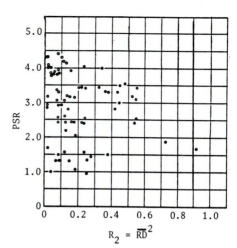

Figure 9.28 PSR vs. $\overline{RD}^2$ for flexible pavements. (After Carey and Irick (1960).)

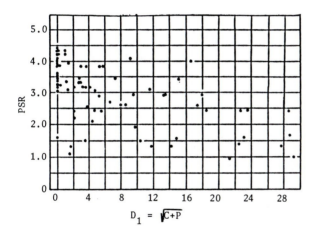

Figure 9.29 PSR vs. $\sqrt{C + P}$ for flexible pavements. (After Carey and Irick (1960).)

the best linearizing functions available. Therefore, the equation for flexible pavements can be written as

$$\begin{aligned} PSI &= A_0 + A_1 R_1 + A_2 R_2 + B_1 D_1 \\ &= A_0 + A_1 \log(1 + \overline{SV}) + A_2 \overline{RD}^2 + B_1 \sqrt{C + P} \end{aligned} \tag{9.6}$$

For rigid pavements, there is no rut depth, so the equation becomes

$$PSI = A_0 + A_1 R_1 + B_1 D_1 = A_0 + A_1 \log(1 + \overline{SV}) + B_1 \sqrt{C + P} \tag{9.7}$$

Determination of Coefficients

The coefficients A_0, A_1, A_2, and B_1 in Eq. 9.6 and A_0, A_1, and B_1 in Eq. 9.7 can be determined by linear multiple regression. From Eq. 9.6, the error between

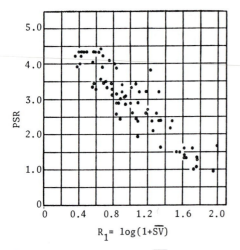

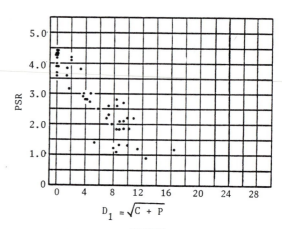

Figure 9.30 PSR vs. log(1 + $\overline{SV}$) for rigid pavements. (After Carey and Irick (1960).)

Figure 9.31 PSR vs. $\sqrt{C + P}$ for rigid pavements. (After Carey and Irick (1960).)

PSR and PSI for all sections can be expressed as

$$E = \sum(PSR - PSI)^2 = \sum(PSR - A_0 - A_1R_1 - A_2R_2 - B_1D_1)^2 \qquad (9.8)$$

in which Σ is the summation over all sections, e.g., the 74 flexible pavement sections in the Road Test. The coefficients A_0, A_1, A_2, and B_1 must be selected so that the least error is obtained. The value of E can be minimized by taking partial derivatives with respect to each of the coefficients and setting them to zero. This results in four equations and four unknown coefficients, which can easily be solved. The derivation of these four equations is presented below:
From Eq. 9.8

$$\frac{\partial E}{\partial A_0} = -2\sum(PSR - A_0 - A_1R_1 - A_2R_2 - B_1D_1) = 0$$

or

$$\sum PSR - nA_0 - A_1\sum R_1 - A_2\sum R_2 - B_1\sum D_1 = 0$$

in which n is the number of pavement sections. Dividing by n and rearranging yields

$$\overline{PSR} = A_0 + A_1\overline{R}_1 + A_2\overline{R}_2 + B_1\overline{D}_1 \qquad (9.9)$$

From Eq. 9.8

$$\frac{\partial E}{\partial A_1} = -2\sum[(PSR - A_0 - A_1R_1 - A_2R_2 - B_1D_1) R_1] = 0$$

Replacing A_0 by Eq. 9.9 yields

$$\sum\{[PSR - \overline{PSR} - A_1(R_1 - \overline{R}_1) - A_2(R_2 - \overline{R}_2) - B_1(D_1 - \overline{D}_1)]R_1\} = 0$$

To make the equation more systematic, a constant term $\overline{R}_1$ can be added:

$$\sum\{[PSR - \overline{PSR} - A_1(R_1 - \overline{R}_1) - A_2(R_2 - \overline{R}_2)$$
$$- B_1(D_1 - \overline{D}_1)] (R_1 - \overline{R}_1)\} = 0 \qquad (9.10)$$

This addition has no effect on the equation because

$$\overline{R}_1\sum[PSR - \overline{PSR} - A_1(R_1 - \overline{R}_1) - A_2(R_2 - \overline{R}_2) - B_1(D_1 - \overline{D}_1)]$$
$$= \overline{R}_1\left[\sum PSR - \sum\overline{PSR} - A_1\left(\sum R_1 - \sum\overline{R}_1\right) - A_2\left(\sum R_2 - \sum\overline{R}_2\right)\right.$$
$$\left. - B_1\left(\sum D_1 - \sum\overline{D}_1\right)\right]$$
$$= 0$$

Equation 9.10 can be simplified to

$$A_1\sum(R_1 - \overline{R}_1)^2 + A_2\sum(R_2 - \overline{R}_2)(R_1 - \overline{R}_1) + B_1\sum(D_1 - \overline{D}_1)(R_1 - \overline{R}_1)$$
$$= \sum(PSR - \overline{PSR})(R_1 - \overline{R}_1) \qquad (9.11)$$

The remaining two equations can be obtained by taking partial derivatives of E with respect to A_2 and B_1, or simply by replacing the multiplier $(R_1 - \overline{R}_1)$ in Eq. 9.11 with $(R_2 - \overline{R}_2)$ and $(D_1 - \overline{D}_1)$, respectively.

$$A_1\sum(R_1 - \overline{R}_1)(R_2 - \overline{R}_2) + A_2\sum(R_2 - \overline{R}_2)^2 + B_1\sum(D_1 - \overline{D}_1)(R_2 - \overline{R}_2)$$
$$= \sum(PSR - \overline{PSR})(R_2 - \overline{R}_2) \qquad (9.12)$$
$$A_1\sum(R_1 - \overline{R}_1)(D_1 - \overline{D}_1) + A_2\sum(R_2 - \overline{R}_2)(D_1 - \overline{D}_1) + B_1\sum(D_1 - \overline{D}_1)^2$$
$$= \sum(PSR - \overline{PSR})(D_1 - \overline{D}_1) \qquad (9.13)$$

Equations 9.9, 9.11, 9.12, and 9.13 can be used to solve the four coefficients.

The final PSI equation obtained by the Road Test based on 74 flexible sections is

$$PSI = 5.03 - 1.91\log(1 + \overline{SV}) - 1.38\overline{RD}^2 - 0.01\sqrt{C + P} \qquad (9.14)$$

in which the units are 10^{-6} rad for $\overline{SV}$, in. for $\overline{RD}$, ft/1000 ft^2 for C, and ft^2/1000 ft^2 for P. The equation based on 49 rigid sections is

$$PSI = 5.41 - 1.71\log(1 + \overline{SV}) - 0.09\sqrt{C + P} \qquad (9.15)$$

Example 9.1

Table 9.3 gives the measurements and ratings on seven sections of rigid pavements. Find the regression equation for PSI.

Solution: Consider $\log(1 + \overline{SV})$ as R_1 and $\sqrt{C + P}$ as D_1. The values of $R_1 - \overline{R}_1$, $D_1 - \overline{D}_1$, and $PSR - \overline{PSR}$ are tabulated in Table 9.4. From Eq. 9.9

$$2.79 = A_0 + 1.30A_1 + 5.69B_1 \qquad (9.16)$$

From Eq. 9.11, $[(0.42)^2 + (-0.42)^2 + (0.07)^2 + (0.13)^2 + (0.39)^2 + (0.12)^2 + (-0.68)^2]A_1 + [0.42 \times 2.11 + (-0.42)(-3.69) + 0.07 \times 1.62 + 0.13 \times 1.61 + 0.39 \times 5.71 + 0.12(-1.69) + (-0.68)(-5.69)]B_1 = 0.42(-0.79) + (-0.42) \times 1.41 + 0.07(-0.19) + 0.13 \times (-0.49) + 0.39(-1.59) + 0.12 \times 0.01 + (-0.68) \times 1.61$, or

$$1.004A_1 + 8.652B_1 = -2.715 \qquad (9.17)$$

TABLE 9.3 MEASUREMENTS AND RATINGS ON RIGID PAVEMENT SECTIONS

Section no.	Slope variance $\overline{SV}$ (10^{-6})	Cracking and patching $C + P$ (ft or ft^2/1000 ft^2)	Present serviceability rating (PSR)
1	52.0	60.8	2.0
2	6.5	4.0	4.2
3	22.2	53.5	2.6
4	26.2	53.3	2.3
5	47.8	130.0	1.2
6	25.5	16.0	2.8
7	3.2	0.0	4.4

Note. 1 ft = 0.305 m.

TABLE 9.4 COMPUTATION OF REGRESSION EQUATION FOR PSI

Section no.	$\log(1 + \overline{SV})$ R_1	$\sqrt{C + P}$ D_1	PSR	$R_1 - \overline{R}_1$	$D_1 - \overline{D}_1$	$PSR - \overline{PSR}$	PSI check
1	1.72	7.80	2.00	0.42	2.11	−0.79	1.92
2	0.88	2.00	4.20	−0.42	−3.69	1.41	3.94
3	1.37	7.31	2.60	0.07	1.62	−0.19	2.43
4	1.43	7.30	2.30	0.13	1.61	−0.49	2.36
5	1.69	11.40	1.20	0.39	5.71	−1.59	1.34
6	1.42	4.00	2.80	0.12	−1.69	0.01	2.93
7	0.62	0.00	4.40	−0.68	−5.69	1.61	4.60
Average	1.30	5.69	2.79				

From Eq. 9.13, $8.652A_1 + [(2.11)^2 + (-3.69)^2 + (1.62)^2 + (1.61)^2 + (5.71)^2 + (-1.69)^2 + (-5.69)^2]B_1 = 2.11(-0.79) + (-3.69) \times 1.41 + 1.62(-0.19) + 1.61 \times (-0.49) + 5.71(-1.59) + (-1.69) \times 0.01 + (-5.69) \times 1.61$, or

$$8.652A_1 + 91.121B_1 = -26.223 \tag{9.18}$$

From Eqs. 9.17 and 9.18, $A_1 = -1.23$ and $B_1 = -0.17$. From Eq. 9.9, $A_0 = 5.36$. The regression equation is

$$PSI = 5.36 - 1.23 \log(1 + \overline{SV}) - 0.17\sqrt{C + P} \tag{9.19}$$

Values of PSI computed by Eq. 9.19 are tabulated in the last column of Table 9.4.

Deficiencies in PSI Equation

The PSI equations developed by the AASHO Road Test have the following shortcomings:

1. They were based on the evaluations of the Road Test rating panel. Whether the public's perception of serviceability is the same today as 30 years ago is

questionable because vehicles, highway characteristics, and travel speeds have changed significantly.

2. They include not only the rideability but also the surface defects. For the management of pavement inventory, it would be better to have separate measures of ride quality and surface defects. Therefore, there is a need to develop a new pavement rating scale to ensure that the objective pavement evaluations are directly and reasonably related to the public's perception of riding quality.

3. The prolifometer used in the Road Test is no longer in use today. Errors are multiplied when the original AASHO PSI equation is used with a different method for measuring roughness. The best way to eliminate this problem is to form a new pavement rating panel and to correlate the result directly with the particular roughness instrument of interest.

Since both pavement distress and ride quality are terms that describe the physical condition of pavements, it is frequently assumed that they will be highly correlated and that a relationship exists allowing reliable estimates of one based on a measure of the other. However, this relationship can rarely be found in the pavement performance data bases. The reason for the lack of correlation between distress and roughness is that in reality they are physically different attributes of pavement conditions.

9.2.2 Roughness

Although the physical measurements used for computing PSI include the distress data of rut depth, cracking, and patching, it is the longitudinal profile or roughness that provides the major correlation variable. The correlation coefficient between PSR and PSI is increased by only about 5% after the addition of distress data (Zaniewski et al., 1985). Because of the relatively small contribution to PSI by physical distress and the difficulty in obtaining the distress data, many agencies rely only on roughness to estimate PSI.

Methods for Measuring Roughness

Either direct or indirect methods can be used to measure roughness. Direct measures of longitudinal profile can be obtained from class I or class II devices. Class I devices include the traditional longitudinal surveys by rod and level or other labor-saving apparatus such as the Face Dipstick, which is walked along the test path on two feet spaced 1 ft (0.305 m) apart. Class II devices include various types of profilometers such as the chloe type rolling straightedge profilometer used in the AASHO Road Test, the Surface Dynamics Profilometer (SDP) designed by General Motors (Spangler and Kelly, 1964), the siometer employed by the Texas State Department of Highways and Public Transportation for measuring the serviceability index (Fernando et al., 1990), and the profilometer and rut measuring device called PRORUT owned by FHWA (1987a). Indirect measures of longitudinal profile are those obtained from response-type road roughness meters (RTRRM) such as the Mays ridemeter. RTRRM systems are referred to as class III devices.

A survey of 48 states shows that the Mays ridemeter is the most popular device for roughness measurements and is used by 22 states (Epps and Monismith, 1986). The RTRRMs measure the relative movement in inches per mile (m/km) between the body of the automobile and the center of the rear axle. They are inexpensive, simple, and easy to operate at speeds of up to 50 mph (80 km/h). The recording systems are portable and can be installed in selected standard automobiles. The RTRRM measurements are sensitive to the type of tires, tire pressure, load, vehicle suspension system, speed of car, and factors that affect vehicle responses. Because of such sensitivity, they need to be calibrated when any of the above factors change significantly.

When the response-type road roughness meters are properly calibrated, the output from all the meters may be placed on a common scale. One primary advantage of this approach is that the subjective interpretation of road roughness can be eliminated and pavement engineers will have a common scale for describing the ride quality of a road. Once the common scale is accepted, each highway agency may define where good, fair, and poor ride quality should be located on the scale. Certainly, PSR values from large panels could be correlated with the roughness scale as a means for intepreting the roughnes data.

International Roughness Index

To provide a common quantitative basis on which the different measures of roughness can be compared, the International Roughness Index (IRI) was developed at the International Road Roughness Experiment held in Brazil in 1982 under the sponsorship of the World Bank. Since the World Bank published guidelines for conducting and calibrating roughness measurements (Sayers et al., 1986a), the IRI has been adopted as a standard for the FHWA Highway Performance Monitoring System (HPMS) data base (FHWA, 1987b).

The IRI summarizes the longitudinal surface profile in the wheelpath and is computed from surface elevation data collected by either a topographic survey or a mechanical profilometer. It is defined by the average rectified slope (ARS), which is a ratio of the accumulated suspension motion to the distance traveled obtained from a mathematical model of a standard quarter car transversing a measured profile at a speed of 50 mph (80 km/h). It is expressed in units of inches per mile (m/km). General methods for use in simulation are also specified in ASTM E 1170 "Standard Practices for Simulating Vehicular Response to Longitudinal Profiles of a Vehicular Traveled Surface."

Since no two RTRRMs are exactly alike, it is necessary to convert the measures to the standard IRI scale using the relationships established through calibration. Calibration can be achieved by measuring the ARS of the RTRRM at a standard speed on special calibration sites of known IRI roughness values. The measured ARS values are plotted against the IRI values, and a line is fitted to the data points and used to estimate IRI from the RTRRM measurements in the field. A detailed discussion of this procedure was presented by Sayer et al. (1986b).

Riding Number

Janoff et al. (1985) investigated the relationship between roughness and serviceability. They found that the riding number RN, as measured with a profilometer in the band of frequencies between 0.125 and 0.63 cycle/ft (10 to 15 Hz at 55 mph), is highly correlated with the mean panel rating MPR for all three types of pavements: asphalt, concrete, and composite. Note that RN, which ranges from 0 to 5, is equivalent to PSI, while MPR is equivalent to PSR. The resulting regression equation that transforms the longitudinal roughness in this frequency band into RN is

$$RN = -1.74 - 3.03 \log(PI) \qquad (9.20)$$

in which RN is the riding number of a given pavement section, which gives the best estimate of MPR, and PI is the profile index, defined as the square root of the mean square of the profile height in the specified frequency band. The MPR of a given test section is also an accurate predictor of the public's subjective perception of whether a specific test section needs repairs. The percentage of the driving public that feels that a given section requires repair is defined by

$$NR = 132.6 - 33.5 MPR \qquad (9.21)$$

or

$$NR = 190.9 + 101.5 \log(PI) \qquad (9.22)$$

in which NR is the percentage of drivers who think that the pavement needs repair. Therefore, based only on longitudinal roughness measures, one can compute both the rideability number and the exact percentage of the driving population that thinks the road should be repaired.

Equations 9.20 to 9.22 were based on data from Ohio only. The data base was later extended to include 282 pavement sections in New Jersey, Michigan, New Mexico, and Louisiana. Based on combined data from all five states, the equations were modified as follows (Janoff, 1988):

$$RN = -1.47 - 2.85 \log(PI) \qquad (9.23)$$

$$NR = 131.7 - 33.9 RN \qquad (9.24)$$

$$NR = 181.5 + 96.6 \log(PI) \qquad (9.25)$$

A nonlinear equation was also developed to compute RN:

$$RN = 5 \exp[-11.72(PI)^{0.89}] \qquad (9.26)$$

Both Eqs. 9.23 and 9.26 are accurate and valid within the range of the data base collected. The nonlinear equation extends the prediction to the entire range of roughness, although it is no more accurate or valid than the linear equation within the range of the data base.

9.3 SURFACE FRICTION

Adequate surface friction must be provided on a pavement so that loss of control does not occur in normally expected situations when the pavement is wet. The skid accident reduction program issued by the Federal Highway Administration (FHWA, 1980) encourages each state highway agency to minimize wet weather skidding accidents by identifying and correcting sections of roadway with high or potentially high skid accident incidence and ensuring that new surfaces have adequate, durable skid-resistant properties.

9.3.1 Factors Affecting Skid Accidents

Pavement characteristics are only one element of a multiple-component system in which each component has a significant effect and interacts with the others to cause skid accidents. The four major elements are the driver, the roadway, the vehicle, and the weather.

Friction demands on a pavement vary greatly, depending on the speed of a vehicle, its design, and the design and condition of its braking system. The skill of the operator also affects the potential for loss of control or skidding. Weather, especially wet pavements and the thickness of water film on the pavement, also affects the available friction. If the water layer on the pavement is very thick and the tire moves at a high speed, hydroplaning occurs. This phenomenon is analogous to water skiing and makes the vehicle uncontrollable. The following are some of the surface conditions that are indicative of potential safety hazards:

1. Bleeding of asphalt, which covers the aggregates and obscures the effectiveness of their skid-resistant qualities.
2. Polished aggregate with smooth microtexture, which reduces friction between the aggregate and the tire.
3. Smooth macrotexture, which lacks suitable channels to facilitate drainage.
4. Rutting, which holds water in the wheelpaths after rain and causes hydroplaning.
5. Inadequate cross slope, which retains water on the pavement for a longer time, reduces friction, and increases the thickness of water layer and the potential for hydroplaning.

9.3.2 Measurement of Surface Friction

Various methods have been used to determine the coefficient of friction between pavement and tire. The coefficient depends not only on pavement properties but also on many other factors, such as tire type, tire wear, inflation pressure, vehicle speed, whether the wheel is rolling or locked, and whether the pavement is wet or dry. To obtain meaningful results, all the factors not related to the pavement must be fixed or clearly defined so that the only variables are the pavement properties.

Definition of Skid Number

Surface friction is defined as the force developed when a tire that is prevented from rotating slides along the pavement surface. Surface friction can be determined by

$$F = \mu W \tag{9.27}$$

in which F is the tractive force applied to the tire at the tire–pavement contact, μ is the coefficient of friction, and W is the dynamic vertical load on the tire. The coefficient of friction determined from Eq. 9.27 is multiplied by 100 to obtain the skid number:

$$SN = 100\mu = 100\left(\frac{F}{W}\right) \tag{9.28}$$

Direct Measurement of Skid Number

Pavement friction is measured most frequently by the locked wheel trailer procedure, as specified in ASTM E 274 "Standard Test Method for Skid Resistance of Paved Surfaces Using a Full-Scale Tire." The full-scale tire is usually a standard tire specified in ASTM E 501 "Standard Specification for Standard Rib Tire for Pavement Skid-Resistance Tests." The test tire is installed on the wheel or wheels of a single- or two-wheel trailer. The trailer is towed at a speed of 40 mph (64 km/h) over the dry pavement and water is applied in front of the test wheel. The braking system is actuated to lock the test tire. Equipment is provided to measure the friction force generated when the tire is locked and the vehicle and trailer are moving at the prescribed speed. The results of the test are reported as a skid number SN.

The ASTM E 274 method is used by more than forty states. Another method, which measures the coefficient of friction between tire and pavement in the yaw mode, is used by at least four states. A simple trailer called Mu Meter is commercially available, which measures the side force developed by two yawed wheels with smooth tires. Since both wheels are yawed at equal but opposite angles, the trailer will travel in a straight line without requiring a restraining mechanism. The use of Mu Meter is specified in ASTM E 670 "Standard Test Method for Side Force Friction on Paved Surfaces Using the Mu-Meter."

The most natural method for determining the skid resistance is to drive an automobile on a pavement, lock the wheels after the desired speed is reached, and measure how far the vehicle slides until it comes to a full stop. This method is specified in ASTM E 445 "Standard Test Method for Stopping Distance on Paved Surfaces Using a Passenger Vehicle Equipped with Full-Scale Tires." The coefficient of friction can then be easily determined from mechanics by

$$\mu = \frac{V^2}{2gS} \tag{9.29}$$

in which V is the initial vehicle speed, g is the acceleration of gravity, and S is the braking distance. If V is in mph and S in ft, then from Eqs. 9.28 and 9.29

$$SDN = 100\left(\frac{V^2}{30S}\right) \tag{9.30}$$

Note that the skid number obtained by the stopping distance method in ASTM E 445 is called the stopping distance number, SDN, instead of SN because values of SDN are numerically larger than those of SN. The reason is that wet pavement friction increases with decreasing speeds and the speed in stopping distance tests decreases from 40 to 0 mph (64 to 0 km/h), whereas the standard skid number is determined at 40 mph (64 km/h). The use of stopping distance method is potentially hazardous, especially at high speeds. The hazard can be reduced by locking two diagonal wheels instead of all four wheels. The results, however, may be questionable and the stopping distance may not be representative of the actual value.

Relating Surface Texture to Skid Number

Many methods have been used for measuring surface texture such as tracing by stylus, measuring the outflow rate, and using putty to fill surface irregularities. One method that has been used frequently is the sand patch method, specified as ASTM E 965 "Standard Test Method for Measuring Surface Macrotexture Depth Using a Sand Volumetric Technique." The test procedure involves spreading a known volume of material on a clean and dry pavement surface, measuring the area covered, and calculating the mean texture depth MTD as a quotient between volume and area. This method is not suited for extensive field use because it requires the diversion of traffic. It is preferable to measure texture by some noncontact method from a vehicle moving at normal speed, as reported by Henry et al. (1984).

The macrotexture provides drainage paths for the water entrapped between pavement surface and tire imprint. After most of the water is expelled, some portion of the tire will make actual contact with the tips and plateaus of the pavement surface. Significant friction can be developed because the small-scale asperities of the aggregate particles penetrate whatever moisture may remain. This fine texture is referred to as microtexture. The division between macrotexture and microtexture is arbitrarily set at a texture depth of 0.5 mm, as defined in ASTM E 867 "Standard Definitions of Terms Relating to Traveled Surface Characteristics."

Unfortunately, no practical method has been developed to directly measure microtexture. A commonly used substitute is to measure low-speed friction by the British Pendulum Tester, as specified in ASTM E 303 "Standard Method for Measuring Surface Friction Properties Using the British Pendulum Tester." The tester is equipped with a standard rubber slider that is positioned to barely come in contact with the test surface prior to conducting the test. The pendulum is raised to a locked position, then released, thus allowing the slider to make contact with the test surface. A drag pointer is used to read the British Pendulum Number, BPN. The greater the friction between the slider and the test surface, the more the swing is retarded and the larger is the BPN reading. The test can be conducted either in the field or in the laboratory.

The relationship between skid number SN and speed V can be expressed by (Leu and Henry, 1978)

$$SN = SN_0 \exp\left[-\left(\frac{PNG}{100}\right)V\right] \qquad (9.31)$$

in which SN_0 is the SN at zero speed and PNG is the percent normalized gradient of the SN versus V curve. Equation 9.31 shows that the skid number is contributed by microtexture in terms of SN_0 and macrotexture in terms of PNG. It can be easily proved from Eq. 9.31 that $PNG = -100(dSN/dV)/SN$, which is the percent gradient, $100(dSN/dV)$, normalized by dividing with SN.

Figure 9.32 shows the relationship between zero intercept skid number SN_0 and the British pendulum number BPN. The linear relationship can be expressed as (Meyer, 1991)

$$SN_0 = 1.32BPN - 34.9 \qquad (9.32)$$

Figure 9.33 shows the relationship between percent normalized gradient PNG and mean texture depth MTD. The relationship can be expressed as (Meyer, 1991)

$$PNG = 0.157(MTD)^{-0.47} \qquad (9.33)$$

in which PNG is in h/mile and MTD is in inches.

It should be pointed out that Eqs. 9.32 and 9.33 are based on limited data and different regression equations were presented by Wambold et al. (1989). Values of PNG and SN_0 can also be obtained from the SN versus speed curve obtained from locked-wheel skid resistance tests at several speeds. From Eq. 9.31

$$\log(SN) = \log(SN_0) - 0.00434(PNG)V \qquad (9.34)$$

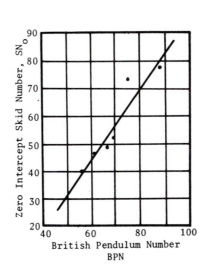

Figure 9.32 Relationship between SN_0 and BPN. (After Meyer (1991).)

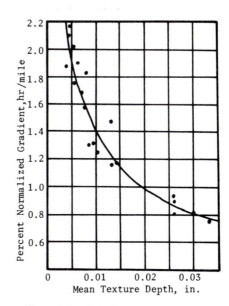

Figure 9.33 Relationship between PNG and MTD (1 h/mile = 0.63 h/km, 1 in. = 25.4 mm). (After Meyer (1991).)

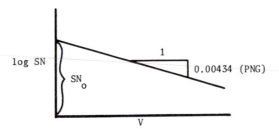

Figure 9.34 Method for determining SN_0 and PNG.

Therefore, a plot of log(SN) versus V should result in straight line, as shown in Figure 9.34. The intercept at zero speed is SN_0 and the slope of the straight line is 0.00434 PNG.

Example 9.2:

The skid numbers obtained from locked-wheel skid tests at various speeds are listed in Table 9.5. Determine the skid number at zero speed SN_0 and the percent normalized gradient PNG. Estimate the probable British pendulum number BPN and the mean texture depth MTD of the pavement surface.

Solution: Plot SN versus V on a semilog paper and draw a best-fit line passing through these four points, as shown in Figure 9.35. The intercept at $V = 0$ is $SN_0 = 58$ and the slope of the straight line is $0.00434(\text{PNG}) = (\log 58 - \log 26)/80 = 0.00436$, or PNG = 1.0%. From Eq. 9.32, BPN = $(58 + 34.9)/1.32 = 70$, and from Eq. 9.33, $1 = 0.157(\text{MTD})^{-0.47}$, or MTD = 0.0195 in. (0.5 mm).

9.3.3 Control of Skid Resistance

The skid resistance of new and existing pavements must be properly controlled so that the minimum requirement can be met.

Minimum Required Skid Resistance

The minimum skid resistance required for a pavement depends on the method of measurement and the traffic speed. Table 9.6 shows the minimum skid numbers, measured according to ASTM E 274, for main rural highways as recommended by the NCHRP Report 37 (Kummer and Meyer, 1967). For interstate highways, lower values of SN than those shown in the table may be sufficient, whereas certain sites may require higher values. Based on a mean traffic speed of 50 mph (80 km/h), NCHRP Report 37 recommended a SN of 37, measured at 40 mph (64 km/h), as the minimum permissible for standard main

TABLE 9.5 SKID NUMBERS MEASURED AT VARIOUS SPEEDS

Speed V (mph)	10	30	50	70
Skid number SN	52	44	34	29

Note. 1 mph = 1.6 km/h.

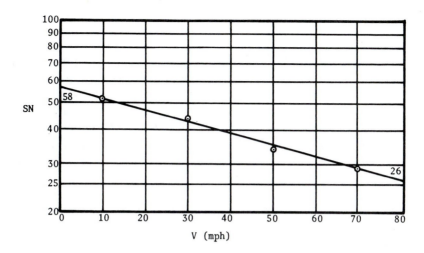

Figure 9.35 Example 9.2 (1 mph = 1.6 km/h).

rural highways. Since there are no definite federal and state standards on the minimum SN required, most highway agencies follow the guidelines recommended by NCHRP Report 37.

Design of New Pavements

The skid resistance of wearing surfaces depends on both the aggregate and the mixture characteristics.

Aggregate Characteristics

The most frequently sought after characteristics for a skid-resistant aggregate are its resistance to polish and wear, texture, shape, and size, as described below:

1. The ability of an aggregate to resist the polish and wear actions of traffic has long been recognized as a most important characteristic. When an aggregate becomes smooth, it will have poor skid resistance.

TABLE 9.6 RECOMMENDED MINIMUM SKID NUMBER FOR MAIN RURAL HIGHWAYS

Traffic speed (mph)	SN measured at traffic speed	SN measured at 40 mph
30	36	31
40	33	33
50	32	37
60	31	41
70	31	46

Note. 1 mph = 1.6 km/h.

Source. After Kummer and Meyer (1967).

2. Both the microtexture and the macrotexture have a great effect on skid resistance. Microtexture describes surface coarseness as governed by the size of individual mineral grains and the matrix in which they are cemented. Macrotexture refers to the angularity of the aggregate particles and the voids and pits in the pavement surface. An aggregate with larger than sand sizes of hard grains and weak cementation of the grains will wear under traffic and expose a continually renewed nonpolished surface. However, if the matrix of aggregate is strong, the individual grains will be tightly held and subsequently may be polished by traffic. For an aggregate to exhibit satisfactory skid resistance properties, it probably should contain at least two mineral constituents of different hardness in order to wear differentially and expose new surfaces.

3. The shape of an aggregate particle significantly affects its skid-resistant properties. Angular particles are more skid resistant as long as they remain angular. The retention of angularity depends on such characteristics as mineralogical composition and the amount of polish and wear produced by traffic. Some minerals will crush into mostly flat and elongated particles, resulting in poor skid resistance.

4. The size of an aggregate has considerable effect on skid resistance. For HMA pavements, the quality of large-size aggregates has more effect on skid resistance than that of small-size aggregates, whereas for PCC pavements, due to the presence of cement mortar on the surface, the sand-size aggregates have more influence.

Mixture Characteristics

Mixture characteristics can be controlled by blending of aggregates, selection of proper aggregate size and gradation, use of appropriate binder content, and application of good construction practice, as described below:

1. Blending of aggregates to achieve the desired skid resistance is resorted to when superior quality aggregates are expensive or in limited supply. Blending may be accomplished by combining natural aggregates with a synthetic aggregate. Most frequently, one of the aggregates will constitute the entire coarse or fine aggregate. To provide good skid resistance, high-quality coarse aggregates should be used for HMA, and high-quality fine aggregates should be used for PCC.

2. The maximum size of aggregate, as well as the mix gradation, may be varied to provide the desired surface texture. The skid resistance of asphalt pavements can be greatly increased by using an open-graded surface course or a porous pavement. In addition to reducing splash and spray and helping to maintain high friction levels between vehicle tires and wet pavements, porous asphalt pavements are recognized for their ability to reduce night reflectance and decrease tire and vehicle noise.

3. The binder content depends on the design criteria to be satisfied. The criteria for asphalt pavements include durability and stability, and those for concrete pavements include durability, strength, and workability. This does not mean that binder content has no effect on skid resistance. For asphalt pavements, too much asphalt will cause bleeding and result in a slippery surface, whereas too little may lead to raveling and deterioration of the pavement surface. For concrete pavements, if the mortar is improperly balanced in the mix, rapid wear and early

deterioration of the concrete surface may occur. The skid resistance will reduce as the pavement deteriorates.

4. The texture of concrete surface can be controlled by the finishing method. The mean texture depth and the initial skid number for some of the finishing methods are shown in Table 9.7.

Improvement of Existing Surface

The correcting measure for a pavement surface of low skid resistance usually consists of modifying the existing surface rather than applying a new surface. The nature of the modification varies with the type of surface and the cause of low skid resistance. For example, grooving may be the answer for a concrete pavement, while bonding aggregates on the surface to form the desired texture may be the answer for a bituminous pavement.

Concrete Pavements

For concrete pavements, surface modification may involve roughing, acid etching, grooving, or bonding a thin layer of aggregate, as described below:

1. Concrete surfaces may be roughened by mechanical means to improve their skid resistance. Steel shots and hand-operated machines have been used to cut into the pavement surface with varying degrees of success. The success of these methods depends on a number of variables, such as the hardness of the cement mortar and the type of aggregate in the concrete.

2. Concrete surfaces can also be roughened by acid etching. The chemicals most frequently used are concentrated hydrochloric or hydrofluoric acids that are diluted with water at the time of application. After the acid has reacted with the minerals of the aggregates and with the cement, the residue is flushed and an improved surface texture is obtained.

3. Grooving is the process of cutting shallow, narrow channels on the concrete surface using a series of rotating diamond saw blades. It is used most often at locations where hydroplaning or wet pavement accident is a problem. Although longitudinal grooving does not improve the drainage in the transverse direction, it is generally preferred over transverse grooving because the longitudi-

TABLE 9.7 MEAN TEXTURE DEPTH AND INITIAL SKID NUMBER OF CONCRETE PAVEMENTS WITH DIFFERENT FINISHING METHODS

Method of finishing	Mean texture depth (in.)	Initial skid number (SN)		
		Minimum	Maximum	Mean
Burlap drag	0.019	38	64	52
Paving broom	0.031	46	72	58
Wire broom	0.042	51	72	61
Fluted float	0.045	40	72	61

Note. 1 in. = 25.4 mm.

Source. After NCHRP (1972).

nal grooving can be produced faster and the process does not require the closing of adjacent lanes.

4. Bonding a thin layer of aggregate to the surface is another method for improving skid resistance. The bonding medium may be an epoxy resin that is applied to the surface after it has been cleaned. The aggregate to be bonded should be extremely hard, highly angular, and free of fine material. Crushed quartz sand, synthetic aggregate, or slag can be used.

Bituminous Pavements

The modification of bituminous surfaces to correct low skid resistance can be accomplished by grooving, surface treatment or thin overlay, and heater planer, as described below:

1. Grooving techniques for asphalt pavements are similar to those for concrete pavements. However, if the asphalt content is high and the weather is hot, grooves in asphalt pavements may flow together and lose their effectiveness.

2. Some types of surface treatment or thin overlay, such as the open-graded friction course, are specially designed to reduce hydroplaning and wet weather accidents. This can be achieved through careful selection of aggregate types and gradation and by improving the transverse slope to facilitate drainage.

3. A heater planer can be used to correct a surface with low skid resistance due to bleeding of the bituminous binder. The equipment consists essentially of a unit to heat the pavement surface and then remove the excess material by cutting or "planing" it away. After planing, stone chips or sand is spread and rolled into the surface while it is still hot. This method is most effective and economical in correcting isolated spots of bleeding, such as at high-volume intersections.

9.4 NONDESTRUCTIVE DEFLECTION TESTING

Deflection measurements have long been used to evaluate the structural capacity of in situ pavements. They can be used to backcalculate the elastic moduli of various pavement components, evaluate the load transfer efficiency across joints and cracks in concrete pavements, and determine the location and extent of voids under concrete slabs. Many devices are being used to perform nondestructive testing (NDT) on pavements. However, only the devices that measure deflections are discussed. Other nondestructive testing involving the use of wave propagation, impact hammer, ground-penetrating radar, and impedance devices are not presented.

9.4.1 Type of NDT Equipment

Based on the type of loading applied to the pavement, NDT deflection testing can be divided into three categories: static or slowly moving loads, steady-state vibration, and impulse loads.

Static or Slowly Moving Loads

The Benkelman beam, California traveling deflectometer, and LaCroix deflectometer are the best known devices in this category.

The Benkelman beam, which was developed by A. C. Benkelman during the WASHO Road Test, is perhaps the best-known deflection measuring device in the world. It consists essentially of a measurement probe hinged to a reference beam supported on three legs, as shown in Figure 9.36. The deflection at one end of the probe is measured by a dial placed at the other end. The probe is placed between the rear dual tires of a loaded truck, normally with an 18-kip (80-kN) single-axle load, and the rebound deflection of the probe is measured after the truck is slowly driven away. By taking several dial readings as the tire passes a series of designated points at different distances from the probe, it is possible to measure the deflection basin using the principle of reciprocal displacement. The test is easy to operate but is very slow. The beam finds limited use on stiff pavements when the legs of the reference beam are located within the influence of the deflection basin.

The traveling deflectometer, developed by the California Division of Highways, is used to measure deflections while a truck, generally with an 18-kip (80-kN) rear axle, is moving. The LaCroix deflectometer was developed in France and is used extensively in Europe. Like the California deflectometer, the system measures deflections under both rear wheels.

The most serious problem with these devices is the difficulty of obtaining an immovable reference for making deflection measurements. This makes the validity of their use on stiff pavements questionable. They also suffer from the disadvantage that the static or slowly moving loads do not represent the transient or impulse loads actually imposed on pavements. Therefore, they cannot be applied directly to any mechanistic method of pavement design and evaluation without extensive empirical correlations.

Steady-State Vibration

Dynaflect and road rater are the best known devices in this category. The deflections are generated by vibratory devices that impose a sinusoidal dynamic

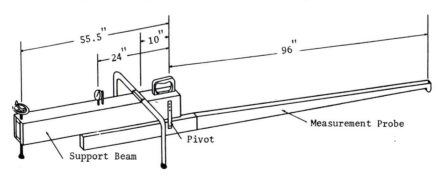

Figure 9.36 Benkelman beam (1 in. = 25.4 mm).

force over a static force. The magnitude of the peak-to-peak dynamic force is less than twice that of the static force, so the vibratory device always applies a compressive force of varying magnitude on the pavement. The deflections are measured by accelerators or velocity sensors. These sensors are placed directly under the center of the load and at specified distances from the center, usually at 1-ft (0.3-m) intervals.

One advantage of this type of equipment over the static equipment is that a reference point is not required. An inertial reference is used so the change in deflection can be compared to the magnitude of the dynamic force. The disadvantages of the method are that the actual loads applied to pavements are not in the form of steady-state vibration and that the use of relatively large static load may have some effect on the behaviors of stress sensitive materials.

Dynaflect

Dynaflect was one of the first commercially available steady-state dynamic devices. It is trailer mounted and can be towed by a standard vehicle. A static weight of 2000 to 2100 lb (8.9 to 9.3 kN) is applied to the pavement through a pair of rigid steel wheels. A dynamic force generator is used to produce a 1000-lb (4.45-kN) peak-to-peak force at a frequency of 8 cycles per second. The dynamic force is superimposed on the static force and the deflections due to the dynamic force are measured by five transducers.

The normal sequence of operations is to move the device to the test location and hydraulically lower the loading wheels and transducers to the pavement surface. A test is then conducted and the data are recorded. If the next test site is nearby, the device can be moved on the loading wheels at speeds up to 6 mph (9.6 km/h). After the last test is completed, the loading wheels and transducers are hydraulically lifted and locked in a secure position. The fixed magnitude and frequency of the loading are the major limitations of the device.

Road Rater

Road raters are commercially available in several models. The models vary primarily in the magnitude of the loads, with the static loads ranging from 2400 to 5800 lb (10.7 to 25.8 kN) and the peak-to-peak dynamic loads ranging from 500 to 8000 lb (2.2 to 35.6 kN). The loading frequency can be varied continuously from 5 to 70 cycles per second. Four sensors are used to measure the deflection basin. Earlier versions of the light model were mounted on vehicles and later versions are all trailer mounted.

In normal operations, the device is moved to the test location and the loading plate and deflection sensors are lowered to the pavement surface. After completing the test at selected loads and frequencies, the loading plate and sensors are lifted from the surface, and the device is ready to move to the next test location. The major limitations of this equipment include the small levels of load for the lighter models and the need for a heavy static load for the heavier models.

Impulse Loads

All devices that deliver a transient force impulse to the pavement surface, such as the various types of falling weight deflectometers (FWD), are included in this category. By varying the amount of weight and the height of drop, different impulse forces can be generated. The normal operation is to move the trailer-mounted device to the test location, lower the loading plate and transducers hydraulically to the pavement surface, complete the test sequence by dropping the weight at each height selected, lift the loading plate and sensors, and tow the device to the next site. The major advantages of the impulse loading device are the ability to accurately model a moving wheel load in both magnitude and duration and the use of a relatively small static load compared to the impulse loading. Three types of FWD are currently commercially available: Dynatest, KUAB, and Phoenix.

Dynatest Falling Weight Deflectometer

The most widely used FWD in the United States is the Dynatest Model 8000 FWD system. The impulse force is created by dropping a weight of 110, 220, 440, or 660 lb (50, 100, 200, or 300 kg) from a height of 0.8 to 15 in. (20 to 381 mm). By varying the drop height and weight, a peak force ranging from 1500 to 24,000 lb (6.7 to 107 kN) can be generated. The load is transmitted to the pavement through a loading plate, 11.8 in. (300 mm) in diameter, to provide a load pulse in the form of a half sine wave with a duration from 25 to 30 ms. The magnitude of load is measured by a load cell.

Deflections are measured by seven velocity transducers mounted on a bar that can be lowered automatically to the pavement surface with the loading plate. One of the transducers is located at the center of the plate, while the remaining six can be placed at locations up to 7.4 ft (2.25 m) from the center. The Dynatest FWD is also equipped with a microprocessor-based control console that can fit on the passenger side of the front seat of a standard automobile.

KUAB Falling Weight Deflectometer

The impulse force is created by dropping a set of two weights from different heights. By varying the drop height and weights, an impulse force ranging from 2698 to 35,000 lb (12 to 156 kN) can be generated. The load is transferred through a loading plate, 11.8 in. (300 mm) in diameter. A two-mass falling weight system is used to create a smoother rise of the force pulse on pavements with both stiff and soft subgrade support. Deflections are measured by five transducers.

Phoenix Falling Weight Deflectometer

The Phoenix FWD is the earlier version of deflectometers commercially available in Europe and in the United States. A single weight is dropped from different heights to develop impact loads from 2248 to 11,240 lb (10 to 50 kN). The load is transferred to the pavement through a plate, 11.8 in. (300 mm) in diameter.

Deflections are measured by three transducers, one at the center of the loading plate and the others at 11.8 and 29.5 in. (300 and 750 mm) from the center.

9.4.2 Factors Influencing Deflections

The major factors that influence deflections include loading, climate, and pavement conditions. These factors must be carefully considered when conducting nondestructive tests.

Loading

The magnitude and duration of loading have a great influence on pavement deflections. It is desirable that the NDT device can apply a load to the pavement similar to the actual design load, e.g., a 9000-lb (40-kN) wheel load. Unfortunately, not every commercially available NDT device is capable of simulating the design load. Some can simulate the magnitude of the design load but not its duration or frequency. It is generally agreed that the falling weight deflectometer is the best NDT device developed so far to simulate the magnitude and duration of actual moving loads (Lytton, 1989).

Due to the nonlinear, or stress-sensitive, properties of most paving materials, pavement deflections are not proportional to load. Test results obtained for light loads must be extrapolated to those for heavy loads. Because extrapolation may lead to a significant error, the use of NDT devices that produce loads approximating those of heavy truck loads is strongly recommended.

Several agencies have developed correlations or regression equations relating the deflection under the lighter load of one device to the heavier load of another device. However, great caution should be exercised in applying these correlations because the data from which these relationships were developed usually contain a large amount of scatter and a large error is always possible. Furthermore, the correlations for one type of pavement structure may not be applicable to a different structure. Even for the same pavement structure, considerable variation has been found between correlations developed by different agencies due to the differences in construction practice and environmental conditions.

Climate

Temperature and moisture are the two climatic factors that affect pavement deflections. For asphalt pavements, higher temperatures cause the asphalt binder to soften and increase deflections. For concrete pavements, temperature in the form of overall change or thermal gradient has a significant influence on the deflections near joints and cracks. The slab expands in warmer temperatures, causing tighter joints and cracks and resulting in greater efficiency of load transfer and smaller deflections. The curling of slab due to temperature gradients can cause a large variation in measured deflections. Measurements taken at night or early morning, when the top of slab is colder than the bottom, will result in higher corner and edge deflections than those taken in the afternoon, when the top of slab is much warmer than the bottom.

The season of the year has a great effect on deflection measurements. In cold regions, four distinct periods can be distinguished. The period of deep frost occurs during the winter season when the pavement is the strongest. The period of spring thaw starts when the frost begins to disappear from the pavement system and the deflection increases greatly. The period of rapid strength recovery takes place in early summer when the excess free water from the melting frost leaves the pavement system and the deflection decreases rapidly. The period of slow strength recovery extends from late summer to fall when the deflection levels off slowly as the water content slowly decreases. For pavements in areas that do not experience freeze–thaw, the deflection generally follows a sine curve, with the peak deflection occurring in the wet season when the moisture contents are high. In relatively dry areas, the period of maximum deflection may occur in the summer when the asphalt surface softens due to the intense solar radiation (Poehl, 1971).

To compare and interpret deflection measurements, the time of the day and the season of the year when the measurements are made must be considered. Generally, deflection measurements are corrected to a standard temperature, say 70°F (21°C), and a critical period condition based on locally developed procedures.

Pavement Conditions

Pavement conditions have significant effects on measured deflections. For asphalt pavements, deflections obtained in areas with cracking and rutting are normally higher than those free of distress. For concrete pavements, voids beneath concrete slabs will cause increased deflections, and the absence or deterioration of load transfer devices will affect the deflections measured on both sides of the joint. Deflections taken near or over a culvert may be much higher, and pavements in cut or fill sections may show significantly different deflections. These conditions must be carefully considered when selecting test locations.

9.4.3 Back-Calculation of Moduli

One of the most useful applications of NDT testing is to back-calculate the moduli of pavement components including the subgrade. The basic procedure is to measure the deflection basin and vary the set of moduli until a best match between the computed and measured deflections is obtained. A layer system program similar to KENLAYER can be used for computing the deflections of flexible pavement, and a finite element program similar to KENSLABS can be used for rigid pavements. If the load is applied in the interior of a slab on a solid or layer foundation, the layer system program can also be used for rigid pavements.

A variety of methods based on layer system programs have been used to back-calculate layer moduli. Unfortunately, none of the methods currently available is guaranteed to give reasonable moduli values for every deflection basin measured. It was reported that two agencies using the same computer program derived very different back-calculated results for the same pavement cross sections (Lytton and Chou, 1988). This is especially true for thin layers because the deflection basin is insensitive to their moduli and a good match between computed and measured deflections can be obtained even if totally unreasonable moduli are derived for these thin layers. Because engineering judgments play an important role in such situations, Chou et al. (1989) developed an expert system to serve as an intelligent preprocessor and postprocessor to the back-calculation program.

Manual Method

Figure 9.37 shows a three-layer system with the surface deflections measured by sensors at five locations. It is arbitrarily assumed that the load is distributed through the various layers according to the broken lines. Because sensors 4 and 5 are outside the stress zone of the HMA and granular base, the deflections at sensors 4 and 5 depend on the modulus of subgrade only and are independent of the moduli of HMA and granular base. Therefore, in applying the layer system program, any reasonable moduli can be assumed for the HMA and granular base, and the modulus of the subgrade can be varied until a satisfactory match between the computed and measured deflections at sensors 4 and 5 is obtained. The deflection at sensor 3 depends on the moduli of the granular base and the subgrade, independent of the modulus of the HMA. Because the modulus of the subgrade has been determined previously, the modulus of the granular base can be varied until a satisfactory match between the computed and measured deflection at sensor 3 is obtained. The same applies to sensors 1 and 2, and the modulus of the HMA can be determined.

The above explanation on how the layer moduli affect surface deflections is over simplified. In reality, the deflection at any given sensor is affected by the moduli of all layers, each to a different degree. However, for a three-layer system, the following general principle always applies: The deflections of the sensors far away from the load are matched by adjusting the modulus of the bottom layer, those at intermediate distances from the load are matched by adjusting the

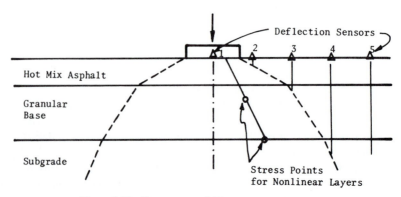

Figure 9.37 Stress zone within pavement structure.

modulus of the middle layer, and those near the load are matched by adjusting the modulus of the top layer.

The problem is more complicated if the granular base and the subgrade are nonlinear. Theoretically, to match the deflection at a given sensor, the stress points for computing the elastic moduli of all nonlinear layers should be located directly beneath that sensor. The use of different stress points for different sensors is not possible because the layer theory limits only one stress point in each layer. The best approach is to assume an average stress distribution with the stress points shown by the smaller circles in Figure 9.37.

MODULUS Program

MODULUS is a microcomputer program for back-calculating layer moduli (Uzan et al., 1988, 1989; Scullion et al., 1990). It can be applied to a two-, three-, or four-layer system with or without a rigid bedrock layer. A linear elastic program is used to generate a data base of deflection bowls by assuming different modulus ratios. Once the data base is generated for a particular pavement, the linear elastic program is not called again, no matter how many bowls are to be analyzed. In the case of a four-layer system, the elastic layer program is automatically run at least 27 times (3 surface × 3 base × 3 subbase modulus ratios). A pattern search routine is used to fit measured and calculated bowls.

The objective function to be minimized can be expressed as

$$\epsilon^2 = \sum_{i=1}^{s} \left(\frac{w_i^m - w_i^c}{w_i^m} \right)^2 = \sum_{i=1}^{s} \left(1 - \frac{w_i^c}{w_i^m} \right)^2 \tag{9.35}$$

in which ϵ^2 is the squared error, w^m is the measured deflection, w^c is the computed deflection, and i is a sensor number from 1 to s. The original equation contains a weighting factor for each sensor i. Because the same weighting factors are usually used, they are not shown in Eq. 9.35. Similar to Eq. 2.14, the computed deflection can be expressed as a function of modulus ratios:

$$w_i^c = \frac{qa}{E_n} f_i \left(\frac{E_1}{E_n}, \frac{E_2}{E_n}, \ldots, \frac{E_{n-1}}{E_n} \right) \tag{9.36}$$

in which q is the contact pressure, a is the contact radius, and E is the elastic modulus with a subscript indicating the layer number. The value of f_i for sensor i can be obtained from the layer system program and used as a data base. By assuming $n - 1$ modulus ratios, w_i^c can be obtained from Eq. 9.36 and ϵ^2 from Eq. 9.35. As in any other back-calculation program, a set of seed moduli is required for the first trial. The following procedure is used to determine the seed modulus of subgrade E_n. When E_n and the modulus ratios are known, the modulus of each layer can be determined.

For a fixed set of modulus ratios, functions f_i can be considered as constants. To minimize the error, the value of E_n can be determined by taking derivative of ϵ^2 with respect to E_n and setting the result to zero:

$$\frac{\partial \epsilon^2}{\partial E_n} = \sum_{i=1}^{s} 2 \left(1 - \frac{w_i^c}{w_i^m} \right) \left(\frac{1}{w_i^m} \right) \left(\frac{\partial w_i^c}{\partial E_n} \right) = 0 \tag{9.37}$$

From Eq. 9.36

$$\frac{\partial w_i^c}{\partial E_n} = -\frac{qaf_i}{E_n^2} = -\frac{w_i^c}{E_n}$$

so Eq. 9.37 can be reduced to

$$\sum_{i=1}^{s} \left(1 - \frac{w_i^c}{w_i^m}\right)\left(\frac{w_i^c}{w_i^m}\right) = 0$$

or

$$\sum_{i=1}^{s} \frac{w_i^c}{w_i^m} = \sum_{i=1}^{s} \left(\frac{w_i^c}{w_i^m}\right)^2 \tag{9.38}$$

When both sides are divided by w_1^c with $w_1^c = qaf_1/E_n$ and $w_i^c/w_1^c = f_i/f_1$, Eq. 9.38 becomes

$$\sum_{i=1}^{s} \frac{f_i}{f_1 w_i^m} = \sum_{i=1}^{s} \frac{qaf_i^2}{f_1(w_i^m)^2 E_n}$$

or

$$E_n = \frac{qaf_1 \sum_{i=1}^{s} \left(\frac{f_i}{f_1 w_i^m}\right)^2}{\sum_{i=1}^{s} \left(\frac{f_i}{f_1 w_i^m}\right)} \tag{9.39}$$

where the functions of f_i are taken from the generated data base. The squared error ϵ^2 is computed using the subgrade modulus from Eq. 9.39 and the modulus ratios of the data base points. A pattern search routine is used to find the optimum set of modulus ratios so that a minimum ϵ^2 is obtained.

Example 9.3:

A pressure of 80 psi (552 kPa) is applied to a pavement surface through a circular area 6 in. (152 mm) in radius. The measured deflections w_i^m at seven sensor locations and the deflection functions f_i generated from the data base for given modulus ratios are shown in Table 9.8. Determine the seed modulus of the subgrade E_n and the squared error ϵ^2.

Solution: Values of $f_i/(f_1 w_i^m)$ are calculated and shown in Table 9.8. If the calculated and measured deflections match exactly, then all points will have the same values of $f_i/(f_1 w_i^m)$. The degree of their variation is an indication of the amount of error.
   From Eq. 9.39, $E_n = 80 \times 6 \times 0.39156 [(22.676)^2 + (25.767)^2 + (28.298)^2 + (23.238)^2 + (24.220)^2 + (21.314)^2 + (31.924)^2]/[22.676 + 25.767 + 28.298 + 23.238 + 24.220 + 21.314 + 31.924] = 4850$ psi (33.5 MPa).
   Values of w_i^c can be determined from Eq. 9.36 and are shown in Table 9.8. Also shown are values of $(w_i^m - w_i^c)/w_i^m$. From Eq. 9.35, $\epsilon^2 = (0.12018)^2 + (-0.09630)^2 + (0.10046)^2 + (0.06087)^2 + (0.17647)^2 + (-0.24390)^2 = 0.1282$.

WESDEF Program

WESDEF (Van Cauwelaert et al., 1989) is the latest back-calculation program developed by the U.S. Army Engineer Waterways Experiment Station (WES). The program is similar to the previous versions called CHVDEF (Bush, 1980), which used the CHEV n-layer program as a subroutine, and BISDEF (Bush and Alexander, 1985), which used the BISAR n-layer program. However, a new,

TABLE 9.8 DEFLECTION DATA

	Point no.						
	1	2	3	4	5	6	7
w_i^m(in.)	0.0441	0.0334	0.0270	0.0219	0.0115	0.0085	0.0041
f_i	0.39156	0.33698	0.29917	0.19927	0.10906	0.07094	0.05125
$f_i/(f_1 w_i^m)$	22.676	25.767	28.298	23.238	24.220	21.314	31.924
w_i^c(in.)	0.0388	0.0334	0.0296	0.0197	0.0108	0.0070	0.0051
$(w_i^m - w_i^c)/w_i^m$	0.12018	0.00000	−0.09630	0.10046	0.06087	0.17647	−0.24390

Note. 1 in. = 25.4 mm.

fast, five-layer program called WESLEA is used in WESDEF to replace the BISAR program.

WESDEF contains a computer optimization routine to determine a set of modulus values that provide the best fit between a measured deflection basin and the computed deflection basin when given an initial estimate of the elastic modulus values and a limiting range of moduli. As in the collocation method described in Section 2.3.2, the number of deflections on the basin must be greater than the number of layer moduli to be determined.

A set of E values is assumed and the deflection is computed corresponding to the measured deflection. Each unknown E is varied individually, and a new set of deflections is computed for each variation. For each layer i and each sensor j, the intercept A_{ji} and the slope S_{ji}, as shown in Figure 9.38, are determined. For multiple deflections and layers, the solution is obtained by developing a set of equations that define the slope and intercept for each deflection and each unknown modulus:

$$\log (\text{deflection}_j) = A_{ji} + S_{ji} (\log E_i) \tag{9.40}$$

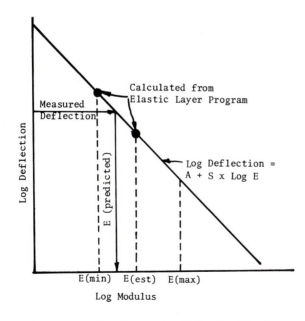

Figure 9.38 Relationship between deflection and modulus.

Normally three iterations of the program can produce a set of modulus values that yield a deflection basin within an accuracy of 3%. Iterations will cease when the absolute sum of the percent differences between computed and measured deflections or the predicted change in modulus values becomes less than 10%.

ILL-BACK Program

In contrast to the optimization programs that require the input of seed moduli, ILL-BACK can determine the moduli of rigid pavement directly without the use of a search routine (Ioannides et al., 1989). The program can be applied to a concrete slab on a liquid or solid foundation with a circular load at the interior. The deflections at four sensors with distances of 0, 12, 24, and 36 in. (0, 305, 610, and 914 mm) from the center are measured and the moduli of the slab and the foundation can be back-calculated.

For liquid foundations, Westergaard's equation for deflection at the center of loading was shown in Eq. 4.21 and can be expressed as

$$d_0 = \frac{w_0 k \ell^2}{P} = f\left(\frac{a}{\ell}\right) \tag{9.41}$$

in which d_0 is the normalized deflection under the center of loading and f is a function of a/ℓ. For a given contact radius a, d_0 is a function of the radius of relative stiffness ℓ only. Similar equations can be obtained for sensor i at a given distance from the center:

$$d_i = \frac{w_i k \ell^2}{P} = f_i(\ell) \tag{9.42}$$

in which i can be 0, 1, 2, or 3. For a given ℓ, the normalized deflection d_i can be determined from theory. If the deflection w_i at any sensor i is measured, the modulus of subgrade reaction k can be computed by Eq. 9.42. Now the crucial question is how to determine ℓ. Once ℓ is known, the deflection measured by any one of the sensors can be used to determine k. A comparison of these k values can provide an evaluation of the agreement between theory and field measurements.

If two deflections, w_0 and w_1, are measured, then from Eq. 9.42

$$d_0 = \frac{w_0 k \ell^2}{P} = f_0(\ell) \tag{9.43a}$$

$$d_1 = \frac{w_1 k \ell^2}{P} = f_1(\ell) \tag{9.43b}$$

Dividing Eq. 9.43b by 9.43a gives

$$\frac{d_1}{d_0} = \frac{w_1}{w_0} = \frac{f_1}{f_0} = f(\ell) \tag{9.44}$$

Equation 9.44 indicates that ℓ can be determined from the deflection ratio, w_1/w_0. If four sensors are used, it is more reliable to use the average deflection of all four sensors, as indicated by the area of deflection basin:

$$\text{AREA} = 6\left[1 + 2\left(\frac{w_1}{w_0}\right) + 2\left(\frac{w_2}{w_0}\right) + \left(\frac{w_3}{w_0}\right)\right] \qquad (9.45)$$

Note that Eq. 9.45 is based on a sensor spacing of 12 in. (305 mm) and the unit of AREA is in inches.

Figure 9.39 shows the relationship between AREA and ℓ for both liquid and solid foundations, as obtained from theory. The solid curves are based on a circular load with a radius of 5.9 in. (150 mm) and the broken curves on a concentrated load with a radius of zero. It can be seen that the effect of radius on the AREA versus ℓ relationship is not very significant.

After ℓ is determined from Figure 9.39, the normalized deflection d_i can be determined from Figure 9.40 for liquid foundations and Figure 9.41 for solid foundations. For liquid foundations, k can be determined from Eq. 9.42:

$$k = \frac{Pd_i}{\ell^2 w_i} \qquad (9.46)$$

It can be shown that E_f for solid foundation can be determined by (Ioannides, 1990)

$$E_f = \frac{2(1 - v_f^2)\, Pd_i}{\ell w_i} \qquad (9.47)$$

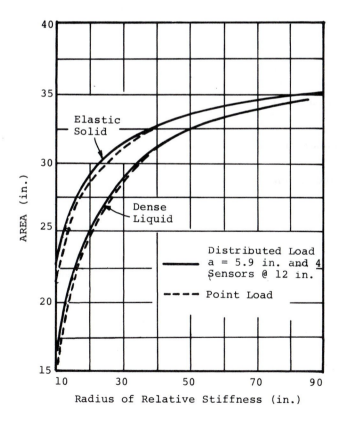

Figure 9.39 Relationship between AREA and ℓ (1 in. = 25.4 mm). (After Ioannides et al. (1989).)

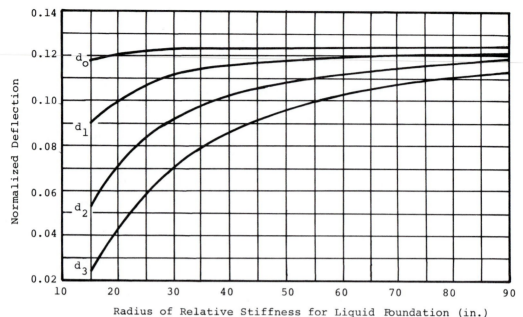

Figure 9.40 Relationship between normalized deflections and ℓ for liquid foundations (1 in. = 25.4 mm). (After Ioannides et al. (1989).)

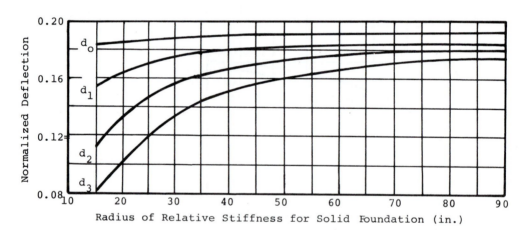

Figure 9.41 Relationship between normalized deflections and ℓ for solid foundations (1 in. = 25.4 mm). (After Ioannides et al. (1989).)

The radius of relative stiffness for liquid foundation is defined by Eq. 4.10, and that for solid foundation is defined by Eq. 5.38. These equations can be used to compute the elastic modulus of concrete E_c if the slab thickness h is known or vice versa.

The elastic modulus of concrete E_c for liquid foundations can be computed from Eq. 4.10:

Pavement Performance   Chap. 9

$$E_c = \frac{12(1 - \nu_c^2) \, k\ell^4}{h^3} \qquad (9.48)$$

For solid foundations, from Eq. 5.38

$$E_c = \frac{6(1 - \nu_c^2) \, E_f \ell^3}{(1 - \nu_f^2) \, h^3} \qquad (9.49)$$

The procedure for back-calculating the moduli of concrete and foundation can be summarized as follows:

1. Measure deflections, w_0, w_1, w_2, and w_3.
2. Compute AREA by Eq. 9.45.
3. Determine ℓ from Figure 9.39.
4. For liquid foundations, determine the normalized deflection from Figure 9.40 and compute k by Eq. 9.46. Four k values, one for each sensor, are obtained and their average is taken. For solid foundations, determine the normalized deflection from Figure 9.41 and compute E_f by Eq. 9.47. The average of the four values is taken as E_f.
5. Based on the average k or E_f, compute E_c by Eq. 9.48 or 9.49.

Example 9.4

A FWD test was conducted on a 10-in. (254-mm) concrete pavement. The radius of the loaded plate is 5.9 in. (300 mm) and the recorded load is 7792 lb (34.7 kN). Sensors were located at 0, 12, 24, and 36 in. (0, 305, 610, and 914 mm) and the corresponding deflections recorded are 0.0030, 0.0028, 0.0024, and 0.0021 in. (0.076, 0.071, 0.061, and 0.053 mm). It is assumed that the concrete has a Poisson ratio of 0.15. (a) If the subgrade is considered as a liquid foundation, determine k and E_c. (b) If the subgrade is considered as a solid foundation with a Poisson ratio of 0.45, determine E_f and E_c.

Solution: From Eq. 9.45, AREA $= 6[1 + 2(2.8/3) + 2(2.4/3) + 2.1/3] = 31.0$ in. (787 mm). From Figure 9.39, $\ell = 39$ in. (991 mm) for liquid foundation and 28 in. (711 mm) for solid foundation.

(a) For a liquid foundation, from Figure 9.40, $d_0 = 0.123$, $d_1 = 0.115$, $d_2 = 0.102$, and $d_3 = 0.085$. From Eq. 9.46 and based on w_0 and d_0, $k = 7792 \times 0.123/[(39)^2 \times 0.0030] = 210$ pci (57.0 MN/m^3). The k values based on the other three sensors are 210, 218, and 207 pci (57.0, 59.1, and 56.2 MN/m^3). The mean of the four k values is 211 pci (57.3 kN/m^3). From Eq. 9.48, $E_c = 12[1 - (0.15)^2] \times 211 \times (39)^4/(10)^3 = 5.7 \times 10^6$ psi (39.3 GPa).

(b) For a solid foundation, from Figure 9.41, $d_0 = 0.188$, $d_1 = 0.174$, $d_2 = 0.152$, and $d_3 = 0.128$. From Eq. 9.47 and based on w_0 and d_0, $E_f = 2[1 - (0.45)^2] \times 7792 \times 0.188/(28 \times 0.0030) = 27,816$ psi (192 MPa). Values of E_f based on the other three sensors are 27,583, 28,111, and 27,055 psi (190, 194, and 187 MPa). The mean E_f is 27,641 psi (191 MPa). From Eq. 9.49, $E_c = 6 \times [1 - (0.15)^2] \times 27,641 \times (28)^3/\{[1 - (0.45)^2] \times (10)^3\} = 4.5 \times 10^6$ psi. Because of the incompatibility between liquid and solid foundations, it is expected that the elastic modulus of concrete based on a solid foundation is somewhat different from that based on a liquid foundation.

Multidepth Deflectometer System

The multi-depth deflectometer (MDD) was developed by the National Institute for Transport and Road Research (NITRR) in South Africa and can be used to back-calculate layer moduli (Basson et al., 1981; DeBeer et al., 1989; Yazdani and Scullion, 1990). The system consists of a number of LVDT modules installed vertically at various depths in the pavement, usually at the interfaces of pavement layers. A maximum of six MDD modules may be placed in a single 1.5-in.-(38-mm)-diameter hole. The modules are clamped against the sides of the hole and the center core is attached to an anchor located approximately 7 ft (2.1 m) below the pavement surface. The MDD can measure either the relative elastic deformation or the total permanent deformation of each layer in the pavement system. By matching the measured deflections and those computed by the layer system program, the modulus of each layer can be determined.

The use of MDD has an advantage over the use of surface deflection basin because the sensors are placed at locations closer to the load, which are the most critical areas for design considerations. If the impulse loading has the same magnitude as the design wheel load, the moduli thus determined represent the actual field conditions and can be used directly with the linear elastic layer program, even though some of the materials may be nonlinear. If the impulse loading is quite different from the design wheel load or the design is based on various load groups with widely different loading magnitudes, then a nonlinear layer program, such as KENLAYER, can be used to adjust the nonlinear coefficients K_1 until a satisfactory match between the measured and computed deflections is obtained.

One disadvantage of using MDD is that a hole must be drilled in the pavement, so the test is actually a destructive test. The presence of the hole may change the stress and deflection regimes around the MDD hole, and the measured deflections may not be representative of the actual deflections when the hole is not present.

9.5 PAVEMENT PERFORMANCE

The evaluation of pavement performance is an important part of pavement design, rehabilitation, and management. It includes the evaluation of distress, roughness, friction, and structure, as described in Sections 9.1 through 9.4, as well as traffic, materials, and drainage. The latter three are presented in Chapters 6, 7, and 8 for pavement design and can also be applied to pavement evaluation.

9.5.1 Expert Systems

Pavement evaluation and rehabilitation are not exact sciences and depend to a large extent on past experiences and engineering judgments. They can be achieved by two different types of knowledge: deterministic and heuristic. Deterministic knowledge is that body of factual information that is widely accepted and can be obtained from textbooks or published literature. Heuristic knowledge is the subjective or private knowledge possessed by each individual, which is largely characterized by beliefs, opinions, and rules of thumb. Organizing and preserving the wealth of heuristic problem-solving knowledge forms the basis of a relatively new type of tool known as knowledge-based systems. These systems are computer programs in which heuristic knowledge is utilized to solve problems that are intractable with a purely deterministic approach. A subset of knowledge-based systems contains expert systems, which employ both the knowledge and reasoning methods of human experts to solve difficult problems in a narrowly defined problem domain.

Figure 9.42 shows the basic architecture of an expert system. The three main components are the knowledge base, the inference engine, and the context. The knowledge base contains all the empirical and factual information for the problem domain. The inference engine searches through the knowledge base or the context to find a conclusion for each subgoal and thus the entire problem. The context is the work space or short-term memory in which currently relevant facts and knowledge are placed successively by the inference engine. Finally, a user-friendly interface contains an explanation module to explain the system's problem-solving strategy to the user and a knowledge-acquisition module to help experts articulate their knowledge in a form acceptable to the system's architecture.

SCEPTRE

SCEPTRE is the acronym for Surface Conditon Expert for Pavement Rehabilitation. It was developed for use by the Washington Department of Transportation to evaluate pavement surface distress and recommend feasible rehabilitation strategies for detailed anaylsis and design (Ritchie et al., 1986). Currently, it is applicable only to flexible pavements.

The major task in building an expert system is to acquire and encode the expertise and knowledge of experts into the knowledge base. The factual and empirical information in the knowledge base can be represented in various ways.

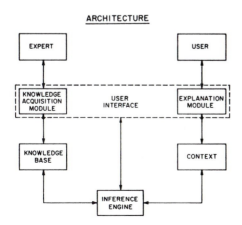

Figure 9.42 Basic architecture of an expert system. (Ritchie et al. (1986).)

The most common is by the "IF THEN" statements; for example, IF transverse cracking exists and crack width is greater than $\frac{1}{8}$ in. (3.2 mm), THEN fill cracks.

General purpose program languages, such as LISP and PROLOG, can be used to build expert systems. However, a much faster route is to use one of several knowledge engineering tool kits, or shells, which comprise an inference engine, empty knowledge base, and context structure. The system developer simply has to enter the rules into the knowledge base. SCEPTRE utilized a shell named EXSYS (EXSYS, 1985), which was developed for use with IBM personal computers and other compatibles. The types of surface distress to be considered include longitudinal cracking in wheelpaths, alligator cracking in wheelpaths, block cracking, transverse and longitudinal cracking outside wheelpaths, and rutting.

The following is a list of the rehabilitation and maintenance strategies (RAMs) from which a subset is drawn by SCEPTRE to form a feasible set for any particular combination of distress types:

1. Do nothing.
2. Fog seal.
3. Chip seal.
4. Double chip seal.
5. Thin HMA overlay, thickness $\leq$ 1.2 in. (31 mm).
6. Medium HMA overlay, 1.2 in. (31 mm) < thickness < 3 in. (76 mm).
7. Thick HMA overlay, thickness $\geq$ 3 in. (76 mm).
8. Friction course.
9. Fill cracks.
10. Reconstruct.
11. Recycled HMA.
12. Level up, mill, and make medium HMA overlay, 1.2 in. (31 mm) < thickness < 3 in. (76 mm).
13. Level up, mill, and make thick HMA overlay, thickness $\geq$ 3 in. (76 mm).

EXPEAR

EXPEAR (Expert System for Pavement Evaluation and Rehabilitation) was developed by the University of Illinois for the Federal Highway Administration and the Illinois Department of Transportation (Hall et al., 1989). It is an advisory program to assist practicing engineers in evaluating a specific pavement section and selecting rehabilitation alternatives. The evaluation part is an extension of COPES (Concrete Pavement Evaluation System) developed earlier (Darter et al., 1985). Currently, the program can be used for only three types of rigid pavement: JPCP, JRCP, and CRCP.

Project level evaluation using EXPEAR begins with the collection of some basic design, construction, traffic, and climate data for the project in question, together with a visual condition survey. Back in the office, the design and condition data are entered into EXPEAR by the engineer. The program extrapolates the overall condition of the project from the distress data for one or more sample units.

EXPEAR evaluates the project in several key problem areas related to specific aspects of performance for that pavement type. For example, the problem areas for JPCP and JRCP are structural adequacy, roughness, drainage, joint deterioration, foundation movement, skid resistance, joint sealant condition, joint construction, concrete durability, load transfer, loss of support, and shoulders. The evaluation is performed using decision trees, which compare the pavement condition to predefined critical levels for key design and distress variables. EXPEAR produces a summary of the deficiencies found, and by interacting with the engineer, formulates a rehabilitation strategy that will correct all the deficiencies. The major rehabilitation options are reconstruction of both lanes, reconstruction of the outer lane and restoration of the inner lane, bonded or unbonded PCC overlay, HMA overlay, crack and seat and HMA overlay, and restoration. Appropriate repair techniques for the shoulders, which are compatible with the rehabilitation strategy of mainline pavements, are also selected.

A large number of predictive models for concrete pavement performance with and without rehabilitation are incorporated into EXPEAR. Some of the models were developed from national data bases of new construction and rehabilitation projects, while others were developed using data from Illinois pavements. Some of these models are presented in Section 9.5.2. The models allow the engineer to predict the performance of the rehabilitation strategy developed. This information is then used, along with rehabilitation unit cost, to compute the cost of the strategy over the predicted life. EXPEAR produces a summary of the project's data file, the evaluation results, recommendations for physical testing, predictions of the pavement's future condition without rehabilitation, rehabilitation techniques, performance predictions, and cost estimates for as many rehabilitation strategies as the engineer wishes to investigate.

9.5.2 Predictive Models

Various equations, mostly based on regression analysis, were developed for predicting pavement performance. These equations illustrate the effect of various factors on pavement performance but their usefulness in practice may be limited by the scope of the data base that was used in their development. Regression equations are valid only under certain conditions and should not be applied when the actual conditions are different. Because the regression constants vary with the units used, all equations are based on the U.S. customary unit, as specified for each variable.

Flexible Pavements

The major failure modes for flexible pavements are fatigue cracking, rutting, and low-temperature cracking.

Fatigue Cracking

The equations used in PDMAP for predicting the allowable number of load repetitions when crackings occur over 10 and 45% of the area in the wheelpath are (Finn et al., 1986)

$$\log N_f (10\%) = 15.947 - 3.291 \log(\epsilon_t/10^{-6}) - 0.854 \log(E_1/10^3) \qquad (9.50a)$$

$$\log N_f (45\%) = 16.086 - 3.291 \log(\epsilon_t/10^{-6}) - 0.854 \log(E_1/10^3) \qquad (9.50b)$$

in which N_f is the allowable number of repetitions for fatigue cracking, ϵ_t is the horizontal tensile strain at the bottom of asphalt layer in in./in., and E_l is the resilient modulus of asphalt layer in psi. Note that Eq. 9.50b is the Asphalt Institute's criterion presented in a different form and that 45% of the area in the wheelpath is equivalent to about 20% of the total area.

Example 9.4

An asphalt pavement, after being subjected to one million load repetitions at a strain level of 10^{-4}, has experienced fatigue cracking over 10% of the area in the wheelpath. At the same strain level, what will be the number of load repetitions when cracking occurs over 45% of the area in the wheelpath? What should be the strain level to cause 45% cracking in the wheelpath after one million repetitions?

Solution: Subtracting Eq. 9.50a from Eq. 9.50b gives $\log[N_f(45\%)] - \log[N_f(10\%)]$ = 16.086 − 15.947 = 0.139, or $\log[N_f(45\%)] = 0.139 + \log(10^6) = 6.139$. Therefore, $N_f(45\%) = 1.38 \times 10^6$, or the number of repetitions increases by 38% when the percentage of area cracked increases from 10 to 45%.

For the same number of repetitions at one million and subtracting Eq. 9.50a from Eq. 9.50b, $0 = 0.139 - 3.291 \log[\epsilon_t(45\%)/\epsilon_t(10\%)]$, or $\epsilon_t(45\%) = 1.102\epsilon_t(10\%)$ = 1.102×10^{-4}. Therefore, a 10% increase in strain causes an increase in area cracked from 10 to 45%.

Rutting

In PDMAP, for conventional construction with HMA less than 6 in.,

$$\log RR = -5.617 + 4.343 \log w_0 - 0.167 \log(N_{18}) - 1.118 \log \sigma_c \qquad (9.51a)$$

For full-depth with HMA equal to or greater than 6 in.,

$$\log RR = -1.173 + 0.717 \log w_0 - 0.658 \log(N_{18}) + 0.666 \log \sigma_c \qquad (9.51b)$$

in which RR is the rate of rutting in microinches (1 μin. = 10^{-6} in.) per axle load repetition, w_0 is the surface deflection in mil (1 mil = 10^{-3} in.), σ_c is the vertical compressive stress under HMA in psi, and N_{18} is the equivalent 18-kip (80-kN) single-axle load in 10^5. For full-depth predictions, all traffic during frozen periods are neglected.

Example 9.5

A full-depth asphalt pavement was constructed on a subgrade with a resilient modulus of 10,000 psi (69 MPa). The asphalt layer is 8 in. (203 mm) thick with a resilient modulus of 500,000 psi (3.45 GPa). Assuming that both HMA and subgrade are incompressible with a Poisson ratio of 0.5, estimate the rut depth after 1 million applications of a 9000-lb (40-kN) wheel load with a contact pressure of 80 psi (552 kPa).

Solution: Contact radius $a = \sqrt{9000/(80\pi)} = 6$ in. (152 mm). With $h/a = \frac{8}{6} = 1.33$ and $E_1/E_2 = 500,000/10,000 = 50$, from Figure 2.17, $F_2 = 0.24$, or $w_0 = 1.5 \times 80 \times 6 \times 0.24/10,000 = 0.0173$ in. (0.44 mm). With $a/h = 0.75$ and $E_1/E_2 = 50$, from Figure 2.15, $\sigma_c/q = 0.075$, or $\sigma_c = 80 \times 0.075 = 6$ psi (41 kPa). From Eq. 9.51b, log RR $= -1.173 + 0.717 \log(17.3) - 0.658 \log(10) + 0.666 \log(6) = -0.425$, or RR $= 0.376$ μin. per one repetition. For one million repetitions, rut depth $= 0.376 \times 10^{-6} \times 10^6 = 0.376$ in.

Low-Temperature Cracking

The cracking index due to low-temperature cracking can be computed by (Hajek and Haas, 1972)

$$10^I = 2.497 \times 10^{30} \times (0.1 S_{\text{bit}})^{(6.7966 - 0.8740h + 1.3388a)} \times (7.054 \times 10^{-3})^d$$

$$\times (3.193 \times 10^{-13})^{(0.1m)} \times d^{(0.06026 s_{\text{bit}})} \qquad (9.52a)$$

or $\qquad I = 30.3974 + (6.7966 - 0.8741h + 1.3388a) \log(0.1 S_{\text{bit}})$

$$- 2.1516d - 1.2496m + 0.06026 S_{\text{bit}} \log d \qquad (9.52b)$$

in which I = cracking index (≥ 0) in terms of the number of full cracks plus one-half of the half-transverse cracks per 500-ft section of two-lane road; cracks shorter than half-width are not due to low-temperature cracking and are not considered in the regression equation;

S_{bit} = stiffness modulus of the original asphalt in kg/cm², as determined from Van der Poel nomograph shown in Figure 7.20, using a loading time of 20,000 s and the winter design temperature; however, values of PI and $T_{\text{R\&B}}$ of bitumen should be determined from penetration at 77°F (25°C) and kinematic viscosity at 275°F (135°C) as suggested by McLeod (1970);

h = total thickness of asphalt layer in inches;

a = pavement age in years;

m = winter design temperature in $-$°C, neglect the negative sign and use positive value only;

d = subgrade type in terms of a dimensionless code with 5 for sand, 3 for loam, and 2 for clay.

Example 9.6

Given thickness of asphalt layer, 4 in. (102 mm); stiffness of original asphalt cement, 205 kg/cm²; winter design temperature, -27°C; and sand subgrade. Determine the cracking index I at 5 and 10 years after construction.

Solution: With $h = 4$ in. (102 mm), $a = 5$ years, $S_{\text{bit}} = 205$ kg/cm², $d = 5$, and $m = 27$°C, from Eq. 9.52b, $I = 30.3974 + (6.7966 - 0.8740 \times 4 + 1.3388 \times 5) \times \log(20.5) - 2.1516 \times 5 - 1.2496 \times 27 + 0.06026 \times 205 \times \log 5 = 7.6$. When $a = 10$ years, from Eq. 9.52b, $I = 16.4$.

Rigid Pavements

A large number of models or regression equations have been developed for predicting pumping, joint faulting, joint deterioration, cracking, and present serviceability rating of both jointed plain concrete pavements (JPCP) and jointed reinforced concrete pavements (JRCP). These equations can be developed using a combination of multiple linear regression and nonlinear regression techniques as included in the SPSS statistical package (Nie et al., 1975). Multiple linear regression is used to determine which independent variables are significant. Nonlinear regression is then used to compute the coefficients and exponents for the final predictive equation. Various forms of equations can be obtained, depending on the extent of data base. The presence or absence of certain variables in the data base will have a significant effect on the equation obtained.

Notable examples of regression models are the AASHTO design equation (AASHTO, 1986), the PEARDARP models (Purdue Economic Analysis of Rehabilitation and Design Alternative for Rigid Pavements, van Wiji, 1985), and the COPES models (Concrete Pavement Evaluation System, Darter et al., 1985). The COPES models were also incorporated in the EXPEAR system, as described in Section 9.5.1, to formulate rehabilitation strategies (Hall et al., 1989). A recent review of these models by ERES Consultants, Inc. (Smith et al., 1990a) indicated that when they were applied to 95 pavement sections located in four major climate zones, none could adequately predict the distress (such as faulting, cracking, joint deterioration, and pumping), serviceability, or roughness. Based on these 95 sections plus the more than 400 sections from the COPES data base, new prediction models were developed and are presented below. It should be emphasized that the models are valid within the data base from which they were derived. Due to the large number of design, materials, climate, and construction variables, it is extremely difficult to develop nationwide models. Regional models or general models calibrated to each state may be required.

Present Serviceability Rating

The best way to predict PSR for a given pavement is using roughness. However, to relate serviceability to physical deterioration, which may be modeled mechanistically, a PSR model based on key distress types is desirable. Models were developed for both JPCP and JRCP. Whereas all types of distress were initially included, only four were found significant.

For JPCP

$$PSR = 4.356 - 0.0182 \text{ TFAULT} - 0.00313 \text{ SPALL}$$
$$- 0.00162 \text{ TCRKS} - 0.00317 \text{ FDR} \qquad (9.53a)$$

Statistics: $R^2 = 0.58$

$$SEE = 0.31$$

$$n = 282$$

in which   PSR = present serviceability rating (from 0 to 5);

TFAULT = cumulative transverse joint faulting in inches per mile;

SPALL = number of deteriorated joints (medium and high severity) per mile;

TCRKS = number of transverse cracks (all severity) per mile;

FDR = number of full-depth repair per mile;

R = correlation coefficient;

SEE = standard error of estimate;

n = number of sample points.

For JRCP

$$\text{PSR} = 4.333 - 0.0539 \text{ TFAULT} - 0.00372 \text{ SPALL}$$

$$- 0.00425 \text{ MHTCRKS} - 0.000531 \text{ FDR} \qquad (9.53b)$$

Statistics: $R^2 = 0.64$

$$\text{SEE} = 0.37$$

$$n = 434$$

in which MHTCRKS is the number of medium and high severity cracks per mile.

Transverse Joint Faulting

Two models, one for pavements with dowels and the other for pavements without dowels, were developed. Because of the mechanisms involved in faulting, it was not possible to combine these two design types into a single model. The models were developed using a combination of mechanistic and empirical approaches.

With dowels

$$\text{FAULT} = \text{ESAL}^{0.5280}[0.1204 + 0.04048(\text{BSTRESS}/1000)^{0.3388} - 0.007353$$

$$\times (\text{AVJSPACE}/10)^{0.6725} - 0.1492(\text{KSTAT}/100)^{0.05911}$$

$$- 0.01868 \text{ DRAIN} - 0.00879 \text{ EDGESUP}$$

$$- 0.00959 \text{ STYPE}] \qquad (9.54a)$$

Statistics: $R^2 = 0.67$

$$\text{SEE} = 0.0571 \text{ in.}$$

$$n = 559$$

in which   FAULT = mean transverse joint faulting in inches;

ESAL = cumulative equivalent 18-kip (80-kN) single-axle loads per lane in millions;

BSTRESS = maximum concrete bearing stress in psi computed by Eq. 4.45, in which the load on each dowel P_t is computed by assuming that the 9000-lb (40-kN) wheel load is distributed over a length ℓ, where ℓ is the radius of relative stiffness, and that 45% of the load is transferred through the joint;

AVJSPACE = average transverse joint spacing in ft;

KSTAT = effective modulus of subgrade reaction on top of base in pci;

DRAIN = index for drainage condition, 0 without edge drain and 1 with edge drain;

EDGESUP = index for edge support, 0 without edge support and 1 with edge support;

STYPE = index for AASHTO subgrade soil classification, 0 for A-4 to A-7 and 1 for A-1 to A-3.

Without dowels

$$\text{FAULT} = \text{ESAL}^{0.2500}[0.000038 + 0.01830 (100 \text{ OPENING})^{0.5585}$$

$$+ 0.000619 \times (100 \text{ DEFLAMI})^{1.7229} + 0.0400 (\text{FI}/1000)^{1.9840}$$

$$+ 0.00565 \text{ BTERM} - 0.00770 \text{ EDGESUP}$$

$$- 0.00263 \text{ STYPE} - 0.00891 \text{ DRAIN}] \tag{9.54b}$$

Statistics: $R^2 = 0.81$

SEE $= 0.028$ in.

$n = 398$

in which   OPENING = average transverse joint opening in inches computed by Eq. 4.36 in which temperature change ΔT is equal to one-half of the annual temperature range;

DEFLAMI = Ioannides' corner deflection in inches (Ioannides et al., 1985) $= P[1.2 - 0.88\sqrt{2}(a/\ell)]/(\text{KSTAT}\ell^2)$, where P is the wheel load and a is the contact radius;

FI = freezing index in degree days;

BTERM = base type factor, which can be computed by

$$\text{BTERM} = 10[\text{ESAL}^{0.2076} (0.04546 + 0.05115 \text{ GB} + 0.007279 \text{ CTB}$$

$$+ 0.003183 \text{ ATB} - 0.003714 \text{ OGB} - 0.006441 \text{ LCB} \tag{9.55}$$

in which GB = 1 for dense-graded aggregate base and 0 for others; CTB = 1 for cement-treated base and 0 for others; ATB = 1 for dense-graded asphalt-treated base and 0 for others; OGB = 1 for open-graded base, either untreated or asphalt-treated, and 0 for others; and LCB = 1 for lean concrete base and 0 for others.

Transverse Joint Spalling

Predicting models were developed separately for JPCP and JRCP. Extensive efforts to develop a single model for joint spalling were not successful. One reason may be that most of the joint spalling for JPCP was of medium severity, while a much greater proportion of joint spalling for JRCP was of high severity.

For JPCP

$$\text{JTSPALL} = \text{AGE}^{2.178}[0.0221 + 0.5494 \text{ DCRACK} - 0.0135 \text{ LIQSEAL}$$
$$- 0.0419 \text{ PREFSEAL} + 0.0000362 \text{ FI} \qquad (9.56a)$$

Statistics: $R^2 = 0.59$

SEE = 15 joints per mile

$n = 262$

in which   JTSPALL = number of medium- and high-severity joint spalls per mile;

AGE = age since original construction in years;

DCRACK = 0 with no "D" cracking and 1 with "D" cracking;

LIQSEAL = 0 if no liquid sealant exists in joint and 1 if liquid sealant exists in joint;

PREFSEAL = 0 if no preformed compression seal exists and 1 if preformed compression seal exists;

FI = freezing index in degree days.

For JRCP

$$\text{JTSPALL} = \text{AGE}^{4.1232}[0.00024 + 0.0000269 \text{ DCRACK}$$
$$+ 0.000307 \text{ REACTAGG} - 0.000033 \text{ LIQSEAL}$$
$$- 0.0003 \text{ PREFSEAL} + 0.00000014 \text{ FI}] \qquad (9.56b)$$

Statistics: $R^2 = 0.47$

SEE = 13 joints per mile

$n = 280$

in which REACTAGG = 0 if no reactive aggregate exists and 1 if reactive aggregate exists.

9.5.3 Effects of Design Features on Rigid Pavement Performance

The performance of rigid pavements is affected by a variety of design features, including slab thickness, base type, joint spacing, reinforcement, joint orientation, load transfer, dowel bar coatings, longitudinal joint design, joint sealant, tied concrete shoulders, and subdrainage. A study was made by ERES Consultants, Inc. under FHWA contract on the effects of these features on rigid pavement performance. Ninety-five pavement sections located in four major climatic regions were thoroughly evaluated. The following conclusions, which provide some revealing insights into pavement performance, are abstracted from the report (Smith et al., 1990a).

Slab Thickness

The effect of slab thickness on pavement performance was significant. It was found that increasing slab thickness reduced transverse and longitudinal cracking in all cases. This effect was much more pronounced for thinner slabs than for thicker slabs. It was not possible to directly compare the performance of the thinner slabs and the thicker slabs, because the thick slabs were all constructed directly on the subgrade and the thinner slabs were all constructed on a base course.

Increasing the thickness of slab did not appear to reduce joint spalling or joint faulting. Thick slabs placed directly on the subgrade, especially in wet climates and exposed to heavy traffic, faulted as much as thin slabs constructed on a base course.

Base Type

Base types, including base/slab interface friction, base stiffness, base erodibility, and base permeability, seemed to have a great effect on the performance of jointed concrete pavements. The major performance indicators, which were affected by variations in base type, were transverse and longitudinal cracking, joint spalling, and faulting.

The worst performing base type was the cement-treated or soil cement bases, which tended to exhibit excessive pumping, faulting, and cracking. This is most likely due to the impervious nature of the base, which traps moisture and yet can break down and contribute to the movement of fines beneath the slab.

The use of lean concrete bases generally produced poor performance. Large curling and warping stresses have been associated with slabs constructed over lean concrete bases. These stresses result in considerable transverse and longitudinal cracking of the slab. The poor performance of these bases can also be attributed to a bathtub design in which moisture is trapped within the pavement cross section.

Dense-graded asphalt-treated base courses ranged in performance from very poor to good. The fact that these types of bases were often constructed as a bathtub design contributed to their poor performance. This improper design often resulted in severe cracking, faulting, and pumping.

The construction of thicker slabs directly on the subgrade with no base resulted in a pavement that performed marginally. These pavements were especially susceptible to faulting, even under low traffic levels.

Pavements constructed over aggregate bases had varied performance, but were generally in the fair to very good category. In general, the more open-graded the aggregate, the better the performance. An advantage of aggregate bases is that they contribute the least to the high curling and warping stresses in the slab. Even though aggregate bases are not open-graded, they are more permeable and have a lower friction factor than stabilized bases.

The best bases in terms of pavement performance were the permeable bases. Typical base courses have permeabilities ranging from 0 to less than 1 ft/day

(0.3 m/day), while good permeable bases have permeabilities up to 1000 ft/day (305 m/day). Specific areas of concern were the high corner deflections and the low load transfer exhibited by the permeable bases. Since these could impact their long-term performance, the use of dowel bars may be required. An unexpected benefit of using permeable bases was the reduction in "D" cracking on pavements susceptible to this type of distress.

Slab Length

For JPCP, the length of slabs investigated ranged from 7.75 to 30 ft (2.4 to 9.1 m). It was found that reducing the slab length decreased both the magnitude of joint faulting and the amount of transverse cracking. On pavements with random joint spacings, slabs with joint spacings greater than 18 ft (5.5 m) experienced more transverse cracking than the shorter slabs.

For JRCP, the length of slabs investigated ranged from 21 to 78 ft (6.4 to 23.9 m). Generally, shorter joint spacings performed better, as measured by the deteriorated transverse cracks, joint faulting, and joint spalling. However, several JRCP with long joint spacings performed quite well. In particular, the long jointed pavements in New Jersey, which were constructed with expansion joints, displayed excellent performance.

An examination of the stiffness of foundation was made through the use of the radius of relative stiffness ℓ. Generally speaking, when the ratio L/ℓ, where L is the length of slab, was greater that 5, transverse cracking occurred more frequently. This factor was further examined for different base types. It was found that stiffer base courses required shorter joint spacings to reduce or eliminate transverse cracking.

Reinforcement

The amount of steel reinforcement appeared to have an effect in controlling the amount of deteriorated transverse cracking. Pavement sections with less than 0.1% reinforcing steel often displayed significant deteriorated transverse cracking. A minimum of 0.1% reinforcing steel is therefore recommended, with larger amounts required for more severe climate and longer slabs.

Joint Orientation

Conventional wisdom has it that skewed joints prevent the application of two wheel loads to the joint at the same time and thus may reduce load-associated distresses. While the results from the limited sample size in this study were ambiguous, all of the nondoweled sections with skewed joints had a lower PSR than similar designs with perpendicular joints. The available data provide no definite conclusions on the effectiveness of skewing transverse joints for nondoweled slabs. Skewed joints are not believed to provide any benefit to doweled slabs.

Load Transfer

Dowel bars were found to be effective in reducing the amount of joint faulting when compared with nondoweled sections of comparable designs. The diameter of dowels had an effect on performance, since larger diameter bars provided better load transfer and control of faulting under heavy traffic than did smaller dowels. It appeared that a minimum dowel diameter of 1.25 in. (32 mm) was necessary to provide good performance.

Nondoweled JPCP slabs generally developed significant faulting, regardless of pavement design or climate. This effect was somewhat mitigated by the use of permeable bases. However, the sections in this group had a much lower number of accumulated ESAL, so no definite conclusions can be drawn yet.

Dowel Bar Coatings

Corrosion-resistant coatings are needed to protect dowels from the adverse effects of moisture and deicing chemicals. While most of the sections in this study did not contain corrosion-resistant dowel bars, those that did generally exhibited enhanced performance. Very little deteriorated transverse cracking was identified on these sections. In fact, one section in New Jersey with stainless steel-clad dowel bars was performing satisfactorily after 36 years of service.

Longitudinal Joint Design

The longitudinal joint design was found to be a critical design element. Both inadequate forming techniques and insufficient depths of joint can contribute to the development of longitudinal cracking. There was evidence on the advantage of sawing the joints over the use of inserts. The depth of longitudinal joints is generally recommended to be one-third of the actual, not designed, slab thickness, and may have to be deeper when stabilized bases are used.

Joint Sealant

Joint sealing appeared to have a beneficial effect on performance. This was particularly true in harsh climates with excessive amounts of moisture. Preformed compression sealants were shown to perform well for more than 15 years under heavy traffic. Except where ''D'' cracking occurred, pavement sections containing preformed sealants generally exhibited little joint spalling and were in good overall conditions. Rubberized asphalt joint sealants showed good performance for 5 to 7 years.

Tied Concrete Shoulders

It is generally believed that tied concrete shoulders can reduce edge stresses and corner deflections by providing more lateral supports to the mainline pavement, thus improving pavement performance. Surprisingly, this study showed that while tied concrete shoulders performed better than asphalt shoulders, many of the tied shoulders were not designed properly and actually contributed to poor performance of the mainline pavement. The tiebars were spaced too far apart, sometimes at a spacing of 40 in. (1016 mm), and not strategically located near slab corners to provide adequate support. In some cases, tied concrete shoulders were constructed over a stabilized dense-graded base in a bathtub design, which resulted in the poor performance of mainline pavement.

Subdrainage

The provision of positive subdrainage, either in the form of longitudinal edge drains or the combination of a drainage layer and edge drains, generally reduced the amount of faulting and spalling related to "D" cracking. With few exceptions, the load-associated distresses, especially faulting and transverse cracking, decreased as the drainage characteristics improved. The overall pavement performance can be improved by using an open-graded base or restricting the percentage of fines. A filter layer must be placed below the permeable base and regular maintenance of the outlets must be provided.

9.6 SUMMARY

Pavement performance is an important factor of pavement design because it provides a framework upon which a judgement on the success or failure of a design procedure or the need for further improvements can be made. This chapter discusses pavement performance, including distress, serviceability, skid resistance, nondestructive testing, and pavement evaluation.

Important Points Discussed in Chapter 9

1. Distress or failure can be divided into structural failure, which is associated with the ability of the pavement to carry the design load, and functional failure, which is concerned mainly with riding comfort and safety. Pavements that exhibit structural failures may finally end up in functional failures, whereas pavements with functional failures may be structurally sound. Immediate attention must be paid to correct any structural failure so that damage to the pavement will not be accelerated.

2. Distress can be classified into load-associated distress and non-load-associated distress. This distinction is important because it identifies the causes of distress so that proper remedial measures can be taken. The non-load-associated distress is caused by climates, materials, or construction. A distress may begin as a non-load-associated distress and later become more severe as the number of load repetitions increases.

3. The types of distress that should be considered in the design of asphalt pavement are fatigue cracking, which is a load-associated structural failure, and rutting, which is a load-associated functional failure. In climates with large temperature variations or very cold temperatures, low-temperature cracking, which is a non-load-associated structural failure, should also be considered.

4. The most common problem with composite pavements is reflection cracking in the asphalt overlay above a joint or crack in the PCC slab. This is a non-load-associated structural failure caused mainly by the movement of the concrete slab beneath the asphalt surface because of thermal or moisture changes.

5. Common types of distress in PCC pavements are pumping, faulting, and cracking. Joint deterioration is a major problem in JRCP due to longer joint spacings, while edge punchout is a major problem in CRCP due to closely spaced cracks.

6. The pavement serviceability–performance concept was developed during the AASHO Road Test. Detailed procedures for developing the AASHO PSI equations relating physical measurements to panel ratings are presented. Due to changes in the public's perception of serviceability and the use of new equipment for physical measurements, it is doubtful that the AASHO equations developed more than 30 years ago are still applicable. One major deficiency of the AASHO equations is that they include both rideability and surface defects. For the management of pavement inventory, it is preferable to have separate measures of ride quality and surface defects. Therefore, there is a need to develop new equations that reflect the public's perception of riding quality.

7. A variety of devices can be used to measure roughness and riding quality. The most popular is the response-type road roughness meter (RTRRM). RTRRMs measure the relative movement in inches per mile (m/km) between the body of the automobile and the center of the rear axle. Because the measurements are sensitive to the type and condition of the vehicle used, regular calibrations are required. To provide a common base for roughness measurements, the use of the international roughness index (IRI) is recommended.

8. Pavement friction is measured most frequently by the locked-wheel trailer procedure, as specified in ASTM E 274. The skid number (SN), which is the measured coefficient of friction multiplied by 100, is determined when the vehicle travels at a speed of 40 mph (64 km/h).

9. Regression equations are available to relate the surface texture to the skid number determined by the ASTM E 274 method. The British pendulum number (BPN) can be used as an indicator for microtexture, which governs the surface friction at very low speeds, while the mean texture depth (MTD) can be used for macrotexture, which affects the surface drainage and friction at higher speeds.

10. Skid resistance of wearing surfaces depends on both the aggregate and the mixture characteristics. The most frequently sought after characteristics for a skid-resistant aggregate are its resistance to polish and wear, texture, shape, and size. The mixture characteristics can be controlled by blending of aggregates, selection of proper aggregate size and gradation, use of appropriate binder content, and application of good construction practice.

11. The surface friction of PCC pavement can be improved by roughing, acid etching, grooving, and bonding a thin layer of aggregate to the surface; while that of bituminous pavement by grooving, surface treatment or thin overlay, and heater planer.

12. Nondestructive deflection tests can be conducted by three types of

equipment using static or slowly moving loads, steady-state vibration, or impulse loads. The falling weight deflectometer, which imposes an impulse load on the pavement surface, can simulate a moving wheel load in both magnitude and duration and is considered as the best device for nondestructive testing.

13. Various computer programs are available to back-calculate the layer moduli. Because the solutions are not unique and depend on the initial moduli assumed, engineering judgments are needed in using these programs. The layer moduli can also be determined by a trial and error method using any layer system computer programs. The deflection of the sensor farthest from the load must be matched first by adjusting the modulus of the lowest layer. The process is then moved progressively inward by adjusting the moduli of the upper layers to match with the deflections of the inner sensors until the innermost sensor is reached. An exception to the above procedures is the ILL-BACK program for directly determining the composite modulus of the foundation and the elastic modulus of the PCC slab.

14. Pavement evaluation and rehabilitation are not exact sciences and depend to a large extent on past experiences and engineering judgments. Until recently, many problems in this domain could be solved only by human experts with extensive practical experience and knowledge. The new technology of knowledge-based expert systems offers a powerful means for acquiring and organizing this human expertise so that it can be preserved and communicated to others. Two of the systems, SCEPTRE for flexible pavements and EXPEAR for rigid pavements, are briefly described.

15. A large number of regression equations can be found in the literature to predict pavement performance. These equations illustrate the effects of various factors on pavement performance, but their usefulness in practice may be limited by the scope of the data base that was used in their development. Some equations are presented for predicting fatigue cracking, rutting, and low-temperature cracking of flexible pavements, as well as present serviceability rating, transverse joint faulting, and transverse joint spalling of rigid pavements.

16. An insight into the effects of various design features on the performance of jointed concrete pavements was presented. These features include slab thickness, base type, joint spacings, reinforcement, joint orientation, load transfer, dowel bar coatings, longitudinal joint design, joint sealant, tied concrete shoulders, and subdrainage.

PROBLEMS

9.1. It is assumed that the present serviceability index PSI of pavements is related to the roughness index RI by

$$PSI = A_0 + A_1 \log(RI)$$

Derive the general equations for determining A_0 and A_1. Using the equations derived, find A_0 and A_1 based on the data shown in Table P9.1. [Answer: PSI = 9.50 − 2.98 log(RI)]

TABLE P9.1

Section no.	1	2	3	4	5
RI (in./mile)	800	300	200	150	80
PSR	1.0	2.0	2.5	3.0	4.0

9-2. Table P9.2 shows the measurements and ratings of five sections of flexible pavement. Develop the equation for the present serviceability index. [Answer: PSI = 5.51 − 1.70 log (1 + $\overline{SV}$) − 38.09$\overline{RD}^2$ − 0.004 $\sqrt{C + P}$]

TABLE P9.2

Section no.	Slope variance $\overline{SV}$ (10^{-6})	Rut depth $\overline{RD}$ (in.)	Cracking and patching $C + P$ (ft or ft²/1000 ft²)	PSR
1	2.8	0.06	0	4.3
2	5.8	0.10	1	3.8
3	10.9	0.11	13	3.2
4	16.8	0.16	23	2.4
5	56.0	0.19	31	1.1

9-3. Derive Eq. 9.29.

9-4. Glass beads are spread over a pavement surface to determine its mean texture depth. The volume of beads is 2 in.³ and the diameter of the patch area is 10 in. If the skid number of the pavement is 40 at a speed of 40 mph, estimate the skid number at speeds of 20 and 60 mph, respectively. [Answer: SN_{20} = 47.7 and SN_{60} = 33.6]

9-5. FWD deflection measurements were made on a pavement, as shown in Figure P9.5. A 9000-lb load was applied uniformly over a circular area with a radius of 6 in. The surface deflections at radial distances of 0, 8, 12, 24, 48, 72, and 96 in. were measured and found to be 23.7, 19.7, 17.1, 10.5, 5.4, 3.5, and 2.6 mil (1 mil = 10^{-3} in.) Assume the layer system to be linear elastic with the thicknesses and Poisson ratios shown in the figure and back-calculate the layer moduli E_1, E_2, and E_3 by KENLAYER.

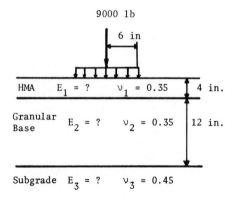

Figure P9.5

9-6. Same as Problem 9-5 except that the granular base is nonlinear elastic with the properties shown in Figure P9.6. Consider the granular base as a single layer with the stress point at the upper quarter as shown and back-calculate E_1, K_1, and E_3 by KENLAYER.

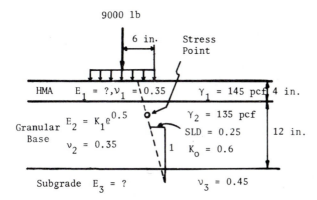

Figure P9.6

9-7. Same as Problem 9-6, but the granular base is divided into 6 layers each 2 in. thick.

9-8. Figure P9.8 is a linear elastic three-layer system with the load, material properties, and layer thicknesses as shown. The deflection sensors are placed at 0, 12, 24, 36, and 48 in. from the axis of symmetry. Determine the functions f_1, f_2, f_3, f_4, and f_5, as shown in Eq. 9.36, by running KENLAYER. [Answer: 0.2971, 0.2427, 0.1907, 0.1530, 0.1125]

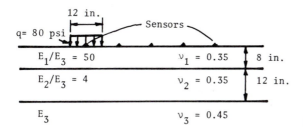

Figure P9.8

9-9. A FWD test was made on a concrete pavement. The radius of the load plate was 5.9 in. and the recorded load was 9000 lb. Sensors were located at 0, 12, 24, and 36 in. from the center of plate and the corresponding deflections recorded were 0.0047, 0.0040, 0.0035, and 0.0026 in. Laboratory tests on the cored samples indicated that the concrete has a modulus of elasticity of 4.7×10^6 psi. The foundation is assumed to be a dense liquid. Determine the modulus of subgrade reaction k and the thickness of the concrete slab. [Answer: 285 pci, 8 in.]

9-10. Same as problem 9-9, but the foundation is considered as an elastic solid with a Poisson ratio of 0.4. [Answer: 37000 psi, 5.5 in.]

9-11. An asphalt pavement has a HMA thickness of 4 in., a stiffness of original asphalt of 200 kg/cm², and a loam type of subgrade. For a winter design temperature of $-20°C$

and an age of 6 years, estimate the length of low-temperature cracks in ft per 1000 ft of two-lane road. [Answer: 40 ft/1000 ft]

9-12. A doweled concrete pavement has a joint spacing of 20 ft and no tied concrete shoulders for edge support. The subgrade soil is A-6 and there is no edge drain installed. The calculated bearing stress between dowel and concrete is 3000 psi and the pavement is placed on a subbase with an effective modulus of subgrade reaction of 100 pci. Estimate the amount of transverse joint faulting after the pavement has been subjected to 20 million repetitions of an 18-kip equivalent single-axle load. [Answer: 0.089 in.]

9-13. No evidence of "D" cracking can be found on a 15-year-old jointed plain concrete pavement. If the freezing index at the pavement site is 1000 degree days, estimate the average number of medium- and high-severity joint spall per mile if (a) liquid joint sealant is used and (b) preformed compressive seal is used. [Answer: 16.3 and 6.0 joints per mile]

10

Reliability

10.1 STATISTICAL CONCEPTS

Due to the variability of design factors, the probabilistic method has been used for the design of both flexible and rigid pavements (Lemer and Moavenzadeh, 1971; Darter et al., 1973b; Kher and Darter, 1973; Darter, 1976). The method was also incorporated in the AASHTO design guide (AASHTO, 1986). The application of the method requires an understanding of some statistical concepts.

10.1.1 Definitions

Two of the most useful properties of a random variable are its expectation, or mean, and its variance. A random variable is a function that takes a defined value for every point in a sample space. For example, consider the sample space generated by the tossing of two fair coins. The random variable x is defined as the total number of heads observed. The sample space consists of four points (0, 0), (0, 1), (1, 0) and (1, 1) and at these points x takes the values 0, 1, 1, and 2. The probability that the random variable x takes these values is known as the probability function of x, $f(x)$. This can be expressed in tabular form:

x	0	1	2
$f(x)$	$\frac{1}{4}$	$\frac{1}{2}$	$\frac{1}{4}$

Note that the sum of $f(x)$ for all possible values of x must be equal to one. If the random variable is not discrete, but rather is continuous within the sample space, as shown in Figure 10.1, $f(x)$ is called the probability density function. For $f(x)$ to

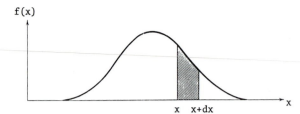

f(x)

x   x+dx   x

Figure 10.1   Probability density function.

be a probability density function, it must satisfy the condition that $f(x) \geq 0$ for all x and that

$$\int_{-\infty}^{\infty} f(x) \, dx = 1 \tag{10.1}$$

Expectation

The expectation, or mean, of a random variable x is defined as follows:

Discrete case $\qquad\qquad E[x] = \sum_{\text{all} x_i} x_i f(x_i) \tag{10.2a}$

Continuous case $\qquad\qquad E[x] = \int_{-\infty}^{\infty} xf(x) \, dx \tag{10.2b}$

If n independent observations of x are taken, each with the same probability of $1/n$, the mean of the observation $\bar{x}$ as obtained from Eq. 10.2a is

$$\bar{x} = \frac{\sum_{i=1}^{n} x_i}{n} \tag{10.3}$$

This value is called the sample mean and is the best estimate of the true or population mean μ.

Example 10.1

Compute $E[x]$ for a random variable x with the following probability function:

x	1	2	3	4
$f(x)$	$\frac{1}{8}$	$\frac{1}{8}$	$\frac{1}{4}$	$\frac{1}{2}$

Solution: By Eq. 10.2a, $E[x] = 1(\frac{1}{8}) + 2(\frac{1}{8}) + 3(\frac{1}{4}) + 4(\frac{1}{2}) = 3.125$.

Example 10.2

Compute $E[x]$ for a random variable x with the following probability density function:

$$f(x) = \begin{cases} \frac{1}{3}x & 0 \leq x \leq 1 \\ \frac{2}{3} & 1 < x \leq 3 \end{cases}$$

Solution: By Eq. 10.2*b*,

$$E[x] = \int_0^1 \tfrac{1}{3}x^2 \, dx + \int_1^3 \tfrac{2}{3}x \, dx$$

$$= \tfrac{1}{9}x^3 \Big|_0^1 + \tfrac{2}{6}x^2 \Big|_1^3 = 2\tfrac{7}{9}$$

If *c* is a constant, then

$$E[c] = \int_{-\infty}^{\infty} cf(x) \, dx = c \int_{-\infty}^{\infty} f(x) \, dx = c \qquad (10.4)$$

Using the definition shown in Eq. 10.2*b*, it can be easily proved that

$$E[cx] = cE[x] \qquad (10.5)$$

$$E[g_1(x) + g_2(x)] = E[g_1(x)] + E[g_2(x)] \qquad (10.6)$$

in which g_1 and g_2 are functions of *x*.

Variance

The variance of a random variable *x* is defined as the expected value of the square of the deviation from its expectation:

$$V[x] = E[(x - E[x])^2] \qquad (10.7a)$$

Discrete case

$$V[x] = \sum_{\text{all }x_i} (x_i - E[x])^2 f(x_i) \qquad (10.7b)$$

Continuous case

$$V[x] = \int_{-\infty}^{\infty} (x - E[x])^2 f(x) \, dx \qquad (10.7c)$$

If $E[x]$ is considered a constant, then Eq. 10.7*a* can be written as

$$V[x] = E[x^2 + (E[x])^2 - 2xE[x]] = E[x^2] + (E[x])^2 - 2(E[x])^2$$

$$= E[x^2] - (E[x])^2 \qquad (10.8)$$

Equation 10.8 indicates that the variance of the random variable *x* is equal to the expectation of the square of *x*, abbreviated as square mean, minus the square of the expectation of *x*, abbreviated as mean square.

Example 10.3

Compute $V[x]$ for the random variable *x* with the probability function shown in Example 10.1.

Solution: The variance can be calculated by either Eq. 10.7*b* or 10.8. From Eq. 10.7*b*, $V[x] = (1 - 3.125)^2(\tfrac{1}{8}) + (2 - 3.125)^2(\tfrac{1}{8}) + (3 - 3.125)^2(\tfrac{1}{4}) + (4 - 3.125)^2(\tfrac{1}{2})$ = 1.109. From Eq. 10.8, $V[x] = 1(\tfrac{1}{8}) + 4(\tfrac{1}{8}) + 9(\tfrac{1}{4}) + 16(\tfrac{1}{2}) - (3.125)^2 = 1.109$.

If *n* independent observations of *x* are taken, the variance of *x* is determined by

$$V[x] = \sum_{i=1}^{n} \frac{(x_i - \bar{x})^2}{n - 1} \qquad (10.9)$$

The sum is divided by $n - 1$ rather than n because $\bar{x}$ is the sample mean rather than the true mean, so the degree of freedom is $n - 1$.

If a, b, and c are constants, it can be easily proved that

$$V[c] = 0 \tag{10.10}$$

$$V[cx] = c^2 \, V[x] \tag{10.11}$$

$$V[a + bx] = b^2 \, V[x] \tag{10.12}$$

The standard deviation s of a random variable x is defined as the square root of the variance:

$$s = \sqrt{V[x]} \tag{10.13}$$

The coefficient of variation is generally used in percentile form but for convenience can be expressed as a decimal by

$$C[x] = \frac{s}{\mu} \tag{10.14}$$

in which μ is the mean.

Example 10.4

Given 10 values of x as 17, 19, 14, 11, 18, 16, 16, 10, 9, and 12, calculate the mean, variance, standard deviation, and coefficient of variation.

Solution: From Eq. 10.3, $\bar{x} = (17 + 19 + 14 + 11 + 18 + 16 + 16 + 10 + 9 + 12)/10 = 14.2$. From Eq. 10.9, $V[x] = [(17 - 14.2)^2 + (19 - 14.2)^2 + (14 - 14.2)^2 + (11 - 14.2)^2 + (18 - 14.2)^2 + (16 - 14.2)^2 + (16 - 14.2)^2 + (10 - 14.2)^2 + (9 - 14.2)^2 + (12 - 14.2)^2]/(10 - 1) = 12.4$. From Eq. 10.13, $s = \sqrt{12.4} = 3.52$. From Eq. 10.14, $C[x] = (3.52/14.2) \times 100 = 24.8\%$.

Covariance

The covariance of two random variables x and y is defined as the expected value of the product of the deviation of x and y from their expected values:

$$\text{Cov}[x, y] = E[(x - E[x])(y - E[y])] \tag{10.15a}$$

Discrete case

$$\text{Cov}[x, y] = \sum_{\text{all } x_i, y_i} (x_i - E[x])(y_i - E[y])f(x_i, y_i) \tag{10.15b}$$

Continuous case

$$\text{Cov}[x, y] = \int_{-\infty}^{\infty} \int_{-\infty}^{\infty} (x - E[x])(y - E[y])f(x, y)\, dx\, dy \tag{10.15c}$$

From Eq. 10.15a

$$\text{Cov}[x, y] = E[xy - yE[x] - xE[y] + E[x]E[y]]$$

$$= E[xy] - E[x]E[y] \tag{10.16}$$

For actual computation of covariance, the use of Eq. 10.16 is more convenient than that of Eq. 10.15a.

If large positive deviations of x are associated with large positive deviations of y, or large negative deviations of x with large negative deviations of y, then the covariance will be positive. If positive deviations of x are associated with negative deviations of y, and vice versa, the covariance will be negative. On the other hand, if positive and negative deviations of x occur as frequently as positive and negative deviations of y, then the covariance will tend to 0. Therefore, the covariance is a measure of correlation between two random variables. It should be noted that variance is a special case of covariance of a random variable with itself:

$$\text{Cov}[x, x] = E[(x - E[x])(x - E[x])] = E[(x - E[x])^2] = V[x] \qquad (10.17)$$

The correlation coefficient between random variables x and y is defined by

$$\rho(x, y) = \frac{\text{Cov}[x, y]}{\sqrt{V[x]\, V[y]}} \qquad (10.18)$$

It can be shown that $-1 \le \rho \le 1$ and that

$$\rho = 1 \qquad \text{when } y = a + bx \qquad (10.19)$$

and

$$\rho = -1 \qquad \text{when } y = a - bx \qquad (10.20)$$

in which a and b are constants. The variance of $x + y$ can be expressed as

$$V[x + y] = E[(x + y - E[x + y])^2] = E[(x - E[x] + y - E[y])^2]$$

$$= E[(x - E[x])^2 + E[(y - E[y])^2] + 2E[(x - E[x])(y - E[y])]$$

$$= V[x] + V[y] + 2\,\text{Cov}[x, y] \qquad (10.21)$$

If x and y are independent, $\text{Cov}[x, y] = 0$.

Example 10.5

Compute $\text{Cov}[x, y]$ and $\rho(x, y)$ for the following probability functions of random variables x and y:

$$f(x, y) = \begin{cases} \frac{1}{4} & x = 1 \quad y = 3 \\ \frac{1}{8} & x = 2 \quad y = 6 \\ \frac{1}{2} & x = 3 \quad y = 9 \\ \frac{1}{8} & x = 4 \quad y = 12 \end{cases}$$

Solution: From Eq. 10.2a

$$E[x] = (1)(\tfrac{1}{4}) + (2)(\tfrac{1}{8}) + (3)(\tfrac{1}{2}) + (4)(\tfrac{1}{8}) = 2.5$$

$$E[y] = (3)(\tfrac{1}{4}) + (6)(\tfrac{1}{8}) + (9)(\tfrac{1}{2}) + (12)(\tfrac{1}{8}) = 7.5$$

From Eq. 10.7b

$$V[x] = (1 - 2.5)^2(\tfrac{1}{4}) + (2 - 2.5)^2(\tfrac{1}{8}) + (3 - 2.5)^2(\tfrac{1}{2}) + (4 - 2.5)^2(\tfrac{1}{8}) = 1$$

$$V[y] = (3 - 7.5)^2(\tfrac{1}{4}) + (6 - 7.5)^2(\tfrac{1}{8}) + (9 - 7.5)^2(\tfrac{1}{2}) + (12 - 7.5)^2(\tfrac{1}{8}) = 9$$

From Eq. 10.15*b*

$$\text{Cov}[x, y] = (1 - 2.5)(3 - 7.5)(\tfrac{1}{4}) + (2 - 2.5)(6 - 7.5)(\tfrac{1}{8})$$
$$+ (3 - 2.5)(9 - 7.5)(\tfrac{1}{2}) + (4 - 2.5)(12 - 7.5)(\tfrac{1}{8}) = 3$$

The covariance can also be determined from Eq. 10.16:

$$\text{Cov}[x, y] = (1)(3)(\tfrac{1}{4}) + (2)(6)(\tfrac{1}{8}) + (3)(9)(\tfrac{1}{2})$$
$$+ (4)(12)(\tfrac{1}{8}) - (2.5)(7.5) = 3$$

From Eq. 10.18

$$\rho(x, y) = \frac{3}{\sqrt{9}} = 1$$

This is as expected because $y = 3x$.

Example 10.6

Compute $\text{Cov}[x, y]$ and $\rho(x, y)$ of the joint density function of random variables x and y:

$$f(x, y) = x + y \qquad \text{when } 0 \le x \le 1 \text{ and } 0 \le y \le 1$$
$$= 0 \qquad \text{otherwise}$$

Solution: To find the expectation involving two random variables, one is considered as a constant and the other as a variable. From Eq. 10.2*b*,

$$E[xy] = \int_{-\infty}^{\infty} \int_{-\infty}^{\infty} xy(x + y)\, dx\, dy = \int_0^1 \int_0^1 (x^2 y + xy^2)\, dx\, dy$$
$$= \int_0^1 \left[\frac{x^3}{3} y + \frac{x^2 y^2}{2} \right]_0^1 dy = \int_0^1 \left[\frac{1}{3} y + \frac{1}{2} y^2 \right] dy = \left[\frac{y^2}{6} + \frac{y^3}{6} \right]_0^1 = \frac{1}{3}$$

$$E[x] = \int_0^1 \int_0^1 x(x + y)\, dx\, dy = \int_0^1 \left[\frac{x^3}{3} + \frac{x^2 y}{2} \right]_0^1 dy = \left[\frac{y}{3} + \frac{y^2}{4} \right]_0^1 = \frac{7}{12}$$

$$E[y] = \int_0^1 \int_0^1 y(x + y)\, dx\, dy = \frac{7}{12}$$

$$E[x^2] = \int_0^1 \int_0^1 x^2(x + y)\, dx\, dy = \int_0^1 \left[\frac{x^4}{4} + \frac{x^3 y}{3} \right]_0^1 dy = \left[\frac{y}{4} + \frac{y^2}{6} \right]_0^1 = \frac{5}{12}$$

$$E[y^2] = \int_0^1 \int_0^1 y^2(x + y)\, dx\, dy = \frac{5}{12}$$

From Eq. 10.8, $V[x] = V[y] = \frac{5}{12} - (\frac{7}{12})^2 = \frac{11}{144}$. From Eq. 10.16, $\text{Cov}[x, y] = \frac{1}{3} - (\frac{7}{12})(\frac{7}{12}) = -\frac{1}{144}$. From Eq. 10.18, $\rho(x, y) = -\frac{1}{11}$.

10.1.2 Taylor's Expansion

Taylor's expansion for a function $f(x, y)$ about the point (a, b) can be expressed as

$$f(x, y) = f(a, b) + f_x(a, b)(x - a) + f_y(a, b)(y - b) + \tfrac{1}{2} [f_{xx}(a, b)(x - a)^2$$
$$+ 2f_{xy}(a, b)(x - a)(y - b) + f_{yy}(a, b)(y - b)^2] + \cdots \qquad (10.22)$$

in which the subscripts x and y indicate the partial differentiation with respect to x and y. Taylor's expansion and its extension into n random variables can be used to determine the mean and variance of a function (Benjamin and Cornell, 1970).

Mean

If a and b are taken as the means of x and y, the expectation of $f(x, y)$ can be obtained from Eq. 10.22 by taking the expectation on both sides. When all terms higher than the second order are neglected, the following relation is obtained:

$$E[f(x, y)] = f(a, b) + \tfrac{1}{2}\{f_{xx}(a, b) V[x] + 2f_{xy}(a, b) \text{Cov}[x, y]$$

$$+ f_{yy}(a, b) V[y]\} \tag{10.23}$$

Note that $E[f_x(a, b)(x - a)] = f_x(a, b)\{E[x] - a\} = 0$ because $E[x] = a$. If $g(x_1, x_2, \ldots, x_n)$ is a function of n random variables x_i and μ is the mean value for each of these random variables, the second-order Taylor series approximation to the expected value of g is

$$E[g] = g(\mu) + \tfrac{1}{2} \sum_{i=1}^{n} \sum_{j=1}^{n} \left(\frac{\partial^2 g}{\partial x_i\, \partial x_j}\right) \text{Cov}[x_i, x_j] \tag{10.24}$$

in which μ is the set of mean values for x_i. If x_i and x_j are independent, then $\text{Cov}[x_i, x_i] = V[x_i]$ and $\text{Cov}[x_i, x_j] = 0$, so Eq. 10.24 becomes

$$E[g] = g(\mu) + \tfrac{1}{2} \sum_{i=1}^{n} \frac{\partial^2 g}{\partial x_i^2} V[x_i] \tag{10.25}$$

The second-order term may be neglected if the coefficients of variations of x_i are small, so

$$E[g] = g(\mu) \tag{10.26}$$

Equation 10.26 indicates that the mean value of g can be obtained simply by substituting the mean value of each random variable into the function.

Example 10.7

The maximum surface deflection of a homogeneous and incompressible half-space under a circular loaded area, as indicated by Eq. 2.8, is

$$w_0 = \frac{1.5qa}{E} \tag{10.27}$$

If the contact pressure q, the contact radius a, and the elastic modulus E are independent with mean values of $q = 100$ psi (690 kPa), $a = 6$ in. (152 mm), and $E = 10,000$ psi (69 MPa) and coefficients of variation of $C[q] = 0.1$, $C[a] = 0.05$, and $C[E] = 0.2$, determine the mean of w_0 by the first- and second-order approximations.

Solution: By the first-order approximation, or Eq. 10.26, $E[w_0] = 1.5 \times 100 \times 6/10,000 = 0.09$ in. (2.29 mm). For the second-order approximation $\partial^2 w_0/\partial q^2 = 0$, $\partial^2 w_0/\partial a^2 = 0$, and $\partial^2 w_0/\partial E^2 = 3qa/E^3$, so the variance of q and a has no effect on $E[w_0]$. From Eqs. 10.13 and 10.14, $V[E] = (0.2 \times 10,000)^2 = 4 \times 10^6$ lb²/in.⁴ (190 MN²/m⁴). From Eq. 10.25, $E[w_0] = 0.09 + 0.5 \times 3 \times 100 \times 6 \times 4 \times 10^6/10^{12} = 0.0936$ in. (2.38 mm).

Variance

The first-order approximation for the variance of $f(x, y)$ can be determined from the first three terms on the right side of Eq. 10.22 by applying Eq. 10.21:

$$V[f(x, y)] = \{f_x(a, b)\}^2 V[x] + \{f_y(a, b)\}^2 V[y]$$

$$+ 2\{f_x(a, b)\}\{f_y(a, b)\}\text{Cov}[x, y] \qquad (10.28)$$

Equation 10.28 can be extended to $g(x_1, x_2, \ldots, x_n)$ by

$$V[g] = \sum_{i=1}^{n} \sum_{j=1}^{n} \left(\frac{\partial g}{\partial x_i}\right)_\mu \left(\frac{\partial g}{\partial x_j}\right)_\mu \text{Cov}[x_i, x_j] \qquad (10.29)$$

If $x_1, x_2, \ldots, x_n$ are independent, there are no cross product terms and Eq. 10.29 becomes

$$V[g] = \sum_{i=1}^{n} \left(\frac{\partial g}{\partial x_i}\right)^2 V[x_i] \qquad (10.30)$$

Example 10.8

Compute $V[w_0]$ for the problem in Example 10.7.

Solution: With $V[q] = (0.1 \times 100)^2 = 100$ lb^2/in.^{4} (4.8 GN^2/m^4), $V[a] = (0.05 \times 6)^2 = 0.09$ in.^{2} (58.1 mm^2) and $V[E] = 4 \times 10^6$ lb^2/in.^{4} (190 MN^2/m^4), and applying Eq. 10.30 to Eq. 10.27, $V[w_0] = (1.5)^2 \{(a/E)^2 V[q] + (q/E)^2 V[a] + (qa/E^2)^2 V[E]\} = 2.25 \{(6/10,000)^2 \times 100 + (100/10,000)^2 \times 0.09 + (100 \times 6/10^8)^2 \times 4 \times 10^6\} = 4.25 \times 10^{-4}$ in.^{2} (0.274 mm^2).

10.1.3 Normal Distribution

The distribution function most frequently used as a probability model is called the normal or Gaussian distribution. Although this symmetrical and bell-shaped distribution is very important, it is not the only type of distribution to be used in the probabilistic method.

Basic Equation

The mathematical equation of normal distribution expressing the frequency of occurrence of the random variable x is

$$f(x) = \frac{1}{s\sqrt{2\pi}} \exp\left[-\frac{1}{2}\left(\frac{x - \mu}{s}\right)^2\right] \qquad (10.31)$$

in which s is the standard deviation and μ is the mean.

Figure 10.2 shows a plot of normal distribution with $s = 1$ and $\mu = 0$ and 4, respectively. Note that both curves are similar except that the x coordinate is displaced by a constant distance. If the x coordinate at the peak is not equal to zero, it can be made to zero by a simple displacement.

The cumulative distribution function for a normally distributed random variable x can be expressed as

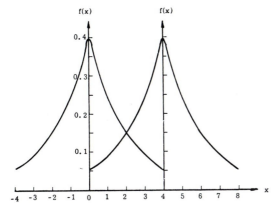

Figure 10.2 Normal distribution.

$$\psi(x) = \int_{-\infty}^{x} \frac{1}{s\sqrt{2\pi}} \exp\left[-\frac{1}{2}\left(\frac{x-\mu}{s}\right)^2\right] dx \qquad (10.32)$$

A simple way to eliminate μ and s in Eq. 10.32 is to introduce a normal deviate defined as

$$u = \frac{x - \mu}{s} \qquad (10.33)$$

Replacing x in Eq. 10.32 by u gives

$$\psi(z) = \frac{1}{\sqrt{2\pi}} \int_{0}^{z} \exp\left(-\frac{u^2}{2}\right) du \qquad (10.34)$$

in which $\psi(z)$ is the area under the standard normal distribution curve $f(u)$ between 0 and z, as indicated in Figure 10.3 and tabulated in Table 10.1. By using Eq. 10.34 and recognizing that the area under half of the standardized normal curve is $\frac{1}{2}$, the probability associated with the value of the random variable being less or greater than any specified value can be determined.

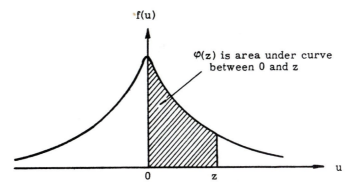

Figure 10.3 Area under normal curve.

TABLE 10.1 AREA ψ(z) UNDER NORMAL CURVE

z	.00	.01	.02	.03	.04	.05	.06	.07	.08	.09
0	0	.003969	.007978	.011966	.015953	.019939	.023922	.027903	.031881	.035856
.1	.039828	.043795	.047758	.051717	.055670	.059618	.063559	.067495	.071424	.075345
.2	.079260	.083166	.087064	.090954	.094835	.098706	.102568	.106420	.110251	.114092
.3	.117911	.121720	.125516	.129300	.133072	.136831	.140576	.144309	.148027	.151732
.4	.155422	.159097	.162757	.166402	.170031	.173645	.177242	.180822	.184386	.187933
.5	.191462	.194974	.198466	.201944	.205401	.208840	.212260	.215661	.219043	.222405
.6	.225747	.229069	.232371	.235653	.238914	.242154	.245373	.248571	.251748	.254903
.7	.258036	.261148	.264238	.267305	.270350	.273373	.276373	.279350	.282305	.285236
.8	.288145	.291030	.293892	.296731	.299546	.302337	.305105	.307850	.310570	.313267
.9	.315940	.318589	.321214	.323814	.326391	.328944	.331472	.333977	.336457	.338913
1.0	.341345	.343752	.346136	.348495	.350830	.353141	.355428	.357690	.359929	.362143
1.1	.364334	.366500	.368643	.370762	.372857	.374928	.376976	.379000	.381000	.382977
1.2	.384930	.386861	.388768	.390651	.392512	.393350	.396165	.397958	.399727	.401475
1.3	.403200	.404902	.406582	.408241	.409877	.411492	.413085	.414657	.416207	.417736
1.4	.419243	.420730	.422196	.423641	.425066	.426471	.427855	.429219	.430563	.431888
1.5	.433193	.434476	.435745	.436992	.438220	.439429	.440620	.441792	.442947	.444083
1.6	.445201	.446301	.447384	.448449	.449497	.450529	.451543	.452540	.453521	.454486
1.7	.455435	.456367	.457284	.458185	.459070	.459941	.460796	.461636	.462462	.463273
1.8	.464070	.464852	.465620	.466375	.467116	.467843	.468557	.469258	.469946	.470621
1.9	.471283	.471933	.472571	.473197	.473610	.474412	.475002	.475581	.476148	.476705

	.00	.01	.02	.03	.04	.05	.06	.07	.08	.09
2.0	.477250	.477784	.478308	.478822	.479325	.479818	.480301	.480774	.481237	.481691
2.1	.482136	.482571	.482997	.483414	.483823	.484222	.484614	.484997	.485371	.485738
2.2	.486097	.486447	.486791	.487126	.487455	.487776	.488089	.488396	.488696	.488989
2.3	.489276	.489556	.489830	.490097	.490358	.490613	.490863	.491106	.491344	.491576
2.4	.491802	.492024	.492240	.492451	.492656	.492857	.493053	.493244	.493431	.493613
2.5	.493790	.493963	.494132	.494297	.494457	.494614	.494766	.494915	.495060	.495201
2.6	.495339	.495473	.495604	.495731	.495855	.495975	.496093	.496207	.496319	.496427
2.7	.496533	.496636	.496736	.496833	.496928	.497020	.497110	.497197	.497282	.497365
2.8	.497445	.497523	.497599	.497673	.497744	.497814	.497882	.497948	.498012	.498074
2.9	.498134	.498193	.498250	.498305	.498359	.498411	.498462	.498511	.498559	.498605
3.0	.498650	.498694	.498736	.498777	.498817	.498856	.498893	.498930	.498965	.498999
3.1	.499032	.499065	.499096	.499126	.499155	.499184	.499211	.499238	.499264	.499289
3.2	.499313	.499336	.499359	.499381	.499402	.499423	.499443	.499462	.499481	.499499
3.3	.499517	.499534	.499550	.499566	.499581	.499596	.499610	.499624	.499638	.499651
3.4	.499663	.499675	.499687	.499698	.499709	.499720	.499730	.499740	.499749	.499758
3.5	.499767	.499776	.499784	.499792	.499800	.499807	.499815	.499822	.499828	.499835
3.6	.499841	.499847	.499853	.499858	.499864	.499869	.499874	.499879	.499883	.499888
3.7	.499892	.499896	.499900	.400904	.499908	.499912	.499915	.499918	.499922	.499925
3.8	.499928	.499931	.499933	.499936	.499938	.499941	.499943	.499946	.499948	.499950
3.9	.499952	.499954	.499956	.499958	.499959	.499961	.499963	.499964	.499966	.499967

Example 10.9

Given a normally distributed random variable x with a mean of 135 and a variance of 125, determine the probability that x is greater than 150.

Solution: From Eq. 10.33, $z = (150 - 135)/\sqrt{125} = 1.34$. From Table 10.1, $\psi(z) = 0.41$, so there is a 0.09 chance, or 9% probability, that x is greater than 150.

Log Normal Distribution

When a variable extends over several orders of magnitude, such as traffic volume, damage ratio, grain size, or permeability, it is necessary to present it in log scale and assume its distribution as log normal, so Eq. 10.33 becomes

$$u = \frac{\log x - \log \mu}{s_{\log x}} \tag{10.35}$$

in which $s_{\log x}$ is the standard deviation based on log normal distribution of x. From Eq. 10.30

$$V[\log x] = \left(\frac{\log e}{x}\right)^2 V[x] = \frac{0.1886}{x^2} V[x] \tag{10.36}$$

$$s_{\log x} = \frac{0.4343}{x} s = 0.4343 C[x] \tag{10.37}$$

Equation 10.37 can be used to relate the standard deviation based on the normal distribution of x to that based on the log normal distribution of x.

Example 10.10

If the distribution of x in Example 10.9 is log normal, determine the probability that x is greater than 150.

Solution: From Eq. 10.37, $s_{\log x} = (0.4343/135) \sqrt{125} = 0.036$. From Eq. 10.35, $u = (\log 150 - \log 135)/0.036 = 1.27$. From Table 10.1, $\psi(z) = 0.398$; so there is a 0.102 chance, or 10.2% probability, that x is greater than 150.

10.2 PROBABILISTIC METHODS

There are two methods of pavement design: deterministic and probabilistic. In the deterministic method, each design factor has a fixed value based on the factor of safety assigned by the designer. Using judgment, the designer usually assigns a higher factor of safety to those factors that are less certain or that have a greater effect on the final design. Application of this traditional approach based on the factors of safety can result in over design or under design, depending on the magnitudes of the safety factors applied and the sensitivity of the design procedures. A more realistic approach is the probabilistic method in which each design factor is assigned a mean and a variance. The factor of safety assigned to each design factor and its sensitivity to the final design are automatically taken care of and the reliability of the design can be evaluated. Reliability is defined as the probability that the design will perform its intended function over its design life.

Traffic, or the number of load repetitions, is one of the most important factors in pavement design. There are two types of load repetitions: the predicted number of load repetitions n and the allowable number of load repetitions N. In the deterministic method, both n and N have fixed values, whereas in the probabilistic method, each has a mean and a variance. If the design is based on a single value of equivalent single-axle load (ESAL), n and N can be compared directly to evaluate the adequacy of the design. If the design is based on a variety of axle loads or on ESAL applications over two or more seasons, the concept of damage ratio, as indicated by Eq. 3.19, may be used.

The use of load repetitions as a failure criterion is only one of the general methods to evaluate the reliability of a design. Other failure criteria can also be used. For example, VESYS (FHWA, 1978) employed the serviceability index as a design criterion in which traffic is treated as one of the many variables that affect the serviceability.

10.2.1 Method Based on Equivalent Single-Axle Load

For convenience, the determination of the predicted number of load repetitions and its variance is called traffic prediction, while the determination of the allowable number of load repetitions and its variance is called performance prediction.

Traffic Prediction

The predicted ESAL during the design period is designated W_T and can be determined from Eq. 6.28 and rewritten as

$$W_T = \left(\sum_{i=1}^{m} p_i F_i \right) (\text{ADT}_0)(T)(A)(G)(D)(L)(365)(Y) \qquad (10.38)$$

in which p_i is the percentage (decimal) of axles in the ith load group, F_i is the equivalent axle load factor (EALF) for the ith load group, m is the number of load groups, ADT_0 is the average daily traffic at the start of the design period, T is the percentage (decimal) of trucks in ADT, A is the average number of axles per truck, D is the percentage (decimal) of ADT in design direction, L is the percentage (decimal) of ADT in design lane, Y is the design period in years, and G is the growth factor, which can be computed by Eq. 6.31 and repeated as

$$G = \tfrac{1}{2}[1 + (1 + r)^Y] \qquad (6.31)$$

in which r is the growth rate (decimal) per year. When log is applied to both sides, Eq. 10.38 becomes

$$\log W_T = \log (\Sigma p_i F_i) + \log (\text{ADT}_0) + \log G + \log T$$
$$+ \log A + \log D + \log L + \log Y + 2.562 \qquad (10.39)$$

From Eq. 10.30, the variance of log W_T can be determined by

$$V[\log W_T] = (\log e)^2 \left\{ \frac{V[\Sigma p_i F_i]}{(\Sigma p_i F_i)^2} + \frac{V[\text{ADT}_0]}{(\text{ADT}_0)^2} + \frac{V[G]}{G^2} \right.$$

$$\left. + \frac{V[T]}{T^2} + \frac{V[A]}{A^2} + \frac{V[D]}{D^2} + \frac{V[L]}{L^2} \right\} \qquad (10.40)$$

Example 10.11

The truck traffic on a flexible pavement is composed of 55% 20-kip (90-kN) single-axle loads and 45% 36-kip (160-kN) tandem-axle loads. The average daily traffic at the start of the period is 5000 per day with a coefficient of variation of 15%. The annual traffic growth rate is 6%, the percentage of trucks is 20%, and the number of axles per truck is 2.5, each having a coefficient of variation of 10%. The directional distribution factor is 50% and the lane distribution factor is 100%, both being considered deterministic with no variations. If the coefficient of variation of $p_i F_i$, which is the product of axle load percentage and its equivalent factor, is 35% and the Asphalt Institute's equivalent factors are used, determine the mean and the variance of log W_T for a design period of 20 years.

Solution: The given data and the computed variances are shown in Table 10.2.

From Table 6.4, the equivalent factors for 20-kip (90-kN) single-axle and 36-kip (160-kN) tandem-axle loads are 1.51 and 1.38, respectively, so $p_i F_i = 0.55 \times 1.51 + 0.45 \times 1.38 = 1.452$ and $V[p_i F_i] = (0.35 \times 1.452)^2 = 0.258$. From Eq. 6.31, $G = 0.5[1 + (1 + 0.06)^{20}] = 2.104$, and $V[G] = \{0.5Y(1 + r)^{Y-1}\}^2 V[r] = \{0.5 \times 20 \times (1.06)^{19}\}^2 \times 0.000036 = 0.033$. From Eq. 10.39, log $W_T = \log 1.452 + \log 5000 + \log 2.104 + \log 0.2 + \log 2.5 + \log 0.5 + \log 1 + \log 20 + 2.562 = 7.445$. From Eq. 10.40, $V[\log W_T] = 0.1886[0.258/(1.452)^2 + 562,500/(5000)^2 + 0.033/(2.104)^2 + 0.0004/(0.2)^2 + 0.0625/(2.5)^2] = 0.1886(0.122 + 0.023 + 0.007 + 0.010 + 0.010) = 0.033$. It can be seen that the most important factor causing the variance is $p_i F_i$.

Performance Prediction

The allowable ESAL during the design period is designated W_t and can be determined from design equations or computer programs. The AASHTO design equation for flexible pavement can be used as an example. Because of the empirical nature of this equation, only the English units will be shown.

$$\log W_t = 9.36 \log (\text{SN} + 1) - 0.2 + \frac{\log [(p_0 - p_t)/(4.2 - 1.5)]}{0.4 + 1094/(\text{SN} + 1)^{5.19}}$$

$$+ 2.32 \log M_R - 8.07 \qquad (10.41)$$

in which p_0 is the initial serviceability index, p_t is the terminal serviceability index, M_R is the resilient modulus in psi, and SN is the structural number, defined as

$$\text{SN} = a_1 D_1 + a_2 D_2 m_2 + a_3 D_3 m_3 \qquad (10.42)$$

in which a_i is the ith layer coefficient, D_i is the ith layer thickness in inches, and m_i is the ith layer drainage coefficient. More about AASHTO design equations is

TABLE 10.2 MEAN, COEFFICIENT OF VARIATION, AND VARIANCE OF TRAFFIC DATA

Design factor	ADT_0 (vehicles per day)	r (%)	T (%)	A (axles per truck)	D (%)	L (%)
Mean	5000	6	20	2.5	50	100
CV	15	10	10	10	0	0
Variance	562,500	0.000036	0.0004	0.0625	0	0

presented in Section 12.3.1.

$$V[\log W_t] = \left(\frac{\partial \log W_t}{\partial SN}\right)^2 V[SN] + \left(\frac{\partial \log W_t}{\partial p_0}\right)^2 V[p_0] + \left(\frac{\partial \log W_t}{\partial M_R}\right)^2 V[M_R]$$

$$= \left\{\frac{9.36 \log e}{SN + 1} + \frac{\log [(p_0 - p_t)/(4.2 - 1.5)][(1094 \times 5.19)/(SN + 1)^{6.19}]}{[0.4 + 1094/(SN + 1)^{5.19}]^2}\right\}^2 V[SN]$$

$$+ \left\{\frac{\log e/(4.2 - 1.5)}{[0.4 + 1094/(SN + 1)^{5.19}][(p_0 - p_t)/(4.2 - 1.5)]}\right\}^2 V[p_0]$$

$$+ \left\{\frac{2.32 \log e}{M_R}\right\}^2 V[M_R] \tag{10.43}$$

$$V[SN] = D_1^2 V[a_1] + a_1^2 V[D_1] + (D_2m_2)^2 V[a_2] + (a_2m_2)^2 V[D_2]$$
$$+ (a_2D_2)^2 V[m_2] + (D_3m_3)^2 V[a_3] + (a_3m_3)^2 V[D_3]$$
$$+ (a_3D_3)^2 V[m_3] \tag{10.44}$$

Example 10.12

Given the mean and coefficient of variation of the initial serviceability index, layer coefficients, drainage coefficients, layer thicknesses, and subgrade resilient modulus, as shown in Table 10.3, determine the mean and variance of the performance prediction for a terminal serviceability of 2.0.

Solution: From Eq. 10.42, SN = 0.42 × 8 + 0.14 × 7 × 1.2 + 0.08 × 11 × 1.2 = 5.592. From Eq. 10.41, log W_t = 9.36 log 6.592 − 0.2 + log[(4.6 − 2.0)/2.7]/[0.4 + 1094/(6.592)^{5.19}] + 2.32 log 5700 − 8.07 = 8.074. From Eq. 10.44, $V[SN]$ = (8)² × (0.42 × 0.1)² + (0.42)² × (8 × 0.1)² + (7 × 1.2)² × (0.14 × 0.143)² + (0.14 ×

TABLE 10.3 MEAN AND COEFFICIENT OF VARIATION OF PAVEMENT VARIABLES

Design factor	p_0	a_1	D_1 (in.)	a_2	D_2 (in.)	m_2	a_3	D_3 (in.)	m_3	M_R (psi)
Mean	4.6	0.42	8.0	0.14	7.0	1.2	0.08	11.0	1.2	5700
CV	6.7	10.0	10.0	14.3	10.0	10.0	18.2	10.0	10.0	15.0

Note. 1 in. = 25.4 mm, 1 psi = 6.9 kPa.

$1.2)^2 \times (7 \times 0.1)^2 + (0.14 \times 7)^2 \times (1.2 \times 0.1)^2 + (11 \times 1.2)^2 \times (0.08 \times 0.182)^2 + (0.08 \times 1.2)^2 \times (11 \times 0.1)^2 + (0.08 \times 11)^2 \times (1.2 \times 0.1)^2 = 0.341$. From Eq. 10.43, $V[\log W_t] = \{9.36 \times 0.4343/6.592 + [(\log 0.963) \times 5677.8/(6.592)^{6.19}]/[0.4 + 1094/(6.592)^{5.19}]^2\}^2 \times 0.341 + \{(0.4343/2.7)/[0.4 + 1094/(6.592)^{5.19}]/0.963\}^2(4.6 \times 0.067)^2 + \{2.32 \times 0.4343/5700\}^2(5700 \times 0.15)^2 = 0.128 + 0.012 + 0.023 = 0.163$.

Reliability

When the mean and variance of log W_T and log W_t are known, the reliability of the design can be determined. Reliability is the probability that log W_T − log W_t < 0, or

$$\text{Reliability} = \text{Probability} (\log D_r = \log W_T - \log W_t < 0) \qquad (10.45)$$

in which $D_r = W_T/W_t$ is the damage ratio. The variance of log D_r can be obtained by

$$V[\log D_r] = V[\log W_T] + V[\log W_t] \qquad (10.46)$$

By assuming D_r to be a log normal distribution, the reliability of the design, as indicated by the shaded area in Figure 10.4, can be determined.

Example 10.13

The results of the traffic prediction are shown in Example 10.11 and those of the performance prediction are shown in Example 10.12. Determine the reliability of the design.

Solution: From Example 10.11, log W_T = 7.445 and $V[\log W_T]$ = 0.033. From Example 10.12, log W_t = 8.074 and $V[\log W_t]$ = 0.163. The mean of log D_r = 7.445 − 8.074 = −0.629 and the variance of log D_r = 0.033 + 0.163 = 0.196, so z = (0 + 0.629)/$\sqrt{0.196}$ = 1.42. From Table 10.1, the probability that log D_r is smaller than 0 is 0.5 + 0.422, or 92.2%.

The above procedure for determining reliability is highly simplified for easy understanding. A more rigorous procedure (Irick et al., 1987) defines reliability as

$$\text{Reliability} = \text{Probability} (\log N_T - \log N_t < 0) \qquad (10.47)$$

in which N_T is the actual traffic and N_t is the actual performance. Both W_T and W_t can be predicted but their actual values, N_T and N_t, can be known only after the

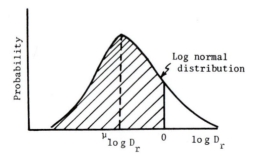

Figure 10.4 Reliability based on damage ratio.

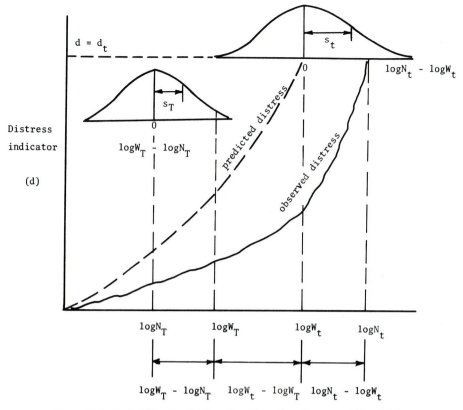

Figure 10.5 Probability distribution of traffic and performance. (After Irick et al. (1987).)

pavement has been constructed and its traffic and performance have been observed. This can be explained by Figure 10.5 in which the performance is evaluated by the distress indicator d. The predicted number of load repetitions to cause the distress indicator to reach the terminal value is W_t, while the actual number is N_t. If the performance prediction is unbias with positive errors for some projects and negative errors for others, the mean of $\log N_t - \log W_t$ should be zero. The same is true for traffic prediction with the mean of $\log W_T - \log N_T$ equal to zero. The standard deviation for performance prediction is s_T and that for the traffic prediction is s_t. Since

$$\log N_t - \log N_T = (\log W_T - \log N_T) + (\log W_t - \log W_T)$$
$$+ (\log N_t - \log W_t) \tag{10.48}$$

and the first and third terms on the right side have a mean of zero, Eq. 10.48 can be reduced to

$$\log N_T - \log N_t = \log W_T - \log W_t \tag{10.49}$$

Equation 10.49 implies that the mean of $\log N_T - \log N_t$ is the same as the mean of $\log W_T - \log W_t$. Because $\log W_t - \log W_T$ is a reliability interval selected by the designer and should be considered as a constant, the variance of $\log N_t - \log N_T$ is

$$s^2 = V[\log N_t - \log N_T] = V[\log W_T - \log N_T]$$

$$+ V(\log N_t - \log W_t) = s_T^2 + s_t^2 \qquad (10.50)$$

which is identical to Eq. 10.46. Therefore, the reliability determined by Eqs. 10.45 and 10.46 is the same as that determined by Eqs. 10.47 and 10.50.

10.2.2 Method Based on Different Axle Load Groups

If the design is based on a number of load groups, the damage ratio, indicated by Eq. 3.19, may be used. In the design of concrete pavements by fatigue principle, a cracking index c, similar to the damage ratio, is usually used and defined as

$$c = \sum_{i=1}^{m} \frac{n_i}{N_i} \qquad (10.51)$$

in which m is the number of axle load groups, n_i is the predicted number of stress repetitions for load i during the design life, and N_i is the allowable number of stress repetitions for load i. The design is considered satisfactory if the cracking index is less than or equal to 1.

Traffic Prediction

The procedure for predicting the mean and variance of n_i is similar to the method based on ESAL except that each load group is considered separately. If $(n_0)_i$ is the number of the ith-axle load group per day at the start of the design period, the predicted number of repetitions n_i for load group i during the design period is

$$n_i = 365(n_0)_i (G)(D)(L)(Y) \qquad (10.52)$$

Note that the growth factor G, the percentage of axles in the design direction D, and the percentage of axles on the design lane L may be different for each load group.

$$V[n_i] = (365 \times Y)^2 \{(GDL)^2 V[(n_0)_i] + [(n_0)_i DL]^2 V[G]$$

$$+ [(n_0)_i GL]^2 V[D] + [(n_0)_i GD]^2 V[L]\} \qquad (10.53)$$

Example 10.14

A rigid pavement is subjected to a combination of single-, tandem-, and tridem-axle loads. The means, coefficients of variation, and variances of the initial repetitions of each axle type, growth factor, directional distribution factor, and lane distribution factor are shown in Table 10.4. Determine the mean and variance of the number of repetitions of each axle type for a period of 20 years.

Solution: For single-axle loads, from Eq. 10.52, $n_1 = 365 \times 500 \times 2 \times 0.5 \times 1 \times 20 = 3.65 \times 10^6$; from Eq. 10.53, $V[n_1] = (365 \times 20)^2[(2 \times 0.5)^2 \times 22{,}500 + (500 \times 0.5)^2 \times 0.16] = 1.73 \times 10^{12}$. For tandem-axle loads, $n_2 = 365 \times 100 \times 2 \times 0.5 \times 1 \times 20 = 7.3 \times 10^5$ and $V[n_2] = (365 \times 20)^2 [(2 \times 0.5)^2 \times 6400 + (100 \times 0.5)^2 \times$

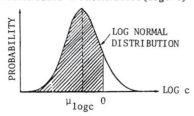

RELIABILITY=1-PROBABILITY(logc>0)

PROBABILITY

LOG NORMAL
DISTRIBUTION

LOG c

$\mu_{\log c}$ 0

Figure 10.6 Reliability of cracking index.

The standard deviation of log c is

$$s_{\log c} = \frac{0.4343}{c} \sqrt{V[c]} \qquad (10.59)$$

When the mean and the standard deviation of log c are known, the reliability of the design, as indicated by the shaded area in Figure 10.6, can be defined as

$$\text{Reliability} = 1 - \text{Probability} (\log c \geq 0) \qquad (10.60)$$

Example 10.16

The results of traffic prediction are shown in Example 10.14 and those of performance prediction in Example 10.15. Determine the reliability of the design.

Solution: The means and variances of predicted and allowable repetitions for the three different axle loads are summarized in Table 10.6.

From Eq. 10.51, $c = 3.65 \times 10^6/(4.67 \times 10^8) + 7.30 \times 10^5/(7.15 \times 10^7) + 2.92 \times 10^5/(2.03 \times 10^7) = 0.0324$. From Eq. 10.57, $V[c] = 1.73 \times 10^{12}/(4.67 \times 10^8)^2 + [3.65 \times 10^6/(4.67 \times 10^8)^2]^2 \times 9.02 \times 10^{18} + 3.62 \times 10^{11}/(7.15 \times 10^7)^2 + [7.30 \times 10^5/(7.15 \times 10^7)^2]^2 \times 2.58 \times 10^{17} + 8.87 \times 10^{10}/(2.03 \times 10^7)^2 + [2.92 \times 10^5/(2.03 \times 10^7)^2]^2 \times 3.07 \times 10^{16} = 0.0235$. From Eq. 10.59, $s_{\log c} = 0.4343\sqrt{0.0235}/0.0324 = 2.055$. With log $c = \log 0.0324 = -1.489$, $z = (0 + 1.489)/2.055 = 0.725$, from Table 10.1, reliability = probability (log $c < 0$) = 0.5 + 0.261 = 0.761, or 76.1%.

10.3 VARIABILITY

To determine the reliability of a design, the variances due to various sources must be known or estimated. This information can be obtained from field observations or laboratory tests or from past experience on similar projects.

TABLE 10.6 MEAN AND VARIANCE OF PREDICTED AND ALLOWABLE REPETITIONS

Axle load	Single	Tandem	Tridem
n_i	3.65×10^6	7.30×10^5	2.92×10^5
$V[n_i]$	1.73×10^{12}	3.62×10^{11}	8.87×10^{10}
N_i	4.67×10^8	7.15×10^7	2.03×10^7
$V[N_i]$	9.02×10^{18}	2.58×10^{17}	3.07×10^{16}

10.3.1 Frequency Distribution

The frequency distribution of some design factors and their coefficients of variation are presented below.

Stiffness Coefficient of the Subgrade

The stiffness coefficient of the subgrade, based on the surface curvature index measured by the Dynaflect, is an indication of the subgrade support. The in situ stiffness coefficients, which vary from about 0.2 for a weak clay to 1.0 for HMA are calculated by means of a computer program. The theory and development of the stiffness coefficient are similar to those of the elastic modulus, with simplifying assumptions, as described by Scrivner and Moore (1968).

Figure 10.7 is a histogram showing typical distribution of the stiffness coefficient of subgrade within a project. The normal distribution curve fit to the plot is shown by the dotted line. The mean μ, standard deviation s, and the number of measurements n are also shown in the figure. The average coefficient of variation within a project, based on a large number of projects, is about 10% (Darter, 1976).

Stiffness Coefficient of Pavement

Figure 10.8 shows a typical frequency distribution of stiffness coefficient for a pavement consisting of hot mix asphalt and cement-treated base. The stiffness coefficient of the pavement is much larger than the subgrade with a greater coefficient of variation.

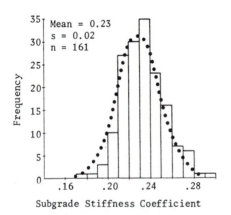

Figure 10.7 Frequency distribution of subgrade stiffness coefficient. (After Darter et al. (1973a).)

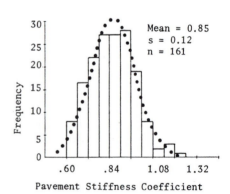

Figure 10.8 Frequency distribution of pavement stiffness coefficient. (After Darter et al. (1973a).)

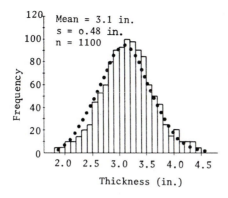

Figure 10.9 Frequency distribution of asphalt surfacing thickness. (1 in. = 25.4 mm). (After Darter et al. (1973a).)

Thickness of Asphalt Surface

Figure 10.9 shows a typical frequency distribution of asphalt surfacing thickness determined by individual cores. The coefficient of variation is not a good representation of variance because it increases with the decrease in thickness. Sherman (1971) presented results from California showing thickness variations in various pavement layers from 1962 to 1969. Based on over 8000 tests for each material, the standard deviations of thickness for four different paving materials are presented in Table 10.7. Using corresponding thicknesses of these layers, an average coefficient of variation of about 10% was obtained.

Thickness of Concrete

Figure 10.10 shows a typical frequency distribution of concrete slab thickness based on the measurement of individual cores. From the thicknesses obtained on 27 projects in four states, the standard deviations for three nominal pavement thicknesses are shown in Table 10.8. As can be seen, the standard deviation is not proportional to, or even dependent on, the average thickness of the pavement.

TABLE 10.7 STANDARD DEVIATIONS OF LAYER THICKNESS FOR FLEXIBLE PAVEMENTS

Material	Standard deviation
Hot mix asphalt	0.41 in.
Cement-treated base	0.68 in.
Aggregate base	0.79 in.
Aggregate subbase	1.25 in.

Note. 1 in. = 25.4 mm.

Source. After Darter et al. (1973a).

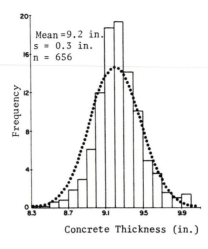

Mean = 9.2 in.
s = 0.3 in.
n = 656

Frequency (y-axis)
Concrete Thickness (in.) (x-axis)

Figure 10.10 Frequency distribution of concrete slab thickness. (1 in. = 25.4 mm). (After Kher and Darter (1973).)

Modulus of Rupture of Concrete

Figure 10.11 shows a typical frequency distribution of concrete modulus of rupture. The coefficient of variation for this particular project is 9.6%. The average coefficient of variation based on 15 projects is about 10.7% (Kher and Darter, 1973).

Initial Serviceability Index

Figure 10.12 shows a typical frequency distribution for the initial serviceability index of a newly constructed flexible pavement and Figure 10.13 shows that of the AASHTO Road Test rigid pavements. Note that rigid pavements have a higher initial serviceability index and a lower coefficient of variation compared with flexible pavements.

10.3.2 Coefficients of Variation

The use of a coefficient of variation is a convenient way to specify the variability of design factors. Experience has shown that the standard deviations of most random variables are proportional to their average values. Therefore, a fixed

TABLE 10.8 STANDARD DEVIATIONS OF CONCRETE THICKNESS FOR RIGID PAVEMENTS

Concrete thickness	Standard deviation	Number of projects
8 in.	0.32 in.	14
9 in.	0.29 in.	8
10 in.	0.29 in.	5

Note. 1 in. = 25.4 mm.

Source. After Kher and Darter (1973).

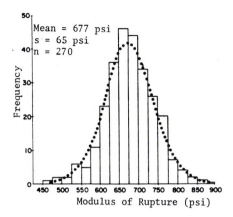

Figure 10.11 Frequency distribution of concrete modulus of rupture. (1 psi = 6.9 kPa). (After Kher and Darter (1973).)

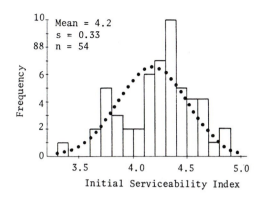

Figure 10.12 Initial serviceability index of flexible pavements (After Darter et al. (1973a).)

coefficient of variation can be applied to a design factor regardless of its magnitude and dimensional unit. Even if the standard deviation is a constant and independent of the mean, such as pavement thickness, a coefficient of variation can still be easily figured out by simply dividing the standard deviation by the mean. Estimates on the coefficients of variation for some of the traffic and design factors are presented below (AASHTO, 1985).

Traffic Prediction

Table 10.9 shows the coefficients of variation for design period traffic prediction. These coefficients fall in line with those originally suggested by Darter et al. (1973a). The overall coefficient of 42% is obtained by $[(35)^2 + (15)^2 + (10)^2 + (10)^2 + (10)^2]^{0.5}$.

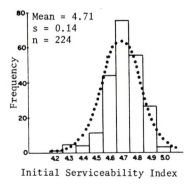

Initial Serviceability Index

Figure 10.13 Initial serviceability index of concrete pavements. (After Kher and Darter (1973).)

Description	Symbol	Coefficient of variation (%)
Summation of EALF over % axle distribution	$\Sigma\, p_i F_i$	35
Initial average daily traffic	ADT_0	15
Traffic growth factor	G	10
Percentage of trucks	T	10
Average no. of axles per truck	A	10
Overall traffic prediction		42

Source. After AASHTO (1985).

Performance Prediction

Table 10.10 shows the coefficients of variation for performance prediction of flexible pavements and Table 10.11 shows those of rigid pavements. The strength factor of each flexible layer is an indicator of its resilient modulus. The coefficients of variation of the resilient modulus vary from 10.0% for the asphalt surface layer to 18.2% for the subbase layer. If the concrete slab is placed on a solid or a layer foundation, a coefficient of variation of 15% may be assumed for the foundation. If the slab is placed on a liquid foundation, a coefficient of variation of 35% may be assumed for the modulus of subgrade reaction k.

10.3.3 Variances

All the variances discussed so far are caused by the variation of traffic and design factors. Other variances also have to be considered. The source and distribution of these variances are presented below.

TABLE 10.10 COEFFICIENTS OF VARIATION FOR
PERFORMANCE PREDICTION OF FLEXIBLE PAVEMENTS

Description	Symbol	Coefficient of variation (%)
Initial serviceability index	p_0	6.7
Surface strength factor	a_1	10.0
Surface thickness	D_1	10.0
Base strength factor	a_2	14.3
Base drainage factor	m_2	10.0
Base thickness	D_2	10.0
Subbase strength factor	a_3	18.2
Subbase drainage factor	m_3	10.0
Subbase thickness	D_3	10.0
Subgrade resilient modulus	M_R	15.0

Source. After AASHTO (1985).

TABLE 10.11 COEFFICIENTS OF VARIATION FOR PERFORMANCE PREDICTION OF RIGID PAVEMENTS

Description	Symbol	Coefficient of variation (%)
Initial serviceability index	p_0	6.7
Slab thickness	D	4.0
Slab elastic modulus	E_c	10.0
Load transfer factor	J	5.0
Drainage factor	C_d	10.0
Modulus of subgrade reaction	k	35.0
Concrete modulus of rupture	S_c	10.0

Source. After AASHTO (1985).

Sources

There are three sources of variance:

1. *Variances due to traffic or design factors.* Sampling errors in all traffic factors or construction errors in all design factors will increase the chance deviation between design and performance.

2. *Unexplained variance.* Many variables other than traffic and design factors give rise to chance deviations. For example, the allowable number of repetitions in concrete pavement design is governed by the stress ratio σ/S_c, according to Eq. 5.35. Even though the variation of σ/S_c caused by all the relevant design factors has been properly considered, the constants in the fatigue equations are actually not deterministic and may exhibit a range of variations. The negligence of these variations will increase the chance deviation between design and performance.

3. *Variance due to inadequacy of design procedure or lack of fit of design equation.* The application of a design procedure or equation always involves some degree of uncertainty due to the use of simplifying assumptions or the negligence of certain design factors.

Distribution

Table 10.12 shows the percentage distribution of overall variance when the AASHTO equations are used for pavement design. It can be seen that the variance due to performance prediction is much greater than that due to traffic prediction. The design equation for rigid pavements has a much smaller variance compared to that for flexible pavements. The large lack of fit variance in the performance prediction is due to the empirical nature of the AASHTO equations. This variance can be greatly reduced if mechanistic–empirical methods are used.

10.4 ROSENBLUETH METHOD

Taylor series expansion has been used for determining the mean and variance of a function. This method requires the knowledge of each term in the function and the

TABLE 10.12 PERCENTAGE DISTRIBUTION OF OVERALL VARIANCE

Type of prediction	Source of variance		Flexible pavement		Rigid pavement
Traffic prediction	Traffic factor		14%		22%
	Unexplained		3%		4%
	Lack of fit		1%		1%
	Total variance	0.0429	18%	0.0429	27%
Performance prediction	Design factor		45%		42%
	Unexplained		5%		8%
	Lack of fit		32%		23%
	Total variance	0.1938	82%	0.1128	73%
Overall variance		0.2367	100%	0.1557	100%

Source. After AASHTO (1985).

existence and continuity of the first and second derivatives. Rosenblueth (1975) developed a simple method similar to the finite different procedure that can be used directly to determine the mean and variance of any function without knowing its formulation.

10.4.1 Description of Method

In the design of concrete pavements by the probabilistic method, the most difficult part is to determine the maximum stress σ and its variance. From Eq. 10.8, the variance can be determined by

$$V[\sigma] = E[\sigma^2] - (E[\sigma])^2 \qquad (10.61)$$

The expectations of σ and σ^2 can be determined by

$$E[\sigma^m] = \frac{1}{2}(\sigma_+^m - \sigma_-^m) \qquad \text{for one variable} \qquad (10.62a)$$

$$E[\sigma^m] = \frac{1}{2^2}(\sigma_{++}^m + \sigma_{+-}^m + \sigma_{-+}^m + \sigma_{--}^m) \qquad \text{for two variables} \qquad (10.62b)$$

$$E[\sigma^m] = \frac{1}{2^3}(\sigma_{+++}^m + \sigma_{++-}^m + \sigma_{+-+}^m + \sigma_{+--}^m + \sigma_{-++}^m + \sigma_{-+-}^m$$
$$+ \sigma_{--+}^m + \sigma_{---}^m) \qquad \text{for three variables} \qquad (10.62c)$$

$$E[\sigma^m] = \frac{1}{2^4}(\sigma_{++++}^m + \sigma_{+++-}^m + \sigma_{++-+}^m + \sigma_{++--}^m + \sigma_{+-++}^m + \sigma_{+-+-}^m$$
$$+ \sigma_{+--+}^m + \sigma_{+---}^m + \sigma_{-+++}^m + \sigma_{-++-}^m + \sigma_{-+-+}^m + \sigma_{-+--}^m$$
$$+ \sigma_{--++}^m + \sigma_{--+-}^m + \sigma_{---+}^m + \sigma_{----}^m) \qquad \text{for four variables} \quad (10.62d)$$

The equation can be extended to any number of variables n by simply adding a plus sign and then a minus sign to the previous equation with a total of 2^n terms. Note that m is either 1 or 2, corresponding to $E[\sigma]$ and $E[\sigma^2]$.

If there are four random variables, it is necessary to determine the stress 16 times and then compute $E[\sigma]$ and $E[\sigma^2]$ by Eq. 10.62 and $V[\sigma]$ by Eq. 10.61.

10.4.2 Comparison with Taylor's Expansion

A comparison of Taylor's expansion and the Rosenblueth method can be best illustrated by a simple closed-form equation used previously by the U.S. Corps of Engineers (Turnbull and Ahlvin, 1957):

$$t = (0.231 \log n + 0.144) \sqrt{\frac{P}{8.1 \text{ CBR}} - \frac{A}{\pi}} \qquad (10.63)$$

in which t is the pavement thickness in inches, n is the number of passes or coverages, P is the wheel load in lb, CBR is the California bearing ratio of subgrade, and A is the tire contact area in in.^{2}. Equation 10.63 can be written as

$$\log n = \frac{4.329t}{\sqrt{P/(8.1 \text{ CBR})} - A/\pi} - 0.623 \qquad (10.64)$$

Given $t = 20$ in., $P = 30{,}000$ lb, CBR $= 4$, $A = 285$ in.^{2}, and that the coefficients of variation, $C[t]$, $C[P]$, $C[\text{CBR}]$, and $C[A]$, are all 0.1, determine the number of repetitions for 75% reliability. It is assumed that $\log n$ is normally distributed and that n is deterministic with no variations.

Taylor's Expansion

$$E[\log n] = \frac{4.329 \times 20}{\sqrt{30{,}000/(8.1 \times 4)} - 285/\pi} - 0.623 = 2.373$$

The variances of $\log n$ due to the variances of t, p, CBR, and A can be computed by the first-order Taylor's expansion:

$$V[\log n]_t = \left(\frac{\partial \log n}{\partial t}\right)^2 V[t] = \left[\frac{4.329}{\sqrt{P/(8.1 \text{ CBR})} - A/\pi}\right]^2 (t \times C[t])^2$$

$$= \frac{(4.329 \times 20 \times 0.1)^2}{30{,}000/(8.1 \times 4) - 285/\pi} = \frac{74.96}{835.21} = 0.0898$$

$$V[\log n]_p = \left\{-\frac{1}{2}\frac{4.329t}{[P/(8.1 \text{ CBR}) - A/\pi]^{1.5}}\frac{1}{8.1 \text{ CBR}}\right\}^2 (P \times C[P])^2 = 0.0276$$

$$V[\log n]_{\text{CBR}} = \left\{\frac{1}{2}\frac{4.329t}{[P/(8.1 \text{ CBR}) - A/\pi]^{1.5}}\frac{P}{8.1 \text{ (CBR)}^2}\right\}^2 (\text{CBR} \times C[\text{CBR}])^2 = 0.0276$$

$$V[\log n]_A = \left\{\frac{1}{2}\frac{4.329t}{[P/(8.1 \text{ CBR}) - A/\pi]^{1.5}}\frac{1}{\pi}\right\}^2 (A \times C[A])^2 = 0.0003$$

$$V[\log n] = 0.0898 + 0.0276 + 0.0276 + 0.0003 = 0.1453$$

$$s_{\log n} = \sqrt{0.1453} = 0.381$$

For 75% reliability, the area between the number of repetitions and the mean is 25%, as indicated in Figure 10.14. From Table 10.1, $z = -0.674$. From Eq. 10.35,

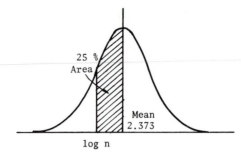

Figure 10.14 Allowable repetitions at 75% reliability.

$-0.674 = (\log n - 2.373)/0.381$, or $\log n = 2.373 - 0.674 \times 0.381 = 2.116$, so the allowable number of repetitions for 75% reliability is $n = 131$.

Rosenblueth's Procedure

This procedure does not take derivatives on Eq. 10.64 with respect to t, P, CBR, and A, but requires the evaluation of $\log n$ for a total of 16 times. It is convenient to evaluate $\log n$ at plus and minus one standard deviation, and the results are presented in Table 10.13. The sum of $\log n$ is 38.1461 and the sum of $(\log n)^2$ is 93.3041. From Eq. 10.62d, $E[\log n] = 38.1461/16 = 2.3841$, which checks with 2.373 derived by Taylor's expansion, and $E[(\log n)^2] = 93.3041/16 = 5.8315$. From Eq. 10.61, $V[\log n] = 5.8315 - (2.3841)^2 = 0.1476$, which checks with 0.1453 derived by Taylor's expansion. For 75% reliability, $\log n = 2.3841 - 0.674 \times \sqrt{0.1476} = 2.125$, or $n = 133$, which checks with 131 derived by Taylor's expansion.

TABLE 10.13 COMPUTATION OF LOG n AND (LOG $n)^2$ BY EQ. 10.64

Term	t (in.)	P (lb)	CBR	A (in.)	$\log n$	$(\log n)^2$
+ + + +	22	33,000	4.4	313.5	2.6905	7.2387
+ + + −	22	33,000	4.4	256.5	2.6547	7.0474
+ + − +	22	33,000	3.6	313.5	2.3418	5.4839
+ + − −	22	33,000	3.6	256.5	2.3161	5.3643
+ − + +	22	27,000	4.4	313.5	3.0904	9.5504
+ − + −	22	27,000	4.4	256.5	3.0402	9.2428
+ − − +	22	27,000	3.6	313.5	2.6905	7.2387
+ − − −	22	27,000	3.6	256.5	2.6547	7.0473
− + + +	18	33,000	4.4	313.5	2.0880	4.3599
− + + −	18	33,000	4.4	256.5	2.0587	4.2384
− + − +	18	33,000	3.6	313.5	1.8027	3.2498
− + − −	18	33,000	3.6	256.5	1.7817	3.1744
− − + +	18	27,000	4.4	313.5	2.4152	5.8332
− − + −	18	27,000	4.4	256.5	2.3742	5.6366
− − − +	18	27,000	3.6	313.5	2.0880	4.3599
− − − −	18	27,000	3.6	256.5	2.0587	4.2384
				Sum	38.1461	93.3041

Note. 1 in. = 25.4 mm and 1 lb = 4.45 N.

Reliability   Chap. 10

10.5 PROBABILISTIC COMPUTER PROGRAMS

Two probabilistic computer programs, one for flexible pavements and the other for rigid pavements, are described. Both programs are based on linear theory and cannot be used for nonlinear analysis.

10.5.1 VESYS Computer Program

VESYS program can be used to predict the structural responses and hence the integrity of flexible pavements. Similar to KENLAYER, the program is based on Burmister's layered theory. However, the program considers the variability of materials and traffic and evaluates the reliability of the design.

Program Development

The original concepts were developed at the Massachusetts Institute of Technology (Moavenzadeh et al., 1974). The work was refined by FHWA and presented at the 4th International Conference on the Structural Design of Asphalt pavements (Kenis, 1977). This widely publicized version was called VESYS-IIM. Major deficiencies of VESYS-IIM are that it can be applied only to a three-layer system subjected to a given single-axle load and the base and subgrade properties cannot be varied throughout the year. Many researchers contributed to the development and improvement of VESYS system. The various versions of VESYS are shown in Figure 10.15.

The Georgia Institute of Technology extended VESYS-IIM to an N-layer system. The modified code, known as VESYS-G (Lai, 1977), can handle up to seven layers and can be easily expanded to any number of layers. Another modification of VESYS-IIM was made by Austin Research Engineers (Rauhut and Jordahl, 1979). This modified version, known as VESYS-A, added capabilities for the seasonal characterization of pavement materials by layers, the prediction of low-temperature cracking, and a discrete representation of axle load distribution.

VESYS-III is a combination of VESYS-A and VESYS-G in which the three-layer system in VESYS-A is replaced with the N-layer system in VESYS-G. To develop zero-maintenance design procedures for flexible and composite pavements, Austin Research Engineers simplified VESYS-III and renamed this version VESYS-IIIA (Von Quintus et al., 1980). A major change was the input of a time-independent modulus for each layer by season, rather than the creep compliances at reference temperature and moisture conditions. The other modifications in VESYS-IIIA allowed for the computation of stress-dependent fatigue damage in a PCC base layer.

Under a contract with FHWA, the Austin Research Engineers combined VESYS-III and VESYS-IIIA to form VESYS-IV, which was a completely modular program that allowed various optional computations to be performed. It also included subroutines for the computation of permanent deformation in each of the N layers. A modified version of VESYS-IV is called VESYS-IVB and was developed by Brent Rauhut Engineering, Inc. (Jordahl and Rauhut, 1983). It is a

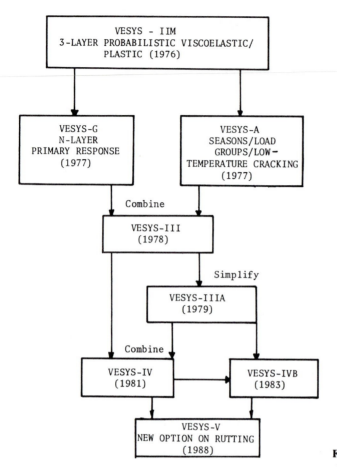

Figure 10.15 Evolution of VESYS.

combination of VESYS-IIIA and VESYS-IV and includes the following features not present in VESYS-IV:

1. Channelization of traffic, which permits the definition of rut depth as used in actual measurements.
2. Analysis involving tandem axles.
3. Allowance for seasons with unequal length.
4. Input of modulus and permanent deformation parameters by load group.
5. Internal calculation of fatigue constants K_1 and K_2 from seasonal temperatures or from seasonal moduli.

The latest modification of VESYS was made by MIT (Brademeyer, 1988) and is now called VESYS-V. Major features include tandem- and tridem-axle analysis, bilinear rutting curves, and "layer formation" for rutting. The layer formation approach is based on the strains in each layer, and the permanent deformation properties of each individual layer are used directly to calculate the rutting in each layer. The total rut depth for the pavement system is then obtained by integration. Thus, strains in the primary response model, rather than deflections, are needed as input to the rutting model.

Although the VESYS program has been under constant change, its major components remain the same as the original VESYS-IIM and are shown in Figure 10.16. The program is composed of four major interactive models: primary response, general response, damage, and performance.

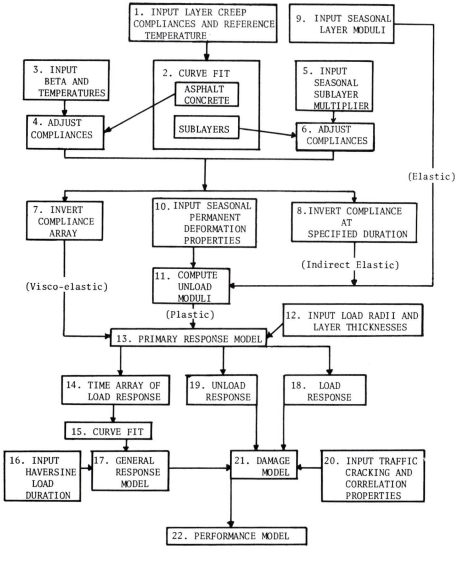

Figure 10.16   Major component of VESYS.

Primary Response Model

This model is used to compute stresses, strains, and displacements under static loading. Three methods are available to determine the responses: viscoelastic, indirect elastic, and elastic. These methods are presented in Chapters 2 and 3, and it will be helpful to review them and see how they are incorporated in VESYS. The viscoelastic method was the only procedure employed in VESYS-IIM and is indicated by steps 1 through 7 in Figure 10.16. The indirect elastic procedure is the same as the viscoelastic method except that step 7 is replaced by step 8. The elastic method needs only the input of seasonal layer moduli and their coefficients of variation, as indicated in step 9. Each step is explained as follows:

1. Input the creep compliances $D(t)$ for each layer, usually at 11 time durations, and the reference temperature T_0 for each asphalt layer. The coefficient of variation of creep compliances for each layer should also be specified.

2. Curve fit the creep compliances with a Dirichlet series, as shown by Eq. 2.38 and explained in Section 2.3.2.

3. Input the seasonal temperature T of each asphalt layer and the β coefficient of HMA.

4. Determine the creep compliances of each asphalt layer at temperature T by Eqs. 2.46 and 2.48.

5. Input the seasonal multiplier for each sublayer, including the untreated base, subbase, and subgrade. These multipliers account for the change in moisture content or freeze–thaw conditions of the material.

6. Multiply the creep compliances specified in step 1 for each sublayer by the seasonal multiplier to obtain the creep compliances for each period.

7. In the viscoelastic method, determine the elastic moduli by inverting the creep compliance at each time duration and obtain the primary responses under a static load at a number of time durations, usually 11.

8. In the direct elastic method, a single time duration commensurate with the vehicle speed is specified. The creep compliance is obtained by substituting this time duration into the curve-fitted Dirichlet series. The reciprocal of the creep compliance is the elastic modulus to be used for the analysis of primary responses.

9. In the elastic method, input the seasonal modulus of each layer and its coefficient of variation. This eliminates the time-fitting procedure and, in contrast to the viscoelastic method, determines the primary responses at one time only, instead of at 11 times.

10. Input seasonal permanent deformation parameters α and μ. Definitions of α and μ are discussed in Section 7.4.1.

11. Compute the unload modulus $E_r(N)$ for each layer under a given load repetition by Eq. 7.47.

12. Input the radius of loaded area for each group and the thickness of each layer.

13. The primary response model determines the means and variances of the

tensile strain at the bottom of asphalt layer and the vertical deflections on the surface and at each layer interface under a static load. The mean is determined from Eq. 10.25 and the variance from Eq. 10.30 by considering the layer moduli as uncorrelated random variables. The variability of material properties is represented by a random variable η with a mean of 1 and a variance $V[\eta]$, which is actually the square of the coefficient of variation of vertical displacement.

General Response Model

This model applies only to the viscoelastic case in which the tensile strains and deflections at 11 time durations are reduced to single values under a moving load. The steps shown in Figure 10.16 are discussed below:

14. The tensile strains and vertical deflections at 11 time durations are obtained from the primary response model together with the coefficients of variation at the first time.
15. Curve fit the strains and deflections with a seven-term Dirichlet series, as indicated by Eqs. 2.49 and 2.50.
16. Input the amplitude q and duration d of a haversine load, as indicated by Eq. 2.55, together with their coefficients of variation.
17. The general response model determines the response under moving load by Boltzmann's superposition principle, as discussed in Section 2.3.3. As shown by Eq. 2.59, the response can be expressed as

$$R = \frac{q\eta\pi^2}{2} \sum_{i=1}^{n} c_i \frac{1 + e^{-u_i}}{\pi^2 + u_i^2} \qquad (10.65)$$

in which $u_i = d/(2T_i)$ and η is a random variable with a mean of 1 and a variance $V[\eta]$ representing the variability of layer moduli. Assuming the random variables q, d, and η to be uncorrelated and applying Eqs. 10.25 and 10.30 to Eq. 10.65 yields

$$E[R] = \frac{q\pi^2}{2} \sum_{i=1}^{n} c_i \left(\frac{1 + e^{-u_i}}{\pi^2 + u_i^2}\right) + \frac{\pi^2 q}{4} V[d] \sum_{i=1}^{n} c_i \left(\frac{u_i}{d}\right)^2 \left[\frac{e^{-u_i}}{\pi^2 + u_i^2} \right.$$

$$\left. + \frac{4u_i e^{-u_i} - 2(1 + e^{-u_i})}{(\pi^2 + u_i^2)^2} + \frac{8u_i^2(1 + e^{-u_i})}{(\pi^2 + u_i^2)^3} \right] \qquad (10.66)$$

$$V[R] = \left[\frac{\pi^2}{2} \sum_{i=1}^{n} c_i \left(\frac{1 + e^{-u_i}}{\pi^2 + u_i^2}\right) \right]^2 \left[V[q] + q^2 \, V[\eta] \right]$$

$$+ \left\{ \frac{\pi^2}{2} \sum_{i=1}^{n} c_i \frac{u_i}{d} \left[\frac{-e^{-u_i}}{\pi^2 + u_i^2} - \frac{1 + e^{-u_i}}{(\pi^2 + u_i^2)^2} (2u_i) \right] \right\}^2 q^2 \, V[d] \qquad (10.67)$$

Example 10.17

The static response of surface deflection is characterized by the Dirichlet series

$$R = 0.0001q \left[1 - \exp\left(-\frac{t}{0.05}\right) \right]$$

The contact pressure q is 80 psi (552 kPa) with a coefficient of variation of 0.3, the duration of the moving load is 0.1 s with a coefficient of variation of 0.2, and $V[\eta]$ is 0.09. Compute the responses under a moving load using Eq. 10.65 for the first-order approximation of surface deflection, Eq. 10.66 for the second-order approximation, and Eq. 10.67 for the variance of surface deflection.

Solution: The Dirichlet series has two terms. The first term has $T_i = \infty$ or $u_i = 0$, and the second term has $T_i = 0.05$ or $u_i = 0.1/(2 \times 0.05) = 1$. Note that $1 + e^{-u_i} = 1 + e^{-1} = 1.368$ and $\pi^2 + u_i^2 = \pi^2 + (1)^2 = 10.87$. From Eq. 10.65, $R = 0.5 \times 80 \ \pi^2 \times 0.0001 \ (2/\pi^2 + 1.368/10.87) = 0.0130$ in. (0.33 mm).

From Eq. 10.66, $E[R] = 0.0130 + 0.25\pi^2 \times 80 \times (0.1 \times 0.2)^2 \times 0.0001 \times (1/0.1)^2[e^{-1}/10.87 + (4 \times 1 \times e^{-1} - 2 \times 1.368)/(10.87)^2 + 8 \times (1)^2 \times 1.368/(10.87)^3] = 0.0130 + 0.0000249 = 0.0130$ in. (0.33 mm). For this particular case, Eq. 10.65 based on the first-order approximation yields practically the same result as Eq. 10.66 based on the second-order approximation.

From Eq. 10.67, $V[R] = (0.5\pi^2 \times 0.0001)^2\{(2/\pi^2 + 1.368/10.87)^2[(80 \times 0.3)^2 + (80)^2 \times 0.090] + (1/0.1)^2[(-e^{-1})/10.87 - 1.368 \times 2 \times 1/(10.87)^2]^2 \times (80)^2 \times (0.1 \times 0.2)^2\} = 2.435 \times 10^{-7} \{0.108 \times 1152 + 0.832\} = 3.05 \times 10^{-5}$ in.^{2} (1.97 $\times$ 10^{-2} mm^2).

Damage Model

The damage model includes steps 18 through 21 in Figure 10.16. It consists of three submodels: cracking, rut depth, and roughness.

Cracking Model

The cracking model includes both fatigue and low-temperature cracking and is shown in Figure 10.17. Each step is explained below:

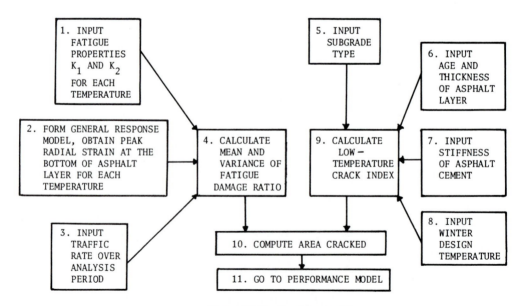

Figure 10.17 Cracking model.

1. Input the fatigue constants K_1 and K_2 for each temperature. The allowable number of repetitions N_f is determined by

$$N_f = K_1 \left(\frac{1}{\epsilon_t}\right)^{K_2} \tag{10.68}$$

2. From the general response model, obtain the peak radial strain ϵ_t at the bottom of the asphalt layer for each temperature. In the elastic and indirect elastic cases, the radial strain is obtained from the primary response model.
3. Input the predicted number of load repetitions n for each period.
4. The mean and variance of fatigue damage ratio can be determined by the following equations. From Eq. 10.68,

$$\frac{1}{N_f} = \frac{(\epsilon_t)^{K_2}}{K_1} \tag{10.69}$$

By assuming that ϵ_t is uncorrelated with K_1 and K_2 but that K_1 and K_2 are correlated, the second-order Taylor series approximation to the expected value of the function $1/N_f$ can be obtained from Eq. 10.24:

$$E\left[\frac{1}{N_f}\right] = \frac{(\epsilon_t)^{K_2}}{K_1} \left\{ 1 + \frac{V[K_1]}{(K_1)^2} + 0.5 (\ln \epsilon_t)^2 V[K_2] - \frac{\ln \epsilon_t}{K_1} \text{Cov}[K_1, K_2] \right.$$

$$\left. + \frac{K_2 (K_2 - 1)}{2(\epsilon_t)^2} V[\epsilon_t] \right\} \tag{10.70}$$

The first-order approximation to the variance of $1/N_f$ can be computed by Eq. 10.29:

$$V\left[\frac{1}{N_f}\right] = (\epsilon_t)^{2K_2} \left\{ \frac{V[K_1]}{(K_1)^4} + \frac{(\ln \epsilon_t)^2 V[K_2]}{(K_1)^2} - \frac{2 \ln \epsilon_t \text{ Cov}[K_1, K_2]}{(K_1)^3} \right\}$$

$$+ \frac{(K_2)^2}{(K_1)^2} (\epsilon_t)^{2K_2-2} V[\epsilon_t] \tag{10.71}$$

It is assumed that the predicted number of load repetitions n follows a Poisson arrival process in which the mean is equal to the variance, or $V[n] = E[n]$, and that the predicted number of repetitions and the allowable number of repetitions N_f are uncorrelated random variables. For cases with variance equal to the mean, the coefficient of variation is large when the mean is very small but becomes negligibly small when the mean is large. The damage ratio is defined as n/N_f so, based on Eqs. 10.25 and 10.30, the mean and variance of damage ratio can be determined by

$$E[c] = E[n] E\left[\frac{1}{N_f}\right] \tag{10.72}$$

$$V[c] = (E[n])^2 V\left[\frac{1}{N_f}\right] + \left(E\left[\frac{1}{N_f}\right]\right)^2 E[n] \tag{10.73}$$

5. The subgrade type is specified in terms of a dimensionless code with 5 for sand, 3 for loam, and 2 for clay.
6. The age is specified in terms of years and the thickness of asphalt layer is in inches.

7. The stiffness of asphalt is specified in terms of kg/cm^2.
8. The winter design temperature is specified in terms of °C.
9. The parameters specified in steps 5 to 8 are used for the analysis of low-temperature cracking. The crack index I, defined as the number of full transverse cracks per 500-ft section of two-lane road, is computed from Eq. 9.52.
10. The fatigue damage ratio and the low-temperature crack index are converted into area cracked per 1000 ft^2. By assuming the damage ratio as a normal distribution with a mean indicated by Eq. 10.72 and a variance by Eq. 10.73, the area cracked per 1000 ft^2, A, can be computed by

$$A = 1000[1 - \text{probability} (c \leq 1)] \tag{10.74}$$

Note that pavements crack when $c > 1$, so the expression within the brackets is the percentage of area that cracks. By assuming each foot of crack is equivalent to 3 ft^2 of area, the area of crack in ft^2 per 1000 ft^2 due to low-temperature cracking is equal to six times the crack index, or $6I$. Note that the assumption of c as a normal distribution is different from many other analyses, which assume c as a log normal distribution. The use of log normal distribution appears to be more reasonable because, similar to traffic, the values of c extend over several orders of magnitude and should be expressed in logarithmic scales.

11. The sum of the area cracked due to both fatigue and low-temperature cracking and their variance are transferred to the performance model.

Example 10.18

Given the fatigue coefficients K_1 of 6.57×10^{-5} with a coefficient of variation of 0.3 and K_2 of 2.69 with a coefficient of variation of 0.04, a correlation coefficient between K_1 and K_2 of -0.867, tensile strain $\epsilon_t = 3.54 \times 10^{-4}$ with a coefficient of variation of 0.175, and the predicted number of load repetitions n of 50,000 with a Poisson arrival process, determine the percentage of area cracked based on the assumption that (a) the fatigue damage ratio is normally distributed, and (b) the fatigue damage ratio is log normally distributed.

Solution: With $V[K_1] = (6.57 \times 10^{-5} \times 0.3)^2 = 3.885 \times 10^{-10}$, $V[K_2] = (2.69 \times 0.04)^2 = 0.0116$, $\text{Cov}[K_1, K_2] = -0.867 \sqrt{3.885 \times 10^{-10} \times 0.0116} = -1.839 \times 10^{-6}$, and $V[\epsilon_t] = (3.54 \times 10^{-4} \times 0.175)^2 = 3.838 \times 10^{-9}$, from Eq. 10.70, $E[1/N_f] = (3.54 \times 10^{-4})^{2.69}/(6.57 \times 10^{-5}) [1 + 3.885 \times 10^{-10}/(6.57 \times 10^{-5})^2 + 0.5(\ln 3.54 \times 10^{-4})^2 \times 0.0116 - (\ln 3.54 \times 10^{-4}) \times (-1.839 \times 10^{-6})/(6.57 \times 10^{-5}) + 0.5 \times 2.69 \times 1.69 \times 3.838 \times 10^{-9}/(3.54 \times 10^{-4})^2 = 7.93 \times 10^{-6}(1 + 0.0900 + 0.3662 - 0.2224 + 0.0696) = 7.93 \times 10^{-6} \times 1.303 = 1.035 \times 10^{-5}$. Note that in this particular case, the difference between Eqs. 10.69 and 10.70 is about 30%. From Eq. 10.71, $V[1/N_f] = (3.54 \times 10^{-4})^{5.38}[3.885 \times 10^{-10}/(6.57 \times 10^{-5})^4 + (\ln 3.54 \times 10^{-4})^2 \times 0.0116/(6.57 \times 10^{-5})^2 + 2 \times (\ln 3.54 \times 10^{-4}) \times 1.839 \times 10^{-6}/(6.57 \times 10^{-5})^3] + (2.69)^2 (3.54 \times 10^{-4})^{3.38} \times 3.838 \times 10^{-9}/(6.57 \times 10^{-5})^2 = 2.714 \times 10^{-19}(2.085 \times 10^7 + 16.969 \times 10^7 - 10.306 \times 10^7) + 1.394 \times 10^{-11} = 3.768 \times 10^{-11}$.

With $E[n] = 50,000$ and $V[n] = 50,000$, from Eq. 10.72, $E[c] = 50,000 \times 1.035 \times 10^{-5} = 0.5175$; from Eq. 10.73, $V[c] = (50,000)^2 \times 3.768 \times 10^{-11} + (1.035 \times 10^{-5})^2 \times 50,000 = 0.0942 + 5.3 \times 10^{-6} = 0.0942$. The coefficient of variation of c is $\sqrt{0.0942}/0.5175 = 0.593$, which can be considered as the coefficient of variation for the area cracked.

(a) If c is normally distributed, $z = (1 - 0.5175)/\sqrt{0.0942} = 1.572$. From Table 10.1, Probability($c \leq 1$) $= 0.5 + 0.442 = 0.942$. From Eq. 10.74, percentage of area cracked $A = 1000(1 - 0.942) = 58$ ft^2/1000 ft^2.

(b) If c is assumed to be a log normal distribution, from Eq. 10.37, $s_{\log c} = 0.4343 \times \sqrt{0.0942}/0.5175 = 0.2575$, or $z = (\log 1 - \log 0.5175)/0.2575 = 1.111$. From Table 10.1, Probability($\log c \leq 0$) $= 0.867$, or $A = 1000(1 - 0.867) = 133$ ft^2/1000 ft^2.

Rut Depth Model

The rut depth model is shown in Figure 10.18. Each step is explained below:

1. The mean and variance of vertical deflections are obtained from the general response model in the viscoelastic case and from the primary response model in the elastic or indirect elastic case.
2. The traffic rate over the analysis period is expressed as the number of repetitions n with a variance equal to the mean.
3. From the primary response model, the permanent deformation parameters of the system, α_{sys} and μ_{sys}, are computed, as explained in Section 7.4.1. The following procedures are used to calculate the mean and the variance of the accumulated permanent deformation. From Eq. 7.49 by replacing N by n,

$$w_p(n) = \mu_{sys} \, w n^{-\alpha_{sys}} \tag{7.49}$$

The accumulated permanent deformation can be obtained by integrating $w_p(n)$ with respect to n:

$$w_p = w\mu_{sys} \frac{n^{1-\alpha_{sys}}}{1 - \alpha_{sys}} \tag{10.75}$$

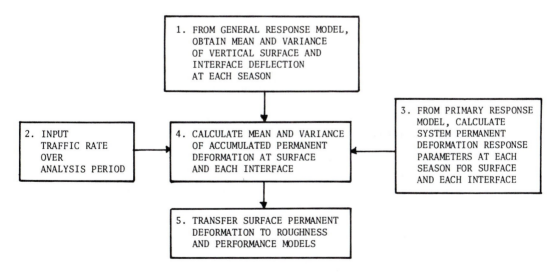

Figure 10.18 Rut depth model.

Let

$$F(n) = \mu_R n^{\alpha_R} \tag{10.76}$$

in which $\alpha_R = 1 - \alpha_{sys}$ and $\mu_R = \mu_{sys}/\alpha_R$. Equation 10.75 can be written as

$$w_p = wF(n) \tag{10.77}$$

The permanent deformation parameter $F(n)$ and the resilient deformation w are assumed to be uncorrelated random variables, so from Eq. 10.25,

$$E[w_p] = E[w] E[F(n)] \tag{10.78}$$

The expected value of general response $E[w]$ is expressed by Eq. 10.66, while $E[F(n)]$ can be derived from Eqs. 10.25 and 10.76 by noting that $V[n] = n$:

$$E[F(n)] = \mu_R \left[n^{\alpha_R} + \frac{\alpha_R(\alpha_R - 1)}{2} (n)^{\alpha_R - 1} \right] \tag{10.79}$$

Applying Eq. 10.30 to Eq. 10.77 gives

$$V[w_p] = \{E[w]\}^2 V[F(n)] + \{E[F(n)]\}^2 V[w] \tag{10.80}$$

The variance of general response $V[w]$ is expressed by Eq. 10.67, while $V[F(n)]$ can be obtained from Eqs. 10.30 and 10.76, or

$$V[F(n)] = \alpha_R^2 \mu_R^2 (n)^{2\alpha_R - 1} \tag{10.81}$$

5. The mean and variance of accumulated permanent deformation, or rut depth, are transferred to the roughness and performace models. To be consistent with the PSI equation, the symbol w_p for permanent deformation is replaced by RD in the roughness and performance models.

Example 10.19

Given $E[w] = 0.0130$ in. (0.33 mm) and $V[w] = 3.05 \times 10^{-5}$ in.^{2} (1.97×10^{-2} mm^2), as obtained from Example 10.17, $E[n] = 50{,}000$ and $V[n] = 50{,}000$, as assumed in Example 10.18, and the permanent deformation parameters $\alpha_{sys} = 0.756$ and $\mu_{sys} = 0.145$, compute $E[w_p]$ and $V[w_p]$.

Solution: With $\alpha_R = 1 - 0.756 = 0.244$ and $\mu_R = 0.145/0.244 = 0.594$, from Eq. 10.79, $E[F(n)] = 0.594[(50{,}000)^{0.244} + 0.5 \times 0.244 \times (0.244 - 1) \times (50{,}000)^{0.244-1}]$ $= 8.324$ in. (211 mm); from Eq. 10.81, $V[F(n)] = (0.244)^2(0.594)^2(50{,}000)^{0.488-1} = 8.250 \times 10^{-5}$ in.^{2} (5.32×10^{-2} mm). From Eq. 10.78, $E[w_p] = 0.0130 \times 8.324 = 0.108$ in. (2.75 mm). From Eq. 10.80, $V[w_p] = (0.0130)^2 \times 8.250 \times 10^{-5} + (8.324)^2 \times 3.05 \times 10^{-5} = 2.113 \times 10^{-3}$ in.^{2} (1.36 mm^2).

Roughness Model

The roughness model is shown in Figure 10.19. Each step is explained below:

1. Pavement roughness properties B and C are specified by the user.
2. From the rut depth model, the mean and the variance of rut depth, $E[RD]$ and $V[RD]$, are obtained.

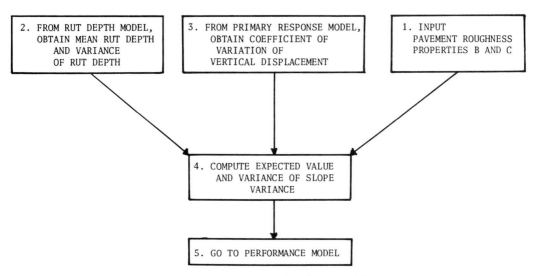

| 2. FROM RUT DEPTH MODEL, OBTAIN MEAN RUT DEPTH AND VARIANCE OF RUT DEPTH | 3. FROM PRIMARY RESPONSE MODEL, OBTAIN COEFFICIENT OF VARIATION OF VERTICAL DISPLACEMENT | 1. INPUT PAVEMENT ROUGHNESS PROPERTIES B AND C |

4. COMPUTE EXPECTED VALUE AND VARIANCE OF SLOPE VARIANCE

5. GO TO PERFORMANCE MODEL

Figure 10.19 Roughness model.

3. From the primary response model, a random variable η with a mean of 1 and a variance $V[\eta]$ is obtained.
4. The slope variance can be expressed as

$$\text{SV} = \frac{2B}{C^2}\, V[\eta]\, (E[\text{RD}])^2 \tag{10.82}$$

The expected value of slope variance is

$$E[\text{SV}] = \frac{2B}{C^2}\, V[\eta]\{(E[RD])^2 + V[RD]\} \tag{10.83}$$

The variance of slope variance is

$$V[\text{SV}] = \left(\frac{4B}{C^2}\, V[\eta]\, E[\text{RD}]\right)^2 V[\text{RD}] \tag{10.84}$$

5. The values of $E[\text{SV}]$ and $V[\text{SV}]$ are used in the performance model.

Example 10.20

Given $E[\text{RD}] = 0.108$ in. (2.75 mm) and $V[\text{RD}] = 2.113 \times 10^{-3}$ in.^{2} (1.36 mm^2), as obtained in Example 10.19, $V[\eta] = 0.09$, and roughness parameters $B = 1$ and $C = 0.06$, determine $E[\text{SV}]$ and $V[\text{SV}]$.

Solution: From Eq. 10.83, $E[\text{SV}] = 2 \times 1 \times 0.09 \times [(0.108)^2 + 2.113 \times 10^{-3}]/(0.06)^2 = 0.689 \times 10^{-6}$ rad. From Eq. 10.84, $V[\text{SV}] = (4 \times 1 \times 0.09 \times 0.108)^2 \times 2.113 \times 10^{-3}/(0.06)^4 = 0.246 \times 10^{-12}$ rad.

Performance Model

The performance model is shown in Figure 10.20. Each step is explained below:

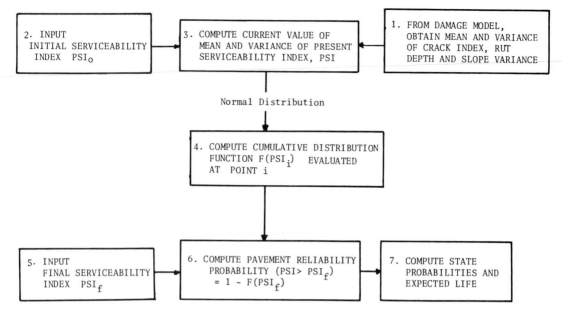

Figure 10.20 Performance model.

1. The mean and variance of area cracked, rut depth, and slope variance at the end of various periods are determined from the cracking, rut depth, and roughness models, respectively.

2. The initial serviceability index PSI_0 is specified.

3. The current value of PSI is computed by

$$PSI = PSI_0 - 1.91 \log(1 + SV) - 1.38RD^2 - 0.01\sqrt{A} \quad (10.85)$$

in which PSI_0 is the initial serviceability index, SV is the slope variance in 10^{-6} rad, RD is the rut depth in inches, and A is the area cracked in $ft^2/1000$ ft^2. Assuming that the initial serviceability index and these three modes of distress are uncorrelated random variables, from Eqs. 10.25, 10.30, and 10.85 yield,

$$E[PSI] = PSI_0 - 1.91\left\{\log(1 + E[SV]) - \frac{\log e}{2(1 + E[SV])^2} V[SV]\right\}$$

$$- 1.38\{E[RD])^2 + V[RD]\}$$

$$- 0.01\left\{(E[A])^{0.5} - \frac{(E[A])^{-1.5}}{8} V[A]\right\} \quad (10.86)$$

$$V[PSI] = V[PSI_0] + \left\{\frac{1.91 \log e}{1 + E[SV]}\right\}^2 V[SV] + \{2.76 E[RD]\}^2 V[RD]$$

$$+ \frac{(0.005)^2}{E[A]} V[A] \quad (10.87)$$

4. PSI is assumed to be normally distributed, and the cumulative distribution function $F(\text{PSI}_i)$ is evaluated at a specified number of points. These probabilities are used later to compute the state probability.

5. The final or terminal serviceability index PSI_f is specified.

6. Reliability is defined as the probability that PSI is greater than PSI_f, or reliability $= 1 - F(\text{PSI}_f)$.

7. The state probability is the probability of PSI being in a given interval $i - 1$ to i:

$$M_i(\text{PSI}) = F(\text{PSI}_i) - F(\text{PSI}_{i-1}) \qquad (10.88)$$

in which $M_i(\text{PSI})$ is the state or marginal probability, which is the probability that the PSI of the pavement will be in a certain specified state, and $F(\text{PSI}_i)$ is the cumulative distribution function of PSI evaluated at PSI_i. The reliability of a pavement at any point in time is the probability that the PSI exceeds the minimum acceptable level. The time for the reliability to fall below a predefined minimum value is taken as the expected life.

Example 10.21

Shown in Table 10.14 are the means and variances of slope variance, rut depth, and areas cracked obtained from the previous examples together with the mean and variance of the initial PSI. Determine (a) the expected value of PSI, $E[\text{PSI}]$, (b) the variance of PSI, $V[\text{PSI}]$, (c) the level of reliability for a terminal serviceability index of 3.0, and (d) the state probability for PSI between 3.0 and 3.5.

Solution: (a) From Eq. 10.86, $E[\text{PSI}] = 4.2 - 1.91\,[\log(1 + 0.689) - 0.5 \times 0.434 \times 0.246/(1 + 0.689)^2] - 1.38\,[(0.108)^2 + 0.002113] - 0.01\,[(58)^{0.5} - (58)^{-1.5} \times 1183/8] = 4.2 - 0.399 - 0.019 - 0.073 = 3.709$.

(b) From Eq. 10.87, $V[\text{PSI}] = 0.06 + (1.91 \times 0.434/1.689)^2 \times 0.246 + (2.76 \times 0.108)^2 \times 0.002113 + 2.5 \times 10^{-5} \times 1183/58 = 0.06 + 0.059 + 0 + 0.001 = 0.120$.

(c) With $E[\text{PSI}] = 3.709$, $V[\text{PSI}] = 0.12$, and $\text{PSI}_f = 3.0$, $z = (3.0 - 3.709)/\sqrt{0.12} = -2.05$. From Table 10.1, the reliability is 98.0%.

(d) The probability for PSI greater than 3.0 is 98.0%. For a PSI of 3.5, $z = (3.5 - 3.709)/\sqrt{0.12} = -0.603$. From Table 10.1, the probability for PSI greater than 3.5 is 72.7%. The state probability that PSI falls between 3.0 and 3.5 is $98.0 - 72.7$, or 25.3%.

10.5.2 PMRPD Computer Program

The PMRPD (Probabalistic Method for Rigid Pavement Design) computer program (Huang and Sharpe, 1989) is a simplification of KENSLABS that considers a

TABLE 10.14 MEAN AND VARIANCE OF PAVEMENT CONDITION DATA

PSI$_0$		SV		RD		A	
Mean	Variance	Mean (10^{-6} rad)	Variance (10^{-12} rad)	Mean (in.)	Variance (in.^{2})	Mean (10^{-3})	Variance (10^{-6})
4.2	0.06	0.689	0.246	0.108	2.113×10^{-3}	58	1183

Note. 1 in. = 25.4 mm, unit of A is ft^2 per 1000 ft^2, or 10^{-3}.

single slab with full subgrade contact. The design is based on fatigue cracking, but without considering the deflection and pumping at the joints and corners. Only one slab is needed because the maximum stress due to wheel loads occurs at the slab edge near the midslab far away from the joints. Consequently, the presence of adjoining slabs should have very little effect on the maximum stress. Only the case of full contact is considered because partial contact due to pumping or temperature curling does not have a significant effect on the edge stress. However, a stress adjustment factor is employed as an input parameter to reflect the effect of curling, pumping, tied shoulder, and the load placement relative to the slab edge.

General Features

The use of probabilistic method requires the input of the coefficients of variation of traffic and design factors. By specifying a series of thicknesses, the program will determine the reliability of the design for each thickness. The minimum thickness that meets the reliability requirement can then be selected.

Limitations

It should be pointed out that the design of concrete pavements is a complex process and involves many factors that are difficult to evaluate. Therefore, some simplified assumptions must be made to obtain a solution.

Due to the difficulty of superimposing the rapidly applied loading stress and the slowly varied curling stress, it is assumed that the effect of curling stress is negligible for fatigue analysis. This assumption is reasonable if relatively short panels of slabs are used. If the slabs are long, the curling stress may become quite large and a larger stress adjustment factor should be used.

It is generally recognized that slab deflection and drainage are important factors in rigid pavement design because they will cause the pumping of subgrade soils or the blowing of subbase materials. However, in PMRPD neither deflection nor drainage is considered, not because they are unimportant but because the state of the art has not advanced to the point where they can be treated theoretically. The difficulty in analyzing pavement deflection is that the most critical deflection occurs at the slab corner and is affected to a large extent by the slab–subgrade contact, which varies with temperature differentials, joint design, type of subbase, and subsurface drainage. Because PMRPD considers only fatigue cracking, a designer using this program should make sure that proper drainage, nonerodible subbase, and adequate load transfer across the joints are provided, so that the pavement will not fail by pumping and erosion.

Advantages of Probabilistic Method

In the deterministic method of design based on fatigue cracking, a slab thickness is selected so that the cracking index is smaller than 1. It is well known that the cracking index is very sensitive to the thickness of the concrete slab. For example, a cracking index of 50% is obtained for a slab thickness of 8 in. (203 mm). If the thickness is increased to 8.5 in. (216 mm), the cracking index is decreased to less than 10%. If the thickness is decreased to 7.5 in. (190 mm), the cracking index is increased to more than 100%. It does not appear reasonable that a slight change in thickness should affect the success or failure of the design to such an extent. In fact, the cracking index depends on a large number of traffic, material, and geometric parameters. It is difficult to select these parameter values because they exhibit a wide range of variations. If average values are used, the reliability of the design is only 50% and is usually not considered sufficient. To be on the safe side, designers tend to use more conservative values. If conservative values are used for every parameter, the final design will become too conservative. The arbitrary selection of these parameter values may result in widely different designs. These difficulties can be overcome if a probabilistic approach is employed.

The probabilistic method of design has several advantages over the deterministic method. First, reliability is not so sensitive to thickness as the cracking index. A gradual increase in thickness should result in a gradual increase in reliability, which makes sense. Second, reliability indicates the probability of success and has a physical meaning. A reliability of 90% implies that 10% of the pavement may fail at or before the end of the design period, while the cracking index does not relate even remotely to pavement conditions. Third, pavement thickness should vary with the type and importance of highways; therefore, different levels of reliability should be assigned to each. It is impractical to assign different cracking indices to different highways because the cracking index is sensitive to thickness and varies over several orders of magnitude. Last, but not least, the thickness required depends to a great extent on workmanship. Poorly constructed pavements with a wide range of variability should be thicker than well-constructed pavements with good quality control.

Capabilities

The PMRPD program has the following capabilities that make it particularly useful as a practical design tool:

1. The program can be applied to a liquid, solid, or layer foundation. With the ever-increasing speed and storage of personal computers, it is no longer necessary to assume the foundation to be a liquid with a fictitious k value. The more realistic solid or layer foundation consisting of a maximum of 6 layers can be used.
2. The program can be applied to mixed traffic consisting of single-, tandem-, and tridem-axle loads. Starting from the heaviest axle load in one group, the analysis will stop or switch to the next group if the stress ratio falls below a minimum value. The program was so written that the stiffness matrix is reduced to an upper triangular matrix only once. After the matrix is reduced, very little time is required to analyze additional loads.
3. The program can determine the reliability of different trial thicknesses in the same run. For solid and layer foundations, this capability can also save some computer time because the stiffness matrix of the foundation obtained for the first thickness can be used directly for the remaining thicknesses.
4. The program can be used for the deterministic method of design by simply specifying all the coefficients of variations as 0.

5. The finite element grid can be either specified by the user or determined automatically by the program based on the axle load data.

Theoretical Development

Similar to KENSLABS, the slab is considered as an elastic plate on a liquid, solid, or layer foundation. The displacements are determined first by solving a system of simultaneous equations. In the numerical solutions, not only the value of each expression but also its variance must be evaluated.

Finite Element Method

After dividing the plate into rectangular finite elements, the relationship between displacements and forces at the nodes can be written as

$$R\,[K(R = 1)]\{\delta\} = q\{F(q = 1)\} \tag{10.89}$$

in which R is the modulus of rigidity of the plate; $[K(R = 1)]$ is the overall stiffness matrix of the plate–foundation system when $R = 1$; $\{\delta\}$ are nodal displacements, including a vertical deflection and two rotations; q is the contact pressure, which is equal to the total load divided by the contact area; and $\{F(q = 1)\}$ are the nodal forces when $q = 1$. Note that

$$R = \frac{Eh^3}{12(1 - v^2)} \tag{5.11}$$

$$[K(R = 1)] = [K_p\,(R = 1)] + \frac{k}{R}\,[K_f\,(k = 1)] \tag{10.90}$$

in which $[K_p\,(R = 1)]$ is the stiffness matrix of the plate when $R = 1$ and $[K_f\,(k = 1)]$ is the stiffness matrix of the foundation when $k = 1$. For a solid foundation, $k = 1$ should be replaced by $E_f = 1$, and for a layer foundation $k = 1$ should be replaced by $E_1 = 1$, where E_1 is the elastic modulus of the uppermost layer. The reason that R, k, and q are factored out from $[K_p]$, $[K_f]$, and $\{F\}$, respectively, is to solve Eq. 10.89 with only one random variable, k/R, or

$$[K(R = 1)]\{\delta(q/R = 1)\} = \{F(q = 1)\} \tag{10.91}$$

in which $[K(R = 1)]$ contains the random variable k/R, as shown in Eq. 10.90 and $\{\delta(q/R = 1)\}$ are displacement factors that must be multiplied by q/R to obtain the nodal displacements. The probabilistic solution will become more difficult if more than one random variable exists in Eq. 10.91.

After the displacement factors are determined, the stress $\{\sigma\}$ can be determined by

$$\{\sigma\} = \frac{6q}{h^2}\,[K_m\,(R = 1)]\{\delta(q/R = 1)\} \tag{10.92}$$

in which $[K_m\,(R = 1)]$ is the moment matrix, which relates nodal moments to nodal displacements, when $R = 1$. To obtain the stresses, $[K_m\,(R = 1)]$ must be multiplied by R and $\{\delta(q/R) = 1)\}$ by q/R. Because R cancels out, it does not appear in Eq. 10.92. In Eq. 10.92, $h^2/6$ is the section modulus of the plate.

Probabilistic Procedure

In the probabilistic approach, every variable has a mean and a variance. Each computer statement is a mathematical expression relating one variable, say X, to other variables, say Y and Z. If all input parameters are based on average values and the coefficients of variations are small, the value of X, Y, and Z can be taken as the mean, as shown by Eq. 10.26. For each deterministic statement giving the mean of X, there is a probabilistic statement for determining the variance, $V[X]$. Consider the more general case that g is a function of $x_1, x_2, \ldots, x_n$. If $x_1, x_2, \ldots, x_n$ are functions of the same random variable r, the variance of g can be determined by Eq. 10.30:

$$V[g] = \left(\frac{\partial g}{\partial x_1} \frac{dx_1}{dr} + \frac{\partial g}{\partial x_2} \frac{dx_2}{dr} + \cdots + \frac{\partial g}{\partial x_n} \frac{dx_n}{dr} \right)^2 V[r] \qquad (10.93)$$

If, in addition to r, g is also a function of random variable s, then similar terms with r replaced by s must be added to Eq. 10.93. Although variance is always positive, the derivatives within the parentheses may be positive or negative. To keep the proper sign, it is more convenient to use derivatives for mathematical manipulation, rather than Eq. 10.93:

$$D[g] = \frac{\partial g}{\partial x_1} D[x_1] + \frac{\partial g}{\partial x_2} D[x_2] + \cdots + \frac{\partial g}{\partial x_n} D[x_n] \qquad (10.94a)$$

in which $\qquad D[x_i] = \sqrt{V[r]} \, \dfrac{dx_i}{dr} \qquad (10.94b)$

At the final step when the variance is required, $V[g]$ can be determined from $D[g]$ by

$$V[g] = (D[g])^2 = \left(\sum_{i=1}^{n} \frac{\partial g}{\partial x_i} D[x_i] \right)^2 \qquad (10.95)$$

Consider a simple example in which $X = Y/Z$ and the means Y and Z and the derivatives $D[Y]$ and $D[Z]$ are known. From Eq. 10.94

$$D[X] = D\left[\frac{Y}{Z} \right] = \frac{1}{Z} D[Y] - \frac{Y}{Z^2} D[Z]$$

and $\qquad V[X] = (D[X])^2$

Example 10.22

Given $x = a + br$ and $y = c - dr$, in which a, b, c, and d are constants and r is a random variable with known $V[r]$, determine the variances of xy and x/y.

Solution: $D[xy] = yD[x] + xD[y]$. Since $D[x] = bD[r]$ and $D[y] = -dD[r]$, $D[xy] = (c - dr) bD[r] + (a + br)(-dD[r]) = (bc - ad - 2bdr) D[r]$, or $V[xy] = (bc - ad - 2bdr)^2 V[r]$.
$\qquad D[x/y] = D[x]/y - xD[y]/y^2 = bD[r]/(c - dr) - (a + br)(-dD[r])/(c - dr)^2 = (bc + ad)D[r]/(c - dr)^2$, or $V[x/y] = (bc + ad)^2 V[r]/(c - dr)^4$.

In PMRPD, the random variables include slab thickness h, contact pressure q, Young's modulus of concrete E, subgrade modulus k, concrete modulus of rupture S_c, and the predicted number of load repetitions n. Let r be a random variable defined as

$$r = \frac{k}{R} = \frac{12(1 - v^2)\, k}{Eh^3} \tag{10.96}$$

By comparing Eqs. 10.96 and 4.10, it can be seen that r is related to the radius of relative stiffness ℓ by $r = 1/\ell^4$. Applying Eq. 10.94 to Eq. 10.96 gives

$$D[r] = 12(1 - v^2)\left\{\left(-\frac{3k}{Eh^4}\right) D[h]\right.$$

$$\left. + \left(-\frac{k}{E^2h^3}\right) D[E] + \left(\frac{1}{Eh^3}\right) D[k]\right\} \tag{10.97}$$

If h, E, and k are assumed to be uncorrelated, then there are no covariance terms, so

$$V[r] = \{12(1 - v^2)\}^2 \left\{\left(-\frac{3k}{Eh^4}\right)^2 V[h]\right.$$

$$\left. + \left(-\frac{k}{E^2h^3}\right)^2 V[E] + \left(\frac{1}{Eh^3}\right)^2 V[k]\right\} \tag{10.98}$$

Because there is only one random variable, r, the variance of any mathematical expression can be determined by Eq. 10.94. The solution of Eq. 10.91 by the Gauss elimination method to determine the mean and variance of $\{\delta\}$ was presented by Huang and Sharpe (1989).

Let the stress at each node be expressed as

$$\sigma = \frac{6q}{h^2} M \tag{10.99}$$

As shown by Eq. 10.92, M is the moment in the slab due to $q = 1$ and $R = 1$. Because M and h are correlated, applying Eq. 10.29 to Eq. 10.99 yields

$$V[\sigma] = 36\left\{\left(\frac{M}{h^2}\right)^2 V[q] + \left(\frac{2q\,M}{h^3}\right)^2 V[h] + \left(\frac{q}{h^2}\right)^2 V[M]\right.$$

$$\left. + 2\left(\frac{2q\,M}{h^3}\right)\left(\frac{q}{h^2}\right) f\, D[M]\, \sqrt{V[h]}\right\} \tag{10.100}$$

in which $D[M]$ is the derivative of M defined by Eq. 10.94 and f is the fraction of $\sqrt{V[r]}$ contributed by $\sqrt{V[h]}$ and can be obtained from Eq. 10.98:

$$f = \frac{36(1 - v^2)\, k}{Eh^4} \sqrt{\frac{V[h]}{V[r]}} \tag{10.101}$$

The first three terms on the right side of Eq. 10.100 indicate that q, M, and h are independent, and the last term is the correlation between h and M because M is a function of h.

Given the stress σ, the modulus of rupture S_c, and their variances $V[\sigma]$ and $V[S_c]$, the allowable number of load repetitions N can be determined from Eq. 5.35 or 5.36 and its variance $V[N]$ from Eqs. 10.54 to 10.56. Given the predicted number of load repetitions n and its variance $V[n]$, the reliability of the design can be determined by Eqs. 10.57 through 10.60.

10.6 SUMMARY

This chapter discusses the concept of reliability and its application to pavement design. Reliability is the probability of success. Depending on how the success or failure is defined, various design criteria, such as equivalent 18-kip single-axle loads, damage ratio or cracking index, and terminal serviceability index, may be used for computing reliability.

Important Points Discussed in Chapter 10

1. There are two methods of pavement design: deterministic and probabilistic. In the deterministic method, each design factor is assigned a fixed value, usually a more conservative value compared to the mean. In the probabilistic method, each design factor has a mean and a variance. If any two of the design factors are correlated, their covariance must also be specified.

2. In the probabilistic method of design, the mean and variance of random variables can be determined by the Taylor series expansion. If the random variables are normally distributed, the second-order terms are small and can be neglected. Thus, Eq. 10.26 can be used to determine the mean and Eq. 10.29 or 10.30 can be used to determine the variance.

3. Various criteria can be used to evaluate the reliability of a design. The reliability can be obtained by comparing the mean value of the criterion with the required criterion. If the mean value is equal to the required value, the reliability of the design is 50%. To increase the reliability, the required criterion must be more stringent depending on the variability of the mean value.

4. The most popular probabilistic method of pavement design is to compare the allowable number of load repetitions with the predicted number of load repetitions, all based on the mean values and each with its own coefficient of variation.

5. It is reasonable to assume axle load repetitions as a log normal distribution. If an equivalent axle load is used for design purposes, the reliability is the probability that $\log D_r = \log W_T - \log W_t \leq 0$, in which D_r is the damage ratio or the ratio between the allowable number of load repetitions W_t and the predicted number of load repetitions W_T. When $\log D_r$ and its standard deviation are known, the probability for $\log D_r$ not greater than 0 can be determined. If the axle loads are divided into a number of groups and a year is divided into several periods, each with a different W_t and W_T, the D_r is the summation of damage ratios for all load groups over all periods.

6. To apply the probabilistic method, it is necessary to know the coefficients of variation of various traffic and design factors. These coefficients depend on the method of construction and quality control and can be obtained only from past experiences and field measurements. In the absence of more definite information, the coefficients of variation presented in Tables 10.9, 10.10, and 10.11 can be used as a guide.

7. The Rosenblueth method can be used to determine the probabilistic solutions from the deterministic solutions. Any deterministic design method can be easily converted to a probabilistic method based solely on a number of deterministic solutions.

8. A mechanistic-based probabilistic computer program for the analysis of flexible pavements was described. The program is called VESYS and has been constantly improved and upgraded through research contracts by the Federal Highway Administration. Particular emphases were placed on the organization of the program into different models and the derivation of some equations for defining means and variances. These basic concepts are very useful for the development of any probabilistic computer programs.

9. A probabilistic computer program named PMRPD for the design of rigid pavements was briefly described. The program is similar to KENSLABS but, for each deterministic expression g in KENSLABS, a probabilistic expression $D[g]$ was included. To simplify the program, several random variables were combined into a single variable so only the derivative with respect to one variable was required.

PROBLEMS

10-1. The probability function of a random variable x is

$$f(x) = \begin{cases} \frac{1}{3} & x = 0 \\ \frac{1}{2} & x = 1 \\ \frac{1}{6} & x = 2 \end{cases}$$

Find the mean and the variance of x. [Answer: 0.833, 0.472]

10-2. If $f(x) = t \exp(-t^2/2)$ when $t \geq 0$ and $f(x) = 0$ when $t < 0$, compute the mean and the variance of x. [Answer: 1.253, 0.429]

10-3. Two random variables, x and y, have the following probability functions:

$$f(x, y) = \begin{cases} \frac{1}{4} & x = 1 \quad y = 2 \\ \frac{1}{4} & x = 2 \quad y = 3 \\ \frac{1}{2} & x = 3 \quad y = 4 \end{cases}$$

Compute Cov[x, y], $\rho(x, y)$, and $V[x + y]$. [Answer: 0.688, 1, 2.75]

10-4. The joint density function of two random variables, x and y, is

$$f(x, y) = 6xy(2 - x - y) \quad \text{when } 0 \leq x \leq 1 \text{ and } 0 \leq y \leq 1$$
$$= 0 \quad \text{otherwise}$$

Compute Cov[x, y] and $\rho(x, y)$. [Answer: $-\frac{1}{144}$, $-\frac{5}{43}$]

10-5. The thickness of a full-depth asphalt pavement is determined by the equation

$$t = (0.25 + 0.125 \log n) \left[\sqrt{\left(\frac{75P}{\pi E}\right)^2 - a^2} \right] \sqrt[3]{\frac{E}{E_p}}$$

in which t is the thickness of asphalt concrete; n is the number of load repetitions; P is the wheel load; E is the modulus of the subgrade; a is the tire contact radius, and E_p is the modulus of asphalt concrete. Assume that $t = 6.5$ in., $n = 100,000$, $P = $

9000 lb, $E = 10,000$ psi, $a = 6$ in., and $E_p = 400,000$ psi and that the coefficients of variation of t, n, P, E, a, and E_p are all 10%. Also assume that t is a normal distribution and the reliability is based on the required t obtained from the above equation versus the designed t of 6.5 in. Determine the reliability of the design based on Taylor series expansion. [Answer: 95.5%]

10-6. Solve Problem 10-5 by the Rosenblueth method. [Answer: 95.0%]

10-7. In Problem 10-5, it is assumed that log n is normally distributed and that the reliability is based on the n obtained from the equation and the predicted n of 100,000. Determine the reliability of the design based on Taylor series expansion. [Answer: 86.8%]

10-8. Solve Problem 10-7 by the Rosenblueth method. [Answer: 87.6%]

10-9. The number of load repetitions to induce the fatigue cracking of a certain hot mix asphalt is given by

$$N_f = f_1(\epsilon_t)^{-f_2}$$

For this problem $f_1 = 0.00462$ with a coefficient of variation of 30%, $f_2 = 2.69$ with a coefficient of variation of 5%, the coefficient of correlation for f_1 and f_2 is -0.867, and the tensile strain ϵ_t is 0.0021 with a coefficient of variation of 10%. Compute the expectation and variance of N_f based on the first-order Taylor series expansion. [Answer: 73,770, 2.28×10^9]

10-10. Rework on Example 10.18 by assuming that the predicted number of load repetitions n does not follow a Poisson arrival process, but with a coefficient of variation of 0.4. [Answer: 96 and 178 ft^2/1000 ft^2]

10-11. The serviceability index of flexible pavements is determined by

$$\text{PSI} = \text{PSI}_0 - 1.91 \log(1 + \text{SV}) - 1.38\text{RD}^2 - 0.01\sqrt{C}$$

in which PSI_0 is the initial serviceability index, SV is the slope variance (10^{-6} rad), RD is the rut depth (in.), and C is the cracked area (sq ft/1000 sq ft). With the data shown in Table P10.11 determine (a) the expected value of PSI using second-order Taylor series expansion, (b) the variance of PSI, and (c) the terminal serviceability index for 95% reliability. [Answer: 3.21, 0.095, 2.7]

TABLE P10.11 INITIAL SERVICEABILITY AND PHYSICAL MEASUREMENTS

PSI$_0$		SV		RD		C
Mean	Variance	Mean (10^{-6} rad)	Variance (10^{-12} rad)	Mean (in.)	Variance (in.^{2})	Cracked area (ft^2/1000 ft^2)
4.2	0.06	1.572	0.32	0.284	0.003	120

10-12. Given slab thickness $h = 10$ in. and $C[h] = 0.04$, slab elastic modulus $E = 4 \times 10^6$ psi and $C[E] = 0.1$, and modulus of subgrade reaction $k = 200$ pci and $C[k] = 0.35$, (a) determine r and $D[r]$ by Eqs. 10.96 and 10.97, respectively, and (b) derive and compute $E[r]$ based on second-order Taylor series expansion. It is assumed that the concrete has a fixed Poisson ratio of 0.15. [Answer: 5.87×10^{-7}, 7.62×10^{-8}, 5.99×10^{-7}]

10-13. Find $V[g]$ when $r = 1$, $C[r] = 0.2$ and

$$g = \frac{1 - 2r}{1 + r} + \frac{2 + r}{3 + 6r + r^2} + \frac{r}{(4 + r^2)(6 + r)}$$

[Answer: 0.01425]

10-14. Find $V[g]$ when $r = V[r] = 10$, $s = V[s] = 20$ and

$$g = \frac{1 - 2r - s}{1 + rs} + \frac{r + s}{3 + 6s + r^2} + \frac{rs}{(r + s)^2}$$

[Answer: 0.00139]

Flexible Pavement Design

11.1 CALIBRATED MECHANISTIC DESIGN PROCEDURE

The design equations presented in the 1986 AASHTO design guide were obtained empirically from the results of the AASHO Road Test. To develop a mechanistic pavement analysis and design procedure suitable for future versions of AASHTO guide, a research project entitled ''Calibrated Mechanistic Structural Analysis Procedures for Pavements'' was awarded to the University of Illinois under NCHRP 1-26. The research included both flexible and rigid pavements, and a two-volume report, hereafter referred to as Report 1-26, was prepared for the National Cooperative Highway Research Program (NCHRP, 1990). Some of the information presented herein was obtained from Report 1-26.

The calibrated mechanistic procedure is a more specific name for the mechanistic–empirical procedure. It contains a number of mechanistic distress models which require careful calibration and verification to ensure that satisfactory agreement between predicted and actual distress can be obtained. The purpose of calibration is to establish transfer functions relating mechanistically determined responses to specific forms of physical distress. Verification involves the evaluation of the proposed models by comparing results to observations in other areas not included in the calibration exercise. This procedure has been used in several design methods, such as the Asphalt Institute method described in Section 11.2.6. However, these existing methods are based on many simplifying assumptions and are not as rigorous as desirable. The recommended procedures presented in this section should be looked upon as the long-range goals to be accomplished in the future.

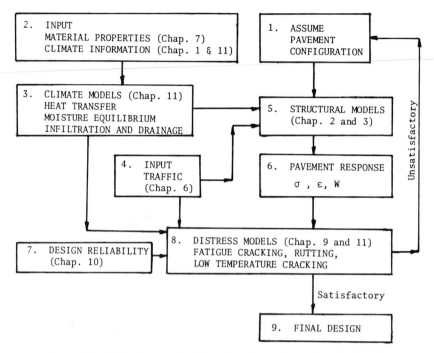

Figure 11.1 Methodology of calibrated mechanistic procedure for flexible pavement design.

11.1.1 General Methodology

Figure 11.1 shows a general methodology for flexible pavement design. In the figure, it is assumed that the materials to be used for the pavement structure are known a priori and that only the pavement configuration is subjected to design iterations. If changing the pavement configuration does not satisfy the design requirements, it may be necessary to change the types and properties of the materials to be used. If so, the iteration should also go through steps 2 and 3, instead of directly from step 1 to 5. Also shown in the figure are the chapters covering the appropriate subjects.

Explanation of Figure 11.1

1. Pavement configuration includes the number of layers, the thickness of each layer, and the type of materials.

2. The basic material properties for the structural models are the resilient moduli of HMA, base, subbase, and subgrade, while those for the distress models involve the various failure criteria, one for each distress. If temperature and moisture at different times of the year vary significantly, it is unreasonable to use the same modulus for each layer throughout the entire year. Each year should be divided into a number of periods, each with a different set of moduli based on the climatic data specified.

3. Climatic models include the heat transfer model (Dempsey and Thompson, 1970) for determining the temperature distribution with respect to space and time, the moisture equilibrium model (Dempsey et al., 1986) for determining the final moisture distribution in the subgrade, and the infiltration and drainage model (Liu and Lytton, 1984) for predicting the degree of saturation of granular bases. These models are described in Section 11.1.2.

4. Traffic should be divided into a number of load groups, each with different load magnitudes and configurations and different numbers of load repetitions. When the design is based on each type of distress, it is unreasonable to use an equivalent single-axle load because the equivalent factor for one type of distress is different from that for the other. The load magnitude and configuration, such as wheel spacing, contact radius, and contact pressure, are used in the structural models, while the number of repetitions is used in the distress models.

5. The finite element models can analyze nonlinear pavement systems more realistically than any other structural model by considering the variation of modulus within each layer. However, due to the many deficiencies of the finite element programs currently available, it appears that the layer system programs are the best and the most practical structural models to be used at the present time. More about structural models is presented in Section 11.1.3.

6. Pavement responses include stresses σ, strains ϵ, and deflections w. Theoretically, only those responses that contribute to each distress should be evaluated by the structural models and used in the distress models. However, if the responses can be related to surface deflections, it is possible to use surface deflections as input to the distress models. The most attractive feature of surface deflections is that they can be easily, rapidly, and inexpensively measured utilizing automatic equipment.

7. Due to the variabilities of materials, climate, traffic, and construction practice, a probabilistic method should be used to evaluate pavement distress. To determine the design reliability, the variances or coefficients of variation of some input parameters on materials, climate, traffic, and even layer thicknesses must also be specified in steps 1 to 4, so the variances of σ, ϵ, and w can be computed and used in the distress models.

8. Distress models include fatigue cracking, rutting, and low-temperature cracking. The roughness or performance model can also be included. If the reliability for a certain distress is less than the minimum level required, the assumed pavement configuration should be changed and steps 5, 6, and 8 repeated until the required level of reliability is met. More about distress models is presented in Section 11.1.4.

9. The final design is selected when the assumed pavement thicknesses satisfy the reliability requirement for each type of distress.

11.1.2 Climate Models

Temperature and moisture are significant climatic inputs for pavement design. The modulus of the HMA depends on pavement temperature, while the moduli of the base, subbase, and subgrade vary appreciably with moisture content. Report 1-26 indicated that the current technology, as utilized in the climatic–materials–

structural (CMS) model developed at the University of Illinois (Dempsey et al., 1986), is adequate for characterizing the pavement temperature regime but the capabilities for moisture modeling are not as advanced as those for temperature modeling. The strength and modulus of cohesive soils as well as granular materials with a high percentage of fines are very sensitive to even a small change in moisture content, say ±1%. Report 1-26 also indicated that the moisture equilibrium model in the CMS model for determining subgrade moisture is a reasonable and practical choice for design purpose.

Heat Transfer Model

The heat transfer model was originally developed at the University of Illinois (Dempsey and Thompson, 1970) for evaluating frost action and temperature regime in multilayered pavement systems. The model applies the finite difference method to solve the following Fourier equation for one-dimensional heat flow:

$$\frac{\partial T}{\partial t} = \alpha \frac{\partial^2 T}{\partial z^2} \tag{11.1}$$

in which T is the temperature, t is the time, z is the depth below surface, and α is the thermal diffusivity, which is related to the thermal conductivity and heat capacity of the pavement materials. Given initial temperature distribution and the two boundary conditions at the pavement surface and at a depth of H below the surface, Eq. 11.1 can be solved.

Convection and radiation play a dominant role in transferring heat between the air and the pavement surface, whereas conduction plays a separate role in transferring heat within the pavement system. The total depth H is a variable input parameter in the heat transfer model. It can be determined from a study of deep soil temperatures at a given geographic location. For example, studies of soil temperatures in northern Illinois have indicated that the ground temperature is about 51°F (10.6°C) at a depth of 12 ft (3.7 m) below surface. The temperature at this depth can be used as a boundary to solve Eq. 11.1.

Inputs to the heat transfer model are the climatic data and the thermal properties of paving materials and soils. The climatic data include maximum and minimum daily air temperatures, percent sunshine, and wind speed. The thermal properties include thermal conductivity, heat capacity, and latent heat of fusion. The model recognizes three sets of thermal properties depending on whether the material is in an unfrozen, freezing, or frozen condition. To facilitate the application of the heat transfer model to various pavement-related problems, a comprehensive climate data base for the state of Illinois was developed by Thompson et al. (1987).

Moisture Equilibrium Model

The moisture equilibrium model in the CMS model (Dempsey et al., 1986) is based on the assumption that the subgrade cannot receive moisture by infiltration through the pavement. Any rainwater will drain out quickly through the drainage

534 Flexible Pavement Design Chap. 11

layer to the side ditch or longitudinal drain, so the only water in the subgrade is the capillary water caused by the water table. Because of the thermodynamic relationship between soil suction and moisture content, a simple way to determine the moisture content in a soil is to determine its suction, which is related to the pore water pressure.

Figure 11.2 shows the suction–moisture curves for five different soils with varying clay contents, as indicated by the numerals in parentheses under each soil title. These curves were obtained in the laboratory by drying tests, in which different levels of vacuum or suction were applied to a wet soil specimen until the moisture was reduced to an equilibrium value. The suction is expressed in pF scale, which is the logarithm of water tension in cm as defined by Schofield (1935). The corresponding values in terms of psi are shown on the left scale.

It can be seen that suction increases as the moisture content decreases or the clay content increases. The increase in suction is due to the smaller menisci formed between soil particles. In the CMS model, empirical relationships were used to define the suction–moisture curve based on the liquid limit, the plasticity index, and the saturated moisture content of the soil.

When there is no loading or overburden pressure, suction is equal to the negative pore pressure. When a load or overburden is applied to an unsaturated soil with a given moisture content or suction, the suction or moisture content remains the same but the pore pressure becomes less negative. The relationship between suction and pore pressure can be expressed as

$$u = S + \alpha p \tag{11.2}$$

in which u is the pore pressure when soil is loaded; S is the soil suction, which is a negative pressure; p is the applied pressure or overburden, which is always positive; and α is the compressibility factor, varying between 0 for unsaturated, cohesionless soils and 1 for saturated soils. For unsaturated cohesive soils, α is related to the plasticity index PI by (Black and Croney, 1957)

$$\alpha = 0.03 \times PI \tag{11.3}$$

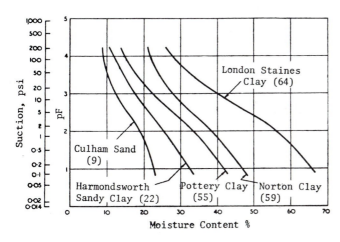

Figure 11.2 Soil suction–moisture content curves for different soils (1 psi = 6.9 kPa). (After Road Research Laboratory (1952).)

The pore pressure in a soil depends solely on its distance above the ground-water table:

$$u = -z\gamma_w \tag{11.4}$$

in which z is the distance above the water table and γ_w is the unit weight of water. This simple fact can be explained by considering soils as a bundle of capillary tubes with varying sizes. Water will rise in these capillary tubes to various elevations depending on the size of the tube. At any distance z above the water table, a large number of menisci will form at the air–water interfaces, thus resulting in tension at that elevation corresponding to the height of capillary rise. Combining Eqs. 11.2 and 11.4 yields

$$S = -z\gamma_w - \alpha p \tag{11.5}$$

The procedures for determining the equilibrium moisture content at any point in a pavement system can be summarized as follows:

1. Determine the distance z from the point to the water table.
2. Determine the loading or overburden pressure p.
3. Determine the compressibility factor α by Eq. 11.3.
4. Determine the suction S by Eq. 11.5.
5. Determine the moisture content from the suction–moisture curve.

Example 11.1

Figure 11.3 shows an 8-in. (203-mm) full-depth asphalt pavement on a subgrade composed of two different materials. The top 16 in. (406 mm) of subgrade is a Culham sand below which is a Norton clay with a PI of 18. The relationship between the suction and moisture content of these two soils is shown in Figure 11.2. The water table is located 12 in. (305 mm) below the top of the clay. The unit weight γ of each material is shown in the figure. Predict the moisture contents at point A on top of the clay, point B at the bottom of the sand layer, and point C on top of the sand layer.

Solution: At point A, the overburden pressure $p = (8 \times 145 + 16 \times 120)/12 = 256.7$ psf. From Eq. 11.3, $\alpha = 0.03 \times 18 = 0.54$. From Eq. 11.5, $S = -1 \times 62.4 - 0.54 \times 256.7 = -201.0$ psf $= -1.40$ psi. From Figure 11.2, the moisture content is 38%.

At point B, $\alpha = 0$ and, from Eq. 11.5, $S = -62.4$ psf $= -0.43$ psi. From Figure 11.2, the moisture content is 21%.

At point C, from Eq. 11.5, $S = -(12 + 16) \times 62.4/12 = -145.6$ psf $= 1.01$ psi. From Figure 11.2, the moisture content is 19.5%.

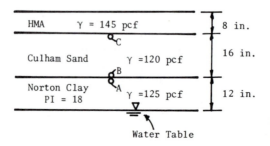

Figure 11.3 Example 11.1 (1 in. = 25.4 mm, 1 pcf = 157.1 N/m³).

Flexible Pavement Design   Chap. 11

Infiltration and Drainage Model

The infiltration and drainage model developed by Texas A&M University (Liu and Lytton, 1984) can be used to evaluate the effects of rainfall on the degree of saturation and the resilient moduli of the base course and subgrade. The degree of saturation for the base and the subgrade is predicted daily by considering the probability distribution of the amount of rainfall, the probabilities of wet and dry days, the infiltration of water into the pavement through cracks and joints, the drainage of the base course, and the wet and dry probabilities of pavement sublayers. Under a FHWA contract, Texas A&M University is in the process of upgrading the infiltration and drainage model. A research study has been initiated by the FHWA to combine the CMS model, the infiltration and drainage model, and the frost model, which is a mathematical model of coupled heat and moisture flow in soils developed by the Cold Regions Research and Engineering Laboratory (Johnson et al., 1986).

11.1.3 Structural Models

Report 1-26 recommended the use of elastic layer programs (ELP) and the ILLI-PAVE finite element program for the development of future AASHTO design guide. It suggested the use of the modulus–depth relationship obtained from ILLI-PAVE to establish the various moduli for the ELP, thus capitalizing on the stress-sensitive feature of ILLI-PAVE and the multiple wheel capability of ELP.

Finite Element Models

The two finite element modes, ILLI-PAVE and MICH-PAVE, are described in Section 3.3.2. As discussed previously, many questions need to be resolved before these models can be used with ELP. The most serious problem is that the linear elastic solutions obtained by both models differ significantly from those obtained by ELP. A basic requirement for any finite element program to be used for pavement design is the capability to check with ELP solutions when the elements in the same layer are assigned the same modulus. Failure to meet this basic requirement clearly indicates that something is wrong with these finite element programs.

Another limitation of ILLI-PAVE and MICH-PAVE is the representation of wheel loading by a single circular area. As shown in Figure 3.24, the use of a single wheel to replace a set of duals is unsafe for thin asphalt pavements because, due to the larger contact radius of a single wheel, the tensile strain at the bottom of the asphalt layer is smaller under a single wheel than under dual wheels. A finite element program capable of analyzing multiple wheels is highly desirable. Although the program will be very complex and require an extensive interpolation and superposition of stresses, there is no reason that such a program cannot be developed. When the iteration method is used for nonlinear analysis, the superposition principle can still be applied. The system is considered linear during each iteration, and the stresses due to various wheel loads can be superimposed to determine the modulus of each element.

Elastic Layer Programs

A large number of elastic layer programs are available. Three programs were used to compare results with KENLAYER, as shown by ELSYM5 in Table 3.5, VESYS in Table 3.7, and DAMA in Tables 3.8 and 3.9. KENLAYER as presented in this book is believed to be the most comprehensive structural model. It considers a variety of cases not available elsewhere and can be applied to layer systems consisting of linear elastic, nonlinear elastic, and viscoelastic materials under single- and multiple-wheel loads.

A major deficiency of KENLAYER is in the nonlinear analysis where a stress point must be specified for each nonlinear layer to compute its modulus based on the state of stresses. The assumption of a uniform modulus throughout each nonlinear layer is not theoretically correct but its effects on the critical pavement responses are believed to be small. Since the most critical responses always occur near the loads and the layer moduli near the loads have the most effect, a stress point directly under the center of a single wheel or between the centers of dual wheels can be reasonably selected. It is hoped that ILLI-PAVE or MICH-PAVE can be further improved so that its linear solutions will check closely with the ELP solutions. Thus, by comparing the nonlinear solutions obtained from KENLAYER with those from the finite element programs, some guidelines on the selection of stress points can be developed.

11.1.4 Distress Models

Distress models are sometimes called transfer functions which relate structural responses to various types of distress. Distress models are the weak link in the mechanistic–empirical methods, and extensive field calibration and verification are needed to establish reliable distress predictions.

Usable transfer functions for HMA fatigue and subgrade rutting are available but those for the rutting of HMA and granular materials are marginal and require further developments. Report 1-26 indicated that the use of "rutting rate" concept advanced by Ohio State University (Majidzadeh et al., 1976) appears to be very promising because it can be applied to all paving materials, including HMA, granular materials, and fine-grained soils. The report also recommended the use of the Shahin–McCullough thermal cracking model (1972) as a checking procedure to assess the thermal cracking potential after the thickness design is completed; if unsatisfactory, a softer asphalt cement should be used. Because HMA rutting is the major cause of permanent deformation on heavily traveled pavements with thicker HMA, the design procedure may be simplified by checking the rutting potential after the thickness design is completed. If unsatisfactory, the selection of different mixture design procedures and practices should be made until the rut depth is reduced to the acceptable limit.

Fatigue Cracking Models

Miner's (1945) cumulative damage concept has been widely used to predict fatigue cracking. It is generally agreed that the allowable number of load repetitions is related to the tensile strain at the bottom of the asphalt layer. The amount of damage is expressed as a damage ratio, which is the ratio between predicted and allowable number of load repetitions. Damage occurs when the sum of damage ratios reaches a value of 1. Because of variabilities, damage will not occur exactly and all at once when the ratio reaches 1. If mean parameter values are used for design, a damage ratio of 1 indicates that the probability of failure is 50%, or 50% of the area will experience fatigue cracking. By assuming the damage ratio to be a log normal distribution, the probability of failure, or the percentage of area cracked, can be computed and checked with field performance.

The major difference in the various design methods is the transfer functions which relate the HMA tensile strains to the allowable number of load repetitions. In the Asphalt Institute and Shell design methods, the allowable number of load repetitions N_f to cause fatigue cracking is related to the tensile strain ϵ_t at the bottom of the HMA and the HMA modulus E_1 by

$$N_f = f_1(\epsilon_t)^{-f_2} (E_1)^{-f_3} \tag{3.6}$$

For the standard mix used in design, the Asphalt Insititute equation for 20% of area cracked is

$$N_f = 0.0796(\epsilon_t)^{-3.291}(E_1)^{-0.854} \tag{11.6}$$

and the Shell equation is

$$N_f = 0.0685(\epsilon_t)^{-5.671}(E_1)^{-2.363} \tag{11.7}$$

Because the exponent f_2 is much greater than f_3, the effect of ϵ_t on N_f is much greater than that of E_1. Therefore, the E_1 term may be neglected:

$$N_f = f_1(\epsilon_t)^{-f_2} \tag{11.8}$$

Equation 11.8 has been used by several agencies as listed below:

Illinois Department of Transportation (Thompson, 1987)

$$N_f = 5 \times 10^{-6} \, (\epsilon_t)^{-3.0} \tag{11.9}$$

Transport and Road Research Laboratory (Powell et al., 1984)

$$N_f = 1.66 \times 10^{-10} \, (\epsilon_t)^{-4..32} \tag{11.10}$$

Belgian Road Research Center (Verstraeten et al., 1982)

$$N_f = 4.92 \times 10^{-14} \, (\epsilon_t)^{-4.76} \tag{11.11}$$

It can be seen that the exponent f_2 of the fatigue equations varies from 3.0 to 5.671, but the coefficient f_1 varies over several orders of magnitude from 5×10^{-6} to 4.92×10^{-14}. The exponents f_2 and f_3 are usually determined from fatigue tests on laboratory specimens, while f_1 must shift from laboratory to field values by calibration. Pell (1987) indicated that the shift factor may range from 5 to 700. Due to differences in materials, test methods, field conditions, and structural models, a large variety of transfer functions are expected. No matter what transfer function is used, it is important to carefully calibrate the function by applying an appropriate shift factor so that the predicted distress can match with field observations.

Rutting Models

Two procedures have been used to limit rutting: one to limit the vertical compressive strain on top of the subgrade and the other to limit the total accumulated permanent deformation on the pavement surface based on the permanent deformation properties of each individual layer. In the Asphalt Institute and Shell design methods, the allowable number of load repetitions N_d to limit rutting is related to the vertical compressive strain ϵ_c on top of the subgrade by

$$N_d = f_4 \, (\epsilon_c)^{-f_5} \tag{3.7}$$

Equation 3.7 is also used by several other agencies with the values of f_4 and f_5 shown in Table 11.1.

As can be seen from Table 11.1, the exponent f_5 falls within a narrow range, but the coefficient f_4 varies a great deal. Both f_4 and f_5 should be calibrated by comparing the predicted performance with field observations.

In the subgrade strain method, it is assumed that if the subgrade compressive strain is controlled, reasonable surface rut depths will not be exceeded. For example, designs by the Asphalt Institute method are not expected to have a rut depth greater than 0.5 in. (12.7 mm) and designs by the TRRL procedure are not expected to have a rut depth of more than 0.4 in. (10.2 mm). The Shell method has a suggested procedure for estimating permanent deformation, as shown by Eq. 7.51. Without such a procedure, HMA rutting can best be addressed by improving the material selection and mixture design.

Unless standard thicknesses and materials are used for design, the evaluation of surface rutting based on the subgrade strain does not appear to be reasonable. Under heavy traffic with thicker HMA, most of the permanent deformation occurs in the HMA, rather than in the subgrade. Because rutting is caused by the accumulation of permanent deformation over all layers, it is more reasonable to determine the permanent deformation in each layer and sum up the results, as described by the direct method in Section 7.4.1.

Many methods are available to determine rut depth. The VESYS method was described in Section 7.4.1. In VESYS-V, a new option similar to the direct method was added. In this section, only the Ohio State model, which was recommended in Report 1-26, is described.

TABLE 11.1 SUBGRADE STRAIN CRITERIA USED BY VARIOUS AGENCIES

Agency	f_4	f_5	Rut depth (in.)
Asphalt Institute	1.365×10^{-9}	4.477	0.5
Shell (revised 1985)			
50% reliability	6.15×10^{-7}	4.0	
85% reliability	1.94×10^{-7}	4.0	
95% reliability	1.05×10^{-7}	4.0	
U.K. Transport & Road Research			
Laboratory (85% reliability)	6.18×10^{-8}	3.95	0.4
Belgian Road Research Center	3.05×10^{-9}	4.35	

Note. 1 in. = 25.4 mm.

Similar to VESYS, the Ohio State model also assumes a linear relationship between permanent strain ϵ_p and number of load repetitions N when plotted in log scales. However, instead of total rutting, the rutting rate is considered, as indicated by

$$\frac{\epsilon_p}{N} = A(N)^{-m} \qquad (11.12)$$

in which A is an experimental constant depending on material type and state of stress and m is an experimental constant depending on material type. Figure 11.4 shows typical straight-line relationships between $\log(\epsilon_p/N)$ and $\log N$ of HMA specimens under different stress and temperature conditions. All lines are nearly parallel, indicating that the value of m, which is the slope of the straight lines, is a constant independent of the levels of stress and temperature.

Equation 11.12 can be written as

$$\epsilon_p = A(N)^{1-m} \qquad (11.13)$$

Compared with Eq. 7.41 in VESYS

$$m = \alpha \qquad (11.14a)$$

and

$$A = \frac{\epsilon\,\mu}{1 - \alpha} \qquad (11.14b)$$

VESYS defines permanent deformation parameters by α and μ, while Ohio State defines them by A and m. To facilitate the computation of permanent deformation without subdividing the system into a large number of layers, VESYS relates the permanent strain ϵ_p to the resilient strain ϵ, while the Ohio State model applies the direct method and defines the parameter A directly as a material property. Khedr (1986) conducted a large number of repeated load tests on HMA specimens and found a straight-line relationship between $\log A$ and $\log(M_R/\sigma_d)$, as shown in Figure 11.5. Therefore, the value of A can be expressed as

$$A = a\left(\frac{M_R}{\sigma_d}\right)^{-b} = a(\epsilon)^b \qquad (11.15)$$

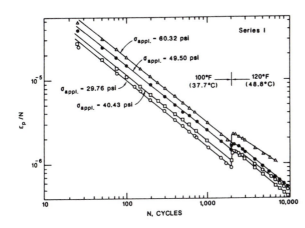

Figure 11.4 Typical relationship between ϵ_p/N and N for HMA specimens (1 psi = 6.9 kPa). (After Khedr (1986).)

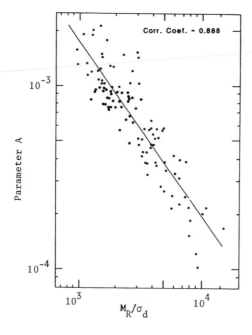

Figure 11.5 Relationship between parameter A and M_R/σ_d. (After Khedr (1986).)

in which a and b are material constants, ϵ is the resilient strain, M_R is the resilient modulus, and σ_d is the applied stress.

A comparison of Eqs. 11.14b and 11.15 shows that VESYS assumes that the parameter A is proportional to ϵ while the Ohio State model assumes that A is proportional to $(\epsilon)^b$. It was reported that rut depths were generally seriously underpredicted by VESYS (Rauhut et al., 1977). The addition of a new material constant, b, certainly gives the Ohio State model more flexibility in predicting rut depths. The final form of Ohio State model is

$$\epsilon_p = a(\epsilon)^b (N)^{1-m} \tag{11.16}$$

Considerable experimental data (Majidzadeh et al., 1976, 1978; Khedr, 1985) indicate that m varies within a narrow range for cohesive soils and granular materials. A typical value is 0.88 for cohesive soils and 0.8 for granular materials. The A term is quite variable and is dependent on material properties, repeated stress states, and environmental conditions. Several equations were developed to relate A to these parameters. When A and m are known for each material in the pavement, the rut depth, or the permanent deformation on the surface, can be determined by the direct method, as described in Section 7.4.1.

Thermal Cracking Models

There are two forms of thermal cracking in asphalt pavements: low-temperature cracking and thermal fatigue cracking. The mechanism of low-temperature cracking is illustrated in Figure 11.6. When the pavement temperature decreases, the tensile stress always increases but the tensile strength increases only to a maximum and then decreases. Low-temperature cracking occurs when

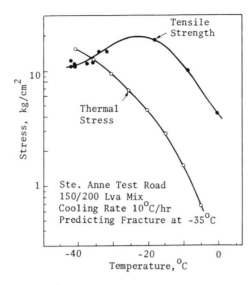

Figure 11.6 Mechanism of low-temperature cracking. (After McLeod (1970).)

the thermal tensile stress in the HMA exceeds its tensile strength. If the tensile stress is smaller than the tensile strength, the pavement will not crack under a single daily temperature cycle but may still crack under a large number of cycles. This is called thermal fatigue cracking, which occurs when the fatigue consumed by daily temperature cycles exceeds the HMA fatigue resistance.

Several mechanistic thermal cracking models are available, such as those developed by Finn et al. (1986), Ruth et al. (1982), Lytton et al. (1983), and Shahin and McCullough (1972). The latter two are the most comprehensive models which examine both low-temperature and thermal fatigue cracking and were recommended by Report 1-26 for further studies. They use the same basic structure to examine the accumulation of damage but with very different approaches. The model developed by Lytton et al. is more theoretical and is based on the principles of viscoelastic fracture mechanics, whereas that by Shahin and McCullough is more phenomenal and much easier to understand. Figure 11.7 shows how thermal cracking is analyzed in the Shahin–McCullough model.

Explanation of Figure 11.7

1. Weather data include mean annual air temperature, annual range in air temperature, daily range in air temperature, mean annual solar radiation, July average solar radiation, mean annual wind velocity, and surface absorbtivity. Asphalt properties include original penetration, original softening point, and percentage of penetration after the thin film oven test. HMA properties include thermal conductivity, specific heat, density, mix composition, coefficient of thermal contraction, maximum tensile strength, and fatigue characteristics.

2. From the mean annual air temperature, the mean annual solar radiation, and the July average solar radiation, the mean daily air temperature and mean daily solar radiation can be estimated. For each day the hourly temperatures at a designated point in the pavement, usually on the pavement surface where the

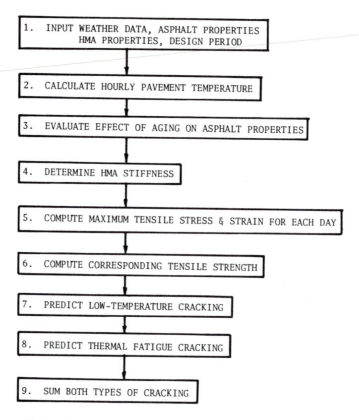

1. INPUT WEATHER DATA, ASPHALT PROPERTIES HMA PROPERTIES, DESIGN PERIOD

2. CALCULATE HOURLY PAVEMENT TEMPERATURE

3. EVALUATE EFFECT OF AGING ON ASPHALT PROPERTIES

4. DETERMINE HMA STIFFNESS

5. COMPUTE MAXIMUM TENSILE STRESS & STRAIN FOR EACH DAY

6. COMPUTE CORRESPONDING TENSILE STRENGTH

7. PREDICT LOW-TEMPERATURE CRACKING

8. PREDICT THERMAL FATIGUE CRACKING

9. SUM BOTH TYPES OF CRACKING

Figure 11.7 Analysis procedures for Shahin–McCullough model.

thermal cracking initiates, can be computed by a modified version of Barber's equation (1957).

3. The effect of aging on penetration and the ring and ball softening point of the asphalt was considered by two empirical equations. The penetration at any time after HMA placement is a function of time duration, original penetration, percentage of void in HMA, and percentage of original penetration after the thin film oven test. The softening point is a function of time duration, original softening point, and the percentage of original penetration after the thin film test. Both equations were used in conjunction with the Van der Poel nomograph to estimate the asphalt stiffness as a function of time.

4. Van der Poel's nomograph, as shown in Figure 7.19, was applied to estimate the stiffness of asphalt, which was then used to determine the stiffness of HMA, based on the volume concentration of the aggregate. Two techniques were established to computerize the nomograph. The first technique, which is more accurate but is limited to three loading times of 0.01 s, 0.02 s, and 1 h, is to convert the nomograph to a computerized form. The second technique, which is less accurate but can be extended to other times of loading, is to develop predictive models through the use of regression.

5. Thermal stresses can be estimated by assuming the temperature drop to be uniform and dividing it into equal intervals ΔT of 1-h duration. Starting from the maximum temperature and going down on a hourly basis to the minimum temperature, estimate the stiffness at the middle of the temperature interval. The increase in thermal stress can be expressed as

$$\Delta\sigma(t, \Delta T) = \alpha_t \Delta T S_m(t, \overline{\Delta T}) \tag{11.17}$$

in which $\Delta\sigma(t, \Delta T)$ is the increase in thermal stress for a given time of loading t and temperature interval ΔT; α_t is the coefficient of thermal contraction; and S_m is the HMA stiffness at a given time of loading and at the mean value of ΔT. The increments of thermal stresses obtained from Eq. 11.17 are accumulated to estimate the maximum stress and strain for that day:

$$\sigma(t, T) = \alpha_t \sum \{\Delta T S_m(t, \overline{\Delta T})\} \tag{11.18}$$

In solving Eq. 11.18, it is assumed that the stress and strain are negligible at the end of each day and that the maximum daily stress occurs at the minimum daily pavement temperature as a result of accumulation of thermal stress increments during the day. Asphalt stiffness is affected strongly by the loading time, which is not the time corresponding to ΔT but depends on the rate of temperature drop. The time of loading was determined by matching the maximum thermal stress measured in a laboratory specimen under the given rate of temperature drop with that computed from Eq. 11.18.

6. The extent of low-temperature cracking is evaluated by comparing the maximum tensile stress σ with the tensile strength H of HMA at the corresponding temperature. The temperature–strength relationship can be specified directly or computed by Heukelom's procedure (1966) based on the maximum HMA tensile strength.

7. Given the coefficients of variation of α_t, T, and S_m, the variance of σ can be determined, which in combination with the variance of H results in the total variance. Assuming both stress and strength as normally distributed random variables, the probability of failure, which indicates the percentage of area cracked, can be determined by

$$\text{Probability (failure)} = \text{Probability } (\sigma - H > 0) \tag{11.19}$$

The area of cracking can be converted to linear cracking by assuming that 1 ft (0.3 m) of linear cracking is equivalent to 5 ft² (0.46 m²) of area cracking. Note that in VESYS 1 ft (0.3 m) of low-temperature cracking is equivalent to 3 ft² (0.28 m²) of area cracking.

8. The analysis of thermal fatigue cracking is similar to that of fatigue cracking under repeated loading. The same fatigue equation, as indicated by Eq. 10.68, can be used:

$$N_f = K_1 \left(\frac{1}{\epsilon_t}\right)^{K_2} \tag{10.68}$$

in which N_f is the allowable number of cycles to cause fatigue cracking under tensile strain ϵ_t, and K_1 and K_2 are fatigue constants based on constant strain tests. Because K_1 and K_2 vary with the HMA stiffness, two sets of values, one at a very low stiffness and the other at a very high stiffness, must be specified, so the values at any intermediate stiffness can be interpolated. Since there is only one thermal cycle per day, the damage ratio can be expressed as

$$c = \sum_{i=1}^{n} \frac{1}{N_i} \qquad (11.20)$$

in which n is the number of days and N_i is the allowable number of repetitions for the ith day. If N is assumed to be a log normal distribution with a given standard deviation, then the probability of failure can be determined by

$$\text{Probability (failure)} = \text{Probability (log } c > 0) \qquad (11.21)$$

For easy understanding, the procedure described above is slightly different from the original report by Shahin and McCullough (1972), but the basic principle is the same. The probability of failure is the percentage of cracked area, which can be converted to linear cracking by multiplying by 0.2.

9. Low-temperature cracking and thermal fatigue cracking can be summed to estimate total cracking for a specified time after construction.

Example 11.2

Given a tensile stress σ of 100 psi (690 kPa) with a coefficient of variation of 0.3 and a tensile strength H of 150 psi (1.04 MPa) with a coefficient of variation of 0.2, estimate the length of crack in ft per 1000 ft^2.

Solution: The variance of $(\sigma - H)$ is $(0.3 \times 100)^2 + (0.2 \times 150)^2 = 1800$ (psi)^2 or the standard deviation of $(\sigma - H) = \sqrt{1800} = 42.4$ psi (293 kPa). The mean of $(\sigma - H)$ is $100 - 150 = -50$, so the normal deviate for $(\sigma - H) = 0$ is $z = [0 - (-50)]/42.4 = 1.18$. From Table 10.1, $\psi(z) = 0.381$, so there is 11.9% probability that $(\sigma - H)$ is greater than 0. The area cracked is $0.119 \times 1000 = 119$ ft^2/1000 ft^2, or the length cracked is $119/5 = 23.8$ ft/1000 ft^2 (78.1 m/1000 m^2).

11.2 ASPHALT INSTITUTE METHOD

From 1954 to 1969, eight editions of Manual Series No. 1 (MS-1) were published by the Asphalt Institute for the thickness design of asphalt pavements. The procedures recommended in these manuals were empirical. The seventh and eighth editions of MS-1 were based on data from the AASHO Road Test, the WASHO Road Test, and a number of British test roads, as well as comparisons with the design procedures of U.S. Army Corps of Engineers and some state agencies. In 1981, the ninth edition of MS-1 was published. Unlike previous editions, the ninth edition is based on mechanistic–empirical methodology, using the mechanistic multilayer theory in conjuction with empirical failure criteria to determine pavement thicknesses. Based on the results from a computer program named DAMA, a series of design charts covering three different temperature regimes were developed. However, only the charts for one regime, which represents a large part of the United States, were included in MS-1. In 1991, a revision of the ninth edition of MS-1 was made, in which charts for all three temperature regimes were included (AI, 1991). Details about the DAMA program are presented in Section 3.3.4.

11.2.1 Design Criteria

As has been explained in previous chapters, two types of strains have frequently been considered most critical for the design of asphalt pavements. One is the horizontal tensile strain ϵ_t at the bottom of the asphalt layer, which causes fatigue cracking, and the other is the vertical compressive strain ϵ_c on the surface of the subgrade, which causes permanent deformation or rutting. These two strains are used as failure criteria in the Asphalt Institute method.

Fatigue Criterion

The fatigue equation employed by the Asphalt Institute is discussed in Section 7.3.1. For a standard mix with an asphalt volume of 11% and air void volume of 5%, the equation is

$$N_f = 0.0796(\epsilon_t)^{-3.291}|E^*|^{-0.854} \tag{7.38}$$

in which N_f is the allowable number of load repetitions to control fatigue cracking and $|E^*|$ is the dynamic modulus of the asphalt mixture. If the asphalt or air void volume is different from 11 or 5%, a correction factor C, shown by Eq. 7.37, must be applied. However, Eq. 7.38, based on the standard mix, was used to develop the design charts. It was reported that the use of Eq. 7.38 would result in fatigue cracking of 20% of the total area, as observed on selected sections of the AASHO Road Test (AI, 1982).

Permanent Deformation Criterion

The allowable number of load repetitions to control permanent deformation can be expressed as

$$N_d = 1.365 \times 10^{-9} (\epsilon_c)^{-4.477} \tag{11.22}$$

Equation 11.22 was used to develop the design charts. As long as good compaction of the pavement components is obtained and the asphalt mix is well designed, the use of Eq. 11.22 should not result in rutting greater than 0.5 in. (12.7 mm) for the design traffic.

11.2.2 Traffic Analysis

Methods for traffic analysis are discussed in Section 6.4. Load repetitions, expressed in terms of an 18-kip (80-kN) single-axle load, are determined from traffic estimates using AASHTO equivalent factors for a structural number SN of 5 and a terminal serviceability index p_t of 2.5, as shown in Table 6.4.

Determination of Design ESAL

The procedure for determining the design ESAL can be summarized as follows:
1. Estimate the number of vehicles of different types, such as passenger

cars, single-unit trucks, including buses, and multiple unit trucks of various types, expected on the proposed facility. In the United States, traffic classification counts are made periodically by state highway and other agencies and should be available for use in pavement design. If the total number of trucks can be estimated but the classification is not known, Table 6.9 can be used as a guide.

2. Determine the number of each type of truck on the design lane during the first year of traffic. For two-lane highways, the design lane may be either lane of the pavement facility. Under some conditions, more trucks may travel in one direction than in the other. In many locations, heavily loaded trucks will travel in one direction and empty trucks in the other direction. In the absence of specific data, Table 6.15 may be used for determining the relative proportion of trucks to be expected on the design lane.

3. Determine a truck factor for each vehicle type. Truck factor is defined as the number of 18-kip (80-kN) axle load applications contributed by one passage of a truck. The computation of the truck factor is described in Section 6.4.2 and illustrated in Table 6.11. Typical truck factors are given in Table 6.10. If only the total number of trucks is known, it is not necessary to divide the number into different types according to Table 6.9. The equivalent 18-kip (80-kN) single-axle load applications can be obtained directly by multiplying the total number of trucks with the truck factor for all trucks shown at the bottom of Table 6.10.

4. For the given design period, select from Table 6.13 a single-growth factor for all trucks or separate factors for each truck type, as appropriate.

5. Multiply the number of trucks of each type by the truck factor and the growth factor and sum the values determined to obtain the design ESAL.

Example 11.3

Table 11.2 shows a four-lane rural highway with the number of trucks during the first year on the design lane shown in column 2 and the truck factor in column 3. If the annual growth rate is 4%, determine ESAL for a design period of 20 years.

TABLE 11.2 EXAMPLE FOR TRAFFIC ANALYSIS

Vehicle type (1)	Number of vehicles (2)	Truck factor (3)	Growth factor (4)	ESAL ($2 \times 3 \times 4$) (5)
Single-unit trucks				
2 axles, 4 tires	78,800	0.003	29.8	7000
2 axles, 6 tires	13,200	0.25	29.8	98,300
3 axles or more	3900	0.86	29.8	100,000
All singles	95,900			205,300
Tractor semitrailers and combinations				
4 axles or less	3900	0.92	29.8	106,900
5 axles	30,200	1.25	29.8	1,125,000
6 axles or more	1300	1.54	29.8	59,700
All tractors, etc.	35,400		Subtotal	1,291,000
All trucks	131,300		Design ESAL =	1,496,900

Solution: With the annual growth rate of 4% and a design period of 20 years, from Table 6.13, growth factor = 29.8, as shown by column 4 in Table 11.2. Column 5 is the product of columns 2, 3, and 4. The sum of column 5 is 1,496,900, which is the design ESAL.

It is interesting to note that column 2 is based on an AADT of 4000, 20% trucks, 45% in the design lane, and an average truck distribution shown in Table 6.9 under "Other Principal" of rural systems, and column 3 is the average truck factor in Table 6.10 under the same category. Actually the ESAL can be obtained directly by multiplying the total number of trucks with the truck factor for all trucks, or ESAL = 131,300 × 0.38 × 29.8 = 1,486,800, which is slightly different from the 1,496,900 shown in Table 11.2 due to round-off error.

Simplified Procedure for Determining Design ESAL

If detailed traffic data are not available, the Asphalt Institute recommends the use of Table 11.3 for estimating the design ESAL (AI, 1981b). This simplified procedure separates traffic into six classes, each associated with a type of highway or street and an average number of heavy trucks expected on the facility during the design period. Heavy trucks are defined as two-axle, six-tire trucks or larger. Pickup, panel, and light four-tire trucks are not included. The ESAL shown for each class can be used for design purposes.

TABLE 11.3 TRAFFIC CLASSIFICATION

Traffic class	Type of street or highway	Range of heavy trucks expected in design period	ESAL
I	Parking lots, driveways Light traffic residential streets Light traffic farm roads	Less than 7000	5×10^3
II	Residential streets Rural farm and residential roads	7000 to 15,000	10^4
III	Urban minor collector streets Rural minor collector roads	70,000 to 150,000	10^5
IV	Urban minor arterial and light industrial streets Rural major collector and minor arterial highways	700,000 to 1,500,000	10^6
V	Urban freeways, expressways, and other principal arterial highways Rural interstate and other principal arterial highways	2,000,000 to 4,500,000	3×10^6
VI	Urban interstate highways Some industrial roads	7,000,000 to 15,000,000	10^7

Note. Whenever possible, more rigorous traffic analysis should be used for roads and streets in traffic category IV or higher.

Source. AI (1981b).

11.2.3 Material Characterization

The material properties to be used for analysis include the resilient moduli and Poisson ratios of subgrade, granular base, and asphalt layer. Poisson ratios can be reasonably assumed as 0.45 for subgrade soils and 0.35 for all other materials.

Subgrade Soils

The resilient modulus used in this design procedure is the normal resilient modulus, which is not representative of times when the subgrade is freezing or thawing. It can be determined from the resilient modulus test or correlated with other tests, such as CBR or R values, as described in Section 7.1.6. To determine a representative resilient modulus, substantial testing of subgrade materials within 2 ft (0.6 m) of the planned subgrade elevation is required. If significant variations are present, random sampling should be made to determine the controlling soil type or the boundaries between different soils. If the soil types are significantly different and each soil covers a sufficiently large area, consideration should be given to subdividing the project for separate designs.

A minimum of six to eight test values are usually used to determine the design subgrade resilient modulus. The design subgrade resilient modulus is defined as the modulus value that is smaller than 60, 75, or 87.5% of all the test values. These percentages are known as percentile values and are related to traffic levels, as shown in Table 11.4.

Example 11.4

The results of eight tests produced the following subgrade resilient modulus values; 6200, 7800, 8800, 9500, 10,000, 11,300, 11,900, and 13,500 psi (42.8, 53.8, 60.7, 65.6, 69.0, 78.0, 82.1 and 93.2 MPa). Determine the design subgrade resilient modulus for ESAL of 10^4, 10^5, and 10^6

Solution: The percentile values are calculated in Table 11.5. From Table 11.4, the percentile values for ESAL of 10^4, 10^5, and 10^6 are 60, 75, and 87.5%, respectively. It can be estimated from Table 11.5 that the design resilient moduli corresponding to 60, 75, and 87.5 percentile values are 9600, 8800, and 7800 psi (66.2, 60.7, and 53.8 MPa), respectively. For more accurate results, it is preferable to plot the percentile values versus the subgrade resilient moduli, as shown in Figure 11.8, and draw a smooth curve to correct any irregularities. Note that at 75 percentile, the resilient modulus determined from Figure 11.8 is 8700 psi (60 MPa), which is slightly different from the original 8800 psi (60.7 MPa).

TABLE 11.4 DESIGN SUBGRADE RESILIENT MODULUS

Traffic level ESAL	Design resilient modulus percentile value (%)
10^4 or less	60.0
Between 10^4 and 10^6	75.0
10^6 or more	87.5

Source. After AI (1981a).

TABLE 11.5 COMPUTATION OF PERCENTILE VALUE

Test value (psi)	Number equal to or greater than	Percent equal to or greater than
13,500	1	12.5
11,900	2	25.0
11,300	3	37.5
10,000	4	50.0
9500	5	62.5
8800	6	75.0
7800	7	87.5
6200	8	100.0

Note. 1 psi = 6.9 kPa.

Untreated Granular Materials

The effect of stresses on the resilient modulus of granular materials is indicated by Eq. 3.8. The coefficient K_1 was selected in the range from 8000 to 12,000 psi (55.2 to 82.8 MPa) and the exponent K_2 was set equal to 0.5. Instead of using the iterative method as in KENLAYER, DAMA applies Eqs. 3.28 and 3.29 to determine the modulus of granular base.

When untreated aggregate base and subbase are used, it is recommended that they comply with ASTM Specification D 2940 "Graded Aggregate Material for Bases or Subbases for Highways and Airports," except that the requirements given in Table 11.6 should apply where appropriate.

Hot Mix Asphalt

The resilient modulus is determined by Eq. 7.27. The design charts were developed using the following parameter values in Eq. 7.27: $P_{200} = 5\%, f = 10$ Hz, $V_a = 4\%$ for the surface course and 7% for the base course, and $V_b = 11\%$ for both courses. Three temperature regimes, representative of New York, South Carolina, and Arizona, were considered with mean annual air temperatures

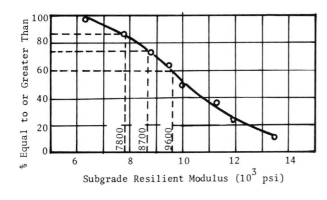

Figure 11.8 Graphical determination of design subgrade resilient modulus (1 psi = 6.9 kPa).

Test	Subbase	Base
CBR, minimum or	20	80
R value, minimum	55	78
Liquid limit, maximum	25	25
Plasticity index, maximum	6	NP
Sand equivalent, minimum	25	35
% passing No. 200, maximum	12	7

Source. After AI (1981a).

(MAAT) of 45, 60, and 75°F (7, 15.5, and 24°C), respectively. The types of asphalt cement and their viscosities for use in the development of design charts are shown in Table 11.7.

Table 11.8 shows the mean monthly air temperatures for the three representative temperature regimes. The temperature T in Eq. 7.27 is the mean pavement temperature M_p, which can be computed from the mean monthly air temperature by Eq. 3.27.

Emulsified Asphalt Mixtures

It is permissible to use emulsified asphalt mixtures for base courses. Depending on aggregate types, three types of mixes are specified:

1. *Type I:* mixes with processed dense graded aggregates, which should be mixed in a plant and have properties similar to HMA.
2. *Type II:* mixes with semiprocessed, crusher run, pit run, or bank run aggregates.
3. *Type III:* mixes with sands or silty sands.

Representative stiffness moduli at the time of placement and after being fully cured were used for each of the mix types based on the results of 32 different mixes tested at 73°F (23°C) and 100°F (38°C). The moduli at other temperatures

TABLE 11.7 ASPHALT GRADES AND VISCOSITY FOR DIFFERENT TEMPERATURE REGIMES

Location	Mean annual air temperature (MAAT)	Asphalt grades	Viscosity λ at 70°F (10^6 poise)
New York	45°F (7°C)	AC-5, AC-10	0.6
South Carolina	60°F (15.5°C)	AC-10, AC-20	1.6
Arizona	75°F (24°)	AC-40	5.0

Source. After AI (1982).

TABLE 11.8 VARIATIONS OF MEAN MONTHLY AIR TEMPERATURES WITH MAAT

MAAT	Mean monthly air temperature (F°)											
	Jan	Feb	Mar	Apr	May	Jun	Jul	Aug	Sep	Oct	Nov	Dec
45°F	24	25	14	27	42	48	61	69	65	55	48	41
60°F	45	38	43	45	56	70	78	81	78	73	58	54
75°F	55	61	61	73	90	91	92	93	93	86	72	55

Note. °F = 32 + 1.8 × °C.

Source. After AI (1982).

can be obtained by a straight-line interpolation. The effect of curing time on the stiffness modulus is represented by

$$E_t = E_f - (E_f - E_i)(RF) \qquad (11.23)$$

in which E_t is the modulus at curing time t, E_f is the modulus in the fully cured state, E_i is the modulus in the uncured or initial state, and RF is the reduction factor representing the amount of cure at time t.

A six-month cure period was used to prepare the design charts, since longer periods of curing up to 30 months do not have a significance influence on the thickness obtained from the design charts. The reduction factor for a six-month cure period is shown in Figure 11.9.

11.2.4 Environmental Effects

In addition to the effect of monthly temperature changes throughout the year on the stiffness moduli of HMA and emulsified asphalt mixtures, the design charts also take into consideration the effect of freezing and thawing on the resilient modulus of the subgrade and granular materials. This was accomplished by using an increased modulus to represent the freezing period and a reduced modulus to

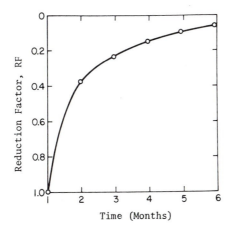

Figure 11.9 Reduction factor during six-month cure period. (After AI (1982).)

represent the thaw period. These adjustments are needed for regions with a MAAT of 45°F (7°C) or 60°F (15.5°C) but not for those of 75°F (24°C).

Subgrade

Figure 11.10 shows the variations of a subgrade resilient modulus throughout a year. The diagram represents four distinct periods: freeze, thaw, recovery, and normal. When the subgrade is completely frozen, a frozen modulus of 50,000 psi (345 MPa) is assumed. The modulus is reduced during the thaw period and reaches a minimum thaw modulus, which is a fraction of the normal modulus. The magnitude of the thaw modulus and the duration of each period are shown in Table 11.9 for the two temperature regimes. When the resilient moduli at the beginning and end of each period are known, those at any month during that period can be interpolated from Figure 11.10. Table 11.10 shows the monthly subgrade moduli used in DAMA for developing the design charts.

Untreated Granular Materials

A similar procedure was used to adjust the moduli of untreated granular sections. The coefficeint K_1 in Eq. 3.28 was increased by a factor of 300% for the frozen condition and reduced during the spring thaw to 25% of the normal value.

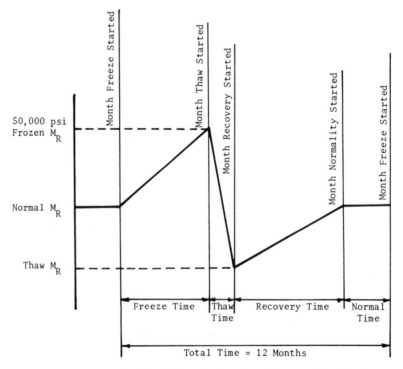

Figure 11.10 Seasonal variations of subgrade resilient modulus (1 psi = 6.9 kPa). (After AI (1982).)

TABLE 11.9 CONDITIONS USED TO REPRESENT FROST EFFECTS ON SUBGRADE

MAAT	Normal modulus (psi)	Thaw modulus		Month freeze started	Duration (month)			
		% Normal	psi		Freeze	Thaw	Recovery	Normal
45°F	4500	20	900	Dec	4	1	5	2
(7°C)	12,000	50	6000	Dec	4	1	5	2
	22,500	70	15,800	Dec	4	1	5	2
60°F	4500	30	1350	Jan	2	1	4	5
	12,000	60	7200	Jan	2	1	4	5
(15.5°C)	22,500	80	18,000	Jan	2	1	4	5

Note. 1 psi = 6.9 kPa.

Source. After AI (1982).

Table 11.11 contains a summary of the monthly K_1 values used. The value of K_2 was maintained constant at 0.5.

11.2.5 Design Procedure

The DAMA computer program was used to determine the minimum thickness required to satisfy both fatigue cracking and rutting criteria. For any given material and environmental conditions, two thicknesses were obtained, one by each criterion, and the larger of the two was used to prepare the design charts. For this reason, many of the design curves represent shapes associated with two different criteria.

Figures 11.11 through 11.20 are the design charts reproduced from the MS-1 manual. These charts are based on a mean annual air temperature of 60°F (15.5°C), which covers a major part of the continental United States. It is assumed that if asphalt cements are selected according to Table 11.7, the resulting HMA modulus will not change significantly to affect the thickness design, even if the temperature is somewhat different. However, in a report (AI, 1982) documenting the research and development of MS-1 manual, charts for mean annual air temperatures of 45°F (7.0°C) and 75°F (24°C) were also presented. These charts were included in the 1991 version of MS-1. If the mean annual air temperature is significantly different from 60°F (15.5°C), it is recommended that these design charts be utilized. The Asphalt Institute has issued a computer program called HWY which can be used to determine the thickness for all three temperature regimes. The program can also be used for the design of overlay, as is discussed in Section 13.3, and of full-depth pavements for parking lots, service stations, and driveways.

Full-Depth HMA

Figure 11.11 is the design chart for full-depth asphalt pavements. Given the subgrade resilient modulus M_R and the equivalent 18-kip single-axle load, ESAL, the total HMA thickness including both surface and base courses can be read directly from the chart.

TABLE 11.10 SUBGRADE MODULUS USED IN DAMA PROGRAM

MAAT	Normal modulus (psi)	Subgrade modulus by month (10^3 psi)											
		Dec	Jan	Feb	Mar	Apr	May	Jun	Jul	Aug	Sep	Oct	Nov
45°F (7°C)	4500	4.5	15.9	27.3	38.7	50.0	0.9	1.62	2.34	3.06	3.78	4.5	4.5
	12,000	12.0	21.5	31.0	40.5	50.0	6.0	7.20	8.40	9.60	10.8	12.0	12.0
	22,500	22.5	29.4	36.3	43.1	50.0	15.8	17.1	18.5	19.8	21.2	22.5	22.5
60°F (15.5°C)	4500	4.5	4.5	27.3	50.0	1.35	2.14	2.93	3.71	4.5	4.5	4.5	4.5
	12,000	12.0	12.0	31.0	50.0	7.2	8.4	9.6	10.8	12.0	12.0	12.0	12.0
	22,500	22.5	22.5	38.3	50.0	18.0	19.1	20.3	21.4	22.5	22.5	22.5	22.5

Note. 1 psi = 6.9 kPa.
Source. After AI (1982).

TABLE 11.11 MONTHLY K_1 VALUES FOR GRANULAR BASES

MAAT	Normal K_1 (psi)	Monthly value for K_1 (10^3 psi)											
		Dec	Jan	Feb	Mar	Apr	May	Jun	Jul	Aug	Sep	Oct	Nov
45°F (7°C)	8000	8.0	12.0	16.0	20.0	24.0	2.0	3.2	4.4	5.6	6.8	8.0	8.0
	12,000	12.0	18.0	24.0	30.0	36.0	3.0	4.8	6.6	8.4	10.2	12.0	12.0
60°F (15.5°C)	8000	8.0	8.0	16.0	24.0	2.0	3.5	5.0	6.5	8.0	8.0	8.0	8.0
	12,000	12.0	12.0	24.0	36.0	3.0	5.25	7.50	9.75	12.0	12.0	12.0	12.0

Note. 1 psi = 6.9 kPa.
Source. After AI (1982).

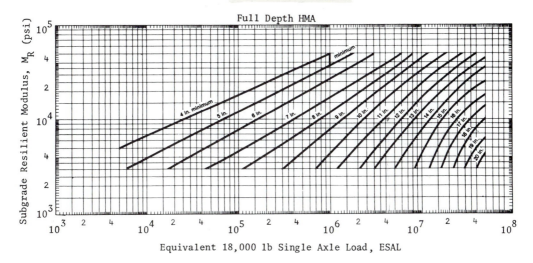

Figure 11.11 Design chart for full-depth HMA (1 psi = 6.9 kPa, 1 in. = 25.4 mm). (After AI (1981a).)

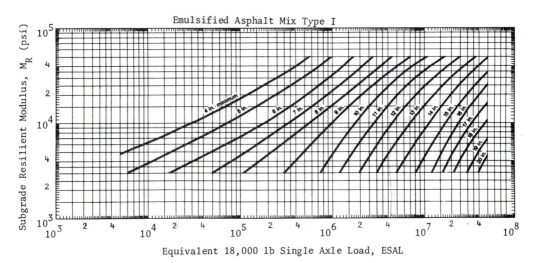

Figure 11.12 Design chart for type I emulsified asphalt mix (1 psi = 6.9 kPa, 1 in. = 25.4 mm). (After AI (1981a).)

Example 11.5

Given M_R = 10,000 psi (69 MPa) and ESAL = 10^6, determine the thickness of HMA for a full-depth asphalt pavement.

Solution: On Figure 11.11, a horizontal line is drawn at M_R = 10^4 psi (69 MPa) and a vertical line from ESAL = 10^6. The intersection of these two lines gives a HMA thickness of 8.5 in. (216 mm).

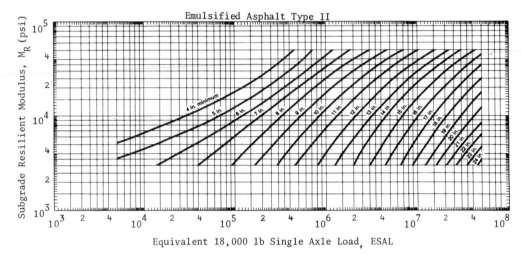

Figure 11.13 Design chart for type II emulsified asphalt mix (1 psi = 6.9 kPa, 1 in. = 25.4 mm). (After AI (1981a).)

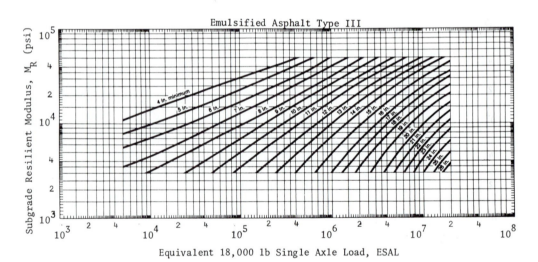

Figure 11.14 Design chart for type III emulsified asphalt mix (1 psi = 6.9 kPa, 1 in. = 25.4 mm). (After AI (1981a).)

HMA Over Emulsified Asphalt Base

Figures 11.12, 11.13, and 11.14 are the design charts for types I, II, and III emulsified asphalt mixes, respectively. The chart gives the combined thickness of HMA surface course and emulsified asphalt base course. The minimum thickness of HMA over the emulsified asphalt base varies with the traffic level and are shown in Table 11.12. The difference between the combined thickness and the HMA thickness is the thickness of emulsified asphalt required.

Untreated Aggregate Base 4.0 in. Thickness

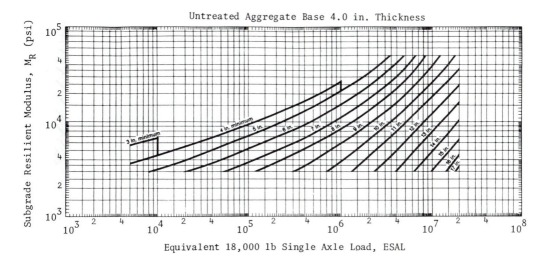

Figure 11.15 Design chart for HMA with 4-in. untreated aggregate base (1 psi = 6.9 kPa, 1 in. = 25.4 mm). (After AI (1981a).)

Untreated Aggregate Base 6.0 in. Thickness

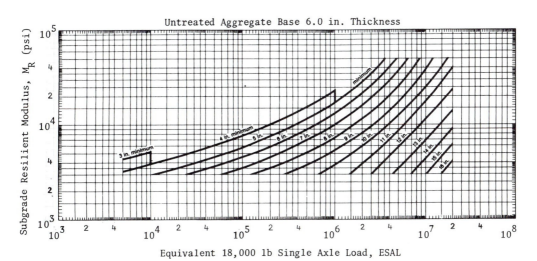

Figure 11.16 Design chart for HMA with 6-in. untreated aggregate base (1 psi = 6.9 kPa, 1 in. = 25.4 mm). (After AI (1981a).)

Example 11.6

Given M_R = 10,000 psi (69 MPa) and ESAL = 10^6, design the thickness of a pavement with a HMA surface over a type II emulsified asphalt base.

Solution: With M_R = 10,000 psi (69 MPa) and ESAL = 10^6, from Figure 11.13, the combined thickness is 10.5 in. (267 mm). From Table 11.12, the minimum HMA thickness is 3 in. (76 mm). If the minimum HMA thickness is used, the thickness of emulsified asphalt base is 10.5 − 3 = 7.5 in. (191 mm). Compared with Example 11.5, the total pavement thickness is increased by 2 in. (51 mm) when the emulsified asphalt base is used.

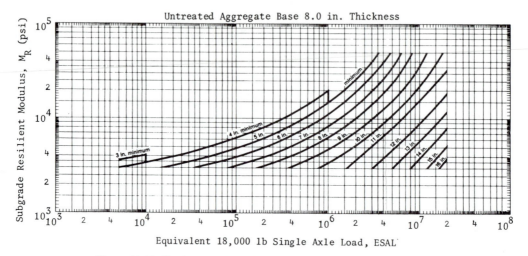

Figure 11.17 Design chart for HMA with 8-in. untreated aggregate base (1 psi = 6.9 kPa, 1 in. = 25.4 mm). (After AI (1981a).)

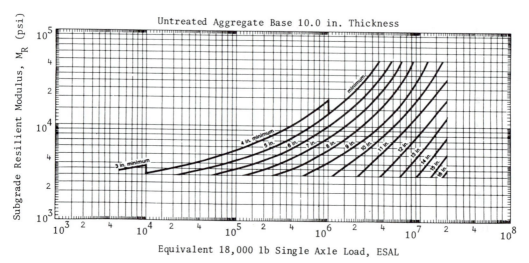

Figure 11.18 Design chart for HMA with 10-in. untreated aggregate base (1 psi = 6.9 kPa, 1 in. = 25.4 mm). (After AI (1981a).)

HMA Over Untreated Aggregate Base

Figures 11.15 through 11.20 are the design charts for HMA surface courses on untreated aggregate base courses of 4, 6, 8, 10, 12, and 18 in. (102, 152, 203, 254, 305, and 457 mm), respectively. Due to the large number of charts needed to cover all three temperature regimes, charts for the 4, 8, 10, and 18 in. (102, 203, 252, and 457 mm) granular base were not included in the 1991 version of MS-1. To use these charts, the designer must first determine what thickness of aggregate

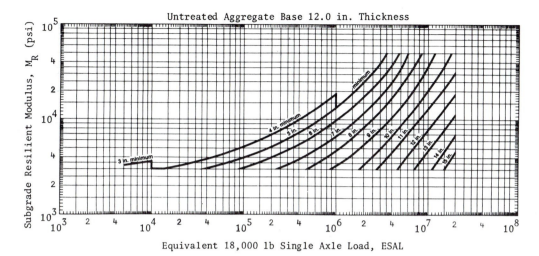

Equivalent 18,000 lb Single Axle Load, ESAL

Figure 11.19 Design chart for HMA with 12-in. untreated aggregate base (1 psi = 6.9 kPa, 1 in. = 25.4 mm). (After AI (1981a).)

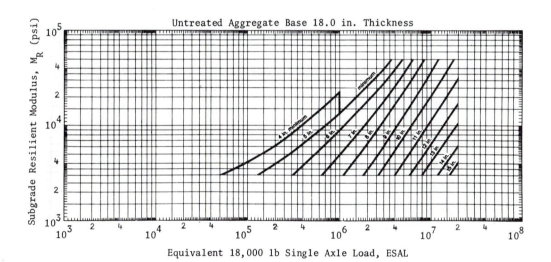

Untreated Aggregate Base 18.0 in. Thickness

Equivalent 18,000 lb Single Axle Load, ESAL

Figure 11.20 Design chart for HMA with 18-in. untreated aggregate base (1 psi = 6.9 kPa, 1 in. = 25.4 mm). (After AI (1981a).)

base is to be used and then select the appropriate design chart from which the HMA thickness can be found.

Example 11.7

Given M_R = 10,000 psi (69 MPa), ESAL = 10^6, and an untreated aggregate base of 8 in. (203 mm), determine the thickness of HMA required.

Solution: With M_R = 10^4 psi (69 MPa) and ESAL = 10^6, from Figure 11.17, the thickness of HMA is 6.5 in. (165 mm). Compared with Example 11.5, the use of an 8-in. (203-mm) untreated base reduces the HMA thickness by 2 in. (51 mm).

Sec. 11.2   Asphalt Institute Method

TABLE 11.12   MINIMUM THICKNESS OF HMA OVER
EMULSIFIED ASPHALT BASES

Traffic level ESAL	HMA thickness for type I mix (in.)	HMA thickness for type II and type III mixes (in.)
10^4	1	2
10^5	1.5	2
10^6	2	3
10^7	2	4
$>10^7$	2	5

Note. 1 in. = 25.4 mm.

Source. After AI (1981a).

HMA and Emulsified Asphalt Mix Over Untreated Aggregate Base

Design charts for pavements consisting of a HMA surface, an emulsified asphalt base, and an untreated base are currently not available. The best alternative is to use the charts for full-depth HMA and emulsified asphalt mix to determine a substitution ratio, which indicates the thickness of emulsified asphalt mix required to substitute a unit thickness of HMA. Then the chart for HMA over untreated aggregate base is applied to determine the thickness of HMA, part of which can be substituted by the emulsified asphalt mix according to the substitution ratio. The following method has been recommended by the Asphalt Institute:

1. Design a full-depth HMA pavement for the appropriate traffic and subgrade conditions. Assume a 2-in. (51-mm) surface course and calculate the corresponding base thickness.
2. Design a pavement for the same traffic and subgrade conditions using the selected emulsified mix type. Assume a 2-in. (51-mm) surface course and calculate the corresponding base thickness.
3. Divide the thickness of emulsified asphalt base in step 2 by the thickness of HMA base in step 1 to obtain a substitution ratio.
4. Design a pavement for the same traffic and subgrade conditions using HMA and untreated base.
5. Select a portion of the HMA thickness to be replaced by the emulsified asphalt mix, based on the minimum HMA thickness specified in Table 11.12.
6. Multiply the above thickness by the substitution ratio determined in step 3 to obtain the thickness of emulsified asphalt mix required.

Example 11.8

Given M_R = 10,000 psi (69 MPa), ESAL = 10^6, and an 8-in. (203-mm) untreated aggregate base, design the thicknesses of HMA surface course and type II emulsified asphalt base course.

Solution: 1. The thickness for full-depth HMA is 8.5 in. (216 mm), as determined in Example 11.5. If the HMA surface is 2 in. (51 mm), then the thickness of the HMA base is 6.5 in. (165 mm).

2. The thickness for emulsified asphalt mix to be placed directly on the subgrade is 10.5 in. (267 mm), as obtained in Example 11.6. If the HMA surface is 2 in. (51 mm), then the thickness of the emulsified asphalt base is 8.5 in. (216 mm).

3. The substitution ratio is 8.5/6.5 = 1.31.

4. The thickness of HMA over an 8 in. (203 mm) untreated aggregate base is 6.5 in. (165 mm), as shown in Example 11.7.

5. From Table 11.12, the minimum HMA thickness is 3 in. (76 mm), so 3.5 in. (89 mm) of HMA base must be replaced by the emulsified asphalt base.

6. The thickness of the emulsified asphalt base is 3.5 $\times$ 1.31 = 4.5 in. (114 mm). The final design consists of 3 in. (76 mm) of HMA, 4.5 in. (114 mm) of emulsified asphalt base, and 8 in. (203 mm) of untreated aggregate base, or a total thickness of 15.5 in (394 mm).

The Asphalt Institute suggested the use of a 2-in. (51-mm) surface course to determine the substitution factor. If the actual thickness is 3 in. (76 mm), as shown in the above example, the substitution factor is (10.5 − 3)/(8.5 − 3) = 1.36, which is not too much different from the 1.31 used.

Planned Stage Construction

Planned stage construction involves the successive applications of HMA layers according to a predetermined time schedule. The procedure is based on the concept of remaining life, which implies that the second stage will be constructed before the first stage shows serious signs of distress. Stage construction is beneficial when funds are insufficient for constructing a pavement with a long design life. This approach is also desirable when there is a great amount of uncertainty in estimating traffic. The pavement can be designed for an initial traffic and the next stage of construction can be designed using traffic projections based on the traffic in service. Finally, stage construction allows weak spots that develop in the first stage to be detected and repaired in the next stage.

If n_1 is the actual ESAL for stage 1 and N_1 is the allowable ESAL for the initial thickness h_1 selected for stage 1, then the damage ratio D_r at the end of stage 1 is

$$D_r = \frac{n_1}{N_1} \tag{11.24}$$

Note that D_r must be smaller than 1 because when $D_r = 1$ the pavement fails. Therefore, the remaining life in the existing pavement at the end of stage 1 is $(1 - D_r)$. The thickness obtained from the design chart is based on a D_r of 1 with no remaining life. To keep some remaining life, the thickness h_1 should be determined from the design chart based on an adjusted design ESAL N_1, which is somewhat greater than the design ESAL n_1, depending on the D_r specified. If h_1 is based on N_1 and

$$N_1 = \frac{n_1}{D_r} \tag{11.25}$$

then the damage ratio after N_1 load applications is 1, so N_1 can be considered to be the allowable number of applications. Since n_1 is smaller than N_1, the damage ratio is smaller than 1 and can be determined from Eq. 12.24.

If n_2 is the design ESAL for stage 2 and N_2 is the allowable or adjusted ESAL for stage 2, then the damage incurred in stage 2 should not exceed the remaining life:

$$\frac{n_2}{N_2} = 1 - D_r \tag{11.26}$$

or

$$N_2 = \frac{n_2}{1 - D_r} \tag{11.27}$$

Note that N_2 is an adjusted design ESAL to permit the selection of a thickness h_2 that will carry traffic n_2 and use the remaining life. The difference between h_2 and h_1 is the additional thickness required in stage 2.

Thus, in a stage construction analysis, the designer is required to select a stage time period and the amount of damage D_r to be incurred during this stage. Given D_r and n_1, the adjusted design ESAL N_1 can be computed by Eq. 11.25 and used for determining h_1. Given n_2 and D_r, the adjusted design ESAL N_2 can be computed by Eq. 11.27 and used for determining h_2. The MS-1 manual recommends the use of 5 to 10 years for the first stage, with a damage ratio of 60% at the end of the stage.

Example 11.9

A full-depth HMA pavement with a subgrade resilient modulus of 10,000 psi (69 MPa) will be constructed in two stages. The first stage is 5 years with 150,000 ESAL repetitions and the second stage is 15 years with 850,000 ESAL repetitions. Limiting the damage ratio to 0.6 at the end of stage 1, determine the thickness of HMA required for the first 5 years and the thickness of overlay required to accommodate the additional traffic expected during the next 15 years.

Solution: Given $n_1 = 150,000$ and $D_r = 0.6$, from Eq. 11.25, $N_1 = 150,000/0.6 = 2.5 \times 10^5$. From Figure 11.11, $h_1 = 6.5$ in. (165 mm)

Given $n_2 = 850,000$, from Eq. 11.27, $N_2 = 850,000/(1 - 0.6) = 2.1 \times 10^6$. From Figure 11.11, $h_2 = 10$ in. (254 mm).

The thickness for the first stage is 6.5 in. (165 mm) and the overlay for the second stage is 3.5 in. (89 mm). If the design is not divided into two stages, the thickness of the pavement is 8.5 in. (216 mm), as shown in Example 11.5. The use of stage construction decreases the thickness in the first stage by 2 in. (51 mm) but increases the total thickness by 1.5 in. (38 mm).

11.2.6 Comparison with Observed Performance

Extensive efforts were devoted by the Asphalt Institute to compare the predicted HMA thickness with the actual thickness. The predicted thickness was determined from the design charts presented in MS-1 based on the estimated subgrade modulus, the thickness of untreated granular base, and the actual ESAL repetitions, which reduced the PSI level to 2.5. The actual thickness was the thickness of HMA incorporated in the pavement. The difference between predicted and actual thicknesses is Δh. If the predicted thickness is greater than the actual thickness, or Δh is positive, the use of the design charts is considered conservative. If the predicted thickness is smaller than the actual thickness, or Δh is negative, the design is unconservative.

Six separate sources of data, representing a total of 394 individual data points, were utilized. The studies include the San Diego Test Road in California, the Brampton Test Road in Canada, California investigation, Minnesota investigation, and the main factorial experiment and special base study of the AASHO Road Test in Illinois. The southern and eastern parts of the United States were not included due to the lack of documented data.

Figure 11.21 is a plot of Δh for the 394 sections. The figure shows that the MS-1 design procedure yields conservative results, especially when ESAL is above 2×10^6. The mean Δh of the 394 sections is 1.61 in. (41 mm) with a standard deviation of 1.24 in. (31.5 mm).

Figure 11.22 shows the distribution of Δh and illustrates that the probability of achieving unconservative design with $\Delta h < 0$ is 12.7%. This percentage is based on the actual data. If the data are assumed to be normally distributed with a mean of 1.61 and a standard deviation of 1.24, then the normal deviate for $\Delta h = 0$ is $(0 - 1.61)/1.24 = 1.298$ and the probability that $\Delta h < 0$ is 40%.

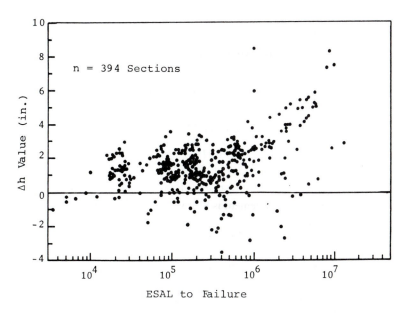

Figure 11.21 Thickness comparison for all pavement sections (1 in. = 25.4 mm). (After AI (1982).)

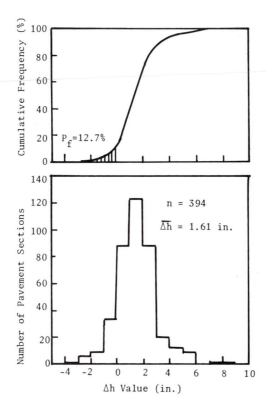

Figure 11.22 Distribution of Δh parameter for all pavement sections. (After AI (1982).)

11.3 AASHTO METHOD

The design procedure recommended by the American Association of State Highway and Transportation Officials (AASHTO) is based on the results of the extensive AASHO Road Test conducted in Ottawa, Illinois, in the late 1950s and early 1960s. The AASHO Committee on Design first published an interim design guide in 1961. It was revised in 1972 and 1981. In 1984–85, the Subcommittee on Pavement Design and a team of consultants revised and expanded the guide under NCHRP Project 20-7/24 and issued the current guide in 1986.

The empirical performance equations obtained from the AASHO Road Test are still being used as the basic models in the current guide but were modified and extended to make them applicable to other regions in the nation. It should be kept in mind that the original equations were developed under a given climatic setting with a specific set of pavement materials and subgrade soils. The climate at the test site is temperate with an average annual precipitation of about 34 in. (864 mm). The average depth of frost penetration is about 28 in. (711 mm). The subgrade soils consists of A-6 and A-7-6 that are poorly drained, with CBR values ranging from 2 to 4.

11.3.1 Design Variables

Several general design variables related to both flexible and rigid pavements are presented in this section. Other variables such as the effective roadbed soil resilient modulus and the structural number are presented in Sections 11.3.3 and 11.3.4, respectively.

Time Constraints

To achieve the best use of available funds, the AASHTO design guide encourages the use of a longer analysis period for high-volume facilities, including at least one rehabilitation period. Thus, the analysis period should be equal to or greater than the performance period, as described below.

Performance Period

The performance period refers to the time that an initial pavement structure will last before it needs rehabilitation or the performance time between rehabilitation operations. It is equivalent to the time elapsed as a new, reconstructed, or rehabilitated structure deteriorates from its initial serviceability to its terminal serviceability. The designer must select the performance period within the minimum and maximum allowable bounds that are established by agency experience and policy. The selection of performance period can be affected by such factors as the functional classification of the pavement, the type and level of maintenance applied, the funds available for initial construction, life cycle costs, and other engineering considerations.

Analysis Period

The analysis period is the period of time that any design strategy must cover. It may be identical to the selected performance period. However, realistic performance limitations may necessitate the consideration of staged construction or planned rehabilitation for the desired analysis period. In the past, pavements were typically designed and analyzed for a 20-year performance period. It is now recommended that consideration be given to longer analysis periods because they may be better suited for the evaluation of alternative long-term strategies based on life cycle costs. Table 11.13 contains general guidelines for the length of the analysis period.

Traffic

The design procedures are based on culmulative expected 18-kip (80-kN) equivalent single-axle load (ESAL). The determination of equivalent axle load factors (EALF) is discussed in Section 6.3.1 for flexible pavements and Section 6.3.2 for rigid pavements. The procedure for converting mixed traffic into ESAL is described in Section 6.4. The use of Table 6.13 for total growth factors and Table 6.16 for lane distribution factors is recommended.

TABLE 11.13 GUIDELINES FOR LENGTH OF
ANALYSIS PERIOD

Highway conditions	Analysis period (years)
High-volume urban	30–50
High-volume rural	20–50
Low-volume paved	15–25
Low-volume aggregate surface	10–20

Source. After AASHTO (1986).

If a pavement is designed for the analysis period without any rehabilitation or resurfacing, all that is required is the total ESAL over the analysis period. However, if stage construction is considered and rehabilitation or resurfacing is anticipated, a graph or equation of cumulative ESAL versus time is needed so that the ESAL traffic during any given stages can be obtained.

Reliability

Reliability concepts for pavement design are discussed in Section 10.2. Basically, realiablity is a means of incorporating some degree of certainty into the design process to ensure that the various design alternatives will last the analysis period. The level of reliability to be used for design should increase as the volume of traffic, difficulty of diverting traffic, and public expectation of availability increase. Table 11.14 presents recommended levels of reliability for various functional classifications.

Application of the reliability concept requires the selection of a standard deviation that is representative of local conditions. It is suggested that standard deviations of 0.45 be used for flexible pavements and 0.35 for rigid pavements. These correspond to variances of 0.2025 and 0.1225, which are smaller than those shown in Table 10.12.

TABLE 11.14 SUGGESTED LEVELS OF
RELIABILITY FOR VARIOUS FUNCTIONAL
CLASSIFICATIONS

Functional classification	Recommended level of reliability	
	Urban	Rural
Interstate and other freeways	85–99.9	80–99.9
Principal arterials	80–99	75–95
Collectors	80–95	75–95
Local	50–80	50–80

Note. Results based on a survey of AASHTO
Pavement Design Task Force.
Source. After AASHTO (1986).

When stage construction is considered, the reliability of each stage must be compounded to achieve the overall reliability:

$$R_{\text{stage}} = (R_{\text{overall}})^{1/n} \qquad (11.28)$$

in which n is the number of stages being considered. For example, if two stages are contemplated and the desired level of overall reliability is 95%, the reliability of each stage must be $(0.95)^{1/2}$, or 97.5%.

Environmental Effects

The AASHO design equations were based on the results of traffic tests over a two-year period. The long-term effects of temperature and moisture on the reduction of serviceability were not included. If problems of swell clay and frost heave are significant in a given region and have not been properly corrected, the loss of serviceability over the analysis period should be estimated and added to that due to cumulative traffic loads. Figure 11.23 shows the serviceability loss versus time curves for a specific location. The environmental loss is a summation of losses from both swelling and frost heave. The chart may be used to estimate the serviceability loss at any intermediate period, for example, a loss of 0.73 at the end of 13 years. Of course, if only swelling or frost heave is considered, there will be only one curve on the graph. The shape of these curves indicates that the serviceability loss due to environment increases at a decreasing rate. This may favor the use of stage construction because most of the loss will occur during the first stage and can be corrected with little additional loss in later stages.

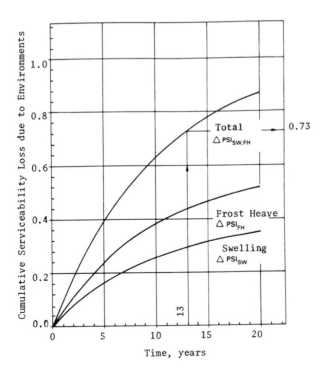

Figure 11.23 Environmental serviceability loss versus time for a specific location. (From the *AASHTO Guide for Design of Pavement Structures*. Copyright 1986. American Association of State Highway and Transportation Officials, Washington, DC. Used by permission.)

The serviceability loss due to roadbed swelling depends on the swell rate constant, the potential vertical rise, and the swell probability; that due to frost heave depends on the frost heave rate, the maximum potential serviceability loss, and the frost heave probability. Methods for evaluating these losses are described in Appendix G of the AASHTO design guide.

Serviceability

Initial and terminal serviceability indexes must be established to compute the change in serviceability, ΔPSI, to be used in the design equations. The initial serviceability index is a function of pavement type and construction quality. Typical values from the AASHO Road Test were 4.2 for flexible pavements and 4.5 for rigid pavements. The terminal serviceability index is the lowest index that will be tolerated before rehabilitation, resurfacing, and reconstruction become necessary. An index of 2.5 or higher is suggested for design of major highways and 2.0 for highways with lower traffic. For relatively minor highways where economics dictate a minimum initial capital outlay, it is suggested that this be accomplished by reducing the design period or total traffic volume, rather than by designing a terminal serviceability index less than 2.0.

11.3.2 Design Equations

The original equations were based purely on the results of the AASHO Road Test but were modified later by theory and experience to take care of subgrade and climatic conditions other than those encountered in the Road Test.

Original Equations

The following are the basic equations developed from the AASHO Road Test for flexible pavements (HRB, 1962):

$$G_t = \beta(\log W_t - \log \rho) \tag{11.29}$$

$$\beta = 0.40 + \frac{0.081 \, (L_1 + L_2)^{3.23}}{(SN + 1)^{5.19} \, L_2^{3.23}} \tag{11.30}$$

$$\log \rho = 5.93 + 9.36 \log(SN + 1) - 4.79 \log(L_1 + L_2)$$
$$+ 4.33 \log L_2 \tag{11.31}$$

in which G_t = logarithm of the ratio of loss in serviceability at time t to the potential loss taken at a point where $p_t = 1.5$, or $G_t = \log[(4.2 - p_t)/(4.2 - 1.5)]$, noting that 4.2 is the initial serviceability for flexible pavements;

β = a function of design and load variables, as shown by Eq. 11.30, that influences the shape of ρ versus W_t curve;

ρ = a function of design and load variables, as shown by Eq. 11.31, that denotes the expected number of load applications to a p_t of 1.5, as can be seen from Eq. 11.29 where $\rho = W_t$ when $p_t = 1.5$;

W_t = axle load application at end of time t;

p_t = serviceability at end of time t;

L_1 = load on one single axle or a set of tandem axles, in kip;

L_2 = axle code, 1 for single axle and 2 for tandem axle;

SN = structural number of pavement, which was computed by

$$SN = a_1 D_1 + a_2 D_2 + a_3 D_3 \qquad (11.32)$$

in which a_1, a_2, and a_3 are layer coefficients for the surface, base, and subbase, respectively; and D_1, D_2, and D_3 are the thicknesses of the surface, base, and subbase, respectively.

The procedure is greatly simplified if an equivalent 18-kip (80-kN) single-axle load is used. By combining Eqs. 11.29, 11.30, and 11.31 and setting $L_1 = 18$ and $L_2 = 1$, the following equation is obtained:

$$\log W_{t18} = 9.36 \log (SN + 1) - 0.20 + \frac{\log[(4.2 - p_t)/(4.2 - 1.5)]}{0.4 + 1094/(SN + 1)^{5.19}} \qquad (11.33)$$

in which W_{t18} is the number of 18-kip (80-kN) single-axle load applications to time t and p_t is the terminal serviceability index. Equation 11.33 is applicable only to the flexible pavements in the AASHO Road Test with an effective subgrade resilient modulus of 3000 psi (20.7 MPa).

Modified Equations

For other subgrade and environmental conditions, Eq. 11.33 is modified to

$$\log W_{t18} = 9.36 \log (SN + 1) - 0.20 + \frac{\log[(4.2 - p_t)/(4.2 - 1.5)]}{0.4 + 1094/(SN + 1)^{5.19}}$$

$$+ 2.32 \log M_R - 8.07 \qquad (11.34)$$

in which M_R is the effective roadbed soil resilient modulus. Note that when $M_R = 3000$ psi (20.7 MPa), Eq. 11.34 is identical to Eq. 11.33. To take local precipitation and drainage conditions into account, Eq. 11.32 was modified to

$$SN = a_1 D_1 + a_2 D_2 m_2 + a_3 D_3 m_3 \qquad (11.35)$$

in which m_2 is the drainage coefficient of base course and m_3 is the drainage coefficient of subbase course.

Equation 11.34 is the performance equation which gives the allowable number of 18-kip (80-kN) single-axle load applications W_{t18} to cause the reduction of PSI to p_t. If the predicted number of applications W_{18} is equal to W_{t18}, the reliability of the design is only 50% because all variables in Eq. 11.34 are based on mean values. To achieve a higher level of reliability, W_{18} must be smaller than W_{t18} by a normal deviate Z_R, as shown in Figure 11.24:

$$Z_R = \frac{\log W_{18} - \log W_{t18}}{S_0} \qquad (11.36)$$

in which Z_R is the normal deviate for a given reliability R, and S_0 is the standard deviation. Z_R can be determined from Table 10.1 and more conveniently from Table 11.15.

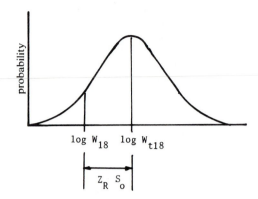

log W_{18} log W_{t18}

Z_R S_o

Figure 11.24 Reliability of design based on ESAL.

Combining Eqs. 11.34 and 11.36 and replacing $(4.2 - p_t)$ by ΔPSI yields

$$\log W_{18} = Z_R S_0 + 9.36 \log(SN + 1) - 0.20 + \frac{\log[\Delta PSI/(4.2 - 1.5)]}{0.4 + 1094/(SN + 1)^{5.19}}$$

$$+ 2.32 \log M_R - 8.07 \tag{11.37}$$

Equation 11.37 is the final design equation for flexible pavements. Figure 11.25 is a nomograph for solving Eq. 11.37. The DNPS86 computer program issued by AASHTO can also be used to solve Eq. 11.37 and perform the design procedure.

Example 11.10

Given $W_{18} = 5 \times 10^6$, $R = 95\%$, $S_0 = 0.35$, $M_R = 5000$ psi (34.5 MPa), and ΔPSI = 1.9, determine SN from Figure 11.25.

Solution: As shown by the arrows in Figure 11.25, starting from $R = 95\%$, a series of lines are drawn through $S_0 = 0.35$, $W_{18} = 5 \times 10^6$, $M_R = 5000$ psi (34.5 MPa), and ΔPSI = 1.9 and finally intersect SN at 5.0, so SN = 5.0.

The chart is most convenient for determining SN because the solution of SN by Eq. 11.37 is cumbersome and requires a trial and error process. If W_{18} is the unknown to be determined, the use of Eq. 11.37 is more accurate.

TABLE 11.15 STANDARD NORMAL DEVIATES FOR VARIOUS LEVELS OF RELIABILITY

Reliability (%)	Standard normal deviate (Z_R)	Reliability (%)	Standard normal deviate (Z_R)
50	0.000	93	-1.476
60	-0.253	94	-1.555
70	-0.524	95	-1.645
75	-0.674	96	-1.751
80	-0.841	97	-1.881
85	-1.037	98	-2.054
90	-1.282	99	-2.327
91	-1.340	99.9	-3.090
92	-1.405	99.99	-3.750

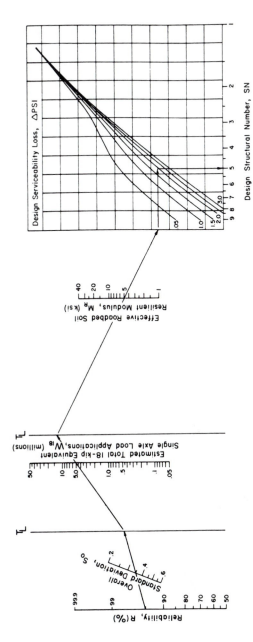

Figure 11.25 Design chart for flexible pavements based on mean values for each input (1 ksi = 6.9 MPa). (From the *AASHTO Guide for Design of Pavement Structures*. Copyright 1986. American Association of State Highway and Transportation Officials, Washington, DC. Used by permission.)

Example 11.11

Given $R = 95\%$, $SN = 5$, $S_0 = 0.35$, $M_R = 5000$ psi (34.5 MPa), $\Delta PSI = 1.9$, determine W_{18} by Eq. 11.37.

Solution: For $R = 95\%$, from Table 11.15, $Z_R = -1.645$. From Eq. 11.37, log $W_{18} = -1.645 \times 0.35 + 9.36 \log(5 + 1) - 0.2 + \log(1.9/2.7)/[0.4 + 1094/(6)^{5.19}] + 2.32 \log(5000) - 8.07 = 6.714$, or $W_{18} = 5.18 \times 10^6$, which checks with 5×10^6 in the previous example.

11.3.3 Effective Roadbed Soil Resilient Modulus

The effective roadbed soil resilient modulus M_R is an equivalent modulus that would result in the same damage if seasonal modulus values were actually used. The equation for evaluating the relative damage to flexible pavements u_f and the method for computing M_R are discussed below.

Relative Damage

From Eq. 11.37, the effect of M_R on W_{18} can be expressed as

$$\log W_{18} = \log C - \log(1.18 \times 10^8 M_R^{-2.32}) \tag{11.38}$$

in which log C is the sum of all but the last two terms in Eq. 11.37. Equation 11.38 can be written as

$$W_{18} = \frac{C}{1.18 \times 10^8 M_R^{-2.32}} \tag{11.39}$$

If W_T is the predicted total traffic, the damage ratio, which is a ratio between predicted and allowable number of load repetitions, can be expressed as

$$D_r = \frac{W_T}{C/(1.18 \times 10^8 M_R^{-2.32})} = \frac{W_T}{C}(1.18 \times 10^8 M_R^{-2.32}) \tag{11.40}$$

If W_T is uniformly distributed over n periods, the cumulative damage ratio is

$$D_r = \sum_{i=1}^{n} \frac{W_T/n}{C/(1.18 \times 10^8 M_{Ri}^{-2.32})} = \frac{W_T}{C}\frac{1}{n}\sum_{i=1}^{n}(1.18 \times 10^8 M_{Ri}^{-2.32}) \tag{11.41}$$

Equating Eq. 11.40 to Eq. 11.41 gives

$$1.18 \times 10^8 M_R^{-2.32} = \frac{1}{n}\sum_{i=1}^{n}(1.18 \times 10^8 M_{Ri}^{-2.32}) \tag{11.42}$$

Equation 11.42 can be used to determine the effective roadbed soil resilient modulus M_R in terms of seasonal moduli M_{Ri}. Although the coefficient 1.18×10^8 can be canceled out to simplify the equation, the AASHTO design guide keeps the coefficient and defines the relative damage u_f as

$$u_f = 1.18 \times 10^8 M_R^{-2.32} \tag{11.43}$$

Computation of Effective Roadbed Soil Resilient Modulus

Figure 11.26 is a worksheet for estimating effective roadbed soil resilient modulus, in which Eq. 11.43 together with a vertical scale for graphical solution of u_f are also shown. A year is divided into a number of periods during which different roadbed soil resilient moduli are specified. The shortest time period is half a month. These seasonal moduli can be determined from correlations with soil moisture and temperature conditions or from nondestructive deflection testing.

Month	Roadbed Soil Modulus, M_R (psi)	Relative Damage, u_f
Jan.	15,900	0.02
Feb.	27,300	0.01
Mar.	38,700	0.00
Apr.	50,000	0.00
May	900	16.52
June	1,620	4.22
July	2,340	1.80
Aug.	3,060	0.97
Sept.	3,780	0.59
Oct.	4,500	0.39
Nov.	4,500	0.39
Dec.	4,500	0.39
Summation: $\Sigma u_f =$		25.30

Average: $\bar{u}_f = \dfrac{\Sigma u_f}{n} = \underline{2.11}$

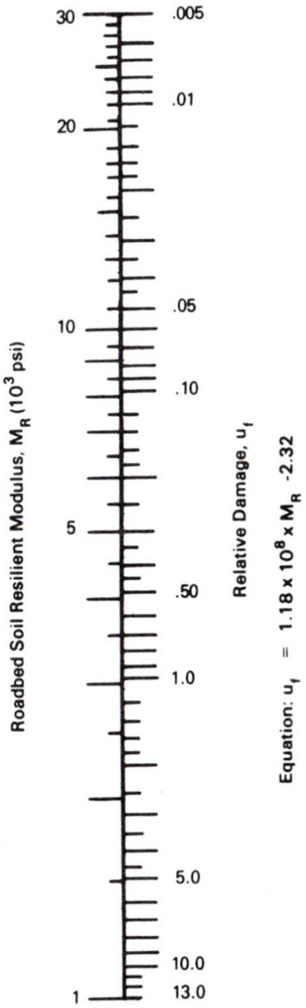

Equation: $u_f = 1.18 \times 10^8 \times M_R^{-2.32}$

Effective Roadbed Soil Resilient Modulus, M_R (psi) $= \underline{2,200}$ (corresponds to $\bar{u}_f$)

Figure 11.26 Worksheet for estimating effective roadbed soil resilient modulus (1 psi = 6.9 kPa).

In the worksheet, the 12 monthly subgrade moduli used in DAMA program for a MAAT of 45°F (7.2°C) and a normal modulus of 4500 psi (31 MPa), as shown in Table 11.10, are used as an example. The relative damage during each month can be obtained from the vertical scale or computed from Eq. 11.43 and a sum of 25.30 is shown at the bottom. The average relative damage = 25.30/12 = 2.11, which corresponds to an effective roadbed resilient modulus of 2200 psi (15.2 MPa).

In the above example, there is a large variation in the monthly resilient modulus. The maximum and minimum values are outside the range of the vertical scale and can only be computed from Eq. 11.43. About 65% of the damage is done in May alone. This is the reason why a very low effective modulus of 2200 psi (15.2 MPa) is obtained, which is much lower than the normal modulus of 4500 psi (31.1 MPa).

11.3.4 Structural Number

Structural number is a function of layer thicknesses, layer coefficients, and drainage coefficients and can be computed from Eq. 11.35.

Layer Coefficient

The layer coefficient a_i is a measure of the relative ability of a unit thickness of a given material to function as a structural component of the pavement. Layer coefficients can be determined from test roads or satellite sections, as was done in the AASHO Road Test, or from correlations with material properties, as shown in Figures 7.13, 7.15, and 7.16. It is recommended that the layer coefficient be based on the resilient modulus, which is a more fundamental material property. The procedure for determining the resilient modulus of a particular material varies with its type. Except for the higher stiffness materials, such as HMA and stabilized bases, that may be tested by the repeated load indirect tensile test (ASTM D-4123), all materials should be tested using the resilient modulus test methods (AASHTO T274). Methods for determining the resilient modulus were described in Section 7.1.

In following the AASHTO design guide, the notation M_R, as used herein, refers only to roadbed soils, while E_1, E_2, and E_3 apply to the HMA, base, and subbase, respectively.

Asphalt Concrete Surface Course

Figure 11.27 is a chart relating the layer coefficient of a dense-graded HMA to its resilient modulus at 70°F (21°C). Caution should be used in selecting layer coefficients with modulus values greater than 450,000 psi (3.1 GPa) because the use of these larger moduli is accompanied by increased susceptiblity to thermal and fatigue cracking. The layer coefficient a_1 for the dense-graded HMA used in the AASHO Road Tests is 0.44, which corresponds to a resilient modulus of 450,000 psi (3.1 GPa).

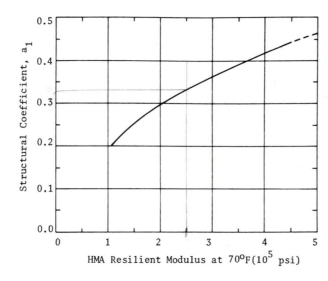

Figure 11.27 Chart for estimating layer coefficient of dense-graded asphalt concrete based on elastic modulus (1 psi = 6.9 kPa). (After Van Til et al. (1972).)

Untreated and Stabilized Base Courses

Figure 7.15 shows the charts that can be used to estimate the layer coefficient a_2 for untreated, bituminous-treated, and cement-treated base courses. In lieu of Figure 7.15a, the following equation can also be used to estimate a_2 for untreated base course from its resilient modulus E_2:

$$a_2 = 0.249(\log E_2) - 0.977 \qquad (11.44)$$

The layer coefficient a_2 for the granular base material used in the AASHO Road Test is 0.14, which corresponds to a base resilient modulus of 30,000 psi (207 GPa).

The resilient modulus of untreated granular materials depends on the stress state θ, as indicated by Eq. 3.8 and rewritten as

$$E_2 = K_1 \, \theta^{K_2} \qquad (11.45)$$

Typical values of K_1 for base materials range from 3000 to 8000 and those of K_2 range from 0.5 to 0.7. Values of K_1 and K_2 for each specific base material should be determined using AASHTO Method T274. In the absence of this information, the values shown in Table 11.16 can be used.

TABLE 11.16 TYPICAL VALUES OF K_1 AND K_2 FOR UNTREATED BASE MATERIALS

Moisture condition	K_1	K_2
Dry	6000–10,000	0.5–0.7
Damp	4000–6000	0.5–0.7
Wet	2000–4000	0.5–0.7

Source. After AASHTO (1986).

TABLE 11.17 TYPICAL VALUES OF
STRESS STATE θ FOR BASE COURSE

Asphalt concrete thickness (in.)	Roadbed soil resilient modulus (psi)		
	3000	7500	15,000
Less than 2	20	25	30
2–4	10	15	20
4–6	5	10	15
Greater than 6	5	5	5

Note. Unit of θ is in psi, 1 in. = 25.4 mm,
1 psi = 6.9 kPa.

Source. After AASHTO (1986).

The resilient modulus of base course is a function of not only K_1 and K_2, but also the stress state θ. Values for the stress state within the base course vary with the roadbed soil resilient modulus and the thickness of the surface layer. Typical values of θ are shown in Table 11.17. Given K_1, K_2, and θ, E_2 can be determined from Eq. 11.45.

Granular Subbase Course

Figure 7.16 provides the chart that may be used to estimate layer coefficient a_3 of granular subbase courses. The relationship between a_3 and E_3 can be expressed as

$$a_3 = 0.227(\log E_3) - 0.839 \tag{11.46}$$

The layer coefficient a_3 for the granular subbase in the AASHO Road Test is 0.11, which corresponds to a resilient modulus of 15,000 psi (104 MPa). Similar to granular base courses, values of K_1 and K_2 for granular subbase courses can be determined from the resilient modulus test (AASHTO T274) or estimated from Table 11.18. Values of K_1, K_2, θ, and E_3 for the subbase in the AASHO Road Test are shown in Table 11.19.

TABLE 11.18 TYPICAL VALUES OF K_1
AND K_2 FOR GRANULAR SUBBASE
MATERIALS

Moisture condition	K_1	K_2
Dry	6000–8000	0.4–0.6
Damp	4000–6000	0.4–0.6
Wet	1500–4000	0.4–0.6

Source. After AASHTO (1986).

TABLE 11.19 VALUES OF RESILIENT MODULUS FOR AASHO ROAD TEST SUBBASE MATERIALS

| Moisture condition | K_1 | K_2 | Stress state θ (psi) | | |
			5	7.5	10
Damp	5400	0.6	14,183	18,090	21,497
Wet	4600	0.6	12,082	15,410	18,312

Note. Resilient modulus is in psi, 1 psi = 6.9 kPa.

Source. After Finn et al. (1986).

Drainage Coefficient

Depending on the quality of drainage and the availability of moisture, drainage coefficients m_2 and m_3 should be applied to granular bases and subbases to modify the layer coefficients, as shown in Eq. 11.35. At the AASHTO Road Test site, these drainage coefficients are all equal to 1, as indicated by Eq. 11.32.

Table 11.20 shows the recommended drainage coefficients for untreated base and subbase materials in flexible pavements. The quality of drainage is measured by the length of time for water to be removed from bases and subbases and depends primarily on their permeability. The percentage of time during which the pavement structure is exposed to moisture levels approaching saturation depends on the average yearly rainfall and the prevailing drainage conditions.

11.3.5 Selection of Layer Thicknesses

Once the design structural number SN for an initial pavement structure is determined, it is necessary to select a set of thicknesses so that the provided SN, as computed by Eq. 11.35, will be greater than the required SN. Note that Eq. 11.35 does not have a single unique solution. Many combinations of layer thicknesses

TABLE 11.20 RECOMMENDED DRAINAGE COEFFICIENTS FOR UNTREATED BASES AND SUBBASES IN FLEXIBLE PAVEMENTS

| Quality of drainage | | Percentage of time pavement structure is exposed to moisture levels approaching saturation | | | |
Rating	Water removed within	Less than 1%	1–5%	5–25%	Greater than 25%
Excellent	2 hours	1.40–1.35	1.35–1.30	1.30–1.20	1.20
Good	1 day	1.35–1.25	1.25–1.15	1.15–1.00	1.00
Fair	1 week	1.25–1.15	1.15–1.05	1.00–0.80	0.80
Poor	1 month	1.15–1.05	1.05–0.80	0.80–0.60	0.60
Very poor	Never drain	1.05–0.95	0.95–0.75	0.75–0.40	0.40

Source. After AASHTO (1986).

are acceptable, so their cost effectiveness along with the construction and maintenance constraints must be considered to avoid the possibility of producing an
impractical design. From a cost-effective viewpoint, if the ratio of costs for HMA
and granular base is less than the corresponding ratio of layer coefficients times
the drainage coefficient, then the optimum economical design is to use a minimum
base thickness by increasing the HMA thickness.

Minimum Thickness

It is generally impractical and uneconomical to use layers of material that are
less than some minimum thickness. Furthermore, traffic considerations may
dictate the use of a certain minimum thickness for stability. Table 11.21 shows the
minimum thicknesses of asphalt surface and aggregate base. Because such mini-
mums depend somewhat on local practices and conditions, they may be changed if
needed.

General Procedure

The procedure for thickness design is usually started from the top, as shown
in Figure 11.28 and described below:

1. Using E_2 as M_R, determine from Figure 11.25 the structural number SN_1
required to protect the base and compute the thickness of layer 1 by

$$D_1 \geq \frac{SN_1}{a_1} \qquad (11.47)$$

2. Using E_3 as M_R, determine from Figure 11.25 the structural number SN_2
required to protect the subbase and compute the thickness of layer 2 by

$$D_2 \geq \frac{SN_2 - a_1 D_1}{a_2 m_2} \qquad (11.48)$$

TABLE 11.21 MINIMUM THICKNESS FOR ASPHALT
SURFACE AND AGGREGATE BASE

Traffic (ESAL)	Asphalt concrete	Aggregate base
Less than 50,000	1.0	4
50,001–150,000	2.0	4
150,001–500,000	2.5	4
500,001–2,000,000	3.0	6
2,000,001–7,000,000	3.5	6
Greater than 7,000,000	4.0	6

Note. Minimum thickness is in in., 1 in. = 25.4 mm.

Source. After AASHTO (1986).

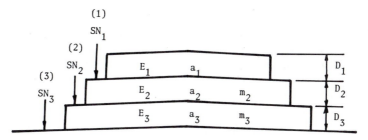

Figure 11.28 Selection of thicknesses.

3. Based on the roadbed soil resilient modulus M_R, determine from Figure 11.25 the total structural number SN_3 required and compute the thickness of layer 3 by

$$D_3 \geq \frac{SN_3 - a_1 D_1 - a_2 D_2 m_2}{a_3 m_3} \qquad (11.49)$$

Example 11.12

Figure 11.29 is a pavement system with the resilient moduli, layer coefficients, and drainage coefficients as shown. If predicted ESAL $= 18.6 \times 10^6$, $R = 95\%$, $S_0 = 0.35$, and $\Delta PSI = 2.1$, select thicknesses D_1, D_2, and D_3.

Solution: With $M_R = E_2 = 30,000$ psi (207 MPa), from Figure 11.25, $SN_1 = 3.2$; from Eq. 11.47, $D_1 \geq 3.2/0.42 = 7.6$ in. (193 mm); use $D_1 = 8$ in. (203 mm).
  With $M_R = E_3 = 11,000$ psi (76 MPa), from Figure 11.25, $SN_2 = 4.5$; from Eq. 11.48, $D_2 \geq (4.5 - 0.42 \times 8)/(0.14 \times 1.2) = 6.8$ in. (173 mm); use $D_2 = 7$ in. (178 mm). Note that a surface thickness of 8 in. (203 mm) and a base thickness of 7 in. (178 mm) meet the minimum thicknesses shown in Table 11.21.
  With $M_R = 5700$ psi (39.3 MPa), from Figure 11.25, $SN_3 = 5.6$; from Eq. 11.49, $D_3 \geq (5.6 - 0.42 \times 8 - 0.14 \times 7 \times 1.20)/(0.08 \times 1.2) = 11.1$ in. (282 mm); use $D_3 = 11.5$ in. (292 mm).

11.3.6 Stage Construction

If the maximum performance period is less than the analysis period, any initial structure selected will require an overlay to last the analysis period. The thickest recommended initial structure is that corresponding to the maximum performance

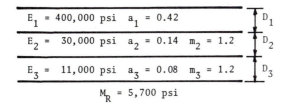

Figure 11.29 Example 11.12 (1 psi = 6.9 kPa).

period. Thinner initial structures, selected for the purpose of life cycle cost analyses, will result in shorter performance periods and require thicker overlays to last the same analysis period. The design of initial structure for stage construction is the same as that for new construction except that the reliability must be compounded over all stages. The design of overlay as a stage construction alternative is presented in Chapter 13.

If the loss of serviceability is caused by traffic loads alone, the length of performance period, which is related to W_{18}, for a given serviceability loss can be determined from Figure 11.25 or directly from Eq. 11.37. However, if the serviceability loss is caused by both traffic loads and the environmental effects of roadbed swelling and frost heave, the performance period for a given terminal serviceability can be determined only by an iterative process, as illustrated by the following example.

Example 11.13

Given the following design inputs, determine the length of performance period required: structural number SN = 5.0, reliability R = 95%, standard deviation S_0 = 0.35, initial serviceability p_0 = 4.3, terminal serviceability p_t = 2.5, effective roadbed soil resilient modulus M_R = 5000 psi (35 MPa), ΔPSI due to both swelling and frost heave as shown in Figure 11.23, and traffic versus time relationship as

$$W_{18} = 10 \times 10^6 [(1.03)^Y - 1] \tag{11.50}$$

or

$$Y = 77.9 \log\left(\frac{W_{18}}{10 \times 10^6} + 1\right) \tag{11.51}$$

Solution: First, assume Y = 13 years. From Figure 11.23, ΔPSI due to environmental effects = 0.73; ΔPSI due to traffic = 4.3 − 2.5 − 0.73 = 1.07. From Eq. 11.37 or Figure 11.25, W_{18} = 1.6 × 10⁶. From Eq. 11.51, Y = 5.1 years, which is much smaller than the 13 years assumed.

Next assume Y as the average of 13 and 5.1 years, or 9.0 years. From Figure 11.23, ΔPSI due to environmental effects = 0.59; ΔPSI due to traffic = 4.3 − 2.5 − 0.59 = 1.21. From Eq. 11.37 or Figure 11.25, W_{18} = 2.1 × 10⁶. From Eq. 11.51, Y = 6.5 years.

Finally, assume Y = (9 + 6.5)/2 = 7.7 years. From Figure 11.23, ΔPSI due to environmental effects = 0.52; ΔPSI due to traffic = 4.3 − 2.5 − 0.52 = 1.28. From Eq. 11.37 or Figure 11.25, W_{18} = 2.4 × 10⁶. From Eq. 11.51, Y = 7.3, which is nearly equal to the 7.7 years assumed. When the difference between the assumed and calculated values is smaller than 1 year, no more iterations are needed and the average of the two values can be used as the performance period. Therefore, the performance period = (7.7 + 7.3)/2 = 7.5 years.

11.3.7 Comparison with Asphalt Institute Method

Due to the difference in design variables, it is difficult to compare two different design methods. The AASHTO design method applies the reliability concept by using average values for all variables including the effective roadbed soil resilient modulus. The Asphalt Institute method does not consider reliability and uses a normal subgrade resilient modulus that is smaller than 60 to 87.5% of all the test

TABLE 11.22 COMPARISON OF THICKNESS BETWEEN AASHTO AND AI METHODS

AASHTO method			AI method
HMA thickness (in.)	SN	ESAL	HMA thickness (in.)
5.0	2.2	5.2×10^3	4.5
10.0	4.4	3.6×10^5	9.0
15.0	6.6	6.8×10^6	14.0

Note. 1 in. = 25.4 mm.

values, depending on the level of traffic. As shown by the example in Figure 11.26, a normal subgrade resilient modulus of 4500 psi (31 MPa) is equivalent to an effective roadbed soil resilient modulus of 2200 psi (15.2 MPa). Because the resilient modulus is based on the average value in the AASHTO method but on the 60 to 87.5 percentile value in the Asphalt Institute method, it is reasonable to assume that an effective roadbed soil resilient modulus of 3000 psi (20.7 MPa) in the AASHTO method is about equivalent to a normal modulus of 4500 psi (31 MPa) in the Asphalt Institute method. Other parameter values to be used in the AASHTO method are reliability $R = 95\%$, standard deviation $S_0 = 0.45$, serviceability loss $\Delta PSI = 1.7$, and layer coefficient for full-depth HMA $a_1 = 0.44$. These values are reasonable for use in the design of major highways.

Table 11.22 shows the thicknesses of full-depth HMA required by the AASHTO method versus those required by the Asphalt Institute method. Three HMA thicknesses of 5, 10, and 15 in. (127, 254, and 381 mm) are assumed. The structural number SN is computed by multiplying the HMA thickness with 0.44. The ESAL based on the AASHTO method is determined from Eq. 11.37. Based on the ESAL thus obtained, the thickness by the Asphalt Institute method is obtained from Figure 11.11. It can be seen that both methods check quite well. The thicknesses obtained by the Asphalt Institute method are 0.5 to 1 in. (127 to 254 mm) thinner than those by the AASHTO method.

11.4 DESIGN OF FLEXIBLE PAVEMENT SHOULDERS

As defined by AASHO (1968), a highway shoulder is the "portion of roadway contiguous with the traveled way for accommodation of stopped vehicles for emergency use, and for lateral support of base and surface courses." Shoulders also provide recovery space for errant vehicles, lateral clearance for signs and guardrails, improved sight distance in cuts, area for maintenance operations, and additional lane for peak hour or detoured traffic.

11.4.1 Current Status of Shoulder Design

During the early years of highway construction, the need for first-class shoulders was of secondary importance. Therefore, the structural design of shoulder pavements has been developed mostly by experience rather than by rational methods.

However, with the tremendous increase in both number and speed of vehicles, the need for adequate shoulders has greatly increased. In this section, a survey of state practices reported by the National Cooperative Highway Research Program (NCHRP, 1979) and a position paper on shoulder design prepard by the AASHTO Joint Task Force on Pavements (AASHTO, 1986) are presented.

Survey of State Practices

A comprehensive survey of shoulder design, construction, and maintenance was conducted during 1967 by the Committee on Shoulder Design of the Highway Research Board and again during 1977 by NCHRP. It was found that nearly half of the states did not make any changes in their shoulder design standards during the ten-year period. Following is a summary of the findings reported by NCHRP:

1. The predominant criterion for selecting shoulder type, thickness, width, and slope is the combination of highway classification and traffic volume.
2. Although surface drainage can be provided by the cross slope of the shoulders, a significant number of states use some form of dikes, catch basins, or gutters to minimize slope erosion. For subsurface drainage, the predominant policy is to use underdrains or a free-draining base.
3. Shoulder conditions are evaluated in nearly all states through visual inspections, usually by a member of the maintenance team.
4. Most states use the same width and slope of shoulders for both rigid and flexible pavements. On interstate highways and freeways, nearly 80% of the states use a 10-ft (3-m) width for the outside shoulder. More than 40% of the states use a 4-ft (1.2-m) width for the median shoulder; the other states use a median shoulder of 3 to 10 ft (0.9 to 3.0 m), depending upon traffic volume and number of lanes. Shoulder slope is usually the same for both outside and median shoulders. The predominant slope is 4% or $\frac{1}{2}$ in./ft.
5. Only five states permit regular use of shoulders for slow-moving vehicles, although ten states permit such use under certain conditions.
6. All states use a 4-in. (102-mm) white reflectorized edge stripe to delineate the outside shoulder and a yellow edge stripe to delineate the median shoulder. However, a number of states supplement the edge stripe with contrasting color, texture, or rumble strips on the shoulder. Edge stripes are usually placed on the travel lane at or near the shoulder, but some states placed them on the paved shoulder.

Position Paper on Shoulder Design

In view of the wide variety of practices related to shoulder design, a position paper was published as Appendix E in the AASHTO design guide. It was mentioned that California has a formal procedure for shoulder design, 14 other states have documented policies but no formal procedure, 28 states have no policy, and 5 states pave their shoulders integrally with the mainline pavements. Therefore, a definite need exists to develop criteria for a unified and widely accepted shoulder design guide.

The paper lists the following points that may need to be considered in shoulder design:

1. Shoulder thickness design as related to load-carrying requirements
2. Cost and service criteria for stabilization of shoulder aggregate base course
3. Problems associated with the pavement edge and shoulder edge interface
4. Abrasive effects of traffic
5. Permeability or degree of imperviousness required for shoulder aggregate base course
6. Relationship between shoulder performance and subgrade support
7. Relationship of shoulder drainage subsystem to overall subsurface drainage system
8. Construction and maintenance methods and operations that result in adverse shoulder performance
9. Effect of environment on shoulder performance
10. Type and texture of shoulder surface for waterproofing and delineating purposes
11. Effects of shoulder geometrics on performance
12. Warrants for paved shoulders

Regarding the specific design of shoulders, the paper lists the following recommendations:

1. Design shoulder thickness based on the criteria reflecting the magnitude and frequency of loads to which the shoulder will be subjected.
2. Integrate shoulder drainage with the overall pavement subdrainage system.
3. Avoid the use of aggregate base courses having a significant amount of fine materials passing through a No. 200 sieve to prevent frost heaving, pumping, clogging of the shoulder drainage system, and base instability.
4. Take advantage of the desirable performance features of plant-mixed bituminous and various stabilized shoulder materials as opposed to bituminous surface-treated, unbound shoulder aggregate bases.
5. Have a definite program of shoulder maintenance.
6. Take advantage of the desirable performance of rigid shoulders adjacent to rigid mainline pavements.
7. Develop criteria for paving shoulders.

The position paper was submitted to the states for comments in 1981. There was no general consensus expressed. However, the following comments are pertinent and reflect the preferences of individual agencies:

1. Design mainline and shoulder pavements as a single unit to allow for future additions of new traveled lanes.
2. Use the same material for both mainline and shoulder pavements and tie concrete shoulders to the mainline concrete pavement.
3. Provide for means to seal the joint between shoulder and traveled way.
4. Develop shoulder design criteria for low-volume roads and investigate the advantage and economy of using a 28- to 30-ft (8.5- to 9.1-m) mainline section with aggregate shoulders.
5. Give proper consideration to a full-depth shoulder alternative.
6. Consider the use of shoulders for detouring traffic or as an extra lane during peak hours.

11.4.2 Prediction of Traffic for Shoulder Design

The factors affecting shoulder design are similar to those affecting mainline pavement design. The major difference is the amount of traffic. Traffic volume on shoulders is lower than on mainline pavements and more difficult to predict. Three types of traffic may be considered in shoulder design: encroaching traffic, parking traffic, and regular traffic. Regular traffic is considered only if the use of shoulder as an additional lane for peak hour or detoured traffic is anticipated. If there is no regular traffic, the sum of encroaching and parking traffic is used to design the inner edge of shoulder adjacent to the mainline pavement; while parking traffic is used to design the outer edge of shoulder. If a uniform thickness is used for shoulders, only the inner edge need be considered. However, this may not be true for rigid pavement shoulders, because the stresses and deflections at the outer edge are much greater than those at the inner edge and may cause more damage even the traffic volume is smaller.

Encroaching Traffic

When there is a paved shoulder and no lateral obstruction within the shoulder area, trucks using the outer traffic lane tend to encroach on the shoulder as much as 12 in. (305 mm) and sometimes even more. In California (1972), shoulder sections are designed for 1% of the mainline traffic in the adjacent lane with a minimum traffic index of 5, which corresponds to approximately 10^4 equivalent 18-kip single-axle load applications. Results of a study in Georgia (Emery, 1975) showed that the use of 1% is low for some traffic flow conditions. In the absence of additional data, Barksdale and Hicks (1979) recommended that for free flow traffic conditions in rural areas of the south, the inner edge of the shoulder should be designed for at least 2 to 2.5% of the truck traffic on the outer lane.

Because local conditions vary significantly, the best way to determine the percentage of encroaching traffic is to make an actual survey on a segment of highway with paved shoulders, which has the same traffic, geometric, and topographic conditions as the design case in question. For the study performed in Georgia, trucks were selected at random and followed by observers for 10 miles (16 km). Those trucks not completing the full 10-mile (16-km) trip were dropped from the analysis. The longitudinal distance for each encroachment is estimated from the prevailing speed and the time during which the truck encroaches on the shoulder. The percent encroaching traffic, which is the ratio of load applications on the shoulder to those on the adjacent lane, can be computed by

$$P_e = \frac{N_e L_e}{N_o L_o} \times 100 \tag{11.52}$$

in which P_e is the percent encroaching traffic, N_e is the total number of encroachments per day, L_e is the average length of each encroachment, N_o is the number of load applications per day on the outside lane, and L_o is the length of observed distance, such as 10 miles (16 km) used in the Georgia study. Note that $N_e L_e$ is the total length of encroachment for all trucks and $N_e L_e / N_o$ is the length of encroachment per truck, so P_e can also be defined as the length of encroachment per truck within an observed distance L_o.

Field observations indicate that P_e usually varies from 1 to 8% of the traffic volume on the adjacent lane. The percentage of parking traffic should be added to P_e because any truck must encroach to park on the shoulder.

Example 11.14

This example is the result of an actual survey on a 10-mile segment of I-75 at Perry, Georgia (Emery, 1975). Given the number of trucks on the outside lane per day $N_o = 2239$, the number of shoulder encroachments in the 10 mile (16 km) stretch $N_e = 7389$, and the average distance of each encroachment $L_e = 384$ ft (117 m), determine the percent encroachment P_e.

Solution: Given $N_e = 7839$, $L_e = 384$ ft (117 m), $N_o = 2239$, and $L_o = 52{,}800$ ft (16,100 m), from Eq. 11.52, $P_e = [(7389 \times 384)/(2239 \times 52{,}800)] \times 100 = 2.4\%$.

Parking Traffic

Parking traffic is the number of load applications for trucks that park on the shoulder for emergencies or other purposes. This information can also be estimated for the design section using traffic counts on an existing pavement with similar traffic and design characteristics. Parking traffic usually varies greatly along a given route, depending on geometric and interchange conditions. Because most trucks park near interchange ramps, it may be necessary to identify separate design sections for areas where parking is likely and for areas where minimum parking is expected. The parking survey should last at least one day and cover the early morning hours during which more parking usually takes place.

Similar to percent encroaching traffic, the percent parking traffic can be computed by

$$P_p = \frac{N_p L_p}{N_o L_o} \times 100 \tag{11.53}$$

in which P_p is the percent parking traffic, which is the ratio between parking traffic and the traffic on the outside lane: N_p is the number of parked trucks per day; L_p is the average distance the trucks drive on the shoulder during a typical stop, which can be determined from a field survey; N_o is the number of trucks traveling on the outside lane per day; and L_o is the length of segment for the parking survey. Field observations indicate that P_p may range from 0.0005 to 0.02 percent. Note that the percent parking traffic is much smaller than the percent encroaching traffic and can usually be neglected.

Example 11.15

Based on a limited field survey on I-80 in Illinois, the average number of trucks that may park on a 2-mile (3.2-km) stretch of shoulder during one day could range from 1 to 25 (Sawan and Darter, 1979). If the number of trucks on the outside lane is 2951 per day and a truck drives an average of 200 ft (61 m) during each parking manuever, determine the range of percent parking traffic.

Solution: Given $N_p = 1$, $L_p = 200$ ft (31 m), $N_o = 2951$, and $L_o = 2 \times 5280 = 10,560$ ft (1573 m), from Eq. 11.53, $P_p = [(1 \times 200)/(2951 \times 10,560)] \times 100 = 0.00064\%$. If $N_p = 25$, $P_p = 0.00064 \times 25 = 0.016\%$. The range of P_p is from 0.00064 to 0.016%.

Regular Traffic

If it is anticipated that the shoulder will be used by regular traffic at any stage of its design life, this additional traffic should be added to the encroaching and parking traffic to form the total shoulder design traffic. The ultimate design is to consider the shoulder as an extra lane with the same traffic and cross section as that of the mainline outer lane.

11.4.3 Thickness Design Concepts

As indicated in the AASHTO position paper, many problems need to be solved before a unified and widely accepted shoulder design procedure can be developed. In this section, some of the design concepts are discussed.

Type of Shoulders

The first decision to be made is the type of shoulders to be used. The Technical Advisory issued by FHWA (1982) recommends the use of similar materials for mainline and shoulder pavements. Therefore, flexible shoulder pavements should be used with flexible mainline pavements, and rigid shoulder pavements should be used with rigid mainline pavements. If asphalt shoulders are placed adjacent to concrete mainline slabs, serious problems will occur at the longitudinal joint between slabs and shoulders because the difference in thermal expansion and contraction between dissimilar materials will introduce additional stresses to joint sealants which are already being stressed by differential vertical deflections across the joint. The problem is further compounded by the fact that few joint sealants are suited to bond dissimilar materials. Longitudinal joints between flexible shoulders and flexible mainline pavements generally do not experience serious problems and can be easily sealed even if some separation takes place.

Configuration of Shoulder Section

The thickness of shoulder can be uniform or nonuniform. Many agencies suggested that the same thickness be used for shoulder and mainline pavements to allow for maximum safety and performance, flexibility in shoulder use during periods of construction or congestion, and the potential for future additions of traffic lanes. Some suggested that the thickness at the inner edge of shoulder pavement should be the same as the thickness of the mainline pavement to avoid problems associated with varying support and volume change on both sides of the joint (ERES, 1987). Barksdale and Hicks (1979) recommended that shoulder sections be designed structurally for the expected amount of traffic due to encroachment using presently accepted mainline design methods. Because virtually all truck encroachments and shoulder failures occur within 2 ft (0.61 m) of the longitudinal joint, Barksdale and Hicks suggested the use of a shoulder design having a variable structural strength, such as tapered sections or special support in the critical areas of heavy loading. On low-volume roads, the use of a 28- to 30-ft (8.5- to 9.1-m) mainline section with granular shoulders has special advantage. The mainline pavement can be striped to provide standard traffic lanes and reduce the premature shoulder deterioration due to encroachment, while a sufficiently wide granular shoulder is provided to permit emergency use.

It can be seen that a variety of opinions exist on shoulder thickness design. If the thickness of the shoulder is the same as that of the mainline pavement and shoulder drainage is integrated with the overall pavement drainage design, as has been practiced by several states, there is no need to design shoulder thickness separately. The reason for using thinner pavements or tapered sections for shoulders is for economy. Since it is difficult to find a consensus on shoulder design, the final selection should be based on local experience, engineering judgment, and the availability of funds.

Environmental Effects

Due to the use of thinner shoulder sections, the environmental effects, such as roadbed swelling, frost heave, and drainage conditions, become more important and may govern the thickness of shoulder pavement required.

First, the foundation soil should be analyzed to determine its susceptibility to frost damage. If the average depth of frost penetration is more than 6 in. (152 mm), the thickness of the base and subbase must be selected to provide frost protection using some appropriate methods, such as the procedure developed by the U.S. Army Corps of Engineers (1961).

Next, the need for proper drainage must be evaluated. A system for evaluating the moisture accelerated distress (MAD) was developed by Carpenter et al. (1981) and may be used for shoulder design. If the MAD index, which is based on the climate of the area and the properties of the subgrade, indicates that drainage is required, a drainage layer is assumed and its adequacy is checked using the drainage evaluation criteria in the MAD system. If the assumed drainage is inadequate, a different design must be assumed and the process repeated until a satisfactory design is obtained. The thickness of the drainage layer should then be compared with the minimum thickness required for frost penetration. If the drainage layer is thinner than required for frost protection, the thickness of drainage layer must be increased or a subbase added and the adequacy of the design reevaluated. The MAD system can also be used to determine the need for underdrains, if needed.

Finally, based on the predicted traffic, material properties, and climate variables, a structural design can be conducted using the method for mainline pavements. The thicknesses thus determined must be checked against those required for frost penetration and drainage and modified as necessary to meet these criteria.

11.5 SUMMARY

This chapter presents several methods for flexible pavement design, including a calibrated mechanistic procedure, the Asphalt Institute method, the AASHTO design guide, and several methods for the design of shoulders.

Important Points Discussed in Chapter 11

1. The most ideal design method still under develepment is the calibrated mechanistic procedure. It consists of a number of response models to evaluate the stresses and strains in the pavement system and a number of distress models using the pertinent stress or strain as an input. These distress models require careful calibration and verification to ensure that satisfactory agreements between predicted and observed distress can be obtained.

2. The response models include the heat transfer model, the moisture equilibrium model, the infiltration and drainage model, and the structural model. The finite element structural model can analyze nonlinear pavement systems more realistically by considering the variation of elastic modulus within each layer. However, due to its inaccuracy in analyzing pavement systems and the large amount of computer time required, it appears that the layer system program is the best and the most practical structural model to be used at the present time.

3. The distress models include the fatigue cracking model, the rutting model, and the low-temperature cracking model. The roughness model and the performance model may also be included, if needed. When the design is based on each type of distress, it is preferable to use the various load groups for damage analysis, rather than an equivalent single-axle load, because the equivalent factor for each type of distress is different from that for the others.

4. From a theoretical view point, the use of the reliability concept to define the extent of distress is a much better approach than the use of deterministic method. Due to the variability of the design factors, a pavement will not fail when the stress is exactly equal to the strength or when the damage ratio is exactly equal to one. Consequently, the extent of distress depends on the probability that the stress will exceed the strength or the damage ratio will exceed one. The thermal cracking model developed by Shahin and McCullough (1972) was described in details because it applies the reliability concept to predict both the low-tem-

perature cracking and the thermal fatigue cracking. It is hoped that the same concept would be used in other distress models.

5. The design method by the Asphalt Institute is based on the equivalent 18-kip single-axle load. Two failure criteria are employed: (a) fatigue cracking based on the horizontal tensile strain at the bottom of the asphalt layer and (b) rutting based on the vertical compressive strain on top of the subgrade. The method applies the cumulative damage concept and assumes that failures occur when the damage ratio is equal to one. The variability of the design factors and the reliability of the design are not considered in the design procedure.

6. A computer program called DAMA was used by the Asphalt Institute to develop a series of design charts for full-depth HMA, HMA over aggregate base, and HMA over emulsified asphalt base. The program considers the asphalt layer and the subgrade as linear elastic and the granular base as nonlinear elastic. Each year was divided into 12 months, and the elastic moduli of asphalt layer, granular base, and subgrade were varied monthly based on assumed climatic conditions. Three temperature regimes representing New York, South Carolina, and Arizona were assumed but only the charts for South Carolina, which represents the climate over a large part of the United States, were included in the 1981 version of MS-1 and reproduced here. Charts representing New York and Arizona were later added in the 1991 version of MS-1.

7. The Asphalt Institute devoted extensive effort to comparing the predicted thickness obtained from the charts with the actual thickness on pavements of known performance. Six separate sources of data were utilized, representing a total of 394 individual data points for pavements from California, Minnesota, Illinois, and Canada. It was found that the predicted thickness was about 1.61 in. (41 mm) greater than the actual thickness, so the use of the design charts is quite conservative. Due to the lack of documented data, the southern and eastern parts of the United States were not included in the comparison.

8. The AASHTO design method is based on the empirical regression equation obtained from the AASHO Road Test in Ottawa, Illinois. The design is based on the equivalent 18-kip single-axle load. The original equation is applicable only to the specific environmental and soil conditions at the test site. To make it applicable to other areas of the nation, the equation was modified by introducing an effective roadbed soil resilient modulus M_R and two drainage coefficients m_2 and m_3 for granular base and subbase, respectively. Although only one type of HMA, granular base, and subbase was used in the Road Test, the equation can be applied to other materials by varying the layer coefficients, which have been correlated empirically with the resilient moduli.

9. The effective roadbed resilient modulus is an equivalent modulus that would result in the same damage if seasonal modulus values were used. The equation for evaluating relative damage was derived from the AASHTO equation based on the cumulative damage concept. It has been demonstrated that most of the damage is done during the short period of spring breakup. As a result, the effective roadbed resilient modulus is much smaller than the normal modulus usually taken during summer or fall.

10. A salient feature of the AASHTO design guide is the inclusion of reliability and standard deviation as design factors. The guide encourages the use

of stage construction for high-traffic facilities by selecting a longer analysis period with an initial performance period plus at least one rehabilitation period. Because the Road Test lasted only about two years, the AASHTO equation does not include the loss of serviceability due to the environmental effects of roadbed swelling and frost heave. The consideration of these effects on stage construction was illustrated.

11. A definite need exists to develop criteria for a unified and widely accepted shoulder design guide. To avoid the difficulty of sealing the longitudinal joint between shoulder and mainline pavements, asphalt shoulders should be used for asphalt pavements but not for PCC pavements.

12. Factors affecting shoulder design are similar to those affecting mainline pavement design. Shoulders should be designed according to the amount of traffic anticipated, including encroaching traffic, parking traffic, and regular peak hour or detoured traffic, if any. When thinner sections are used for shoulders, the environmental effects, such as roadbed swelling, frost heave, and drainage conditions, become more important and may govern the thickness of shoulder required.

PROBLEMS

11-1. Figure P11.1a shows a thin pavement laid on a subgrade consisting of a layer of sand, 5 ft thick, underlain by a clay. The water table is 10 ft below the surface. The mass unit weight of sand is 120 pcf and that of clay is 100 pcf. The compressibility factor for the clay is 0.5. The soil suction–moisture content curves for each soil is shown in Figure P11.1b. Neglecting the weight of the pavement, determine the moisture content of the subgrade at the following two points: (a) point A, which is 7 ft above the water table, and (b) point B, which is 3 ft above the water table. [Answer; 10%, 21%]

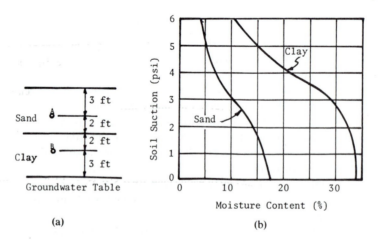

(a)   (b)

Figure P11.1

11-2. Based on the test results shown in Figures 11.4 and 11.5, estimate the coefficients of permanent deformation in the Ohio State model, as indicated by a, b, and m in Eq. 11.16. [Answer: 0.89, 0.91, 0.79]

11-3. The fatigue equation for an asphalt pavement is $N_f = 5 \times 10^{-6}(\epsilon_t)^{-3}$, in which ϵ_t is the tensile strain. If the pavement is subjected to 5000 thermal cycles with a maximum tensile strain of 0.0005 in./in. and the coefficient of variation of N_f, $C[N_f]$, is 0.8, estimate the percent area cracked due to thermal fatigue cracking. [Answer: 0.5%]

11-4. Figure P11.4 shows an asphalt pavement with the thicknesses and the resilient moduli of the granular base and subgrade as indicated. The hot mix asphalt has a viscosity of 2.5×10^6 poise at 70°F, a bitumen volume of 11%, and a fine content of 5% passing No. 200, and is subjected to a loading frequency of 8 Hz. The air void content is 4% for the surface course and 7% for the binder course. The mean monthly air temperature is 68°F. By the Asphalt Institute procedure, as used in DAMA program, determine (a) temperatures for surface and binder courses using Eq. 3.27, (b) dynamic moduli of surface and binder courses using Eq. 7.27, and (c) the modulus of the untreated base using Eqs. 3.28 and 3.29. [Answer: (a) 81.3 and 78.3°F, (b) 4.2×10^5 and 3.8×10^5 psi, and (c) 20,700 psi]

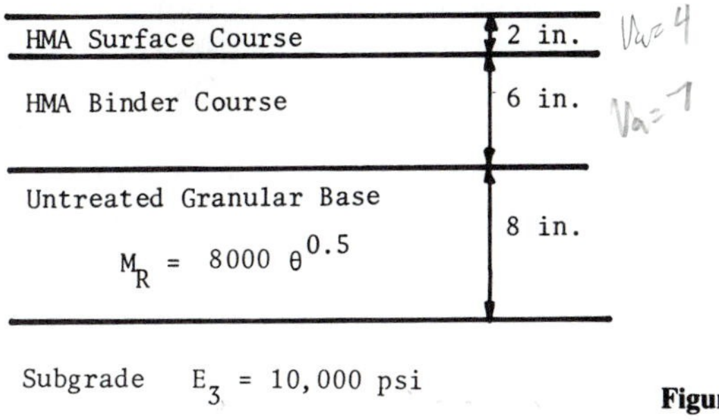

HMA Surface Course	2 in. $V_w = 4$
HMA Binder Course	6 in. $V_a = 7$
Untreated Granular Base $M_R = 8000 \, \theta^{0.5}$	8 in.
Subgrade $E_3 = 10,000$ psi	

Figure P11.4

11-5. A six-lane (three in each direction) rural interstate highway has a truck count of 1885 per day (including two-axle, four-tire panel and pickup trucks) and an annual growth rate of 4%. The HMA will be laid on an untreated granular base, 8 in. thick, which is placed on a subgrade with a resilient modulus of 10,000 psi. (a) Referring to Tables 6.9 and 6.10, estimate ESAL for a design period of 20 years by the Asphalt Institute method. (b) Determine the HMA thickness required by the Asphalt Institute method. (c) If part of the HMA is replaced by emulsified asphalt mix type I, determine the thicknesses of HMA and emulsified asphalt required. [Answer: (a) 4.26×10^6, (b) 10 in., (c) 3.5 in. and 7 in.]

11-6. A stage construction is planned for a full-depth asphalt pavement with a subgrade resilient modulus of 5000 psi. The design period is 30 years and is divided into three stages of 5, 10, and 15 years each. The first year traffic on the design lane is 20,000 equivalent 18-kip single-axle load applications and the growth rate is 3.5%. If the damage ratios at the end of each stage are 0.5, 0.75, and 1.0, respectively, determine the thicknesses of HMA to be placed in each stage by the Asphalt Institute method. [Answer: 8.5, 2.5, and 1.5 in.]

11-7. Based on the properties given in Problem 11-4, estimate the structural number of the pavement shown in Figure P11.4, assuming that the drainage coefficient of base equals to 1. [Answer: 4.6]

11-8. For a mean annual air temperature of 45°F and a normal modulus of 22,500 psi, determine the effective roadbed soil resilient modulus based on the monthly moduli shown in Table 11.10 [Answer: 22,200 psi]

11-9. A 12-in. full-depth asphalt pavement is placed on a subgrade with an effective roadbed resilient modulus of 10,000 psi. Assuming a layer coefficient of 0.44 for the hot mix asphalt, a drop in PSI from 4.2 to 2.5, an overall standard deviation of 0.5, and a predicted ESAL of 3×10^7, determine the reliability of the design by the AASHTO equation and check the result by the AASHTO design chart. [Answer: 88%]

11-10. An interstate highway pavement composed of a HMA surface course, a cement-treated base course, and a sand–gravel subbase is to be designed for an ESAL of 1.2×10^6. The quality of drainage is considered fair because water can be removed from the subbase within a week. However, due to the large amount of precipitation, more than 25% of the time the pavement will be exposed to moisture levels approaching saturation. The material properties are summarized below: effective roadbed soil resilient modulus = 5500 psi, resilient modulus of subbase = 15,000 psi, unconfined compressive strength of cement-treated base at 7 days = 500 psi (see Figure 7.15c for correlation), and resilient modulus of HMA = 4.3×10^5 psi. Assuming a minimum thickness for HMA, determine the thicknesses of the surface, base, and subbase courses required. [Answer: 3 in., 6 in., and 9 in.]

11-11. The design ESAL for a mainline pavement is 5×10^6. The encroaching traffic on the shoulder is 2.5% and the parking traffic is 0.02%. The subgrade resilient modulus is 8000 psi and a full-depth construction is used for both mainline and shoulder pavements. By the Asphalt Institute method, determine the thicknesses of mainline and shoulder pavements required. [Answer: 12 in., 6.5 in.]

11-12. Given ESAL = 5×10^6, P_e = 2.5%, P_p = 0.02%, ΔPSI = 1.7, and M_R = 5000 psi, determine the structural number SN of the mainline and shoulder pavement required for a reliability of 50%. [Answer: 4.2, 2.4]

$R = 85\%$ $\Delta PSI = 2.5$ $S_o = 0.35$

$E_1 = 4.3 \times 10^5 \, psi$

$a_1 = 0.44$

$E_3 = 15,000 \, psi$

$M_R = 5500 \, psi$

Rigid
Pavement Design

12.1 CALIBRATED MECHANISTIC DESIGN PROCEDURE

Similar to flexible pavements, the calibrated mechanistic design procedure involves the application of structural models to calculate pavement responses, the development of distress models to predict pavement distress from structural responses, and the calibration of the predicted distress with the observed distress on in-service pavements. Figure 12.1 shows the general methodology for rigid pavement design. This figure is similar to Figure 11.1 for flexible pavements except for step 5 on structural models and step 8 on distress models.

The structural models for rigid pavement analysis are more advanced than the distress models. Several finite element programs can be used as structural models but most of the distress models are regression equations derived empirically with a large scatter of data. The major types of distress to be modeled include fatigue cracking, pumping, faulting, and joint deterioration for jointed concrete pavements and punchouts for continuous reinforced concrete pavements. Some steps in Figure 12.1 are described in Sections 11.1.1 and 11.1.2; only the steps involving these new models are discussed in this section. Most of the models presented herein were developed by the University of Illinois and described in Report 1-26 (NCHRP, 1990).

12.1.1 Structural Models

To accurately analyze rigid pavement systems, Report 1-26 (NCHRP, 1990) indicated that the structural models used must have the following minimum capabilities:

1. To analyze slab of any arbitrary dimensions.

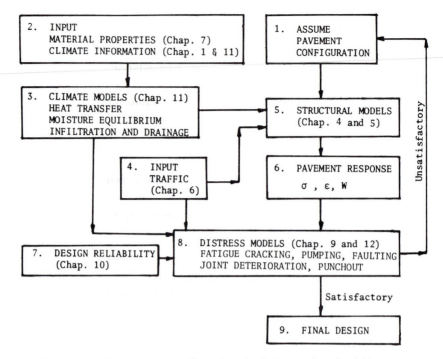

Figure 12.1 Methodology of calibrated mechanistic procedure for rigid pavement design.

2. To analyze systems with two layers (slab and subbase), either bonded or unbonded, with the same or different material properties.
3. To analyze slab systems on either liquid or solid subgrade.
4. To analyze slab systems with either uniform or nonuniform support so that the loss of support due to erosion or other causes can be taken into account.
5. To analyze multiple slabs with load transfer across the joints or cracks.
6. To consider slab warping and curling simultaneously with load responses.
7. To analyze slabs with variable crack spacings for CRCP design.
8. To analyze slabs with any arbitrary loading conditions including single or multiple wheels, variable tire pressures, and loads applied at arbitrary assigned distances from cracks, joints, or slab edges.
9. To analyze pavement systems with arbitrary shoulder conditions including asphalt shoulders, tie concrete shoulders, and extended driving lanes with asphalt or concrete shoulders beyond the extended slab.
10. To analyze systems with nonuniform slab or shoulder thicknesses.

After reviewing several finite element models, Report 1-26 recommended the use of ILLI-SLAB as the basic model for the analysis of rigid pavements. The KENSLABS program presented in this book also meets all the above requirements. In addition to liquid and solid foundations, KENSLABS can be applied to a layer subgrade consisting of up to six layers as well.

Report 1-26 also indicated that the finite element programs require a large computer core capacity, and conventional personal computers with a core capacity of 640 K are not adequate to obtain reliable and accurate results. This is true only for multiple slabs on solid or layer foundation but not necessarily for liquid foundation, which is the most popular and even the only foundation in use today. The very complex cases of four slabs on liquid foundation, as shown in Figure 5.25, and two slabs on solid foundation, as shown in Figure 5.37, have been solved satisfactorily by KENSLABS and requires a core smaller than 600 K. This is made possible by an iteration method to reduce the storage required, as explained in Section 5.1.3. Pavement design is not an exact science and, with a judicious selection of proper mesh size, many practical problems can be solved by PC without the service of a main-frame computer.

One way to reduce the computer time and storage is to develop algorithms or regression equations to replace structural models for calculating pavement responses (Zollinger and Barenberg, 1989). These algorithms can be developed by running a finite element program thousands of times and fitting the results with regression equations. Tabulated values of the regression coefficients can be provided on a disk and the algorithms run on a PC. However, this method can be used only for typical designs with a limited number of variables that can be changed, such as PCC thickness, modulus of subgrade reaction, magnitude of single-axle load, and concrete modulus of elasticity. For unusual situations, such as axle loads with special configurations, voids under the slabs, two layers of slabs, and slabs of nonuniform thickness, the finite element program must still be used.

12.1.2 Fatigue Cracking Models

The fatigue of PCC was described in Section 7.3.2. Similar to flexible pavements, the accumulation of fatigue damage can be expressed as a summation of damage ratios defined as the ratio between predicted and allowable number of load repetitions. However, instead of relating to tensile strain, the allowable number of load repetitions is related to the stress ratio, which is the ratio between the flexural stress and the modulus of rupture. The same probability concept used to define percent area cracked can be used to define percent of slabs cracked.

Truck Load Placement

The fatigue of concrete can cause both transverse cracking, which initiates at the pavement edge midway between transverse joints, and longitudinal cracking, which initiates in the wheelpaths at transverse joints, usually at the wheelpath nearest the slab centerline. Figure 12.2 shows the most critical loading and stress locations to be considered for fatigue analysis. Transverse cracking is caused by the midslab edge loading, and longitudinal cracking is caused by the joint loading.

Due to the lateral distribution of traffic, wheel loads are not applied at the same location, so only a fraction of the load repetitions need be considered for fatigue damage. Report 1-26 suggested the use of an equivalent damage ratio, EDR, for each critical loading position. EDR is the ratio of the traffic applied at

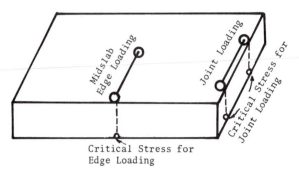

Critical Stress for
Edge Loading

Figure 12.2 Critical loading and stress locations for fatigue analysis.

the same critical location that will produce the same accumulated fatigue damage as the total traffic distributed over all locations. It was demonstrated in Report 1-26 that an EDR of 0.05 to 0.06 can be used for the midslab edge loading with asphalt shoulders, and an EDR of 0.25 to 0.28 can be used for joint loading. For edge loading with tied concrete shoulders, the EDR ranges from 0.12 to 0.34. Therefore, the truck load placement, which is not a factor in flexible pavement design, must be carefully considered in rigid pavement design.

Curling Stress

Report 1-26 suggested the use of combined loading and curling stresses for determining the stress ratio and thus the allowable number of load repetitions. In addition to the number of periods and load groups, a new loop indicating curling conditions must be added to Eq. 3.19:

$$D_r = \sum_{i=1}^{p} \sum_{k=1}^{3} \sum_{j=1}^{m} \frac{n_{i,k,j}}{N_{i,k,j}} \tag{12.1}$$

in which D_r is the accumulated damage ratio over the design period at the critical location, i is the counter for periods or subgrade support values, p is the total number of periods, k is the counter for three curing conditions (day, night, and zero temperature gradient), j is the counter for load groups, m is the total number of load groups, $n_{i,k,j}$ is the predicted number of load repetitions for the jth load group, kth curling condition, and ith period, and $N_{i,k,j}$ is the allowable number of load repetitions for the jth load groups, kth curling condition, and ith period. The inclusion of curling stress complicates the computation because the traffic has to be divided into three time periods, each with a different temperature gradient. It does not appear reasonable to combine loading and temperature stresses because they do not occur at the same frequency. A pavement may be subject to thousands of load repetitions per day due to traffic but the number of repetitions due to temperature curling is mostly only once a day. If curling stresses cannot be ignored and longer panel lengths have significant effects on fatigue cracking because of higher curling stresses, it is more reasonable to consider the damage ratios due to loading and curling separately and then combined, as illustrated by the Shahin–McCullough model for flexible pavements presented in Section 11.1.4.

Curling may not affect the fatigue life significantly because the curling stress may be subtracted from or added to the loading stress, thus neutralizing the effect.

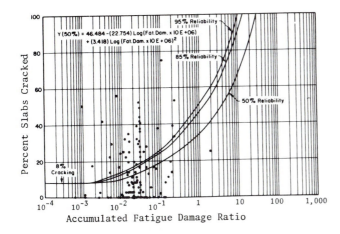

Figure 12.3 Calibrated performance curves based on Illinois COPES data. (After NCHRP (1990).)

The edge stress is further reduced by moisture warping because the moisture contents at the bottom of slab are more frequently higher than those at the top. The curling stress should be much reduced when new pavements are to be constructed with reasonably short panel lengths. The calibration of the model can further minimize the effect of curling stress. For example, Figure 12.3 shows a plot of calibrated performance curves for jointed concrete pavements relating the percent slabs cracked to the accumulated damage ratio. The stress ratio used in calculating the fatigue relationships shown in the figure included both loading and curling stresses. If curling stresses were eliminated from this calculation, different performance curves would be obtained. However, the percent slabs cracked should not be significantly affected if the same procedure, either including or excluding the curling stress, is used in both design and calibration processes.

The performance curves shown in Figure 12.3 were based on field calibration. For 50% reliability, the theoretical percent slabs cracked at a damage ratio of 1 should be 50%, but the percentage shown in the figure is only 27%. One possible cause for the discrepancy is the difficulty of determining the concrete modulus of rupture during the entire evaluation period from the initial loading to the time of evaluation. Additional research needs to be done on the best method for estimating concrete strength in existing pavements and how the observed cracking can be correlated with the damage ratio and the probability of cracking.

12.1.3 Pumping and Erosion Models

There is an important mode of distress in addition to fatigue cracking that needs to be addressed in the design of rigid pavements. This is the pumping and erosion of material beneath and beside the slab. In fact, most of the failures in the Maryland and AASHO road tests were the result of pumping.

Factors that influence pumping and erosion include the presence of water, the rate at which water is ejected under the slab, the erodibility of the subbase material, the magnitude and number of repeated loads, and the amount of deflection. No mechanistic models currently available take all of the above factors into account. The only available model is the one developed by PCA, which is described in Section 12.2. The PCA model was based primarily on the results of AASHO Road Test and only the corner deflection was taken directly into consideration. Because the subbase materials used in the AASHO Road Test are highly erodible and are currently not used by any of the highway agencies, the application of the model appears to be limited.

Attempts have been made to correlate erosion with rate of water ejection, traffic loads, and pavement deflection through an energy model (Dempsey, 1983; Phu et al., 1986). The thrust of this approach is to calculate the amount of energy involved in the deflection of a pavement system and establish a correlation between the total energy absorbed for given levels of traffic and erosion. These attempts have been moderately successful for specified conditions, but there are other factors affecting erodibility that have not been duly considered. Additional work is needed before these models can be incorporated into a mechanistic-based pavement design procedure.

Report 1-26 also presented an empirically based pumping model that considers all factors believed to influence erosion and pumping of rigid pavements. The model, which can be applied to both JPCP and JRCP by including the type of load transfer as a variable, was developed using nonlinear regression techniques from a data base of 927 pavement sections from seven states in every continental climatic zone. Because some of the regression constants in the model do not appear reasonable, the model is not presented here. Instead, the COPES pumping models (Darter et al., 1985), which include separate equations for JPCP and JRCP, are presented below. Note that the erodibility of the base course, which is an important factor contributing to pumping and was considered in Report 1-26, was not included in the COPES models.

Jointed Plain Concrete Pavements

$$PI = (N_{18})^{0.443} [-1.479 + 0.255(1 - S) + 0.0605(P)^{0.5}$$
$$+ 52.65(H)^{-1.747} + 0.0002269(FI)^{1.205}] \qquad (12.2a)$$

$$\text{Statistics:} \qquad R^2 = 0.68$$

$$\text{SEE} = 0.42$$

$$n = 289$$

in which   PI = pumping index rated on a scale of 0 to 3, with 0 for no pumping, 1 for low-severity pumping, 2 for medium-severity pumping, and 3 for high-severity pumping

N_{18} = number of equivalent 18-kip single-axle loads in millions

S = soil type based on AASHTO classification, with 0 for coarse-grained soils (A-1 to A-3) and 1 for fine-grained soils (A-4 to A-7)

P = annual precipitation in cm

H = slab thickness in inches

FI = freezing index in degree days

Jointed Reinforced Concrete Pavements

$$PI = (N_{18})^{0.670} [-22.82 + 26,102.2(H)^{-5.0} - 0.129(D)$$

$$- 0.118(S) + 13.224(P)^{0.0395} + 6.834(FI + 1)^{0.00805} \quad (12.2b)$$

Statistics: $R^2 = 0.57$

$$SEE = 0.52$$

$$n = 481$$

in which D is the indicator for the presence of subdrainage systems with 0 for no subdrainage system and 1 for subdrainage system.

A pumping prediction model was developed by Purdue University and incorporated in the PEARDARP computer program (Van Wiji, 1985; Van Wiji et al., 1989). The model is based on field data from different sources and the amount of deformation energy in the pavement structure. The volume of pumped material is calculated as a function of the deformation energy produced by traffic. This volume is then adjusted to take into account the variations in subbase type, drainage conditions, load transfer adequacy, subgrade conditions, and climate. The model can be used in the optimum design of rigid pavements by predicting the effect of different design alternatives on the development of pumping. As is true with any regression models, the model is valid only within the data base from which it was derived. It is hoped that more mechanistic-based models can be developed to replace the regression models.

12.1.4 Faulting Models

Faulting at transverse joints is a serious problem that can lead to severe roughness in jointed concrete pavements. The mechanisms of faulting distress in doweled pavements are quite different from those in undoweled pavements. Therefore, these two pavements are discussed separately.

Doweled Pavements

Faulting of doweled pavements is caused by the erosion of concrete around the dowels under repeated loading. Because the design of dowels is based on the bearing stress between dowel and concrete, as described in Section 4.4.1, it is natural to assume that faulting is due to excessive bearing stress. It was found that if the bearing stress is kept below approximately 1500 psi (10.4 MPa), faulting can be maintained at an acceptable level.

The maximum bearing stress on dowel bars can be obtained directly by the finite element computer programs. As an alternative, Report 1-26 recommended the use of Eq. 4.45 to compute the bearing stress. The procedure is the same as described in Example 4.12, Section 4.4.1, except that the load is distributed over an effective length of 1.0ℓ, instead of 1.8ℓ, and the load transferred through the joint is assumed to be $0.45W$ instead of $0.5W$, where W is the total load. Using the above method to calculate the bearing stress and assuming a modulus of dowel support K of 1.5×10^6 psi (10.4 GPa), the following regression equation based on 280 pavement sections in the COPES data base was obtained:

$$F = (N_{18})^{0.5377} [2.2073 + 0.002171(S)^{0.4918}$$

$$+ 0.0003292(JS)^{1.0793} - 2.1397(k)^{0.01305}] \qquad (12.3)$$

in which F = pavement faulting in inches
N_{18} = number of equivalent 18-kip single-axle loads in millions
S = maximum bearing stress in psi
JS = transverse joint spacing in ft
k = estimated modulus of subgrade reaction on the top of the subbase in pci

Several climatic variables, such as the precipitation and freezing indexes, were introduced originally but did not show any statistical significance and, therefore, were not included in the model. This is probably due to the limited number of climatic zones in the data base. Because of insufficient data, many other variables such as permeable base, subgrade type, edge support, and subdrainage are not included. This model must not be used to predict faulting by extrapolation beyond the data range used in its generation. This is particularly true for open-graded drainable bases which were not included. Figure 12.4 is a plot of predicted faulting versus actual faulting for these 280 sections of doweled pavements. Note that Eq. 12.3 is quite different from Eq. 9.54a, which is a later model with a more extended data base and considers the effect of drainage condition, edge support, and soil type.

Figure 12.5 shows the effect of bearing stress on faulting, as obtained from Eq. 12.3 based on $N_{18} = 10$ million. Two different joint spacings and subgrade moduli were used. It can be seen that bearing stress has the most significant effect on faulting, while the joint spacing has the least effect.

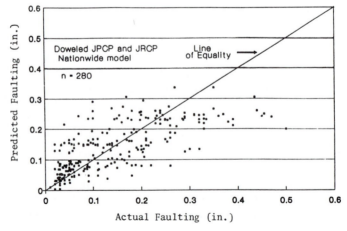

Figure 12.4 Predicted faulting versus observed faulting for doweled pavements (1 in. = 25.4 mm). (After NCHRP (1990).)

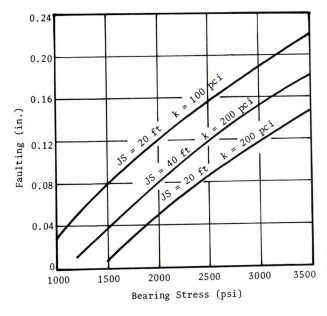

Figure 12.5 Effect of bearing stress on faulting. (1 in. = 25.4 mm, 1 ft = 0.305 m, 1 psi = 6.9 kPa, 1 pci = 271.3 kN/m³).

Undoweled Pavements

To date no mechanistic-based analyses have been attempted for undoweled pavements. The following regression equation was derived from 186 pavement sections in the COPES data base and presented in Report 1-26:

$$F = (N_{18})^{0.3157} [0.4531 + 0.3367(z)^{0.3322} - 0.5376(100w)^{-0.008437}$$

$$+ 0.0009092(\text{FI})^{0.5998} + 0.004654(B) - 0.03608(\text{ES})$$

$$- 0.01087(S) - 0.009467(D)] \tag{12.4}$$

in which F = faulting in inches

N_{18} = number of equivalent 18-kip single-axle loads in millions

z = joint opening in inches, which can be determined from Eq. 4.36

w = corner deflection in inches, which was determined from Eq. 4.16 based on a 9000-lb (40-kN) load with a contact pressure of 90 psi (621 kPa) applied at a free corner

FI = mean air freezing index in degree days

B = erodibility factor for subbase materials, with 0.5 for lean concrete subbase, 1.0 for cement-treated granular subbase, 1.5 for cement-treated nongranular subbase, 2.0 for asphalt-treated subbase, 2.5 for untreated granular subbase

ES = edge support condition, with 0 for no edge support and 1 for tied edge beam or tied concrete shoulder

S = subgrade soil type, with 0 for A-4 to A-7 and 1 for A-1 to A-3

D = drainage index, with 0 for no edge drains and 1 for edge drains

Figure 12.6 is a plot of predicted faulting versus observed faulting for these 186 sections of undoweled pavements. Note that Eqs. 12.4 and 9.54b are of the same

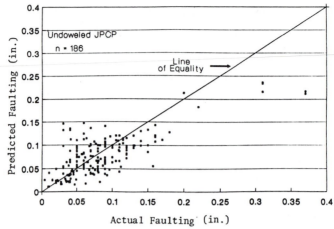

Figure 12.6 Predicted faulting versus observed faulting for undoweled pavements (1 in. = 25.4 mm). (After NCHRP (1990).)

form and contain the same variables, but the regression constants are completely different due to the scope of the data base.

Table 12.1 shows a sensitivity analysis of faulting in undoweled pavements based on Eq. 12.4. The parameter values shown in column 2 are used as the standard case. In the other seven cases, only the parameter shown in column 1 is changed in value from column 2 to column 3, while the other values remain the same as in column 2. For each case, the amount of faulting and its ratio to the standard case are tabulated.

A review of Table 12.1 indicates that, unless the freezing index is extremely high, edge support and the joint opening, which are directly related to the load transfer across the joint and edge, have the most effect on faulting, while corner deflection, which is related to slab thickness, is not very sensitive to faulting.

TABLE 12.1 SENSITIVITY ANALYSIS OF UNDOWELED PAVEMENTS

Case	Change		Faulting	
	From	To	in.	Ratio
(1)	(2)	(3)	(4)	(5)
Standard case			0.118	1.00
ESAL, N_{18} (10^6)	10	20	0.147	1.25
Joint opening z (in.)	0.04	0.08	0.180	1.53
Corner deflection w (in.)	0.01	0.04	0.131	1.11
Freezing index FI (degree-days)	100	2000	0.268	2.27
Edge support ES	0	1	0.043	0.36
Base erodibility B	2.5	0.5	0.099	0.84
Subgrade soil type S	0	1	0.095	0.81
Drainage index D	0	1	0.098	0.83

Note. 1 in. = 25.4 mm.

12.1.5 Joint Deterioration

Joint deterioration includes spalling and general breakup of the concrete near the joints. No mechanistic methods have been developed to analyze joint deterioration. The most important factor that causes joint deterioration is the "D" cracking of concrete, as described in Section 9.1.2. Therefore, the use of aggregates that do not cause "D" cracking is the first requisite to prevent joint deterioration.

A statistically based model was developed using the COPES data base. All pavements showing "D" cracking and reactive aggregate distress were removed from the data base and nonlinear regression techniques were used on the remaining pavement sections. With over 501 pavement sections around the country, the following regression equation was obtained:

$$DETJT = (AGE)^{2.3503}(N_{18})^{0.62974} [-0.0021443 + 3.6239 \times 10^{-6} (FI)$$

$$+ 7.08597 \times 10^{-5} (JS) + 3.5307 \times 10^{-5} (SCTE)] \qquad (12.5)$$

in which $DETJT$ = percentage of deteriorated joints
AGE = time in years since construction started
N_{18} = number of 18-kip equivalent single-axle loads in millions
FI = mean air freezing index in degree days
JS = transverse joint spacing in ft
$SCTE$ = Thornthwaite summer concentration of thermal energy

Table 12.2 shows a sensitivity analysis of Eq. 12.5. It can be seen that age and joint spacing have the most effect on joint deterioration, while the summer concentration of thermal energy has the least effect.

12.1.6 Punchout Models

Punchouts are the primary mode of distress in CRCP, as described in Section 9.1.2. If this type of distress could be eliminated, CRCP would have an outstanding performance record. A comprehensive study was made by Zollinger (1989) on punchout distress. He described four modes of distress that eventually lead to punchout, as illustrated in Figure 12.7.

TABLE 12.2 SENSITIVITY ANALYSIS OF JOINT DETERIORATION

Case (1)	Change From (2)	Change To (3)	Joint deterioration % (4)	Joint deterioration Ratio (5)
Standard case			7.426	1.00
AGE (year)	10	30	98.209	13.23
ESAL N_{18} (10^6)	10	30	14.833	2.00
FI (degree-day)	100	2000	14.002	1.89
JS (ft)	100	15	1.674	0.23
SCTE	70	35	6.246	0.84

Note. 1 ft = 0.305 m.

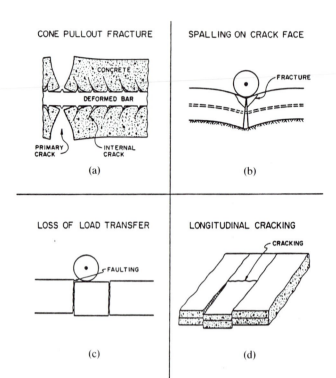

CONE PULLOUT FRACTURE

CONCRETE

DEFORMED BAR

PRIMARY
CRACK
INTERNAL
CRACK

(a)

SPALLING ON CRACK FACE

FRACTURE

(b)

LOSS OF LOAD TRANSFER

FAULTING

(c)

LONGITUDINAL CRACKING

CRACKING

(d)

Figure 12.7 Failure modes leading to punchout distress. (After Zollinger and Barenberg (1990).)

When the concrete slab cracks, the tensile stress in the steel reinforcement causes the fracture of surrounding concrete, as shown in (a). The fracture of concrete reduces the stiffness of the slab and results in spalling on the crack surface, as shown in (b). The spalling on the crack surface makes the crack open wide and results in the loss of load transfer across the crack, as shown in (c). Without the load transfer, the slab between two cracks acts as a cantilever beam and the tensile stress in the transverse direction due to the edge loading causes the slab to crack at the top, as shown in (d).

Figure 12.8 shows the stresses at two critical locations in a CRCP under an 18-kip (80-kN) single-axle load. The stresses are presented in terms of load transfer efficiency (LTE) and crack spacing. Based on the analysis by ILLI-SLAB, the figure shows that the maximum longitudinal tensile stress σ_B at the bottom of the slab is highly sensitive to crack spacing but is relatively unaffected by LTE across the crack. The transverse tensile stress σ_A on the top of the slab is highly sensitive to LTE whenever the crack spacing is less than 4 ft. These results indicate that if the LTE can be maintained above 90%, the maximum transverse stress will remain relatively low even when the crack spacing is small. When the LTE drops to 70% or lower, the maximum transverse stress becomes increasingly large as the LTE and crack spacing decrease. When the tensile stress exceeds the tensile strength, a longitudinal crack will occur. The combination of poor load transfer, voids around the reinforcing steel, longitudinal cracks, and so on, results in a punchout.

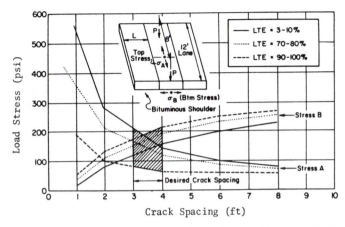

Figure 12.8 Comparison of σ_A and σ_B in CRCP for a 10-in. slab (1 psi = 6.9 kPa, 1 ft = 0.305 m). (After Zollinger and Barenberg (1990).)

A mechanistic method of design to prevent punchout requires a knowledge of LTE, which depends on the level of shear stress across the crack and the crack width. The level of shear stress can be obtained from the finite element computer programs, but the determination of crack width requires an accurate prediction of crack spacing. Under normal construction conditions, crack spacings vary so widely that a reliable design procedure cannot be achieved. For the design approach to be reliable, some means of external control on the crack spacing is needed.

12.2 PORTLAND CEMENT ASSOCIATION METHOD

The Portland Cement Association's (PCA) thickness design procedure for concrete highways and streets was published in 1984, superseding that published in 1966. The procedure can be applied to JPCP, JRCP, and CRCP. A finite element computer program called JSLAB (Tayabji and Colley, 1986) was employed to compute the critical stresses and deflections, which were then used in conjunction with some design criteria to develop the design tables and charts. The design criteria are based on general pavement design, performance, and research experience, including relationships to performance of pavements in the AASHO Road Test and to studies of pavement faulting. Design problems can be worked out by hand with tables and charts presented herein or by a microcomputer program available from PCA.

12.2.1 Design Criteria

One aspect of the new design procedure is the inclusion of an erosion analysis, in addition to the fatigue analysis. Fatigue analysis recognizes that pavements can fail by fatigue of concrete, while in erosion analysis pavements fail by pumping, erosion of foundation, and joint faulting.

Fatigue Analysis

Fatigue analysis is based on the edge stress midway between the transverse joints with the most critical loading position shown in Figure 12.9. Because the load is near the midslab far away from the joints, the presence of the joints has practically no effect on the edge stress. When a concrete shoulder is tied onto the mainline pavement, the magnitude of the critical stress is reduced considerably.

The cumulative damage concept, as described in Section 12.1.2, is used for fatigue analysis. However, only an average modulus of subgrade reaction k is used for the entire design period, and the stresses due to warping and curling are not considered. Warping and curling are excluded because the moisture content and temperature are usually higher at the bottom of the slab than at the top; thus, the combined effect of warping and curling stresses are subtractive from the loading stresses. Consequently, Eq. 12.1 is reduced to

$$D_r = \sum_{i=1}^{m} \frac{n_i}{N_i} \tag{12.6}$$

in which D_r is the damage ratio accumulated over the design period due to all load groups, m is the total number of load groups, n_i is the predicted number of repetitions for the ith load group, and N_i is the allowable number of repetitions for the ith load group, which can be determined from Eq. 5.36. The accumulated damage ratio at the end of design period should be smaller than one.

The placement of outside wheels at the pavement edge produces a critical stress higher than that at other locations. As the truck placement moves inward a few inches from the edge, the stress decreases significantly. At increasing distances inward from the edge, the frequency of load applications increases but the magnitude of edge stress decreases. Theoretically, the distribution of load placement across the traffic lane must be known, so that the damage ratio at the pavement edge caused by each load pacement can be computed and summed to obtain the total damage. This procedure is too cumbersome for design purposes but was analyzed by PCA to develop a more easily applied method. In this method, fatigue at the pavement edge was computed by placing the load incrementally at different distances inward from the slab edge for typical distributions of truck placement. It was found that the same fatigue damage can be obtained by considering the edge loading only and placing 6% of the total number of load repetitions at the pavement edge. If the total number of repetitions are used for design, the edge stress must be reduced to obtain the same fatigue consumption.

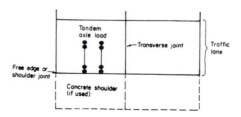

Figure 12.9 Critical loading position for fatigue analysis.

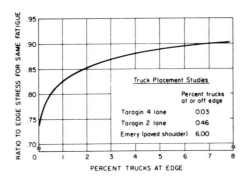

Figure 12.10 Equivalent stress factor versus percentage of trucks at edge. (After PCA (1984).)

For 6% truck encroachment, the edge stress must be multiplied by an adjusting factor of 0.894, as shown in Figure 12.10. This factor was used in preparing the design tables.

Erosion Analysis

Pavement distresses such as pumping, erosion of foundation, and joint faulting are related more to pavement deflections than to flexural stresses. The most critical pavement deflection occurs at the slab corner when an axle load is placed at the joint near to the corner, as shown in Figure 12.11.

The principal mode of failure in the AASHO Road Test was pumping or erosion of the granular subbase from under the slabs. However, satisfactory correlations between corner deflections and the performance of these pavements could not be obtained. It was found that to be able to predict their performance, different values of deflection criteria would have to be applied, depending on the slab thickness and, to a small extent, on the modulus of subgrade reaction. A better correlation was obtained by relating the performance to the rate of work defined as the product of corner deflection w and pressure p at the slab–foundation interface divided by the length of deflection basin, which is a function of the radius of relative stiffness ℓ. The concept is that a thin slab with a shorter deflection basin receives a faster load punch than a thicker slab. The following equations were developed to compute the allowable load repetitions:

$$\log N = 14.524 - 6.777(C_1 P - 9.0)^{0.103} \tag{12.7}$$

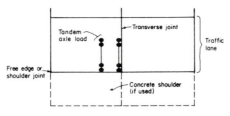

Figure 12.11 Critical loading position for erosion analysis.

in which N is the allowable number of load repetitions based on a PSI of 3.0, C_1 is an adjustment factor with a value of 1 for untreated subbases and 0.9 for stabilized subbases, and P is the rate of work or power defined by

$$P = 268.7\frac{p^2}{hk^{0.73}} \tag{12.8}$$

in which p is the pressure on the foundation under the slab corner in psi, which is equal to kw for a liquid foundation, h is the thickness of slab in inches, and k is the modulus of subgrade reaction in pci. The equation for erosion damage is

$$\text{Percent erosion damage} = 100 \sum_{i=1}^{m} \frac{C_2 n_i}{N_i} \tag{12.9}$$

in which $C_2 = 0.06$ for pavements without concrete shoulders and 0.94 for pavements with tied concrete shoulders. With a concrete shoulder, the corner deflection is not significantly affected by the truck load placement, so a large C_2 should be used. The percent erosion damage should be less than 100%.

Example 12.1

The sensitivity analysis of corner deflection presented in Table 5.14 shows that for the standard case with $h = 8$ in. (203 mm) and $k = 100$ pci (27.1 MN/m³), the corner deflection is 0.0353 in. (0.90 mm) under an 18-kip (80-kN) single-axle load and 0.0458 in. (1.16 mm) under a 36-kip (160-kN) tandem-axle load. If the predicted number of load repetitions is 5×10^6, compute the percent erosion damage under the single- and tandem-axle loads, respectively.

Solution: For the case of an 18-kip (80-kN) single-axle load, $p = kw = 100 \times 0.0353 = 3.53$ psi (24.4 kPa). From Eq. 12.8, $P = 268.7 \times (3.53)^2/[8 \times (100)^{0.73}] = 14.512$. Assuming that $C_1 = 1.0$, from Eq. 12.7, log $N = 14.524 - 6.777(14.512 - 9.0)^{0.103} = 6.444$, or $N = 2.78 \times 10^6$. With $C_2 = 0.06$, from Eq. 12.9, percent erosion damage $= 100 \times 0.06 \times 5 \times 10^6/(2.78 \times 10^6) = 10.8\%$.

For the case of a 36-kip (160-kN) tandem-axle load, $p = 100 \times 0.0458 = 4.58$ psi (31.6 kPa); $P = 268.7 \times (4.58)^2/[8 \times (100)^{0.73}] = 24.429$; log $N = 14.524 - 6.777(24.429 - 9.0)^{0.103} = 5.541$; $N = 3.47 \times 10^5$; and percent erosion damage $= 100 \times 0.06 \times 5 \times 10^6/(3.47 \times 10^5) = 86.5\%$.

12.2.2 Design Factors

After deciding whether doweled joints and concrete shoulders are to be used, the thickness design is governed by four design factors: concrete modulus of rupture, subgrade and subbase support, design period, and traffic. These factors are discussed below.

Concrete Modulus of Rupture

The flexural strength of concrete is defined by the modulus of rupture, which is determined at 28 days using the method specified by ASTM in "C78-84 Standard Test Method for Flexural Strength of Concrete Using Simple Beam with

Third Point Loading." The 28-day flexural strength is used as the design strength. The variability of strength and the gain in strength with age should be considered in the fatigue analysis.

In view of the fact that the variations in modulus of rupture have far greater effect on thickness design than the usual variations in other material properties, it is recommended that the modulus of rupture be reduced by one coefficient of variation. A coefficient of variation of 15%, which represents good to fair quality control, was assumed and incorporated into the design charts and tables. Also incorporated was the effect of strength gain after 28 days.

Subgrade and Subbase Support

Subgrade and subbase support is defined by the modulus of subgrade reaction k. Methods for determining the modulus of subgrade reaction are described in Section 7.5.1. Figure 7.36 can be used to correlate k values with other soil properties.

The PCA method does not consider the variation of k values over the year. The contention is that the reduced subgrade support during thaw periods has very little or no effect on the required thickness of concrete pavements, as evidenced by the results of AASHO Road Test. This is true because the brief periods when k values are low during spring thaws are more than offset by the longer freezing periods when k values are much higher than the design value. To avoid the tedious method of considering seasonal variations in k values, normal summer or fall k values can be used as reasonable mean values for design purposes.

It is not economical to use untreated subbases for the sole purpose of increasing k values. However, there are many reasons to use subbases, as described in Section 1.2.2. Where a subbase is used, there will be an increase in k that should be considered in the thickness design. If the subbase is composed of untreated granular materials, the approximate increase in k values can be taken from Table 12.3. For a cement-treated subbase, the design k values can be obtained from Table 12.4.

TABLE 12.3 EFFECT OF UNTREATED SUBBASE ON k VALUES

Subgrade k value (pci)	Subbase k values (pci)			
	4 in.	6 in.	9 in.	12 in.
50	65	75	85	110
100	130	140	160	190
200	220	230	270	320
300	320	330	370	430

Note. 1 in. = 25.4 mm, 1 pci = 271.3 kN/m³.

Source. After PCA (1984).

TABLE 12.4 DESIGN k VALUES FOR
CEMENT-TREATED BASES

Subgrade k value (pci)	Subbase k values (pci)			
	4 in.	6 in.	8 in.	10 in.
50	170	230	310	390
100	280	400	520	640
200	470	640	830	—

Note. 1 in. = 25.4 mm, 1 pci = 271.3 kN/m³.
Source. After PCA (1984).

Design Period

The term "design period" should not be confused with the term "pavement life," which is not subject to precise definition. Design period is more synonymous to the term "traffic analysis period." Because traffic probably cannot be predicted with much accuracy for a longer period, a design period of 20 years has been commonly used in pavement design. However, there are cases where the use of a shorter or longer design period may be economically justified. For example, a special haul road that will be used only a few years requires a much shorter design period, while a premium facility that provides a high level of performance for a long time with little or no pavement maintenance may require a design period up to 40 years.

Traffic

The information presented in Section 6.4, such as Eq. 6.26, can be used to determine the design traffic. The growth factor can be determined from Table 6.12 and the lane distribution factor for multilane highways from Figure 6.8. Information on the average daily truck traffic (ADTT) and the axle load distribution is needed in using the PCA design procedure. The ADTT includes only trucks with six tires or more and does not include panel and pickup trucks or other vehicles with only four tires.

Axle Load Distribution

Data on the axle load distribution of truck traffic is needed to compute the number of single and tandem axles of various weights expected during the design period. These data can be obtained from special traffic studies to establish the loadometer data for the specific project or from the W-4 table of a loadometer station representing truck weights and types that are expected to be similar to the project under design. If axle load distribution data are not available, the simplified design procedure described in Section 12.2.4 can be used.

Table 12.5 illustrates how the information in a W-4 table for a loadometer station can be used to determine the number of various axles based on the total number of trucks. In the W-4 table, axle loads are grouped by 2-kip (8.9-kN)

TABLE 12.5 AXLE LOAD DISTRIBUTION FOR A GIVEN
FACILITY

Axle load kip (1)	Axles per 1000 trucks (2)	Adjusted axles per 1000 trucks (3)	Axles in design period (4)
Single axles			
28–30	0.28	0.58	6310
26–28	0.65	1.35	14,690
24–26	1.33	2.77	30,140
22–24	2.84	5.92	64,410
20–22	4.72	9.83	106,900
18–20	10.40	21.67	235,800
16–18	13.56	28.24	307,200
14–16	18.64	38.83	422,500
12–14	25.89	53.94	586,900
10–12	81.05	168.85	1,837,000
Tandem axles			
48–52	0.94	1.98	21,320
44–48	1.89	3.94	42,870
40–44	5.51	11.48	124,900
36–40	16.45	34.27	372,900
32–36	39.08	81.42	885,800
28–32	41.06	85.54	930,700
24–28	73.07	152.23	1,656,000
20–24	43.45	90.52	984,900
16–20	54.15	112.81	1,227,000
12–16	59.85	124.69	1,356,000

Note. 1 kip = 4.45 kN.

Source. After PCA (1984).

increments for single axles and 4-kip (17.8-kN) increments for tandem axles, as shown in column 1. The axles per 1000 trucks shown in column 2 are obtained from the W-4 table. The table also shows 13,215 total trucks counted with 6918 two-axle, four-tire trucks, which constitute 52% of total trucks. Because the trucks in column 2 include two-axle, four-tire trucks, which should be excluded from consideration, the data in column 2 must be divided by $(1 - 0.52)$ to obtain the adjusted axles per 1000 trucks shown in column 3. Column 4 is the number of load repetitions to be used for the sample problem in Figure 12.15 and can be obtained by the following formula: Column 4 = Column 3 × (Trucks on design lane in design period)/1000. The number of trucks on the design lane in the design period is 10,880,000, as will be determined later in the sample problem.

Load Safety Factors

In the design procedure, the axle load must be multiplied by a load safety factor (LSF). The recommended load safety factors are as follows:

1. For interstate highways and other multilane projects where there will be uninterrupted traffic flow and high volumes of truck traffic, LSF = 1.2.

2. For highways and arterial streets where there will be moderate volumes of truck traffic, LSF = 1.1.

3. For roads, residential streets, and other streets that will carry small volumes of truck traffic, LSF = 1.0.

In special cases, the use of a load safety factor as high as 1.3 may be justified for a premium facility to maintain a higher than normal level of pavement serviceability throughout the design period.

12.2.3 Design Procedure

The method presented in this section can be used when detailed axle load distributions have been determined or estimated, as described in Section 12.2.2. If the axle load data are not available, the simplified method presented in Section 12.2.4 should be used.

Design Tables and Charts

Separate sets of tables and charts are used to evaluate fatigue and erosion damages. The following parameter values are used in their development: elastic modulus of concrete = 4×10^6 psi (28 GPa), Poisson ratio of concrete = 0.15, diameter of dowels = $\frac{1}{8}$ in./in. of slab, spacing of dowels = 12 in. (305 mm), modulus of dowel support = 2×10^6 pci (543 GN/m^3), spring constant for aggregate interlock joints = 5000 psi (34.5 MPa), spring constant for tie concrete shoulder = 25,000 psi (173 MPa).

Fatigue Damage

Fatigue damage is based on the edge stress. Because the edge stress on mainline pavements without concrete shoulders is much greater than that with tied concrete shoulders, two different tables are needed: Table 12.6 for slabs without concrete shoulders and Table 12.7 for slabs with concrete shoulders. The equivalent stresses shown in these tables are the edge stresses multiplied by a factor of 0.894. It is not known what axle load was used to generate these stresses. Based on the levels of stress, it appears that an 18-kip (80-kN) load was used for single axles and a 36-kip (160-kN) load was used for tandem axles. Both tables show that the equivalent stresses under 36-kip (160-kN) tandem-axle loads are smaller than those under 18-kip (80-kN) single-axle loads, which is as expected.

After the equivalent stress is determined, the stress ratio factor can be computed by dividing the equivalent stress with the design modulus of rupture, so the allowable number of load repetitions can be obtained from Figure 12.12. Note that the reduction in the modulus of rupture by 15% and the increase in the modulus of rupture with age have been incorporated in the chart, so the user simply inputs the 28-day strength as the design modulus of rupture. Figure 12.12 can be applied to pavements both with and without concrete shoulders. If the allowable repetitions fall outside the range of the chart, the allowable number of repetitions is considered to be unlimited.

TABLE 12.6   EQUIVALENT STRESSES FOR SLABS WITHOUT CONCRETE SHOULDERS

Slab thickness (in.)	\multicolumn{7}{c}{k of Subgrade–subbase (pci)}						
	50	100	150	200	300	500	700
4	825/679	726/585	671/542	634/516	584/486	523/457	484/443
4.5	699/586	616/500	571/460	540/435	498/406	448/378	417/363
5	602/516	531/436	493/399	467/376	432/349	390/321	363/307
5.5	526/461	464/387	431/353	409/331	379/305	343/278	320/264
6	465/416	411/348	382/316	362/296	336/271	304/246	285/232
6.5	417/380	367/317	341/286	324/267	300/244	273/220	256/207
7	375/349	331/290	307/262	292/244	271/222	246/199	231/186
7.5	340/323	300/268	279/241	265/224	246/203	224/181	210/169
8	311/300	274/249	255/223	242/208	225/188	205/167	192/155
8.5	285/281	252/232	234/208	222/193	206/174	188/154	177/143
9	264/264	232/218	216/195	205/181	190/163	174/144	163/133
9.5	245/248	215/205	200/183	190/170	176/153	161/134	151/124
10	228/235	200/193	186/173	177/160	164/144	150/126	141/117
10.5	213/222	187/183	174/164	165/151	153/136	140/119	132/110
11	200/211	175/174	163/155	154/143	144/129	131/113	123/104
11.5	188/201	165/165	153/148	145/136	135/122	123/107	116/98
12	177/192	155/158	144/141	137/130	127/116	116/102	109/93
12.5	168/183	147/151	136/135	129/124	120/111	109/97	103/89
13	159/176	139/144	129/129	122/119	113/106	103/93	97/85
13.5	152/168	132/138	122/123	116/114	107/102	98/89	92/81
14	144/162	125/133	116/118	110/109	102/98	93/85	88/78

Note. Number at left is for single axle and number at right is for tandem axle (single/tandem);
1 in. = 25.4 mm, 1 pci = 271.3 kN/m^3.

Source. After PCA (1984).

Example 12.2

The sensitivity analysis presented in Table 5.13 shows an average edge stress of about 283 psi (1.95 MPa) under a 36-kip (160-kN) tandem-axle load when $h = 8$ in. (203 mm) and $k = 100$ pci (27.1 MN/m^3). Assuming a concrete modulus of rupture of 500 psi (4.5 MPa) and using the PCA fatigue equation, or Eq. 5.36, determine the allowable number of load repetitions. Compare the result with that obtained from Table 12.6 and Figure 12.12.

Solution: Because only a small portion of wheel loads is applied at the pavement edge, the edge stress must be multiplied by an adjustment factor of 0.894, so the actual stress is $\sigma = 0.894 \times 283 = 253$ psi (1.75 MPa). With $\sigma/S_c = 253/500 = 0.506$, from Eq. 5.36$b$, $N_f = [4.2577/(0.506 - 0.43250)]^{3.268} = 5.8 \times 10^5$.

From Table 12.6, the equivalent stress for a tandem-axle load with a k value of 100 pci (27.1 MN/m^3) and a slab thickness of 8 in. (203 mm) is 249 psi (1.72 MPa), which is slightly smaller than the 253 psi (1.75 MPa) obtained in the sensitivity analysis. This is as expected because the dual and tandem spacings as well as the tire contact pressure assumed by PCA is not known and may be quite

TABLE 12.7 EQUIVALENT STRESSES FOR SLABS WITH CONCRETE SHOULDERS

Slab thickness (in.)	k of Subgrade–subbase (pci)						
	50	100	150	200	300	500	700
4	640/534	559/468	517/439	489/422	452/403	409/388	383/384
4.5	547/461	479/400	444/372	421/356	390/338	355/322	333/316
5	475/404	417/349	387/323	367/308	341/290	311/274	294/267
5.5	418/360	368/309	342/285	324/271	302/254	276/238	261/231
6	372/325	327/277	304/255	289/241	270/225	247/210	234/203
6.5	334/295	294/251	274/230	260/218	243/203	223/188	212/180
7	302/270	266/230	248/210	236/198	220/184	203/170	192/162
7.5	275/250	243/211	226/193	215/182	201/168	185/155	176/148
8	252/232	222/196	207/179	197/168	185/155	170/142	162/135
8.5	232/216	205/182	191/166	182/156	170/144	157/131	150/125
9	215/202	190/171	177/155	169/146	158/134	146/122	139/116
9.5	200/190	176/160	164/146	157/137	147/126	136/114	129/108
10	186/179	164/151	153/137	146/129	137/118	127/107	121/101
10.5	174/170	154/143	144/130	137/121	128/111	119/101	113/95
11	164/161	144/135	135/123	129/115	120/105	112/95	106/90
11.5	154/153	136/128	127/117	121/109	113/100	105/90	100/85
12	145/146	128/122	120/111	114/104	107/95	99/86	95/81
12.5	137/139	121/117	113/106	108/99	101/91	94/82	90/77
13	130/133	115/112	107/101	102/95	96/86	89/78	85/73
13.5	124/127	109/107	102/97	97/91	91/83	85/74	81/70
14	118/122	104/103	97/93	93/87	87/79	81/71	77/67

Note. Number at left is for single axle and number at right is for tandem axle (single/tandem); 1 in. = 25.4 mm, 1 pci = 271.3 kN/m³.

Source. After PCA (1984).

different from those in the sensitivity analysis. Hence, the stress ratio factor = 249/500 = 0.498.

From Figure 12.12, a straight line is drawn starting from the 18-kip (80-kN) single-axle load on the left axis, which is the same as the 36-kip (160-kN) tandem-axle load, passing through a point on the middle axis with a stress ratio factor of 0.498, and ending at the allowable load repetitions on the right axis. The allowable number of repetitions read from the right axis is 7×10^5, which is also slightly larger than the 5.8×10^5 obtained previously. Note that in applying Eq. 5.36b, no adjustment was made on the modulus of rupture, while in developing the chart, the modulus of rupture was reduced by 15% for variability and then increased by a certain amount for strength gain with age. It is not known how much gain in strength was assumed by PCA, so a discrepancy between the two results is expected.

Erosion Damage

Because erosion damage occurs at the pavement corner and is affected by the type of joints, separate tables for doweled and aggregate interlock joints are

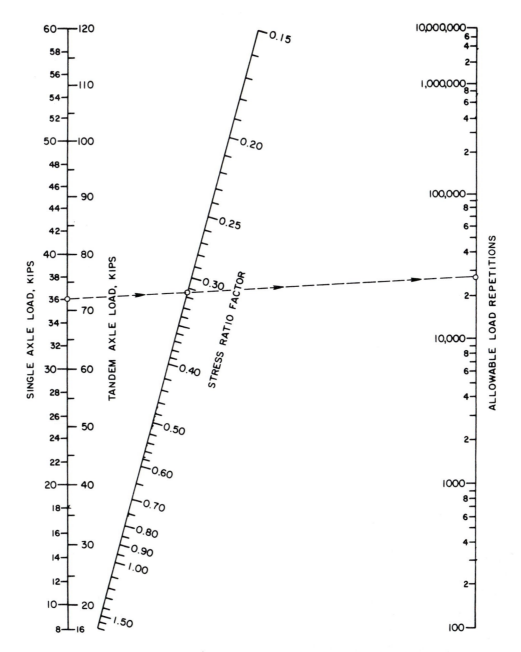

Figure 12.12 Stress ratio factors versus allowable load repetitions both with and without concrete shoulders (1 kip = 4.45 kN). (After PCA (1984).)

Slab thickness (in.)	k of Subgrade–subbase (pci)					
	50	100	200	300	500	700
4	3.74/3.83	3.73/3.79	3.72/3.75	3.71/3.73	3.70/3.70	3.68/3.67
4.5	3.59/3.70	3.57/3.65	3.56/3.61	3.55/3.58	3.54/3.55	3.52/3.53
5	3.45/3.58	3.43/3.52	3.42/3.48	3.41/3.45	3.40/3.42	3.38/3.40
5.5	3.33/3.47	3.31/3.41	3.29/3.36	3.28/3.33	3.27/3.30	3.26/3.28
6	3.22/3.38	3.19/3.31	3.18/3.26	3.17/3.23	3.15/3.20	3.14/3.17
6.5	3.11/3.29	3.09/3.22	3.07/3.16	3.06/3.13	3.05/3.10	3.03/3.07
7	3.02/3.21	2.99/3.14	2.97/3.08	2.96/3.05	2.95/3.01	2.94/2.98
7.5	2.93/3.14	2.91/3.06	2.88/3.00	2.87/2.97	2.86/2.93	2.84/2.90
8	2.85/3.07	2.82/2.99	2.80/2.93	2.79/2.89	2.77/2.85	2.76/2.82
8.5	2.77/3.01	2.74/2.93	2.72/2.86	2.71/2.82	2.69/2.78	2.68/2.75
9	2.70/2.96	2.67/2.87	2.65/2.80	2.63/2.76	2.62/2.71	2.61/2.68
9.5	2.63/2.90	2.60/2.81	2.58/2.74	2.56/2.70	2.55/2.65	2.54/2.62
10	2.56/2.85	2.54/2.76	2.51/2.68	2.50/2.64	2.48/2.59	2.47/2.56
10.5	2.50/2.81	2.47/2.71	2.45/2.63	2.44/2.59	2.42/2.54	2.41/2.51
11	2.44/2.76	2.42/2.67	2.39/2.58	2.38/2.54	2.36/2.49	2.35/2.45
11.5	2.38/2.72	2.36/2.62	2.33/2.54	2.32/2.49	2.30/2.44	2.29/2.40
12	2.33/2.68	2.30/2.58	2.28/2.49	2.26/2.44	2.25/2.39	2.23/2.36
12.5	2.28/2.64	2.25/2.54	2.23/2.45	2.21/2.40	2.19/2.35	2.18/2.31
13	2.23/2.61	2.20/2.50	2.18/2.41	2.16/2.36	2.14/2.30	2.13/2.27
13.5	2.18/2.57	2.15/2.47	2.13/2.37	2.11/2.32	2.09/2.26	2.08/2.23
14	2.13/2.54	2.11/2.43	2.08/2.34	2.07/2.29	2.05/2.23	2.03/2.19

Note. Number at left is for single axle and number at right is for tandem axle (single/tandem);
1 in. = 25.4 mm, 1 pci = 271.3 kN/m³.

Source. After PCA (1984).

needed. The erosion criteria also require two separate charts for slab with and without concrete shoulders.

Table 12.8 shows the erosion factors for slabs with doweled joints and no concrete shoulders, and Table 12.9 shows the erosion factors for slabs with aggregate interlock joints and no concrete shoulders. After the erosion factor is found, the allowable number of load repetitions can be obtained from Figure 12.13. Note that the allowable load repetitions shown in Figure 12.13 have already been divided by a C_2 of 0.06, so it is not necessary to multiply the predicted repetitions by C_2, as indicated by Eq. 12.9.

Example 12.3

Same as Example 12.1 except that the percent erosion damages for the 18-kip (80-kN) single-axle and 36-kip (160-kN) tandem-axle loads are to be determined by Table 12.8 and Figure 12.13.

Solution: For the 18-kip (80-kN) single-axle load with a k value of 100 pci (27.1 MN/m³) and a slab thickness of 8 in. (203 mm), from Table 12.8, erosion factor =

TABLE 12.9 EROSION FACTORS FOR SLABS WITH AGGREGATE INTERLOCK JOINTS AND NO CONCRETE SHOULDERS

Slab thickness (in.)	k of Subgrade–subbase (pci)					
	50	100	200	300	500	700
4	3.94/4.03	3.91/3.95	3.88/3.89	3.86/3.86	3.82/3.83	3.77/3.80
4.5	3.79/3.91	3.76/3.82	3.78/3.75	3.71/3.72	3.68/3.68	3.64/3.65
5	3.66/3.81	3.63/3.72	3.60/3.64	3.58/3.60	3.55/3.55	3.52/3.52
5.5	3.54/3.72	3.51/3.62	3.48/3.53	3.46/3.49	3.43/3.44	3.41/3.40
6	3.44/3.64	3.40/3.53	3.37/3.44	3.35/3.40	3.32/3.34	3.30/3.30
6.5	3.34/3.56	3.30/3.46	3.26/3.36	3.25/3.31	3.22/3.25	3.20/3.21
7	3.26/3.49	3.21/3.39	3.17/3.29	3.15/3.24	3.13/3.17	3.11/3.13
7.5	3.18/3.43	3.13/3.32	3.09/3.22	3.07/3.17	3.04/3.10	3.02/3.06
8	3.11/3.37	3.05/3.26	3.01/3.16	2.99/3.10	2.96/3.03	2.94/2.99
8.5	3.04/3.32	2.98/3.21	2.93/3.10	2.91/3.04	2.88/2.97	2.87/2.93
9	2.98/3.27	2.91/3.16	2.86/3.05	2.84/2.99	2.81/2.92	2.79/2.87
9.5	2.92/3.22	2.85/3.11	2.80/3.00	2.77/2.94	2.75/2.86	2.73/2.81
10	2.86/3.18	2.79/3.06	2.74/2.95	2.71/2.89	2.68/2.81	2.66/2.76
10.5	2.81/3.14	2.74/3.02	2.68/2.91	2.65/2.84	2.62/2.76	2.60/2.72
11	2.77/3.10	2.69/2.98	2.63/2.86	2.60/2.80	2.57/2.72	2.54/2.67
11.5	2.72/3.06	2.64/2.94	2.58/2.82	2.55/2.76	2.51/2.68	2.49/2.63
12	2.68/3.03	2.60/2.90	2.53/2.78	2.50/2.72	2.46/2.64	2.44/2.59
12.5	2.64/2.99	2.55/2.87	2.48/2.75	2.45/2.68	2.41/2.60	2.39/2.55
13	2.60/2.96	2.51/2.83	2.44/2.71	2.40/2.65	2.36/2.56	2.34/2.51
13.5	2.56/2.93	2.47/2.80	2.40/2.68	2.36/2.61	2.32/2.53	2.30/2.48
14	2.53/2.90	2.44/2.77	2.36/2.65	2.32/2.58	2.28/2.50	2.25/2.44

Note. Number at left is for single axle and number at right is for tandem axle (single/tandem); 1 in. = 25.4 mm, 1 pci = 271.3 kN/m^3.

Source. After PCA (1984).

2.82, and from Figure 12.13, allowable load repetitions = 3×10^7. Given predicted repetitions = 5×10^6, percent erosion damage = $100 \times (5 \times 10^6)/(3 \times 10^7)$ = 16.7%, compared with 10.8% in Example 12.1.

For the 36-kip (160-kN) tandem-axle load, from Table 12.8, erosion factor = 2.99; from Figure 12.13, allowable load repetitions = 6.2×10^6; percent erosion damage = $100 \times (5 \times 10^6)/(6.2 \times 10^6)$ = 80.6%, compared with 86.5% in Example 12.1. Since allowable repetitions are very sensitive to corner deflections, the agreement between the two examples is considered exceedingly good.

Similar tables and charts for slabs with concrete shoulders are shown in Tables 12.10 and 12.11 and Figure 12.14.

Sample Problem

Figure 12.15 is the worksheet on a sample problem. The design involves a four-lane interstate pavement with doweled joints and no concrete shoulders. A 4-in. (102-mm) untreated subbase will be placed on a clay subgrade with a k value

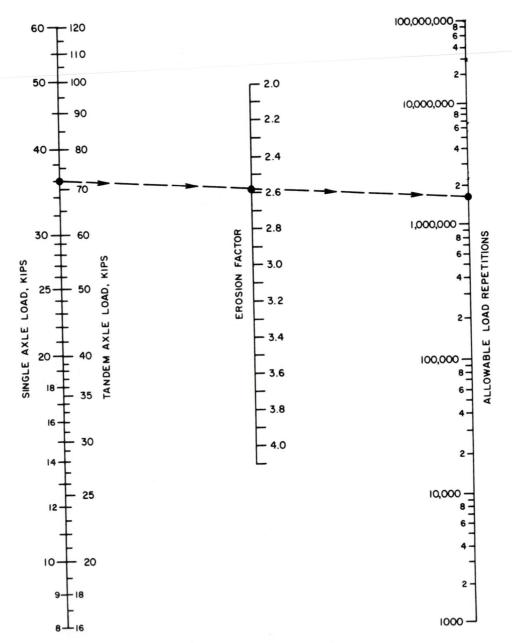

Figure 12.13 Erosion factors versus allowable load repetitions without concrete shoulders (1 kip = 4.45 kN). (After PCA (1984).)

Rigid Pavement Design   Chap. 12

TABLE 12.10 EROSION FACTORS FOR SLABS WITH DOWELED JOINTS AND CONCRETE SHOULDERS

Slab thickness (in.)	k of Subgrade–subbase (pci)					
	50	100	200	300	500	700
4	3.28/3.30	3.24/3.20	3.21/3.13	3.19/3.10	3.15/3.09	3.12/3.08
4.5	3.13/3.19	3.09/3.08	3.06/3.00	3.04/2.96	3.01/2.93	2.98/2.91
5	3.01/3.09	2.97/2.98	2.93/2.89	2.90/2.84	2.87/2.79	2.85/2.77
5.5	2.90/3.01	2.85/2.89	2.81/2.79	2.79/2.74	2.76/2.68	2.73/2.65
6	2.79/2.93	2.75/2.82	2.70/2.71	2.68/2.65	2.65/2.58	2.62/2.54
6.5	2.70/2.86	2.65/2.75	2.61/2.63	2.58/2.57	2.55/2.50	2.52/2.45
7	2.61/2.79	2.56/2.68	2.52/2.56	2.49/2.50	2.46/2.42	2.43/2.38
7.5	2.53/2.73	2.48/2.62	2.44/2.50	2.41/2.44	2.38/2.36	2.35/2.31
8	2.46/2.68	2.41/2.56	2.36/2.44	2.33/2.38	2.30/2.30	2.27/2.24
8.5	2.39/2.62	2.34/2.51	2.29/2.39	2.26/2.32	2.22/2.24	2.20/2.18
9	2.32/2.57	2.27/2.46	2.22/2.34	2.19/2.27	2.16/2.19	2.13/2.13
9.5	2.26/2.52	2.21/2.41	2.16/2.29	2.13/2.22	2.09/2.14	2.07/2.08
10	2.20/2.47	2.15/2.36	2.10/2.25	2.07/2.18	2.03/2.09	2.01/2.03
10.5	2.15/2.43	2.09/2.32	2.04/2.20	2.01/2.14	1.97/2.05	1.95/1.99
11	2.10/2.39	2.04/2.28	1.99/2.16	1.95/2.09	1.92/2.01	1.89/1.95
11.5	2.05/2.35	1.99/2.24	1.93/2.12	1.90/2.05	1.87/1.97	1.84/1.91
12	2.00/2.31	1.94/2.20	1.88/2.09	1.85/2.02	1.82/1.93	1.79/1.87
12.5	1.95/2.27	1.89/2.16	1.84/2.05	1.81/1.98	1.77/1.89	1.74/1.84
13	1.91/2.23	1.85/2.13	1.79/2.01	1.76/1.95	1.72/1.86	1.70/1.80
13.5	1.86/2.20	1.81/2.09	1.75/1.98	1.72/1.91	1.68/1.83	1.65/1.77
14	1.82/2.17	1.76/2.06	1.71/1.95	1.67/1.88	1.64/1.80	1.61/1.74

Note. Number at left is for single axle and number at right is for tandem axle (single/tandem); 1 in. = 25.4 mm, 1 pci = 271.3 kN/m³.

Source. After PCA (1984).

of 100 pci (27.2 MN/m³). Other information include concrete modulus of rupture = 650 psi (4.5 MPa), design period = 20 years, current ADT = 12,900, annual growth rate = 4%, and ADTT = 19% of ADT.

On the worksheet, a trial thickness of 9.5 in. (241 mm) is selected. For a subgrade k value of 100 pci (27.1 MN/m³) and a subbase thickness of 4 in. (102 mm), from Table 12.3, the k value for subbase–subgrade combination is 130 pci (35.3 MN/m³). For an interstate highway, a load safety factor of 1.2 is recommended. With a thickness of 9.5 in. (241 mm) and a k value of 130 pci (35.3 MN/m³), an equivalent stress of 206 psi (731 kPa) for single axles and 192 psi (1.32 MPa) for tandem axles can be obtained from Table 12.6 and entered as items 8 and 11 on the worksheet. The stress ratio factor is the ratio between the equivalent stress and the modulus of rupture, and ratios of 0.317 for single axles and 0.295 for tandem axles are entered as items 9 and 12. Erosion factors of 2.59 for single axles and 2.79 for tandem axles are obtained from Table 12.8 and entered as items 10 and 13. Each column, as numbered from 1 to 7 in the worksheet, is explained on page 622.

TABLE 12.11 EROSION FACTORS FOR SLABS WITH AGGREGATE INTERLOCK JOINTS AND CONCRETE SHOULDERS

Slab thickness (in.)	k of Subgrade–subbase (pci)					
	50	100	200	300	500	700
4	3.46/3.49	3.42/3.39	3.38/3.32	3.36/3.29	3.32/3.26	3.28/3.24
4.5	3.32/3.39	3.28/3.28	3.24/3.19	3.22/3.16	3.19/3.12	3.15/3.09
5	3.20/3.30	3.16/3.18	3.12/3.09	3.10/3.05	3.07/3.00	3.04/2.97
5.5	3.10/3.22	3.05/3.10	3.01/3.00	2.99/2.95	2.96/2.90	2.93/2.86
6	3.00/3.15	2.95/3.02	2.90/2.92	2.88/2.87	2.86/2.81	2.83/2.77
6.5	2.91/3.08	2.86/2.96	2.81/2.85	2.79/2.79	2.76/2.73	2.74/2.68
7	2.83/3.02	2.77/2.90	2.73/2.78	2.70/2.72	2.68/2.66	2.65/2.61
7.5	2.76/2.97	2.70/2.84	2.65/2.72	2.62/2.66	2.60/2.59	2.57/2.54
8	2.69/2.92	2.63/2.79	2.57/2.67	2.55/2.61	2.52/2.53	2.50/2.48
8.5	2.63/2.88	2.56/2.74	2.51/2.62	2.48/2.55	2.45/2.48	2.43/2.43
9	2.57/2.83	2.50/2.70	2.44/2.57	2.42/2.51	2.39/2.43	2.36/2.38
9.5	2.51/2.79	2.44/2.65	2.38/2.53	2.36/2.46	2.33/2.38	2.30/2.33
10	2.46/2.75	2.39/2.61	2.33/2.49	2.30/2.42	2.27/2.34	2.24/2.28
10.5	2.41/2.72	2.33/2.58	2.27/2.45	2.24/2.38	2.21/2.30	2.19/2.24
11	2.36/2.68	2.28/2.54	2.22/2.41	2.19/2.34	2.16/2.26	2.14/2.20
11.5	2.32/2.65	2.24/2.51	2.17/2.38	2.14/2.31	2.11/2.22	2.09/2.16
12	2.28/2.62	2.19/2.48	2.13/2.34	2.10/2.27	2.06/2.19	2.04/2.13
12.5	2.24/2.59	2.15/2.45	2.09/2.31	2.05/2.24	2.02/2.15	1.99/2.10
13	2.20/2.56	2.11/2.42	2.04/2.28	2.01/2.21	1.98/2.12	1.95/2.06
13.5	2.16/2.53	2.08/2.39	2.00/2.25	1.97/2.18	1.93/2.09	1.91/2.03
14	2.13/2.51	2.04/2.36	1.97/2.23	1.93/2.15	1.89/2.06	1.87/2.00

Note. Number at left is for single axle and number at right is for tandem axle (single/tandem); 1 in. = 25.4 mm, 1 pci = 271.3 kN/m³.

Source. After PCA (1984).

Explanation of Worksheet

1. Single-axle loads are incremented at 2-kip (8.9-kN) intervals, and tandem-axle loads are incremented at 4-kip (17.8-kN) intervals. The largest load in the single- or tandem-load group should be entered first. If the allowable number of repetitions for a given load is unlimited, it is not necessary to compute the damage for the remaining loads in the same group.

2. The axle loads in column 1 are multiplied by a load safety factor of 1.2.

3. The predicted or expected repetitions are obtained from Table 12.5. To be on the conservative side, the upper limit of the load in the range is used to represent the range. For example, all axle loads between 28 and 30 kip (125 and 134 kN) are considered as 30 kip (134 kN). With an annual growth rate of 4% and a design period of 20 years, from Table 6.12, growth factor $G = 1.5$. Design ADT = 12,900 × 1.5 = 19,350, or 9675 in one direction. ADTT = 19,350 × 0.19 = 3680, or 1840 in one direction. For an ADT of 9675 in one direction, from Figure 6.8, lane distribution factor $L = 0.81$. Therefore, the total number of trucks on the design lane during the design period is 1840 ×

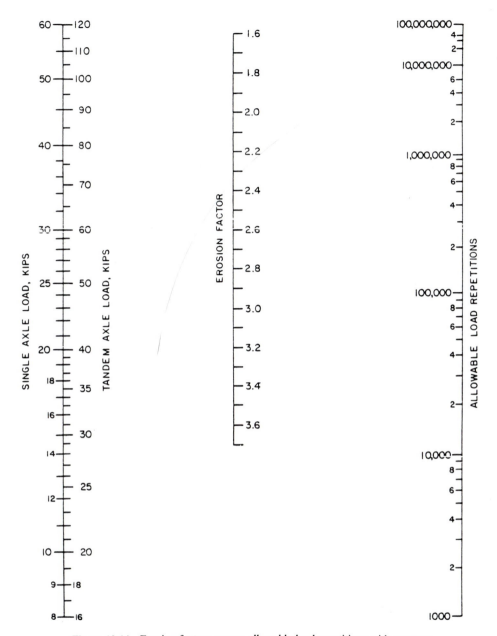

Figure 12.14 Erosion factors versus allowable load repetitions with concrete shoulders (1 kip = 4.45 kN). (After PCA (1984).)

0.81 × 365 × 20 = 10,880,000, which was used to obtain the axle load distribution in Table 12.5.

4. The allowable repetitions in fatigue analysis are obtained from Figure 12.12 based on a stress ratio factor of 206/650, or 0.317, for single axles and 192/650, or 0.295, for tandem axles.

Calculation of Pavement Thickness

Project _Design 1A, four-lane Interstate, rural_

Trial thickness _9.5_ in.

Subbase-subgrade k _130_ pci

Modulus of rupture, MR _650_ psi

Load safety factor, LSF _1.2_

Doweled joints: yes ✓ no ____

Concrete shoulder: yes ____ no ✓

Design period _20_ years

4-in. untreated subbase

Axle load, kips	Multiplied by LSF 1.2	Expected repetitions	Fatigue analysis		Erosion analysis	
			Allowable repetitions	Fatigue, percent	Allowable repetitions	Damage, percent
1	2	3	4	5	6	7

8. Equivalent stress _206_   10. Erosion factor _2.59_

9. Stress ratio factor _0.317_

Single Axles

30	36.0	6,310	27,000	23.3	1,500,000	0.4
28	33.6	14,690	77,000	19.1	2,200,000	0.7
26	31.2	30,140	230,000	13.1	3,500,000	0.9
24	28.8	64,410	1,200,000	5.4	5,900,000	1.1
22	26.4	106,900	Unlimited	0	11,000,000	1.0
20	24.0	235,800	"	0	23,000,000	1.0
18	21.6	307,200	"	0	64,000,000	0.5
16	19.2	422,500			Unlimited	0
14	16.8	536,900			"	0
12	14.4	1,837,000			"	0

11. Equivalent stress _192_   13. Erosion factor _2.79_

12. Stress ratio factor _0.295_

Tandem Axles

52	62.4	21,320	1,100,000	1.9	920,000	2.3
48	57.6	42,870	Unlimited	0	1,500,000	2.9
44	52.8	124,900	"	0	2,500,000	5.0
40	48.0	372,900	"	0	4,600,000	8.1
36	43.2	885,800			9,500,000	9.3
32	38.4	930,700			24,000,000	3.9
28	33.6	1,656,000			92,000,000	1.8
24	28.8	984,900			Unlimited	0
20	24.0	1,227,000			"	0
16	19.2	1,356,000				
				Total 62.8		Total 38.9

Figure 12.15 Worksheet for sample problem (1 in. = 25.4 mm, 1 psi = 6.9 kPa, 1 pci = 271.3 kN/m³). (After PCA (1984).)

5. The fatigue percentages are obtained by dividing column 3 with column 4 and multiplying by 100. The sum of fatigue percentages over all single- and tandem-axle loads is entered at the bottom.

6. The allowable repetitions in erosion analysis are obtained from Figure 12.13 based on an erosion factor of 2.59 for single axles and 2.79 for tandem axles.

7. The erosion percentages are obtained by dividing column 3 with column 6 and multiplying by 100. The sum of erosion percentages over all single- and tandem-axle loads is entered at the bottom.

Figure 12.15 shows that damages caused by fatigue and erosion are 62.8 and 38.9, respectively. Since both are less than 100%, the use of 9.5 in. (241 mm) slab is quite adequate. Separate calculations showed that a slab of 9.0 in. (229 mm) was not adequate because the fatigue damage would increase to 142%. Therefore, this design is controlled by the fatigue analysis.

The fatigue analysis will usually control the design of pavements subjected to light traffic, whether the joints are doweled or not, and medium traffic with doweled joints. The erosion analysis will usually control the design of pavements subjected to medium and heavy traffic with aggregate interlock joints and heavy traffic with doweled joints. For pavements carrying a normal mix of axle weights, single-axle loads will cause more fatigue cracking, while tandem-axle loads will cause more erosion damage.

12.2.4 Other Features

The PCA design manual also includes a simplified design procedure to be used when axle load data are not available, an analysis of concrete pavements with a lean concrete subbase, and an analysis of tridem-axle loads. These features are described in this section.

Simplified Design Procedure

A series of tables were developed by PCA to select the pavement thickness when specific axle load data are not available. The factors to be considered are traffic, subgrade–subbase strength, and the modulus of rupture of concrete.

Traffic Category

Traffic is divided into four axle load categories, as shown in Table 12.12. The ADT and ADTT values should not be used as the primary criteria for selecting the axle load category. More reliance should be placed on word descriptions or the expected maximum axle loads. The axle load distributions used to prepare the simplified design tables for each traffic category are shown in Table 12.13. Each of these is the average of several W-4 tables representing pavement facilities in the appropriate category.

Subgrade–Subbase Strength

Subgrade–subbase strength is characterized by the descriptive terms low, medium, high, and very high. These terms are related to the modulus of subgrade

TABLE 12.12 AXLE LOAD CATEGORIES FOR SIMPLIFIED DESIGN PROCEDURE

Axle load category	Description	Traffic			Maximum axle loads (kips)	
		ADT	ADTT		Single axles	Tandem axles
			%	Per day		
1	Residential streets Rural and secondary roads (low to medium)	200–800	1–3	Up to 25	22	36
2	Collector streets Rural and secondary roads (high) Arterial streets and primary roads (low)	700–5000	5–18	40–1000	26	44
3	Arterial streets and primary roads (medium) Expressways and urban and rural interstate highways (low to medium)	3000–12,000 2 lanes 3000–50,000 + 4 lanes or more	8–30	500–5000 +	30	52
4	Arterial streets, primary roads, expressways (high) Urban and rural interstate highways (medium to high)	3000–20,000 2 lanes 3000–150,000 + 4 lanes or more	8–30	1500–8000 +	34	60

Note. The descriptors high, medium, or low refer to the relative weights of axle loads for the type of street or road; ADTT does not include two-axle, four-tire trucks; 1 kip = 4.45 kN.

Source. After PCA (1984).

TABLE 12.13 AXLE LOAD DISTRIBUTION FOR FOUR TRAFFIC CATEGORIES

Axle load (kips)	Axles per 1000 trucks			
	Category 1	Category 2	Category 3	Category 4
Single axles				
4	1693.31			
6	732.28			
8	483.10	233.60		
10	204.96	142.70		
12	124.00	116.76	182.02	
14	56.11	47.76	47.73	
16	38.02	23.88	31.82	57.07
18	15.81	16.61	25.15	68.27
20	4.23	6.63	16.33	41.82
22	0.96	2.60	7.85	9.69
24		1.60	5.21	4.16
26		0.07	1.78	3.52
28			0.85	1.78
30			0.45	0.63
32				0.54
34				0.19
Tandem axles				
4	31.90			
8	85.59	47.01		
12	139.30	91.15		
16	75.02	59.25	99.34	
20	57.10	45.00	85.94	
24	39.18	30.74	72.54	71.16
28	68.48	44.43	121.22	95.79
32	69.59	54.76	103.63	109.54
36	4.19	38.79	56.25	78.19
40		7.76	21.31	20.31
44		1.16	8.01	3.52
48			2.91	3.03
52			1.19	1.79
56				1.07
60				0.57

Note. 1 kip = 4.45 kN; all two-axle, four-tire trucks are excluded.
Source. After PCA (1984).

reaction k, as shown in Table 12.14. When a subbase is used, the increase in k value can be determined from Table 12.3 or 12.4, depending on whether the subbase is untreated or stabilized.

Design Tables

The PCA design manual contains a series of tables showing the allowable ADTT for pavements with either doweled or aggregate interlock joints. Separate tables were developed for each axle load category. To illustrate the method, only the table for axle load category 3 with doweled joints is reproduced here, as shown in Table 12.15.

TABLE 12.14 SUBGRADE SOIL TYPES AND APPROXIMATE *k* VALUES

Type of soil	Support	*k* Values (pci)
Fine-grained soils in which silt and clay-size particles predominate	Low	75–120
Sands and sand–gravel mixtures with moderate amounts of silt and clay	Medium	130–170
Sands and sand–gravel mixtures relatively free of plastic fines	High	180–220
Cement-treated subbases	Very high	250–400

Note. 1 pci = 271.3 kN/m³.

Source. After PCA (1984).

Three different moduli of rupture can be specified. The two values of 650 and 600 psi (4.5 and 4.1 MPa) on the upper portion of the tables are for good concrete with normal aggregates and are recommended for general design use; the 550 psi (3.8 MPa) on the bottom portion is for a special case where high-quality aggregates are not available.

The allowable ADTT is based on a 20-year design period and does not include any two-axle, four-tire trucks. If the design period is not 20 years, the predicted ADTT must be changed proportionately. Incorporated in the tables are load safety factors of 1.0, 1.1, 1.2, and 1.2 for axle load categories of 1, 2, 3, and 4, respectively. The tables were developed by first assuming an ADTT and then determining the percentages of fatigue and erosion damage based on the given slab thickness, concrete modulus of rupture, and subgrade–subbase *k* value. The allowable ADTT was then computed by

$$\text{Allowable ADTT} = \frac{100 \times (\text{assumed ADTT})}{\% \text{ fatigue or erosion damage}}$$

Example 12.4

The following information is given for a concrete pavement: arterial street, doweled joints, curb and gutter, design ADT = 6200, total trucks per day = 1440, ADTT = 630, concrete modulus of rupture = 650 psi (4.5 MPa), and 4 in. (102 mm) of untreated granular subbase on a subgrade with *k* = 150 pci (40.7 MN/m³). Determine slab thickness by the simplified method.

Solution: Both ADT and ADTT fit well with axle load category 3. From Table 12.3, the *k* value of the subgrade and subbase combined is about 170 pci (46.1 MN/m³), so the subgrade–subbase support is classified as medium according to Table 12.14. From Table 12.15, a 7.5-in. (191-mm) slab gives an allowable ADTT of 1200; while a 7 in. (178 mm) slab gives only 220. The predicted ADTT is 630, so the use of 7.5 in. (191 mm) is adequate.

Lean Concrete Subbase

The finite element computer program can be used to analyze two layers of slab, either bonded or unbonded. If the bottom layer is a hardened lean concrete

TABLE 12.15 Allowable ADTT[a] FOR AXLE LOAD CATEGORY 3 WITH DOWELED JOINTS

| | No concrete shoulder or curb | | | | | Concrete shoulder or curb | | | |
| Slab thickness (in.) | Subgrade–subbase support | | | | Slab thickness (in.) | Subgrade–subbase support | | | |
	Low	Medium	High	Very high		Low	Medium	High	Very high
MR = 650 psi									
7.5				250	6.5			83	320
8		130	350	1300	7	52	220	550	1900
8.5	160	640	1600	6200	7.5	320	1200	2900	9800
9	700	2700	7000	11,500[b]	8	1600	5700	13,800	
9.5	2700	10,800			8.5	6900	23,700[b]		
10	9900								
MR = 600 psi									
8			73	310	6.5				67
8.5		140	380	1500	7			120	440
9	160	640	1700	6200	7.5		270	680	2300
9.5	630	2500	6500		8	370	1300	3200	10,800
10	2300	9300			8.5	1600	5800	14,100	
10.5	7700				9	6600			
MR = 550 psi									
8.5			70	300	7				82
9		120	340	1300	7.5			130	480
9.5	120	520	1300	5100	8	67	270	670	2300
10	460	1900	4900	19,100	8.5	330	1200	2900	9700
10.5	1600	6500	17,400		9	1400	4900	11,700	
11	4900				9.5	5100	18,600		

Note. [a] ADTT excludes two-axle, four-tire trucks.
[b] Erosion controls the design; otherwise fatigue controls. 1 in. = 25.4 mm, 1 psi = 6.9 kPa.
Source. After PCA (1984).

on which a layer of normal concrete is placed, the layers can be considered unbonded. If the two layers are built monolithically with the joints sawed deep enough to induce cracking through both layers, the case of two bonded layers applies. Design charts were developed by PCA for both bonded and unbonded cases. However, only the chart for the more popular unbonded case, which involves a normal concrete slab on a lean concrete subbase, is presented here. In the finite element analysis, the two layers of slab were assumed to have the same width. Because the lean concrete subbase is usually built at least 2 ft (0.61 m) wider than the pavement on each side to support the tracks of the slipform paver, the assumption of equal width provides additional margin of safety to the design.

Figure 12.16 is the design chart for concrete pavements with lean concrete subbases. To use the design chart, the slab thickness required for a conventional pavement without a lean concrete subbase must be determined by the procedure described previously. For a given thickness of lean concrete subbase, the thickness of concrete slab can be reduced, depending on the moduli of rupture of the two concrete materials. For example, if the moduli of rupture are 650 psi (4.5 MPa) for normal concrete and 200 psi (1.4 MPa) for lean concrete, the design

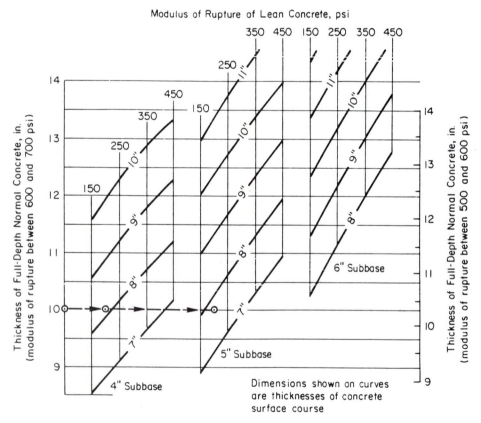

Figure 12.16 Design chart for concrete pavements with lean concrete subbase (1 in. = 25.4 mm, 1 psi = 6.9 kPa). (After PCA (1984).)

equivalent to the 10-in. (254-mm) pavement can be either a 7.7-in. (196-mm) concrete slab on a 5-in. (127-mm) lean concrete subbase or an 8.1-in. (206-mm) concrete slab on a 4-in. (102-mm) lean concrete subbase, as shown by the dashed line in Figure 12.16.

The normal practice has been to select a surface thickness about twice the subbase thickness. Therefore, either an 8-in. (203-mm) slab on a 5-in. (127-mm) subbase or an 8.5-in. (216-mm) slab on a 4-in. (127-mm) subbase can be used for practical design. The use of the design chart will ensure that the fatigue and erosion damage in the two layers of concrete does not exceed that in the conventional pavement. The use of a very low modulus of rupture of 200 psi (1.4 MPa) is recommended to minimize reflection cracking from the unjointed subbase through the concrete surface. If, contrary to current practice, joints are placed in the subbase at the same location as in the concrete surface, higher moduli of rupture for lean concrete may be used.

Tridem-Axle Loads

Three more tables, one for equivalent stresses and two for erosion factors, were developed by PCA for tridem axles. One of the tables that can be used to determine erosion factors for slabs with doweled joints is shown in Table 12.16 for illustrative purposes. The procedure is similar to that for single and tandem axles. After the equivalent stress or erosion factor is obtained from the tables, Figure 12.12, 12.13, or 12.14 can be used to determine the allowable number of load repetitions. Although tridem-axle loads are not shown in these figures, the scale for single-axle loads can be used by dividing the tridem-axle load by three.

Example 12.5

Given a concrete pavement with a thickness of 8 in. (203 mm), a k value of 100 pci (27.1 MN/m^3), doweled joints, and no concrete shoulders, determine the allowable repetitions under a 54-kip (240-kN) tridem-axle load based on erosion criteria.

Solution: From Table 12.16, erosion factor = 3.14. With a tridem-axle load of 54 kip (240 kN), or a single-axle load of 18 kip (80 kN), from Figure 12.13, the allowable number of repetitions is 2.3×10^6.

Some comments on the PCA method are in order. From Examples 12.3 and 12.5, the allowable repetitions based on erosion criteria are 3×10^7 for the 18-kip (80-kN) single-axle load, 6.2×10^6 for the 36-kip (160-kN) tandem-axle load, and 2.3×10^6 for the 54-kip (240-kN) tridem-axle load. Because erosion was the major distress in the AASHO Road Test, these allowable repetitions can be used to determine the equivalent axle load factor (EALF). The EALF for the 36-kip (160-kN) tandem-axle load is $(3 \times 10^7)/(6.2 \times 10^6)$, or 4.84, and that for the 54-kip (240-kN) tridem-axle load is $(3 \times 10^7)/(2.3 \times 10^6)$, or 13.0. These equivalent factors are much larger than the AASHTO factors of 2.42 and 4.0, as presented in Table 6.7. This is not surprising because EALF valves of 4.58 and 15.5 are also shown in Table 6.8 based on the fatigue analysis by KENLAYER. Under a certain combination of slab thickness and subgrade support, an extremely high or low EALF may be obtained.

TABLE 12.16 EROSION FACTORS FOR SLABS WITH DOWELED JOINTS UNDER TRIDEM AXLES

Slab thickness (in.)	k of Subgrade–subbase (pci)					
	50	100	200	300	500	700
4	3.89/3.33	3.82/3.20	3.75/3.13	3.70/3.10	3.61/3.05	3.53/3.00
4.5	3.78/3.24	3.69/3.10	3.62/2.99	3.57/2.95	3.50/2.91	3.44/2.87
5	3.68/3.16	3.58/3.01	3.50/2.89	3.46/2.83	3.40/2.79	3.34/2.75
5.5	3.59/3.09	3.49/2.94	3.40/2.80	3.36/2.74	3.30/2.67	3.25/2.64
6	3.51/3.03	3.40/2.87	3.31/2.73	3.26/2.66	3.21/2.58	3.16/2.54
6.5	3.44/2.97	3.33/2.82	3.23/2.67	3.18/2.59	3.12/2.50	3.08/2.45
7	3.37/2.92	3.26/2.76	3.16/2.61	3.10/2.53	3.04/2.43	3.00/2.37
7.5	3.31/2.87	3.20/2.72	3.09/2.56	3.03/2.47	2.97/2.37	2.93/2.31
8	3.26/2.83	3.14/2.67	3.03/2.51	2.97/2.42	2.90/2.32	2.86/2.25
8.5	3.20/2.79	3.09/2.63	2.97/2.47	2.91/2.38	2.84/2.27	2.79/2.20
9	3.15/2.75	3.04/2.59	2.92/2.43	2.86/2.34	2.78/2.23	2.73/2.15
9.5	3.11/2.71	2.99/2.55	2.87/2.39	2.81/2.30	2.73/2.18	2.68/2.11
10	3.06/2.67	2.94/2.51	2.83/2.35	2.76/2.26	2.68/2.15	2.63/2.07
10.5	3.02/2.64	2.90/2.48	2.78/2.32	2.72/2.23	2.64/2.11	2.58/2.04
11	2.98/2.60	2.86/2.45	2.74/2.29	2.68/2.20	2.59/2.06	2.54/2.00
11.5	2.94/2.57	2.82/2.42	2.70/2.26	2.64/2.16	2.55/2.05	2.50/1.97
12	2.91/2.54	2.79/2.39	2.67/2.23	2.60/2.13	2.51/2.02	2.46/1.94
12.5	2.87/2.51	2.75/2.36	2.63/2.20	2.56/2.11	2.48/1.99	2.42/1.91
13	2.84/2.48	2.72/2.33	2.60/2.17	2.53/2.08	2.44/1.96	2.39/1.88
13.5	2.81/2.46	2.68/2.30	2.56/2.14	2.49/2.05	2.41/1.93	2.35/1.86
14	2.78/2.43	2.65/2.28	2.53/2.12	2.46/2.03	2.38/1.91	2.32/1.83

Note. Number at left is without concrete shoulder and number at right is with concrete shoulder (without concrete shoulder/with concrete shoulder); 1 in. = 25.4 mm, 1 pci = 271.3 kN/m³.
Source. After PCA (1984).

12.3 AASHTO METHOD

The design guide for rigid pavements was developed at the same time as that for flexible pavements and was published in the same manual. The design is based on the empirical equations obtained from the AASHO Road Test with further modifications based on theory and experience. In this section, only the thickness design is presented. The design of steel reinforcements and tiebars is similar to that discussed in Section 4.3.2 and is not presented here.

12.3.1 Design Equations

The basic equations developed from the AASHO Road Test for rigid pavements are in the same form as those for flexible pavements but with different values for the regression constants. The equations were then modified to include many variables originally not considered in the AASHO Road Test.

Original Equations

Similar to flexible pavements, the regression equations are

$$G_t = \beta(\log W_t - \log \rho) \tag{11.29}$$

$$\beta = 100 + \frac{3.63(L_1 + L_2)^{5.20}}{(D + 1)^{8.46}L_2^{3.52}} \tag{12.10}$$

$$\log \rho = 5.85 + 7.35 \log(D + 1) - 4.62 \log (L_1 + L_2)$$
$$+ 3.28 \log L_2 \tag{12.11}$$

in which $G_t = \log [(4.5 - p_t)/(4.5 - 1.5)]$, where 4.5 is the initial serviceability for rigid pavement at the AASHO Road Test, which is different from the 4.2 for flexible pavements, and p_t is the serviceability at time t;

D = slab thickness in inches, which replaces SN for flexible pavements.

Using an equivalent 18-kip (80-kN) single-axle load with $L_1 = 18$ and $L_2 = 1$ and combining Eqs. 11.29, 12.10, and 12.11 yields

$$\log W_{t18} = 7.35 \log(D + 1) - 0.06 + \frac{\log[(4.5 - p_t)/(4.5 - 1.5)]}{1 + 1.624 \times 10^7/(D + 1)^{8.46}} \tag{12.12}$$

in which W_{t18} is the number of 18-kip (80-kN) single-axle load applications to time t. Equation 12.12 is applicable only to the pavements in the AASHO Road Test with the following conditions: modulus of elasticity of concrete $E_c = 4.2 \times 10^6$ psi (29 GPa), modulus of rupture of concrete $S_c = 690$ psi (4.8 MPa), modulus of subgrade reaction $k = 60$ pci (16 MN/m³), load transfer coefficient $J = 3.2$, and drainage coefficient $C_d = 1.0$.

Modified Equations

To account for conditions other than those that existed in the road test, it is necessary to modify Eq. 12.12 using experience and theory. After comparing stresses calculated from strain measurements on the Road Test pavements with theoretical solutions, the Spangler equation for corner loading (Spangler, 1942) was selected for its simplicity by AASHTO (1972) to extend Eq. 12.12 to other conditions. The Spangler equation is given as

$$\sigma = \frac{JP}{D^2}\left(1 - \frac{a_1}{\ell}\right) \tag{12.13}$$

in which σ is the maximum tensile stress in concrete in psi, J is the load transfer coefficient, P is the wheel load in lb, a_1 is the distance from corner of slab to center of load, and ℓ is the radius of relative stiffness defined by Eq. 4.10 and rewritten as

$$\ell = \left[\frac{ZD^3}{12(1 - v^2)}\right]^{0.25} \tag{12.14}$$

in which $Z = E_c/k$ and ν is Poisson ratio of concrete. Assuming that $a_1 = 10$ in. (254 mm) and $\nu = 0.2$, and substituting Eq. 12.14 into 12.13 gives

$$\sigma = \frac{JP}{D^2}\left(1 - \frac{18.42}{Z^{0.25}D^{0.75}}\right) \tag{12.15}$$

Stresses were calculated for different combinations of Road Test variables using Eq. 12.15. The ratio between the calculated stresses and the modulus of rupture, σ/S_c, was subsequently compared with axle load applications. These comparisons indicated that for any given load and terminal serviceability level p_t, the following relationship similar to the general fatigue equation exists:

$$\log W_t = a - (4.22 - 0.32p_t)\log \frac{\sigma}{S_c} \tag{12.16}$$

Assuming the same form of equation for other pavements with W_t', σ', and S_c' yields

$$\log W_t' = a - (4.22 - 0.32p_t)\log \frac{\sigma'}{S_c'} \tag{12.17}$$

Combining Eqs. 12.16 and 12.17 and using the equivalent 18-kip (80-kN) single-axle load gives

$$\log W_{t18}' = \log W_{t18} + (4.22 - 0.32p_t)\log\left(\frac{S_c'}{S_c}\frac{\sigma}{\sigma'}\right) \tag{12.18}$$

From Eq. 12.15

$$\frac{\sigma}{\sigma'} = \frac{J[1 - 18.42/(Z^{0.25}D^{0.75})]}{J'[1 - 18.42/(Z'^{0.25}D^{0.75})]} \tag{12.19}$$

Combining Eqs. 12.12, 12.18, and 12.19 results in

$$\log W_{t18}' = 7.35 \log(D + 1) - 0.06 + \frac{\log[(4.5 - p_t)/(4.5 - 1.5)]}{1 + 1.624 \times 10^7/(D + 1)^{8.46}}$$

$$+ (4.22 - 0.32p_t)\log\left[\left(\frac{S_c'J}{S_cJ'}\right)\left(\frac{D^{0.75} - 18.42/Z^{0.25}}{D^{0.75} - 18.42/Z'^{0.25}}\right)\right] \tag{12.20}$$

Letting $Z = E_c/k = 4.2 \times 10^6/60 = 70{,}000$, $S_c = 690$, and $J = 3.2$; adding a drainage coefficient C_d and a reliability term $Z_R S_o$; replacing the term $(4.5 - p_t)$ by ΔPSI; and removing the primes for simplicity; the final design equation for rigid pavements becomes

$$\log W_{18} = Z_R S_o + 7.35 \log(D + 1) - 0.06 + \frac{\log[\Delta PSI/(4.5 - 1.5)]}{1 + 1.624 \times 10^7/(D + 1)^{8.46}}$$

$$+ (4.22 - 0.32p_t)\log\left\{\frac{S_c C_d(D^{0.75} - 1.132)}{215.63J[D^{0.75} - 18.42/(E_c/k)^{0.25}]}\right\} \tag{12.21}$$

Figure 12.17 is a nomograph for solving Eq. 12.21. Note that p_t does not appear in the nomograph because it was assumed that $p_t = 4.5 - \Delta PSI$. The DNPS86 computer program can also be used to solve Eq. 12.21 and perform the design procedure.

Example 12.6

Given $k = 72$ pci (19.5 MN/m^3), $E_c = 5 \times 10^6$ psi (34.5 GPa), $S_c = 650$ psi (4.5 MPa), $J = 3.2$, $C_d = 1.0$, $\Delta PSI = 4.2 - 2.5 = 1.7$, $R = 95\%$, $S_o = 0.29$, and $W_t = 5.1 \times 10^6$, determine thickness D from Figure 12.17.

Solution: The required thickness D can be determined by the following steps:

1. Starting from Figure 12.17a with $k = 72$ pci (19.5 MN/m^3), a series of lines, as indicated by the arrows, are drawn through $E_c = 5 \times 10^6$ psi (34.5 GPa), $S_c = 650$ (4.5 MPa), $J = 3.2$, and $C_d = 1.0$ until a scale of 74 is obtained at the match line.

2. Starting at 74 on the match line in Figure 12.17b, a line is drawn through $\Delta PSI = 1.7$ until it intersects the vertical axis.

3. From the scale with $R = 95\%$, a line is drawn through $S_o = 0.29$ and then through $W_{18} = 5.1 \times 10^6$ until it intersects the horizontal axis.

4. A horizontal line is drawn from the last point in steps 2 and a vertical line from that in step 3. The intersection of these two lines gives a D of 9.75 in. (246 mm), which is rounded to 10 in. (254 mm).

Example 12.7

Same as Example 12.6 except that D is given as 9.75 in. (246 mm). Determine W_{18} by using Eq. 12.21.

Solution: For $R = 95\%$, from Table 11.15, $Z_R = -1.645$. From Eq. 12.21, $\log W_{18} = -1.645 \times 0.29 + 7.35 \log(9.75 + 1) - 0.06 + \log(1.7/2.7)/[1 + 1.624 \times 10^7/(9.75 + 1)^{8.46}] + (4.22 - 0.32 \times 2.5) \log\{[(650 \times 1.0)/(215.63 \times 3.2)][(9.75)^{0.75} - 1.132]/[(9.75)^{0.75} - 18.42/(5 \times 10^6/72)^{0.25}]\} = -0.477 + 7.581 - 0.06 - 0.195 - 0.088 = 6.761$, or $W_{18} = 5.8 \times 10^6$, which checks well with the 5.2×10^6 obtained from the chart.

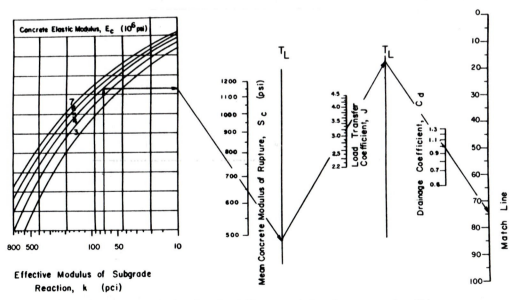

Figure 12.17(a)   Design chart for rigid pavements based on mean values (1 in. = 25.4 mm, 1 psi = 6.9 kPa, 1 pci = 271.3 kN/m^3). (From the *AASHTO Guide for Design of Pavement Structures*. Copyright 1986. American Association of State Highway and Transportation Officials, Washington, DC. Used by permission.)

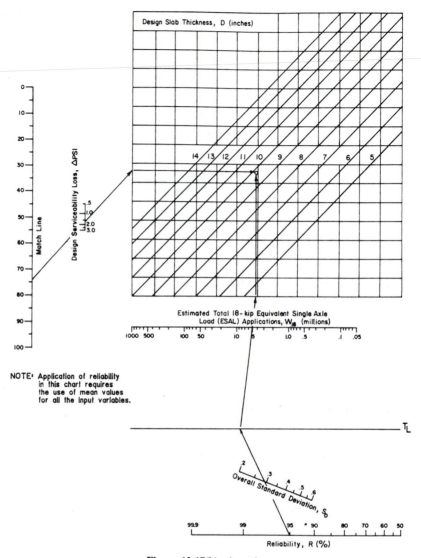

Figure 12.17(b) (cont.)

12.3.2 Modulus of Subgrade Reaction

The property of roadbed soil to be used for rigid pavement design is the modulus of subgrade reaction k, rather than the resilient modulus M_R. It is therefore necessary to convert M_R to k. Similar to M_R, the values of k also vary with the season of the year and the relative damage caused by the change of k needs to be evaluated.

Correlation with Resilient Modulus

As discussed in Section 5.1.1, there is no unique correlation between the modulus of subgrade reaction for liquid foundation and the resilient modulus for solid foundation. Any relationship between k and M_R is arbitrary and depends on whether stresses or deflections are to be compared or whether the loads are applied at the interior, edge, or corner of the slab.

Without Subbase

If the slab is placed directly on the subgrade without a subbase, AASHTO suggested the use of the following theoretical relationship based on an analysis of plate bearing test:

$$k = \frac{M_R}{19.4} \tag{12.22}$$

in which k is in pci and M_R is in psi. This equation gives a k value that is too large. For example, the resilient moduli equivalent to a k value of 100 pci (27.1 MN/m^3) are about 4000 psi (27.6 kPa) for an 8-in. (203-mm) slab and 4720 psi (32.6 MPa) for a 10-in. (254-mm) slab, based on the edge stress as shown in Table 5.13, while those based on the corner deflection are 6400 and 7560 psi (44.2 and 52.2 MPa), as shown in Table 5.14. For a k value of 100 pci (27.1 MPa), the resilient modulus obtained from Eq. 12.22 is only 1940 psi (13.4 MPa).

Equation 12.22 was based on the definition of k using a 30-in. (762-mm) plate. The deflection w_0 of a plate on a solid foundation can be determined by Eq. 2.10. The modulus of subgrade reaction, which is defined as the ratio between an applied pressure q and the deflection w_0, can be expressed as

$$k = \frac{q}{w_0} = \frac{2M_R}{\pi(1 - v^2)a} \tag{12.23}$$

in which v is Poisson ratio of the foundation and a is the radius of the plate. If $v = 0.45$ and $a = 15$ in. (381 mm), then Eq. 12.23 becomes

$$k = \frac{M_R}{18.8} \tag{12.24}$$

Equation 12.24 is a more exact solution compared to Eq. 12.22, which is an approximate solution using the average surface deflection under a flexible loaded area as w_0 (AASHTO, 1985). Equations 12.22 and 12.24 give a k value that is too large because k is inversely proportional to a, as indicated by Eq. 12.23. To correlate k with M_R, a very large plate should be used. The use of a 30-in. (762-mm) plate is arbitrary and is the only practical way to obtain a given value of k because tests with larger plates will be more expensive and difficult to perform. Therefore, the use of Eq. 12.22 or 12.24 based on a 30-in. (762-mm) plate to compute k from M_R is misleading and will result in stresses and deflections that are too small compared with those based on M_R.

With Subbase

If a subbase exists between the slab and the subgrade, the composite modulus of subgrade reaction can be determined from Figure 12.18. The modulus is based on a subgrade of infinite depth and is denoted by k_∞. The chart was developed using the same method as for a homogeneous half-space except that the 30-in. (762-mm) plate is applied on a two-layer system. Therefore, the k values obtained from the chart are too large and do not represent what actually occurs in the field.

Example 12.8

Given a subbase thickness D_{SB} of 6 in. (152 mm), a subbase resilient modulus E_{SB} of 20,000 psi (138 MPa), and a roadbed soil resilient modulus M_R of 7000 psi (48 MPa), determine the composite modulus of subgrade reaction k_∞.

Solution: The composite modulus of subgrade reaction can be determined as follows:

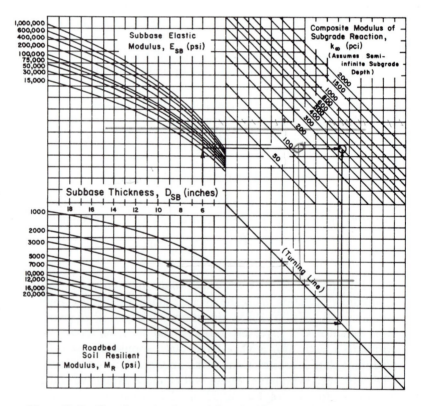

Figure 12.18 Chart for estimating modulus of subgrade reaction (1 in. = 25.4 mm, 1 psi = 6.9 kPa, 1 pci = 271.3 kN/m³). (From the *AASHTO Guide for Design of Pavement Structures*. Copyright 1986. American Association of State Highway and Transportation Officials, Washington, DC. Used by permission.)

Rigid Pavement Design   Chap. 12

1. In Figure 12.18, a vertical line is drawn upward from the horizontal scale with a subbase thickness E_{SB} of 6 in. (152 mm) until it reaches a point with a subbase modulus E_{SB} of 20,000 psi (138 MPa).

2. The same line is drawn downward until it intersects the curve with a roadbed soil resilient modulus M_R of 7000 psi (48 MPa), and then the line is turned horizontally until it intersects the turning line.

3. A horizontal line is drawn from the point in step 1 and a vertical line from the point on the turning line in step 2. The intersection of these two lines gives a k_∞ of 400 pci (108 MN/m³).

Rigid Foundation at Shallow Depth

Equation 12.22 and Figure 12.18 are based on a subgrade of infinite depth. If a rigid foundation lies below the subgrade and the subgrade depth to rigid foundation D_{SG} is smaller than 10 ft (3 m), then the modulus of subgrade reaction must be modified by the chart shown in Figure 12.19. The chart can be applied to slabs either with or without a subbase.

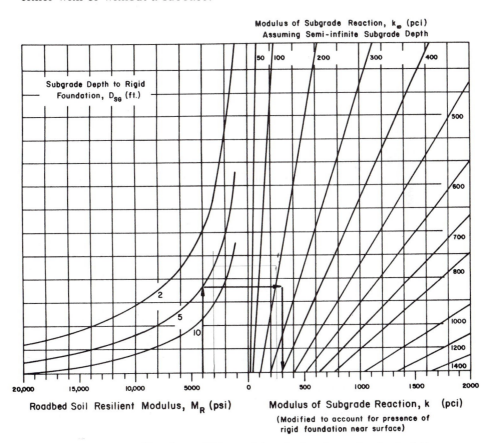

Figure 12.19 Chart for modifying modulus of subgrade reaction due to rigid foundation near surface (1 ft = 0.305 m, 1 psi = 6.9 kPa, 1 pci = 271.3 kN/m³). (From the *AASHTO Guide for Design of Pavement Structures*. Copyright 1986. American Association of State Highway and Transportation Officials, Washington, DC. Used by permission.)

Example 12.9

Given M_R = 4000 psi (27.6 MPa), D_{SG} = 5 ft (1.52 m), and k_∞ = 230 pci (62.4 MN/m³), determine k.

Solution: In Figure 12.19, a vertical line is drawn from the horizontal scale with a M_R of 4000 psi (27.6 MPa) until it intersects the curve with a D_{SG} of 5 ft. The line is turned horizontally until it reaches a point with a k_∞ of 230 pci (62.4 MN/m³) and then vertically until a k of 300 pci (81.4 MN/m³) is obtained.

Effective Modulus of Subgrade Reaction

The effective modulus of subgrade reaction is an equivalent modulus that would result in the same damage if seasonal modulus values were used throughout the year. The equation for evaluating the relative damage and the method for computing the effective k are discussed below.

Relative Damage

From Eq. 12.21, the effect of k on W_{18} can be expressed as

$$\log W_{18} = \log C - \log\left[\left(D^{0.75} - \frac{18.42k^{0.25}}{E_c^{0.25}}\right)^{(4.22-0.32p_t)}\right]$$

or

$$W_{18} = \frac{C}{(D^{0.75} - 18.42k^{0.25}/E_c^{0.25})^{(4.22-0.32p_t)}} \qquad (12.25)$$

in which C is the sum of all terms except for those related to k. Because of the relatively small variation in E_c and p_t, Eq. 12.25 can be simplified by assuming that $E_c = 5 \times 10^6$ psi (34.5 GPa) and $p_t = 2.5$:

$$W_{18} = \frac{C}{(D^{0.75} - 0.39k^{0.25})^{3.42}} \qquad (12.26)$$

If W_T is the predicted total traffic, then the damage ratio can be expressed as

$$D_r = \frac{W_T}{C}(D^{0.75} - 0.39k^{0.25})^{3.42} \qquad (12.27)$$

If W_T is distributed uniformly over n periods, then the cumulative damage ratio is

$$D_r = \frac{W_T}{C}\frac{1}{n}\sum_{i=1}^{n}(D^{0.75} - 0.39k_i^{0.25})^{3.42} \qquad (12.28)$$

Equating Eq. 12.27 to Eq. 12.28 gives

$$(D^{0.75} - 0.39k^{0.25})^{3.42} = \frac{1}{n}\sum_{i=1}^{n}(D^{0.75} - 0.39k_i^{0.25})^{3.42} \qquad (12.29)$$

Equation 12.29 can be used to determine the effective modulus of subgrade reaction k in terms of seasonal moduli k_i. The relative damage to rigid pavements u_r is defined as

$$u_r = (D^{0.75} - 0.39k^{0.25})^{3.42} \qquad (12.30)$$

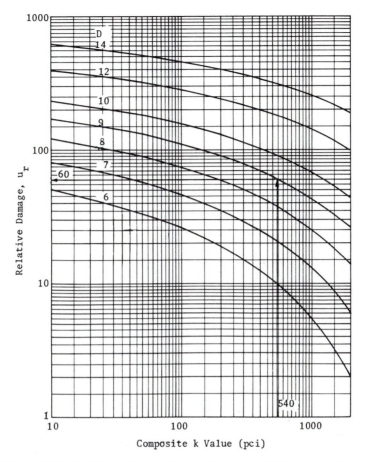

Figure 12.20 Chart for estimating relative damage to rigid pavements (1 in. = 25.4 mm, 1 pci = 271.3 kN/m³). (From the *AASHTO Guide for Design of Pavement Structures*. Copyright 1986. American Association of State Highway and Transportation Officials, Washington, DC. Used by permission.)

Figure 12.20 is a chart for solving Eq. 12.30

Example 12.10

Given $D = 9$ in. (229 mm) and $k = 540$ pci (147 MN/m³), determine u_r by Eq. 12.30 and compare the result with Figure 12.20.

Solution: From Eq. 12.30, $u_r = [(9)^{0.75} - 0.39(540)^{0.25}]^{3.42} = 60.3$, which checks with a relative damage of 60 obtained from the chart.

Computation

Table 12.17 shows the computation of the effective modulus of subgrade reaction for a slab thickness of 9 in. (229 mm). It is assumed that the slab is placed directly on the subgrade having the monthly resilient moduli shown in the table.

TABLE 12.17 COMPUTATION OF
EFFECTIVE MODULUS OF SUBGRADE
REACTION

Month (1)	Roadbed modulus M_R (psi) (2)	k Value (pci) (3)	Relative damage u_r (%) (4)
Jan	4500	232	85.7
Feb	27,300	1407	34.2
Mar	50,000	2577	20.5
Apr	1350	70	121.4
May	2140	110	108.1
Jun	2930	151	98.7
Jul	3710	191	91.6
Aug	4500	232	85.7
Sep	4500	232	85.7
Oct	4500	232	85.7
Nov	4500	232	85.7
Dec	4500	232	85.7

Average $\bar{u}_r = \dfrac{\Sigma u_r}{n} = 82.4$ $\Sigma u_r = 988.7$

Effective modulus of subgrade reaction,
$k = 263$ pci

Note. 1 psi = 6.9 kPa, 1 pci = 271.3 kN/m³.

Explanation of Columns in Table 12.17

1. Each year is divided into 12 months, each with different subgrade moduli.
2. The roadbed resilient moduli are the same as those used in the DAMA program for a MAAT of 60°F (15.5°C) and a normal modulus of 4500 psi (31 MPa), as shown in Table 11.10. The maximum modulus is 50,000 psi (345 MPa) and occurs in March when the subgrade is frozen.
3. The k values are obtained from Eq. 12.22. Because no corrections for rigid foundation are needed, these k values can be used for computing relative damage.
4. The relative damage can be obtained from Eq. 12.30 or Figure 12.20. The sum of relative damage is 988.7 and the average over the 12 months is 82.4, which is equivalent to an effective modulus of 263 pci (71.4 MN/M³).

It is interesting to note the significant difference in behaviors between flexible and rigid pavements under freeze–thaw conditions. For the flexible pavement analyzed in Figure 11.26, the damage caused by the spring breakup in May constitutes about 65% of the total damage, and the effective roadbed soil modulus is 2200 psi (15 MPa) versus the normal value of 4500 psi (31 MPa). For the rigid pavement analyzed in Table 12.17, the damage caused by the spring breakup in April constitutes only 12% of the total damage, and the effective modulus of subgrade reaction is 263 pci (71.4 MN/m³) versus the normal value of 232 pci (62.9

MN/m^3). The fact that the effective modulus is even greater than the normal modulus substantiates the claim by PCA (1984) that the normal summer or fall k values can be used for design purposes to avoid the tedious method of considering seasonal variations.

Loss of Subgrade Support

To account for the potential loss of support by foundation erosion or differential vertical soil movements, the effective modulus of subgrade reaction must be reduced by a factor, LS. Figure 12.21 shows a chart for correcting the effective modulus of subgrade reaction due to the loss of foundation support. For example, if the effective modulus of subgrade reaction for full contact, LS = 0, is 540 pci (147 MN/m^3), the effective modulus of subgrade reaction for partial contact with LS = 1 is 170 pci (46 MN/m^3).

Figure 12.21 was developed by computing the maximum principal stress under a single-axle load for four different contact conditions with LS = 0, 1, 2, and 3. The best case is LS = 0 when the slab and foundation are assumed to be in full contact. The worst case is LS = 3 when an area of slab, 9 ft (2.7 m) long and 7.25 ft (2.2 m) wide along the pavement edge, is assumed not to be in contact with the subgrade. The area assumed not to be in contact for LS = 2 is smaller than that for LS = 3 but greater than that for LS = 1. Since the result of AASHO Road Test indicates that the stresses produced in a concrete pavement are proportional to the number of load applications it can carry, the equivalent k value for partial

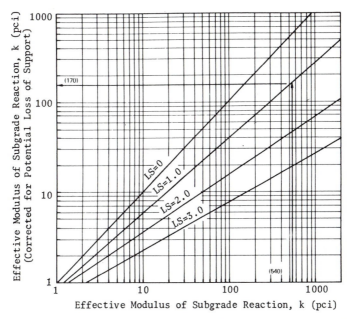

Figure 12.21 Correction of effective modulus of subgrade reaction due to loss of foundation contact (1 pci = 271.3 kN/m^3). (After McCullough and Elkins (1979).)

TABLE 12.18 TYPICAL RANGES OF LS FACTORS FOR VARIOUS TYPES OF MATERIALS

Type of material	Loss of support (LS)
Cement-treated granular base ($E = 1 \times 10^6$ to 2×10^6 psi)	0.0 to 1.0
Cement aggregate mixtures ($E = 500,000$ to 1×10^6 psi)	0.0 to 1.0
Asphalt-treated bases ($E = 350,000$ to 1×10^6 psi)	0.0 to 1.0
Bituminous-stabilized mixture ($E = 40,000$ to $300,000$ psi)	0.0 to 1.0
Lime-stabilized materials ($E = 20,000$ to $70,000$ psi)	1.0 to 3.0
Unbound granular materials ($E = 15,000$ to $45,000$ psi)	1.0 to 3.0
Fine-grained or natural subgrade materials ($E = 3000$ to $40,000$ psi)	2.0 to 3.0

Note. E in this table refers to the general symbol of the resilient modulus.
Source. After AASHTO (1986).

contact can be obatined by varying the *k* values until the maximum principal stress for full contact is equal to that for partial contact.

Table 12.18 provides some suggested ranges of LS for different types of subbase and subgrade materials. In the selection of LS factor, consideration should be given to differential vertical soil movements that may result in voids beneath the pavement. Even though a nonerosive subbase is used, LS values of 2.0 to 3.0 may still be used for active swelling clays or excessive frost heave.

12.3.3 Design Variables

The design variables presented in Sections 11.3.1 and 11.3.6 for flexible pavements, such as time constraints, traffic, reliability, environmental effects, serviceability, stage construction, and the analysis of swelling and frost heave, are the same as those for rigid pavements and are not repeated here. Since the effective modulus of subgrade reaction is discussed in Section 12.3.2, only the elastic modulus of concrete E_c, the concrete modulus of rupture S_c, the load transfer coefficient J, and the drainage coefficient C_d are presented in this section.

Elastic Modulus of Concrete

The elastic modulus of concrete can be determined according to the procedure described in ASTM C469 or correlated with the compressive strength. The following is a correlation recommended by the American Concrete Institute:

$$E_c = 57,000 \ (f'_c)^{0.5} \tag{12.31}$$

in which E_c is the concrete elastic modulus in psi and f'_c is the concrete compressive strength in psi as determined by AASHTO T22, T140, or ASTM C39.

Concrete Modulus of Rupture

The modulus of rupture required by the design procedure is the mean value determined after 28 days using third-point loading, as specified in AASHTO T97 or ASTM C78. If center-point loading is used, a correlation should be made between the two tests.

Load Transfer Coefficient

The load transfer coefficient J is a factor used in rigid pavement design to account for the ability of a concrete pavement structure to transfer a load across joints and cracks. The use of load transfer devices and tied concrete shoulders increases the amount of load transfer and decreases the load transfer coefficient. Table 12.19 shows the recommended load transfer coefficients for various pavement types and design conditions. The AASHO Road Test conditions represent a J value of 3.2, since all joints were doweled and there were no tied concrete shoulders.

Drainage Coefficient

The drainage coefficient C_d has the same effect as the load transfer coefficient J. As indicated by Eq. 12.21, an increase in C_d is equivalent to a decrease in J, both causing an increase in W_{18}. Table 12.20 provides the recommended C_d values based on the quality of drainage and the percentage of time during which the pavement structure would normally be exposed to moisture levels approaching saturation. Similar to flexible pavements, the percentage of time is dependent on the average yearly rainfall and the prevailing drainage conditions.

12.3.4 Comparison with PCA Method

It is difficult to compare the results between AASHTO and PCA methods because the AASHTO method is based on reliability, using mean values for all variables, while the PCA method does not consider reliability, but incorporates load safety factors and more conservative material properties. The AASHTO method is based on the equivalent 18-kip (80-kN) single-axle load applications and does not distinguish the type of distress, while the PCA method considers both fatigue cracking and foundation erosion using actual single- and tandem-axle loads. In view of the fact that fatigue cracking is more critical under single-axle loads and foundation erosion is more critical under tandem-axle loads, it is unreasonable to use ESAL for rigid pavement design because the conversion of a tandem-axle load

TABLE 12.19 RECOMMENDED LOAD TRANSFER COEFFICIENT FOR VARIOUS PAVEMENT TYPES AND DESIGN CONDITIONS

Type of shoulder	Asphalt		Tied PCC	
Load transfer devices	Yes	No	Yes	No
JPCP and JRCP	3.2	3.8–4.4	2.5–3.1	3.6–4.2
CRCP	2.9–3.2	N/A	2.3–2.9	N/A

Source. After AASHTO (1986).

TABLE 12.20 RECOMMENDED VALUES OF DRAINAGE COEFFICIENTS C_d FOR RIGID PAVEMENTS

Quality of drainage		Percentage of time pavement structure is exposed to moisture levels approaching saturation			
Rating	Water removed within	Less than 1%	1–5%	5–25%	Greater than 25%
Excellent	2 hours	1.25–1.20	1.20–1.15	1.15–1.10	1.10
Good	1 day	1.20–1.15	1.15–1.10	1.10–1.00	1.00
Fair	1 week	1.15–1.10	1.10–1.00	1.00–0.90	0.90
Poor	1 month	1.10–1.00	1.00–0.90	0.90–0.80	0.80
Very poor	Never drain	1.00–0.90	0.90–0.80	0.80–0.70	0.70

Source. After AASHTO (1986).

to an equivalent single-axle load actually changes the failure mode from the erosion at the joint to the fatigue at midslab. In the following comparison, it is assumed that the pavement is subjected to only one type of axle load, that is, the same 18-kip (80-kN) single-axle load, so the predominant mode of distress is fatigue cracking.

In applying the AASHTO method, the following parameter values are assumed: reliability $R = 95\%$, standard deviation $S_o = 0.35$, serviceability loss $\Delta PSI = 2.0$ ($p_t = 2.5$), drainage coefficient $C_d = 1.0$, load transfer coefficient $J = 3.2$ without concrete shoulders and 2.5 with concrete shoulders, concrete modulus of rupture $S_c = 650$ psi (4.5 MPa), and modulus of subgrade reaction $k = 100$ pci (27.1 MN/m³). In the PCA method, a load safety factor of 1.2 and the same S_c of 650 psi (4.5 MPa) and k of 100 pci (27.1 MN/m³) are assumed. The PCA design chart has already taken into account the variation of S_c and the increase of S_c with time, so only the input of S_c at 28 days is required, which is the same as the S_c specified by AASHTO. The same k values are used in both methods because the normal summer or fall modulus of subgrade reaction specified by PCA is not too much different from the effective modulus of subgrade reaction specified by AASHTO, as illustrated by the example in Table 12.17. Table 12.21 shows a comparison of thickness design between the AASHTO and PCA methods.

In Table 12.21, comparisons are made for pavements both with and without tied concrete shoulders. The allowable ESAL for five different slab thicknesses are shown. The AASHTO ESAL was computed by the AASHTO design equation (Eq. 12.21), and the PCA ESAL was obtained from the PCA design tables and charts based on the fatigue criterion (Table 12.6 or 12.7 and Figure 12.12) using a single-axle load of 21.6 kip (96 kN), which is the 18-kip (80-kN) load multiplied by a load safety factor of 1.2. It was found that when the joints are doweled, the use of the erosion criterion is not as critical as the use of the fatigue criterion. The thickness by PCA, as shown in Table 12.21, is the slab thickness determined by the PCA method based on the AASHTO ESAL.

As can be seen from Table 12.21, a large difference in ESAL exists between the two methods. For slabs with a thickness smaller than 8 in. (203 mm) and no concrete shoulder or slabs with a thickness smaller than 7 in. (178 mm) and a tied

TABLE 12.21 COMPARISON OF THICKNESS BETWEEN AASHTO AND PCA METHODS

Slab thickness (in.)	No concrete shoulder			With concrete shoulder		
	AASHTO ESAL	PCA ESAL	Thickness by PCA (in.)	AASHTO ESAL	PCA ESAL	Thickness by PCA (in.)
5	1.2×10^5	<100	7.5	2.8×10^5	350	6.6
6	3.2×10^5	500	7.9	7.4×10^5	4.0×10^4	6.9
7	7.4×10^5	3.0×10^4	8.1	1.7×10^6	1.0×10^6	7.1
8	1.6×10^6	5.0×10^5	8.2	3.7×10^6	$>10^7$	7.2
9	3.3×10^6	$>10^7$	8.3	7.7×10^6	$>10^7$	7.3

Note. 1 in. = 25.4 mm.

concrete shoulder, the ESAL determined by the AASHTO method is one or more orders of magnitude greater than that obtained by the PCA method. Therefore, the use of AASHTO equation for thin pavements, such as concrete shoulders, is less conservative. For example, with an ESAL of 2.8×10^5 and a tied concrete shoulder, a thickness of 5 in. (122 mm) is sufficient by the AASHTO equation but the thickness required by the PCA method is 6.6 in. (168 mm).

It appears that the ESAL obtained by the PCA method is more reasonable, at least for thinner pavements. Consider the case of a 5-in. (127-mm) slab without a concrete shoulder. A run of KENSLABS shows that the critical edge stress under an 18-kip (80-kN) single-axle load is 641 psi (4.4 MPa), which is nearly equal to the concrete modulus of rupture. Thus, it is not possible that the pavement can withstand 120,000 applications of an 18-kip (80-kN) single-axle load as computed by the AASHTO equation.

12.4 CONTINUOUS REINFORCED CONCRETE PAVEMENTS

In both PCA and AASHTO design methods, the procedure for determining the thickness of CRCP is the same as that for JPCP and JRCP. According to PCA (1984), the analysis of CRCP by the JSLAB computer program indicated that the critical stresses and deflections in CRCP were about the same as those in conventional pavements, sometimes slightly larger and sometimes slightly smaller depending on the crack spacing, so the use of the same thickness is recommended. The difference in thickness design between CRCP and conventional pavements in the AASHTO method is that CRCP allows the use of a slightly smaller load transfer coefficient, as shown in Table 12.19. If the load transfer coefficient in Example 12.6 is reduced from 3.2 to 2.9, the thickness of the slab can be reduced from 9.75 in. (248 mm) to 9 in. (229 mm).

In this section, only the design of longitudinal steel recommended by AASHTO (1986) is presented. The design of transverse steel is similar to that discussed in Example 4.10 and is not presented here. The AASHTO method for designing longitudinal reinforcements is the same as the method published by Austin Research Engineers, Inc. (McCullough and Elkins, 1979) and later adopted as a design procedure by the Associated Reinforcing Bar Producers–CRSI (McCullough and Cawley, 1981).

12.4.1 Design Variables

The input variables for the design of longitudinal reinforcement are discussed below.

Concrete Tensile Strength

The concrete indirect tensile strength at 28 days, as determined by AASHTO T198 or ASTM C496, should be used for the design of longitudinal steel reinforcement. The indirect tensile strength f_t can be assumed as 86% of the modulus of rupture S_c that is used for thickness design.

Concrete Shrinkage

Concrete shrinkage at 28 days Z is a significant factor in the reinforcement design. Shrinkage is strongly related to strength because both depend on the water-cement ratio. Table 12.22 may be used as a guide in selecting a shrinkage value corresponding to the indirect tensile strength.

Concrete Thermal Coefficient

The thermal coefficient of expansion for PCC, α_c, depends on many factors such as the water:cement ratio, concrete age, richness of the mix, relative humidity, and the type of aggregate in the mix. However, the type of coarse aggregate has the most influence. Recommended values of the thermal coefficient of PCC as a function of aggregate types is presented in Table 12.23.

Bar or Wire Diameter

Typical No. 5 and No. 6 deformed bars are used for longitudinal reinforcement in CRCP. The nominal diameter of a reinforcing bar in inches is simply the bar number divided by 8. The No. 6 bar is the largest practical size that should be used to meet bond requirements and control crack widths. The relationship between longitudinal and transverse wire should conform to manufacturers' recommendations.

TABLE 12.22 APPROXIMATE RELATIONSHIP BETWEEN SHRINKAGE AND INDIRECT TENSILE STRENGTH OF PCC

Indirect tensile strength (psi)	Shrinkage (in./in.)
300 or less	0.0008
400	0.0006
500	0.00045
600	0.0003
700 or greater	0.0002

Source. After AASHTO (1986).

Type of coarse aggregate	Concrete thermal coefficient (10^{-6}/°F)
Quartz	6.6
Sandstone	6.5
Gravel	6.0
Granite	5.3
Basalt	4.8
Limestone	3.8

Note. Value/°F must be multiplied by 1.8 to obtain value/°C.

Source. After AASHTO (1986).

Steel Thermal Coefficient

Unless specific knowledge of the thermal coefficient is known, a value of 5×10^{-6}/°F (9×10^{-6}/°C) may be assumed for design purposes.

Design Temperature Drop

The design temperature drop is the difference between the average concrete curing temperature and a design minimum temperature and can be computed by

$$DT_D = T_H - T_L \tag{12.32}$$

in which DT_D is the design temperature drop, T_H is the average daily high temperature during the month the pavement is constructed, and T_L is the average daily low temperature during the coldest month of the year.

Wheel Load Tensile Stress

Figure 12.22 can be used to estimate the wheel load tensile stress developed during initial loading of the constructed pavement by either construction equipment or truck traffic. This stress depends on the design slab thickness, the wheel load magnitude, and the effective modulus of subgrade reaction, as shown by the example in Figure 12.22.

12.4.2 Limiting Criteria

Three limiting criteria must be considered for the design of longitudinal steel: crack spacing, crack width, and steel stress.

Crack Spacing

The limits on crack spacing are based on the possibility of spalling and punchout. To minimize spalling, the maximum spacing between consecutive cracks should be limited to 8 ft (2.4 m). To minimize the potential for punchout, the minimum desirable spacing is 3.5 ft (1.1 m).

Sec. 12.4   Continuous Reinforced Concrete Pavements   **649**

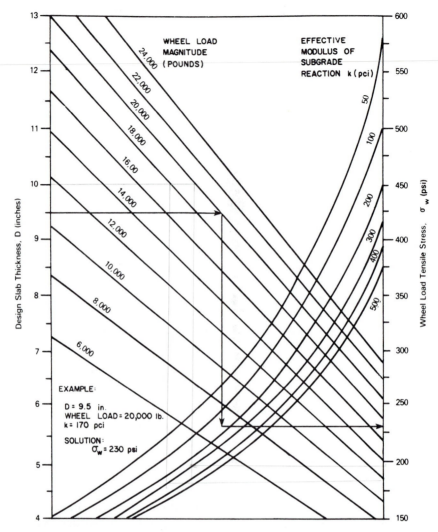

Figure 12.22 Chart for estimating wheel load tensile stress (1 in. = 25.4 mm, 1 lb = 4.45 N, 1 psi = 6.9 kPa, 1 pci = 271.3 kN/m³). (From the *AASHTO Guide for Design of Pavement Structures*. Copyright 1986. American Association of State Highway and Transportation Officials, Washington, DC. Used by permission.)

Crack Width

The limit on crack width is based on a consideration of spalling and water infiltration. The allowable crack width should not exceed 0.04 in. (1.0 mm). The crack width should be reduced as much as possible through the selection of a higher steel percentage or smaller diameter reinforcing bars.

TABLE 12.24 ALLOWABLE STEEL WORKING STRESS

Indirect tensile strength of concrete at 28 days (psi)	Reinforcing bar size[a]		
	No. 4	No. 5	No. 6
300 or less	65,000	57,000	54,000
400	67,000	60,000	55,000
~500	67,000	61,000	56,000
600	67,000	63,000	58,000
700	67,000	65,000	59,000
800 or greater	67,000	67,000	60,000

Note. Steel stress is in psi, 1 psi = 6.9 kPa.

[a] Proportional adjustments may be made for DWF based on wire diameter to bar diameter.

Source. After AASHTO (1986).

Steel Stress

A limiting stress of 75% of the ultimate tensile strength is recommended. Table 12.24 shows the mean steel working stress as a function of reinforcing bar size and concrete strength. The indirect tensile strength of concrete should be determined according to AASHTO T198 or ASTM C496. The limiting steel working stresses in Table 12.24 are for Grade 60 steel meeting ASTM A615 specifications. Guidance for the allowable steel stress for other types of steel is provided by Majidzadeh (1978).

12.4.3 Design Nomographs and Equations

Figures 12.23, 12.24, and 12.25 show the nomographs for determining the percentage of longitudinal reinforcement to satisfy the criteria of crack spacing, crack width, and steel stress, respectively. Also shown in the figures are the equations for developing the nomographs. To facilitate the computation of percent steel P from the limiting criteria, the equations can be rewritten as presented below.

Crack Spacing

Given the crack spacing, the percent steel can be determined by

$$P = \frac{1.062(1 + f_t/1000)^{1.457}(1 + \alpha_s/2\alpha_c)^{0.25}(1 + \phi)^{0.476}}{(\overline{X})^{0.217}(1 + \sigma_w/1000)^{1.13}(1 + 1000Z)^{0.389}} - 1 \quad (12.33)$$

in which P is the amount of longitudinal steel in percent, $\overline{X}$ is the crack spacing in ft, f_t is the indirect tensile strength in psi, α_s/α_c is the thermal coefficient ratio, ϕ is the reinforcing bar or wire diameter in inches, σ_w is the wheel load stress in psi, and Z is the concrete shrinkage at 28 days in in./in. For the example shown in Figure 12.23, the percent steel computed by Eq. 12.33 is $P = [1.062(1.55)^{1.457}$

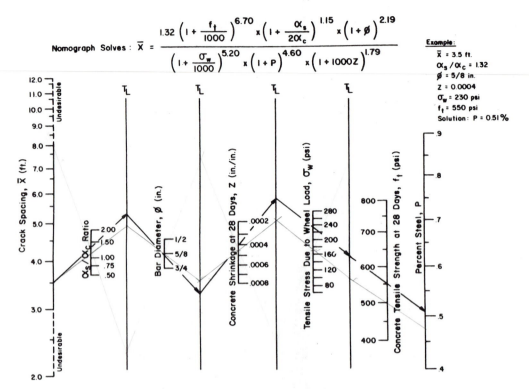

Nomograph Solves:
$$\bar{X} = \frac{1.32\left(1+\frac{f_t}{1000}\right)^{6.70}\times\left(1+\frac{\alpha_s}{2\alpha_c}\right)^{1.15}\times\left(1+\phi\right)^{2.19}}{\left(1+\frac{\sigma_w}{1000}\right)^{5.20}\times\left(1+P\right)^{4.60}\times\left(1+1000Z\right)^{1.79}}$$

Example:
$\bar{X}$ = 3.5 ft.
α_s / α_c = 1.32
ϕ = 5/8 in.
Z = 0.0004
σ_w = 230 psi
f_t = 550 psi
Solution: P = 0.51%

Figure 12.23 Percent steel to satisfy crack spacing criterion (1 ft = 0.305 m, 1 in. = 25.4 mm, 1 psi = 6.9 kPa). (From the *AASHTO Guide for Design of Pavement Structures*. Copyright 1986. American Association of State Highway and Transportation Officials, Washington, DC. Used by permission.)

$(1.66)^{0.25} (1.625)^{0.476}]/[(3.5)^{0.217} (1.23)^{1.13} (1.4)^{0.389}] - 1 = 0.52$, which checks with 0.51% obtained from the chart.

Crack Width

Given the crack width, the percent steel can be computed by

$$P = \frac{0.358(1 + f_t/1000)^{1.435}(1 + \phi)^{0.484}}{(CW)^{0.220}(1 + \sigma_w/1000)^{1.079}} - 1 \tag{12.34}$$

in which CW is the crack width in inches. For the example shown in Figure 12.24, the percent steel computed by Eq. 12.34 is $P = [0.358 (1.55)^{1.435} (1.75)^{0.484}]/[(0.04)^{0.22} (1.23)^{1.079}] - 1 = 0.43$, which is the same as obtained from the chart.

Steel Stress

Given the steel stress, the percent of steel can be computed by

$$P = \frac{50.834(1 + DT_D/100)^{0.155}(1 + f_t/1000)^{1.493}}{(\sigma_s)^{0.365}(1 + \sigma_w/1000)^{1.146}(1 + 1000Z)^{0.180}} - 1 \tag{12.35}$$

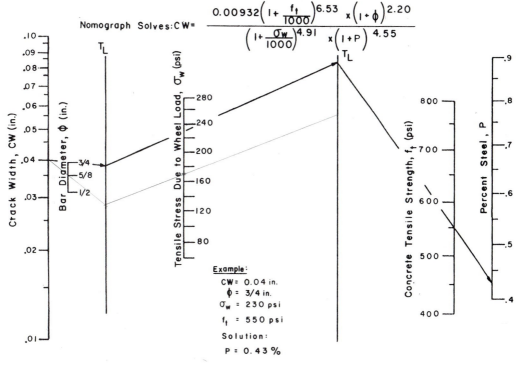

Nomograph Solves: $CW = \dfrac{0.00932\left(1+\dfrac{f_t}{1000}\right)^{6.53} \times \left(1+\phi\right)^{2.20}}{\left(1+\dfrac{\sigma_w}{1000}\right)^{4.91} \times \left(1+P\right)^{4.55}}$

Crack Width, CW (in.)

Bar Diameter, φ (in.)

Tensile Stress Due to Wheel Load, σ_w (psi)

Concrete Tensile Strength, f_t (psi)

Percent Steel, P

Example:
CW = 0.04 in.
φ = 3/4 in.
σ_w = 230 psi
f_t = 550 psi

Solution:
P = 0.43 %

Figure 12.24 Minimum percent steel to satisfy crack width criterion (1 in. = 25.4 mm, 1 psi = 6.9 kPa). (From the *AASHTO Guide for Design of Pavement Structures*. Copyright 1986. American Association of State Highway and Transportation Officials, Washington, DC. Used by permission.)

in which σ_s is the steel stress in psi and DT_D is the design temperature drop in °F. For the example shown in Figure 12.25, the percent steel computed by Eq. 12.35 is $P = [50.834\ (1.55)^{0.155}\ (1.55)^{1.493}]/[(57,000)^{0.365}\ (1.23)^{1.146}\ (1.4)^{0.18}] - 1 = 0.43$, which is slightly smaller than the 0.47% obtained from the chart.

12.4.4 Design Procedure

The following design procedures may be used to determine the amount of longitudinal reinforcement required.

1. Determine the required amount of steel reinforcement to satisfy each limiting criterion using Figures 12.23, 12.24, and 12.25 or Eqs. 12.33, 12.34, and 12.35. The minimum amount of steel P_{min} is selected as the largest among the three criteria: crack spacing of 8.0 ft (2.4 m), crack width, and steel stress; the maximum amount P_{max} is based solely on a crack spacing of 3.5 ft (1.1 m).

2. If P_{max} is less than P_{min}, the design is unsatisfactory and some of the inputs must be revised until P_{max} is greater than P_{min}.

3. Determine the range in the number of reinforcing bars or wires required

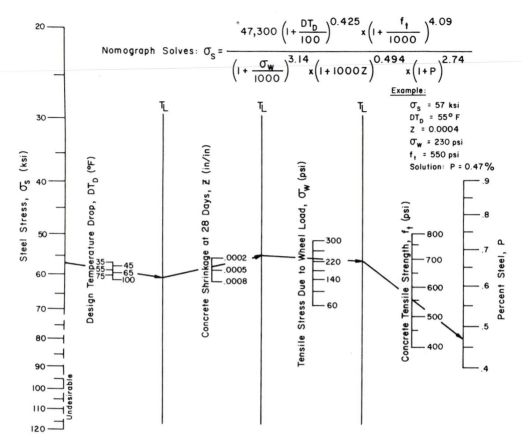

$$\text{Nomograph Solves: } \sigma_s = \frac{47{,}300\left(1+\dfrac{DT_D}{100}\right)^{0.425} \times \left(1+\dfrac{f_t}{1000}\right)^{4.09}}{\left(1+\dfrac{\sigma_w}{1000}\right)^{3.14} \times \left(1+1000Z\right)^{0.494} \times \left(1+P\right)^{2.74}}$$

Example:

σ_s = 57 ksi
DT_D = 55° F
Z = 0.0004
σ_w = 230 psi
f_t = 550 psi
Solution: P = 0.47 %

Figure 12.25 Minimum percent steel to satisfy steel stress criterion (1 psi = 6.9 kPa, 1°F = 0.56°C). (From the *AASHTO Guide for Design of Pavement Structures.* Copyright 1986. American Association of State Highway and Transportation Officials, Washington, DC. Used by permission.)

by

$$N_{\min} = 0.01273 P_{\min} W_s D/\phi^2 \qquad (12.36)$$

$$N_{\max} = 0.01273 P_{\max} W_s D/\phi^2 \qquad (12.37)$$

in which $N_{\min}$ and $N_{\max}$ are the minimum and maximum number of reinforcing bars or wires required, $P_{\min}$ and $P_{\max}$ are the minimum and maximum amount of steel in percent, W_s is the total width of pavement section in inches, D is the thickness of the concrete slab in inches, and ϕ is the reinforcing bar or wire diameter in inches, which may be increased if loss of cross section due to corrosion is anticipated.

    4. Determine the final steel design by selecting the total number of bars or wires N such that N is a whole number between $N_{\min}$ and $N_{\max}$. The appropriateness of the final design may be checked by converting N to P and working backward through the design charts, or by using the equation at the top of each chart to estimate crack spacing, crack width, and steel stress.

Example 12.11

Given the following information, determine the number of longitudinal steel bars per 12 ft (3.66 m) lane for a continuously reinforced concrete pavement: thickness of slab $D = 9.5$ in. (241 mm), magnitude of wheel load due to construction traffic = 20,000 lb (89 kN), effective modulus of subgrade reaction $k = 170$ pci (46.1 MN/m³), concrete indirect tensile strength $f_t = 550$ psi (3.8 MPa), concrete shrinkage $Z = 0.0004$ in./in., thermal coefficient of steel $\sigma_s = 5 \times 10^{-6}$ in./in./°F (9×10^{-6} mm/mm/°C), thermal coefficient of concrete (limestone aggregate) $\alpha_c = 3.8 \times 10^{-6}$ in./in./°F (6.8×10^{-6} mm/mm/°C), and design temperature drop $DT_D = 55$°F (31°C).

Solution: Table 12.25 is a worksheet for longitudinal reinforcement design. The design inputs are entered at the top part of the table. It is assumed that No. 5 bars with a diameter of $\frac{5}{8}$ in. (16 mm) are used. The thermal coefficient ratio is $\alpha_s/\alpha_c = (5 \times 10^{-6})/(3.8 \times 10^{-6}) = 1.32$. The wheel load stress σ_w is 230 psi (1.6 MPa) and can be obtained from Figure 12.22.

The percentages of steel based on the limiting criteria are shown at the bottom part of Table 12.25. The percent steel for a crack spacing of 3.5 ft (1.1 m) is 0.51%, as shown in Figure 12.23, and that for a crack spacing of 8 ft (2.4 m) is less than 0.4% and outside the range of the chart (Eq. 12.33 gives P = 0.27%). The percent steel for

TABLE 12.25 WORKSHEET FOR LONGITUDINAL REINFORCEMENT DESIGN

Design inputs			
Input variable	Value	Input variable	Value
Reinforcing Bar/Wire Diameter ϕ (in.)	5/8 (No. 5)	Thermal Coefficient Ratio α_s/α_c	1.32
Concrete Shrinkage Z (in./in.)	0.0004	Design Temperature Drop DT_D(°F)	55
Concrete Tensile Strength, f_t(psi)	550	Wheel Load Stress σ_w(psi)	230

Design criteria and required steel percentage				
	Crack Spacing, $\bar{x}$ (ft)	Allowable Crack Width, CW_{max} (in.)	Allowable Steel Stress, $(\sigma_s)_{max}$ (ksi)	Steel Range[b]
Value of Limiting Criteria	Max: 8.0 Min: 3.5	0.04	62	
Minimum Required Steel Percentage	<0.40%	<0.40%	<0.40%	0.40% (P_{min})[a]
Maximum Allowable Steel Percentage	0.51%			0.51% p_{max}

Note. [a] Enter the largest percentage across line.

[b] If $p_{max} < p_{min}$, reinforcement criteria are in conflict, design not feasible

1 in. = 25.4 mm, 1 ft = 0.305 m, 1 psi = 6.9 kPa.

a crack width of 0.04 in. (1 mm) can be determined from Figure 12.24 or computed by Eq. 12.34. A steel of 0.43%, as shown in Figure 12.24, is for $\frac{3}{4}$-in. bars. For $\frac{5}{8}$-in. bars, the percent steel computed by Eq. 12.34 is 0.38%, which is less than the minimum of 0.4% required. For a No. 5 bar with a f_t of 550 psi (3.8 MPa), the allowable steel stress σ_s is 62,000 psi (428 MPa), as shown in Table 12.24. The example in Figure 12.25 is for No. 6 bars with a σ_s of 57 ksi (393 MPa). For No. 5 bars with a σ_s of 62 ksi (428 MPa), the percent steel obtained from Eq. 12.35 is 0.39%, which is smaller than 0.4% and outside the range of the chart. With $P_{min} = 0.4\%$ and $P_{max} = 0.51\%$, from Eq. 12.36, $N_{min} = 0.01273 \times 0.4 \times 12 \times 12 \times 9.5/(0.625)^2 = 17.8$; from Eq. 12.37, $N_{max} = 0.01273 \times 0.51 \times 12 \times 12 \times 9.5/(0.625)^2 = 22.7$. Using 18 No. 5 bars per lane, the longitudinal reinforcing bar spacing is 8 in. (203 mm).

From the equation at the top of each chart, the predicted crack spacing is 5.12 ft (1.56 m), which is between the required 3.5 to 8 ft (1.1 to 2.4 m), the predicted crack width is 0.037 in. (0.94 mm), which is smaller than the allowable 0.04 in. (1 mm), and the predicted steel stress is 60,200 psi (415 MPa), which is smaller than the allowable 62,000 psi (428 MPa).

12.5 DESIGN OF RIGID PAVEMENT SHOULDERS

Most of the information presented in Section 11.4 on the design of flexible pavement shoulders is also applicable to the design of rigid pavement shoulders. Some of the features of rigid pavement shoulders that are different from those of flexible pavement shoulders are presented here.

PCC shoulders have been used in urban expressways for many years, but their use on rural highways began only in mid 1960s. Due to the good performance of these pavements, it is now the standard practice of many agencies to utilize PCC shoulders for rigid pavements.

12.5.1 Advantages of Tied Concrete Shoulders

Concrete shoulders must be tied to the mainline concrete pavements. The advantages of tied concrete shoulders are as follows:

1. The placement of a tied concrete shoulder next to the mainline pavement can substantially increase the load-carrying capacity of the pavement. The tied concrete shoulder provides support to the edge of the pavement and reduces stresses and deflections in the mainline slab. The shoulder is also benefited by receiving support from the mainline slab, so the damage due to encroaching traffic can be greatly reduced.

2. A tied longitudinal joint between mainline and shoulder pavements can be easily sealed to reduce the amount of surface runoff infiltrating into the pavement structure. Field studies conducted in Georgia and Illinois showed that sealing the longitudinal joint greatly reduced the amount of inflow from rainfall into the pavement structure (Dempsey et al. 1982).

3. Pumping beneath the mainline slab is reduced due to the reduction of edge and corner deflections, the reduction of water infiltration through the longitudinal joint, and the draining of water far away from the traffic lane.

4. Tied concrete shoulders can reduce differential movements at the longitudinal shoulder joint and do not experience the lane/shoulder dropoff type of distress that occurs so frequently in flexible shoulders.

12.5.2 Types of Rigid Pavement Shoulders

Similar to mainline pavements, three types of shoulder pavements are available: jointed plain concrete pavement (JPCP), jointed reinforced concrete pavement (JRCP), and continuous reinforced concrete pavement (CRCP). Generally, the type of shoulder should match the type of mainline pavement. However, some exceptions may be accepted:

1. For mainline JPCP, only JPCP shoulders with the same joint spacings as the mainline pavement are recommended because of their low cost. If JRCP shoulders with longer joint spacings are used, the excessive joint movements may cause problems in the adjacent mainline slabs. All transverse joints should be provided with an adequate reservoir and sealed similarly to the mainline joints.

2. For mainline JRCP, either JRCP shoulders that match the mainline pavement in design or JPCP shoulders with closer joint spacings may be used. The use of JPCP shoulders is more cost effective because no steel reinforcement is needed. They can be placed at the same time as the JRCP mainline pavement by leaving out the reinforcing steel and cutting transverse joints at shorter intervals.

3. For mainline CRCP, either CRCP shoulders that match the mainline pavement in design or JPCP shoulders with short joint spacings may be used. The use of short joint spacing for JPCP shoulders will reduce potential movements of the joints to cause cracking in the mainline CRCP. The elimination of steel reinforcement in the JPCP shoulders can save construction cost.

12.5.3 Design of Longitudinal Shoulder Joint

Adequate load transfer across the longitudinal shoulder joint must be provided to reduce the stresses and deflections in both mainline and shoulder slabs. Tied and keyed joints have been used most frequently to ensure a high degree of load transfer. Colley et al. (1978) investigated load transfers in laboratory slabs constructed with keyed, tied and keyed, and tied butt joints and concluded that all three were equally effective in reducing load-induced strains and deflections. However, the use of a keyed joint without tiebars was not recommended due to the possibility of shoulder joint separation. The excellent performance of the tied butt joint suggests that this type of construction is feasible and can reduce costs. Malleable tiebars of No. 4 or No. 5 size spaced at 18 to 24 in. (457 to 610 mm) are preferable to stiffer short bars spaced at larger intervals. This will substantially reduce stress concentration and the possibility of joint spall in the vicinity of the bar.

When a PCC shoulder is to be constructed adjacent to an existing pavement, tiebars can be installed by drilling holes in the edge of the existing slab. This can be done by using a tractor-mounted drill that can drill several holes at one time. Tiebars are installed in the holes by using epoxy or cement grout. The bar should be inserted into the slab over such a length as to develop sufficient bond. To avoid spalling over the base, a minimum insertion of 9 in. (229 mm) is required.

In the case of new construction, tiebars can be inserted into the plastic concrete near the rear of the slip form paver. Bent bars can be installed manually or by mechanical means. The bent portion can be straightened later to tie the shoulder to the mainline pavement. In addition to tiebars, a keyway can be formed to provide additional load transfer capability.

The longitudinal joint between the traffic lane and the shoulder should be provided with a sealant reservoir and sealed with an effective sealant. This will minimize the possibility of foreign materials collecting inside the joint to cause joint spall and reduce the amount of water and deicing salts entering into the joint and corroding the tiebars.

12.5.4 Shoulder Thickness Design

The thickness design concepts presented in Section 11.4.3 for flexible pavement shoulders are also applicable to rigid pavement shoulders. One major difference is that the inner edge is always more critical for flexible shoulders due to encroaching traffic but the outer edge may be more critical for rigid shoulders due to parking traffic. There is also a question of whether a separately designed shoulder is really needed. Lokken (1973) reviewed the performance of 16 projects located in 12 states and recommended the use of a 6-in. (152-mm) slab with an alternative tapered slab varying from roadway pavement depth at the longitudinal joint to 6 in. (152 mm) at the outside edge of the shoulder. Slavis (1981) reported on the performance review of these same projects in 1980 and indicated that the vast majority performed extremely well. The only notable deficiency identified in the field investigation was the faulting in one project due to inadequately covered tiebars. Since it is impossible to place the tiebars at the middepth of a 6-in. (152-mm) shoulder and a thicker mainline pavement, it was recommended in the 1980 review that the shoulder thickness be equal to the mainline slab at the longitudinal joint. This thickness can be used for the entire width of the shoulder or tapered to 6 in. (152 mm) at the outside edge. The use of the same thickness for both mainline and shoulder pavements is not only easier to construct, especially in installing the longitudinal joint, but has the further advantages of improving drainage by the elimination of bathtub trench and reducing differential frost heave.

If it is necessary to use thinner shoulder sections due to economic or other reasons, the thickness of inner edge can be designed based on the encroaching and parking traffic combined, while that of the outer edge can be based on the parking traffic alone. The design method used for the mainline pavement can also be used for the shoulder except that the traffic on the shoulder is much lighter. The following example illustrates how the PCA method can be used for determining the thickness of shoulder. In applying the PCA method to real situations, various weights of single- and tandem-axle loads must be analyzed separately because each has a different effect on the mode of failure. However, for simplicity, only the 18-kip (80-kN) single-axle loads will be used in the example.

Example 12.12

The outside lane on a heavily traveled highway is subjected to 10 million applications of an 18-kip (80-kN) single-axle load during the design life. A JPCP shoulder with aggregate interlock transverse joints is tied onto the traffic lane. Assuming an encroaching traffic of 3.5%, a parking traffic of 0.02%, a load safety factor of 1.2, a concrete modulus of rupture of 650 psi (4.5 MPa), and a modulus of subgrade reaction of 100 pci (27.1 MN/m³), determine the thickness of tied concrete shoulder by the PCA design method.

Solution: The outer edge of shoulder slab should be designed as aggregate interlock joints with no concrete shoulder, and the inner edge as aggregate interlock joints with concrete shoulder. Parking traffic on the outer edge = 10,000,000 × 0.0002 = 2000. Total traffic on the inner edge including both encroaching and parking traffic = 10,000,000 × 0.0352 = 352,000. Based on both fatigue and erosion analyses, the allowable repetitions for several assumed thicknesses are tabulated in Table 12.26.

In the fatigue analysis, the equivalent stress was found from Table 12.6 for the outer edge with no concrete shoulder and Table 12.7 for the inner edge with concrete shoulder. The stress ratio was computed by dividing the equivalent stress with 650, which is the concrete modulus of rupture. The allowable number of repetitions was obtained from Figure 12.12. In the erosion analysis, the erosion factor was found from Table 12.9 for the outer edge and Table 12.11 for the inner edge. The allowable number of repetitions was obtained from Figure 12.13 for the outer edge and Figure 12.14 for the inner edge. The single-axle load to be used with the charts is 1.2 × 18, or 21.6 kip (96 kN).

It can be seen from Table 12.26 that fatigue is more critical for 6 and 6.5 in. (152 and 165 mm) slabs, as indicated by the smaller allowable load repetitions compared with the erosion analysis, but erosion is more critical for the 7-in. (178-mm) slab. The required thickness is 6.5 in. (165 mm) for the outer edge and 7.0 in. (178 mm) for the inner edge. That fatigue prevails in thin pavements and erosion in thick pavements can be explained by the fact that the edge stress decreases more rapidly than the corner deflection as the thickness increases. Separate calculations also indicate that the thickness for the mainline slab with aggregate interlock joints is 8 in. (203 mm) based on fatigue analysis, but 9 in. (229 mm) based on erosion analysis.

The thickness of shoulder can be designed in several ways. The best and most expensive method is to use a uniform slab of 9 in. (229 mm). Another method is to use

TABLE 12.26 COMPUTATION OF ALLOWABLE LOAD REPETITIONS BY PCA METHOD

Assumed thickness (in.)	Fatigue analysis			Erosion analysis	
	Equivalent stress	Stress ratio	Allowable repetitions	Erosion factor	Allowable repetitions
Outer edge					
6.0	411	0.63	640	3.40	120,000
6.5	367	0.56	5000	3.30	240,000
Inner edge					
6.0	327	0.50	52,000	2.95	160,000
6.5	294	0.45	200,000	2.86	320,000
7.0	266	0.41	1,500,000	2.77	650,000

Note. 1 in. = 25.4 mm.

9 in. (229 mm) at the longitudinal joint and taper to 6.5 in. (165 mm) at the outer edge. The last resort is to use a uniform thickness of 7 in. (178 mm). It is not worth the effort to taper the section from 7 to 6.5 in. (18 to 165 mm) because the saving is too small.

12.6 SUMMARY

This chapter presents several methods for rigid pavement design, including a calibrated mechanistic procedure, the Portland Cement Association method, and the AASHTO method, as well as methods for the design of shoulders.

Important Points Discussed in Chapter 12

1. The most ideal design method, still under development, is the calibrated mechanistic procedure. Similar to flexible pavements, it consists of a number of response and distress models. The environmental response models, such as heat transfer, moisture equilibrium, and infiltration and drainage, can be applied to both flexible and rigid pavements. The structural response model based on the finite element plate theory and the fatigue cracking model based on the cumulative damage concept are quite advanced but the other distress models are not adequate and need to be improved.

2. The distress models include fatigue cracking, erosion, faulting, and joint deterioration for JPCP and JRCP, and punchout for CRCP. One major problem of fatigue cracking is how temperature warping should be included in the analysis. The prospects of developing mechanistic erosion and punchout models are promising but more work needs to be done to put them into practical use. The faulting and joint deterioration models currently available are not mechanistic; they are empirical regression equations valid only within the data base from which they were derived.

3. The PCA design method considers the fatigue of concrete caused by the edge stress at midslab and the erosion of foundation caused by corner deflection. The erosion criterion was developed primarily from the results of the AASHTO Road Test using a specific subbase and may not be applicable to situations where different types of subbase are used. Generally, the fatigue analysis controls the design of thinner pavements under lighter traffic, while the erosion analysis controls the design of thicker pavements under heavier traffic. For pavements carrying a normal mix of axle weights, single-axle loads will cause more fatigue cracking, and tandem-axle loads will cause more erosion damage.

4. The PCA method considers each load group individually and sums up the damage according to the cumulative damage concept. A slab thickness is assumed and the damage ratios due to both fatigue and erosion are determined. If either ratio is greater than 1, the thickness is increased every 0.5 in. (127 mm) until both ratios are smaller than 1. The smallest thickness with both damage ratios smaller than 1 is the thickness to be used. The design variables include modulus of subgrade reaction, concrete modulus of rupture, load safety factors, types and weights of axle load, types of joint, and types of shoulder.

5. Design tables and charts were developed by PCA using a finite element computer program called JSLAB. The program was also used to develop the design charts for lean concrete lower course and the tables for equivalent stress under tridem-axle loads. A simplified design procedure was also recommended by PCA when specific axle load data are not available.

6. The AASHTO design method is based on the empirical equation obtained from the AASHO Road Test, which is similar in form to the flexible pavement equation but with different regression constants. However, two additional terms are added, one relating to reliability and standard deviation and the other to several design variables not included in the original equation, such as concrete modulus of rupture, concrete modulus of elasticity, drainage coefficient, load transfer coefficient, and modulus of subgrade reaction. A comparison of results with the PCA method, as well as with the KENSLABS computer program, indicates that the AASHTO equation is unsafe for thin pavements and gives a performance traffic that is too large.

7. The effective modulus of subgrade reaction k is an equivalent modulus that would result in the same damage if seasonal k values are used. The equation for evaluating relative damage was derived from the AASHTO equation based on the cumulative damage concept. Unlike flexible pavements, the effective k values are not much different from the normal summer or fall k values. To avoid the tedious method of dividing a year into 12 months, each having a different k value, a mean normal k value may be used for design purposes. The correlations between the resilient moduli and k values, as recommended by AASHTO and shown in Eq. 12.22 and Figure 12.18, are questionable and result in a k value that is too large and does not check with theory.

8. The thickness of CRCP can be designed by the same methods used for JPCP and JRCP. Due to the stiffness of steel reinforcement which prevents the concrete from shrinkage, CRCP will always experience a large number of cracks. The amount of steel reinforcement should be selected so that the crack spacing lies between 3.5 and 8 ft (1.1 and 2.4 m), the crack width is less than 0.04 in. (1.0 mm), and the steel stress does not exceed 75% of the ultimate tensile strength. Empirical equations and design nomographs are available to predict crack spacing, crack width, and steel stress and can be used for design.

9. Concrete shoulders tied onto mainline slabs should be used for concrete pavements. It is preferable that the thickness of shoulders be the same as that of the mainline pavement for ease of construction, to facilitate drainage, and to provide additional lanes for possible future use. This is similar to the use of same thickness for both inside and outside lanes in the mainline pavement, even though the traffic on the inside lane is much lighter. If this cannot be done, at least the thickness at the longitudinal joint should be the same so that tiebars can be installed at the middepth of both mainline and shoulder slabs. The thickness can be tapered to a minimum of 6 in. (152 mm) at the outer edge. The thickness at the outside edge can be designed according to the parking traffic.

10. If thinner shoulder sections have to be used due to economic or other reasons, shoulder slabs should be designed according to the traffic they will carry. The design methods for mainline pavement can also be applied to the design of

shoulder pavement. The inner edge of the shoulder slab should be designed for encroaching and parking traffic, and the outer edge for parking traffic alone. If the shoulder may be used as an additional lane for peak hour or detoured traffic, this traffic should also be estimated and added to the encroaching and parking traffic at the inner edge and the parking traffic at the outer edge.

PROBLEMS

12-1. A 10-in. concrete pavement without concrete shoulders is placed on an untreated subbase with a k value of 150 pci. Estimate the allowable corner deflection by the PCA erosion criterion if the pavement is subjected to 2 million applications of a given axle load. [Answer: 0.0468 in.]

12-2. Same pavement as in Problem 12-1 but with concrete shoulders. Estimate the allowable corner deflection. [Answer: 0.0318 in.]

12-3. Determine the thickness of a concrete pavement for a two-lane highway by the PCA method. The pavement has doweled joints and no concrete shoulders. The modulus of subgrade reaction is 200 pci and the concrete modulus of rupture is 650 psi. Assume a load safety factor of 1.1 and a design period of 20 years. The average daily traffic during the design period is 2500, of which 35% are trucks. Truck weight distribution data for single (S) and tandem (T) loads are tabulated in Table P12.3. [Answer: 8.0 in.]

TABLE P12.3

Axle loads (kip)	No. axles per 1000 trucks	Axle loads (kip)	No. axles per 1000 trucks
16 S	130.9	24 T	80.2
18 S	110.8	28 T	34.4
20 S	65.4	32 T	24.0
22 S	15.6	36 T	17.2
24 S	2.3	40 T	16.8
26 S	1.9	44 T	10.5
28 S	0.9	48 T	9.6

12-4. A concrete pavement has doweled joints and tied concrete shoulders. Given a concrete modulus of rupture of 650 psi, a modulus of subgrade reaction of 150 pci, a load safety factor of 1.2, a design period of 20 years, an average daily truck traffic of 3460 (excluding all two-axle, four-tire trucks) on the design lane during the design life, and an axle load distribution shown by category 3 in Table 12.13. Determine the slab thickness by the PCA method using the procedure similar to the worksheet shown in Figure 12.15. Check the result by the simplified procedure using Table 12.15. [Answer: 8.5 in. versus 8.0 in.]

12-5. Same as Problem 12-4 but with no concrete shoulders. Determine the slab thickness by the regular PCA procedure and check the result with the simplified procedure. [Answer: 9.5 in.]

12-6. The predicted traffic W_{18} and the serviceability loss due to frost heave ΔPSI_{FH} as a function of time can be expressed by

$$W_{18} = 5 \times 10^6 [(1.04)^Y - 1]$$

$$\Delta PSI_{FH} = 0.08(Y)^{0.6}$$

in which Y is the time in years. Determine the performance period for a rigid pavement with the following information: $R = 90\%$, $S_0 = 0.4$, $D = 8$ in., $\Delta PSI = 4.5 - 2.5 = 2.0$, $S_c = 600$ psi, $C_d = 1.05$, $J = 3.2$, $E_c = 5 \times 10^6$, and $k = 100$ pci. [Answer: 6.5 years]

12-7. Figure P12.7 shows a concrete pavement with the thicknesses and elastic moduli as indicated. Assume a modulus of rupture of 650 psi, a load transfer coefficient of 3.2, a serviceability loss from 4.5 to 2.0, and poor drainage with 5% of time near saturation. Determine the performance traffic by Eq. 12.21 without the reliability term and check the result by the AASHTO design chart. [Answer: 8.5×10^6]

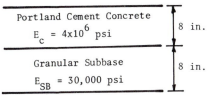

Figure P12.7

12-8. A 8.5-in. (216-mm) concrete slab is placed directly on a subgrade. The relationship between the resilient modulus and k value is indicated by Eq. 12.22. The monthly subgrade resilient moduli from January to December are 15,900, 27,300, 38,700, 50,000, 900, 1620, 2340, 3060, 3780, 4500, 4500, and 4500 psi. Determine the effective modulus of subgrade reaction. [Answer: 305 pci]

12-9. Same as Problem 12-8 except that the relationship between the resilient modulus and k value is indicated by Eq. 5.7. Assuming that the concrete has an elastic modulus of 4,000,000 psi and a Poisson ratio of 0.15 and the subgrade has a Poisson ratio of 0.45, determine the effective modulus of subgrade reaction. [Answer: 105 pci]

12-10. A continuous reinforced concrete pavement using limestone as coarse aggregate is subjected to 10^7 ESAL. Assume $k = 75$ pci, $E_c = 4 \times 10^6$ psi, $S_c = 650$ psi, $C_d = 1.05$, $J = 2.9$, $R = 95\%$, $S_0 = 0.3$, $\Delta PSI = 2.0$, $DT_D = 60°F$, and wheel load due to construction traffic $= 18,000$ lb. Determine the thickness of concrete and the number of No. 5 bars per 12-ft lane required. [Answer: 9.5 in. and 18 to 24 bars]

12-11. The mainline pavement and traffic are the same as Problem 12-4. If dowels are also used for the shoulders and the parking traffic is 2% of the mainline traffic, determine the thickness of outer pavement edge by the PCA method. [Answer: 8.5 in.]

12-12. The mainline pavement and traffic are the same as Problem 12-4. If dowels are also used for the shoulders and the encroaching and parking traffic combined is 2% of the mainline traffic, determine the thickness of inner shoulder edge by the PCA method. [Answer: 7 in.]

13

Design
of Overlays

13.1 TYPES OF OVERLAYS

As the nation's highways age and deteriorate, some type of treatment is eventually required to provide a safe and serviceable facility for the users. The types of treatments can range from simple maintenance to complete reconstruction, depending on the circumstances. For pavements subjected to moderate and heavy traffic, the most prevalent treatment is to place an overlay on the existing pavement. Depending on the types of overlay and existing pavement, four possible designs may exist: HMA overlays on asphalt pavements, HMA overlays on PCC pavements, PCC overlays on asphalt pavements, and PCC overlays on PCC pavements.

13.1.1 HMA Overlays on Asphalt Pavements

HMA overlay is the predominant type of resurfacing on asphalt pavements. Design methods ranging from engineering judgment to mechanistic–empirical procedures have been used. If the existing pavement is adequately repaired prior to overlay, a satisfactory design can usually be achieved. Due to the flexible types of materials used in the existing pavement and the overlay, the elastic layer programs can be applied for design purposes.

13.1.2 HMA Overlays on PCC Pavements

Although HMA overlays have been used extensively on PCC pavements, this type of overlay is most difficult to analyze mechanistically because it involves two different types of materials. Theoretically, the finite element plate programs can be used by considering HMA as the top layer and PCC as the bottom layer. However, with cracks in the existing concrete slabs, it is difficult to model the bottom slab. This type of overlay can also be analyzed by the elastic layer program if the stress adjustment factors for edge and corner loads are known. These factors were obtained by the Austin Research Engineers using the discrete element program called SLAB49 and incorporated in the Pavement Overlay Design (POD) program developed for the Federal Highway Administration (Treybig et al., 1977).

A major problem in the design of HMA overlays on PCC pavements is reflection cracking, defined as the fractures in an overlay or surface that reflect the crack or joint pattern in the underlying layer. It is imperative that such cracking be prevented or controlled to provide a smooth riding surface, maintain the structural integrity of the overlay, and prevent the intrusion of water into the pavement system. It is generally agreed that the primary mechanisms leading to the development of reflection cracks in a HMA overlay on a PCC pavement are the horizontal movement due to temperature and moisture changes and the differential vertical movement due to traffic loadings, both occurring at the joints and cracks in the PCC pavement, with the horizontal movement being considered more critical.

Several methods can be used to minimize the reflection cracking in HMA overlays on PCC pavements:

1. Design a thicker HMA overlay.
2. Crack and seat the existing PCC slab into smaller sections.
3. Use a crack relief layer with drainage system.
4. Saw and seal joints in a HMA overlay.
5. Use a stress-absorbing membrane interlayer with an overlay.
6. Incorporate a fabric membrane interlayer with an overlay.

Method 1 may be used if the thickness of overlay to alleviate reflection cracking is less than 9 in. (229 mm). Normally, when an overlay approaches the 8- to 9-in. (203- to 229-mm) range, other methods should be used. Methods 5 and 6 are being used in some areas and appear to be effective in reducing reflection cracking. However, the available documentation on the performance of these interlayers is insufficient for indicating the proper thickness of overlay. Therefore, if method 5 or 6 is considered, the manufacturers of the interlayer materials should be consulted toward establishing the required overlay thickness. Methods 2, 3, and 4 are described below.

Cracking and Seating

The crack and seat procedure, sometimes also called break and seat procedure, involves cracking the PCC slab into small segments, seating the segments with heavy rollers to eliminate underlying voids, and overlaying the PCC slab with HMA. The purpose is to create small pieces of concrete so slab movement by thermal or other causes is minimal, thereby reducing the reflection cracking in the HMA overlay. The segments, usually 4 to 6 ft² (0.37 to 0.56 m²) in size, are still large enough to have some structural integrity due to aggregate interlock.

In the past, a major problem during the breaking of JRCP slabs has been the difficulty in rupturing the reinforcing steel. If the fragmented pieces are held together by the steel, not only is an effective seating not possible but also horizontal movements cannot be reduced. This problem has been solved recently with the availability of new cracking devices that can break the bond between the concrete and the steel.

There have been some controversies on the effectiveness of using cracking and seating. A report by FHWA (1987c) based on a review of 22 projects indicated that cracking and seating could reduce reflection cracking during the first few years, but after 4 to 5 years the cracking increased and was about the same as that in the control section without cracking and seating. Since the structural capacity of the existing pavement is reduced by cracking and since the condition of the crack and seat sections and the control sections seemed to be the same after some period of time, the report raised questions on the justification for using the cracking and seating method. A recent study by Kilareski and Stoffels (1990) concurred with the above report that cracking and seating only delayed, rather than eliminated, the reflection cracking, but disagreed that cracking and seating reduced the structural capacity. Some of their conclusions are listed below:

1. Over the past 30 years, 24 states have experimented with the crack and seat method of overlay and reported experiences that range from poor to excellent.
2. The crack and seat sections in California exhibited significantly less reflection cracking than the control sections. The use of a fabric interlayer further reduced the quantity of reflection cracking.
3. The crack and seat sections showed significantly less roughness than their corresponding control sections.
4. Based on the analysis of FWD tests, there was no significant loss of structural support or decrease in the modulus of elasticity on the crack and seat sections.
5. The crack and seat sections experienced significant increases in cracking with age.
6. The crack and seat sections displayed more medium and high severity cracking but less total cracking than the corresponding control sections.

The thickness design of HMA overlays on cracked and seated slabs is considered in the Asphalt Institute and AASHTO procedures and is discussed in Sections 13.3.2 and 13.5.3.

Crack Relief Layer

The crack relief layer is designed specifically to minimize the reflection cracking from the old PCC pavement to the new asphalt overlay. It is placed as the first course of an overlay system. Typically, the crack relief layer is a 3.5-in. (90-mm) layer of coarse open-graded HMA, containing 25 to 35% interconnecting voids and made up of 100% crushed material. Due to the large amount of interconnecting voids, the crack relief layer provides a medium through which differential movements of the underlying slab are not readily transmitted. The crack relief layer should be connected to a drainage system so that water can be drained out freely.

Before placing the crack relief layer, the existing pavement surface should be prepared so that it is as structurally sound and clean as possible and a tack coat is generally required. After placing the crack relief layer, a well-graded intermediate leveling or binder course from 2 to 4 in. (51 to 102 mm) thick is laid, followed by a dense-graded surface course about 1.5 in. (38 mm) thick. Figure 13.1 shows a typical cross section of overlay with a crack relief layer.

Table 13.1 gives the gradation recommended for the crack relief layer. Depending on the availability of the size of aggregate, three different gradings may be used. The use of a particular mix also should depend on the characteristics of the material used in the existing PCC pavement. Highly expansive PCC pavements, such as those constructed by washed silica gravel or with joint spacings exceeding 20 ft (6.1 m), should use mix A or B. Less expansive PCC pavements can be overlaid with mix C.

Sawing and Sealing of Joints

Because of the difficulty in eliminating reflection cracking, one method is to control the cracking rather than eliminate it. This is effected by sawing a joint above each existing transverse joint immediately after the overlay. The joints are then sealed and subsequently maintained as typical pavement joints. A recent study by Kilareski and Bionda (1990) on the sawing and sealing of joints resulted in the following conclusions:

1. A total of 12 states, mostly in the northeastern part of the country, have used saw and seal HMA overlay as a routine rehabilitation procedure. Several states have prepared specifications and standards for the saw and seal overlay procedure. States that documented their experiments with sawing and sealing have reported marginal to good results with the technique.

2. The overall documented experience with saw and seal overlay is extremely limited. Information is generally lacking concerning measured field performance, traffic, existing pavement condition, and characterization of the

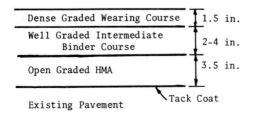

Figure 13.1 Typical cross section of overlay with crack relief layer (1 in. = 25.4 mm).

TABLE 13.1 RECOMMENDED GRADING LIMITS FOR CRACK RELIEF LAYERS

Sieve	Percent passing		
	Mix A	Mix B	Mix C
3 in.	100	—	—
2½ in.	95–100	100	—
2 in.	—	—	100
1½ in.	30–70	35–70	75–90
¾ in.	3–20	5–20	50–70
⅜ in.	0–5	—	—
No. 4	—	—	8–20
No. 8	—	0.5	—
No. 100	—	—	0.5
No. 200	—	0.3	—

Asphalt cement content 1.5–3.0%
(AC-40, 40-50 pen., or AR-16000)

Note. 1 in. = 25.4 mm.

Source. After AI (1983).

existing pavement in terms of joint width, load transfer efficiency, crack spacing, joint and crack opening under known temperature changes, and load deflection.

3. An important step in the construction process is to locate properly the saw cut above the existing joint. Secondary reflection cracking can occur unless the saw cut is made within 1 in. (25 mm) from the existing joint.

4. Saw and seal sections with thick overlay, say 5 in. (127 mm), performed better than sections with thin overlay, say 2.5 in. (64 mm), and had less roughness and reflection cracking.

5. If properly constructed, sawed and sealed joints in a HMA overlay of jointed PCC pavement can reduce the adverse effect of reflection cracking and extend the pavement life. A comparison between control sections and saw and seal sections for 15 pavement projects showed that saw and seal sections reduced pavement roughness by 20% and transverse reflection cracking by 64%.

13.1.3 PCC Overlays on Asphalt Pavements

The use of PCC overlays on asphalt pavements is somewhat uncommon. However, they have been used very successfully in the United States and other countries. This method can be used if the vertical clearance does not pose a problem. It may be cost effective if the asphalt pavement is severely distressed and can be used only as a foundation for the PCC overlay. The design procedure is similar to that of new pavements and uses the existing pavement as the foundation. The finite element plate programs can be used for the mechanistic method of design. To prevent reflection cracking, all cracks of high severity in the existing asphalt pavement should be repaired and sealed. Because the existing asphalt pavement can be considered a nonerodible subbase, only fatigue cracking need be considered for determining the thickness of overlay required.

13.1.4 PCC Overlays on PCC Pavements

Three types of PCC overlay may be used for PCC pavements: unbonded, bonded, and partially bonded. Mechanistic designs can be carried out using the finite element plate programs in which somewhat lower elastic modulus should be specified for the old concrete, depending on the conditions of the existing pavement.

Unbonded or Separated Overlay

Unbonded overlays are typically placed on pavements that are badly cracked. Prior to the overlay, the surface of the existing pavement must be cleaned of debris and excess sealing materials. A separation layer, usually consisting of HMA or sand asphalt of less than 2 in. (51 mm) thick, is then placed between the new overlay and the existing pavement to prevent reflection cracking. The separation layer can also serve as a leveling course so a more uniform thickness of concrete can be obtained. In applying the finite element plate programs, the plate is considered as two unbonded layers and the effect of separation layer is ignored.

Unbonded overlay may be plain, reinforced, continuously reinforced, or prestressed concrete. It is not necessary to match the location or type of joints in the overlay with those in the existing pavement. The minimum thickness of unbonded overlay is 6 in. (152 mm).

Bonded or Monolithic Overlay

Bonded overlays should be used only when the existing pavement is in good condition or when serious distress has been repaired. To achieve a fully bonded overlay, it is necessary to carefully prepare the existing surface before placing the overlay. All oil, grease, paint, and surface contaminants must be removed by cold milling, sandblasting, or waterblasting. Then a thin layer of cement grout should be placed on the cleaned dry surface just in front of the concrete paver. A liquid epoxy resin of low viscosity may also be used as a bonding agent. The finite element plate programs with two bonded layers can be used for the design of bonded overlays.

Irrespective of the type of existing PCC pavements, plain concrete is the most commonly used bonded overlays, although steel reinforcements may be used in thicker overlays to supplement the steel in the existing pavements. The location of the joints in the overlay must match that in the existing pavement. Use of dowel bars in bonded and partially bonded overlays is not recommended because it may produce localized failures in the overlay directly above the dowel and may also cause the overlay to debond. For use over continuously reinforced pavements, the bonded overlay does not require the installation of joints.

Partially Bonded or Direct Overlay

Partially bonded overlays are obtained when the fresh concrete is placed directly on relatively sound and clean existing slabs. Unless steps are taken to prevent bond, some degree of bonding can be assumed, so the overlay can be designed slightly thinner than unbonded overlays to take advantage of the resulting stress reduction.

If the existing pavement is jointed, a joint should be placed immediately above the existing joint to prevent reflection cracking. It is also important to keep the joint spacing of a partially bonded overlay as short as possible by providing additional joints in the overlay to minimize temperature stresses caused by the stiff underlying slabs.

13.2 DESIGN METHODOLOGIES

A variety of methods have been used by various agencies for the design of overlays. Prior to 1960, most agencies relied heavily on engineering judgment and experience in determining the thickness and type of overlay required. Since 1960, the use of nondestructive deflection testing has gained wide acceptance, and more rational methods based on deflection measurements to evaluate the in situ pavement conditions have been gradually developed. Similar to the design of new pavements, most agencies now have their own methods for the design of overlays. Usually the procedure for the design of the overlay is similar to that of new pavement except that the condition or remaining life of the existing pavement at the time of overlay is taken into consideration. By considering the existing pavement as new or having 100% remaining life, the overlay design method can be applied to the design of new composite pavements as well. In this section, the basic concepts on overlay design are presented. Details about AI, PCA, and AASHTO design methods are described in subsequent sections.

Three methods can be used for overlay design: the effective thickness approach, the deflection approach, and the mechanistic–empirical approach. No matter what method is used, it is important to make a pavement condition survey and subdivide the proposed project into one or more homogeneous analysis sections based on age, traffic, design, and pavement conditions. Major distress in the existing pavement must be properly repaired before placing the overlay. If more than one analysis section is identified, practical construction and cost considerations must be used to decide whether separate overlay designs should be developed for each analysis section or which sections should be combined.

13.2.1 Effective Thickness Approach

The basic concept of this method is that the required thickness of the overlay is the difference between the thickness required for a new pavement and the effective thickness of the existing pavement:

$$h_{OL} = h_n - h_e \tag{13.1}$$

in which h_{OL} is the thickness of the overlay, h_n is the thickness of the new pavement, and h_e is the effective thickness of the existing pavement. The procedure assumes that as the pavement deteriorates and uses part of its total life, it behaves as if it were an increasing thinner pavement, i.e., its effective thickness accounts less and less for the expended portion of the total life. Because the effective thickness is based on the type, condition, and thickness of each component layer, this method is also called the component analysis procedure. All thicknesses of new and existing materials must be converted into an equivalent thickness of HMA or PCC based on their types and conditions. If the overlay is HMA, all thicknesses in Eq. 13.1 are expressed in terms of HMA. If the overlay is PCC, all thicknesses are expressed in terms of PCC. This method is used by the Asphalt Institute, as described in Section 13.3.1, and by AASHTO, as described in Section 13.5.

A modified version of Eq. 13.1 was developed by the U.S. Corps of Engineers for the design of PCC overlays on PCC airport pavements. The equation, which was based on the results of full-scale traffic tests, is shown by Eq. 13.2 and was also incorporated in the AASHTO design guide:

$$h_{OL}^n = h_n^n - Ch_e^n \tag{13.2}$$

in which h_e is the thickness of the existing pavement, which is different from the effective thickness shown in Eq. 13.1; $n = 2$ for unbonded overlays, 1.4 for directly placed overlays, and 1 for fully bonded overlays; and C is a condition factor with suggested values as follows:

$C = 1.0$ for existing pavement in good overall structural conditions with little or no cracking.

$C = 0.75$ for existing pavements with initial transverse and corner cracking due to loading but without progressive structural distress or recent cracking.

$C = 0.35$ for existing pavement that is badly cracked or shattered structurally.

For bonded overlays with $n = 1$ and $C = 1$, EQ. 13.2 is identical to Eq. 13.1 in form. Because h_e in Eq. 13.2 is the existing thickness rather than the effective thickness, the application of the Eq. 13.2 for bonded overlays with $n = 1$ and $C = 1$ implies that the concrete in the existing pavement has suffered no fatigue or other damage due to traffic or weathering and is therefore as strong as the concrete in a new pavement. This assumption is contrary to the remaining life concept currently being used for the design of highway overlays.

13.2.2 Deflection Approach

The basic concept of this method is that larger pavement surface deflections imply weaker pavement and subgrade and thus require thicker overlays. The overlay must be thick enough to reduce the defection to a tolerable amount. Usually only the maximum deflection directly under the load is measured. This approach should not be confused with the estimate of in situ material properties from multisensor deflection measurements for use in the effective thickness and the mechanistic–empirical methods. The deflection method is based on the empirical relationship between pavement deflection and overlay thickness and has been used by the Asphalt Institute (AI, 1983), the California Department of Transportation (1979), the Roads and Transportation Association of Canada (1977), and the Transport and Road Research Laboratory of Great Britain (Kennedy and Lister, 1978). Only the Asphalt Institute method is presented in Section 13.3.

13.2.3 Mechanistic–Empirical Approach

Similar to the design of new pavements, this method requires the determination of critical stress, strain, or deflection in the pavement by some mechanistic methods and the prediction of resulting damages by some empirical failure criteria. First, the condition or remaining life of the existing pavement must be evaluated. Based on the pavement condition or remaining life, the thickness of the overlay is then determined so that the damages in either the existing pavement or the new overlay will be within the allowable limits. This mechanistic design procedure has been used by PCA for the design of PCC overlays on PCC pavements, as described in Section 13.4.

The failure criteria that have been used most frequently for the design of flexible overlays are fatigue cracking and permanent deformation, while only fatigue cracking is considered for rigid overlays. The erosion criterion is usually not considered for the design of PCC overlays because of the large increase in pavement thickness and the use of existing pavement as a nonerodible foundation. The following procedure can be used if the existing pavement has remaining fatigue life, as indicated by

$$\sum_{i=1}^{m} \frac{n_i}{N_i} < 1 \qquad (13.3)$$

in which n_i is the actual number of load repetitions for the ith load group, N_i is the allowable number of load repetitions for the ith load group, and m is the number of load groups. If the existing pavement has no remaining life, the overlay should be considered as the top layer of a two-layer system with a modulus of subgrade depending on the conditions of existing pavement.

Fatigue Cracking

For ease of illustration, it is assumed that the design is based on the equivalent single-axle load. The method can be equally applied to various axle loads by using the cumulative damage concept and summing the damage ratios over all load groups. First, the remaining life of existing pavement is estimated by

$$\frac{n_r}{N_a} = 1 - \frac{n_e}{N_a} \qquad (13.4)$$

in which n_r is the additional number of load repetitions that can be applied to the existing pavement after the overlay, N_a is the allowable number of load repetitions on the existing pavement before overlay, and n_e is the actual number of load repetitions on the existing pavement before overlay. Next, based on the tensile stress or strain at the underside of the existing layer after the overlay, the allowable number of load repetitions on the overlaid pavement is determined. Because the existing pavement is not new and has a remaining life of n_r/N_a, the allowable repetitions must be multiplied by the ratio n_r/N_a to obtain the allowable repetitions after the overaly.

If the overlay and existing pavements are not bonded, the fatigue damage in the new overlay should also be checked.

Permanent Deformation

It is assumed that the existing rut, if any, will be filled and the development of rutting has nothing to do with the previous traffic but will only be a function of the additional traffic to be applied on the new pavement structure with the overlay. If the vertical compressive strain at the top of subgrade is used to control the permanent deformation, the strain on the subgrade prior to the overlay is ignored and only the strain after the overlay is used to determine the additional traffic allowed.

Example 13.1

An asphalt overlay is placed on an existing asphalt pavement that has been subjected to an ESAL of 7×10^6. The horizontal tensile strain at the bottom of the asphalt layer is 1×10^{-4} before the placement of overlay and 7×10^{-5} after the overlay. By using the Asphalt Institute fatigue criteria and assuming an elastic modulus of 5×10^5 psi (3.5 GPa) for the HMA, determine the allowable number of ESAL on the overlaid pavement.

Solution: From Eq. 11.6, the allowable number of repetitions on the existing pavement $N_a = 0.0796 \times (0.0001)^{-3.291} \times (500,000)^{-0.854} = 15,677,000$. From Eq. 13.4, $n_r/N_a = 1 - 7,000,000/15,677,000 = 0.553$. The allowable number of repetitions of the overlaid pavement $= 0.553 \times 0.0796 \times (0.00007)^{-3.291} \times (500,000)^{-0.854} = 2.8 \times 10^7$.

13.3 ASPHALT INSTITUTE METHOD

Separate design procedures are used for asphalt overlay on asphalt pavement and asphalt overlay on concrete pavement. The design procedures are illustrated by a number of examples. Most of these examples are the same as those presented by the Asphalt Institute (AI, 1983) with only slight modifications.

13.3.1 Asphalt Overlay on Asphalt Pavement

Two methods are available: the effective thickness method and the deflection method. The effective thickness method is based on the condition of existing pavement at the time of overlay without the necessity of conducting deflection tests.

Effective Thickness Method

To determine the effective thickness of existing pavement in terms of HMA thickness, one or more conversion factors must be found. If the existing pavement is full depth, method 1, based on the Present Serviceability Index (PSI) of the existing pavement, can be used to determine the conversion factor. Otherwise, method 2, based on the condition of each individual layer, should be used to determine the conversion factor of each layer.

Method 1

Figure 13.2 gives the conversion factors C for full-depth asphalt pavements based on the PSI of existing pavement at the time of overlay. The two curves in the figure reflect the difference in performance after the placement of overlay. The upper curve, line A, represents pavements with a reduced rate of change in PSI, compared to their rate of change before overlay. The lower curve, line B, represents a projected rate of change in PSI about the same as before overlay and therefore is somewhat more conservative. The choice between the two curves is largely a matter of judgment and experience.

The conversion factors shown in Figure 13.2 are applicable only for HMA. If emulsifed asphalt mixes are used, the equivalency factors shown in Table 13.2 should be used. The effective thickness of each existing layer is calculated by multiplying the actual thickness of each layer by the conversion factor and the appropriate equivalency factor. The total effective thickness is obtained by summing the individual effective thicknesses of all pavement layers:

$$h_e = \sum_{i=1}^{n} h_i C_i E_i \tag{13.5}$$

in which h_i, C_i, and E_i are the thickness, conversion factor, and equivalency factor of layer i, and n is the total number of layers.

Example 13.2

A full-depth asphalt pavement consisting of a 2-in. (51-mm) HMA and a 6-in. type II emulsified asphalt base course is to be overlaid. Even though there are cracks on the surface, the cracks are not open and the pavement appears to be stable. The pavement has a PSI of 2.0. Determine its effective thickness.

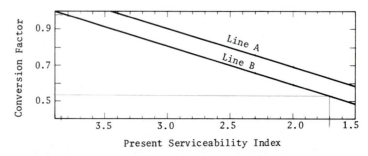

Figure 13.2 Conversion factors for full depth pavements. (After AI (1983).)

TABLE 13.2 EQUIVALENT FACTORS FOR EMULSIFIED ASPHALT BASES

Material type	Equivalency factor (E)
Hot mix asphalt	1.00
Type I emulsified asphalt base	0.95
Type II emulsified asphalt base	0.83
Type III emulsified asphalt base	0.57

Source. After AI (1983).

Solution: From Figure 13.2, $C = 0.7$ based on line A and 0.6 based on line B. From Table 13.2, $E = 0.83$. From Eq. 13.5, $h_e = 2 \times 0.7 \times 1 + 6 \times 0.7 \times 0.83 = 4.9$ in. (124 mm) based on line A, and $h_e = 2 \times 0.6 \times 1 + 6 \times 0.6 \times 0.83 = 4.2$ in. (107 mm) based on line B. If average C is used, $h_e = 4.6$ in. (117 mm).

Method 2

In this method the condition of each individual layer is evaluated and the appropriate conversion factor C is selected from Table 13.3. Similar to Eq. 13.5, the effective thickness is computed by

$$h_e = \sum_{i=1}^{n} h_i C_i \tag{13.6}$$

Method 2 can also be used for full-depth pavements. If the PSI is known, it is suggested that both methods 1 and 2 be used and compared. Although the ranges in values shown in Table 13.3 are based on subjective analysis, experience has shown that they are reasonable and useful for overlay design.

Example 13.3

Determine the effective thickness of a pavement consisting of a 4-in. (102-mm) HMA surface, a 6-in. (152-mm) cement-stabilized base, and a 4-in. (102-mm) untreated crushed gravel subbase. The surface shows numerous transverse cracks and considerable alligator cracking in the wheelpath. The cement-stabilized base shows signs of pumping and loss of stability along the pavement edges, and the crushed gravel subbase is in good condition with a plasticity index smaller than 6.

Solution: The material classification for HMA surface is V(a) in Table 13.3. Because there are considerable transverse and alligator cracks, a lower bound C of 0.5 is selected. The cement-stabilized base is classified as IV(c). Because there are signs of pumping and loss of stability, a lower bound C of 0.3 is selected. The crushed gravel subbase is classified as II. Because it is in good condition with PI less than 6, an upper bound C of 0.2 is selected. From Eq. 13.6, $h_e = 4 \times 0.5 + 6 \times 0.3 + 4 \times 0.2 = 4.6$ in. (117 mm).

Calculation of Overlay Thickness

The required overlay thickness is the difference between the required thickness of a new full-depth pavement and the effective thickness of the existing pavement, as shown by Eq. 13.1. The thickness of the new pavement can be

TABLE 13.3 CONVERSION FACTORS FOR DETERMINING EFFECTIVE THICKNESS

Classification of material	Description of material	Conversion factors[a]
I	(a) Native subgrade in all cases. (b) Improved subgrade, predominantly granular materials, may contain some silt and clay but have P.I. of 10 or less. (c) Lime-modified subgrade constructed from high-plasticity soils, P.I. greater than 10.	0.0
II	Granular subbase or base, reasonably well-graded, hard aggregates with some plastic fines and CBR not less that 20. Use upper part of range if P.I. is 6 or less; lower part of range if P.I. is more than 6.	0.1–0.2
III	Cement or lime-fly ash stabilized subbases and bases constructed from low plasticity soils, P.I. of 10 or less.	0.2–0.3
IV	(a) Emulsified or cutback asphalt surfaces and bases that show extensive cracking, considerable raveling or aggregate degradation, appreciable deformation in the wheelpaths, and lack of stability. (b) Portland cement concrete pavements (including those under asphalt surfaces) that have been broken into small pieces 2 ft (0.6 m) or less in maximum dimension prior to overlay construction. Use upper part of range when subbase is present, lower part of range when slab is on subgrade. (c) Cement or lime-fly ash stabilized bases that have developed pattern cracking, as shown by reflected surface cracks. Use upper part of range when cracks are narrow and tight, lower part of range with wide cracks, pumping, or evidence of instability.	0.3–0.5
V	(a) Asphalt concrete surface and base that exhibit appreciable cracking and crack patterns. (b) Emulsified or cutback asphalt surface and bases that exhibit some fine cracking, some raveling or aggregate degradation, and slight deformation in the wheelpaths but remain stable. (c) Appreciably cracked and faulted portland cement concrete pavement (including such under asphalt surfaces) that cannot be effectively undersealed. Slab fragments, ranging in size from approximately 10 to 40 ft^2 (1 to 4 m^2), and have been well seated on the subgrade by heavy pneumatic-tired rolling.	0.5–0.7
VI	(a) Asphalt concrete surfaces and bases that exhibit some fine cracking, have small intermittent cracking patterns and slight deformation in the wheelpaths but remain stable. (b) Emulsified or cutback asphalt surface and bases that are stable, generally uncracked, show no bleeding, and exhibit little deformation in the wheelpaths. (c) Portland cement concrete pavements (including such under asphalt surfaces) that are stable and undersealed, have some cracking but contain no pieces smaller than about 10 ft^2 (1 m^2).	0.7–0.9
VII	(a) Asphalt concrete, including asphalt concrete base, generally uncracked, and with little deformation in the wheelpaths. (b) Portland cement concrete pavement that is stable, undersealed, and generally uncracked. (c) Portland cement concrete base, under asphalt surface, that is stable, nonpumping, and exhibits little reflected surface cracking.	0.9–1.0

[a] Originally meeting minimum strengths and compaction requirements specified by most state highway departments.

Source. After AI (1983).

determined from the design chart for full-depth pavement presented in Figure 11.11.

Example 13.4

The existing pavement shown in Example 13.3 has a subgrade resilient modulus M_R of 10,000 psi (6.9 MPa). Determine the thickness of overlay required to carry an ESAL of 3×10^6.

Solution: From Example 13.3, $h_e = 4.6$ in. (117 mm). With $M_R = 10,000$ psi (6.9 MPa) and ESAL $= 3 \times 10^6$, from Figure 11.11, $h_n = 10.6$ in. (269 mm). From Eq. 13.1, the thickness of overlay required is $h_{OL} = 10.6 - 4.6 = 6$ in. (152 mm).

Deflection Method

Pavement deflections are measured with the Benkelman beam using a rebound test procedure. Condition survey and deflection data are used to establish the analysis section. At least ten deflection measurements should be made for each analysis section or a minimum of 20 measurements per mile (13 measurements/km). Pavement temperatures are measured at the time of deflection measurements so that deflections can be adjusted to a standard temperature of 70°F (21°C). A random sampling technique is recommended for selecting the locations of deflection measurements.

Representative Rebound Deflection

When the deflection tests on the analysis section are completed, the recorded pavement rebound deflections are used to determine a representative rebound deflection (RRD):

$$\delta_{rrd} = (\bar{\delta} + 2s)\,Fc \tag{13.7}$$

in which δ_{rrd} is the representative rebound deflection, $\bar{\delta}$ is the mean deflection, s is the standard deviation, F is the temperature adjustment factor, and c is the critical period adjustment factor. The use of two standard deviations above the mean implies that 97% of the measurements are smaller than δ_{rrd}. Locations within the analysis section with deflections greater than δ_{rrd} are recommended for special treatment. It is suggested that additional deflections be measured to determine the extent of such weak areas. These locations may require replacement before overlay.

Figure 13.3 gives the temperature adjustment factor curves for various thicknesses of dense-graded aggregate bases. A base thickness of 0 in. corresponds to a full-depth asphalt pavement. Temperatures have most effect on full-depth pavements and the effect decreases as the thickness of granular base increases.

The critical period is the interval during which the pavement is most likely to be damaged by heavy loads. In frost areas, this will normally be the period during or directly after the spring thaw. If the deflection measurements are made during the critical period, the adjustment factor c is equal to 1. If deflections are not measured in the critical period, c is greater than 1 and can be determined from a continuous record of measured deflections for a similar pavement. If no record of

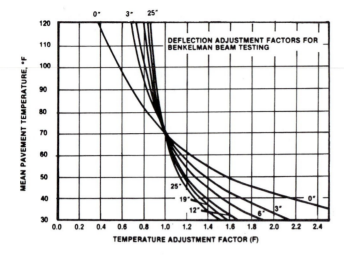

Figure 13.3 Temperature adjustment factors for various thicknesses of untreated granular bases (1 in. = 25.4 mm, 1°F = 0.56°C). (After AI (1983).)

comparable deflection data is available, c should be selected using engineering judgment.

Example 13.5

A series of deflection measurements by a Benkelman beam were made on a full-depth pavement during a critical period with a pavement temperature of 55°F (13.3°C). The measured deflections are 0.033, 0.035, 0.040, 0.025, 0.028, 0.026, 0.020, 0.035, 0.034, and 0.027 in. (0.84, 0.89, 1.02, 0.64, 0.71, 0.66, 0.51, 0.89, 0.86, and 0.69 mm). Compute the representative rebound deflection δ_{rrd}.

Solution: $\bar{\delta}$ = (0.033 + 0.035 + 0.040 + 0.025 + 0.028 + 0.026 + 0.020 + 0.035 + 0.034 + 0.027)/10 = 0.030 in. (0.77 mm). From Eq. 10.9, $V[\delta]$ = [(0.033 − 0.030)² + (0.035 − 0.030)² + (0.040 − 0.030)² + (0.025 − 0.030)² + (0.028 − 0.030)² + (0.026 − 0.030)² + (0.020 − 0.030)² + (0.035 − 0.030)² + (0.034 − 0.030)² + (0.027 − 0.030)²]/(10 − 1) = 3.66 × 10⁻⁵ in.² (0.024 mm²), or s = 0.006 in. (0.15 mm). With a pavement temperature of 55°F (13.3°C) and no untreated aggregate base, from Figure 13.3, temperature adjustment factor F = 1.38. During the critical period, c = 1. From Eq. 13.7, δ_{rrd} = (0.030 + 2 × 0.006) × 1.38 × 1 = 0.058 in. (1.47 mm).

Deflection After Overlay

In developing the design procedure, the overlaid pavement was considered as a two-layer system with the HMA overlay as layer 1 and the existing pavement as layer 2 (Finn and Monismith, 1984). The representative rebound deflection δ_{rrd} was used to determine the modulus of layer 2. By assuming the existing pavement to be a homogeneous half-space with a Poisson ratio of 0.5, from Eq. 2.8

$$E_2 = \frac{1.5qa}{\delta_{rrd}} \tag{13.8}$$

in which q is the contact pressure assumed to be 70 psi (483 kPa) and a is the radius of the equivalent single loaded area to represent the load on dual tires, assumed to be 6.4 in. (163 mm). The expected deflection after the overlay is called the design rebound deflection δ_d and can be determined by the following approximate equation

$$\delta_d = \frac{1.5qa}{E_2}\left(\left\{1 - \left[1 + 0.8\left(\frac{h_1}{a}\right)^2\right]^{-0.5}\right\}\frac{E_2}{E_1} \right.$$
$$\left. + \left\{1 + \left[0.8\frac{h_1}{a}\left(\frac{E_1}{E_2}\right)^{1/3}\right]^2\right\}^{-0.5}\right) \tag{13.9}$$

in which h_1 is the thickness of the overlay and E_1 is the modulus of the overlay, assumed to be 500,000 psi (3.5 GPa).

Overlay Thickness Design

It is assumed that there is a unique relationship between design rebound deflection in inches and the allowable ESAL, as shown in Figure 13.4 and represented by

$$\delta_d = 1.0363\,(\text{ESAL})^{-0.2438} \tag{13.10}$$

Given the ESAL for the overlay, δ_d can be determined from Eq. 13.10. Given the representative rebound deflection δ_{rrd}, E_2 can be obtained from Eq. 13.8. With δ_d and E_2 known and values of q, a, and E_1 assumed, the thickness of overlay h_1 can be computed from Eq. 13.9. Figure 13.5 shows the design chart relating ESAL and δ_{rrd} to overlay thickness.

Example 13.6

By the Asphalt Institute method, determine the allowable ESAL for a 2-in. (51-mm) overlay on an asphalt pavement with a representative rebound deflection δ_{rrd} of 0.062 in. (1.57 mm).

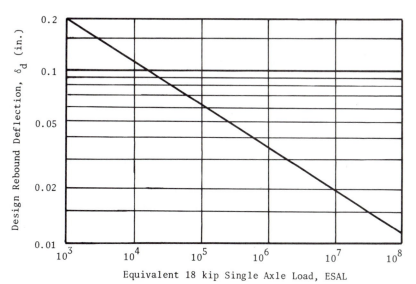

Figure 13.4 Relationship between design rebound deflection and ESAL (1 in. = 25.4 mm). (After AI (1983).)

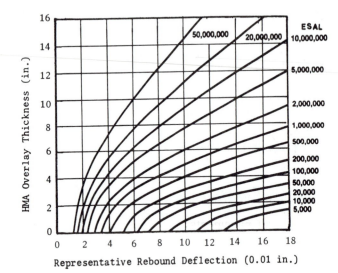

Figure 13.5 Overlay thickness design chart based on representative rebound deflection and design ESAL (1 in. = 25.4 mm). (After AI (1983).)

The chart axes:
- Vertical axis: HMA Overlay Thickness (in.)
- Horizontal axis: Representative Rebound Deflection (0.01 in.)
- ESAL curves: 50,000,000; 20,000,000; 10,000,000; 5,000,000; 2,000,000; 1,000,000; 500,000; 200,000; 100,000; 50,000; 20,000; 10,000; 5,000

Solution: Based on Eq. 13.8 and the parameters assumed by the Asphalt Institute, $E_2 = 1.5 \times 70 \times 6.4/0.062 = 10,840$ psi (74.8 MPa). From Eq. 13.9

$$\delta_d = \frac{1.5 \times 70 \times 6.4}{10,840}\left(\left\{1 - \left[1 + 0.8\left(\frac{2}{6.4}\right)^2\right]^{-0.5}\right\}\frac{10,840}{500,000}\right.$$

$$\left. + \left\{1 + \left[0.8 \times \frac{2}{6.4}\left(\frac{500,000}{10,840}\right)^{1/3}\right]^2\right\}^{-0.5}\right) = 0.04621 \text{ in. (1.17 mm)}$$

From Eq. 13.10, $0.04621 = 1.0363 (\text{ESAL})^{-0.2438}$, or ESAL = 347,000, which checks with the value obtained from Figure 13.5.

Remaining Life

Figure 13.4 or Eq. 13.10 can also be used to estimate the remaining life of an existing pavement or how much time remains before an overlay is needed. The procedure is as follows:

1. Determine the representative rebound deflection δ_{rrd}.
2. Obtain remaining life (ESAL)$_r$ from Figure 13.4 by assuming δ_{rrd} as the design rebound deflection δ_d. A more convenient method is to use Eq. 13.10, which can be rewritten as

$$(\text{ESAL})_r = \left(\frac{1.0363}{\delta_{rrd}}\right)^{4.1017} \tag{13.11}$$

in which δ_{rrd} is in inches.

3. Estimate the design ESAL for the current year (ESAL)$_0$ and determine the growth factor by

$$\text{Growth factor} = \frac{(\text{ESAL})_r}{(\text{ESAL})_0} \tag{13.12}$$

4. Estimate the traffic growth rate in percent and find the design period corresponding to the growth factor from Table 6.13. The design period is the estimated number of years before an overlay is needed.

Example 13.7

Given $(ESAL)_0 = 68,000$ for the current year, $\delta_{rrd} = 0.042$ in. (1.07 mm), and traffic growth rate = 4%, estimate number of years before overlay.

Solution: From Eq. 13.11, $(ESAL)_r = (1.0363/0.042)^{4.1017} = 513,000$. From Eq. 13.12, growth factor = 513,000/68,000 = 7.54. From Table 6.13, estimated time before overlay = 6.5 years.

13.3.2 Asphalt Overlay on PCC Pavement

Similar to asphalt overlays on asphalt pavements, both effective thickness and deflection methods are available for asphalt overlays on PCC pavements.

Effective Thickness Method

The procedure is the same as for asphalt overlays on asphalt pavements and the conversion factors for existing PCC pavements are also shown in Table 13.3. However, there are certain distresses peculiar to rigid pavements that should be considered in design. Among the signals of distress to look for are pumping, cracking, spalling, faulting, and slab movement under traffic. Also, if the pavement is to be undersealed or cracked and seated prior to overlay, this should be considered in selecting the proper conversion factor from Table 13.3. This component analysis procedure also can be applied to PCC pavements that have been previously overlaid with asphalt.

Example 13.8

Given a design ESAL of 2×10^6, design an asphalt overlay for a pavement consisting of 8 in. (203 mm) of jointed PCC slabs and 4 in. (102 mm) of sand–gravel subbase. The rigid pavement is appreciably cracked and faulted but the slab fragments have been broken and well seated on the subbase by heavy rollers prior to overlay. The sand–gravel subbase is of average quality. The subgrade has a resilient modulus of 8000 psi (55 MPa).

Solution: From Table 13.3, the PCC is classified as V(c) with a conversion factor of 0.5 and the sand–gravel subbase is classified as II with a conversion factor of 0.15. From Eq. 13.6, $h_e = 8 \times 0.5 + 4 \times 0.15 = 4.6$ in. (117 mm). With ESAL = 2×10^6 and $M_R = 8000$ psi (55 MPa), from Figure 11.11, a 10.2-in. (259-mm) full-depth pavement is required. Therefore, thickness of overlay = $10.2 - 4.6 = 5.6$ in. (142 mm), and the use of 5.5 in. (140 mm) should be adequate.

Deflection Method

The deflection procedure presented by the Asphalt Institute emphasizes the identification and repair of distress areas prior to overlay. Pavement condition surveys in conjunction with deflection testing are helpful in identifying distress conditions. For two-lane highways, deflection measurements are made on the outside edge on both sides of the center line. For divided highways, deflections should be measured on the outmost edge only. Additional deflections measurements should be made at corners, joints, cracks, and deteriorated pavement areas to determine load transfer capability.

Allowable Deflections

For JPCP and JRCP, vertical deflections at the joints are most important. As measured by the Benkelman beam, the differential deflection, $w_1 - w_r$, should be less than 0.002 in. (0.05 mm) and the mean deflection, $(w_1 + w_r)/2$ should be less than 0.014 in. (0.36 mm). Measurements of these deflections are shown in Figure 13.6a. Stabilization and undersealing are required if these criteria are not met.

Studies on CRCP indicate that Dynaflect deflections of 0.0006 to 0.0009 in. (15 to 23 μm) or greater lead to excessive cracking and deterioration. If the critical deflection w, shown in Figure 13.6b, exceeds 0.0006 in. (15 μm), undersealing or stabilization is required.

Dense-graded HMA can reduce deflections by 5%/in. (0.2%/mm). However, this varies with the mix type and the environmental conditions and may be as high as 10 to 12%/in. (0.4 to 0.5%/mm). If the deflection analysis indicates a desired reduction of 50% or more, overlays should not be used alone. It is more economical to reduce deflections by applying undersealing in conjunction with the overlay.

Overlay Thickness Design Chart

Figure 13.7 is a design chart for selecting the thickness of an asphalt overlay over PCC pavements. The thicknesses of HMA were selected to minimize the reflection cracking by considering the effect of both the horizontal tensile strains and vertical shear stresses. In this procedure, overlay thicknesses are related to slab length and mean annual temperature differential, which is the difference between the highest normal daily maximum temperature and the lowest normal daily minimum temperature for the hottest and coldest months, based on a 30-year average. Maximum and minimum daily temperatures at locations throughout the United States can be found in MS-17 (AI, 1983). An abbreviated table for maximum annual temperature differential is presented in Table 13.4.

The overlay thickness chart is divided into three sections, designated as A, B, and C, according to slab lengths and temperature differentials. In section A, a

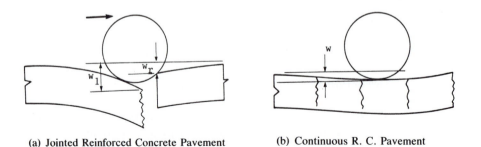

(a) Jointed Reinforced Concrete Pavement   (b) Continuous R. C. Pavement

Figure 13.6 Deflections to be considered in overlay thickness design.

Figure 13.7 Thickness design chart for asphalt overlays on PCC pavements. (After AI (1983).)

Slab Length (Ft)	TEMPERATURE DIFFERENTIAL (°F) 30	40	50	60	70	80	Slab Length (m)
10 or Less	100mm (4 in.)	100mm (4 in.)	100mm (4 in.)	100mm (4 in.)	100mm (4 in.)	100mm (4 in.)	3
15	100mm (4 in.)	100mm (4 in.)	100mm (4 in.)	100mm (4 in.)	100mm (4 in.)	100mm (4 in.)	4.5
20	100mm (4 in.)	100mm (4 in.)	100mm (4 in.)	100mm (4 in.)	125mm (5 in.)	140mm (5.5 in.)	6
25	100mm (4 in.)	100mm (4 in.)	100mm (4 in.)	125mm (5 in.)	150mm (6 in.)	175mm (7 in.)	7.5
30	100mm (4 in.)	100mm (4 in.)	125mm (5 in.)	150mm (6 in.)	175mm (7 in.)	200mm (8 in.)	9
35	100mm (4 in.)	115mm (4.5 in.)	150mm (6 in.)	175mm (7 in.)	215mm (8.5 in.)	Use Alternative 2 or 3	10.5
40	100mm (4 in.)	140mm (5.5 in.)	175mm (7 in.)	200mm (8 in.)	Use Alternative 2 or 3	Use Alternative 2 or 3	12
45	115mm (4.5 in.)	150mm (6 in.)	190mm (7.5 in.)	225mm (9 in.)	Use Alternative 2 or 3	Use Alternative 2 or 3	13.5
50	125mm (5 in.)	175mm (7 in.)	215mm (8.5 in.)	Use Alternative 2 or 3	Use Alternative 2 or 3	Use Alternative 2 or 3	15
60	150mm (6 in.)	200mm (8 in.)	Use Alternative 2 or 3	Use Alternative 2 or 3	Use Alternative 2 or 3	Use Alternative 2 or 3	18
TEMPERATURE DIFFERENTIAL (°C)	17	22	28	33	39	44	

minimum thickness of 4 in. (100 mm) is recommended. This thickness should reduce the deflection by an estimated 20%. The overlay thickness of sections B and C may be used as given. However, the overlay thicknesses may be reduced if the PCC slabs are shortened by cracking and seating to reduce temperature effects. In section C, the thicknesses range upward to 8.5 in. (215 mm). Normally, when an overlay approaches the 8- to 9-in. (200- to 225-mm) range, other alternatives should be considered. Note that alternative 2 indicated in Figure 13.7 is crack and seat and alternative 3 is a crack relief layer, both described in Section 13.1.2.

Example 13.9

A JRCP has a joint spacing of 40 ft (12.2 m) and a temperature differential of 70°F (21°C). Measured Benkelman beam deflections are w_1 = 0.035 in. (0.89 mm) and w_r = 0.025 in. (0.64 mm). Determine the thickness of overlay required.

Solution: *Alternative 1: Thick Overlay.* With a slab length of 40 ft (12.2 m) and a temperature differential of 70°F (21°C), from Figure 13.7, the required overlay thickness is greater than 9 in. (229 mm); therefore, use alternative 2 or 3.

Alternative 2: Reduce Slab Length.
   1. Break the slab into 20-ft (6.1-m) sections. With a slab length of 20 ft (6.1 m) and a temperature differential of 70°F (21°C), from Figure 13.7, the required overlay is 5 in. (125 mm).

Sec. 13.3   Asphalt Institute Method

683

TABLE 13.4 MAXIMUM ANNUAL TEMPERATURE DIFFERENTIAL

State	City	Temperature differential (°F)	State	City	Temperature differential (°F)
AL	Mobile	49	NV	Reno	73
AK	Juneau	46	NH	Concord	73
AZ	Phoenix	67	NJ	Atlantic City	61
AR	Little Rock	64	NM	Albuquerque	69
CA	Los Angeles	30	NY	Albany	71
	Sacramento	56		Buffalo	62
	San Francisco	32		New York	59
CO	Denver	71	NC	Charlotte	56
CT	Hartford	68		Raleigh	58
DE	Wilmington	62	ND	Bismark	87
DC	Washington	61	OH	Cincinnati	64
FL	Jacksonville	46		Cleveland	61
	Miami	31		Columbus	64
GA	Atlanta	53	OK	Okla City	67
HI	Honolulu	22	OR	Portland	47
ID	Boise	69	PA	Philadelphia	62
IL	Chicago	68		Pittsburgh	62
	Peoria	70	RI	Providence	61
IN	Indianapolis	66	SC	Columbia	58
IA	Des Moines	74	SD	Soiux Falls	81
KS	Wichita	71	TN	Memphis	60
KY	Louisville	63		Nashville	61
LA	New Orleans	47	TX	Dallas–Fort Worth	62
ME	Portland	67		El Paso	64
MD	Baltimore	62		Houston	53
MA	Boston	59	UT	Salt Lake City	74
MI	Detroit	64	VT	Burlington	73
	Sault Ste. Marie	69	VA	Norfolk	54
MN	Duluth	77		Richmond	61
	Minneapolis–St. Paul	79	WA	Seattle–Tacoma	42
MS	Jackson	57		Spokane	65
MO	Kansas City	69	WV	Charleston	60
	St. Louis	66	WI	Milwaukee	69
MT	Great Falls	72	WY	Cheyenne	69
NE	Omaha	76	PR	San Juan	19

Source. After AI (1983).

2. Check vertical mean deflection. Mean deflection before overlay = (0.025 + 0.035)/2 = 0.03 in. (0.76 mm). Assuming that each inch of HMA will reduce deflection by 5%, the mean deflection reduced by overlay = 5 × 0.03 × 0.05 = 0.0075 in. Mean deflection after overlay = 0.030 − 0.0075 = 0.0225 (0.56 mm) > 0.014 in. (0.36 mm) allowed. Therefore, undersealing is required.

3. Check differential deflections. Differential deflection before overlay = 0.035 − 0.025 = 0.01 (0.25 mm). Differential deflection reduced by overlay = 5 × 0.01 × 0.05 = 0.0025 in. (0.064 mm). Differential deflection after overlay = 0.01 − 0.0025 = 0.0075 in. (0.19 mm) > 0.002 in. (0.051 mm) allowed. Therefore, undersealing is required.

Alternative 3: Crack Relief Layer. Use the layer thicknesses recommended in Figure 13.1. The design consists of a 3.5-in. (89-mm) mix A crack relief layer, a 2-in. (51 mm) dense-graded HMA leveling course, and a 1.5-in. (38-mm) dense-graded HMA surface. The total thickness is 7 in. (178 mm). A check of deflection indicates that undersealing is required.

13.4 PORTLAND CEMENT ASSOCIATION METHOD

There are three types of PCC overlays on PCC pavement: unbonded, bonded, and partially bonded. In this section, the PCA method of overlay design including the evaluation of existing pavements and the design of unbonded and bonded overlays is presented. Depending on the amount of bonding, the design of partially bonded overlay can be interpolated between the two.

13.4.1 Evaluation of Existing Pavement

As a part of the design process, a comprehensive evaluation of the existing pavement must be made, including pavement condition survey, nondestructive deflection testing, and in situ material evaluation.

Pavement Condition Survey

The condition survey should identify the type, extent, and severity of pavement distress. These distresses are described in Section 9.1.2. The pavement is divided into survey sections based on the differences in design, construction, traffic, and location. For small projects, the entire length of each section should be surveyed. For larger projects, each section may be divided into sample units consisting of about 10 to 20 slabs. Sample units from each section are then randomly selected. The use of 25 to 50% of the sample is recommended. The exact sampling frequency should be based on site conditions. For each sample unit, the type, severity, and extent of distress are recorded. For high severity distress, the cause of distress should be identified and corrective measures taken prior to the overlay.

Nondestructive Deflection Testing

The need for deflection testing is based on site conditions and the information obtained from the condition survey. If the condition survey indicates the existence of, or potential for, load-associated distress, then deflection testing should be conducted to determine the severity of the problem. Deflection measurements should be taken at joints and cracks to determine if loss of support exists and if load transfer across the joints and cracks is adequate. The test should be performed using a NDT device that delivers an 8000- to 10,000-lb (35.6- to 44.5-kN) load to the pavement. The use of lighter loads is not recommended.

In Situ Material Evaluation

An in situ material evaluation should be performed to identify the subgrade and subbase materials and establish the design modulus of subgrade reaction at the top of the subbase to be used in different sections of the project. For bonded overlay projects, the strength related properties of the concrete must also be determined. The modulus of subgrade reaction at the top of the subbase may also be back-calculated from results of nondestructive testing, as described in Section 9.4.3.

The design of bonded overlay requires a knowledge of the modulus of rupture and the modulus of elasticity of the concrete in the existing pavement. Because it is not practical to obtain beam specimens from the pavement, it is recommended that splitting tensile tests be made on cored specimens. The test should be conducted according to ASTM C 496 ''Splitting Tensile Strength of Cylindrical Concrete Specimens.'' One core should be taken every 300 to 500 ft (91 to 152 m) at midslab and about 2 ft (0.6 m) from the edge of outside lane. Core diameter should not be less than 4 in. The bottom of cores may be trimmed, but not more than $\frac{1}{2}$ in. (13 mm).

For each section, the effective splitting tensile strength is determined by

$$f_{te} = \overline{f}_t - 1.65s \qquad (13.13)$$

in which f_{te} is the effective splitting tensile strength, $\overline{f}_t$ is the average value of splitting tensile strength, and s is the standard deviation of splitting tensile strength. Equation 13.13 implies that only 5% of the samples in a section has a tensile strength less than the effective values. The modulus of rupture S_c is then obtained using the relationship

$$S_c = ABf_{te} \qquad (13.14)$$

in which A is a regression constant based on local experience and B is the damage factor equal to 0.9 to relate the strength of concrete specimens taken 2 ft (0.61 m) from the edge of outside lane to that near or at the edge. Values of A reported in the literature range from 1.35 to about 1.55 and a reasonable value based on local experience should be selected for design purposes. In the absence of local information, an A value of 1.45 is suggested.

The modulus of elasticity of the concrete in the existing pavement may be determined by testing concrete cores in accordance with ASTM C 469 ''Static Modulus of Elasticity and Poisson Ratio of Concrete in Compression'' or by the following approximate relationship:

$$E_c = DS_c \qquad (13.15)$$

in which D is a constant $= 6000$ to 7000.

13.4.2 Design of Unbonded Overlay

The JSLAB computer program was used by PCA to analyze both bonded and unbonded overlays.

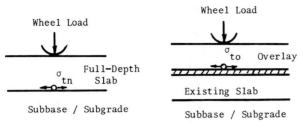

Figure 13.8 Stress equivalent concept for unbonded overlay.

$$\sigma_{to} \leqslant \sigma_{tn}$$

Basic Concept

The design concept is to provide an overlaid pavement that is structurally equivalent to a new full-depth pavement placed on the same subbase and subgrade. The premise of structural equivalency is based on fatigue cracking by ensuring that the edge stress in the unbonded overlay σ_{to} is equal to or less than the edge stress σ_{tn} in the new pavement, as indicated in Figure 13.8.

Cracks in the existing pavement were modeled using soft elements. Wheel load was placed at the edge directly over a crack in the existing pavement, as shown in Figure 13.9. The effect of crack spacing in the existing pavement was determined for spacings of 5, 7, and 10 ft (1.5, 2.1, and 3.0 m). It was found that the edge stress is larger when a crack is present in the existing pavement but additional cracking does not further increase the edge stress.

Design Charts

The edge stresses obtained by JSLAB were used to prepare design charts for determining the thickness of unbonded overlays. The following assumptions were

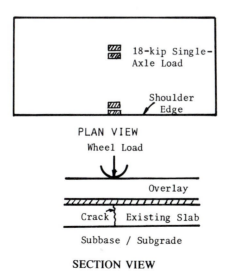

Figure 13.9 Loading position for stress analysis.

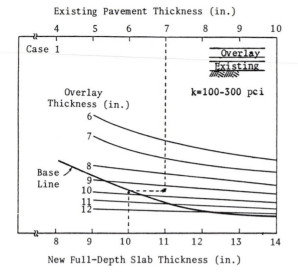

Figure 13.10 Design chart for case 1 condition of existing pavement (1 in. = 25.4 mm, 1 pci = 271.3 kN/m³). (After Tayabji and Okamoto (1985).)

made in the analysis: loading type, 18-kip (80-kN) single-axle load; modulus of elasticity, 5 × 10⁶ psi (35 GPa) for overlays and 3 × 10⁶ and 4 × 10⁶ psi (21 and 28 GPa) for existing pavements; slab length, 20 ft (6.1 m).

Design charts for three cases of existing pavement conditions are shown in Figures 13.10, 13.11, and 13.12. These charts are applicable to existing concrete pavements having modulus of elasticity from about 3 × 10⁶ to 4 × 10⁶ psi (21 to 28 GPa). The three cases are described below:

Case 1. Existing pavement exhibiting a large amount of midslab and corner cracking; poor load transfer across cracks and joints.

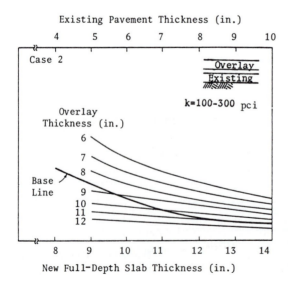

Figure 13.11 Design chart for case 2 condition of existing pavement (1 in. = 25.4 mm, 1 pci = 271.3 kN/m³). (After Tayabji and Okamoto (1985).)

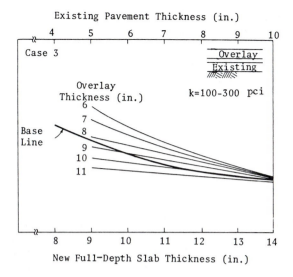

Existing Pavement Thickness (in.)

Case 3

Overlay
Existing

k=100-300 pci

Overlay
Thickness (in.)
6
7
Base 8
Line 9
10
11

New Full-Depth Slab Thickness (in.)

Figure 13.12 Design chart for case 3 condition of existing pavement (1 in. = 25.4 mm, 1 pci = 271.3 kN/m³). (After Tayabji and Okamoto (1985).)

Case 2. Existing pavement exhibiting a small amount of midslab as well as corner cracking; reasonably good load transfer across cracks and joints; localized repairs performed to correct distressed existing slabs.

Case 3. Existing pavement exhibiting a small amount of midslab cracking; good load transfer across cracks and joints; loss of support corrected by undersealing.

The design chart for case 1 was based on the results when a crack was placed in the existing pavement directly under the edge load. The design chart for case 3 was based on the results when no cracking occurred in the existing pavement. The design chart for case 2 was interpolated between cases 1 and 3. The thicknesses of the overlays are based on concrete pavements without tied concrete shoulders. If a tied shoulder is used in conjunction with the overlay, the thickness of the overlay may be reduced by 1 in. (25 mm) provided a minimum thickness of 6 in. (25 mm) is obtained.

Overlay Thickness

The first step in the design process is to determine the thickness of a new full-depth concrete pavement to carry the anticipated traffic, as described in Section 12.2. Based on the condition of the existing pavement, the proper design chart is selected and the overlay thickness corresponding to a new pavement thickness is determined. Use of the design chart is illustrated in Figure 13.10 for an existing pavement of 7 in. (178 mm) and a new full-depth pavement of 10 in. (254 mm). A vertical line is drawn from the 10-in. (254-mm) thickness to intersect the base line and then turn horizontally until it meets the vertical line through the 7-in. (178-mm) existing thickness. The intersection of these two lines gives an overlay thickness of 9.5 in. (241 mm). The minimum overlay thickness provided

TABLE 13.5 REPRESENTATIVE VALUES
OF UNBONDED OVERLAY THICKNESS

h_n (in.)	h_e (in.)	Overlay thickness (in.)		
		Case 1	Case 2	Case 3
8	8	6.8	6.0[a]	6.0[a]
	7	7.2	6.0	6.0
	6	7.6	6.8	6.0
10	9	9.0	7.0	6.0[a]
	8	9.2	7.8	6.0
	7	9.4	8.8	8.0
12	9	11.5	10.4	9.0
	8	11.7	10.8	10.0
	7	11.8	11.2	10.8

Note. [a] indicates minimum thickness;
1 in. = 25.4 mm.

Source. After Tayabji and Okamoto (1985).

by the design charts is 6 in. (152 mm). The use of thinner overlays is not recommended.

Representative values of overlay thickness determined from the charts are summarized in Table 13.5 in terms of new pavement thickness h_n and existing pavement thickness h_e. To compare the results with Eq. 13.2 for the unbonded case with $n = 2$, overlay thicknesses were computed for C values of 0.35, 0.6, and 0.8. These results are presented in Table 13.6. It can be seen that the overlay thicknesses listed in Table 13.5 agree closely with those listed in Table 13.6. This indicates that cases 1, 2, and 3 are equivalent to C values of 0.3 to 0.5, 0.5 to 0.7, and 0.7 to 0.9, respectively.

TABLE 13.6 THICKNESSES OF UNBONDED
OVERLAY BASED ON CORPS OF ENGINEERS'
EQUATION

h_n (in.)	h_e (in.)	Overlay thickness (in.)		
		$C = 0.35$	$C = 0.6$	$C = 0.8$
8	8	6.5	6.0[a]	6.0[a]
	7	6.8	6.0[a]	6.0[a]
	6	7.2	6.5	6.0
10	9	8.5	7.2	6.0
	8	8.5	7.9	7.0
	7	9.1	8.4	7.8
12	9	10.8	9.8	8.9
	8	11.0	10.3	9.6
	7	11.3	10.7	10.2

Note. [a] Indicates minimum thickness;
1 in. = 25.4 mm.

Source. After Tayabji and Okamoto (1985).

13.4.3 Design of Bonded Overlay

Bonded overlay is used when the existing pavement is still in good condition but does not have the structural capacity to carry the anticipated future traffic.

Basic Concept

The basic concept for bonded overlay is the same as that for unbonded overlay, but the location of the edge stress to be compared is different. In the unbonded case, the edge stress at the bottom of the full-depth slab is compared with that at the bottom of overlay, as shown in Figure 13.8. Since both are made of new concrete with the same modulus of rupture S_c, a comparison of edge stress is sufficient. If the edge stress σ_t is the same, the stress ratio σ_t/S_c will be the same, and so will the fatigue life. In the bonded case, the edge stress at the bottom of the full-depth slab is compared with that at the bottom of the existing slab, as shown in Figure 13.13. Because the new full-depth slab and the old existing slab have different moduli of rupture, the equivalency should be based on the stress ratio rather than the stress alone. For both pavements to be equivalent, the stress ratio in the existing overlaid pavement must be equal to or smaller than that in the new full-depth pavement:

$$\frac{\sigma_{te}}{S_{ce}} \leq \frac{\sigma_{tn}}{S_{cn}} \tag{13.16}$$

in which σ_{te} is the critical edge stress in the existing overlaid pavement, σ_{tn} is the critical edge stress in the new full-depth pavement, S_{ce} is the modulus of rupture of existing concrete, and S_{cn} is the modulus of rupture of new concrete.

Design Chart

The JSLAB computer program was used to determine the critical tensile stresses due to edge loading in a full-depth concrete pavement and in an existing pavement with bonded overlay. The computed stresses together with the moduli of rupture were then used to prepare the design chart, as shown in Figure 13.14. In the design chart, three different curves representing different moduli of rupture of the existing concrete are shown. Curve 1 is applicable for modulus of

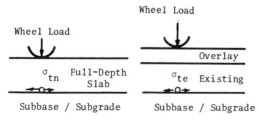

$$\frac{\sigma_{te}}{S_{ce}} \leq \frac{\sigma_{tn}}{S_{cn}}$$

Figure 13.13 Equivalent concept for bonded overlay.

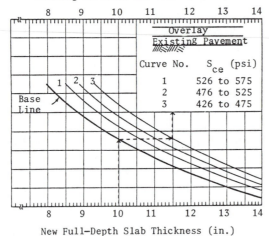

Combined Thickness of
Existing Pavement and Overlay (in.)

New Full—Depth Slab Thickness (in.)

Figure 13.14 Design chart for bonded overlay (1 in. = 25.4 mm, 1 psi = 6.9 kPa). (After Tayabji and Okamoto (1985).)

rupture of 526 to 575 psi (3.6 to 4.0 MPa), curve 2 of 476 to 525 psi (3.3 to 2.6 MPa), and curve 3 of 426 to 475 psi (2.9 to 3.3 MPa). Given the thickness of the full-depth pavement and the curve number, the combined thickness of the existing pavement and the overlay can be read from the chart. The difference between the combined thickness and the thickness of the existing pavement is the overlay thickness required.

To develop the design chart, the moduli of elasticity of the existing concrete corresponding to the three curves were computed by Eq. 13.15 with D values from 6000 to 7000. The moduli of elasticity of the new concrete were assumed from 4×10^6 to 5×10^6 psi (28 to 25 GPa) and the moduli of rupture from 600 to 650 psi (4.1 to 4.5 MPa).

The chart cannot be applied when the modulus of rupture of the existing concrete is lower than 426 psi (2.9 MPa) or higher than 575 psi (4.0 MPa). When the modulus of rupture is lower than 425 psi (2.9 MPa), the very large thickness required may not warrant the use of a bonded overlay. Furthermore, an existing pavement with such a low modulus of rupture may be in such a distressed condition that the use of an unbonded or directly placed overlay, rather than a bonded overlay, should be considered. For the case when the modulus of rupture of the existing concrete is greater than 575 psi (4.0 MPa), the existing pavement is considered as good as the new pavement, so the overlay thickness required is simply the difference between the full-depth pavement thickness and the existing pavement thickness plus the depth of surface milling or scarification.

It should be noted that the maximum bonded overlay thickness recommended is 5 in. (127 mm). When the bonded overlay thickness exceeds 5 in. (127 mm), the use of a directly placed or unbonded overlay may be more cost effective.

Example 13.10

An existing concrete pavement has a thickness of 7.5 in. (191 mm) after milling. Tests on a number of cored samples taken from the existing pavement give an average splitting tensile strength of 440 psi (3.0 MPa) and a standard deviation of 50 psi (345 kPa). If the required thickness for a new full-depth pavement is 10 in. (254 mm) and the conditions used to develop the design chart are all satisfied, determine the thickness of the bonded overlay required.

Solution: From Eq. 13.13, $f_{te} = 440 - 1.65 \times 50 = 358$ psi (2.5 MPa). Based on Eq. 13.14 and a value of 1.45 for constant A, $S_c = 1.45 \times 0.9 \times 358 = 467$ psi (3.2 MPa). From Figure 13.14, the combined thickness of existing pavement and overlay is 11.5 in. (381 mm), so the thickness of bonded overlay = 11.5 − 7.5 = 4 in. (102 mm).

13.5 AASHTO METHOD

The 1986 AASHTO design guide contains the most comprehensive procedure for the design of overlays. The procedure is based on the remaining life concept and can be applied to any type of overlay. The use of nondestructive testing is recommended as part of overlay rehabilitation to evaluate the in situ moduli and structural capacity of subgrade and various layers.

13.5.1 Fundamental Concept

Some terms relating to overlay design are explained first, followed by the introduction of a basic equation for design purposes.

Nomenclature

Many symbols are used in the AASHTO design guide to describe the overlay design procedure. These symbols are listed below alphabetically.

a_{OL} structural layer coefficient of HMA overlay.
a_{bs} structural layer coefficient of PCC layer after it has been cracked by the break and seat approach.
a_{2r} structural layer coefficient of the existing cracked PCC pavement layer.
C condition factor.
C_v visual layer condition factor.
C_x overall condition factor at x defined as a ratio between SC_{xeff} and SC_o.
D_0 existing PCC layer thickness.
D_{OL} required thickness of PCC overlay.
D_{xeff} effective thickness of in situ PCC layer reflecting its reduced modulus value.
D_y PCC thickness required to support the overlay traffic over existing subgrade and subbase conditions.
E_{pcc} PCC effective layer modulus reflecting cracking and loss of support, etc.
E_r modulus ratio between cracked PCC and a modulus of 5×10^6 psi (34.5 GPa).

F_{RL} remaining life factor accounting for damage of the existing pavement as well as the desired degree of damage to the overlay at the end of overlay traffic. It is always less than or equal to a value of 1.

h_{OL} required thickness of HMA overlay.

N number of load repetitions.

N_{fx} total number of ESAL repetitions necessary for the original pavement to reach failure or a p_f of 2.0.

N_{fy} total number of ESAL repetitions necessary for the overlaid pavement to reach failure or a p_f of 2.0.

n an exponent in the basic equation which is a constant dependent upon the type of pavement system used in the analysis.

p_{01} initial serviceability of original pavement when constructed.

p_{02} initial serviceability of pavement after overlay.

p_{t1} terminal serviceability of existing pavement immediately prior to overlay.

p_{t2} desired terminal serviceability of overlaid pavement after the overlaid traffic has been applied.

p_f ultimate failure serviceability for any pavement type corresponding to a completely failed pavement and can be taken as 2.0.

R_{Lx} remaining life of existing pavement prior to overlay.

R_{Ly} remaining life of overlaid pavement after the overlay traffic has been reached.

r annual traffic growth rate of highway section being considered.

SC_0 initial structural capacity of existing pavement.

SC_{OL} required structural capacity of overlay.

SC_{xeff} effective structural capacity of existing pavement immediately prior to the time of overlay and reflecting the damage to that point in time.

SC_y total structural capacity required to support the overlay traffic over existing subgrade or foundation conditions.

SC_{yeff} effective structural capacity of the overlaid pavement at the end of overlay period.

SN_{OL} required structural number of overlay.

SN_{xeff} total effective structural number of the existing pavement structure above the subgrade prior to overlay.

$SN_{xeff-rp}$ effective structural capacity of all of the remaining pavement layers above the subgrade except for the existing PCC layer.

SN_y total structural number required to support the overlay traffic over existing subgrade conditions.

T_f best estimate of the probable time in years that the particular pavement type typically lasts before an overlay is required.

t time in years for which pavement section has been in service prior to overlay.

x actual ESAL repetitions on the original pavement up to the desired time for overlay.

y future ESAL repetitions within the overlay period.

Note that structural capacity SC is a general term that can be applied to both flexible and rigid pavements. For flexible pavements, SC is expressed as a structural number SN, whereas for rigid pavements, it is expressed as the PCC thickness D. Some of the above nomenclature is shown in Figure 13.15 in terms of serviceability, structural capacity, and the condition factor.

In Figure 13.15, the original pavement has an initial serviceability of p_{01}, a structural capacity of SC_0, and a condition factor of 1.0. As the number of load application increases, the pavement gradually deteriorates and its serviceability, structural capacity, and condition factor decrease. After an x number of load

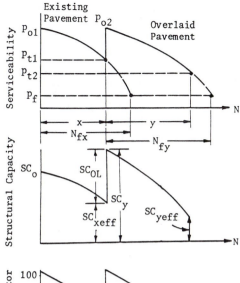
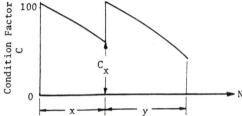

Figure 13.15 Effect of traffic on serviceability, structural capacity, and condition factor. (From the *ASSHTO Guide for Design of Pavement Structures*. Copyright 1986. American Association of State Highway and Transportation Officials, Washington, DC. Used by permission.)

repetitions, the pavement reaches the acceptable serviceability level of p_{t1}. At this time, the pavement reaches a structural capacity of SC_{xeff} and a condition factor of C_x, which is defined as

$$C_x = \frac{SC_{xeff}}{SC_0} \tag{13.17}$$

To upgrade the pavement condition to a serviceability p_{o2}, additional structural capacity SC_{OL} in the form of an overlay is required. The sum of this added structural capacity and the structural capacity of the existing pavement SC_{xeff} is equivalent to the structural capacity of SC_y of a new pavement designed for the existing roadbed modulus and layer properties and for the projected overlay traffic y. The terminal serviceability index for the overlay is p_{t2}, which need not be equal to p_{t1}. The values of N_{fx} and N_{fy} represent the number of load repetitions required to deteriorate the pavement to a failure condition with a serviceability p_f of 2.0.

Basic Equation

The overlay design is based on the remaining life concept that the structural capacity of a pavement decreases with load applications. For a pavement that has been overlaid, the structural capacity of the original pavement is a function of the loads applied before overlay as well as those applied after overlay. The basic equation for overlay design is

$$(SC_{OL})^n = (SC_y)^n - F_{RL}(SC_{xeff})^n \qquad (13.18)$$

in which F_{RL} is the remaining life factor and n is a constant depending on the type of pavement system used in the analysis. For all types of pavement, except for partially bonded or unbonded rigid overlays on rigid pavements, $n = 1$. For partially bonded rigid overlays on rigid pavements, $n = 1.4$; for unbonded rigid overlays on rigid pavements, $n = 2$. The specific form of overlay equation for four possible types of overlays is shown in Table 13.7.

13.5.2 Development of Design Input Factors

Before discussing the design technology for various types of overlay, six common steps should be taken to develop the necessary design input factors. These steps are described below.

Analysis Unit Delineation

The first step in the overlay design process is the delineation of basic analysis units. The objective is to divide the project into statistically homogeneous pavement units having uniform pavement cross sections, subgrade support, construction histories, and pavement conditions. The steps to be taken should be based on the availability of historic data.

Case I: Accurate Historic Data Unavailable

In the absence of historic data, a NDT deflection study should be conducted in the outer wheelpath of the lane adjacent to the shoulder. When prior informa-

TABLE 13.7 SPECIFIC DESIGN EQUATIONS FOR FOUR TYPES OF OVERLAY

Type overlay	Type existing pavement	Specific equation	Conditions/remarks
Flexible	Flexible	$SN_{OL} = SN_y - F_{RL}SN_{xeff}$	$SC = SN$; $n = 1.0$
Flexible	Rigid	$SN_{OL} = SN_y - F_{RL}SN_{xeff}$	$SC = SN$; $n = 1.0$
Rigid	Flexible	$D_{OL} = D_y$ (see remarks)	Treat overlay analysis as new rigid pavement design using existing flexible pavement as new foundation (subgrade)
Rigid	Rigid	$D_{OL} = D_y - F_{RL}(D_{xeff})$	$SC = D$; $n = 1.0$ (bonded overlay)
		$D_{OL}^{1.4} = D_y^{1.4} - F_{RL}(D_{xeff})^{1.4}$	$SC = D$; $n = 1.4$ (partial bond overlay)
		$D_{OL}^2 = D_y^2 - F_{LR}(D_{xeff})^2$	$SC = D$; $n = 2.0$ (unbonded overlay)

Source. From the *AASHTO Guide for Design of Pavement Structures*. Copyright 1986. American Association of State Highway and Transportation Officials, Washington, DC. Used by permission.

tion concerning unit boundary is not available, deflection measurements should be made at equal intervals of 300 to 500 ft (91 to 152 m). The deflection data thus obtained should be analyzed to delineate the boundaries of the units. An evaluation of the practicality of these unit lengths should then be made. A general guideline for a minimum unit length is 0.5 miles (0.8 km).

The analysis units thus selected through the combined use of NDT and engineering judgment are then used as the basis for conducting any destructive tests for determining the thickness and properties of various layers. This information can then be used to verify or modify the units previously established.

Case II: Accurate Historic Data Available

When accurate historic information is available, a general idea on unit boundaries can be formed prior to any field testing. Nondestructive deflection testing is then conducted at 10 to 15 test points randomly selected in each unit to verify or modify the preliminary units selected. If necessary, a destructive sampling plan can be developed to further examine the appropriateness of the analysis units selected.

Traffic Analysis

The purpose of traffic analysis is to determine the 18-kip (80-kN) ESAL repetitions along a pavement length from the date the pavement was originally opened to traffic through the end of the anticipated overlay period. The data actually to be used for design are the x and y repetitions, as shown in Figure 13.15. While calculation of both x and y traffic values is desirable, the determination of y takes priority because knowledge of traffic repetitions anticipated within the overlay design period is absolutely necessary to determine SC_y. The value of x is used to determine the remaining life of the existing pavement. In the absence of other preferred methods, such as NDT, historic traffic data may also be used to estimate the effective structural capacity of the pavement in situ. Methods for traffic computations are presented in Chapter 6.

Materials and Environmental Study

The primary material property of concern is the elastic or resilient modulus of pavement layers and the subgrade. The effects of environment, such as temperature and moisture, on the resilient modulus must also be carefully considered. AASHTO strongly recommends the use of NDT to estimate the in situ resilient modulus of each layer within the pavement structure. Methods for back-calculating the moduli of various layers are described in Section 9.4.3. The reduced moduli predicted will reflect the in situ influence of cracking in stabilized layers or excessive moisture in unbound layers and can therefore be used to determine the effective structural capacity SC_{xeff}. The layer moduli predicted from NDT deflection basin analysis must be compared with previous experience on similar materials to ensure that the results are reasonable. The use of limited destructive testing is encouraged to provide spot verification of the modulus values obtained from the NDT.

The selection of the appropriate design modulus and strength parameters for HMA and PCC to be used in the overlay is similar to that of new materials. This topic is discussed in Chapter 7.

Effective Structural Capacity Analysis

The material properties derived from the previous step can be used to determine the effective structural capacity SC_{xeff}. The analysis procedure for flexible pavements is different from that for rigid pavements. When existing flexible pavements are evaluated, SC_{xeff} is equivalent to the effective structural number SN_{xeff}. When rigid pavements are evaluated, SC_{xeff} is equal to the effective thickness D_{xeff}.

Flexible Pavements

Only the NDT method is recommended for determining the effective structural capacity of existing flexible pavements. Two methods, designated as method 1 and method 2, are available. Method 1 is used to back-calculate the resilient modulus of each pavement layer. The resilient modulus is then related to the layer coefficient. When the thickness, layer coefficient, and drainage coefficient of each layer are known, the structural number of the existing pavement can be determined, as described in Section 11.3.4. Method 2 is used to determine SN_{xeff} directly based on an estimate of the in situ subgrade modulus, the total pavement thickness above the subgrade, and the maximum deflection at the center of the load plate. Details of method 2 are described in the AASHTO design guide (AASHTO, 1986) and in Appendix PP of the design guide (AASHTO, 1985).

Rigid Pavements

Both NDT and non-NDT procedures can be used to determine the effective structural capacity of rigid pavements. Only NDT method 1, which gives the resilient modulus of each layer, is applicable and NDT method 2 cannot be used. The predicted PCC modulus E_{pcc}, which reflects cracking and loss in support, can be used to determine D_{xeff}, as shown by Figure 13.16. The predicted subbase and subgrade moduli can be used to determine the composite modulus of subgrade reaction, as shown in Figure 12.18.

In the event that it is physically impossible to utilize NDT measurements for overlay evaluation of existing rigid pavement systems, three alternative techniques can be used:

1. *Visual Condition Factor Approach.* Figure 13.17 shows the approximate relationship between the visual condition factor C_v and the cracked PCC modulus ratio E_r. Values of C_v can be estimated from Table 13.8. Multiplication of E_r by an uncracked PCC modulus of 5×10^6 psi (35 GPa) gives the in situ PCC modulus E_{pcc}. When E_{pcc} is known, D_{xeff} can be obtained from Figure 13.16. For example, if a 12-in. (305-mm) slab is rated with a C_v of 0.55, the E_r value will be 0.46, or $E_{pcc} = 0.46 \times 5 \times 10^6 = 2.3 \times 10^6$ psi (15.9 GPa). From Figure 13.16, $D_{xeff} = 9.2$ in. (234 mm).

2. *Nominal Size of PCC Slab Fragments.* Figure 13.18 shows the approximate relationship between the E_r ratio and the nominal size of cracked slab

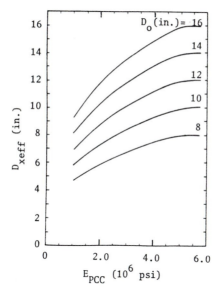

Figure 13.16 Determination of D_{xeff} from NDT derived PCC modulus (1 in. = 25.4 mm, 1 psi = 6.9 kPa). (From the *AASHTO Guide for Design of Pavement Structures*. Copyright 1986. American Association of State Highway and Transportation Officials, Washington, DC. Used by permission.)

fragments. The approach is similar to that of visual condition factor described previously. For example, if an 11-in. (279-mm) slab is cracked into nominal size of 8 ft (2.4 m), then $E_r = 0.7$, $E_{pcc} = 3.5 \times 10^6$ psi (24 GPa), and $D_{xeff} = 9.7$ in. (246 mm).

3. *Remaining Life Approach*. Several techniques can be used for predicting the remaining life R_{Lx} of the existing pavement. If an estimate is made of R_{Lx}, Figure 13.19 can be used to determine the pavement condition factor C_x. Once C_x is known, D_{xeff} can be computed by

$$D_{xeff} = C_x D_0 \tag{13.19}$$

As an example, if a 9-in. (229-mm) PCC pavement is estimated to have a remaining life of 25%, from Figure 13.19, $C_x = 0.79$ and by Eq. 13.19, $D_{xeff} = 0.79 \times 9 = 7.1$ in. (180 mm).

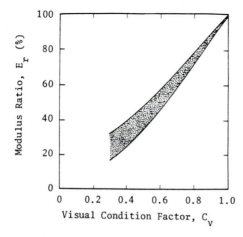

Figure 13.17 Relationship between visual condition factor and PCC modulus ratio. (From the *AASHTO Guide for Design of Pavement Structures*. Copyright 1986. American Association of State Highway and Transportation Officials, Washington, DC. Used by permission.)

TABLE 13.8 VISUAL CONDITION FACTOR FOR PCC PAVEMENTS

Pavement condition	Range of C_v
Uncracked, stable, and undersealed, with no evidence of pumping	0.9–1.0
Stable and undersealed with no evidence of pumping but with initial, tight, and nonworking cracks	0.7–0.9
Appreciably faulted and cracked into 10 to 40 ft² (1 to 4 m²) fragments with signs of progressive crack deterioration and pumping	0.5–0.7
Badly cracked or shattered into fragments with maximum size of 2 to 3 ft (0.6 to 0.9 m)	0.3–0.5

The curve shown in Figure 13.19 was plotted from the relationship that

$$C_x = 1 - 0.7e^{-(R_{Lx}+0.85)^2} \qquad (13.20)$$

Equation 13.20 was derived from an analysis of AASHTO design equations based on a PSI of 1.5 to 2 at failure.

Future Overlay Structural Capacity Analysis

The major objective of this step is to determine the total structural capacity of a new pavement required to carry y load repetitions during the overlay design period, as shown in Figure 13.15. In other words, this step is simply a new pavement design for either a flexible or rigid pavement system based on the existing subgrade or foundation conditions. Consequently, the design procedure for new pavements, as discussed in Section 11.3 or 12.3, can be used. For flexible overlays, the structural capacity of new pavement is expressed in terms of SN_y; for rigid overlays, it is expressed in terms of D_y, as shown in Table 13.7. For rigid overlays on flexible pavements, the existing flexible pavement is treated as a foundation for the new rigid pavement.

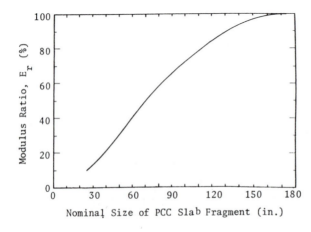

Figure 13.18 Relationship between slab fragment size and PCC modulus ratio (1 in. = 25.4 mm). (From the *AASHTO Guide for Design of Pavement Structures*. Copyright 1986. American Association of State Highway and Transportation Officials, Washington, DC. Used by permission.)

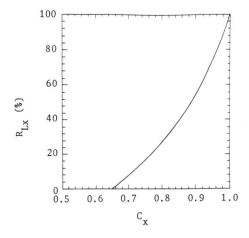

Figure 13.19 Relationship between pavement condition factor and remaining life. (From the *AASHTO Guide for Design of Pavement Structures.* Copyright 1986. American Association of State Highway and Transportation Officials, Washington, DC. Used by permission.)

Remaining Life Factor Determination

The remaining life factor F_{RL} is an adjustment factor applied to the effective structural capacity SC_{xeff} to reflect a more realistic assessment of the weighted effective capacity during the overlay period. This factor depends on the remaining life value of the existing pavement prior to the overlay R_{Lx} and the remaining life of the overlaid pavement system at the end of the overlay period R_{Ly}. Therefore, both R_{Lx} and R_{Ly} must be known to determine F_{RL}.

Remaining Life of the Existing Pavement

The remaining life of the existing pavement is difficult to determine accurately. There are five possible methods to estimate the R_{Lx} value. While these approaches are theoretically equivalent, they rarely yield the same result. The procedure utilizing NDT deflection studies often results in a better quantitative assessment and should be relied on more heavily than other approaches. It was recommended in the AASHTO design guide that the engineer should assess the reasonableness of several results rather than blindly rely on one approach. These five approaches are described as follows:

1. *NDT Approach.* Nondestructive testing can be used to determine SC_{xeff}, as described previously. When SC_{xeff} and the initial structural capacity SC_0 are known, the condition factor C_x can be computed by Eq. 13.17 and R_{Lx} by Figure 13.19.

2. *Traffic Approach.* If reasonably accurate historical traffic x is available, R_{Lx} can be determined by

$$R_{Lx} = \frac{N_{fx} - x}{N_{fx}} \tag{13.21}$$

Note that N_{fx} can be determined from the AASHTO design equations or charts for a p_f of 2.0.

3. *Time Approach.* If specific traffic information is not available, R_{Lx} can be determined by Figure 13.20 based on the time t that the existing pavement has

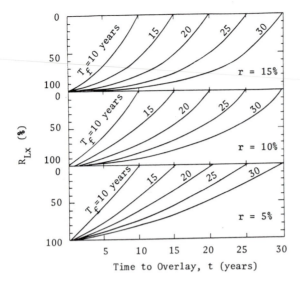

Figure 13.20 Remaining life based on time considerations for various traffic growth rates. (From the *AASHTO Guide for Design of Pavement Structures*. Copyright 1986. American Association of State Highway and Transportation Officials, Washington, DC. Used by permission.)

been in service prior to the overlay, the best estimate of the probable time T_f that the pavement can last before an overlay is required, and the annual traffic growth rate, r. For example, if $r = 5\%$, $T_f = 20$ years, and $t = 15$ years, from Figure 13.20, $R_{Lx} = 35\%$. It can be easily proved from Eq. 6.33 that this time approach is similar to the traffic approach and the value of R_{Lx} can also be computed by

$$R_{Lx} = \frac{T_f - t}{T_f} = \frac{(1 + r)^{T_f} - (1 + r)^t}{(1 + r)^{T_f} - 1} \qquad (13.22)$$

4. *Visual Condition Survey Approach.* The structural condition factor for each layer can be obtained from the visual condition factor using Table 13.9. An overall pavement condition factor C_x can be computed by a weighting procedure based on the layer thickness of each material:

$$C_x = \frac{h_1 C_{x1} + h_2 C_{x2} + \cdots + h_{xn} C_{xn}}{h_1 + h_2 + \cdots + h_n} \qquad (13.23)$$

After the C_x value has been determined, R_{Lx} can be obtained from Figure 13.19.

5. *Serviceability Approach.* If the present serviceability index of a pavement is known, Figure 13.21 can be used to estimate R_{Lx}. Different curves are shown for flexible and rigid pavements as well as for different initial structural capacities.

Remaining Life of the Overlaid Pavement

The remaining life of the overlaid pavement R_{Ly} can be determined by

$$R_{Ly} = \frac{N_{fy} - y}{N_{fy}} \qquad (13.24)$$

For a given y, p_{o2}, and p_{t2}, SC_y can be determined from the AASHTO design charts. The value of N_{fy} can then be determined for a p_f of 2.0.

Design of Overlays   Chap. 13

TABLE 13.9 SUMMARY OF VISUAL AND STRUCTURAL CONDITION FACTORS

Layer type	Pavement condition	C_v visual condition factor range	C_x structural condition factor value
Asphaltic	1. Asphalt layers that are sound, stable, uncracked and have little or no deformation in the wheelpaths.	0.9–1.0	.95
	2. Asphalt layers that exhibit some intermittent cracking with slight to moderate wheelpath deformation but are still stable.	0.7–0.9	.85
	3. Asphalt layers that exhibit some moderate to high cracking, have raveling or aggregate degradation and show moderate to high deformations in wheelpath.	0.5–0.7	.70
	4. Asphalt layers that show very heavy (extensive) cracking, considerable raveling or degradation and very appreciable wheelpath deformations.	0.3–0.5	.60
PCC	1. PCC pavement that is uncracked, stable and undersealed, exhibiting no evidence of pumping.	0.9–1.0	.95
	2. PCC pavement that is stable and undersealed but shows some initial cracking (with tight, nonworking cracks) and no evidence of pumping.	0.7–0.9	.85
	3. PCC pavement that is appreciably cracked or faulted with signs of progressive crack deterioration: slab fragments may range in size from 10 to 40 ft^3 (1 to 4 m^2), pumping may be present.	0.5–0.7	.70
	4. PCC pavement that is very badly cracked or shattered into fragments 2–3 ft (0.6–0.9 m) in maximum size.	0.3–0.5	.60
Pozzolanic base/ subbase	1. Chemically stabilized bases (CTB, LCF) that are relatively crack-free, stable, and show no evidence of pumping.	0.9–1.0	.95
	2. Chemically stabilized bases (CTB, LCF) that have developed very strong pattern of fatigue cracking, with wide and working cracks that are progressive in nature; evidence of pumping or other causes of instability may be present.	0.3–0.5	.60
Granular base/ subbase	1. Unbound granular layers showing no evidence of shear or densification distress, reasonably identical physical properties as when constructed and existing at the same "normal" moisture-density conditions as when constructed.	0.9–1.0	.95
	2. Visible evidence of significant distress within layers (shear or densification), aggregate properties have changed significantly due to abrasion, intrusion of fines from subgrade or pumping, and/or significant change in in situ moisture caused by surface infiltration or other sources.	0.3–0.5	.60

Special notes

1. The visual condition factor C_v is related to the structural condition factor C_x by $C_v = C_x^2$

2. The structural condition factor, C_x and *not* the C_v value, is the variable used in the structural overlay design equation (for all overlay-existing pavement types). It is defined by $SC_{xeff} = C_x SC_0$

Source. From the *AASHTO Guide for Design of Pavement Structures.* Copyright 1986. American Association of State Highway and Transportation Officials, Washington, DC. Used by permission.

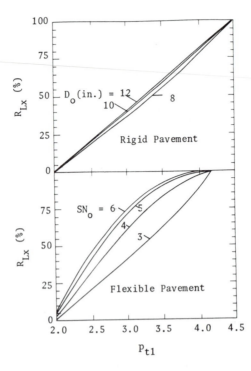

Figure 13.21 Remaining life based on present serviceability index and initial structural capacity (1 in. = 25.4 mm). (From the *AASHTO Guide for Design of Pavement Structures*. Copyright 1986. American Association of State Highway and Transportation Officials, Washington, DC. Used by permission.)

Calculation of the Remaining Life Factor

Having obtained estimates of both R_{Lx} and R_{Ly} as outlined above, the remaining life factor can be obtained from Figure 13.22. Figure 13.22 was obtained from the following equations:

$$F_{RL} = \frac{C_{yx}}{C_x C_y} \qquad (13.25)$$

Note that C_x is the condition factor for the existing pavement prior to overlay and can be related to R_{Lx} by Eq. 13.20. The same equation can be applied to C_y, which is the condition factor for the entire overlaid pavement, as well as to C_{yx}, which is the condition factor of the existing pavement after an overlay has been applied. By assuming that the remaining life R_{Lyx} of the existing pavement after the overlay is equal to the remaining life R_{Lx} at the time of overlay minus the damage, $1 - R_{Ly}$, done during the period of overlay, R_{Lyx} can be expressed as

$$R_{Lyx} = R_{Lx} + R_{Ly} - 1 \qquad (13.26)$$

Example 13.11

Given $R_{Lx} = 0.6$ and $R_{Ly} = 0.4$, compute the remaining life factor F_{RL} by equations and check the results with Figure 13.22.

Solution: From Eq. 13.26, $R_{Lyx} = 0.4 + 0.6 - 1 = 0$. From Eq. 13.20, $C_x = 1 - 0.7e^{-(0.6+0.85)^2} = 0.914$, $C_y = 1 - 0.7e^{-(0.4+0.85)^2} = 0.853$, and $C_{yx} = 1 - 0.7e^{-(0+0.85)^2} = 0.660$. From Eq. 13.25, $F_{RL} = 0.660/(0.914 \times 0.853) = 0.85$, which checks with the value obtained from Figure 13.22.

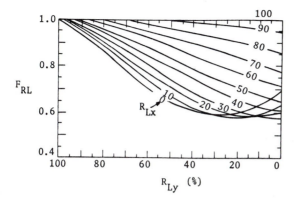

Figure 13.22 Remaining life factor as a function of remaining life of existing and overlaid pavements (From the *AASHTO Guide for Design of Pavement Structures.* Copyright 1986. American Association of State Highway and Transportation Officials, Washington, DC. Used by permission.)

A general trend whereby F_{RL} decreases with both R_{Lx} and R_{Ly} can be noted in Figure 13.22. This is reasonable because if the existing pavement has less remaining life, its effective thickness should be reduced and the required thickness of overlay increases. However, when both R_{Lx} and R_{Ly} are small, some anomalies exist, as indicated by the reversed slope and the crossing of the curves. It has been suggested that when a point is located on the reversed slope, the minimum value of F_{RL} by assuming that the curve is horizontal should be used (ERES Consultant, Inc., 1987). Elliott (1989) revealed some inconsistencies in overlay design using the AASHTO remaining life factor and recommended that the design approach be revised to exclude remaining life considerations by assuming F_{RL} to be unity. Fwa (1991) expressed the need to incorporate a remaining life concept in overlay design and presented a new expression for the remaining life factor which can remove the inconsistencies in the AASHTO design method.

13.5.3 Overlay Design Methodology

Six steps leading to the overlay design methodology are described in the previous section. In this section, the design procedures for various types of overlays on any existing pavements are summarized and examples are given to illustrate the application of the various methods. These examples are essentially the same as those presented in the AASHTO design guides, with only slight modifications.

Flexible Overlays on Flexible Pavements

The basic equation for determining the required SN_{OL} for flexible overlays on flexible pavements is shown in Table 13.7. The required thickness of asphalt overlay h_{OL} can be determined by

$$h_{OL} = \frac{SN_{OL}}{a_{OL}} = \frac{SN_y - F_{RL}SN_{xeff}}{a_{OL}} \tag{13.27}$$

in which a_{OL} is the structural layer coefficient of the overlay material. All the input factors can be determined by following the steps discussed in Section 13.5.2.

Example 13.12

Given the following information, determine the required thickness of asphalt overlay on the existing flexible pavement.

Existing pavement: thickness of HMA surface h_1 = 4 in. (102 mm), thickness of crushed stone base h_2 = 14 in. (356 mm), base drainage coefficient m_2 = 1.0, initial structural number SN_0 = 3.8, initial PSI p_{01} = 4.3, PSI prior to overlay p_{t1} = 2.5, and ESAL to date x = 1.4 × 10⁵. The resilient moduli obtained from NDT are 340,000 psi (2.3 GPa) for the HMA, 25,000 psi (173 MPa) for the crushed stone base, and 5000 psi (35 MPa) for the subgrade. The deflection data were taken during early summer, so only an adjustment in the resilient modulus of roadbed soil is deemed necessary. By following the procedure described in Section 11.3.3, the effective resilient modulus M_R was found to be 3000 psi (21 MPa).

Overlaid pavement: initial PSI p_{02} = 4.3, PSI at end of overlay period p_{t2} = 2.5, future overlay ESAL y = 1.0 × 10⁶, and modulus of the HMA overlay E_{OL} = 375,000 psi (2.6 GPa) at 70°F (21°C).

Other variables: design reliability R = 90%, standard deviation S_0 = 0.5.

Solution: As shown by Eq. 13.27, the parameters to be determined are a_{OL}, SN_{xeff}, SN_y, and F_{RL}. Given E_{OL} = 375,000 psi (2.6 GPa), from Figure 11.27, a_{OL} = 0.40.

First, determine SN_{xeff}. Given a HMA modulus of 340,000 psi (2.3 GPa), from Figure 11.27, a_1 = 0.39. Given a base modulus of 25,000 psi (173 Mpa), from Eq. 11.44, a_2 = 0.249 × log(25,000) − 0.977 = 0.12. From Eq. 11.35, SN_{xeff} = 0.39 × 4 + 0.12 × 14 × 1.0 = 3.24.

Next, determine SN_y, which is the SN of a new pavement. With R = 90%, S_0 = 0.5, $W_{18} = y$ = 1.0 × 10⁶, M_R = 3000 psi (21 MPa), and ΔPSI = 4.3 − 2.5 = 1.8. From Figure 11.25, SN_y = 5.0.

Finally, determine F_{RL}, which depends on R_{Lx} and R_{Ly}. Several methods can be used to determine R_{Lx}. If the traffic approach is used, the value of N_{fx} to reduce the PSI from 4.3 to 2.0, or ΔPSI = 4.3 − 2.0 = 2.3, can be determined from Figure 11.25 or computed more accurately by Eq. 11.37 as follows: $\log N_{fx}$ = $\log W_{18}$ = −1.282 × 0.5 + 9.36 × log(3.8 + 1) − 0.2 + log(2.3/2.7)/[0.4 + 1094/ (3.8 + 1)^{5.19}] + 2.32 log(3000) − 8.07 = 5.435, or N_{fx} = 275,000. Note that a Z_R of −1.282 corresponds to a reliability of 90% and can be found from Table 11.15. From Eq. 13.21, R_{Lx} = (275,000 − 140,000)/275,000 = 0.49 or 49%. If the serviceability approach is used, from Figure 13.21, R_{Lx} = 30%. If the NDT approach is used, from Eq. 13.17, C_x = 3.24/3.80 = 0.85, and from Figure 13.19, R_{Lx} = 40%. It can be seen that the three approaches result in R_{Lx} values of 49, 30, and 40%. These values must be evaluated and the best estimate be made on the basis of engineering judgment. In this instance, an R_{Lx} of 30% has been selected.

The R_{Ly} value is determined in a manner identical to that developed by the traffic method for R_{Lx}. The N_{fy} value to reduce the PSI from 4.3 to 2.0, or ΔPSI = 4.3 − 2.0 = 2.3, can be determined from Figure 11.25 or computed more accurately by Eq. 11.37 as follows: $\log N_{fy}$ = −1.282 × 0.5 + 9.36 × log(5.0 + 1) − 0.2 + log(2.3/2.7)/[0.4 + 1094/(5.0 + 1)^{5.19}] + 2.32 × log(3000) − 8.07 = 6.300, or N_{fy} = 2.0 × 10⁶. From Eq. 13.24, R_{Ly} = (2.0 × 10⁶ − 1.0 × 10⁶)/(2.0 × 10⁶) = 0.50 or 50%. Using R_{Lx} = 30% and R_{Ly} = 50%, from Figure 13.22, F_{RL} = 0.74.

From Eq. 13.27, the overlay thickness required is h_{OL} = (5.0 − 0.74 × 3.24)/ 0.40 = 6.5 in. (165 mm).

Flexible Overlays on Rigid Pavements

Depending on the conditions of the existing rigid pavement, either a normal structural overlay or a break and seat overlay can be used. The overlay equations to be used are shown in Table 13.10.

Normal Structural Overlay

Three different approaches can be used to determine the required thickness of asphalt overlay:

1. *NDT Method 1.* The elastic modulus of each layer is determined by NDT using the back-calculation technique. When E_{pcc} and D_0 are known, D_{xeff} can be obtained from Figure 13.16. The structural number $SN_{xeff-rp}$ for the subbase and the remaining pavement layers above the subgrade, if any, can be computed from the respective thickness, layer coefficient, and drainage coefficient. The overlay thickness required is determined by

$$h_{OL} = \frac{SN_{OL}}{a_{OL}} = \frac{SN_y - F_{RL}(0.8D_{xeff} + SN_{xeff-rp})}{a_{OL}} \tag{13.28}$$

2. *NDT Method 2.* The value of SN_{xeff} is determined directly based on the subgrade modulus, the total thickness of all layers above the subgrade, and the maximum NDT deflection. The required overlay thickness is determined by Eq. 13.27.

3. *Visual Condition Factor.* If it is physically impossible to utilize NDT, the visual condition factor C_v defined in Table 13.8 or 13.9 can be used to determine the structural layer coefficient a_{2r}, as shown in Figure 13.23. However, the structural number of the remaining layers $SN_{xeff-rp}$ still has to be estimated. The value of $SN_{xeff-rp}$ is usually much smaller than D_{xeff}, so a rough estimation of $SN_{xeff-rp}$ is sufficient. The required overlay thickness is determined by

$$h_{OL} = \frac{SN_{OL}}{a_{OL}} = \frac{SN_y - F_{RL}(a_{2r}D_0 + SN_{xeff-rp})}{a_{OL}} \tag{13.29}$$

TABLE 13.10 OVERLAY EQUATIONS FOR FLEXIBLE OVERLAY ON RIGID PAVEMENTS

Major overlay condition	Specific method used	SN_{OL} equation
Normal structural overlay	NDT method 1	$SN_{OL} = SN_y - F_{RL}(0.8D_{xeff} + SN_{xeff-rp})$
	NDT method 2	$SN_{OL} = SN_y - F_{RL}SN_{xeff}$
	Visual condition factor	$SN_{OL} = SN_y - F_{RL}(a_{2r}D_0 + SN_{xeff-rp})$
Break–seat overlay	Estimating nominal crack spacing	$SN_{OL} = SN_y - 0.7(0.4D_0 + SN_{xeff-rp})$
	Postcracking NDT	
	(a) NDT method 1	$SN_{OL} = SN_y - 0.7(a_{bs}D_0 + SN_{xeff-rp})$
	(b) NDT method 2	$SN_{OL} = SN_y - 0.7SN_{xeff}$

Source. From the *AASHTO Guide for Design of Pavement Structures.* Copyright 1986. American Association of State Highway and Transportation Officials, Washington, DC. Used by permission.

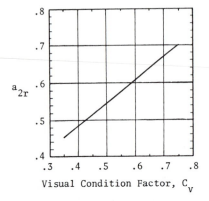

Figure 13.23 Structural layer coefficients of cracked PCC as a function of visual condition factor. (From the *AASHTO Guide for Design of Pavement Structures*. Copyright 1986. American Association of State Highway and Transportation Officials, Washington, DC. Used by permission.)

The overlay thickness obtained by one of the above approaches must be checked with the minimum thickness developed by the Asphalt Institute, as shown in Figure 13.7, to ensure that reflection cracking is kept to a minimum.

Example 13.13

Given the following information, determine the thickness of normal structural overlay required on an existing rigid pavement.

Existing pavement: thickness of PCC slab D_0 = 8.0 in. (203 mm), thickness of granular subbase D_{SB} = 6.0 in. (152 mm), subbase drainage coefficient m_3 = 1.0, and PSI at time of overlay p_{t1} = 2.2. The initial PSI, p_{01}, and the traffic to date x are not known. The moduli obtained from NDT are 1.75×10^6 psi (12 GPa) for PCC slab, 25,000 psi (173 MPa) for subbase, and 7500 psi (52 MPa) for subgrade. After analyzing the roadbed conditions throughout the year, an effective modulus M_R of 6000 psi (41 MPa) has been selected.

Overlaid pavement: initial PSI p_{02} = 4.2, PSI at end of overlay period p_{t2} = 2.5, ESAL during overlay period y = 15×10^6, and modulus of HMA for new overlay E_{OL} = 420,000 psi (2.9 GPa).

Other variables: design reliability R = 85%, standard deviation S_0 = 0.5, slab length = 40 ft (12.2 m), and maximum annual temperature differential = 46°F (26°C).

Solution: As shown by Eq. 13.28, the parameters to be determined are a_{OL}, D_{xeff}, $SN_{xeff-rp}$, SN_y, and F_{RL}. Given E_{OL} = 420,000 psi (2.9 GPa), from Figure 11.27, a_{OL} = 0.43.

With an E_{pcc} of 1.75×10^6 obtained from NDT and a D_0 of 8 in. (203 mm), from Figure 13.16, D_{xeff} = 5.66. From Eq. 11.46, the layer coefficient of subbase a_3 = $0.227 \times \log(25,000) - 0.839 = 0.16$. The value of $SN_{xeff-rp}$ is related only to the subbase, or $SN_{xeff-rp}$ = $0.16 \times 6.0 \times 1.0 = 0.96$. For a new flexible pavement with R = 85%, S_0 = 0.5, W_{18} = y = 15×10^6, M_R = 6000 psi (41 MPa), and ΔPSI = 4.2 − 2.5 = 1.7, SN_y = 5.6, as can be obtained from Figure 11.25.

Because no information regarding previous traffic is available, the NDT results will be used to estimate R_{Lx}. The structural condition factor C_x = D_{xeff}/D_0 = 5.66/8.0 = 0.71. From Figure 13.19, R_{Lx} = 8%. If Figure 13.21 based on a D_0 of 8 in. (203 mm) and a p_{t1} of 2.2 were used, nearly the same R_{Lx} would be obtained. The value of N_{fy} to reduce the PSI from 4.2 to 2.0, or ΔPSI = 4.2 − 2.0 = 2.2, can be determined from Figure 11.25 or computed more accurately by Eq. 11.37 as follows:

$\log N_{fy} = -1.037 \times 0.5 + 9.36 \times \log(5.6 + 1) - 0.2 + \log(2.2/2.7)/[0.4 + 1094/(5.6 + 1)^{5.19}] + 2.32 \log(6000) - 8.07 = 7.455$, or $N_{fy} = 28 \times 10^6$. From Eq. 13.24, $R_{Ly} = (28 \times 10^6 - 15 \times 10^6)/(28 \times 10^6) = 0.46$ or 46%. For $R_{Lx} = 8\%$ and $R_{Ly} = 46\%$, from Figure 13.22, $F_{RL} = 0.64$.

From Eq. 13.28, the overlay thickness required is $h_{OL} = (5.6 - 0.64 \times (0.8 \times 5.66 + 0.96)/0.43 = 4.86$ in. (123 mm). Use 5 in. (127 mm) of asphalt overlay.

The thickness determined by the above normal approach should be checked against the minimum thickness for reflection cracking. For a slab length of 40 ft (12.2 m) and a maximum annual temperature differential of 46°F (26°C), the minimum thickness of asphalt overlay is 6.5 in. (165 mm), as shown in Figure 13.7. Therefore, a thickness of 6.5 in. (165 mm), rather than the 5 in. (127 mm) based on the normal structural design, should be used.

It should be noted that if the slab length were 30 ft (9.1 m), the minimum overlay thickness for reflection cracking would be 4.6 in. (117 mm), so that an overlay thickness of 5 in. (127 mm) based on the normal structural design would control the final design.

Example 13.14

Same as Example 13.13 except that a visual condition factor C_v of 0.5 is assumed and used to determine the effective structural capacity.

Solution: In this example, Eq. 13.29 should be used to determine the overlay thickness. The values of $SN_y = 5.6$, $F_{RL} = 0.64$, and $SN_{xeff-rp} = 0.96$, as obtained in Example 13.13, are still applicable. The only value to be determined is a_{2r}. For a C_v of 0.5, from Figure 13.23, $a_{2r} = 0.55$. From Eq. 13.29, $h_{OL} = [5.6 - 0.64 \times (0.55 \times 8 + 0.96)]/0.43 = 5.0$ in. (127 mm).

Break and Seat Overlay

There are also three approaches for the design of break and seat overlay, one based on nominal crack spacing and the other two based on the NDT results after the pavement has been cracked. Because the broken pavement is transformed into a similar damaged state, a remaining life factor F_{RL}, of 0.7 is assumed.

1. *Nominal Crack Spacing.* This approach assumes that a nominal slab fragment size of approximately 30 in. (0.76 m) will be obtained. The effective structural number of the broken PCC layer is assumed to be $0.4D_o$. The information obtained from NDT method 1 is then used to back-calculate the in situ moduli of other pavement layers so $SN_{xeff-rp}$ can be determined. The required overlay thickness is obtained by

$$h_{OL} = \frac{SN_{OL}}{a_{OL}} = \frac{SN_y - 0.7(0.4D_o + SN_{xeff-rp})}{a_{OL}} \qquad (13.30)$$

2. *Postcracking NDT Method 1.* In some cases, it may be desirable to conduct NDT after the breaking operation. The modulus of the cracked PCC, E_{pcc}, can then be used to determine the in situ structural layer coefficient a_{bs} of the broken layer by Figure 13.24, while the moduli of the other layers can be used to determine $SN_{xeff-rp}$. The required overlay thickness is obtained by

$$h_{OL} = \frac{SN_{OL}}{a_{OL}} = \frac{SN_y - 0.7(a_{bs}D_o + SN_{xeff-rp})}{a_{OL}} \qquad (13.31)$$

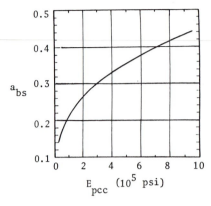

Figure 13.24 Structural layer coefficient of cracked PCC as a function of in situ elastic modulus (1 psi = 6.9 kPa). (From the *AASHTO Guide for Design of Pavement Structures.* Copyright 1986. American Association of State Highway and Transportation Officials, Washington, DC. Used by permission.)

3. *Postcracking NDT Method 2*. If NDT method 2 is used, the SN_{xeff} of the entire broken pavement is found and the required overlay thickness is obtained by

$$h_{OL} = \frac{SN_{OL}}{a_{OL}} = \frac{SN_y - 0.7SN_{xeff}}{a_{OL}} \tag{13.32}$$

Example 13.15

Same as Example 13.13 except that the break and seat method is used by breaking the slab into nominal fragments of approximately 30 in. (0.8 m) in size.

Solution: The thickness of the overlay can be determined by Eq. 13.30. All of the unknown variables have already been determined in Example 13.13 with $SN_y = 5.6$, $D_o = 8.0$ in. (203 mm), $SN_{xeff\text{-}rp} = 0.96$, and $a_{OL} = 0.43$. From Eq. 13.30, $h_{OL} = [5.6 - 0.7 \times (0.4 \times 8.0 + 0.96)]/0.43 = 6.25$ in. (159 mm). It can be seen that an asphalt overlay thickness of 6.25 (159 mm) would be required if the break and seat technique were used, compared to a direct overlay of 6.5 in. (165 mm).

Rigid Overlays on Rigid Pavements

For rigid overlays on rigid pavements, three types of overlay may be considered: fully bonded, partially bonded, and unbonded. The fully bonded overlay can be used only when the structural condition factor is greater than 0.75 and all defects can be repaired. The equations for determining the required PCC overlay thickness over an existing rigid pavement are shown in Table 13.7 and repeated below:

Bonded $\qquad\qquad D_{OL} = D_y - F_R(D_{xeff}) \tag{13.33}$

Partially bonded $\qquad D_{OL}^{1.4} = D_y^{1.4} - F_{RL}(D_{xeff})^{1.4} \tag{13.34}$

Unbonded $\qquad\quad D_{OL}^2 = D_y^2 - F_{RL}(D_{xeff})^2 \tag{13.35}$

D_{xeff} can be determined from the in situ PCC modulus obtained from NDT method 1 or from other non-NDT methods.

Example 13.16

The design variables are the same as those in Examples 13.13 and 13.14 except that an unbonded PCC overlay with a 2-in. (51-mm) HMA leveling course is used in place of the flexible overlay. Assuming that the new overlay has an elastic modulus E_c of 5×10^6 psi (35 GPa), a modulus of rupture S_c of 700 psi (4.8 MPa), an initial serviceability p_{o2} of 4.5, and a joint transfer coefficient J of 3.2, determine the thickness of rigid overlay required.

Solution: The existing pavement is in poor condition with a remaining life of only 8% and a visual condition factor of 0.5, so the only alternative is to use an unbonded overlay. To achieve the bond break, a 2-in. (51-mm) asphalt leveling course is to be placed on the existing rigid pavement. The appropriate equation to be used is Eq. 13.35.

To design a new rigid pavement on the existing subbase, it is necessary to determine the effective modulus of subgrade reaction. Using Figure 12.18 with $E_{SB} = 25,000$ psi (173 MPa), $M_R = 6000$ psi (41 MPa), and $D_{SB} = 6.0$ in. (152 mm), a k value of 350 pci (95 MN/m³) is obtained. Assuming that the roadbed soil is very deep and there is no loss of subgrade support, a k value of 350 pci (95 MN/m³) is selected for design purposes.

A D_{xeff} of 5.66 has already been determined from Example 13.13. A new overlay thickness D_y of 10.2 in. (259 mm) can be determined from Figure 12.17 based on the following design input values: $k = 350$ pci (95 MN/m³), $E_c = 5 \times 10^6$ (35 GPa), $S_c = 700$ psi (4.8 MPa), $J = 3.2$, $C_d = 1.0$, ΔPSI $= 4.5 - 2.5 = 2.0$, $R = 85\%$, $S_o = 0.5$, and ESAL $= 15 \times 10^6$.

To determine the remaining life factor F_{RL}, the values of R_{Lx} and R_{Ly} must be known. A R_{Lx} of 8% has already been determined from Example 13.13. The value of N_{fy} to reduce the PSI from 4.5 to 2.0, or ΔPSI $= 4.5 - 2.0 = 2.5$, can be determined from Figure 12.17 or computed more accurately by Eq. 12.21 as follows: $\log N_{fy} = -1.037 \times 0.5 + 7.35 \times \log(10.2 + 1) - 0.06 + \log(2.5/3.0)/[1 + 1.624 \times 10^7/ (10.2 + 1)^{8.46}] + (4.22 - 0.32 \times 2) \times \log\{[(700 \times 1.0)/(215.63 \times 3.2)] [(10.2)^{0.75} - 1.132]/[(10.2)^{0.75} - 18.42/(5 \times 10^6/350)^{0.25}]\} = -0.519 + 7.712 - 0.06 - 0.078 + 0.223 = 7.278$, or $N_{fy} = 19 \times 10^6$. From Eq. 13.24, $R_{Ly} = (19 \times 10^6 - 15 \times 10^6)/(19 \times 10^6) = 0.21$ or 21%. For $R_{Lx} = 8\%$ and $R_{Ly} = 21\%$, from Figure 13.22, $F_{RL} = 0.58$.

The D_{xeff} value of 5.66 does not include the effect of the 2-in. (51-mm) asphalt leveling course. By assuming a 2:1 substitution ratio for asphalt to concrete, the effective thickness can be increased by 1 in. (51 mm), or $D_{xeff} = 5.66 + 1.0 = 6.7$ in. From Eq. 13.35, $D_{OL}^2 = (10.2)^2 - 0.58 \times (6.7)^2 = 78.0$, or $D_{OL} = 8.83$ in. (224 mm). Use 9 in. (229 mm) unbonded PCC overlay on 2 in. (51 mm) of HMA leveling course. Compared with Example 13.13, the thickness of rigid overlay is much greater than that of the flexible overlay.

Example 13.17

A recently constructed jointed plain concrete pavement has experienced surface distress. Condition surveys indicate that large areas have medium to high severity levels of crazing, map cracking, and scaling. The rehabilitation strategy is to surface grind 1.5 in. (38 mm) of PCC layer and then apply a bonded PCC overlay. Because the pavement is relatively new, no in situ testing is contemplated. Given the following information, determine the thickness of bonded overlay required.

Existing pavement: thickness of PCC slab D_o = 11 in. (279 mm), combined modulus of subgrade reaction k = 300 pci (81 MN/m³), initial serviceability p_{o1} = 4.5, serviceability at time of overlay p_{t1} = 4.25, ESAL to date x = 0.5 × 10⁶, concrete modulus of elasticity E_c = 6 × 10⁶ psi (41 GPa), concrete modulus of rupture S_c = 700 psi (4.8 MPa), load transfer coefficient J = 3.2, drainage coefficient = 1.0, design reliability R = 95%, and standard deviation S_o = 0.5.

Overlaid pavement: initial serviceability p_{o2} = 4.5, serviceability at end of overlay p_{t2} = 2.5, and future overlay ESAL y = 15 × 10⁶. All material properties and design variables are the same as those of the existing pavement.

Solution: The thickness of bonded overlay depends on D_y, F_R, and D_{xeff}, as shown by Eq. 13.33. Without NDT testing, D_{xeff} can be determined from the structural condition factor C_x by estimating the remaining life R_{Lx} of the existing pavement. To compute R_{Lx}, the value of N_{fx} to reduce the PSI from 4.5 to 2.0, or ΔPSI = 4.5 − 2.0 = 2.5, must be determined. The N_{fx} value can be obtained from Figure 12.17 or more accurately by Eq. 12.21 as follows: $\log N_{fx}$ = −1.645 × 0.5 + 7.35 × log(11.0 + 1) − 0.06 + log(2.5/3.0)/[1 + 1.624 × 10⁷/(11.0 + 1)^{8.46}] + (4.22 − 0.32 × 2) × log{[(700 × 1.0)/(215.63 × 3.2)] [(11.0)^{0.75} − 1.132]/[(11.0)^{0.75} − 18.42/(6 × 10⁶/300)^{0.25}]} = −0.823 + 7.932 − 0.06 − 0.078 + 0.160 = 7.131, or N_{fx} = 13.5 × 10⁶. From Eq. 13.21, R_{Lx} = (13.5 × 10⁶ − 0.5 × 10⁶)/(13.5 × 10⁶) = 0.963. From Figure 13.19, C_x = 0.99. If Figure 13.21 based on the serviceability approach were used to determine the remaining life, R_{Lx} = 0.925 and, from Figure 13.19, C_x = 0.985. The D_{xeff} value is simply $C_x D_o$. However, since it is the rehabilitation strategy to grind 1.5 in. (38 mm) of the PCC layer, the D_{xeff} at the time of the overlay is D_{xeff} = 0.99 × (11 − 1.5) = 9.4 in. (239 mm).

The structural capacity D_y is determined by considering the pavement as a new design. The input is as previously noted except that p_{t2} = 2.5 and y = 15 × 10⁶. Using Figure 12.17, a D_y of 11.7 in. (297 mm) is obtained.

The F_{RL} value can be determined from values of R_{Lx} and R_{Ly}. Although R_{Lx} has been determined previously, it can be used to relate only to the condition factor C_x and not to the remaining life factor F_{RL}, because the existing pavement will be reduced 1.5 in. (38 mm) by diamond grinding. Therefire, R_{Lx} should be recalculated based on a new N_{fx}, defined as

$$N_{fx} = x + N'_{fx} \tag{13.36}$$

in which N'_{fx} = ESAL repetitions to reduce the PSI of the ground pavement from 4.5 to 2.0. The values of N'_{fx} can be determined from Figure 12.17 using an effective slab thickness D_{xeff} of 9.4 in. (239 mm) or computed more accurately by Eq. 12.21 as follows: $\log N'_{fx}$ = −1.645 × 0.5 + 7.35 × log(9.4 + 1) − 0.06 + log(2.5/3.0)/ [1 + 1.624 × 10⁷/(9.4 + 1)^{8.46}] + (4.22 − 0.32 × 2) × log{[(700 × 1.0)/ (215.63 × 3.2)][(9.4)^{0.75} − 1.132]/[(9.4)^{0.75} − 18.42/(6 × 10⁶/300)^{0.25}]} = −0.823 + 7.475 − 0.06 − 0.076 + 0.183 = 6.699, or N'_{fx} = 5 × 10⁶. From Eq. 13.36, N_{fx} = 0.5 × 10⁶ + 5 × 10⁶ = 5.5 × 10⁶. From Eq. 13.21, R_{Lx} = (5.5 × 10⁶ − 0.5 × 10⁶)/(5.5 × 10⁶) = 0.91 or 91%. The value of N_{fy} can be determined similarly by simply replacing the 9.4-in. (239-mm) slab with an 11.7-in. (297-mm) slab and was found to be 20 × 10⁶. From Eq. 13.24, R_{Ly} = (20 × 10⁶ − 15 × 10⁶)/(20 × 10⁶) = 0.25 or 25%. With R_{Lx} = 91% and R_{Ly} = 25%, from Figure 13.22, F_{Ly} = 0.98.

The required bonded overlay is D_{OL} = 11.7 − 0.98 × 9.4 = 2.5 in. (64 mm).

Rigid Overlays on Flexible Pavements

The design of rigid overlays on flexible pavements is straightforward because the existing flexible pavement can be viewed as the composite foundation support for a new rigid pavement, as indicated in Table 13.7. The design is simply to determine a composite modulus of subgrade reaction k, so the design procedure for new PCC pavements, as discussed in Section 12.3, can be used. Nondestructive deflection testing may be used to evaluate the k value, which is described in the AASHTO design guide but is not presented here.

13.6 SUMMARY

This chapter presents several methods for overlay design, including the Asphalt Institute method, the Portland Cement Association method, and the AASHTO method.

Important Points Discussed in Chapter 13

1. Either HMA or PCC can be used as an overlay on an existing flexible or rigid pavement. If both the overlay and the existing pavement are flexible, then the layer elastic programs can be used as a mechanistic design model. If either the overlay or the existing pavement is rigid, then the finite element plate programs should be used. By applying stress adjustment factors for edge and corner loading, the layer elastic program has also been used for the design of HMA overlays on PCC pavements.

2. A major problem in the design of HMA overlays on PCC pavements is the occurrence of reflection cracking. The most common methods to minimize reflection cracking include using a thicker overlay, cracking and seating the existing PCC slabs, incorporating a crack relief layer, and sawing and sealing joints on the HMA overlay. However, there has been some controversy on the effectiveness of cracking and seating. It was reported that cracking and seating only delayed rather than eliminated the reflection cracking. Also the AASHTO design method assumes that cracking and seating cause a decrease in the structural capacity of the existing pavement, but a recent study conducted by Pennsylvania State University indicated that there was no significant loss of structural support or decrease in the modulus of elasticity on the crack and seat pavement sections.

3. Three general methods can be used for overlay design: the effective thickness approach, the deflection approach, and the mechanistic–empirical approach. In the effective thickness approach, the thickness of the existing pavement is reduced to an effective thickness depending on the conditions of its component layers. The difference between the required thickness of a new pavement and the effective thickness of the existing pavement is the thickness of overlay required. In the deflection approach, the thickness of overlay should be sufficient to reduce the pavement deflection to a tolerable amount, usually derived empirically based on the amount of traffic. In the mechanistic–empirical ap-

proach, the critical stresses or strains in the existing and the overlaid pavements are determined by mechanistic models and the damage due to traffic loads and environmental causes is evaluated to ensure that failures will not occur.

4. The Asphalt Institute employs both the effective thickness and the deflection methods for the design of HMA overlays on both flexible and rigid pavements. In the effective thickness method, the conversion factors based on the conditions listed in Table 13.3 must be applied to each layer to obtain an effective thickness equivalent to the thickness of new HMA. The difference between the required thickness of a new full-depth pavement and the effective thickness of the existing pavement is the thickness of overlay required. If the existing pavement is full depth, the PSI of the existing pavement at the time of overlay can also be used directly for determining the conversion factor, as shown in Figure 13.2. In the deflection method, separate procedures are used depending on whether the existing pavement is flexible or rigid. For flexible pavements, the overlay thickness can be obtained from Figure 13.5 as a function of the representative rebound deflection, RRD, and the ESAL repetitions during the overlay period. For rigid pavements, unless special techniques described in Section 13.1.2 are used to minimize reflection cracking, the thickness of HMA overlay depends on the slab length and the maximum annual temperature differential, as shown in Figure 13.7. Prior to the overlay, deflection measurements by Benkelman beam should be made at the joints to determine both the mean deflection and the differential deflection. After the overlay, the mean deflection must be reduced to less than 0.014 in. (0.36 mm) and the differential deflection to less than 0.002 in. (0.05 mm). The decrease in deflection after the overlay can be estimated by assuming that each inch of HMA overlay will reduce the deflection by 5%. If the mean and differential deflections cannot be reduced below the above values, undersealing or stabilization of the subgrade will be required.

5. The Portland Cement Association has published design charts for both unbonded and bonded PCC overlays on existing rigid pavements. The method is considered mechanistic–empirical because the JSLAB finite element computer program was used to prepare the design charts. Because of the many arbitrary assumptions employed, the method is more empirical than mechanistic. For example, the modulus of existing PCC pavement in the unbonded pavement system is arbitrarily assumed to be 3×10^6 to 4×10^6 psi (21 to 28 GPa), regardless of the actual fatigue damage caused by the load repetitions prior to the overlay. Furthermore, the concept of providing an overlaid pavement that is structurally equivalent to a new full-depth pavement is approximate at best because the equivalency depends on a large number of factors, such as the k value and type of loading. Another inconsistency is the use of different load groups to design a new pavement but the equivalent 18-kip (80-kN) single-axle load to relate the overlay thickness to the new pavement thickness. However, the PCA method does provide a simple procedure for the design of both unbonded and bonded overlays. Design charts for unbonded overlays are presented in Figures 13.10, 13.11, and 13.12 for three different cases. Some judgment on the condition of existing pavement is needed to select the appropriate chart for use. The design chart for bonded overlays is shown in Figure 13.14.

6. The AASHTO Design Guide provides the most comprehensive overlay design procedure based on the effective thickness approach. The basic equation is

$$(SC_{OL})^n = (SC_y)^n - F_{RL}(SC_{xeff})^n \tag{13.18}$$

in which SC_y is the structural capacity of a new pavement to carry y repetitions of ESAL during the overlay period, SC_{xeff} is the effective structural capacity of the existing pavement after x repetitions of ESAL prior to the overlay, F_{RL} is the remaining life factor, and n is an exponent with a value of 1 for most overlays except for the unbonded or partially bonded PCC overlays on PCC pavements. The determination of SC_y for new pavements is described in Sections 11.3 and 12.3. The value of SC_{xeff} for existing flexible pavements can be determined only by nondestructive testing to back-calculate the resilient modulus of each layer, while that for rigid pavements can be determined by either the NDT or the non-NDT method. The three non-NDT procedures are visual condition factor, nominal size of PCC slab fragments, and remaining life. The value of F_{RL} depends on the remaining life value of the existing pavement R_{Lx} and the remaining life value of the overlaid pavement R_{Ly}. The value of R_{Lx} can be determined from NDT, traffic, time, serviceability, or visual condition survey, but only the traffic approach can be used to determine R_{Ly}. Equation 13.18 can be applied to all four types of overlay systems as described below. The major difference in procedure is how to determine SC_{xeff}.

7. For flexible overlays on flexible pavements, the overlay thickness is determined by

$$h_{OL} = \frac{SN_{OL}}{a_{OL}} = \frac{SN_y - F_{RL}SN_{xeff}}{a_{OL}} \tag{13.27}$$

SN_{xeff} is computed from the resilient modulus of each layer obtained by NDT.

8. For flexible overlays on rigid pavements, the design methods can be divided into two categories: the normal structural overlay method and the break and seat overlay method. The overlay thickness obtained by the normal structural overlay method must be checked against the minimum thickness shown in Figure 13.7 to minimize reflection cracking.

9. In the normal structural overlay method, three different approaches can be used to determine SC_{xeff}: NDT method 1, NDT method 2, and visual condition factor. When NDT method 1 is used, the overlay thickness is determined by

$$h_{OL} = \frac{SN_{OL}}{a_{OL}} = \frac{SN_y - F_{RL}(0.8D_{xeff} + SN_{xeff-rp})}{a_{OL}} \tag{13.28}$$

The value of SC_{xeff} is separated into two parts, one due to the PCC effective thickness D_{xeff} and the other due to the effective structural number of the remaining layers $SN_{xeff-rp}$. Both D_{xeff} and $SN_{xeff-rp}$ are determined from the resilient modulus of each layer obtained from NDT method 1. When NDT method 2 is used, the overlay thickness is determined by

$$h_{OL} = \frac{SN_{OL}}{a_{OL}} = \frac{SN_y - F_{RL}SN_{xeff}}{a_{OL}} \tag{13.27}$$

The value of SN_{xeff} is determined directly from NDT method 2 based on the subgrade resilient modulus, the maximum deflection under load, and the total thickness of pavement above the subgrade. When the visual condition factor is used, the overlay thickness is determined by

$$h_{OL} = \frac{SN_{OL}}{a_{OL}} = \frac{SN_y - F_{RL}(a_{2r}D_0 + SN_{xeff\text{-}rp})}{a_{OL}} \tag{13.29}$$

Similar to Eq. 13.28, SC_{xeff} is separated into two parts. However, the structural capacity of PCC slab is determined from the visual condition factor, rather than from NDT method 1, and the value of $SN_{xeff\text{-}rp}$ has to be estimated.

10. In the break and seat method, three approaches are available: one based on nominal crack spacing and the other two based on the NDT results after the pavement has been cracked. When the nominal crack spacing is used, the overlay thickness is determined by

$$h_{OL} = \frac{SN_{OL}}{a_{OL}} = \frac{SN_y - 0.7(0.4D_0 + SN_{xeff\text{-}rp})}{a_{OL}} \tag{13.30}$$

The value of $SN_{xeff\text{-}rp}$ can be determined from NDT method 1 prior to breaking and seating. When postcracking NDT method 1 is used, the thickness of overlay is determined by

$$h_{OL} = \frac{SN_{OL}}{a_{OL}} = \frac{SN_y - 0.7(a_{bs}D_0 + SN_{xeff\text{-}rp})}{a_{OL}} \tag{13.31}$$

The value of $SN_{xeff\text{-}rp}$ is determined from NDT method 1 after the pavement has been cracked. When postcracking NDT method 2 is used, the overlay thickness is determined by

$$h_{OL} = \frac{SN_{OL}}{a_{OL}} = \frac{SN_y - 0.7SN_{xeff}}{a_{OL}} \tag{13.32}$$

The value of SN_{xeff} is obtained from NDT method 2 after the pavement has been cracked.

11. For rigid overlays on rigid pavements, three types of overlays are available: fully bonded, partially bonded, and unbonded. The fully bonded overlay can be applied only when the existing slabs is in a very good condition and all surface and structural defects have been repaired. The equations to be used for these three types of overlay are

Bonded $\qquad\qquad\qquad D_{OL} = D_y - F_{RL}(D_{xeff}) \tag{13.33}$

Partially Bonded $\qquad\quad D_{OL}^{1.4} = D_y^{1.4} - F_{RL}(D_{xeff})^{1.4} \tag{13.34}$

Unbonded $\qquad\qquad\quad D_{OL}^2 = D_y^2 - F_{RL}(D_{xeff})^2 \tag{13.35}$

D_{xeff} can be determined from the in situ PCC modulus obtained from NDT method 1 or from other non-NDT methods such as visual condition factor, nominal size of slab fragments, and remaining life.

12. The design of rigid overlays on flexible pavements is straightforward because the existing flexible pavement can be viewed as the composite foundation support for a new rigid pavement. The design is simply to determine a composite

modulus of subgrade reaction k, so the design procedure for new PCC pavements can be used. Nondestructive deflection testing may be used to evaluate the composite k value, as described in the AASHTO design guide.

PROBLEMS

13-1. A flexible pavement is composed of a 4-in. HMA surface and a 12-in. crushed stone base. Deflection measurements were made in the spring using a Benkelman beam. The mean deflection was found to be 0.034 in. with a standard deviation of 0.0041 in. The mean pavement temperature during the test was 80°F. Determine the thickness of HMA overlay required for a design ESAL of 10 million using the Asphalt Institute's equations and compared with the result by the design chart. [Answer: 4.5 in.]

13-2. The representative rebound deflection on a flexible pavement is 0.055 in. If the traffic growth rate is 6% and the pavement is to be used for five years before the placement of a new overlay, what is the maximum allowable ESAL during the current year? [Answer: 30,100]

13-3. A CRCP has a crack spacing of 10 ft and a temperature differential of 56°F. Measured edge deflection by dynaflect is 0.0007 in. By the Asphalt Institute procedure, determine the required thickness of asphalt overlay and check whether undersealing is required. [Answer: 4 in., undersealing not required]

13-4. An existing 8-in. PCC slab is in fairly good condition but does not have sufficient structural capacity to carry the anticipated traffic. Twelve core samples were taken from the existing pavement at midslab about 2 ft from the edge and the following splitting tensile strengths were found: 500, 453, 554, 450, 513, 488, 512, 468, 532, 520, 420, and 556 psi. The thickness of a new slab to carry the anticipated traffic is 11 in. Determine the thickness of overlay required by the PCA method for both bonded and unbonded cases. [Answer: 3.5 and 9 in.]

13-5. A 10-year-old pavement with a PCC thickness of 8 in. has been subjected to 2×10^6 ESAL repetitions on the design lane. It is estimated that the pavement can take another 3.3×10^6 ESAL before it is considered a failure with $p_f = 2.0$. Determine the remaining life R_{Lx} of the pavement. If the remaining life based on the above traffic approach is also valid for all other approaches, estimate the structural condition factor C_x, the effective thickness D_{xeff}, the in situ modulus of PCC, E_{pcc}, the visual condition factor C_v, the nominal size of slab fragment, the serviceability p_{t1} of the pavement at the end of 10 years, and the probable life T_f for a traffic growth rate of 5%. [Answer: $R_{Lx} = 62.3\%$, $C_x = 0.92$, $D_{xeff} = 7.36$, $E_{pcc} = 3.8 \times 10^6$ psi, $C_v = 0.8$, fragment size = 106 in., $p_{t1} = 3.7$, and $T_f = 20$ years]

13-6. Nondestructive tests were conducted on an existing flexible pavement consisting of a 4-in. HMA, an 8-in. crushed stone base, and a 10-in. gravel subbase. The average resilient moduli obtained from the NDT tests during early summer are 405,000 psi for HMA, 27,000 psi for base, 11,000 psi for subbase, and 7500 psi for subgrade. After considering the varying subgrade conditions throughout the year, an effective roadbed soil resilient modulus was determined to be 5000 psi. The quality of drainage is fair for the crushed stone base and poor for the gravel subbase with more than 25% of the time under saturation. The initial structural number SN_0 of the existing pavement is 3.3 and its initial serviceability p_{01} is 4.2. The serviceability is 2.6 prior to overlay and 4.3 after the overlay. The modulus of new HMA is 450,000 psi at 70°F and the design ESAL for the overlay based on a terminal serviceability p_{t2} of 2.5 is 5 million. Assuming a reliability of 90% and a standard deviation of 0.35, determine the thickness of asphalt overlay required by the AASHTO procedure. [Answer: 6.5 in.]

13-7. NDT tests were conducted on an existing JPCP consisting of a 9-in. slab and a 6-in. granular subbase. The average back-calculated PCC moduli are 2×10^6 psi near working cracks and 5×10^6 psi on uncracked areas. The subbase resilient modulus is 15,000 psi and the effective roadbed soil resilient modulus is 5000 psi. The drainage quality of the subbase is fair with more than 25% of time under saturation. The slab length is 20 ft and the maximum annual temperature differential is 60°F. The dynamic modulus of new HMA is 450,000 psi at 70°F and the overlay design ESAL is 10×10^6. The initial serviceability p_{02} of the overlaid pavement is 4.2 and the terminal serviceability p_{t2} is 2.5. Assuming a design reliability of 85% and a standard deviation of 0.35, determine the thickness of asphalt overlay required by the AASHTO method. [Answer: 4 in.]

13-8. Same as Problem 13-7, except that the crack and seat method is used by breaking the slab into nominal fragments of approximately 30 in. in size. Determine the thickness of asphalt overlay required. [Answer: 6 in.]

13-9. Same as Problem 13-7 except that an unbonded PCC overlay with a 1.5 in. HMA leveling course is to be used. Determine the thickness of unbonded overlay required. To find the required thickness of a new rigid pavement, the following input values are assumed: $E_c = 5 \times 10^6$ psi, $S_c = 650$ psi, $p_{02} = 4.5$, $p_{t2} = 2.5$, $C_d = 0.9$, and $J = 3.2$. [Answer: 8.5 in.]

13-10. Same as Problem 13-7 except that the existing PCC has an elastic modulus of 4.5×10^6 psi and that a bonded PCC overlay is used. Determine the thickness of bonded overlay required. To find the required thickness of a new rigid pavement, the following input values are assumed: $E_c = 5 \times 10^6$ psi, $S_c = 650$ psi, $p_{02} = 4.5$, $p_{t2} = 2.5$, $C_d = 0.9$, and $J = 3.2$. [Answer: 2.5 in.]

Theory
of Viscoelasticity

This appendix is a supplement to Section 2.3 on viscoelastic solutions and Section 7.2.1 on dynamic complex modulus. It discusses the use of differential operators to characterize viscoelastic materials, the application of the elastic–viscoelastic correspondence principle by using the Laplace transforms, the conversion from creep compliance to the mechanical model by the method of successive residuals, and the change from the complex modulus to creep compliance by Fourier transforms. Readers with only basic calculus and no prior knowledge of these transforms should still have no difficulty in following the presentation. This appendix is presented to give an insight into the classical theory of viscoelasticity.

A.1 DIFFERENTIAL OPERATORS

To facilitate the application of Laplace transforms, the mechanical models may be converted to differential operators. Let $\partial/\partial t = D = $ differential operators.

For an elastic spring $\qquad\qquad\qquad \sigma = E\epsilon \qquad$ or $\dfrac{\sigma}{\epsilon} = E \qquad\qquad$ (A.1)

For a viscous dashpot $\qquad\qquad\quad \sigma = \lambda\dfrac{\partial\epsilon}{\partial t} \qquad$ or $\dfrac{\sigma}{\epsilon} = \lambda D \qquad\qquad$ (A.2)

For a Maxwell model with a spring and a dashpot in series, the total strain is equal to the sum of the two:

$$\epsilon = \frac{\sigma}{E_0} + \frac{\sigma}{\lambda_0 D}$$

$$\frac{\sigma}{\epsilon} = \frac{1}{1/E_0 + 1/(\lambda_0 D)} = \frac{E_0 T_0 D}{T_0 D + 1} \tag{A.3}$$

in which $T_0 = \lambda_0/E_0$. Note that E_0 and $\lambda_0 D$ are the stress–strain ratio of the spring and dashpot, respectively. When two models are connected in series, the stress–strain ratio is the reciprocal of the sum of the reciprocal of the two stress–strain ratios. For a Kelvin model with a spring and a dashpot in parallel, the total stress is equal to the sum of the two:

$$\sigma = E_1 \epsilon + \lambda D \epsilon$$

or
$$\frac{\sigma}{\epsilon} = E_1 + \lambda_1 D = E_1(T_1 D + 1) \tag{A.4}$$

For a generalized model with one Maxwell model and n Kelvin models in series, by the above reciprocal principle

$$\frac{\sigma}{\epsilon} = \frac{1}{\dfrac{T_0 D + 1}{E_0 T_0 D} + \displaystyle\sum_{i=1}^{n} \dfrac{1}{E_i(T_i D + 1)}} \tag{A.5}$$

Example A.1:

Express the stress–strain ratio of the model shown in Figure 2.35a as differential operators.

Solution: Given $E_0 = 2$, $T_0 = 5$, $E_1 = 10$, $T_1 = 10$, $E_2 = 5$, $T_2 = 1$, $E_3 = 1$, and $T_3 = 0.1$, from Eq. A.5

$$\frac{\sigma}{\epsilon} = \frac{1}{\dfrac{5D + 1}{10D} + \dfrac{1}{10(10D + 1)} + \dfrac{1}{5(D + 1)} + \dfrac{1}{0.1D + 1}}$$

or

$$E = \frac{\sigma}{\epsilon} = \frac{10D^4 + 111D^3 + 111D^2 + 10D}{5D^4 + 158.6D^3 + 197.9D^2 + 29.1D + 1} \tag{A.6}$$

It can be seen that Young's modulus of viscoelastic materials can be expressed as the quotient of two polynomials in terms of the differential operator D. The use of differential operations to describe the stress–strain relationship is for convenience only. Otherwise, Eq. A.6 should be written as

$$5\frac{\partial^4 \sigma}{\partial t^4} + 158.6\frac{\partial^3 \sigma}{\partial t^3} + 197.9\frac{\partial^2 \sigma}{\partial t^2} + 29.1\frac{\partial \sigma}{\partial t} + \sigma$$

$$= 10\frac{\partial^4 \epsilon}{\partial t^4} + 111\frac{\partial^3 \epsilon}{\partial t^3} + 111\frac{\partial^2 \epsilon}{\partial t^2} + 10\frac{\partial \epsilon}{\partial t} \tag{A.7}$$

Example A.2

Figure A.1 shows a viscoelastic model. Derive the governing differential equation of the model by using differential operators. Integrate the equation for the following two cases: (a) find the strain–time relationship when a constant stress σ_0 is applied to the model and (b) find the stress–time relationship when a constant strain ϵ_0 is applied to the model.

Solution: The stress–strain relationship of an elastic spring is indicated by Eq. A.1 and that of a Maxwell model by Eq. A.3. When both are connected in parallel, as shown in Figure A.1

$$\frac{\sigma}{\epsilon} = E_1 + \frac{E_2 T_2 D}{T_2 D + 1} = \frac{T_2(E_1 + E_2)D + E_1}{T_2 D + 1} \qquad (A.8)$$

The governing differential equation is

$$T_2 \frac{\partial \sigma}{\partial t} + \sigma = T_2(E_1 + E_2) \frac{\partial \epsilon}{\partial t} + E_1 \epsilon \qquad (A.9)$$

(a) When a constant stress σ_0 is applied, the model experiences an instantaneous strain, $\sigma_0/(E_1 + E_2)$, due to the stretching of the two springs. If $\sigma = \sigma_0$, from Eq. A.9

$$\sigma_0 = T_2(E_1 + E_2) \frac{d\epsilon}{dt} + E_1 \epsilon$$

$$\int_{\sigma_0/(E_1+E_2)}^{\epsilon} \frac{d\epsilon}{\sigma_0 - E_1 \epsilon} = \frac{1}{T_2(E_1 + E_2)} \int_0^t dt$$

$$-\frac{1}{E_1} \ln (\sigma_0 - E_1 \epsilon) \Big|_{\sigma_0/(E_1+E_2)}^{\epsilon} = \frac{t}{T_2(E_1 + E_2)}$$

$$\epsilon = \frac{\sigma_0}{E_1} \left\{ 1 - \frac{E_2}{E_1 + E_2} \exp \left[- \frac{E_1 t}{T_2(E_1 + E_2)} \right] \right\} \qquad (A.10)$$

(b) When a constant strain ϵ_0 is applied, the model experiences an instantaneous stress, $(E_1 + E_2) \epsilon_0$. If $\epsilon = \epsilon_0$, from Eq. A.9

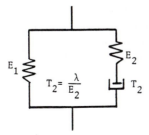

Figure A.1 Example A.2.

$$\int_{(E_1+E_2)\epsilon_0}^{\sigma} \frac{d\sigma}{E_1\epsilon_0 - \sigma} = \frac{1}{T_2} \int_0^t dt$$

$$- \ln (E_1\epsilon_0 - \sigma)\Big|_{(E_1+E_2)\epsilon_0}^{\sigma} = \frac{t}{T_2}$$

$$\sigma = \epsilon_0 \left[E_1 + E_2 \exp\left(-\frac{t}{T_2}\right) \right] \qquad (A.11)$$

A.2 ELASTIC–VISCOELASTIC CORRESPONDENCE PRINCIPLE

Instead of using a Dirichlet series to fit the responses at various times, as presented in Section 2.3.2, the elastic–viscoelastic principle can be applied to obtain the viscoelastic solutions from the elastic solutions. In this traditional method, Laplace transform is applied to remove the time variable t with a transformed variable p, thus changing a viscoelastic problem into an associated elastic problem. The Laplace inversion of the associated elastic problem from the transformed variable p to the time variable t results in the viscoelastic solutions. For simple problems, this method can give directly the closed form solutions. However, for viscoelastic layered systems, it is not possible to obtain Laplace inversion and an approximate method of collocation must be used. Because the collocation with respect to the transformed variable p does not offer any advantage over that with respect to time t, the Laplace transform was not used in the development of KENLAYER.

A.2.1 Laplace Transform

Instead of stresses, strains, and loads, the Laplace transform of stresses, strains, and loads is considered in viscoelastic analysis. When the Laplace transform is applied, the time variable t is removed and replaced by the transformed variable p, thus changing the viscoelastic problem into an elastic problem. The Laplace inversion of p to t results in the viscoelastic solution. The Laplace transform is defined as

$$L[F(t)] = \int_0^\infty F(t)e^{-pt}\,dt \qquad (A.12)$$

in which $F(t)$ is a function of time t, and p is a transformed variable. After integrating Eq. A.12 and substituting the limits of t, F becomes a function of p.

If the applied load q is a constant independent of time, the Laplace transform of q is

$$\bar{q} = L(q) = \int_0^\infty qe^{-pt}\,dt = -\frac{q}{p}e^{-pt}\Big|_0^\infty = \frac{q}{p} \qquad (A.13)$$

in which p is a transformed variable. A bar on top of a variable indicates the Laplace transform of the variable.

The stress–strain relationship of viscoelastic materials can be expressed in the form shown by Eq. A.7. To remove the time variable t, Laplace transforms are applied to the stress or strain on each side of the equation. Take the first derivative $d\sigma/dt$, for example:

$$L\left(\frac{d\sigma}{dt}\right) = \int_0^\infty \frac{d\sigma}{dt}\, e^{-pt}\, dt = \int_0^\infty e^{-pt}\, d\sigma$$

$$= \sigma\, e^{-pt}\Big|_0^\infty - \int_0^\infty \sigma\, d(e^{-pt})$$

If the material is initially undisturbed, or $\sigma = 0$ at $t = 0$, the first term is equal to zero, or

$$L\left(\frac{d\sigma}{dt}\right) = p \int_0^\infty \sigma\, e^{-pt}\, dt = p\bar{\sigma} \tag{A.14}$$

Equation A.14 indicates that the Laplace transform of $\partial\sigma/\partial t$, or $D\sigma$, is simply the replacement of D by p and σ by $\bar{\sigma}$. For the second derivative

$$L\left(\frac{d^2\sigma}{dt^2}\right) = \int_0^\infty \frac{d^2\sigma}{dt^2}\, e^{-pt}\, dt = \int_0^\infty e^{-pt}\, d\left(\frac{d\sigma}{dt}\right)$$

$$= \frac{d\sigma}{dt}\, e^{-pt}\Big|_0^\infty - \int_0^\infty \frac{d\sigma}{dt}\, d(e^{-pt})$$

If the material is initially undisturbed, or $d\sigma/dt = 0$ at $t = 0$, the first term is equal to zero, or

$$L\left(\frac{d^2\sigma}{dt^2}\right) = p \int_0^\infty \frac{d\sigma}{dt}\, e^{-pt}\, dt = p^2\, \bar{\sigma} \tag{A.15}$$

As can be seen, the Laplace transform for the second derivative requires the replacement of D^2 by p^2. The same can be applied to higher derivatives by replacing D^n by p^n. Consequently, the Laplace transform of Young's modulus, as indicated by Eq. A.6, becomes

$$E = \frac{\bar{\sigma}}{\bar{\epsilon}} = \frac{10p^4 + 111p^3 + 111p^2 + 10p}{5p^4 + 158.6p^3 + 197.9p^2 + 29.1p + 1} \tag{A.16}$$

When a viscoelastic material is characterized by a generalized model and a computer is used for the analysis, it is more convenient to apply Eq. A.5 directly:

$$\bar{E} = \frac{1}{\dfrac{T_0p + 1}{E_0T_0p} + \displaystyle\sum_{i=1}^{n} \dfrac{1}{E_i(T_ip + 1)}} \tag{A.17}$$

The Laplace transforms of $e^{-t/T}$ is also of interest and can be determined by

$$L\left(e^{-t/T}\right) = \int_0^\infty e^{-t/T} \, e^{-pt} \, dt = \int_0^\infty e^{-[(Tp + 1)/T]t} \, dt$$

$$= -\frac{T}{Tp + 1} \, e^{-[(Tp + 1)/T]t} \bigg|_0^\infty = \frac{T}{Tp + 1} \qquad (A.18)$$

Example A.3

Figure A.2 shows a viscoelastic homogeneous half-space under a circular loaded area. The half-space has a Young's modulus characterized by a Kelvin model and a constant Poisson ratio v. Derive an expression for the vertical deflection at the center of the loaded area as a function of time.

Solution: The elastic solution can be obtained from Eq. 2.8

$$w_0 = \frac{2(1 - v^2)q \, a}{E} \qquad (2.8)$$

Consider the Laplace transforms of deflection, load, and Young's modulus

$$\overline{w}_0 = \frac{2(1 - v^2)\overline{q} \, a}{\overline{E}} \qquad (A.19)$$

From Eq. A.13, $\overline{q} = q/p$ and from Eq. A.17, with only one Kelvin element, $\overline{E}_1 = E_1(T_1p + 1)$, so Eq. A.19 becomes

$$\overline{w}_0 = \frac{2(1 - v^2)q \, a}{E_1 p(T_1p + 1)} = \frac{2(1 - v^2)q \, a}{E_1}\left(\frac{1}{p} - \frac{T_1}{T_1p + 1}\right) \qquad (A.20)$$

Equation A.20 can be easily inverted from p to t because $1/p$ can be inverted to 1, as indicated by Eq. A.13, and $T_1/(T_1p + 1)$ can be inverted to e^{-t/T_1}, as indicated by Eq. A.18:

$$w_0 = \frac{2(1 - v^2)q \, a}{E_1}\left[1 - \exp\left(-\frac{t}{T_1}\right)\right] \qquad (A.21)$$

When the Poisson ratio is a constant, Eq. A.21 indicates that the viscoelastic deflection is equal to the elastic deflection multiplied by $(1 - e^{-t/T_1})$, which is the expression for a Kelvin model. Because the stresses in a homogeneous half-space are independent of E, the same stresses will be obtained whether the half-space is elastic

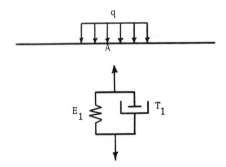

Figure A.2 Example A.3.

or viscoelastic. It can be shown that if the half-space is characterized by the generalized model shown in Figure 2.33f, the surface deflection is

$$w_0 = 2(1 - v^2)q\, a\left[\frac{1}{E_0}\left(1 + \frac{t}{T_0}\right) + \sum_{i=1}^{n}\frac{1}{E_i}(1 - e^{-t/T_i})\right] \qquad (A.22)$$

Example A.4

If the model shown in Figure A.1 is used to characterize a homogeneous half-space subjected to a circular load of intensity q and radius a, as shown in Figure A.3, express the surface deflection at the center of the loaded area in terms of q, a, v, E_1, E_2, and T_2 by the elastic–viscoelastic correspondence principle.

Solution: From Eq. A.8

$$\bar{E} = \frac{\bar{\sigma}}{\bar{\epsilon}} = \frac{T_2(E_1 + E_2)p + E_1}{T_2 p + 1} \qquad (A.23)$$

Substituting Eq. A.23 into Eq. A.19 gives

$$\bar{w}_0 = \frac{2(1 - v^2)qa(T_2 p + 1)}{p[T_2(E_1 + E_2)p + E_1]} \qquad (A.24)$$

Equation A.24 can be fractioned into two terms

$$\bar{w}_0 = 2(1 - v^2)qa\left[\frac{x}{p} + \frac{y}{T_2(E_1 + E_2)p + E_1}\right] \qquad (A.25)$$

When Eqs. A.24 and A.25 are compared for constant and p terms, it can be easily shown that $E_1 x = 1$, or $x = 1/E_1$ and $T_2(E_1 + E_2)x + y = T_2$, or $y = T_2 - T_2(E_1 + E_2)/E_1 = -T_2 E_2/E_1$. Therefore, Eq. A.25 can be written as

$$\bar{w}_0 = \frac{2(1 - v^2)q\, a}{E_1}\left[\frac{1}{p} - \frac{T_2 E_2}{T_2(E_1 + E_2)p + E_1}\right] \qquad (A.26)$$

Laplace inversion of Eq. A.26 results in

$$w_0 = \frac{2(1 - v^2)q\, a}{E_1}\left\{1 - \frac{E_2}{E_1 + E_2}\exp\left[-\frac{E_1 t}{T_2(E_1 + E_2)}\right]\right\} \qquad (A.27)$$

Note that the terms in the braces are the same as those of Eq. A.10, as is expected.

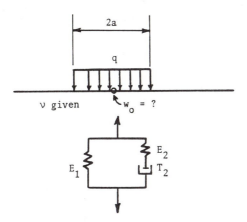

Figure A.3 Example A.4.

A.2.2 Collocation Method

The problem involving a layered system is not as simple as that of a homogeneous half-space. First, each layer has its own elastic modulus and, due to the complex interaction among these moduli, the deflection cannot be expressed as a function of p in closed forms, as shown by Eq. A.20. Second, even if the deflection can be expressed as a polynomial of p, it is of such a high degree that it is difficult to invert. However, the Laplace inversion can be most conveniently accomplished by a collocation method (Huang, 1967).

In Section 2.3.2, it was shown that the viscoelastic response R can be expressed as a Dirichlet series:

$$R = \sum_{i=1}^{7} c_i \exp\left(-\frac{t}{T_i}\right) \tag{2.49}$$

The Laplace transform of Eq. 2.49 is

$$\overline{R} = \sum_{i=1}^{7} \frac{c_i T_i}{T_i p + 1} \tag{A.28}$$

Multiplying Eq. A.28 by p

$$p\overline{R} = \sum_{i=1}^{n} \frac{c_i}{1 + 1/(T_i p)}$$

For any given value of p, $p\overline{R}$ is simply the elastic solution having the elastic modulus expressed as a function of p, as indicated by Eq. A.17. If retardation times T_i of 0.01, 0.03, 0.1, 1, 10, 30, and ∞ are assumed, the coefficients c_i can be obtained by solving the following simultaneous equations:

$$
\begin{bmatrix}
\frac{1}{1+\frac{0.01}{0.01}} & \frac{1}{1+\frac{0.01}{0.03}} & \frac{1}{1+\frac{0.01}{0.1}} & \frac{1}{1+\frac{0.01}{1}} & \frac{1}{1+\frac{0.01}{10}} & \frac{1}{1+\frac{0.01}{30}} & 1 \\
\frac{1}{1+\frac{0.03}{0.01}} & \frac{1}{1+\frac{0.03}{0.03}} & \frac{1}{1+\frac{0.03}{0.1}} & \frac{1}{1+\frac{0.03}{1}} & \frac{1}{1+\frac{0.03}{10}} & \frac{1}{1+\frac{0.03}{30}} & 1 \\
\frac{1}{1+\frac{0.1}{0.01}} & \frac{1}{1+\frac{0.1}{0.03}} & \frac{1}{1+\frac{0.1}{0.1}} & \frac{1}{1+\frac{0.1}{1}} & \frac{1}{1+\frac{0.1}{10}} & \frac{1}{1+\frac{0.1}{30}} & 1 \\
\frac{1}{1+\frac{1}{0.01}} & \frac{1}{1+\frac{1}{0.03}} & \frac{1}{1+\frac{1}{0.1}} & \frac{1}{1+\frac{1}{1}} & \frac{1}{1+\frac{1}{10}} & \frac{1}{1+\frac{1}{30}} & 1 \\
\frac{1}{1+\frac{10}{0.01}} & \frac{1}{1+\frac{10}{0.03}} & \frac{1}{1+\frac{10}{0.1}} & \frac{1}{1+\frac{10}{1}} & \frac{1}{1+\frac{10}{10}} & \frac{1}{1+\frac{10}{30}} & 1 \\
\frac{1}{1+\frac{30}{0.01}} & \frac{1}{1+\frac{30}{0.03}} & \frac{1}{1+\frac{30}{0.1}} & \frac{1}{1+\frac{30}{1}} & \frac{1}{1+\frac{30}{10}} & \frac{1}{1+\frac{30}{30}} & 1 \\
1 & 1 & 1 & 1 & 1 & 1 & 1
\end{bmatrix}
\begin{Bmatrix}
c_1 \\ c_2 \\ c_3 \\ c_4 \\ c_5 \\ c_6 \\ c_7
\end{Bmatrix}
=
\begin{Bmatrix}
(p\overline{R})_{p=1/0.01} \\
(p\overline{R})_{p=1/0.03} \\
(p\overline{R})_{p=1/0.1} \\
(p\overline{R})_{p=1} \\
(p\overline{R})_{p=1/10} \\
(p\overline{R})_{p=1/30} \\
(p\overline{R})_{p=10^{10}}
\end{Bmatrix}
\tag{A.29}
$$

The above procedure is very similar to that presented in Section 2.3.2. If the viscoelastic property is characterized by the creep compliance, the use of the method presented in Section 2.3.2 is more direct. If the elastic–viscoelastic correspondence principle is employed, it is necessary to convert the creep compliance into a mechanical model with the elastic modulus indicated by Eq. A.17. For this reason, the elastic–viscoelastic correspondence principle and the Laplace transform were not used in KENLAYER.

A.3 METHOD OF SUCCESSIVE RESIDUALS

This method is used to determine directly the constants E_i and T_i of a viscoelastic material from the creep curve, rather than arbitrarily assuming T_i and then computing E_i by the collocation method, as described in Section 2.3.2. First, the creep compliances D due to retarded strains are determined by deducting the instantaneous and viscous strains from the total strains, as shown in Figure A.4. The actual number of Kelvin models required is not known at this time but can be determined later. For illustration, it is assumed that three Kelvin models are needed to describe retarded strains:

$$D = \frac{1}{E_1}\left[1 - \exp\left(-\frac{t}{T_1}\right)\right] + \frac{1}{E_2}\left[1 - \exp\left(-\frac{t}{T_2}\right)\right] + \frac{1}{E_3}\left[1 - \exp\left(-\frac{t}{T_3}\right)\right] \quad (A.30)$$

Let

$$b = \frac{1}{E_1} + \frac{1}{E_2} + \frac{1}{E_3}$$

So Eq. A.30 can be written as

$$S_1 = b - D = \frac{1}{E_1}\exp\left(-\frac{t}{T_1}\right) + \frac{1}{E_2}\exp\left(-\frac{t}{T_2}\right) + \frac{1}{E_3}\exp\left(-\frac{t}{T_3}\right) \quad (A.31)$$

If T_1 is much greater than T_2 and T_3, then, after a sufficient period of time, the last two terms on the right side of Eq. A.31 vanish:

$$S_1 = \frac{1}{E_1}\exp\left(-\frac{t}{T_1}\right) \quad (A.32)$$

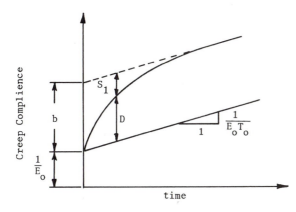

Figure A.4 Separation of creep compliances.

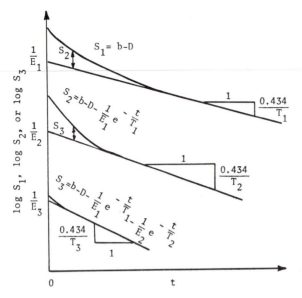

Figure A.5 Method of successive residuals.

Equation A.32 shows that a plot of log S_1 versus t results in a straight line, as indicated by Eq. A.33 and Figure A.5:

$$\log S_1 = \log \left(\frac{1}{E_1}\right) - \frac{0.434t}{T_1} \tag{A.33}$$

The slope of the straight line can be used to determine T_1, and the intercept at $t = 0$ can be used to determine E_1. After E_1 and T_1 are found, Eq. A.31 can be written as

$$S_2 = b - D - \frac{1}{E_1} \exp\left(-\frac{t}{T_1}\right) = \frac{1}{E_2} \exp\left(-\frac{t}{T_2}\right) + \frac{1}{E_3} \exp\left(-\frac{t}{T_3}\right) \tag{A.34}$$

in which S_2 is the vertical intercept between the curve and the straight line. If T_2 is much greater than T_3, a plot of log S_2 versus t should also finally become a straight line, so T_2 and E_2 can be determined. The process is continued until the intercept becomes negligibly small.

Example A.5:

The creep compliances of a viscoelastic material are shown in Table 2.4. Develop a mechanical model and determine its viscoelastic constants.

Solution: The generalized model is represented by Eq. 2.36. When $t = 0$, $D = 1/E_0$. From Table 2.4, $D = 0.5$ when $t = 0$, so $E_0 = 2$. At long loading times, only the viscous strains exist. The rate of change in compliance due to viscous strains is $1/(E_0T_0)$, as can be seen from Eq. 2.29 or 2.36. At $t = 40$, $D = 5.798$ and at $t = 50$, $D = 6.799$, so the change in compliance per unit time is $(6.799 - 5.798)/10 = 0.1$, and $E_0T_0 = 10$, or $T_0 = 5$.

TABLE A.1 COMPUTATION OF SUCCESSIVE RESIDUALS

	Compliance			$\exp(-t/T_1)$		$\exp(-t/T_2)$	
	Dashed						
Time	line	Total	S_1	E_1	S_2	E_2	S_3
(1)	(2)	(3)	(4)	(5)	(6)	(7)	(8)
0	1.799	0.500	1.299	0.100	1.199	0.200	0.999
0.05	1.804	0.909	0.895	0.099	0.796	0.190	0.606
0.1	1.809	1.162	0.647	0.099	0.548	0.181	0.367
0.2	1.819	1.423	0.396	0.098	0.298	0.164	0.134
0.4	1.839	1.592	0.247	0.096	0.151	0.135	0.016
0.6	1.859	1.654	0.205	0.094	0.111	0.111	0.000
0.8	1.879	1.697	0.182	0.092	0.090		
1.0	1.899	1.736	0.163	0.090	0.073		
1.5	1.949	1.819	0.130	0.085	0.045		
2	1.999	1.891	0.108	0.081	0.027		
3	2.099	2.016	0.083	0.073	0.010		
4	2.199	2.129	0.070	0.066	0.004		
5	2.299	2.238	0.061	0.059	0.002		
10	2.799	2.763	0.036	0.035	0.001		
20	3.799	3.786	0.013				
30	4.799	4.795	0.004				
40	5.799	5.798	0.001				
50	6.799	6.799	0.000				

Table A.1 shows the procedure for computing successive residuals. Column 2 is the compliance of the dashed line shown in Figure A.4 and can be computed by $6.799 - (50 - t) \times 0.1$. Column 3 is given in Table 2.4. Column 4 is the difference between columns 2 and 3. A plot of $\log S_1$ versus t is shown in Figure A.6 and results in a straight line. The slope of the straight line is

$$\frac{0.434}{T_1} = \frac{\log 0.1 - \log 0.01}{22} = 0.0455$$

or $T_1 = 9.54$. The intercept at $t = 0$ is $1/E_1 = 0.1$, or $E_1 = 10$. Column 5 can be calculated by $0.1 \exp(-t/9.54)$. Column 6 is the difference between columns 4 and 5. A plot of S_2 versus t results in a straight line. The slope of the straight line is

$$\frac{0.434}{T_2} = \frac{\log 0.2 - \log 0.0015}{5} = 0.426$$

or $T_2 = 1.02$. The intercept at $t = 0$ is $1/E_2 = 0.2$, or $E_2 = 5$. Column 7 can be calculated by $0.2 \exp(-t/1.02)$. Column 8 is the difference between columns 6 and 7. A plot of S_3 versus t results in a straight line. The slope of the straight line is

$$\frac{0.434}{T_3} = \frac{\log 1 - \log 0.017}{0.4} = 4.424$$

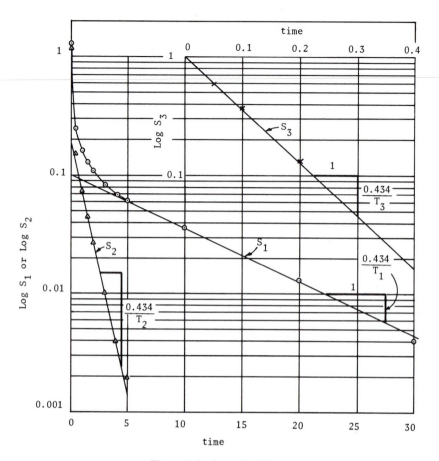

Figure A.6 Example A.5.

or $T_3 = 0.098$. The intercept at $t = 0$ is $1/E_3 = 1$ or $E_3 = 1$. Because all points on S_3 practically lie on a straight line, three Kelvin models are sufficient to describe the creep compliance curve. The equation for predicting the creep compliance is

$$D(t) = \frac{1}{2}\left(1 + \frac{t}{5}\right) + \frac{1}{10}\left[1 - \exp\left(-\frac{t}{9.54}\right)\right] + \frac{1}{5}\left[1 - \exp\left(-\frac{t}{1.02}\right)\right]$$

$$+ \left[1 - \exp\left(-\frac{t}{0.098}\right)\right] \qquad (A.35)$$

Note that the values of E are the same as the original model shown in Figure 2.35a but the values of T are slightly different due to plotting error.

It can be seen that the stress–strain relationship of viscoelastic material can be characterized by a mechanical model or a creep curve. When one is known, the other can be determined.

A.4 COMPLEX MODULUS

The theory of complex modulus can be illustrated by the use of mechanical models. Figure A.7 shows a Kelvin model subjected to a sinusoidal loading. The sinusoidal loading can be represented by a complex number

$$\sigma = \sigma_0 \cos(\omega t) + i\sigma_0 \sin(\omega t) = \sigma_0 e^{i\omega t} \tag{A.36}$$

in which σ_0 is the stress amplitude and ω is the angular velocity, which is related to the frequency f by

$$\omega = 2\pi f \tag{A.37}$$

By assuming that the inertia effect is negligible, the governing differential equation can be written as

$$\lambda_1 \frac{\partial \epsilon}{\partial t} + E_1 \epsilon = \sigma_0 e^{i\omega t} \tag{A.38}$$

The solution of Eq. A.38 can be expressed as

$$\epsilon = \epsilon_0 e^{i(\omega t - \phi)} \tag{A.39}$$

in which ϵ is the strain amplitude and ϕ is the phase angle by which the strain lags the stress, as shown in Figure A.8. Substituting Eq. A.39 into Eq. A.38 gives

$$i\lambda_1 \epsilon_0 \omega e^{i(\omega t - \phi)} + E_1 \epsilon_0 e^{i(\omega t - \phi)} = \sigma_0 e^{i\omega t} \tag{A.40}$$

After canceling $e^{i\omega t}$ on both sides of Eq. A.40 and equating the real terms to σ_0 and the imaginary terms to zero, the following two equations are obtained to solve ϵ_0 and ϕ:

$$\lambda_1 \omega \epsilon_0 \sin \phi + E_1 \epsilon_0 \cos \phi = \sigma_0 \tag{A.41a}$$

$$\lambda_1 \omega \epsilon_0 \cos \phi - E_1 \epsilon_0 \sin \phi = 0 \tag{A.41b}$$

The solutions of Eq. A.41 are

$$\epsilon_0 = \frac{\sigma_0}{\sqrt{E_1^2 + (\lambda_1 \omega)^2}} \tag{A.42a}$$

$$\tan \phi = \frac{\lambda_1 \omega}{E_1} \tag{A.42b}$$

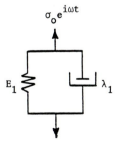

Figure A.7 Kelvin model under sinusoidal loading.

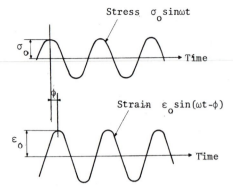

Stress $\sigma_o \sin\omega t$

Time

Strain $\epsilon_o \sin(\omega t - \phi)$

Time

Figure A.8 Lag of strain behind stress.

It can be seen from Eq. A.42b that for elastic materials $\lambda_1 = 0$, so $\phi = 0$; while for viscous materials $E_1 = 0$, so $\phi = \pi/2$. Therefore, the phase angle for viscoelastic materials ranges from 0 to $\pi/2$. The complex modulus E^* is defined as

$$E^* = \frac{\sigma}{\epsilon} = \frac{\sigma_0 e^{i\omega t}}{\epsilon_0 e^{i(\omega t - \phi)}}$$

or

$$E^* = \frac{\sigma_0}{\epsilon_0} \cos \phi + i \frac{\sigma_0}{\epsilon_0} \sin \phi \qquad (A.43)$$

The dynamic modulus is the absolute value of the complex modulus

$$|E^*| = \sqrt{\left(\frac{\sigma_0}{\epsilon_0} \cos \phi\right)^2 + \left(\frac{\sigma_0}{\epsilon_0} \sin \phi\right)^2} = \frac{\sigma_0}{\epsilon_0} \qquad (A.44)$$

It is interesting to note that the real part of Eq. A.43 is actually equal to the elastic stiffness E_1 and the imaginary part to the internal damping $\lambda_1 \omega$.

Example A.6:

A sinusoidal loading with an amplitude of 50 psi (345 kPa) and a frequency of 5 Hz is applied to a specimen characterized by the Kelvin model shown in Figure A.9. Determine the strain amplitude ϵ_0, phase angle ϕ, and dynamic modulus $|E^*|$.

Solution: Given $f = 5$ Hz, from Eq. A.37, $\omega = 2\pi \times 5 = 31.416$ rad/s. $\lambda_1 = T_1 E_1 = 0.05 \times 10^5 = 5000$ lb-s/in.^{2} (35 MPa-s). From Eq. A.42a, $\epsilon_0 = 50/[(10^5)^2 + (5000 \times 31.416)^2]^{0.5} = 2.69 \times 10^{-4}$; from Eq. A.42b, $\tan \phi = 5000 \times 31.416/10^5 = 1.57$, or $\phi = 57°$; and from Eq. A.44, $|E^*| = 50/(2.69 \times 10^{-4}) = 1.85 \times 10^5$ psi (1.38 MPa).

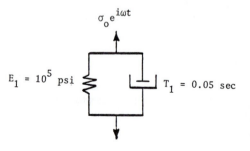

$\sigma_o e^{i\omega t}$

$E_1 = 10^5$ psi

$T_1 = 0.05$ sec

Figure A.9 Example A.6 (1 psi = 6.9 kPa).

Theory of Viscoelasticity   App. A

Most viscoelastic materials cannot be simulated by a single Kelvin model; rather they require a series of Kelvin models. The complex modulus of these models can be more easily obtained by the use of Fourier transform similar to the Laplace transform discussed in Section A.2.1. The Fourier transform is defined as

$$L[F(t)] = \int_{-\infty}^{\infty} F(t)e^{-i\omega t}\, dt \qquad (A.45)$$

For simplicity, the same notations are used for the Fourier transform as for the Laplace transform. The Fourier transform of the first derivative $d\sigma/dt$ is

$$L\left(\frac{d\sigma}{dt}\right) = \int_{-\infty}^{\infty} \frac{d\sigma}{dt}\, e^{-i\omega t}\, dt = \int_{-\infty}^{\infty} e^{-i\omega t}\, d\sigma$$

$$= \left. \sigma e^{-i\omega t} \right|_{-\infty}^{\infty} - \int_{-\infty}^{\infty} \sigma d(e^{-i\omega t})$$

If the material is initially undisturbed, or $\sigma = 0$ when $t = -\infty$, then the first term is equal to 0, or

$$L\left(\frac{d\sigma}{dt}\right) = i\omega \int_{-\infty}^{\infty} \sigma e^{-i\omega t}\, dt = i\omega\bar{\sigma} \qquad (A.46)$$

It can be easily proved that the Fourier transform of $\partial^n\sigma/\partial t^n$, abbreviated as $D^n\sigma$, is obtained by replacing D with $i\omega$ and σ with $\bar{\sigma}$, similar to the Laplace transform where D is replaced by p:

$$L(D^n\sigma) = (i\omega)^n\, \bar{\sigma} \qquad (A.47)$$

The application of the Fourier transform to determine the dynamic modulus of a viscoelastic material characterized by a generalized model can be best illustrated by the following example.

Example A.7:

Figure A.10 shows a generalized model composed of one Maxwell model and two Kelvin models. If the model is subjected to a sinusoidal loading with a frequency of 10 Hz, determine the dynamic modulus of the material.

Solution: From Eq. A.5, the stress–strain ratio of the model can be expressed as

$$\frac{\sigma}{\epsilon} = \left[\frac{0.05D + 1}{2 \times 10^5 \times 0.05D} + \frac{1}{5 \times 10^5(0.05D + 1)} + \frac{1}{10^5(0.1D + 1)}\right]^{-1} \qquad (A.48)$$

From Eq. A.37, $\omega = 2\pi \times 10 = 62.83$. The complex modulus can be obtained by replacing D in Eq. A.48 by $62.83i$.

$$E^* = \frac{10^5}{\dfrac{(3.142i + 1)}{6.283i} + \dfrac{1}{5(3.142i + 1)} + \dfrac{1}{(6.283i + 1)}}$$

$$= \frac{10^5}{\dfrac{(3.142 - i)}{6.283} + \dfrac{(1 - 3.142i)}{54.361} + \dfrac{(1 - 6.283i)}{40.476}} = \frac{10^5}{0.543 - 0.372i}$$

$$= \frac{10^5(0.543 + 0.372i)}{0.433} = 10^5(1.254 + 0.859i)$$

$$|E^*| = 10^5\sqrt{(1.254)^2 + (0.859)^2} = 1.52 \times 10^5 \text{ psi (1.05 GPa)}$$

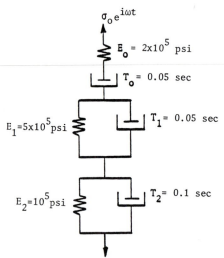

$\sigma_0 e^{i\omega t}$

$E_0 = 2 \times 10^5$ psi

$T_0 = 0.05$ sec

$E_1 = 5 \times 10^5$ psi

$T_1 = 0.05$ sec

$E_2 = 10^5$ psi

$T_2 = 0.1$ sec

Figure A.10 Example A.7 (1 psi = 6.9 kPa).

It was shown in Section 2.3.1 that the stress–strain relationship of viscoelastic materials can be represented by a mechanical model or a creep compliance curve. If one is known, the other can be determined. Since the complex modulus can also be obtained from the mechanical model, the creep compliance and complex modulus should also be related. As long as the material is ideally viscoelastic, any type of test can be used. However, to describe the viscoelastic behavior over a large time range, various frequencies should be used to determine the complex modulus.

APPENDIX

More about Kenlayer

This appendix is a supplement to Chapter 3 and describes in detail the KEN-LAYER computer program. The first part explains the layered theory, sub-routines, and flowcharts and is especially useful to those who would like to modify the program or develop a similar program for their special needs. The latter part contains six illustrative examples for which the data files are supplied with the diskette. Readers can print out these files, study their contents, use them to run KENLAYER, and inspect the computer printout. Instructions on the use of the software are presented in Appendix D.

B.1 LAYERED THEORY

The equations incorporated in KENLAYER for determining the stresses and displacements in a multilayer systems under a circular loaded area are presented in this section.

B.1.1 Basic Equations

As in the classical theory of elasticity, a stress function ϕ that satisfies the governing differential equation

$$\nabla^4\phi = 0 \tag{B.1a}$$

735

is assumed for each of the layers. For systems with an axially symmetrical stress distribution

$$\nabla^4 = \left(\frac{\partial^2}{\partial r^2} + \frac{1}{r}\frac{\partial}{\partial r} + \frac{\partial^2}{\partial z^2}\right)\left(\frac{\partial^2}{\partial r^2} + \frac{1}{r}\frac{\partial}{\partial r} + \frac{\partial^2}{\partial z^2}\right) \qquad (B.1b)$$

in which r and z are the cylindrical coordinates for radial and vertical directions, respectively. After the stress function is found, the stresses and displacements can be determined by

$$\sigma_z = \frac{\partial}{\partial z}\left[(2 - v)\nabla^2\phi - \frac{\partial^2\phi}{\partial z^2}\right] \qquad (B.2a)$$

$$\sigma_r = \frac{\partial}{\partial z}\left(v\nabla^2\phi - \frac{\partial^2\phi}{\partial r^2}\right) \qquad (B.2b)$$

$$\sigma_t = \frac{\partial}{\partial z}\left(v\nabla^2\phi - \frac{1}{r}\frac{\partial\phi}{\partial r}\right) \qquad (B.2c)$$

$$\tau_{rz} = \frac{\partial}{\partial r}\left[(1 - v)\nabla^2\phi - \frac{\partial^2\phi}{\partial z^2}\right] \qquad (B.2d)$$

$$w = \frac{1 + v}{E}\left[(1 - 2v)\nabla^2\phi + \frac{\partial^2\phi}{\partial r^2} + \frac{1}{r}\frac{\partial\phi}{\partial r}\right] \qquad (B.2e)$$

$$u = -\frac{1 + v}{E}\left(\frac{\partial^2\phi}{\partial r \partial z}\right) \qquad (B.2f)$$

Because Eq. B.1 is a fourth-order differential equation, the stresses and displacements thus determined will consist of four constants of integration which must be determined from the boundary and continuity conditions.

Let $\rho = r/H$ and $\lambda = z/H$, in which H is the distance from the surface to the upper boundary of the lowest layer, as shown in Figure 3.1. It can be easily verified by substitution that

$$\phi_i = \frac{H^3 J_0(m\rho)}{m^2}[A_i e^{-m(\lambda_i - \lambda)} - B_i e^{-m(\lambda - \lambda_{i-1})}$$
$$+ C_i m\lambda e^{-m(\lambda_i - \lambda)} - D_i m\lambda e^{-(\lambda - \lambda_{i-1})}] \qquad (B.3)$$

is a stress function for the ith layer which satisfies Eq. B.1, in which J_0 is a Bessel function of the first kind and order 0; m is a parameter; and A, B, C, D are constants of integration to be determined from boundary and continuity conditions. The subscript i varies from 1 to n and refers to the quantities corresponding to the ith layer.

Substituting Eq. B.3 into B.2 gives

$$(\sigma_z^*)_i = -mJ_0(m\rho)\{[A_i - C_i(1 - 2v_i - m\lambda)]e^{-m(\lambda_i - \lambda)}$$
$$+ [B_i + D_i(1 - 2v_i + m\lambda)]e^{-m(\lambda - \lambda_{i-1})}\} \qquad (B.4a)$$

$$(\sigma_r^*)_i = \left[mJ_0(m\rho) - \frac{J_1(m\rho)}{\rho}\right]\{[A_i + C_i(1 + m\lambda)]e^{-m(\lambda_i - \lambda)} + [B_i - D_i(1 - m\lambda)]$$
$$\times e^{-m(\lambda - \lambda_{i-1})}\} + 2v_i mJ_0(m\rho)[C_i e^{-m(\lambda_i - \lambda)} - D_i e^{-m(\lambda - \lambda_{i-1})}] \qquad (B.4b)$$

$$(\sigma_t^*)_i = \frac{J_1(m\rho)}{\rho} \{[A_i + C_i(1 + m\lambda)] \, e^{-m(\lambda_i - \lambda)} + [B_i - D_i(1 - m\lambda)] \, e^{-m(\lambda - \lambda_{i-1})}\}$$
$$+ \, 2\nu_i \, mJ_0(m\rho) \, [C_i e^{-m(\lambda_i - \lambda)} - D_i \, e^{-m(\lambda - \lambda_{i-1})}] \tag{B.4c}$$

$$(\tau_{rz}^*)_i = mJ_1(m\rho) \{[A_i + C_i(2\nu_i + m\lambda)] \, e^{-m(\lambda_i - \lambda)}$$
$$- \, [B_i - D_i(2\nu_i - m\lambda)] \, e^{-m(\lambda - \lambda_{i-1})}\} \tag{B.4d}$$

$$(w^*)_i = -\frac{1 + \nu_i}{E_i} J_0(m\rho) \{[A_i - C_i(2 - 4\nu_i - m\lambda)] \, e^{-m(\lambda_i - \lambda)}$$
$$- [B_i + D_i(2 - 4\nu_i + m\lambda)] \, e^{-m(\lambda - \lambda_{i-1})} \tag{B.4e}$$

$$(u^*)_i = \frac{1 + \nu_i}{E_i} J_1(m\rho) \{[A_i + C_i(1 + m\lambda)] \, e^{-m(\lambda_i - \lambda)}$$
$$+ \, [B_i - D_i(1 - m\lambda)] \, e^{-m(\lambda - \lambda_{i-1})}\} \tag{B.4f}$$

in which σ_z is the stress in the vertical or z direction; σ_r is the stress in the radial or r direction; σ_t is the stress in the tangential or t direction; τ_{rz} is the shear stress; w is the displacement in the vertical or z direction; u is the displacement in the radial or r direction; and J_1 is a Bessel function of the first kind and order one. The subscript i outside the parentheses indicates the ith layer. A star superscript is placed on these stresses and displacements because they are not the actual stresses and displacements due to a uniform load q distributed over a circular area of radius a, but are those due to a vertical load of $-mJ_0(m\rho)$, as can be seen from Eq. B.4a when the terms within the braces are set to 1.

To find the stresses and displacements due to a constant load q distributed over a circular area of radius a, the Hankel transform method is employed. The Hankel transform of such load is

$$\bar{f}(m) = \int_0^\alpha q\rho J_0(m\rho) \, d\rho = \frac{q\alpha}{m} J_1(m\alpha) \tag{B.5}$$

in which $\alpha = a/H$. The Hankel inversion of $\bar{f}(m)$ is

$$q(\rho) = \int_0^\infty \bar{f}(m)mJ_0(m\rho) \, dm = q\alpha \int_0^\infty J_0(m\rho) J_1(m\alpha) \, dm \tag{B.6}$$

If R^* is the stress or displacement in Eq. B.4 due to loading $-mJ_0(m\rho)$ and R is that due to load q, and tension is considered negative, then

$$R = q\alpha \int_0^\infty \frac{R^*}{m} J_1(m\alpha) \, dm \tag{B.7}$$

The analysis of layered systems can therefore be summarized into the following steps:

1. Assign successive values of m, from 0 to a rather large positive number until R in Eq. B.7 converges.
2. For each value of m, determine the constants of integration, A_i, B_i, C_i, and D_i, from the boundary and continuity conditions, which are discussed later.
3. Substitute these constants into Eq. B.4 to obtain R^*.
4. Determine R from Eq. B.7 by numerical integration.

In the numerical integration, the zeros of $J_0(m\rho)$ and $J_1(m\alpha)$ are determined and the integral between these two zeros is evaluated by a four-point Gaussian formula. In view of the fact that the first cycle of integration has to be more finely divided, especially when ρ is large, the interval between 0 and 2.40483, which is the first zero of J_0, is subdivided into 6 intervals and that between 2.40483 and 3.83171, which is the first zero of J_1, is subdivided into 2 intervals. The integral for each subinterval is also evaluated by the four-point formula.

B.1.2 BOUNDARY AND CONTINUITY CONDITIONS

At the upper surface, $i = 1$ and $\lambda = 0$, the boundary conditions

$$(\sigma_z^*)_1 = -mJ_0(m\rho) \tag{B.8a}$$

$$(\tau_{rz}^*)_1 = 0 \tag{B.8b}$$

result in two equations

$$\begin{bmatrix} e^{-m\lambda_1} & 1 \\ e^{-m\lambda_1} & -1 \end{bmatrix} \begin{Bmatrix} A_1 \\ B_1 \end{Bmatrix} + \begin{bmatrix} -(1 - 2v_1)\,e^{-m\lambda_1} & 1 - 2v_1 \\ 2v_1 e^{-m\lambda_1} & 2v_1 \end{bmatrix} \begin{Bmatrix} C_1 \\ D_1 \end{Bmatrix} = \begin{Bmatrix} 1 \\ 0 \end{Bmatrix} \tag{B.9}$$

All the solutions of layered systems presented in Chapter 2 are based on the assumption that the layers are fully bonded with the same vertical stress, shear stress, vertical displacement, and radial displacement at every point along the interface. Therefore, when $\lambda = \lambda_i$, the continuity conditions

$$(\sigma_z^*)_i = (\sigma_z^*)_{i+1} \tag{B.10a}$$

$$(\tau_{rz}^*)_i = (\tau_{rz}^*)_{i+1} \tag{B.10b}$$

$$(w^*)_i = (w^*)_{i+1} \tag{B.10c}$$

$$(u^*)_i = (u^*)_{i+1} \tag{B.10d}$$

result in four equations:

$$\begin{bmatrix} 1 & F_i & -(1 - 2v_i - m\lambda_i) & (1 - 2v_i + m\lambda_i)\,F_i \\ 1 & -F_i & 2v_i + m\lambda_i & (2v_i - m\lambda_i)\,F_i \\ 1 & F_i & 1 + m\lambda_i & -(1 - m\lambda_i)\,F_i \\ 1 & -F_i & -(2 - 4v_i - m\lambda_i) & -(2 - 4v_i + m\lambda_i)\,F_i \end{bmatrix} \begin{Bmatrix} A_i \\ B_i \\ C_i \\ D_i \end{Bmatrix}$$

$$= \begin{bmatrix} F_{i+1} & 1 & -(1 - 2v_{i+1} - m\lambda_i)F_{i+1} & 1 - 2v_{i+1} + m\lambda_i \\ F_{i+1} & -1 & (2v_{i+1} + m\lambda_i)F_{i+1} & 2v_{i+1} - m\lambda_i \\ R_i F_{i+1} & R_i & (1 + m\lambda_i)R_i F_{i+1} & -(1 - m\lambda_i)R_i \\ R_i F_{i+1} & -R_i & -(2 - 4v_{i+1} - m\lambda_i)R_i\,F_{i+1} & -(2 - 4v_{i+1} + m\lambda_i)R_i \end{bmatrix} \begin{Bmatrix} A_{i+1} \\ B_{i+1} \\ C_{i+1} \\ D_{i+1} \end{Bmatrix} \tag{B.11}$$

in which

$$F_i = e^{-m(\lambda_i - \lambda_{i-1})} \tag{B.12a}$$

$$R_i = \frac{E_i}{E_{i+1}} \frac{1 + v_{i+1}}{1 + v_i} \tag{B.12b}$$

Since the stresses and displacements must vanish as λ approaches infinity, it can be concluded from Eq. B.3 that for the lowest layer with $i = n$

$$A_n = C_n = 0 \tag{B.13}$$

For an n-layer system, there are $4n$ constants of integration. With $A_n = C_n = 0$, the remaining $4n - 2$ constants can be determined from the $4n - 2$ equations, two from Eq. B.9 and $4(n - 1)$ from Eq. B.11.

To save computer times in solving simultaneous equations, a procedure was developed by which only two equations need be solved, instead of $4n - 2$ equations. This can be achieved by transforming Eq. B.11 to

$$\begin{Bmatrix} A_i \\ B_i \\ C_i \\ D_i \end{Bmatrix} = \begin{bmatrix} 4 \times 4 \\ \text{matrix} \end{bmatrix} \begin{Bmatrix} A_{i+1} \\ B_{i+1} \\ C_{i+1} \\ D_{i+1} \end{Bmatrix} \tag{B.14}$$

By successive multiplications, the constants for the first layer can be related to those of the last layer by

$$\begin{Bmatrix} A_1 \\ B_1 \\ C_1 \\ D_1 \end{Bmatrix} = \begin{bmatrix} 4 \times 2 \\ \text{matrix} \end{bmatrix} \begin{Bmatrix} B_n \\ D_n \end{Bmatrix} \tag{B.15}$$

When Eq. B.15 is substituted into Eq. B.9, two equations with two unknowns, B_n and D_n, are obtained. After B_n and D_n are solved, they can be substituted into Eq. B.14, with $A_n = C_n = 0$, to obtain the constants for the $(n - 1)$th layer. The procedure is repeated until the constants for every layer up to the first layer are obtained.

If the ith interface, or $\lambda = \lambda_i$, is unbonded or frictionless, the continuation of shear stress and radial displacement must be replaced by zero shear stress on both sides of the interface:

$$(\sigma_z^*)_i = (\sigma_z^*)_{i+1} \tag{B.16a}$$

$$(w^*)_i = (w^*)_{i+1} \tag{B.16b}$$

$$(\tau_{rz}^*)_i = 0 \tag{B.16c}$$

$$(\tau_{rz}^*)_{i+1} = 0 \tag{B.16d}$$

Hence, Eq. B.11 must be replaced by

$$\begin{bmatrix} 1 & F_i & -(1 - 2v_i - m\lambda_i) & (1 - 2v_i + m\lambda_i)\,F_i \\ 1 & F_i & 1 + m\lambda_i & -(1 - m\lambda_i)\,F_i \\ 1 & -F_i & 2v_i + m\lambda_i & (2v_i - m\lambda_i)\,F_i \\ 0 & 0 & 0 & 0 \end{bmatrix} \begin{Bmatrix} A_i \\ B_i \\ C_i \\ D_i \end{Bmatrix}$$

$$= \begin{bmatrix} F_{i+1} & 1 & -(1 - 2v_{i+1} - m\lambda_i)F_{i+1} & 1 - 2v_{i+1} + m\lambda_i \\ R_i F_{i+1} & R_i & (1 + m\lambda_i)R_i F_{i+1} & -(1 - m\lambda_i)R_i \\ 0 & 0 & 0 & 0 \\ F_{i+1} & -1 & (2v_{i+1} + m\lambda_i)F_{i+1} & 2v_{i+1} - m\lambda_i \end{bmatrix} \begin{Bmatrix} A_{i+1} \\ B_{i+1} \\ C_{i+1} \\ D_{i+1} \end{Bmatrix} \tag{B.17}$$

Both Eqs. B.11 and B.17 are used in KENLAYER. If any interface is bonded, Eq. B.11 applies. If any interface is frictionless, Eq. B.17 applies. The program is more efficient when all interfaces are bonded because only two equations are to be solved. If one or more interfaces are frictionless, the program is less efficient because more time is needed to solve the $4n - 2$ equations. KENLAYER was originally developed for bonded interfaces only. When the frictionless interfaces were added, an existing subroutine for collocation was used to solve the $4n - 2$ equations. Because frictionless interfaces are very rarely used, the requirement of more computer time is not of major concern.

B.2 SUBROUTINES

KENLAYER consists of one main program and 18 subroutines. All variables are transferred from the main program to the subroutines through arguments and no common statements are used. Therefore, the program can be easily modified if needed.

The main program is relatively short because its main purpose is to call the various subroutines and conduct a damage analysis, if desired. The subroutines can be divided into six groups: data input, layered system, superposition and principal stresses, nonlinear analysis, viscoelastic analysis, and output.

B.2.1 Data Input

There are four subroutines for data input: ELAINP, NONINP, VISINP, and DAMINP.

ELAINP   reads and writes input data for an elastic layered system under single or multiple wheels. For multiple wheels, the distance and direction cosines from each specified point to each of the wheels are also computed for later use. If a layer is nonlinear, the elastic modulus is the assumed elastic modulus to be used for the first iteration. If a layer is viscoelastic, the elastic modulus may be assigned 0 or any value.

NONINP   reads and writes input data for nonlinear analysis. In the case of single wheel, values of XPTNOL and YPTNOL for locating the stress point may be assigned 0 or any value. For multiple wheels, the radial coordinate, RCNOL, for locating the stress point may be assigned 0 or any value.

VISINP   reads and writes input data for viscoelastic analysis.

DAMINP   reads and writes input data for damage analysis.

B.2.2 Layer System

The heart of KENLAYER is the subroutine called LAYERS which analyzes the multilayer system under a circular loaded area. The subroutine is called to obtain solutions at various points with different radial and vertical coordinates, and the results will be used for later analysis. In the nonlinear or viscoelastic analysis, LAYERS is called repeatedly whenever the moduli of the layers change. The following subroutines are called by LAYERS: BESJ, SOL, COEF, and NOFRIC.

LAYERS   computes the vertical displacement, four components of stress, and four components of strains at different radial and vertical distances under a single wheel.

BESJ   computes Bessel function for a given radial distance and order.

SOL   determines the two constants of integration, B_n and D_n, for the bottom-most layer. This subroutine is called if all interfaces are bonded.

COEF   determines the constants of integration for each specified layer. This subroutine is called if all interfaces are bonded.

NOFRIC   determines the constants of integration for an n-layer system by solving $4n - 2$ equations using a subroutine named GELG, which is described later in the viscoelastic subroutines. This subroutine is called if any interface is frictionless.

B.2.3 Superposition and Principal Stresses

The stresses under multiple-wheel loads are obtained by superimposing those under single-wheel loads. After the six components of normal and shear stresses under multiple-wheel loads are determined, the three principal stresses and strains are then computed. SUPPRI is the only subroutine to accomplish these tasks.

SUPPRI   superimposes the stresses due to single-wheel loads and computes principal stresses and strains under multiple wheels.

B.2.4 Nonlinear Analysis

Three subroutines, NONLAY, SUPERP, and ENOLIN, are used for analyzing nonlinear layers.

NONLAY   computes the vertical, radial, and tangential stresses caused by single wheels at given stress points in nonlinear layers. This subroutine is similar to LAYERS except that only the three normal stresses, σ_r, σ_t, and σ_z, are computed.

MULTIP   calculates the three normal stresses, σ_x, σ_y, and σ_z, under multiple wheels by superimpsoing the stresses obtained from NONLAY. This subroutine is similar to SUPPRI except that only the three normal stresses are determined and no principal stresses and strains are computed.

ENOLIN   computes and prints new elastic moduli during each iteration cycle, checks the convergence, and terminates iterations when the desired accuracy is achieved.

B.2.5 Viscoelastic Analysis

Three subroutines, COLLOC, AFTCOL, and GELG, are used for viscoelastic analysis.

COLLOC   applies the collocation method to determine the coefficients of Dirichlet series for displacement, stresses, and strains. This subroutine calls GELG for solving simultaneous equations.

AFTCOL   computes the displacement, stresses, and strains under a moving load using the Dirichlet series or those under a stationary load for different time durations.

GELG   solves a system of simultaneous equations by Gauss elimination. This subroutine is called in the main program for determining the coefficients of creep compliance in Dirichlet series and also in subroutines NOFRIC and COLLOC.

B.2.6 Output

Separate subroutines, SINOUT and MULOUT, are used for single and multiple wheels.

SINOUT   prints the vertical displacement, vertical stress, radial stress, tangential stress, shear stress, vertical strain, radial strain, tangential strain, and shear strain under a single-wheel load and determines the most critical strains for damage analysis. The stresses and strains are positives when in compression and negative in tension.

MULOUT   prints the vertical displacement, vertical stress, major principal stress, intermediate principal stress, minor principal stress, vertical strain, major principal strain, minor principal strain, and horizontal principal strain under multiple wheels and determines the most critical strains for damage analysis.

B.3 FLOWCHARTS

Instead of drawing a large flowchart, it is more convenient to explain each individual case by separate flowcharts using the various subroutines.

B.3.1 Linear Elastic Systems

Figure B.1 shows the flowcharts for linear elastic systems under single and multiple wheels. The major difference between single and multiple wheels is that the former prints out the results at the same time for every point by r and z coordinates, while the latter requires the superposition of two or more wheels and prints out the results point by point as soon as the results at each point are obtained. In the case of multiple wheels, IPT is the index of each point defined by x and y coordinates, and NPT is the total number of points. The portion of flowcharts marked by SE (single elastic) and ME (multiple elastic) is used later in Figure B.2 to conserve space.

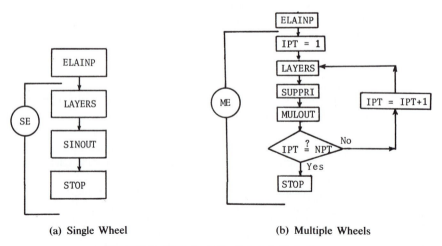

(a) Single Wheel   (b) Multiple Wheels

Figure B.1   Flowcharts for linear elastic system.

B.3.2 Nonlinear Elastic Systems

Figure B.2 shows the flowcharts for nonlinear elastic systems. An iterative process is required until the elastic moduli of nonlinear layers converge and ITE is the iteration number. For the first iteration, the assumed elastic moduli are used for nonlinear layers. When the moduli converge to a specified tolerance, the system is considered linearly elastic and switched to SE or ME, as shown in Figure B.1.

B.3.3 Linear Viscoelastic Systems

Figure B.3 shows the flowcharts for linear viscoelastic systems. This system is linear and no iteration is required. However, it is necessary to obtain elastic solutions at 11 different time durations and IT is the duration number. After the elastic solutions for all 11 durations are obtained, COLLOC is called to evaluate the coefficients of the Dirichlet series, and then AFTCOL is called to evaluate the solutions due to moving loads. If the load is stationary, AFTCOL will evaluate the solutions at different times, instead of at a single duration under moving loads.

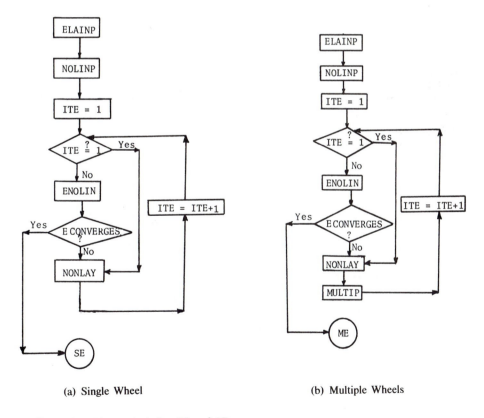

(a) Single Wheel   (b) Multiple Wheels

Note: See Figure B.1 for SE and ME.

Figure B.2 Flowcharts for nonlinear elastic system.

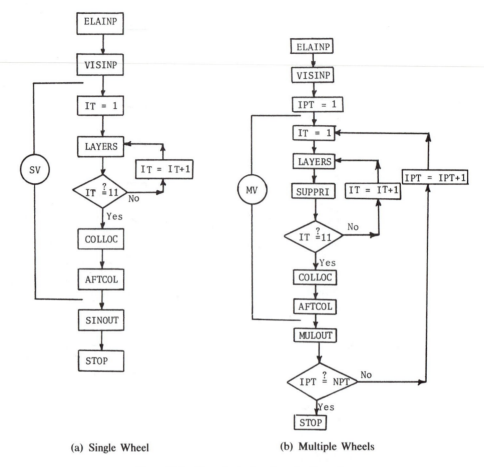

(a) Single Wheel   (b) Multiple Wheels

Figure B.3 Flowcharts for viscoelastic system.

The portion of flowcharts marked SV (single viscoelastic) and MV (multiple viscoelastic) are repeated in Figure B.4.

B.3.4 Nonlinear Elastic and Linear Viscoelastic Systems

Figure B.4 shows the flowcharts for nonlinear elastic and linear viscoelastic systems. This is the most complicated but also the most realistic system for flexible pavements because it is well recognized that under moving loads the hot mix asphalt is linear viscoelastic while the granular bases and subbases are nonlinear elastic. A comparison of Figure B.4 with Figure B.2 shows that NON-LAY in Figure B.2 for single wheel is replaced by SV in Figure B.4 and that NONLAY and MULTIP for multiple wheels are replaced by MV. In other words, during each iteration some of the layers are considered viscoelastic instead of elastic. Due to the interactions among layers, the behavior of nonlinear layers is affected by the viscoelastic layers, and vice versa. Whenever the elastic moduli of

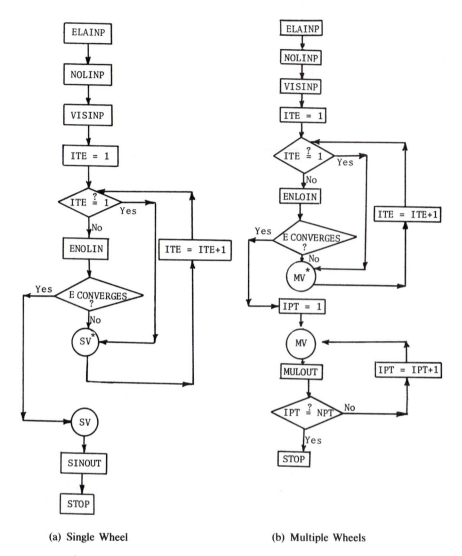

(a) Single Wheel

(b) Multiple Wheels

Note : SV*, MV* LAYERS and SUPPRI should be replaced by NONLAY and MULTIP.
See Figure B.3 for SV and MV.

Figure B.4 Flowcharts for nonlinear elastic and linear viscoelastic system.

nonlinear layers change, the response of viscoelastic layers will also change, and the stresses in the nonlinear layers will be affected. Consequently, iterations continue until the stresses become stable as indicated by the convergence of elastic moduli. Due to the effect of viscoelastic layers, the stresses in nonlinear layers are also time dependent under stationary loads. To obtain a fixed set of stresses for evaluating elastic moduli, it is recommended that this system be used for moving loads only. If stationary loads are specified, the stresses at the last time duration will be used to determine the elastic moduli of nonlinear layers. In

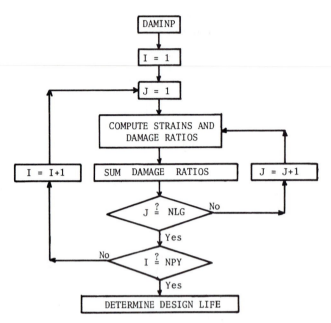

Figure B.5 Flowchart for damage analysis.

other words, the last time duration of stationary loads can be considered as a moving load with a square waveform, instead of a haversine function.

B.3.5 Damage Analysis

Damage analysis is performed in the main program after the critical strains are determined. Starting from the first period, the damage ratios for fatigue cracking and permanent deformation are computed for each load group and summed over all NLG groups. Then the computation continues for the next period until all NPY periods are completed. The reciprocal of the damage ratio is the design life. Figure B.5 is a flowchart for damage analysis.

B.4 ILLUSTRATIVE EXAMPLES

Six examples are given to illustrate the input and output of KENLAYER. Data files for these examples, designated as LAY1.DAT to LAY6.DAT, are contained in the diskette. Each line of the data file is identified by a line number in parentheses, which corresponds to the line number in Section 3.2.3.

The first example is an elastic three-layer system under a set of dual-tandem wheels. The second example is a nonlinear elastic two-layer system under a circular loaded area. The third example is a viscoelastic two-layer system under a stationary circular loaded area. The fourth example is a three-layer viscoelastic system subjected to damage analysis with four periods per year and one load group. The fifth example is an elastic two-layer system subjected to damage analysis with one period per year and three different load groups. The sixth example is a nonlinear elastic and linear viscoelastic three-layer system subjected to damage analysis with two periods per year and two load groups. To show the different forms of output, no damage analysis is made for the first three examples with NSTD = 1 for Example 1, NSTD = 9 for Example 2 and NSTD = 5 for Example 3. The purpose of these examples is to illustrate the capabilities of KENLAYER.

The approximate time for running these examples is shown in Table B.1. It can be seen that the running time for a PC without a math coprocessor is more than 17 times longer than that for a PC with a math coprocessor.

B.4.1 Example 1: Linear Elastic

Figure B.6 shows an elastic three-layer system under a set of dual-tandem tires. Each tire has a contact radius of 4.52 in. (115 mm) and a contact pressure of 70 psi (483 kPa). Layer 1 has an elastic modulus of 7.4×10^5 psi (5.1 GPa), a Poisson ratio of 0.4, and a thickness of 6 in. (152 mm); layer 2 has an elastic modulus of 2.3×10^4 psi (159 MPa), a Poisson ratio of 0.35, and a thickness of 8 in. (203 mm); and layer 3 has an elastic modulus of 1.1×10^4 psi (77 MPa) and a Poisson ratio of 0.45. Compute the surface and interface deflections at the 12 points shown in the figure and determine the maximum surface and interface deflections.

The results obtained by KENLAYER show that the maximum surface deflection is 0.02044 in. (0.52 mm) and occurs at point 3, the maximum deflection at the first interface is 0.02030 in. (0.52 mm) and also occurs at point 3, while the maximum deflection at the second interface is 0.01820 in. (0.46 mm) and occurs at point 6. The movement of maximum deflections from point 3 to point 6 as the depth increases is reasonable because the effect of tandem loads is more pronounced at greater depths.

B.4.2 Example 2: Nonlinear Elastic

Figure B.7 shows a nonlinear elastic two-layer system under a circular loaded area. The loaded area has a radius of 6 in. (152 mm) and a contact pressure of 100 psi (690 kPa). Layer 1 is 12 in. (305 mm) thick and consists of a granular material with a unit weight of 135 pcf (21.2 kN/m³), a Poisson ratio of 0.35, a nonlinear coefficient K_1 of 10,000 psi, and a nonlinear exponent K_2 of 0.5. Layer 2 consists

TABLE B.1 APPROXIMATE EXECUTION TIME FOR EXAMPLE PROBLEMS BY KENLAYER

Example no.	PC with math coprocessor	PC without math coprocessor
1	14.7 min	4.6 h
2	32.0 min	9.9 h
3	0.4 min	7.0 min
4	4.4 min	75.0 min
5	11.4 min	3.5 h
6	87.0 min	24.0 h

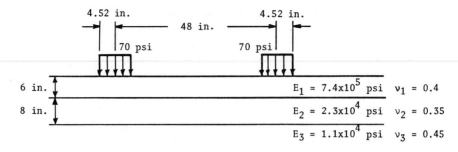

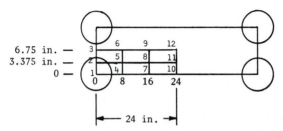

Figure B.6 Example 1 (1 in. = 25.4 mm, 1 psi = 6.9 kPa).

of a fine-grained soil classified as "soft" with a unit weight of 115 pcf (18.1 kN/m³), a Poisson ratio of 0.45, and the nonlinear properties shown in Figure 3.6. Layer 1 is divided into six layers, each 2 in. (51 mm) thick, with the stress points located at the midheight of each layer. The stress point of the subgrade is located 6 in. (152 mm) below the top of the subgrade. Other input data are shown in the data file. Determine the vertical deflection on the surface and the vertical stress on the top of subgrade, both under the center of the loaded area.

The results obtained by KENLAYER show a vertical deflection of 0.05542 in. (1.41 mm) on the surface and a vertical stress of 7.97654 psi (55 kPa) on the top of subgrade.

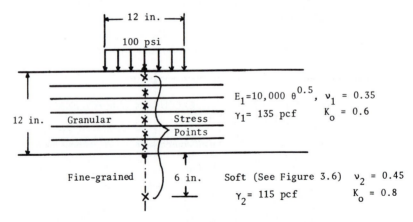

Figure B.7 Example 2 (1 in. = 25.4 mm, 1 psi = 6.9 kPa, 1 pcf = 157.1 N/m³).

More about Kenlayer   App. B

B.4.3 Example 3: Linear Viscoelastic

Figure B.8 shows a viscoelastic two-layer system under a stationary circular loaded area with a contact radius of 10 in. (254 mm) and a contact pressure of 100 psi (690 kPa). Both layers are incompressible with the elastic moduli characterized by the models shown in the figure. The thickness of layer 1 is 10 in. (254 mm). Determine the radial tensile strain at the bottom of layer 1 and the vertical compressive strain on the top of layer 2 at the following loading times: 0.01, 0.03, 0.1, 0.3, 1, 3, and 10 s.

To use KENLAYER, the mechanical models must be replaced by creep compliances:

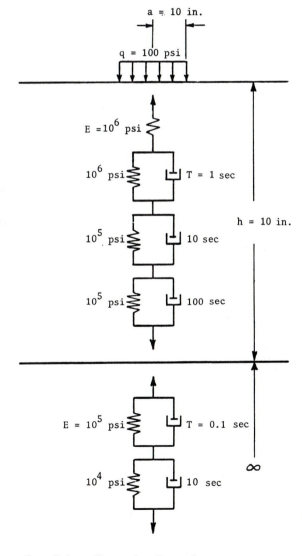

Figure B.8 Example 3 (1 in. = 25.4 mm, 1 psi = 6.9 kPa).

For layer 1 $\quad D(t) = 10^{-6} [1 + (1 - e^{-t}) + 10 (1 - e^{-0.1t}) + 10 (1 - e^{-0.01t})]$

$$= 10^{-6} [22 - e^{-t} - 10 (e^{-0.1t} + e^{-0.01t})] \qquad \text{(B.18)}$$

For layer 2 $\quad D(t) = 10^{-6} [10 (1 - e^{-10t}) + 100 (1 - e^{-0.1t})]$

$$= 10^{-6} (110 - 10e^{-10t} - 100e^{-0.1t}) \qquad \text{(B.19)}$$

The creep compliances at 7 time durations are computed by Eqs. B.18 and B.19 and tabulated in Table B.2.

It is interesting to note that the radial and vertical strains at the bottom of layer 1 are the same, or nearly the same, as those at the top of layer 2 and that the magnitude of vertical strains is twice as large as that of radial strains. These results are expected because the two layers are bonded and both are incompressible with a Poisson ratio of 0.5.

B.4.4 Example 4: Damage Analysis with Four Periods and One Load Group

Figure B.9 shows a three-layer system subjected to a moving circular load with a contact radius of 6 in. (152 mm) and a contact pressure of 80 psi (552 kPa). Layer 1 is linear viscoelastic and layers 2 and 3 are elastic. For illustrative purposes, a year is divided into four periods for damage analysis. The creep compliances shown in the data file were determined at a reference temperature of 70°F (21.1°C). It is assumed that the temperatures of the viscoelastic layer during the four seasons are 49.7, 59.5, 81.6, and 70.1°F (9.8, 15.3, 27.6, and 21.1°C), but the moduli of the elastic layers remain the same throughout the year. The horizontal tensile strain at the bottom of layer 1 and the vertical compression strain at the top of layer 3 on the axis of symmetry, as indicated by the black dots in Figure B.9, are computed and used for damage analysis. Determine the design life based on load repetitions of 500 per day and the Asphalt Institute failure criteria.

A run of KENLAYER shows that the design life is 11.83 years and is controlled by fatigue cracking. To reduce the printout for damage analysis, the first load group is printed in detail, while the other load groups contain a summary only. Detailed information for these later load groups, if needed, can be obtained by rerunning KENLAYER and changing NDAMA to 0.

B.4.5 Example 5: Damage Analysis with One Period and Three Load Groups

Figure B.10 shows a two-layer elastic system subjected to three different load groups: 18-kip (80-kN) single axles, 36-kip (160-kN) tandem axles, and 54-kip

TABLE B.2 CREEP COMPLIANCES AT DIFFERENT TIME DURATIONS

Loading Time (s)	0.01	0.03	0.1	0.3	1	3	10
Layer 1 (10^{-5} in.²/lb)	0.102	0.106	0.121	0.159	0.268	0.484	0.927
Layer 2 (10^{-5} in.²/lb)	0.105	0.289	0.732	1.250	1.950	3.590	7.320

Note. 1 in.²/lb = 0.145 m²/kN.

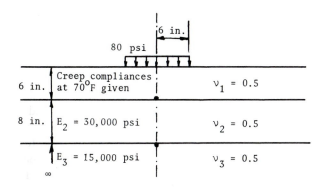

Figure B.9 Example 4 (1 in. = 25.4 mm, 1 psi = 6.9 kPa).

(240-kN) tridem axles. Each year has only one period and the number of repetitions in each year is 50,000 for each load group. The strains at three points, as indicated by the black dots in the figure, are determined and the most critical strains used for computing damage ratios and the design life.

A run by KENLAYER shows that the design life is 19.50 years and is controlled by permanent deformation. Note that multiple axles are analyzed twice. First determine the primary damage ratio by considering all axles and then determine the additional damage ratio by placing the response point halfway between two axles.

B.4.6 Example 6: Damage Analysis with Two Periods and Two Load Groups

Figure B.11 shows a three-layer system subjected to both single- and dual-wheel loads. Each year is divided into two seasons, a cold season with the asphalt temperature at 60°F (15.6°C) and a warm season with the asphalt temperature at

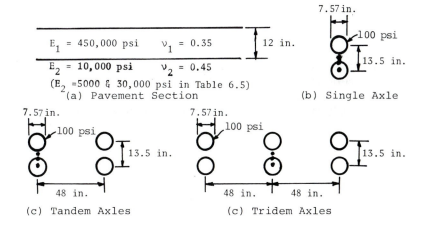

Figure B.10 Example 5 (1 in. = 25.4 mm, 1 psi = 6.9 kPa).

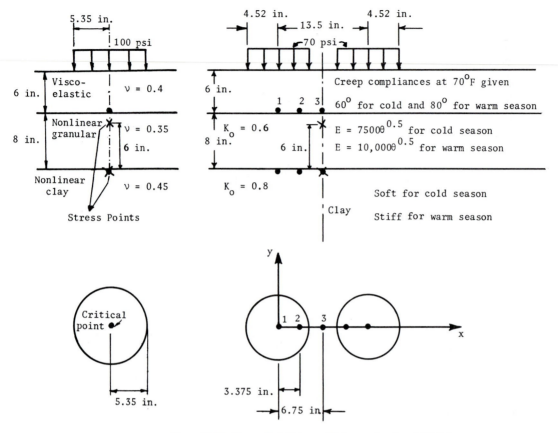

Figure B.11 Example 6 (1 in. = 25.4 mm, 1 psi = 6.9 kPa).

80°F (26.7°C). Layer 1 is viscoelastic with the creep compliances at 70°F (21.1°C) the same as those of Example 4. Layers 2 and 3 are nonlinear elastic with different properties for each season, as shown in the figure. For single wheels, the stresses and strains under the center of the loaded area are used for damage analysis; for dual wheels, the stresses and strains at the three points, shown by the black dots in the figure, are computed and the largest among the three are used. The stress points are located at the upper quarter of layer 2 and on the top of layer 3, as indicated by the crosses. The number of load repetitions per day is 70 for single wheels and 100 for dual wheels. Determine the design life based on the Asphalt Institute failure criteria.

A run of KENLAYER shows that the design life is 25.27 years and is controlled by permanent deformation. Note that only the first load group in the first period is printed in detail.

More About Kenslabs

This appendix is a supplement to Chapter 5 and describes the KENSLABS computer program in detail. It contains additional information on layered theory with concentrated loads, subroutines, flowchart, and five illustrative examples.

C.1 LAYER FOUNDATION UNDER CONCENTRATED LOAD

In Burmister's theory, the load is applied over a circular area. In the finite element method, the load is concentrated at each node. To find the stiffness of foundation, it is necessary to determine the deflection at a given node i due to a concentrated load at a given node j. This can be obtained from Eq. B.7 by reducing the contact radius to zero, or

$$R = \frac{1}{\pi a^2} \frac{a}{H} \int_0^\infty \frac{R^*}{m} J_1(m\alpha) \, dm = \frac{1}{\pi H^2} \int_0^\infty \frac{J_1(m\alpha)}{m\alpha} R^* \, dm \qquad (C.1)$$

Since

$$\lim_{\alpha \to 0} \frac{J_1(m\alpha)}{m\alpha} = 0.5 \qquad (C.2)$$

Eq. C.2 becomes

$$R = \frac{1}{2\pi H^2} \int_0^\infty R^* \, dm \qquad (C.3)$$

in which R is the dimensionless deflection and R^* is expressed by Eq. B.4e and involves the Bessel function $J_0(m\rho)$. To obtain more accurate results when ρ is large, a 32-point Gaussian quadrature formula is used for the first cycle of integration and an 8-point formula is used for the remaining cycles until the specified maximum number of iterations is reached.

C.2 SUBROUTINES

KENSLABS consists of one short main program and 32 subroutines. Similar to KENLAYER, all variables are transferred from the main program to the subroutines through arguments. Subroutines starting with LIQ, BOU, and BUR are used exclusively for liquid, Boussinesq, and Burmister foundations, respectively. Subroutines starting with SOL are used for both solid and layer foundations. The subroutines are listed below in alphabetical order for easy reference:

BESJ computes Bessel function of order 0 at a given radial distance (called by LAYSLA).

BOUDIA determines the flexibility matrix of diagonal elements for Boussinesq foundation by Gaussian quadrature (called by SOLFLE).

BOUINP reads and writes input data for Boussinesq foundation.

BOUOFF determines the flexibility matrix of off diagonal elements for Boussinesq foundation (called by SOLFLE).

BURDEF computes the flexibility matrix of layer foundation at 21 different radial distances (called by BURDIA and BUROFF).

BURDIA determines the flexibility matrix of diagonal elements for Burmister foundation by interpolation and Gaussian quadrature (called by SOLFLE).

BURINP reads and writes input data for Burmister foundation.

BUROFF determines the flexibility matrix of off-diagonal elements for Burmister foundation by interpolation (called by SOLFLE).

COEF determines the constants of integration for the first layer in Burmister foundation (called by LAYSLA).

DISCHE checks the convergence of displacements and iterates until desired accuracy is obtained.

DOWGEO computes number of dowels at each node, average distance between nodes, and equivalent spring constant for doweled joint.

DOWINP reads and writes input data for doweled joints or spring constants.

DOWOUT prints the shear and moment at all nodes along the joints and the shear and bearing stresses of dowels, if any.

DOWSTI adds the stiffness of joints to the overall stiffness matrix.

IDENTI computes local coordinates, temperature curling, nodal forces due to applied loads, modulus of rigidity of slabs, and several parameters for identifying slabs and joints.

GESB applies the Gauss elimination method to solve simultaneous equations without iterations.

LAYSLA computes vertical displacements on the surface of Burmister foundation due to a unit load applied at a given radial distance (called by BURDEF).

LIQCON checks contact conditions of liquid foundation, computes amount of gaps and precompression, and stops iteration when the same contact condition is obtained. This subroutine also sums up the applied loads and the subgrade reactions. If the iteration method is used to solve simultaneous equations and the difference between applied loads and subgrade reactions is larger than the tolerance, more iterations will be made.

LIQINP reads and writes input data for liquid foundation.

LIQMOD modifies nodal forces for liquid foundation due to contact conditions, gaps at joints, and half-band width.

LOADMA computes displacements of slab based on the upper triangular matrix.

MATINV inverts the flexibility matrix to form the stiffness matrix of foundation (called by SOLSTI).

SLAINP reads and writes input data common to all types of foundation.

SOL determines the two constants of integration for the bottommost layer (called by LAYSLA).

SOLCON applies to solid and layer foundations in the same way that LIQCON applies to a liquid foundation.

SOLCOR computes global coordinates for solid or layer foundation.

SOLFLE generates the flexibility matrix of solid or layer foundation.

SOLMOD modifies nodal forces for solid or layer foundation due to contact conditions, half-band width, and gaps at joints.

SOLSTI determines the stiffness matrix of solid or layer foundation.

STIFFN computes the stiffness of slabs and modifies the stiffness due to symmetry.

STROUT computes stresses in the slabs at designated nodes including normal stress in x direction, normal stress in y direction, shear stress in xy plane, major principal stress, minor principal stress, and maximum shear stress.

TRIANG applies the Gauss elimination method to form an upper triangular banded matrix to be used repeatedly during iterations.

C.3 FLOWCHARTS

Figure C.1 shows a simplified flowchart for KENSLABS without damage analysis. Each step shown in Figure C.1 is explained below:

1. Subroutine SLAINP is called first. Depending on the type of foundation, LIQINP, BOUINP, or BURINP is then called. If there is more than one slab, DOWINP should also be called.

2. Based on the externally applied loads, the force vector $\{F\}$ in Eq. 5.15 is formed by calling IDENTI. Any node not subject to external loads is assigned a zero force.

3. If the foundation is a liquid, it is not necessary to form the flexibility matrix of foundation because the stiffness matrix can be determined directly from Eq. 5.2, so steps 4, 5, and 6 can be skipped.

4. If the foundation is a solid, the flexibility matrix can be determined directly from Eq. 5.3, so step 5 can be skipped.

5. If the foundation is layered, the flexibility matrix can be determined from Eq. C.3 based on Burmister's layered theory. To save the computer time, the flexibility coefficients are evaluated at 21 different distances by calling BESJ, SOL, COEF, LAYSLA, and BURDEF. These 21 coefficients are interpolated in step 6 to form the flexibility matrix.

6. The flexibility matrix of solid foundation is formed by calling BOUOFF, BOUDIA, and SOLFLE, and that of layer foundation by calling BUROFF, BURDIA, and SOLFLE.

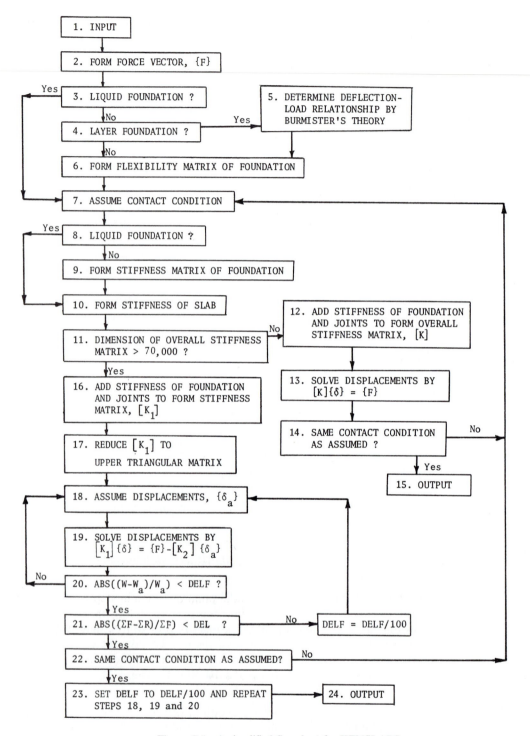

Figure C.1 A simplified flowchart for KENSLABS.

More about Kenslabs   App. C

7. The contact condition between slab and foundation must be assumed before the stiffness of foundation is obtained by inverting the flexibility matrix. Any nodal point not in contact must be removed from the flexibility matrix and not be used for matrix inversion. For the first cycle of iterations, full contact is usually assumed.

8. If the foundation is a liquid, the stiffness matrix is known and no matrix inversion is needed, so step 9 can be skipped.

9. The stiffness matrix of solid and layer foundations is formed by calling MATINV and SOLSTI.

10. The stiffness matrix of the slab has a narrow band and can be formed by calling STIFFN.

11. Depending on the amount of storage required for the overall stiffness matrix, two procedures are used for solving the simultaneous equations. If the matrix has a dimension smaller than 70,000, the direct method of Gauss elimination is used to solve Eq. 5.15, as indicated by steps 12 through 15. If the dimension is greater than 70,000, a maximum allowable half-band width is selected and the iterative method is used to solve Eq. 5.27, as indicated by steps 16 through 24. If there is more than one load group, the iteration method will be used automatically because the reduced matrix can be used repeatedly for other load groups to save the computer time. To provide storage spaces for the reduced matrix, the dimension of stiffness matrix is limited to 35,000, instead of 70,000.

12. The overall stiffness matrix $[K]$ is formed by superimposing the stiffness matrices of foundation and joints to the stiffness matrix of slab. The superposition of liquid foundation is effected by calling LIQSUP; that of solid and layer foundations is effected by calling SOLSUP. The superposition of joints is effected by calling DOWGEO and DOWSTI.

13. The force vector $\{F\}$ is modified for temperature curling and contact conditions by calling LIQMOD or SOLMOD. The simultaneous equations is solved by calling GESB.

14. After the nodal displacements are obtained, the contact conditions at each nodal point is evaluated by calling LIQCON or SOLCON. The contact conditions thus evaluated are compared with the conditions assumed in step 7. If a discrepancy exists at any nodal point, the contact condition at that point should be changed and the process repeated.

15. If the contact condition at every nodal point obtained in step 14 is the same as that assumed in step 7, the final outputs including the stresses and displacements in the slab and at the joints are printed by calling DOWOUT and STROUT.

16. This step is similar to step 12 except that only part of the stiffness matrix is formed, as indicated by $[K_1]$ in Eq. 5.27.

17. The matrix $[K_1]$ is reduced to an upper triangular matrix by calling TRIANG. This triangular matrix is used repeatedly in the iterative process until new contact conditions are assumed.

18. At the first iteration, the displacements $\{\delta_a\}$ are assumed to be zeros. The displacements obtained in step 19 are used as the assumed displacements in later iterations.

19. The force vector $\{F\}$ is modified for bandwidth, temperature curling, and contact conditions by calling LIQMOD or SOLMOD. The simultaneous equations can be solved by calling LOADMA.

20. The convergence of displacements $\{\delta\}$ is checked by calling DISCHE. If $ABS[(w - w_a)/w] > DELF$, iterations will continue until the difference between the assumed and computed deflections converges. Note that w and w_a are the computed and assumed vertical deflections at a designated node and DELF is the tolerance for iterations initially set equal to DEL but later reduced when needed.

21. A check is made on the sum of applied forces ΣF and the sum of subgrade reactions ΣR by calling LIQCON or SOLCON. If $ABS[(\Sigma F - \Sigma R)/\Sigma F] > DEL$, more iterations will be made by reducing DELF to $0.01 \times DELF$. Note that whether more iterations are required is finally decided by DEL, which always remains the same, but the value of DELF continues to decrease until the convergence on DEL is met.

22. After ΣR converges, LIQCON or SOLCON will check the contact conditions. If the contact condition at any nodal point is not the same as the assumed condition, the process is repeated until the same contact conditions are obtained.

23. To obtain more accurate results at the final contact conditions, a few more iterations are run by reducing DELF to $0.01 \times DELF$.

24. When $ABS[(w - w_a)/w] < DELF$, the final results are printed by calling DOWOUT and STROUT.

Damage analysis is based on the fatigue cracking of PCC. Each year can be divided into a number of periods and each period can have a number of different load groups. Damage is defined by the sum of cracking index, which is similar to the damage ratio expressed by Eq. 3.19. The sum of cracking index should not be greater than 1. Figure C.2 shows a simplified flowchart for damage analysis.

C.4 ILLUSTRATIVE EXAMPLES

Five examples are given to illustrate the practical application of KENSLABS. The first four examples are realistic cases in which the slabs are divided into finite elements of appropriate sizes, so the results obtained should be quite accurate. To minimize the amount of printout, MDPO is specified as 0 except in Example 5. Example 5 is the most complicated case involving four slabs on layer foundation with partial contact. To reduce the printout, more detailed output is specified only in the second step and each slab is divided into four elements, which are certainly not sufficient to obtain meaningful results. Therefore, the purpose of this example is merely to illustrate the input and output.

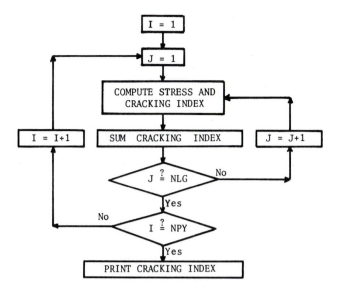

Figure C.2 Simplified flowchart for damage analysis by KENSLABS.

The approximate time for running these examples is shown in Table C.1. It can be seen that the running time for a PC without a math coprocessor is 8 to 13 times longer than that with a math coprocessor.

In all the examples presented, it is assumed that the concrete has an elastic modulus of 4×10^6 psi (27.6 GPa) and a Poisson ratio of 0.15. All other parameter values are shown in the data file and are not repeated in the text.

C.4.1 Example 1: Warping of Slab on Liquid Foundation with Full Contact

This example is the same as Example 4.1 in which the slab is 25 ft (7.62 m) long, 12 ft (3.66 m) wide, and 8 in. (203 mm) thick and is subjected to a temperature differential of 20°F (11.1°C). The manual method based on liquid foundation results in a maximum curling stress of 238 psi (1.64 MPa) in the interior and 214 psi (1.38 MPa) at the edge. It is now desirable to determine these stresses by KENSLABS. The slab is divided into rectangular finite elements as shown in Figure C.3. Due to symmetry, only a quarter of the slab need be used for analysis.

TABLE C.1 APPROXIMATE EXECUTION TIME FOR EXAMPLE PROBLEMS BY KENSLABS

Example no.	PC with math coprocessor	PC without math coprocessor
1	0.4 min	4.0 min
2	1.1 min	11.0 min
3	14.2 min	110.0 min
4	29.2 min	332.0 min
5	11.0 min	143.0 min

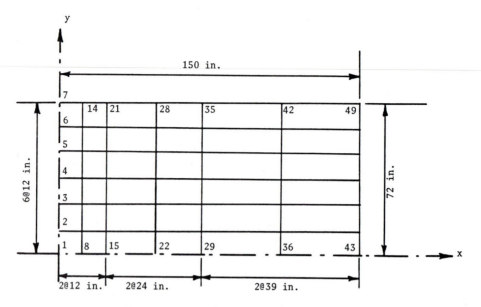

Figure C.3 Examples 1 and 2 (1 in. = 25.4 mm).

The results obtained by KENSLABS show that the stress in the x direction at the center of slab, or node 1, is 237.59 psi (1.64 MPa), which checks with the 238 psi (1.64 MPa) obtained in Example 4.1, while the stress at the edge of slab, or node 7, is 216.22 psi (1.49 MPa), which also checks closely with the 214 psi (1.48 MPa) obtained in Example 4.1.

C.4.2 Example 2: Warping of Slab on Liquid Foundation with Partial Contact

This example is the same as Example 1 except that the spring under each nodal point cannot take tension. If only upward reactions can exist, the forces applied to the slab will not be in equilibrium unless downward forces due to the weight of slab are considered. Similar to Example 1, the slab is divided into finite elements as shown in Figure C.3 and the only differences in input data are a change in NCYCLE from 1 to 10 and NWT from 0 to 1.

The results obtained by KENSLABS show that when the slab curls up, the interior curling stress at node 1 based on partial contact is 230.77 psi (1.59 MPa), which is about 3% smaller than that based on full contact, and the edge curling stress at node 7 is 211.99 psi (1.46 MPa), which is about 2% smaller.

C.4.3 Example 3: Two Slabs on Solid Foundation with Full Contact

Figure C.4 shows a two-slab system on a solid foundation subjected to a 32,000-lb (142.4-kN) tandem-axle load with a tire pressure of 90 psi (621 kPa). One axle load is applied near the doweled joint with the exterior tire on the pavement edge. The

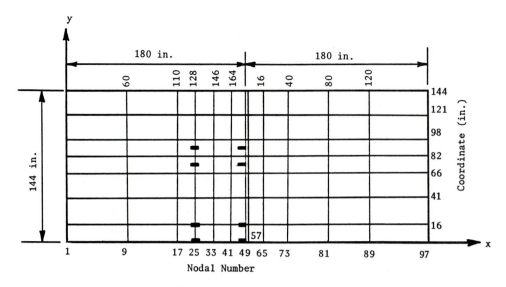

Figure C.4 Example 3 (1 in. = 25.4 mm).

concrete slab is 8 in. (203 mm) thick and the subgrade has an elastic modulus of 5000 psi (34.5 MPa) and a Poisson ratio of 0.45. Assuming that the slab and subgrade are in full contact, determine the maximum stress and deflection in the slabs.

The results obtained by KENSLABS show that the maximum stress is 226.27 psi (1.56 MPa) and occurs at node 25 on the pavement edge. The maximum deflection is 0.05386 in. (1.37 mm) and occurs on node 49 at the pavement corner.

C.4.4 Example 4: Damage Analysis with Two Periods and Three Load Groups

Figure C.5 shows a concrete slab on a solid foundation subject to three different load groups consisting of 18,000-lb (80-kN) single axles, 36,000-lb (160-kN) tandem axles, and 54,000-lb (240-kN) tridem axles, each with 10 repetitions per day. Due to symmetry, only one-half of the slab need be considered. The concrete slab has a thickness of 8 in. (203 mm) and a modulus of rupture of 600 psi (4.9 kPa). The subgrade has an elastic modulus of 5000 psi (69 MPa) and a Poisson ratio of 0.4. Based on a contact pressure of 100 psi (690 kPa), the size of each contact area is 8.08 in. × 5.57 in. (205 mm × 141 mm). Each year is divided into two periods with foundation seasonal adjustment factors, FSAF, of 1.0 and 0.8 for each period. Determine the design life based on PCA fatigue criteria.

As obtained by KENSLABS, the sum of cracking index over two periods, each with three load groups, is 0.057236 and the design life based on fatigue cracking is 17.47 years. In applying KENSLABS for damage analysis, the stresses at three edge points (NPRINT = 3), one under each axle, must be computed. Based on the stresses at these three points, KENSLABS will determine the type of axle load and make the required analysis accordingly. Note also that only the

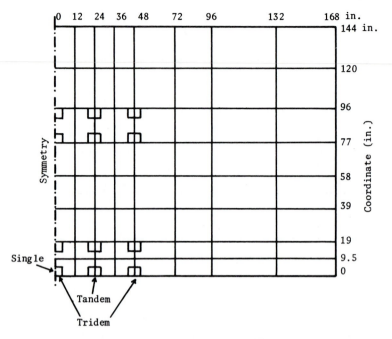

Figure C.5   Example 4 (1 in. = 25.4 mm).

first load group in the first period is printed in detail. Detailed information on all load groups in all periods, if needed, can be obtained by rerunning KENSLABS and changing NDAMA to 0.

C.4.5 Example 5: Four Slabs on Layer Foundation with Partial Contact

Figure C.6 shows a four-slab system subjected to temperature curling, pumping, and two wheel loads. Due to symmetry with respect to the horizontal axis, the problem can be analyzed by considering the upper or lower half of the slabs by specifying NSX = 1. However, all four slabs are used in this example. The results shown by the computer printout are all symmetrical with respect to the horizontal axis, which is as expected. In addition to the more complex situation of four slabs on layer foundations, the following options are invoked:

1. The problem is solved in two steps, or NPROB = 2. First, determine the gaps and precompressions due to the weight of slab, temperature curling, and initial gaps by specifying NGAP = 8, INPUT = 0, NTEMP = 1, and NCYCLE = 10. Then, determine the stresses and deflections under wheel loads by specifying NGAP = 0, INPUT = 1, NTEMP = 0, and NCYCLE = 10.

2. There are two layers of unbonded slabs as indicated by NLAYER = 2 and NBOND = 0.

3. Dowel bars are used for transverse joints and spring constants for longitudinal joints. There are no moment transfers across all joints.

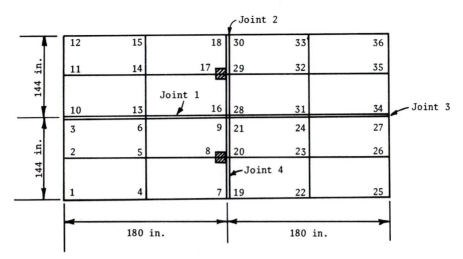

Figure C.6 Example 5 (1 in. = 25.4 mm).

As obtained by KENSLABS, the maximum stresses in step 1 are 43.38 psi (299 kPa) for layer 1 and 13.25 psi (91 kPa) for layer 2, both located at nodes 8 and 17. The maximum stresses in step 2 are 125.17 psi (864 kPa) for layer 1 and 39.68 psi (274 kPa) for layer 2, both also located at nodes 8 and 17. Due to the use of only a very limited number of finite elements, these stresses are not realistic and are cited only for checking purposes.

APPENDIX

Input Programs

D.1 GENERAL FEATURES

The input program to create and edit the data file for **KENLAYER** is called **LAYERINP** and that for **KENSLABS** is called **SLABSINP**. The features common to both programs are discussed in this section.

D.1.1 Use of Keys

There are only a few keys to be used to input or edit the data. The arrow keys, up (↑) or down (↓), can be used to move the cursor to any item on the menu. After pressing the return or enter key, the data entry form for that item will be shown on the screen. Most of the forms have many fields which are filled initially with a zero or a default value. The field in which the cursor resides is highlighted. You can either type in your data to override the zero or default value or keep the value as it appears on the screen. The cursor is then moved to the next field by pressing the return key. You can also use the arrow keys, up (↑), down (↓), left (←), or right (→), to move the cursor to any field and make the necessary change without using the return key. For lengthy forms with more than one screen, you can use the PgDn key to scroll to the next page or the PgUp key to scroll to the previous page. After all the fields have been properly filled, press the Esc key to return to the original menu.

You can use the letter I key to insert a line or the letter D key to delete a line. For example, when using LAYERINP, you input the number of z coordinates (NZ) as 3 in the "General Information" screen. When you select the "Z Coordinates for Analysis" screen, three lines identified as points 1, 2, and 3 appear on the screen for you to enter the z coordinate (ZC). You can delete any point by moving the cursor to that line and pressing the D key. The number of points on the screen is reduced to 2 instantaneously and the number of z coordinates on the "General Information" screen is also changed to 2 automatically. If you wish to add a point, say between points 2 and 3, move the cursor to the line for point 2 and press the I key. An additional line identified as point 3 will appear on the screen. The original point 3 will be changed to point 4 and the total number of z coordinates on the "General Information" screen will also be increased to 4 automatically.

You may have to use the letter X key to reach a screen for editing the data. For example, when you input the number of response points (NR or NPT) on the "Load Information" screen and press the return key, an auxiliary screen for the coordinates of response points will appear automatically. However, if you want to edit the coordinates later, you have to enter the "Load Information" screen, move the cursor to the line on which the number of response points is shown, and press the X key. Of course, you can enter the coordinates screen by retyping the number of response points and pressing the return key. A summary of the keys and their functions is listed in Table D.1.

D.1.2 General Operation

The data entry forms can be reached through a series of menus, main screens, and auxiliary screens. When a control menu consisting of several items appears on the screen, move the cursor to the item you want to enter and press the return key. Another menu with several items or a data entry form with one or more fields will appear on the screen. If another menu appears, repeat the process until a data entry form is obtained. After the form has been filled out properly, press the Esc key one or more times until the screen returns to the control menu. You can then select another item to enter or edit.

The data entry form may have a main screen and an auxiliary screen. The main screen will switch automatically to the auxiliary screen when you input your data into a certain field and press the return key. After filling out the auxiliary screen, press the Esc key to return to the main screen. You can also enter an auxiliary screen by pressing the X key as explained previously. Figure D.1 shows all possible ways of getting into and out of a data field. Note that some data entry forms have auxiliary screens while others have none. If the message "Press X to

TABLE D.1 KEYS AND THEIR FUNCTIONS

Key	Function
Return	Selects an item on menu or moves cursor to next field
Arrow	Moves cursor to an item on menu or moves cursor to any field
Esc	Returns to main or control menu or to previous screen
D	Deletes a line on the screen
I	Inserts a line on the screen
X	Enters into an auxiliary screen for editing
PgDn	Moves down a page when the data have more than one screen
PgUp	Moves up a page when the data have more than one screen

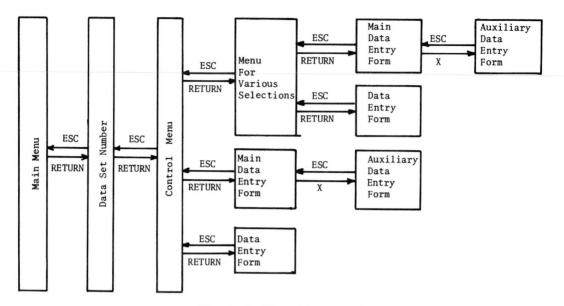

Figure D.1 All possible ways of data entry.

edit'' appears at the bottom of the screen, the existence of auxiliary screens is indicated.

If a data field has more than one option, the option you select may dictate the screens to appear later. If you forget to input the number of z coordinates on the "General Information" screen but try to get into the "Z Coordinates for Analysis" screen, a message that the number of z coordinates is zero will appear on the screen. You should press the return key, get into the "General Information" screen, and input the number of z coordinates. You can then reenter the "Z Coordinates for Analysis" screen and input the necessary data.

D.1.3 Main Menu

After typing LAYERINP, a title screen will be displayed. You can then press the return or enter key and a main menu, as shown by Figure D.2, will appear. The main menu for SLABSINP is the same except that the name LAYERINP is replaced by SLABSINP.

If you do not have a data file, you should move the cursor to "Create an Input Data File" and press the return key. If you have a data file that you want to edit, leave the cursor at "Load an Input Data File Previously Created" and press the return key. The program will ask for a file name. After typing the file name, move the cursor to "Edit the Current Data File" and press the return key. When the data entry or editing is completed, press the Esc key repeatedly until the main menu is displayed. Then move the cursor to "Save the Current Data File" and press the return key. The program will ask for the file name to be saved. After saving the file, move the cursor to "Exit the Program" and press the return key. A

```
Load an Input Data File Previously Created
Edit the Current Data File ................ (Edit Mode )
Save the Current Data File
Create an Input Data File ................ (Input Mode)
Exit the Program
```

Figure D.2 Main menu for LAYERINP.

message "Do you wish to exit (Y/N) ?" will appear on the screen. Type Y to exit from the input program.

D.1.4 Number of Data Sets

After selecting "Edit the Current Data File" or "Create an Input Data File" on the main menu, a screen containing a number of data sets is displayed, as shown in Figure D.3. A total of 10 data sets can be created for KENLAYER and run at the same time, but due to storage limitations only two sets can be created for KENSLABS. Before inputting the data, the status of each data set is not defined because there is no data in the file. After entering the data, the status will be changed to defined. Starting from data set No. 1, you can create as many data sets as you want until the limit of 10 or 2 is reached.

D.2 LAYERINP

The operation of LAYERINP is quite straightforward. By following the instructions on the screens, you should have no problem inputting all the data. Because the number of screens or data entry forms are too numerous to be displayed here,

DATA SETS

Problem number		Status
1	=	not defined
2	=	not defined
3	=	not defined
4	=	not defined
5	=	not defined
6	=	not defined
7	=	not defined
8	=	not defined
9	=	not defined
10	=	not defined

Figure D.3 Screen showing the status of data sets.

only the control menu, the general information, and the load information screens are presented.

D.2.1 Control Menu

After selecting the data set or problem number, a control menu, as shown in Figure D.4, will appear on the screen for each data set. The data are grouped into 11 items. You can move the cursor to any item and press the return key for further information.

D.2.2 General Information

After selecting "GENERAL INFORMATION" on the control menu, the "General Information" screen, as shown in Figure D.5, is displayed. This is the original screen before the entry of any data. All nonzeros are truly default values, while some zeros, such as the number of layers (NL), are not default values and must be replaced by new input data.

D.2.3 Load Information

The load information screen is shown in Figure D.6. This screen is used as an example because it is more complex and consists of an auxiliary screen for each load group, as indicated by the statement "Press X to edit coordinates of response points."

When the number of load groups (NLG) on the general information screen is specified as 6, six lines, one for each load group, is shown in Figure D.6. The number of response points in the last column is represented by NR for a single contact area (LOAD = 0) or NPT for multiple contact areas (LOAD > 0). For example, load group No. 1 is assigned LOAD = 0 and NR = 3. When 3 is typed in the last column and the return key is pressed, an auxiliary screen, as shown in Figure D.7, appears. When the radial coordinates (RC) of all three points are entered, press the Esc key to return to the load information screen. You can then

DATA SET NUMBER 1 CONTROL MENU

```
=    TITLE
=    GENERAL INFORMATION
=    Z COORDINATES FOR ANALYSIS
=    LAYER THICKNESSES, POISSON RATIOS AND UNIT WEIGHTS
=    TYPE OF INTERFACE
=    LAYER MODULI
=    LOAD INFORMATION
=    NONLINEAR LAYERS
=    VISCOELASTIC LAYERS
=    DAMAGE ANALYSIS
=    NUMBER OF LOAD REPETITIONS
```

Figure D.4 Control menu for LAYERINP.

DATA SET NUMBER 1—GENERAL INFORMATION

Type of materials (MATL)	1
Damage analysis (1 = yes, 0 = no), (NDAMA)	0
Number of periods per year (NPY)	1
Number of load groups (NLG)	1
Tolerance for integration (DEL)	.001
Number of layers (NL)	0
Number of z coordinates (NZ)	0
Maximum cycles of integration (ICL)	80
Type of responses (NSTD)	9
All layer interfaces bonded? (1 = yes, 0 = no), (NBOND)	1
Number of layers for bottom tension (NLBT)	1
Number of layers for top compression (NLTC)	1

Figure D.5 General information for LAYERINP.

DATA SET NUMBER 1—LOAD INFORMATION

Load grp. no.	Number of axles (LOAD)	Contact radius (CR)	Contact pressure (CP)	Dual spacing (YW)	Axle spacing (XW)	No. of response points (NR or NPT)
1	0	0	0	0	0	0
2	0	0	0	0	0	0
3	0	0	0	0	0	0
4	0	0	0	0	0	0
5	0	0	0	0	0	0
6	0	0	0	0	0	0

Press X to edit coordinates of response points.

Figure D.6 Main screen of load information for LAYERINP.

LOAD GROUP NO. 1

Response point no.	Radial coordinate (RC)
1	0
2	0
3	0

Figure D.7 Auxiliary screen for single contact area.

continue entering the data for load group No. 2. If load group No. 2 is assigned LOAD = 1 and NPT = 3, an auxiliary screen, as shown in Figure D.8, is displayed. After the x and y coordinates of all three points are entered, press the Esc key to return to the load information screen. The process is repeated until the data for all six load groups are entered.

LOAD GROUP NO. 2

Point no.	X coordinate (XPT)	Y coordinate (YPT)
1	0	0
2	0	0
3	0	0

Figure D.8 Auxiliary screen for multiple contact areas.

You can enter the auxiliary screen of any load group by moving the cursor to the line indicating that load group and pressing the X key. Of course, you can always enter the auxiliary screen by retyping the number of response points and pressing the return key.

D.2.4 Flowchart

Figure D.9 shows a flowchart of LAYERINP. This chart can be used as a handy reference to find the location of input parameters on various screens.

D.3 SLABSINP

As with LAYERINP, only the control menu, general information, and loaded areas and contact pressures are presented.

D.3.1 Control Menu

The control menu has 15 items, as shown in Figure D.10. You can move the cursor to any item for which you wish to input or edit data.

D.3.2 General Information

The general information screen is shown in Figure D.11. All the data shown on the screen are default values. Be sure to override them if they are not applicable to your case.

D.3.3 Loaded Areas and Contact Pressures

When selecting "LOADED AREAS AND CONTACT PRESSURES" on the control menu, the screen shown in Figure D.12 will appear. Because the number of load groups (NLG) on the general information screen is 6, six load groups are shown. The statement "Press X to edit the loaded areas data" indicates the existence of an auxiliary screen for each load group.

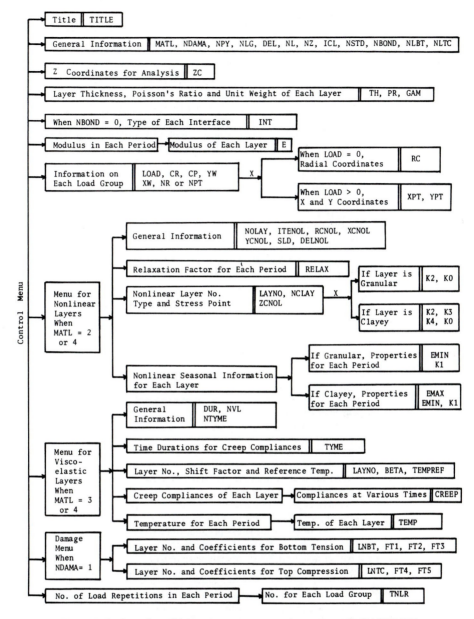

Figure D.9 Location of input parameters on various screens in LAYERINP.

DATA SET NUMBER 1 CONTROL MENU

=	TITLE
=	GENERAL INFORMATION
=	SLAB ARRANGEMENTS
=	CURLING AND CONTACT CONDITIONS
=	SLAB LAYERS AND FATIGUE
=	X COORDINATES OF FINITE ELEMENT MESH
=	Y COORDINATES OF FINITE ELEMENT MESH
=	THICKNESSES, POISSON RATIOS AND MODULI OF SLAB LAYERS
=	LOADED AREAS AND CONTACT PRESSURES
=	CONCENTRATED LOADS AT DESIGNATED NODES
=	OTHER OPTIONAL INFORMATION
=	FOUNDATION SEASONAL ADJUSTMENT FACTORS
=	FOUNDATION INFORMATION
=	JOINT INFORMATION
=	NUMBER OF LOAD REPETITIONS

Figure D.10 Control menu for SLABSINP.

DATA SET NUMBER 1—GENERAL INFORMATION

Type of foundation (NFOUND)	0
Damage analysis (NDAMA), (*)	0
Number of periods per year (NPY)	1
Number of load groups (NLG)	1
Number of slab layers (NLAYER)	1
Bond between two layers (NBOND, 1 = bonded, 0 = unbonded)	0
Number of slabs (NSLAB)	1
Number of joints (NJOINT)	0
Nodal No. for checking deflection (NNCK)	1
Number of nodes for stress printout (NPRINT)	0
No. of nodes on X axis of symmetry (NSX)	0
No. of nodes on Y axis of symmetry (NSY)	0
More detailed printout (1 = yes, 0 = no), (MDPO)	0

* Note: 0 = No damage analysis, 1 = PCA fatigue criteria, and 2 = User specified fatigue coefficients FI and F2.

Figure D.11 General information for SLABSINP.

DATA SET NUMBER 1—LOADED AREAS AND CONTACT PRESSURES

Load grp. no.	No. of uniformly distributed loaded areas (NUDL)
1	0
2	0
3	0
4	0
5	0
6	0

Press X to edit the loaded areas data.

Figure D.12 Main screen of loaded areas for SLABSINP.

LOAD GROUP NO. 2

Load sequence	Slab no. (LS)	X Coordinates		Y Coordinates		Contact pressure (QQ)
		Left (XL1)	Right (XL2)	Bottom (YL1)	Top (YL2)	
1	0	0	0	0	0	0
2	0	0	0	0	0	0
3	0	0	0	0	0	0
4	0	0	0	0	0	0

Figure D.13 Auxiliary screen for uniformly distributed loaded areas.

If load group No. 2 has four loaded areas and you type 4 in the field and press the return key, the screen shown in Figure D.13 is displayed. After entering all data, press the Esc key to return to the main screen. You can enter the auxiliary screen of any load group by moving the cursor to that load group and pressing the X key.

D.3.4 Flowchart

Figure D.14 shows a flowchart of SLABSINP. This chart can be used as a handy reference to find the location of input parameters on various screens.

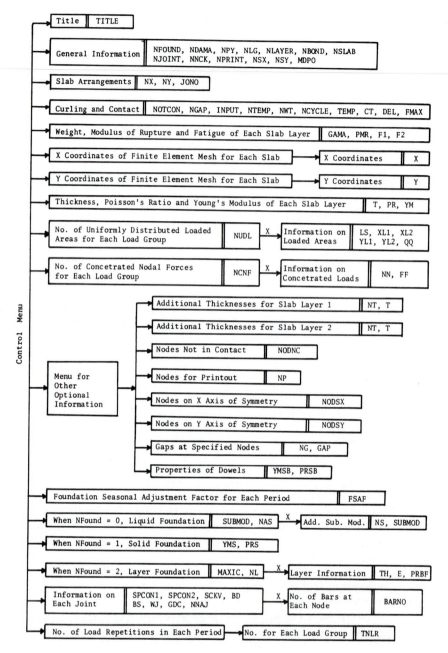

Figure D.14 Location of input parameters on various screens in SLABSINP.

APPENDIX

List of Symbols

A a/h_2 of a three-layer system; or area of dowel; or average number of axles per truck; or temperature susceptibility of bitumen; or cross-section area normal to direction of flow; or area of pipe; or area cracked; or permanent deformation parameter in the Ohio State model; or regression constant relating effective indirect tensile strength to modulus of rupture of concrete

AC   asphalt concrete

ADT   average daily traffic

$(ADT)_0$ average daily traffic at start of design period

ADTT   average daily truck traffic

AOS   apparent opening size of fabric

A_c tire contact area

A_i constant of integration for layer i; or coefficients of PSI equation

A_{ji} intercept of log deflection versus log E

A_s area of steel per unit width of slab

A_1 area of one tie bar at longitudinal joint; or regression constant in stiffness modulus test

a contact radius; or half-length of rectangular finite element; or a constant; or distance from support to third point of beam; or pavement age in years

a' modified contact radius

a_{OL} structural layer coefficient of asphalt overlay

a_T time–temperature shift factor

a_{bs} structural layer coefficient after break and seat

a_d contact radius of dual tires

a_s contact radius of single tire

a_1 distance from corner of slab to center of load

a_1, a_2, a_3 layer coefficients for asphalt surface, base, and subbase courses, respectively

a_{2r} structural layer coefficient of cracked PCC pavement

B a coefficient for determining AOS; or a pavement roughness property; or a damage factor

BPN   British pendulum number

B_i constant of integration for layer i; or coefficients of PSI equation

b total compliance due to retarded strain; or half-width of rectangular element; or a constant; or width of beam

C conversion factor for dual or dual-tandem wheels; or correction factor for warping stress in finite slab with length L; or coefficient of variation; or area cracked; or adjustment factor due to slab–subbase friction; or cohesiometer value; or fatigue correction factor due to different mixes; or a pavement roughness property; or a constant; or pavement condition factor

CBR California Bearing Ratio

CRCP continuous reinforced concrete pavement

CTB cement-treated base

CW crack width

C_d drainage factor for rigid pavements

C_i constant of integration for layer i

C_k Hazen's coefficient relating effective size to permeability

C_m correction factor for dynamic effect

Cov covariance

C_s spacing of transverse cracks or joints

C_u uniformity coefficient

C_v visual layer condition factor

C_w shear spring constant of joint

C_x correction factor for finite slab with length L_x; condition factor of existing pavement before overlay

C_y correction factor for finite slab with length L_y; condition factor of the entire overlaid pavement

C_{yx} condition factor of existing pavement after overlay

C_1 an adjustment factor for erosion, 1 for untreated subbases and 0.9 for stabilized subbases

C_2 equivalent damage ratio, 0.06 for edge loading and 0.94 for corner loading

C_θ moment spring constant of joint

c side length of a square contact area; or initial curling of slab; or a constant; or cracking index; or critical period correction factor

c_i coefficients of Dirichlet series for viscoelastic response; or initial curling of slab at node i

c_1 fatigue coefficient of laboratory specimen

D creep compliance; or differential operator $\partial/\partial t$; or depth of sealant reservoir; or thickness of concrete slab by AASHTO method; or variance before square; or percentage of ADT in design direction; or diameter of pipe; or constant relating modulus of rupture to modulus of elasticity of concrete

DETJT percentage of deteriorated joints

DT_D design temperature drop

D_{OL} required thickness of PCC overlay

D_{SB} thickness of subbase under concrete slab

D_{SG} thickness of subgrade above a rigid foundation

D_a, D_b grain sizes corresponding to a and b % passing

D_i creep compliance at time i; or constant of integration for layer i; or functions of surface deterioration in PSI equation

D_0 existing PCC thickness

D_r damage ratio at end of year; or damage ratio at the end of design period

D_x grain size corresponding to x% passing

D_{xeff} effective thickness of in situ PCC layer prior to overlay

D_y PCC thickness required to support overlay traffic

D_1, D_2, D_3 thickness of hot mix asphalt, base, and subbase by AASHTO method

D_2 amount of displacement in stabilometer test

D_{15}, D_{50}, D_{85} grain size corresponding to 15, 50, and 85% passing, respectively

d distance from neutral axis to bottom of slab; or duration of moving load; or clearance between dual tires; or diameter of steel bar; or distance to the center of slab where curling is zero; or a constant; or distress indicator; or subgrade type

$d_{i,j}$ distance between node i and node j

d_s deformation of saturated specimen

d_u deformation of unsaturated specimen

E elastic modulus; or elastic modulus of concrete; or expectation; or error between PSR and PSI; or equivalent factor for emulsified asphalt base

$\overline{E}$ Laplace transform of E

E^* complex modulus

$|E^*|$ dynamic modulus

EALF   equivalent axle load factor

EDR   equivalent damage ratio

ELP   elastic layer program

ESAL   equivalent single-axle load, which is the total number of repetitions of a standard 18-kip axle load during the design period

E_{SB} resilient modulus of subbase for concrete pavement

(ESAL)$_r$ remaining life in ESAL

(ESAL)$_0$ ESAL for the current year

ESWL   equivalent single-wheel load

E_c elastic modulus of concrete

E_d Young's modulus of dowel

E_f elastic modulus of solid foundation; or modulus in fully cured state

E_i elastic modulus of Kelvin model; or modulus in uncured or initial state

E_o initial elastic modulus for nonlinear soil; or elastic modulus of Maxwell model; or stiffness modulus when $\sigma = 0$

E_{pcc} PCC effective layer modulus reflecting cracking and loss of support

E_r (N) elastic modulus due to unloading at Nth repetition

E_r modulus ratio between cracked concrete and uncracked concrete of 5×10^6 psi

E_s stiffness modulus

E_t modulus at curing time t

E_1 elastic modulus of layer 1 or HMA layer; or elastic modulus of Kelvin model

E_2 elastic modulus of layer 2 or base course

E_3 elastic modulus of layer 3, which may be a subbase or a subgrade

E_4 elastic modulus of subgrade for a four-layer system

E_{1a}, E_{1b} elastic moduli of 1st and 2nd asphalt layers, respectively

e critical tensile strain at the bottom of layer 1

F deflection factor; or a function; or a force vector; or tractive force at tire–pavement contact; or a probability function; or faulting; or temperature adjustment factor

F_{RL} remaining life factor

FWD   falling weight deflectometer

F_d deflection factor under dual wheels

F_e strain factor

F_i EALF for ith load group

F_s deflection factor under single wheels

$F(t)$ function of t

F_w vertical force

$F_{\theta x}$ moment about x axis

List of Symbols   **777**

$F_{\theta y}$ moment about y axis

F_2 deflection factor for two-layer system used in Figure 2.17

f flexural stress; or probability function; or fraction of $\sqrt{V[r]}$ contributed by $\sqrt{V[h]}$

f_a average coefficient of friction between slab and subgrade

f_b allowable bearing stress between dowel and concrete

f'_c ultimate compressive strength of concrete

f_s allowable stress in steel

f_t concrete indirect or splitting tensile strength

f_{te} effective indirect tensile strength of concrete

f_1, f_2, f_3 coefficients of fatigue criterion

f_4, f_5 coefficients of permanent deformation criterion

G shear modulus of dowel; or traffic growth multiplier; or specific gravity

G_b specific gravity of bitumen

G_g specific gravity of aggregate

G_i coefficient of Dirichlet series for creep compliance

G_m specific gravity of mixture

G_s specific gravity of soil solid

G_t function of terminal serviceability index p_t

g function of random variables; or acceleration of gravity

$g_{i,j}$ flexibility coefficient for deflection at i due to force at j

H h_1/h_2 of three-layer system; or thickness of all layers combined; or initial height of groundwater table above impervious layer; or thickness of drainage layer; or tensile strength of HMA

HMA   hot mix asphalt

H_o vertical distance between bottom of drainage layer and impervious layer

h thickness; or thickness of concrete slab; or thickness of beam

h_{OL} thickness of overlay

h_e effective thickness; or thickness of existing pavement

h_i thickness of slab at node i

h_n thickness of new pavement

h_1 thickness of layer 1 or asphalt surface course

h_2 thickness of layer 2 or granular base

h_{1a}, h_{1b} thickness of 1st and 2nd asphalt layers, respectively

h'_1 modified thickness of layer 1

I moment of inertia; or intercept of log ϵ_p versus log N curve at $N = 1$; or low-temperature cracking index

IRI   international roughness index

I_c composite moment of inertia of two layers; or crack infiltration rate

I_d moment of inertia of dowel

i an index for Kelvin models, Dirichlet series, load groups, loading times, layer numbers, design periods, or nodal points; or hydraulic gradient

J load transfer factor for rigid pavements

JPCP   jointed plain concrete pavement

JRCP   jointed reinforced concrete pavement

JS   joint spacing

J_0 Bessel function of first kind and order 0

J_1 Bessel function of first kind and order 1

K modulus of dowel support; or overall stiffness of slab system

K_f stiffness matrix of foundation

K_m moment matrix relating nodal moments to nodal displacements in a slab

K_0 coefficient of earth pressure at rest

K_p stiffness matrix of slab or plate

K_1 nonlinear coefficient of granular materials; or breakpoint resilient modulus of nonlinear fine-grained soils; or stiffness matrix within assigned half band; or coefficient of fatigue equation

K_2 nonlinear exponent of granular materials; or deviator stress at breakpoint for nonlinear fine-grained soils; or stiffness matrix outside the assigned half-band; or exponent of fatigue equation

K_3 slope of first straight line for nonlinear fine-grained soils

K_4 slope of second straight line for nonlinear fine-grained soils

k modulus of subgrade reaction; or permeability

k_i modulus of subgrade reaction at node i

k_p permeability of pavement surface

k_s service or saturated modulus of subgrade reaction

k_u unsaturated modulus of subgrade reaction

k_1 E_1/E_2 of a three-layer system

k_2 E_2/E_3 of a three-layer system

k_∞ modulus of subgrade reaction when D_{SG} is greater than 10 ft

L length of tire contact area; or load factor defined as $2P_d/P_s$; or Laplace transform; or length of slab; or average nodal spacing at joint; or percentage of ADT in design lane; or weight of lead shot in cohesiometer test; or length of beam between supports; or length of drainage layer; or Fourier transform

L' distance from longitudinal joint to the free edge of slab

LS   factor indicating loss of subgrade support

LSF   load safety factor

LTE   load transfer efficiency

L_e average length of each encroachment

L_i distance of groundwater influence

L_o distance between outlets; or length of observed distance

L_p average distance traveled during parking

L_s load in kip on standard axles

$L(t)$ moving load function

L_x load in kip on one single axle, one set of tandem axles, or one set of tridem axles; or length of slab in x direction

L_y length of slab in y direction

L_1 load factor for $a = 6$ in. (152 mm); or distance measured on grain size curve; or load on one single axle or a set of tandem or tridem axles

L_2 load factor for $a = 16$ in. (406 mm); or axle code, 1 for single axle, 2 for tandem axles, and 3 for tridem axles; or distance measured on grain size curve

M moment

MAAT   mean annual air temperature

MAD   moisture accelerated distress

MDD   multidepth deflectometer

MPR   mean panel rating

MTD   mean texture depth

M_a mean monthly air temperature

M_p mean monthly pavement temperature

M_R resilient modulus; or effective roadbed soil resilient modulus

m number of axle load groups; or parameter for layered system; or number of periods in a year; or permanent deformation exponent in the Ohio State model

List of Symbols   **779**

m_1, m_2 drainage coefficients for base and subbase courses, respectively

N allowable number of load repetitions; or number of blocks in influence chart; or number of load repetitions; or number of traffic lanes

NDT   nondestructive testing

NR   percentage of drivers indicating that pavement needs repair

N_T actual traffic during design period

N_a allowable number of load repetitions on existing pavement before overlay

N_c number of longitudinal cracks

N_d allowable number of load repetitions to limit permanent deformation

N_e encroaching traffic; or number of load repetitions on existing pavement before overlay

N_f allowable number of load repetitions to prevent fatigue cracking

N_{fx} total number of repetitions for original pavement to reach failure

N_{fy} total number of repetitions for overlaid pavement to reach failure

N_i allowable number of load repetitions due to ith axle load; or allowable number of load repetitions during ith period; or allowable number of load repetitions during ith stage

N_{max} maximum number of steel bars per traffic lane

N_{min} minimum number of steel bars per traffic lane

N_o number of load applications on the outside lane

N_p number of parked trucks per day

N_t actual performance traffic

N_{18} number of equivalent 18-kip single-axle loads

n number of Kelvin models; or number of terms in Dirichlet series; or number of layers; or predicted number of load repetitions; or direction of moment in influence chart; or number of observations; or number of random variables; or roughness coefficient in Manning formula; or porosity; or number of construction stages; or an exponent in overlay equation

n_e effective porosity

n_i number of passes of ith axle load; or predicted number of load repetitions during ith period; or predicted number of repetitions during ith stage

$(n_0)_1$, $(n_0)_2$, $(n_0)_3$ number of single-, tandem-, and tridem-axle loads per day at start of design period, respectively

n_r additional number of load repetitions that can be applied to existing pavement after overlay

P a concentrated load; or percent by weight; or area patched; or rate of work for erosion; or percentage of steel

PCC   portland cement concrete

PI   penetration index; or profile index; or plasticity index; or pumping index

PMS   pavement management system

PNG   percent normalized gradient

PSI   present serviceability index

PSI_f final or terminal serviceability index

PSI_i present serviceability index at point i

PSI_0 initial serviceability index

PSR   present serviceability rating

P_b bitumen content

P_d load on each of dual wheels

P_e percent encroaching traffic

P_h transmitted horizontal pressure in stabilometer test

P_j force at node j

P_{max} maximum percent steel

P_{min} minimum percent steel

P_p percent parking traffic

P_s load on single wheel

P_t load on one dowel

P_v applied vertical pressure in stabilometer test

P_{200} percent passing 200 sieve

P_{77F} penetration at 77°F

p transform variable; or unit pressure; or shear force per length of joint; or number of periods per year

p_f ultimate failure serviceability

p_i ith transform variable; or percentage of axles in ith load group

p_0 initial serviceability index

p_{01} initial serviceability of existing pavement

p_{02} initial serviceability of overlaid pavement

p_t terminal serviceability index

p_{t1} terminal serviceability of existing pavement

p_{t2} terminal serviceability of overlaid pavement

Q discharge

q uniform contact pressure; or discharge

$\overline{q}$ Laplace transform of contact pressure

q_L total inflow to longitudinal collector

q_d contact pressure of dual tire; or design discharge

q_g inflow to drainage layer from groundwater

q_i inflow to drainage layer from surface infiltration

q_m inflow to drainage layer from melting of ice lenses

q_s contact pressure of single tire

$q(r)$ pressure under a rigid plate as a function of r

q_1 inflow to longitudinal collector from groundwater

q_2 inflow to drainage layer from groundwater

R response due to q expressed as stress, strain, or displacement; or radius of a warped slab; or modulus of rigidity; or resistance value; or hydraulic radius; or level of reliability

R^* response due to $-mJ_0(m\rho)$

$\overline{RD}$ rut depth

$\overline{RD}$ mean rut depth

RF   reduction factor indicating the degree of curing

R_{Lx} remaining life of existing pavement prior to overlay

R_{Ly} remaining life of overlaid pavement

R_{Lyx} remaining life of existing pavement after overlay

RN   roughness number

RR   rate of rutting

RR1   radial stress factor at bottom of layer 1

RR2   radial stress factor at bottom of layer 2

RTRRM   response-type road roughness meter

R_i functions of profile roughness in PSI equation

r radial coordinate; or a random variable; or traffic growth rate

S slope of straight line in permanent deformation test; or slope of pipe; or slope of drainage layer; or slope of load distribution for nonlinear analysis; or braking distance; or pavement slope measured by a profilometer; or soil suction

$\overline{S}$ mean of all slopes measured by a profilometer

SC_{OL} required structural capacity of overlay

SC_0 initial structural capacity of existing pavement

SC_{xeff} effective structural capacity of existing pavement

SC_y total structural capacity required to support overlay traffic

SC_{yeff} effective structural capacity at the end of overlay period

SDN   stopping distance number

SN   structural number; or skid number

SN_{OL} required structural number of overlay

SN_0 skid number at zero speed

SN_{xeff} total effective structural number of existing pavement prior to overlay

$SN_{xeff-rp}$ effective structural capacity of remaining pavement layers excluding PCC

SN_y total structural number to support overlay traffic

SN_1, SN_2, SN_3 structural number needed to protect base, subbase, and subgrade, respectively

SV   slope variance

$\overline{SV}$ mean slope variance

S_b stiffness modulus of bitumen

S_{bit} stiffness modulus of original asphalt for analyzing low-temperature cracking

S_c modulus of rupture of concrete

S_{ce} modulus of rupture of existing concrete

S_{cn} modulus of rupture of new pavement

S_d dual spacing

S_f slope factor

S_{ji} slope of log deflection versus log E

S_m stiffness modulus of mixture

S_{mix} stiffness modulus of mixture determined directly from creep test

S_o overall standard deviation in AASHTO design guide

S_t tandem spacing

S_1, S_2, S_3 vertical intercepts for use in method of successive residuals

s vehicle speed; or standard deviation; or a random variable; or shear strength; or number of sensors

s_T standard deviation for traffic prediction

s_b spacing of dowel bars

s_i gap or precompression at node i, gap is positive and precompression is negative

s_t standard deviation for performance prediction

T temperature; or percentage of truck in ADT

T_H average daily high temperature during the month the pavement is constructed

T_L average daily low temperature during the coldest month of the year

$T_{R\&B}$ temperature at ring and ball softening point

T_f truck factor; or time factor for drainage; or time that a pavement can last before an overlay is required

T_i retardation time of ith Kelvin model

T_0 relaxation time of Maxwell model; or reference temperature

T_1 relaxation time of Kelvin model

T_1, T_2 two temperatures at which penetrations are measured

t time; or length of tie bar; or thickness of specimen

t_T time to obtain creep compliance at temperature T

t_{T_0} time to obtain creep compliance at reference temperature T_0

t_i time at ith increment

t_{50} time for 50% drainage

U degree of drainage

u radial displacement; or normal deviate $(x - \mu)/s$; or ratio $d/2T$; or pore pressure

u_f relative damage to flexible pavement

u_r relative damage to rigid pavement

V variance; or vehicle speed

V_a volume of void

V_b volume of bitumen

V_g volume of aggregate

v discharge velocity

W wheel load; or width of slab; or width of specimen; or total weight of bituminous mixture; or width of roadway; or vertical load on tire

W_T predicted ESAL during design period

W_c length of transverse cracks or joints

W_p width of pavement

W_s width of slab

W_t allowable ESAL during design period

W_{tx} number of x axle load applications at end of time t

W_{t18} number of 18–kip axle load applications at end of time t

W_{18} allowable 18–kip single-axle load applications for a given reliability

w vertical deflection; or elastic deformation; or width of sealant reservoir

w_a assumed vertical deflection at a designated point on the slab

w^c computed deflection

w_d difference in deflection between left and right slab

w_g amount of gap in spring or dowel at joint

w_i vertical deflection at node i

$w_{i,j}$ deflection of node i due to force at node j

w_l deflection of left slab

w^m measured deflection

w_0 surface deflection

$\overline{w}_0$ Laplace transform of surface deflection

w_p permanent or plastic deformation

w_r deflection of right slab; or rebound deflection

$\overline{X}$ crack spacing

x coordinate axis; or distance from a diagonal cross section to the corner of slab; or a random variable; or actual load repetitions on existing pavement prior to overlay

$\overline{x}$ average of x

Y design period in years

y coordinate axis; or a random variable; future load repetitions during the overlay period

y_0 deformation of dowel at the face of joint

Z ratio between concrete modulus and modulus of subgrade reaction; or concrete shrinkage

Z_R normal deviate for a given reliability R

ZZ1   vertical stress factor at interface 1

ZZ2   vertical stress factor at interface 2

z vertical coordinate; or distance below surface; or thickness of pavement considered as homogeneous half-space; or width of joint; or normal deviate; or distance above water table

α contact area/total thickness, a/H; or angle between a line and x axis; or permanent deformation parameter; or thermal diffusivity; or compressibility factor

α_R $1 - \alpha_{sys}$

α_c coefficient of thermal expansion for concrete

α_s coefficient of thermal expansion for steel

α_{sys} permanent deformation parameter of pavement system

α_t coefficient of thermal expansion

β soil constant indicating the increase in E per unit increase in θ; or slope of log a_T versus temperature; or relative stiffness of a dowel embedded in concrete

β_x function of SN, L_x, and L_2

β_1 to β_5 constants to evaluate stiffness or dynamic modulus of HMA

β_{18} value of β_x when L_x is equal to 18 and L_2 to 1

γ unit weight; or shear strain

γ_c unit weight of concrete

γ_d dry unit weight of soil

γ_i unit weight of layer i

γ_w unit weight of water

Δ deflection at beam center; or deflection of plate loading test

ΔL joint opening due to temperature change ΔT and drying shrinkage

ΔPSI serviceability loss

ΔS shear deformation of dowel

ΔT temperature range, which is the temperature at placement minus the lowest mean monthly temperature; or temperature increment

Δ_c corner deflection

Δ_e deflection due to edge loading

Δ_i deflection due to interior loading

Δt temperature differential between top and bottom of slab

$\Delta\sigma$ stress increment

δ displacement vector of slab; or recoverable deformation in indirect tension specimen

δ_{rrd} representative rebound deflection

δ' displacement vector of foundation

δ_a assumed displacement vector of slab by the iterative method

δ_d design rebound deflection

ℓ length of hypotenuse; or radius of relative stiffness

ϵ normal strain; or drying shrinkage coefficient of concrete; or elastic strain

$\bar{\epsilon}$ Laplace transform of strain

ϵ_a strain due to load a

ϵ_b strain due to load b

ϵ_c vertical compressive strain on the surface of subgrade

ϵ_0 initial strain in fatigue test; or strain amplitude

ϵ_p permanent strain

ϵ_r radial strain; or recoverable strain

ϵ_t tangential strain; or tensile strain at bottom of asphalt layer

ϵ_x tensile strain due to x axle load; or strain in x direction

ϵ_y strain in y direction

ϵ_z vertical strain

ϵ^2 squared error

ϵ_1, ϵ_2, ϵ_3 major, intermediate and minor principal strains

ϵ_{18} tensile strain due to 18-kip single-axle load

η a random variable with a mean of 1

θ stress invariant

θ_x rotation about x axis

θ_y rotation about y axis

λ viscosity; or dimensionless vertical distance, z/H

λ_0 viscosity of Maxwell model

λ_1 viscosity of Kelvin model

μ allowable bond stress between concrete and tie bar; or mean; or permanent deformation parameter; or coefficient of friction

μ_R μ_{sys}/α_R

μ_{sys} permanent deformation parameter of pavement system

ν Poisson ratio; or Poisson ratio of concrete

ν_c composite Poisson ratio of two layers

ν_d Poisson ratio of dowel

ν_f Poisson ratio of foundation

ν_1 Poisson ratio of layer 1

ν_2 Poisson ratio of layer 2

ρ dimensionless radial distance, r/H; or correlation coefficient

ΣF sum of applied forces on slab

ΣR sum of subgrade reactions on slab

Σo bar perimeter

σ normal stress; or tensile stress in concrete slab; or tensile stress in beam test

$\bar{\sigma}$ Laplace transform of σ

σ_a stress due to load a

σ_b bearing stress between dowel and concrete; or stress due to load b

σ_c vertical compressive stress on subgrade; or stress in concrete slab due to corner loading; or tensile stress in concrete slab due to friction

σ_d deviator stress $\sigma_1 - \sigma_3$

σ_e stress in concrete slab due to edge loading

σ_i stress in concrete slab due to interior loading

σ_0 initial stress on Maxwell model; or initial stress in fatigue test; or stress amplitude

σ_p consolidation pressure on subgrade

σ_r radial stress

σ_{r1} radial stress at bottom of layer 1

σ'_{r1} radial stress at top of layer 2

σ_{r2} radial stress at bottom of layer 2

σ'_{r2} radial stress at top of layer 3

σ_s steel stress

σ_t tangential stress; edge stress

σ_{te} edge stress in existing overlaid pavement

σ_m edge stress in new pavement

σ_{to} edge stress in new overlay

σ_v vertical stress

σ_w wheel load stress

σ_x stress in x direction

σ_y stress in y direction

σ_z vertical stress

σ_{z1} vertical stress at interface 1

σ_{z2} vertical stress at interface 2

$\sigma_1, \sigma_2, \sigma_3$ three principal stresses

σ'_1 maximum allowable major principal stress

τ shear stress

τ_{rz} shear stresses on r plane in z direction

ψ cumulative distribution functions

ϕ stress function; or angle of internal friction; or reinforcing bar or wire diameter; or phase angle

ϕ_i stress function for layer i

ω angular velocity

APPENDIX

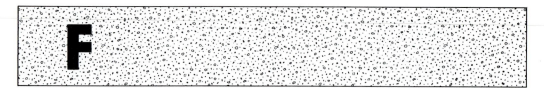

References

AASHO, 1968. *AASHO Highway Definitions*, Special Committee on Nomenclature, American Association of State Highway officials.

AASHTO, 1972. *AASHTO Interim Guide for Design of Pavement Structures*, American Association of State Highway and Transportation Officials.

AASHTO, 1985. *Proposed AASHTO Guide for Design of Pavement Structures*, NCHRP Project 20-7/24, Vol. 2, American Association of State Highway and Transportation Officials.

AASHTO, 1986. *Guide for Design of Pavement Structures*, American Association of State Highway and Transportation Officials.

AASHTO, 1989. *Standard Specifications for Transportation Materials and Methods of Sampling and Testing*, Part II, *Methods of Sampling and Testing*, American Association of State Highway Officials.

ADEDIMILA, A. S., and T. W. KENNEDY, 1976. "Repeated-Load Indirect Tensile Fatigue Characteristics of Asphalt Mixtures," *Transportation Research Record 595*, Transportation Research Board, pp. 25–33.

AHLVIN, R. G., and H. H. ULERY, 1962. "Tabulated Values for Determining the Complete Pattern of Stresses, Strains and Deflections Beneath a Uniform Circular Load on a Homogeneous Half Space," *Bulletin 342*, Highway Research Board, pp. 1–13.

AI, 1973. *Full-Depth Asphalt Pavements for General Aviation Aircraft, IS-154*, Asphalt Institute.

AI, 1981a. *Thickness Design—Asphalt Pavements for Highways and Streets*, Manual Series No. 1, Asphalt Institute.

AI, 1981b. *Asphalt Pavement Thickness Design*, Information Series No. 181, Asphalt Institute.

AI, 1982. *Research and Development of The Asphalt Institute's Thickness Design Manual (MS-1)*, 9th ed., *Research Report 82-2*, Asphalt Institute.

AI, 1983. *Asphalt Overlays for Highway and Street Rehabilitation*, Manual Series No. 17, Asphalt Institute.

AI, 1984. *Drainage of Asphalt Pavement Structures*, Manual Series No. 15, Asphalt Institute.

AI, 1987. *Thickness Design, Asphalt Pavements for Air Carrier Airports*, Manual Series No. 11, Asphalt Institute.

AI, 1989. *The Asphalt Handbook*, Manual Series No. 4, Asphalt Institute.

AI, 1991. *Thickness Design-Asphalt Pavements for Highways & Streets*, Manual Series No. 1, Asphalt Institute.

ALLEN, J. J., 1973. *The Effect of Non-Constant Lateral Pressures on the Resilient Response of Granular Materials*, Ph. D. Dissertation, University of Illinois at Urbana–Champaign.

ALLEN, D. L., and R. C. DEEN, 1986. "A Computerized Analysis of Rutting Behavior of Flexible Pavement," *Transportation Research Record 1095*, Transportation Research Board, pp. 1–10.

ASTM, 1989a. *Annual Books of ASTM Standards*,

Concrete and Aggregates, Vol. 04.02, American Society for Testing and Materials.

ASTM, 1989b. *Annual Books of ASTM Standards, Road and Paving Materials: Traveled Surface Characteristics*, Vol. 04.03, American Society for Testing and Materials.

BAKER, R. F., 1973. "New Jersey Composite Pavement Project," *Highway Research Record 434*, Highway Research Board, pp. 16–23.

BALL, C. G., and L. D. CHILDS, 1975. *Tests of Joints for Concrete Pavements*, Bulletin RD026.01P, Portland Cement Association.

BALLINGER, C. A., 1971. "Cumulative Fatigue Damage Characteristics of Plain Concrete," *Highway Research Record 370*, Highway Research Board, pp. 48–60.

BARBER, E. S., 1946. "Application of Triaxial Compression Test Results to the Calculation of Flexible Pavement Thickness," *Proceedings, Highway Research Board*, pp. 26–39.

BARBER, E. S., 1957. "Calculation of Maximum Pavement Temperatures from Weather Report," *Bulletin 168*, Highway Research Board, pp. 1–8.

BARBER, E. S., and C. L. SAWYER, 1952. "Highway Subdrainage," *Proceedings, Highway Research Board*, pp. 643–666.

BARKSDALE, R. G., 1971. "Compressive Stress Pulse Times in Flexible Pavements for Use in Dynamic Testing," *Highway Research Record 345*, Highway Research Board, pp. 32–44.

BARKSDALE, R. G., and R. G. HICKS, 1979. *Improved Pavement-Shoulder Joint Design*, NCHRP Report 202, Transportation Research Board.

BASSON, J. E. B., O. J. WIJNBERGER, and J. SKULTETY, 1981. *The Multidepth Deflectometer: A Multistage Sensor for the Measurement of Deflections and Permanent Deformations at Various Depths in Road Pavements*, Technical Report RP/3/81, Institute of Transportation and Road Research, Pretoria, South Africa.

BENJAMIN, J. R., and C. A. CORNELL, 1970. *Probability, Statistics and Decision for Civil Engineers*, McGraw-Hill, New York.

BETRAM, G. E., 1940. *An Experimental Investigation of Protective Filters*, Publication No. 267, Graduate School of Engineering, Harvard University.

BLACK, W. P. M., and D. CRONEY, 1957. "Pore Water Pressure and Moisture Content Studies under Experimental Pavements," *Proceedings, 4th International Conference on Soil Mechanics and Foundation Engineering*, Vol. 2, pp. 94–103.

BONNAURE, F., G. GEST, A. GRAVOIS, and P. UGE, 1977. "A New Method of Predicting the Stiffness of Asphalt Paving Mixtures," *Proceedings, Association of Asphalt Paving Technologists*, Vol. 46, pp. 64–100.

BONNAURE, F., A. GRAVOIS, and J. UDRON, 1980. "A New Method for Predicting the Fatigue Life of Bituminous Mixes," *Proceedings, Association of Asphalt Paving Technologists*, Vol. 49, pp. 499–524.

BOUSSINESQ, J., 1885. *Application des Potentiels a l'etude de l'equilibre et du Mouvement des Solids Elastiques*, Gauthier-Villars, Paris.

BOYCE, J. R., S. F. BROWN, and P. S. PELL, 1976. "The Resilient Behavior of a Granular Material Under Repeated Loading," *Proceedings*, Australian Road Research Board.

BOYD, W. K., and C. R. FOSTER, 1950. "Design Curves for Very Heavy Multiple-Wheel Assemblies, Development of CBR Flexible Pavement Design Methods for Airfield, A Symposium," *Transactions, ASCE*, Vol. 115, pp. 534–546.

BRADBURY, R. D., 1938. *Reinforced Concrete Pavements*, Wire Reinforcement Institute, Washington, DC.

BRADEMEYER, B., 1988. *VESYS Modification*, Final Report to FHWA, Work Order DTFH61-87-P-00441.

BROWN, S. F., 1973. "Determination of Young's Modulus for Bituminous Materials in Pavement Design," *Highway Research Record 431*, Highway Research Board, pp. 38–49.

BROWN, S. F., P. S. PELL, and A. F. STOCK, 1977. "The Application of Simplified, Fundamental Design Procedures for Flexible Pavement," *Proceedings, 4th International Conference on the Structural Design of Asphalt Pavements*, Vol. 1, pp. 321–341.

BRUNNER, R. J., 1975. "Prestressed Pavement Demonstration Project," *Transportation Research Record 535*, Transportation Research Board, pp. 62–72.

BURMISTER, D. M., 1943. "The Theory of Stresses and Displacements in Layered Systems and Applications to the Design of Airport Runways," *Proceedings, Highway Research Board*, Vol. 23, pp. 126–144.

BURMISTER, D. M., 1945. "The General Theory of Stresses and Displacements in Layered Soil Systems," *Journal of Applied Physics*, Vol. 16, pp. 84–94, 126–127, 296–302.

BURMISTER, D. M., 1958. "Evaluation of Pavement Systems of the WASHO Road Test by Layered Systems Method," *Bulletin 177*, Highway Research Board, pp. 26–54.

BUSH, A. J., 1980. *Nondestructive Testing of Light*

Aircraft Pavements, Phase II, Development of the Nondestructive Evaluation Methodology, Report No. FAA-RD-80-9-II, Federal Aviation Administration.

BUSH, A. J., and D. R. ALEXANDER, 1985. "Pavement Evaluation Using Deflection Basin Measurements and Layered Theory," *Transportation Research Record 1022,* Transportation Research Board, pp. 16–28.

CALIFORNIA DEPARTMENT OF PUBLIC WORKS, 1972. *Highway Design Manual.*

CALIFORNIA DEPARTMENT OF TRANSPORTATION, 1979. *Asphalt Concrete Overlay Design Manual.*

CAREY, W. N., and P. E. IRICK, 1960. *The Pavement Serviceability Performance Concept,* Bulletin 250, Highway Research Board.

CARPENTER, S. H., M. I. DARTER, and B. J. DEMPSEY, 1981. *A Pavement Moisture Accelerated Distress Identification System, User's Manual,* Vol. 2, Report No. FHWA/RD-81/080, Federal Highway Administration.

CASAGRANDE, A., and W. L. SHANNON, 1952. "Base Course Drainage for Airport Pavements," *Transactions, American Society of Civil Engineering,* pp. 792–814.

CEDERGREN, H. R., 1974. *Drainage of Highway and Airfield Pavements,* Wiley, New York.

CEDERGREN, H. R., 1977. *Seepage, Drainage, and Flow Nets,* Wiley, New York.

CEDERGREN, H. R., 1988. "Why All Important Pavements Should Be Well Drained," *Transportation Research Record 1188,* Transportation Research Board, pp. 56–62.

CEDERGREN, H. R., J. A. ARMAN, and K. H. O'BRIEN, 1972. *Guidelines for the Design of Subsurface Drainage Systems for Highway Pavement Structural Sections,* Report No. FHWA-RD-72-30, Federal Highway Administration.

CEDERGREN, H. R., J. A. ARMAN, and K. H. O'BRIEN, 1973. *Development of Guidelines for the Design of Subsurface Drainage Systems for Highway Pavement Structural Sections,* Report No. FHWA-RD-73-14, Federal Highway Administration.

CHEUNG, Y. K., and O. C. ZIENKIEWICZ, 1965. "Plates and Tanks on Elastic Foundations—An Application of Finite Element Method," *International Journal of Solids and Structures,* Vol. 1, pp. 451–461.

CHOU, Y. J., J. UZAN, and R. L. LYTTON, 1989. "Backcalculation of Layer Moduli from Nondestructive Pavement Deflection Data Using the Expert System Approach," *STP 1026,* American Society for Testing and Materials, pp. 341–354.

CHOU, Y. T., 1981. *Structural Analysis Computer Programs for Rigid Multicomponent Pavement Structures with Discontinuities—WESLIQID and WESLAYER,* Technical Report GL-81-6, U.S. Army Engineer Waterways Experiment Station, Reports 1, 2, and 3.

CHOU, Y. T., 1987. *Probabilistic and Reliability Design Procedures for Flexible Airfield Pavements—Elastic Layered Method,* Technical Report GL-82-24, U.S. Army Engineer Waterways Experiment Station.

CHOU, Y. T., and Y. H. HUANG, 1979. "A Computer Program for Slabs with Discontinuities," *Proceedings, International Air Transportation Conference,* Vol. 1, pp. 121–136.

CHOU, Y. T., and Y. H. HUANG, 1981. "A Computer Program for Slabs with Discontinuities on Layered Elastic Solids," *Proceedings, 2nd International Conference on Concrete Pavement Design,* Purdue University, pp. 78–85.

CHOU, Y. T., and Y. H. HUANG, 1982. "A Finite Element Method for Concrete Pavements," *Proceedings, International Conference on Finite Element Methods,* Shanghai, China, pp. 348–353.

CHOU, Y. T., and H. G. LAREW, 1969. "Stresses and Displacements in Viscoelastic Pavement Systems Under a Moving Load," *Highway Research Record 282,* Highway Research Board, pp. 25–40.

CHRISTISON, J. T., D. W. MURRAY, and K. O. ANDERSON, 1972. "Stress Prediction and Low Temperature Fracture Susceptibility of Asphalt Concrete Pavements," *Proceedings, Association of Asphalt Paving Technologists,* Vol. 41, pp. 494–523.

CHUA, K. M., and R. L. LYTTON, 1985. "Load Rating of Light Pavement Structures," *Transportation Research Record 1043,* Transportation Research Board, pp. 89–102.

CLAROS, G., W. R. HUDSON, and K. H. STOKOE, 1990. "Modifications to Resilient Modulus Testing Procedure and Use of Synthetic Samples for Equipment Calibration," *Transportation Research Record 1278,* Transportation Research Board, pp. 51–62.

CLAUSSEN, A. I. M., J. M. EDWARDS, P. SOMMER, and P. UGE, 1977. "Asphalt Pavement Design—The Shell Method," *Proceedings, 4th International Conference on the Structural Design of Asphalt Pavements,* Vol. 1, pp. 39–74.

CLEMMER, H. F., 1923. "Fatigue of Concrete," *Proceedings, ASTM,* Vol. 22, II, pp. 408–419.

COLLEY B. E., C. G. BALL, and P. ARRIYAVAT, 1978. "Evaluation of Concrete Pavements with Tied Shoulders or Widened Lanes," *Transportation Re-*

search *Record 666,* Transportation Research Board, pp. 39–45.

CRAUS, J., R. YUCE, and C. L. MONISMITH, 1984. "Fatigue Behavior of Thin Asphalt Concrete Layers in Flexible Pavement Structures," *Proceedings, Association of Asphalt Paving Technologists,* Vol. 53, pp. 559–582.

DARTER, M. I., 1976. "Application of Statistical Methods to the Design of Pavement Systems," Transportation Research Record 575, Transportation Research Board, pp. 39–55.

DARTER, M. I., and E. J. BARENBERG, 1977. *Design of Zero-Maintenance Plain Jointed Concrete Pavement,* Report No. FHWA-RD-77-111, Vol. 1, Federal Highway Administration.

DARTER, M. I., J. M. BECKER, M. B. SNYDER, and R. E. SMITH, 1985. *Portland Cement Concrete Pavement Evaluation System (COPES),* NCHRP Report 277, Transportation Research Board.

DARTER, M. I., W. R. HUDSON, and J. L. BROWN, 1973a. "Statistical Variations of Flexible Pavement Properties and Their Consideration in Design," *Proceedings, Association of Asphalt Paving Technologists,* Vol. 42, pp. 589–613.

DARTER, M. I., B. F. MCCULLOUGH, and J. L. BROWN, 1973b. "Reliability Concepts Applied to the Texas Flexible Pavement System," *Highway Research Record 407,* Highway Research Board, pp. 180–190.

DEACON, J. A., 1965. *Fatigue of Asphalt Concrete,* Doctoral Dissertation, University of California, Berkeley.

DEACON, J. A., 1969. "Load Equivalency in Flexible Pavements," *Proceedings, Association of Asphalt Paving Technologists,* Vol. 38, pp. 465–491.

DE BEER, M., E. HORAK, and A. T. VISSER, 1989. "The Multidepth Deflectometer (MDD) System for Determining the Effective Elastic Moduli of Pavement Layers," *STP 1026,* American Society for Testing and Materials, pp. 70–89.

DE JONG, D. L., M. G. F. PEATZ, and A. R. KORSWAGEN, 1973. *Computer Program Bisar Layered Systems Under Normal and Tangential Loads,* Konin Klijke Shell-Laboratorium, Amsterdam, External Report AMSR.0006.73.

DEMPSEY, B. J., 1983. "Laboratory and Field Studies of Channeling and Pumping in Pavement Systems," Paper presented at PIARC Seminar, Paris, France.

DEMPSEY, B. J., M. I. DARTER, and S. H. CARPENTER, 1982. *Improving Subdrainage and Shoulders of Existing Pavements, State of the Art,* Report No. FHWA-RD-81-077, Federal Highway Administration.

DEMPSEY, B. J., W. A. HERLACHE, and A. J. PATEL, 1986. "Climatic–Materials–Structural Pavement Analysis Program," *Transportation Research Record 1095,* Transportation Research Board, pp. 111–123.

DEMPSEY, B. J., and M. R. THOMPSON, 1970. "A Heat Transfer Model for Evaluating Frost Action and Temperature Related Effects in Multilayered Pavement Systems," *Highway Research Record 342,* Highway Research Board, pp. 39–56.

DORMON, G. M., and C. T. METCALF, 1965. "Design Curves for Flexible Pavements Based on Layered System Theory," *Highway Research Record 71,* Highway Research Board, pp. 69–84.

DUNCAN, J. M., C. L. MONISMITH, and E. L. WILSON, 1968. "Finite Element Analysis of Pavements," *Highway Research Record 228,* Highway Research Board, pp. 18–33.

ELLIOTT, J. F., and F. MOAVENZADEH, 1971. "Analysis of Stresses and Displacements in Three-Layer Viscoelastic Systems," *Highway Research Record 345,* Highway Research Board, pp. 45–57.

ELLIOTT, R. P., 1989. "An Examination of the AASHTO Remaining Life Factor," *Transportation Research Record 1215,* Transportation Research Board, pp. 53–59.

EMERY, D. K., 1975. "A Preliminary Report on the Paved Shoulder Encroachment and Transverse Lane Displacement for Design Trucks on Rural Freeways," *Proceedings, ASCE Specialty Conference on Pavement Design for Practicing Engineers,* pp. 6:10–6:15.

EPPS, J. A., and C. L. MONISMITH, 1986. "Equipment for Obtaining Pavement Condition and Traffic Loading Data," *NCHRP Synthesis of Highway Practice No. 126,* Transportation Research Board.

ERES Consultant, Inc., 1987. *Pavement Design, Principles and Practices, A Training Course Participant Notebook,* Federal Highway Administration.

EXSYS, Inc., 1985. *EXSYS Expert System Development Package, User's Manual,* Albuquerque, NM.

FAA, 1988. *Airport Pavement Design and Evaluation,* Consolidated Reprint, AC 150/5320-6C, Federal Aviation Administration.

FERNANDO, E. G., R. S. WALKER, and R. L. LYTTON, 1990. "Evaluation of the Siometer as a Device for Measurement of Pavement Profiles," *Transportation Research Record 1260,* Transportation Research Board, pp. 112–124.

FHWA, 1978. *Predictive Design Procedures, VESYS Users Manual,* Report No. FHWA-RD-77-154, Federal Highway Administration.

FHWA, 1980. *Skid Accident Reduction Program,*

Technical Advisory T-5040.17, Federal Highway Administration.

FHWA, 1982. *Pavement Shoulders,* Technical Advisory T5040.18, Federal Highway Administration.

FHWA, 1986. *Longitudinal Edge Drains in Rigid Pavement Systems,* Report No. FHWA-TS-86-208, Federal Highway Administration.

FHWA, 1987a. *Reference Manual for the UMTRI/ FHWA Road Profiling (PRORUT) System,* Report FHWA/RD-87/042, Federal Highway Administration.

FHWA, 1987b. *Highway Performance Monitoring System Field Manual for the Continuing Analytical and Statistical Database, Appendix J,* FHWA Order M5600.1A, OMB No. 2125–0028, Federal Highway Administration.

FHWA, 1987c. *Crack and Seat Performance,* Review Report, Demonstration Projects Division and Pavement Division, Federal Highway Administration.

FHWA, 1989. *Geotextile Engineering Manual, Course Text,* Publication No. FHWA-HI-89-050, Federal Highway Administration.

FHWA, 1990. *Highway Statistics,* Federal Highway Administration.

FINN, F. N., and C. L. MONISMITH, 1984. *Asphalt Overlay Design Procedures,* Synthesis of Highway Practice 116, National Cooperative Highway Research Program.

FINN, F. N., C. L. MONISMITH, and N. J. MARKEVICH, 1983. "Pavement Performance and Asphalt Concrete Mix Design, *Proceedings, Association of Asphalt Paving Technologists,* Vol. 52, pp. 121–144.

FINN, F., C. L. SARAF, R. KULKARNI, K. NAIR, W. SMITH, and A. ABDULLAH, 1986. *Development of Pavement Structural Subsystems,* NCHRP Report 291, Transportation Research Board.

FOSTER, C. R., and R. G. AHLVIN, 1954. "Stresses and Deflections Induced by a Uniform Circular Load," *Proceedings, Highway Research Board,* Vol. 33, pp. 467–470.

FOSTER, C. R., and R. G. AHLVIN, 1958. "Development of Multiple-Wheel CBR Design Criteria," *Journal of the Soil Mechanics and Foundations Division,* ASCE, Vol. 84, No. SM2, May, pp. 1647–1 to 1647–12.

FRANCHEN, L., and J. VERSTRAETEN, 1974. "Methods for Predicting Modulus and Fatigue Laws of Bituminous Mixes under Repeated Bending," *Transportation Research Record 515,* Transportation Research Board, pp. 114–123.

FREEME and MARAIS, 1973. "Thin Bituminous Surfaces: Their Fatigue Behavior and Prediction," *Special Report 140,* Highway Research Board, pp. 158–179.

FRIBERG, B. F., 1940. "Design of Dowels in Transverse Joints of Concrete Pavements," *Transactions, ASCE,* Vol. 105, pp. 1076–1095.

FWA, T. F., 1991. "Remaining-Life Consideration in Pavement Overlay Design," *Journal of Transportation Engineering,* Vol. 117, No. 6, pp. 585–601.

GAGE, R. B., 1932. "Discussion on Joints in Concrete Pavements," *Proceedings, Highway Research Board,* Part I, pp. 137–141.

GERRARD, C. M., and W. J. HARRISON, 1970. "A Theoretical Comparison of the Effects of Dual-Tandem and Dual-Wheel Assemblies on Pavements," *Proceedings, Fifth Conference, Australian Road Research Board.*

GOLDBECK, A. T., 1919. "Thickness of Concrete Slabs," *Public Roads,* pp. 34–38.

GOMEZ-ACHECAR M., and M. R. THOMPSON, 1986. "ILLI-PAVE-Based Response Algorithms for Full-Depth Asphalt Concrete Flexible Pavements," *Transportation Research Record 1095,* Transportation Research Board, pp. 11–18.

HAAS, R., and W. R. HUDSON, 1978. *Pavement Management Systems,* McGraw-Hill, New York.

HAJEK, J. J., and R. C. G. HAAS, 1972. "Predicting Low-Temperature Cracking Frequency of Asphalt Concrete Pavements," *Transportation Research Record 407,* Transportation Research Board, pp. 39–54.

HALIBURTON, T. A., and J. D. LAWMASTER, 1981. *Training Manual on Use of Engineering Fabrics,* Contract No. DTFH 61-80-C-0094, Federal Highway Administration.

HALL, K. T., J. M. CONNOR, M. I. DARTER, and S. H. CARPENTER, 1989. *Rehabilitation of Concrete Pavements, Vol. 3, Concrete Pavement Evaluation and Rehabilitation System,* Report No. FHWA-RD-88-073, Federal Highway Administration.

HANNA, A. M., P. J. NUSSBAUM, P. ARRIYAVAT, J. J. TSENG, and B. F. FRIBERG, 1976. *Technological Review of Prestressed Pavements,* Report No. FHWA-RD-77-8, Federal Highway Administration.

HARICHANDRAN, R. S., G. Y. BALADI, and M. YEH, 1989. *Development of a Computer Program for Design of Pavement Systems Consisting of Bound and Unbound Materials,* Department of Civil and Environmental Engineering, Michigan State University.

HEALEY, K. A., and R. P. LONG, 1972. "Prefabricated Filter Fin for Subsurface Drains," *Journal of Irrigation and Drainage Division,* ASCE, Vol. IR4, pp. 543–552.

HEINRICHS, K. W., M. J. LIU, M. I. DARTER, S. H. CARPENTER, and A. M. IOANNIDES, 1989. *Rigid Pavement Analysis and Design*, Report No. FHWA-RD-88-068, Federal Highway Administration.

HENRY, J. J., J. C. WAMBOLD, and H. XUE, 1984. *Evaluation of Pavement Texture*, Report No. FHWA-RD-84-016, Federal Highway Administration.

HEUKELOM, W., 1966. "Observations on the Rheology and Fracture of Bitumens and Asphalt Mixes," *Proceedings, Association of Asphalt Paving Technologists*, Vol. 35, pp. 358–399.

HEUKELOM, W., and A. J. G. KLOMP, 1962, "Dynamic Testing as a Means of Controlling Pavements During and After Construction," *Proceedings, (1st) International Conference on the Structural Design of Asphalt Pavements*, pp. 667–685.

HICKS, R. G., 1970. *Factors Influencing the Resilient Properties of Granular Materials*, Ph. D. Dissertation, University of California, Berkeley.

HICKS, R. G., and F. N. FINN, 1970. "Analysis of Results from the Dynamic Measurements Program on the San Diego Test Road," *Proceedings, Association of Asphalt Paving Technologists*, Vol. 39, pp. 153–185.

HOFFMAN, M. S., and M. R. THOMPSON, 1982. "Comparative Study of Selected Nondestructive Testing Devices," *Transportation Research Record 852*, Transportation Research Board, pp. 32–41.

HOGENTOGLER, C. A., and C. TERZAGHI, 1929. "Interrelationship of Load, Road and Subgrade," *Public Roads*, May, pp. 37–64.

HRB, 1945. "Report of Commitee on Classification of Materials for Subgrades and Granular Type Roads," *Proceedings, Highway Research Board*, Vol. 25, pp. 376–384.

HRB, 1952. *Final Report on Road Test One MD*, Special Report 4, Highway Research Board.

HRB, 1955. *The WASHO Road Test*, Part 2; *Test Data Analysis and Findings*, Special Report 22, Highway Research Board.

HRB, 1962. *The AASHO Road Test*, Report 5; *Pavement Research;* Report 6; *Special Studies;* and Report 7: *Summary Report*, Special Reports 61E, 61F, and 61G, Highway Research Board.

HUANG, Y. H., 1967. "Stresses and Displacements in Viscoelastic Layered Systems Under Circular Loaded Areas," *Proceedings, 2nd International Conference on the Structural Design of Asphalt Pavements*, pp. 225–244.

HUANG, Y. H., 1968a. "Stresses and Displacements in Nonlinear Soil Media," *Journal of the Soil Mechanics and Foundation Division, ASCE*, Vol. 94, No. SM1, pp. 1–19.

HUANG, Y. H., 1968b. "Chart for Determining Equivalent Single-Wheel Loads," *Journal of the Highway Division, ASCE*, Vol. 94, No. HW2, pp. 115–128.

HUANG, Y. H., 1969a. "Finite Element Analysis of Nonlinear Soil Media," *Proceedings, Symposium on Application of Finite Element Methods in Civil Engineering*, Vanderbilt University, Nashville, TN, pp. 663–690.

HUANG, Y. H., 1969b. "Influence Charts for Two-Layer Elastic Foundations," *Journal of the Soil Mechanics and Foundation Division, ASCE*, Vol. 95, No. SM2, March, pp. 709–713.

HUANG, Y. H., 1969c. "Computation of Equivalent Single-Wheel Loads Using Layered Theory," *Highway Research Record 291*, Highway Research Board, pp. 144–155.

HUANG, Y. H., 1971. "Deflection and Curvature as Criteria for Flexible Pavement Design and Evaluation," *Highway Research Record 345*, Highway Research Board, pp. 1–11.

HUANG, Y. H., 1972. "Strain and Curvature as Factors for Predicting Pavement Fatigue," *Proceedings, 3rd International Conference on the Structural Design of Asphalt Pavements*, Vol. 1, pp. 622–628.

HUANG, Y. H., 1973a. "Critical Tensile Strain in Asphalt Pavements," *Transportation Engineering Journal, ASCE*, Vol. 99, No. TE3, pp. 553–569.

HUANG, Y. H., 1973b. "Stresses and Strains in Viscoelastic Multilayer Systems Subjected to Moving Loads," *Highway Research Record 457*, Highway Research Board, pp. 60–71.

HUANG, Y. H., 1974a. "Finite Element Analysis of Slabs on Elastic Solids," *Transportation Engineering Journal, ASCE*, Vol. 100, No. TE2, pp. 403–416.

HUANG, Y. H., 1974b. "Analysis of Symmetrically Loaded Slabs on Elastic Solid," *Transportation Engineering Journal, ASCE*, Vol. 100, No. TE2, pp. 537–541.

HUANG, Y. H., 1985. "A Computer Package for Structural Analysis of Concrete Pavements," *Proceedings, 3rd International Conference on Concrete Pavement Design and Rehabilitation*, Purdue University, pp. 295–307.

HUANG, Y. H., and Y. T. CHOU, 1978. "Discussion on Finite Element Analysis of Jointed or Cracked Pavements," *Transportation Research Record 671*, Transportation Research Board, pp. 17–18.

HUANG, Y. H., and X. DENG, 1982. "A Simplified Method for Analyzing Concrete Pavements Composed of Jointed Slabs," *Proceedings, International*

References   **791**

Conference on Finite Element Methods, Shanghai, China, pp. 376–381.

HUANG, Y. H., C. LIN, X. DENG, and J. G. ROSE, 1984b. *KENTRACK: A Computer Program for Hot-Mix Asphalt and Conventional Ballast Railway Trackbeds,* Report No. RR-84-1, The Asphalt Institute, College Park, Maryland.

HUANG, Y. H., C. LIN, X. DENG, and J. G. ROSE, 1986b. *KENTRACK, A Finite Element Computer Program for the Analysis of Railroad Tracks, User's Manual,* Department of Civil Engineering, University of Kentucky, Lexington, KY.

HUANG, Y. H., C. LIN, and J. G. ROSE, 1984a. "Asphalt Pavement Design: Highway versus Railroad," *Journal of Transportation Engineering, ASCE,* Vol. 110, pp. 276–282.

HUANG, Y. H., J. G. ROSE, and C. J. KHOURY, 1986a. "Hot-Mix Asphalt Railroad Trackbeds," *Transportation Research Record 1095,* pp. 102–110.

HUANG, Y. H., J. G. ROSE, and C. J. KHOURY, 1987. "Thickness Design for Hot-Mix Asphalt Railroad Trackbeds," *Proceedings, Association of Asphalt Paving Technologists,* Vol. 56, pp. 427–451.

HUANG, Y. H., J. G. ROSE, and C. LIN, 1985. "Structural Design of Hot Mix Asphalt Underlayments for Railroad Trackbeds," *Proceedings, Association of Asphalt Paving Technologists,* Vol. 54, pp. 502–528.

HUANG, Y. H., and G. W. SHARPE, 1989. "Thickness Design of Concrete Pavements by Probabilistic Method," *Proceedings, 4th International Conference on Concrete Pavement Design and Rehabilitation,* Purdue University, pp. 251–265.

HUANG, Y. H., and S. T. WANG, 1973. "Finite Element Analysis of Concrete Slabs and Its Implications for Rigid Pavement Design," *Highway Research Record 466,* Highway Research Board, pp. 55–69.

HUANG, Y. H., and S. T. WANG, 1974. "Finite-Element Analysis of Rigid Pavements with Partial Subgrade Contact," *Transportation Research Record 485,* Transportation Research Board, pp. 39–54.

HUDSON, W. R., and H. MATLOCK, 1966. "Analysis of Discontinuous Orthotropic Pavement Slabs Subjected to Combined Loads," Highway Research Record 131, Highway Research Board, pp. 1–48.

HWANG, D., and M. W. WITCZAK, 1979. *Program DAMA (Chevron), User's Manual,* Department of Civil Engineering, University of Maryland.

IOANNIDES, A. M., 1990. "Dimensional Analysis in NDT Rigid Pavement Evaluation," *Journal of Transportation Engineering,* Vol. 116, No. 1, pp. 23–36.

IOANNIDES, A. M., E. J. BARENBERG, and J. A.

LARY, 1989. "Interpretation of Falling Weight Deflectometer Results Using Principles of Dimensional Analysis," *Proceedings, 4th International Conference on Concrete Pavement Design and Rehabilitation,* Purdue University, pp. 231–247.

IOANNIDES, A. M., M. R. THOMPSON, and E. J. BARENBERG, 1985. "Westergaard Solutions Reconsidered," *Transportation Research Record 1043,* Transportation Research Board, pp. 13–23.

IRICK, P., W. R. HUDSON, and B. F. McCULLOUGH, 1987. "Application of Reliability Concepts to Pavement Design," *Proceedings, 6th International Conference on the Structural Design of Asphalt Pavements,* Vol. 1, pp. 163–179.

IRWIN, L. H., and B. M. GALLAWAY, 1974. "Influence of Laboratory Test Method on Fatigue Test Results for Asphalt Concrete," *STP 561,* American Society for Testing and Materials, pp. 12–46.

JANOFF, M. S., 1988. *Pavement Roughness and Rideability Field Evaluation,* NCHRP Report 308, Transportation Research Board.

JANOFF, M. S., J. B. NICK, and P. S. DAVIT, 1985. *Pavement Roughness and Rideability,* NCHRP Report 275, Transportation Research Board.

JIMENEZ, R. A., 1972. "Fatigue Testing of Asphaltic Concrete Slabs," *STP 508,* American Society for Testing and Materials, pp. 3–17.

JIMENEZ, R. A., and B. M. GALLAWAY, 1962. "Behavior of Asphalt Concrete Diaphragms to Repetitive Loadings," *Proceedings, (1st) International Conference on the Structural Design of Asphalt Pavements,* pp. 339–344.

JOHNSON, T. C., R. L. BERG, E. J. CHAMBERLAIN, and D. M. COLE, 1986. *Frost Action Predictive Techniques for Roads and Airfields,* Report DOT/FAA/PM-85-23, Federal Aviation Administration.

JONES, A., 1962. "Tables of Stresses in Three-Layer Elastic Systems," *Bulletin 342,* Highway Research Board, pp. 176–214.

JORDAHL, P. R., and J. B. RAUHUT, 1983. *Flexible Pavement Model VESYS IV-B,* Report prepared for Federal Highway Administration under Contract DTFH61-C-00175.

KALCHEFF, I. V, and R. G. HICKS, 1973. "A Test Program for Determining the Resilient Properties of Granular Materials," *Journal of Testing and Evaluation, ASTM,* Vol. 1, No. 6, pp. 472–479.

KALLAS, B. F., 1970. "Dynamic Modulus of Asphalt Concrete in Tension and Tension-Compression," *Proceedings, Association of Asphalt Paving Technologists,* Vol. 39, pp. 1–20.

KALLAS, B. F., and V. P. PUZINAUSKAS, 1972. "Flexure Fatigue Tests on Asphalt Paving Mixtures," *STP 508*, American Society for Testing and Materials, pp. 47–65.

KANSAS STATE HIGHWAY COMMISSION, 1947. *Design of Flexible Pavement Using the Triaxial Compression Test*, Bulletin 8, Highway Research Board.

KENIS, W. J., 1977. "Predictive Design Procedures, a Design Method for Flexible Pavements Using the VESYS Structural Subsystem," *Proceedings, 4th International Conference on the Structural Design of Asphalt Pavements*, Vol. 1, pp. 101–147.

KENNEDY, C. K., and N. W. LISTER, 1978. *Prediction of Pavement Performance and the Design of Overlay*, TRRL Report No. 833, Transport and Road Research Laboratory, Great Britain.

KERKHOVEN, R. E., and G. M. DORMON, 1953. *Some Considerations on the California Bearing Ratio Method for the Design of Flexible Pavement*, Shell Bitumen Monograph No. 1.

KESLER, C. E., 1953. "Effect of Speed of Testing on Flexural Strength of Plain Concrete," *Proceedings, Highway Research Board*, Vol. 32, pp. 251–258.

KHEDR, S. A., 1985. "Deformation Characteristics of Granular Base Course in Flexible Pavements," *Transportation Research Record 1043*, Transportation Research Board, pp. 131–138.

KHEDR, S. A., 1986. "Deformation Mechanism in Asphalt Concrete," *Journal of Transportation Engineering, ASCE*, Vol 112, No. 1, pp. 29–45.

KHER, R. K., and M. I. DARTER, 1973. "Probabilistic Concepts and Their Applications to AASHTO Interim Guide for Design of Rigid Pavements," *Highway Research Record 466*, Highway Research Board, pp. 20–36.

KILARESKI, W. P., and R. A. BIONDA, 1990. *Structural Overlay Strategies for Jointed Concrete Pavements*, Vol. 1, *Sawing and Sealing of Joints in AC Overlay of Concrete Pavements*, Report No. FHWA-RD-89-142.

KILARESKI, W. P., and S. M. STOFFELS, 1990. *Structural Overlay Strategies for Jointed Concrete Pavements*, Vol. II, *Cracking and Seating of Concrete Slabs Prior to AC Overlay*, Report No. FHWA-RD-89-143, Federal Highway Administration.

KOPPERMAN, S., G. TILLER, and M. TSENG, 1986. *ELSYM5, Interactive Microcomputer Version, User's Manual*, Report No. FHWA-TS-87-206, Federal Highway Administration.

KUMMER, H. W., and W. E. MEYER, 1967. *Tentative Skid-Resistance Requirements for Main Rural Highways*, NCHRP Report 37, Highway Research Board.

LAI, J. S., 1977. *VESYS-G, a Computer Program for Analysis of N-layered Flexible Pavement*, Report

No. FHWA-77-117, Federal Highway Administration.

LEMER, A. C., and F. MOAVENZADEH, 1971. "Reliability of Highway Pavements," *Highway Research Record 362*, Highway Research Board, pp. 1–8.

LEU, M. C., and J. J. HENRY, 1978. "Prediction of Skid Resistance as a Function of Speed from Pavement Texture," *Transportation Research Record 666*, Transportation Research Board, pp. 38–43.

LIU, S. J., and R. L. LYTTON, 1984. "Rainfall Infiltration, Drainage, and Load-Carrying Capacity of Pavements," *Transportation Research Record 993*, Transportation Research Board, pp. 28–35.

LOKKEN, E. C., 1973. "What We Have Learned to Date from Experimental Concrete Shoulder Projects," *Highway Research Record 434*, Highway Research Board, pp. 43–53.

LOVERING, W. R., and H. R. CEDERGREN, 1962. "Structural Section Drainage," *Proceedings, (1st) International Conference on the Structural Design of Asphalt Pavements*, pp. 773–784.

LYTTON, R. L., 1989. "Backcalculation of Pavement Layer Properties," *STP 1026*, American Society for Testing and Materials, pp. 7–38.

LYTTON, R. L., and Y. J. CHOU, 1988. "Modulus Backcalculation Exercise," *Informal report to TRB Committee A2B05, Strength and Deformation Characteristics*, Transportation Research Board.

LYTTON, R. L., F. P. GERMANN, Y. J. CHOU, and S. M. STOFFELS, 1990. *Determining Asphalt Concrete Pavement Structural Properties by Nondestructive Testing*, NCHRP Report 327, Transportation Research Board.

LYTTON, R. L., U. SHANMUGHAM, and B. D. GARRETT, 1983. *Design of Asphalt Pavements for Thermal Fatigue Cracking*, Research Report No. 284-4, Texas Transportation Institute, Texas A&M University.

MAHONEY, J. P., and N. C. JACKSON, 1990. "Guidelines on When to Apply and Remove Seasonal Load Restrictions—Development Through Implementation," *Proceedings, 3rd International Conference on Bearing Capacity of Roads and Airfields*, The Norwegian Institute of Technology, Trondheim, Vol. 1, pp. 75–84.

MAJIDZADEH, K., 1978. *Observations of Field Performance of Continuously Reinforced Concrete Pavements in Ohio*, Report No. Ohio-DOT-12-77, Ohio Department of Transportation.

MAJIDZADEH, K., F. BAYOMY, and S. KHEDR, 1978. "Rutting Evaluation of Subgrade Soils in Ohio," *Transportation Research Record 671*, Transportation Research Board, pp. 75–84.

References

MAJIDZADEH, K., G. J. ILVES, and H. SKLYUT, 1984. *Mechanistic Design of Rigid Pavements,* Vol. 1, *Development of the Design Procedure,* Report No. FHWA-RD-86-124, Vol. 2, *Design and Implementation Manual,* Report No. FHWA-RD-86–235, Federal Highway Administration.

MAJIDZADEH, K., S. KHEDR, and H. GUIRGUIS, 1976. "Laboratory Verification of a Mechanistic Subgrade Rutting Model," *Transportation Research Record 616,* Transportation Research Board, pp. 34–37.

MAMLOUK, M. S., 1987. "Dynamic Analysis of Multilayer Pavement Structures—Theory, Significance and Verification," *Proceedings, 6th International Conference on the Structural Design of Asphalt Pavements,* Vol. 1, pp. 466–474.

McCULLOUGH, B. F., and M. L. CAWLEY, 1981. "CRCP Design Based on Theoretical and Field Performance," *Proceedings, 2nd International Conference on Concrete Pavement Design,* Purdue University, pp. 239–251.

McCULLOUGH, B. F., and G. E. ELKINS, 1979. *CRC Pavement Design Manual,* Austin Research Engineers.

McLEAN, D. B., 1974. *Permanent Deformation Characteristics of Asphalt Concrete,* Ph. D. Dissertation, University of California, Berkeley.

McLEOD, N. W., 1953. "Some Basic Problems in Flexible Pavement Design," *Proceedings, Highway Research Board,* pp. 91–118.

McLEOD, N. W., 1970. "Influence of Hardness of Asphalt Cement on Low Temperature Transverse Pavement Cracking," *Proceedings, Canadian Good Roads Association.*

MEYER, W. E., 1991. "Pavement Texture Significance and Measurement," *Standardization News, ASTM,* February, pp. 28–31.

MILLER, E. J., H. H. RICHTER, and D. A. PURKIS, 1986. "Computer Assisted Pavement," *Civil Engineering,* September, pp. 75–77.

MINER, M. A., 1945. "Cumulative Damage in Fatigue," *Transactions of the ASME,* Vol. 67, pp. A159–A164.

MOAVENZADEH, F., J. E. SOUSSOU, H. K. FINDAKLY, and B. BRADEMEYER, 1974. *Synthesis for Rational Design of Flexible Pavements,* Part 3, *Operating Instructions and Program Documentation,* Reports prepared for Federal Highway Administration under Contract FH 11-776.

MONISMITH, C. L., J. SOUSA, and J. LYSMER, 1988. "Modern Pavement Design Technology Including Dynamic Loading Conditions," *Vehicle/Pavement Interaction, Where the Truck Meets the Road,*

SP-765, Society of Automotive Engineers, Inc., pp. 33–52.

MONISMITH, C. L., and M. W. WITCZAK, 1982. "Moderator's Report, Section 1," *Proceedings, Fifth International Conference on the Structural Design of Asphalt Pavements,* Vol. 2, pp. 2–59.

MOORE, J. H., 1956. "Thickness of Concrete Pavements," with discussions by E. C. Sutherland and W. Harwood, *Transactions, ASCE,* Vol. 121, pp. 1125–1152.

MOREELL, B., 1958. "Prestressing Promises Nearly Joint-Free Highways," *Civil Engineering,* Vol. 28, No. 8, pp. 34–37.

MOULTON, L. K., 1980. *Highway Subdrainage Design,* Report No. FHWA-TS-80-224, Federal Highway Administration.

NCHRP, 1972. *Skid Resistance,* Synthesis of Highway Practice 14, National Cooperative Highway Research Program.

NCHRP, 1979. *Design and Use of Highway Shoulders,* Synthesis of Highway Practice 63, National Cooperative Highway Research Program.

NCHRP, 1990. *Calibrated Mechanistic Structural Analysis Procedures for Pavement,* NCHRP 1–26, Vol. 1, *Final Report;* Vol. 2, *Appendices,* University of Illinois at Urbana–Champaign.

NIE, N. H. et al., 1975. *Statistical Package for the Social Sciences,* 2d ed., McGraw-Hill, New York.

NUSSBAUM, P. J., and E. C. LOKKEN, 1978. *Portland Cement Concrete Pavements, Performance Related to Design–Construction–Maintenance,* Report No. FHWA-TS-78-202, Prepared by PCA for Federal Highway Administration.

OLDER, C., 1924. "Highway Research in Illinois," *Transactions, ASCE,* Vol. 87, pp. 1180–1222.

PACKARD, R. G., and S. D. TAYABJI, 1985. "New PCA Thickness Design Procedure for Concrete Highway and Street Pavements," *Third International Conference on Concrete Pavement Design and Rehabilitation,* Purdue University, pp. 225–236.

PAGEN, C. A., 1965. "Rheological Response of Bituminous Concrete," *Highway Research Record 67,* Highway Research Board, pp. 1–26.

PAPAZIAN, H. S., 1962. "The Response of Linear Viscoelastic Materials in the Frequency Domain with Emphasis on Asphaltic Concrete," *(1st) International Conference on the Structural Design of Asphalt Pavements,* pp. 454–463.

PASKO, T. J., 1972. "Prestressed Concrete Pavement at Dulles Airport for Transpo 72," *Journal of the Prestressed Concrete Institute,* Vol. 17, No. 2, pp. 46–54.

PCA, 1951. *Concrete Pavement Design*, Portland Cement Association.

PCA, 1955. *Design of Concrete Airport Pavement*, Portland Cement Association.

PCA, 1966. *Thickness Design for Concrete Pavements*, Portland Cement Association.

PCA, 1969. *Load Stresses at Pavement Edge, a Supplement to Thickness Design for Concrete Pavements*, Portland Cement Association.

PCA, 1975. *Join Design for Concrete Highway and Street Pavements*, Portland Cement Association.

PCA, 1984. *Thickness Design for Concrete Highway and Street Pavements*, Portland Cement Association.

PCA, 1991. *Design and Construction of Joints for Concrete Highways*, Concrete Paving Technology, Portland Cement Association.

PEATTIE, K. R., 1962. "Stress and Strain Factors for Three-Layer Elastic Systems," *Bulletin 342*, Highway Research Board, pp. 215–253.

PELL, P. S., 1962. "Fatigue Characteristics of Bituminous Mixes," *(1st) International Conference on the Structural Design of Asphalt Pavements*, pp. 310–323.

PELL, P. S., 1987. "Keynote Lecture—Pavement Materials," *Sixth International Conference on the Structural Design of Asphalt Pavements*, Vol. 2, pp. 36–70.

PERLOFF, W. H., and F. MOAVENZADEH, 1967. "Deflection of Viscoelastic Medium Due to Moving Load," *Proceedings, 2nd International Conference on the Structural Design of Asphalt Pavements*, University of Michigan, pp. 269–276.

PHU, N. C, J. P. CHRISTORY, and J. RAY, 1986. *The Hydromechanics of Pumping in Concrete Pavements: Modeling and Prevention*, Appendix I, *Combatting Concrete Pavement Pumping: State-of-the-Art and Recommendations*, PIARC Technical Committee on Concrete Roads.

PICKETT, G., and S. BADARUDDIN, 1956. "Influence Chart for Bending of a Semi-infinite Pavement Slab," *Proceedings, Ninth International Congress on Applied Mechanics*, Vol. 6, pp. 396–402.

PICKETT, G., M. E. RAVILLE, W. C. JANES, and F. J. McCORMICK, 1951. "Deflections, Moments and Reactive Pressures for Concrete Pavements," *Bulletin No. 65*, Engineering Experiment Station, Kansas State College.

PICKETT, G., and G. K. RAY, 1951. "Influence Charts for Concrete Pavement," *Transactions, ASCE*, Vol. 116, pp. 49–73.

POEHL, R., 1971. *Seasonal Variations of Pavement Deflections in Texas*, Research Report 136-1, Texas Transportation Institute, Texas A & M University.

PORTER, O. J., 1950. "Development of the Original Method for Highway Design: Symposium on Development of CBR Flexible Pavement Design Method for Airfields," *Transactions, ASCE*, pp. 461–467.

POWELL, W. D., et al., 1984. *The Structural Design of Bituminous Pavements*, TRRL Laboratory Report 1132, Transportation and Road Research Laboratory, U.K.

RAAD, L. and J. L. FIGUEROA, 1980. "Load Response of Transportation Support Systems," *Transportation Engineering Journal, ASCE*, Vol. 106, No. TE1, pp. 111–128.

RADA, G., and M. W. WITCZAK, 1981. "Comprehensive Evaluation of Laboratory Resilient Moduli Results for Granular Materials," *Transportation Research Record 810*, Transportation Research Board, pp. 23–33.

RAITHBY, K. D., and J. W. GALLOWAY, 1974. "Effects of Moisture Condition, Age, and Rate of Loading on Fatigue of Plain Concrete," *SP-41*, American Concrete Institute, pp. 15–34.

RAUHUT, J. B., R. C. G. HAAS, and T. W. KENNEDY, 1977. "Comparison of VESYS IIM Predictions to Brampton/AASHO Performance Measurements," *Proceedings, 4th International Conference on the Structural Design of Asphalt Pavements*, Vol. 1, pp. 131–147.

RAUHUT, J. B., and P. R. JORDAHL, 1979. *Effects on Flexible Pavements of Increased Legal Vehicle Weights Using VESYS-IIM*, Report No. FHWA-77-116, Federal Highway Administration.

RIDGEWAY, H. H., 1976. "Infiltration of water through the Pavement Surface," *Transportation Research Record 616*, Transportation Research Board, pp. 98–100.

RIDGEWAY, H. H., 1982. *Pavement Subsurface Drainage Systems*, Synthesis of Highway Practice 96, Transportation Research Board.

RITCHIE, S. G., C. YEH, J. P. MAHONEY, and N. C. NEWTON, 1986. "Development of an Expert System for Pavement Rehabilitation Decision Making," *Transportation Research Record 1070*, Transportation Research Board, pp. 90–103.

ROAD RESEARCH LABORATORY, 1952. *Soil Mechanics for Road Engineers*, Her Majesty Stationary Company.

ROADS AND STREETS, 1971. "Prestressed Concrete Pavement in Service," Vol 114, No. 10, p. 56.

ROAD AND TRANSPORTATION ASSOCIATION OF CANADA, 1977. *Pavement Management Guide*.

ROSENBLUETH, E., 1975. "Point Estimates for Probability Moments," *Proceedings, National Academy of Sciences, Mathematics*, Vol. 72, No. 10, pp. 3812–3814.

RUTH, B. E., L. A. K. BLOY, and A. A. AVITAL, 1982. "Prediction of Pavement Cracking at Low Temperatures," *Proceedings, Association of Asphalt Paving Technologists,* Vol. 51, pp. 53–90.

RYELL, J., and J. T. CORKILL, 1973. "Long Term Performance of an Experimental Composite Pavement," *Highway Research Record 434,* Highway Research Board, pp. 1–15.

SAAL, R. N. J., and P. S. PELL, 1960. *Kolloid-Zeitschrift MI,* Heft 1, pp. 61–71.

SAWAN, J. S., and M. I. DARTER, 1979. "Structural Design of PCC Shoulders," *Transportation Research Record 725,* Transportation Research Board, pp. 80–88.

SAXENA, S. K., 1973. "Pavement Slabs Resting on Elastic Foundation," *Highway Research Record 466,* Highway Research Board, pp. 163–178.

SAYERS, M. W., T. D. GILLESPIE, and W. D. PATERSON, 1986a. *Guidelines for the Conduct and Calibration of Road Roughness Measurements,* Technical Paper 46, The World Bank, Washington, DC.

SAYERS, M. W., T. D. GILLESPIE, and C. A. QUEIROZ, 1986b. *The International Road Roughness Experiment: Establishing Correlation and a Calibration Standard for Measurements,* Technical Paper 46, The World Bank, Washington, DC.

SCHOFIELD, R. K., 1935. "The PF of the Water in Soil," *Transactions, 3rd International Congress on Soil Science,* Oxford, 1935, Vol. 2, pp. 37–48.

SCRIVNER, F. H., and W. M. MOORE, 1968. *Standard Measurements for Satellite Road Test Program,* NCHRP Report No. 59, National Cooperative Highway Research Program.

SCRIVNER, F. H., R. POEHL, W. M. MOORE, and M. B. PHILLIPS, 1969. *Detecting Seasonal Changes in Load-Carrying Capacities of Flexible Pavements,* NCHRP Report No. 76, National Cooperative Highway Research Program.

SCULLION, T., J. UZAN, and M. PAREDES, 1990. "Modulus: A Microcomputer-Based Backcalculation System," *Transportation Research Record 1260,* Transportation Research Record, pp. 180–191.

SEBAALY, B. E., M. S. MAMLOUK, and T. G. DAVIES, 1986. "Use of Falling Weight Deflectometer Data," *Transportation Research Record 1070,* Transportation Research Board, pp. 63–68.

SELIG, E. T., H. LIN, L. J. DOYER, S. DUANN, and H. TZENG, 1986. *Layered System Performance Evaluation, Phase II,* Final Report, Report No. DOT/OST/P-34/86/047.

SHAHIN, M. Y., and B. F. McCULLOUGH, 1972. *Prediction of Low-Temperature and Thermal-Fatigue Cracking in Flexible Pavements,* Report No. CFHR 1-8-69-123-14, The University of Texas.

SHELL, 1978. *Shell Pavement Design Manual—Asphalt Pavements and Overlays for Road Traffic,* Shell International Petroleum, London.

SHERARD, J. L., R. J. WOODWARD, S. F. GIZIENSKI, and W. A. CLEVENGER, 1963. *Earth and Earth-Rock Dams,* Wiley, New York.

SHERMAN, G. B., 1971. "In Situ Materials Variability," *Special Report 126,* Highway Research Board, pp. 180–188.

SHOOK, J. F., F. N. FINN, M. W. WITCZAK, and C. L. MONISMITH, 1982. "Thickness Design of Asphalt Pavements—The Asphalt Institute Method," *Proceedings, 5th International Conference on the Structural Design of Asphalt Pavements,* Vol. 1, pp. 17–44.

SLAVIS, C., 1981. "Portland Cement Concrete Shoulder Performance in the United States," *Proceedings, 2nd International Conference on Concrete Pavement Design,* Purdue University, pp. 331–341.

SMITH, B. E., and M. W. WITCZAK, 1981. "Equivalent Granular Base Moduli: Prediction," *Transportation Engineering Journal, ASCE,* Vol. 107, No. TE6, pp. 635–652.

SMITH, K. D., M. I. DARTER, J. B. RAUHUT, and K. T. Hall, 1987. *Distress Identification Manual for the LTPP Studies,* Strategic Highway Research Program, National Research Council.

SMITH, K. D., D. G. PESHKIN, M. I. DARTER, A. L. MUELLER, and S. H. CARPENTER, 1990a. *Performance of Jointed Concrete Pavements,* Vol. 1, *Evaluation of Concrete Pavement Performance and Design Features,* Report No. FHWA-RD-89-136, Federal Highway Administration.

SMITH, K. D., D. G. PESHKIN, M. I. DARTER, and A. L. MUELLER, 1990b. *Performance of Jointed Concrete Pavements,* Vol. 2, *Evaulation and Modification of Concrete Pavement Design and Analysis Models,* Report No. FHWA-RD-89-137, Federal Highway Administration.

SMITH, K. D., D. G. PESHKIN, M. I. DARTER, A. L. MUELLER, and S. H. CARPENTER, 1990c. *Performance of Jointed Concrete Pavements,* Vol. 5, *Appendix B: Data Collection and Analysis Procedures,* Report No. FHWA-RD-89-140, Federal Highway Administration.

SMITH, R. E., M. I. DARTER, and S. M. HERRIN, 1979. *Highway Pavement Distress Identification Manual for Highway Condition and Quality of*

Highway Construction Survey, Contract DOT-FH-11-9175/NCHRP 1-19, Federal Highway Administration.

SPANGLER, E. B., and W. J. KELLY, 1964. *GMR Road Profilometer—A Method for Measuring Road Profiles,* Research Publication GMR-452, General Motor Corporation.

SPANGLER, M. G., 1942. *Stresses in the Corner Region of Concrete Pavements,* Bulletin 157, Iowa State College.

STEELE, D. J., 1945. "Application of the Classification and Group Index in Estimating Desirable Subbase and Total Pavement Thickness," (a discussion), *Proceedings, Highway Research Board,* pp. 388–392.

TABATABAIE, A. M., 1977. *Structural Analysis of Concrete Pavement Joints,* Ph.D. Thesis, University of Illinois.

TABATABAIE, A. M., and E. J. BARENBERG, 1979. *Longitudinal Joint Systems in Slip-Formed Rigid Pavements,* Vol. 3; *User's Manual,* Report FAA-RD-79-4, III, U.S. Department of Transportation.

TABATABAIE, A. M., and E. J. BARENBERG, 1980. "Structural Analysis of Concrete Pavement Systems," *Transportation Engineering Journal, ASCE,* Vol. 106, No. TE5, pp. 493–506.

TAYABJI, S. D., and B. E. COLLEY, 1986. *Analysis of Jointed Concrete Pavement,* Report No. FHWA-RD-86-041, Federal Highway Administration.

TAYABJI, S. D., and P. A. OKAMOTO, 1985. "Thickness Design of Concrete Resurfacing, *Proceedings, 3rd International Conference on Concrete Pavement Design and Rehabilitation,"* Purdue University, pp. 367–379.

TELLER, L. W., and E. C. SUTHERLAND, 1935–1943. *The Structural Design of Concrete Pavements,* Reprints from *Public Roads,* Vols. 16, 17, and 23.

TERZAGHI, K., 1943. *Theoretical Soil Mechanics,* Wiley, New York.

THOMPSON, M. R., 1987. "ILLI-PAVE Based Full-Depth Asphalt Concrete Pavement Design Procedure," *Proceedings, Sixth International Conference on Structural Design of Asphalt Pavements,* Vol. 1, pp. 13–22.

THOMPSON, M. R., B. J. DEMPSEY, H. HILL, and J. VOGEL, 1987. "Characterizing Temperature Effects for Pavement Analysis and Design," *Transportation Research Record 1121,* Transportation Research Board, pp. 14–22.

THOMPSON, M. R., and R. P. ELLIOT, 1985. "ILLI-PAVE-Based Response Algorithms for Design of Conventional Flexible Pavements," *Transportation Research Record 1043,* Transportation Research Board, pp. 50–57.

TIMOSHENKO, S., and I. N. GOODIER, 1951. *Theory of Elasticity,* McGraw-Hill, New York.

TREYBIG, H. C., B. F. McCULLOUGH, P. SMITH, H. VONQUINTUS, and P. JORDAHL, 1977. *Flexible and Rigid Pavement Overlay Design Procedures,* Report No. FHWA-RD-77-133, Austin Research Engineers, Inc., Austin, TX.

TURNBULL, W. J., and R. G. AHLVIN, 1957. "Mathematical Expression of the CBR (California Bearing Ratio) Relations," *Proceedings, 4th International Conference on Soil Mechanics and Foundation Engineering,* Vol. 2, pp. 178–180.

ULLIDTZ, P., 1987. *Pavement Analysis,* Elsevier Science, New York.

ULLIDTZ, P., and R. N. STUBSTAD, 1985. "Analytical-Empirical Pavement Evaluation Using the Falling Weight Deflectometer," *Transportation Research Record 1022,* Transportation Research Board, pp. 36–44.

U.S. ARMY CORPS OF ENGINEERS, 1955. *Drainage and Erosion Control—Subsurface Drainage Facilities for Airfields,* Part XIII, Chapter 2, *Engineering Manual,* Military Construction.

U.S. ARMY CORPS OF ENGINEERS, 1961. *Revised Method of Thickness Design of Flexible Highway Pavements at Military Installations,* Technical Report No. 3-582, Waterways Experimental Station.

U.S. BUREAU OF RECLAMATION, 1973. *Design of Small Dams,* U.S. Government Printing Office, Washington, DC.

U.S. NAVY, 1953. Airfield Pavement, Bureau of Yards and Docks, Technical Publication, NAVDOCKS TP-PW-4.

UZAN, J., R. L. LYTTON, and F. P. GERMANN, 1989. "General Procedure for Backcalculating Layer Moduli," *STP 1026,* American Society for Testing and Materials, pp. 217–228.

UZAN, J., T. SCULLION, C. H. MICHALEK, M. PAREDES, and R. L. LYTTON, 1988. *A Microcomputer Based Procedure for Backcalculating Layer Moduli from FWD Data,* Research Report 1123-1, Texas Transportation Institute.

VAN CAUWELAERT, F. J., D. R. ALEXANDER, T. D. WHITE, and W. R. BAKER, 1989. Multilayer Elastic Program for Backcalculating Layer Moduli in Pavement Evaluation, *STP 1026,* American Society for Testing and Materials, pp. 171–188.

VAN DE LOO, P. J., 1974. "Creep Testing, a Simple

Tool to Judge Asphalt Mix Stability,'' *Proceedings, Association of Asphalt Paving Technologists,* Vol. 43, pp. 253–281.

VAN DE LOO, P. J., 1978. ''The Creep Test: A Key Tool in Asphalt Mix Design and in the Prediction of Pavement Rutting,'' *Proceedings, Association of Asphalt Paving Technologists,* Vol. 47, pp. 522–554.

VAN DER POEL, C., 1954. A General System Describing the Visco-Elastic Properties of Bitumens and Its Relation to Routine Test Data, *Journal of Applied Chemistry,* Vol. 4, 1954, pp. 221–236.

VAN TIL, C. J., B. F. McCULLOUGH, B. A. VALLERGA, and R. G. HICKS, 1972. *Evaluation of AASHO Interim Guides for Design of Pavement Structures,* NCHRP 128, Highway Research Board.

VAN WIJI, A. J., 1985. *Purdue Economic Analysis of Rehabilitation and Design Alternatives for Rigid Pavements: A User's Manual for PEARDARP,* Final Report, FHWA Contract DTFH61-82-C-00035.

VAN WIJI, A. J., J. LARRALDE, C. W. LOVELL, and W. F. CHEN, 1989. ''Pumping Prediction Model for Highway Concrete Pavements,'' *Journal of Transportation Engineering,* Vol. 115, No. 2, pp. 161–175.

VERSTRAETEN, J., V. VEVERKA, and L. FRANCKEN, 1982. ''Rational and Practical Designs of Asphalt Pavements to Avoid Cracking and Rutting,'' *Proceedings, Fifth International Conference on the Structural Design of Asphalt Pavements.*

VESIC, A. S., and L. DOMASCHUK, 1964. *Theoretical Analysis of Structural Behavior of Road Test Flexible Pavements,* NCHRP Report No. 10, Highway Research Board.

VESIC, A. S., and K. SAXENA, 1974. *Analysis of Structural Behavior of AASHO Road Test Rigid Pavements,* NCHRP Report No. 97, Highway Research Board.

VON QUINTUS, H. L., F. N. FINN, W. R. HUDSON, and F. L. ROBERTS, 1980. *Flexible and Composite Structures for Premium Pavements,* Report Nos. FHWA-RD-81-154 and -155 Vols. 1 and 2, Federal Highway Administration.

WAMBOLD, J. C., J. J. HENRY, C. E. ANTLE, J. W. BUTTON, and D. A. ANDERSON, 1989. *Pavement Friction Measurement Normalized for Operational, Seasonal, and Weather Effects,* Report No. FHWA-RD-88-069, Federal Highway Administration.

WARREN, H., and W. L. DIECKMANN, 1963. *Numerical Computation of Stresses and Strains in a Multiple-Layer Asphalt Pavement System,* Internal Report, Chevron Research Corporation, Richmond, CA.

WESTERGAARD, H. M., 1926a. ''Analysis of Stresses in Concrete Pavement Due to Variations of Temperature,'' *Proceedings, Highway Research Board,* Vol. 6, pp. 201–215.

WESTERGAARD, H. M., 1926b. ''Stresses in Concrete Pavements Computed by Theoretical Analysis,'' *Public Roads,* Vol. 7, pp. 25–35.

WESTERGAARD, H. M., 1927. ''Theory of Concrete Pavement Design,'' *Proceedings, Highway Research Board,* Part I, pp. 175–181.

WESTERGAARD, H. M., 1933. ''Analytical Tools for Judging Results of Structural Tests of Concrete Pavements,'' *Public Roads,* Vol. 14, No. 10, pp. 185–188.

WESTERGAARD, H. M., 1939. ''Stresses in Concrete Runways of Airports,'' *Proceedings, Highway Research Board,* Vol. 19, pp. 197–202.

WESTERGAARD, H. M., 1943. ''Stress Concentrations in Plates Loaded over Small Areas,'' *Transactions, ASCE,* Vol. 108, pp. 831–856.

WESTERGAARD, H. M., 1948. ''New Formulas for Stresses in Concrete Pavements of Airfields,'' *Transactions, ASCE,* Vol. 113, pp. 425–444.

WHITLOW, R., 1990. *Basic Soil Mechanics,* Wiley, New York.

WINTER, G., and A. H. NILSON, 1979. *Design of Concrete Structures,* McGraw-Hill, New York.

WITCZAK, M. W., and R. E. ROOT, 1974. ''Summary of Complex Modulus Laboratory Test Procedures and Results,'' *STP 561,* American Society for Testing and Materials, pp. 67–94.

WRI, 1975. *Jointed Concrete Pavements Reinforced with Welded Wire Fabric,* Wire Reinforcement Institute, Inc.

YAZDANI, J. I., and T. SCULLION, 1990. ''Comparing Measured and Theoretical Depth Deflections Under a Falling Weight Deflectometer Using a Multidepth Deflectometer,'' *Transportation Research Record 1260,* Transportation Research Board, pp. 216–225.

YODER, E. J., 1959. *Principles of Pavement Design,* Wiley, New York.

YODER, E. J., and M. W. WITCZAK, 1975. *Principles of Pavement Design,* Wiley, New York.

ZANIEWSKI, J. P., W. R. HUDSON, R. HIGH, and S. W. HUDSON, 1985. *Pavement Rating Procedures,* Contract No. DTFH61-83-C-00153, Federal Highway Administration.

ZIENKIEWICZ, O. C., and Y. K. CHEUNG, 1967. *The*

Finite Element Method in Structural and Continuum Mechanics, McGraw-Hill, New York.

ZOLLINGER, D. G., 1989. *Investigation of Punchout Distress of Continuously Reinforced Concrete Pavements,* Ph.D. Thesis, Department of Civil Engineering, University of Illinois at Champaign-Urbana.

ZOLLINGER, D. G., and E. J. BARENBERG, 1989. *Proposed Mechanistic Based Design Procedure for Jointed Concrete Pavements,* Civil Engineering Studies, Transportation Engineering Series No. 57, Illinois Cooperative Highway Research Program.

ZOLLINGER, D. G., and E. J. BARENBERG, 1990. ''Mechanistic Design Considerations for Punchout Distress in Continuously Reinforced Concrete Pavement,'' *Transportation Research Record 1286,* Transportation Research Board, pp. 25–37.

Index

AUTHOR INDEX

Friberg, 194, 195
Fwa, T. F., 705

Gage, R. B., 8
Gallaway, B. M., 337, 347
Galloway, J. W., 355
Gerrard, C. M., 283
Goldbeck, A. T., 6, 175
Gomez-Achecar, M., 5

Haas, R., 39, 465
Hajek, J. J., 465
Haliburton, T. A., 387
Hall, K. T., 3, 462, 466
Hanna, A. M., 19
Harichandran, R. S., 5, 37, 122
Harrison, W. J., 283
Healey, K. A., 386
Heinrichs, K. W., 196, 250
Henry, J. J., 440
Heukelom, W., 330, 545
Hick, R. G., 108, 586, 589
Hogentogler, C. A., 2
HRB, 2, 21, 23, 26
Huang, Y. H., 8, 9, 42, 43, 56,
    57, 60, 62, 65, 68, 70–73, 93,
    217, 220, 222, 283, 289, 521
Hudson, W. R., 7, 39
Hwang, D., 4, 344

Ioannides, A. M., 176, 456
Irick, P., 5, 426, 495
Irwin, L. H., 337

Janoff, M. S., 437
Jimenez, R. S., 347
Johnson, T. C., 537
Jones, A., 75, 77
Jordahl, P. R., 5, 509

Kalcheff, I. V., 108
Kallas, B. F., 335, 336
Kansas State Highway Commis-
    sion, 2, 48
Kelly, W. J., 435
Kenis, W. J., 137, 509
Kennedy, T. W., 348
Kerkhoven, R. E., 3
Kesler, C. E., 355
Khedr, S. A., 541
Kher, R. K., 8, 479, 502, 503
Kilareski, W. P., 666, 667
Klomp. A. J. G., 330
Kopperman, S. G., 4, 121
Kummer, H. W., 442

Lai, J. S., 509
Larew, H. G., 93
Lawmaster, J. D., 387
Lemer, A. C., 5, 479
Leu, M. C., 440
Liu, S. J., 533, 537
Lokken, E. C., 17, 658
Long, R. P., 386
Lovering, W. R., 379
Lytton, R. L., 134, 450, 452,
    533, 537, 543

Majidzadeh, K., 8, 538, 542
Mamlouk, M. S., 5
Marais, C. P., 147
Matlock, H., 7
McCullough, B. F., 538, 543,
    643, 647
McLean, D. B., 318
McLeod, N. W., 2, 465, 543
Metcalf, C. T., 3
Meyer, W. E., 441, 442
Miller, E. J., 12
Miner, M. A., 42, 538
Moavenzadeh, F., 5, 93, 479,
    509
Monismith, C. L., 5, 436, 678
Moore, W. M., 500
Moreell, B., 18
Moulton, L. K., 377, 381, 383,
    392, 393

NCHRP, 445, 531, 584, 595
Nie, N. H., 466
Nilson, A. H., 187
Nussbaum, P. J., 17

Okamoto, P. A., 688, 689, 692
Older, C., 6, 175

Packard, R. G., 235
Pagen, C. A., 91
Papazian, H. S., 334
Pasko, T. J., 18
Peattie, K. R., 78, 79
Pell, P. S., 3, 348, 539
Perloff, W. H., 93
Phu, N. C., 600
Pickett, G., 7, 248
Poehl, R., 451
Porter, O. J., 2
Powell, W. D., 539
Puzinauskas, V. P., 336

Raad, L., 5, 122
Rada, G., 108
Raithby, K. D., 355

Rauhut, J. B., 5, 509, 542
Ridgeway, H. H., 35, 376, 391,
    398
Ritchie, S. G., 461
Road Research Laboratory, 535
Roads and Streets, 18
Roads and Transportation Asso-
    ciation of Canada, 672
Root, R. E., 335, 337
Rosenblueth, E., 506
Ruth, B. E., 543
Ryell, J. T., 19

Saal, R. N. J., 3
Sawan, J. S., 588
Sawyer, C. L., 380, 397, 399
Saxena, S. K., 7, 211
Sayers, M. W., 436
Schofield, R. K., 535
Scrivner, F. H., 500
Scullion, T., 453, 460
Selig, E. T., 125
Shahin, M. Y., 538, 543
Shannon, W. L., 397
Sharpe, G. W., 9, 521
Shell Petroleum International,
    37, 339, 344
Sherard, J. L., 383
Sherman, G. B., 501
Shook, J. F., 4, 37, 104, 108
Slavis, C., 658
Smith, B. E., 139
Smith, K. D., 406, 469
Smith, R. E., 406
Spangler, E. B., 435
Spangler, M. G., 633
Steele, D. J., 2
Stoffels, S. M., 666
Sutherland, E. C., 6

Tabatabaie, A. M., 8, 250
Tayabji, S. D., 8, 607, 688, 689,
    692
Teller, L. W., 6
Terzaghi, K., 2
Thompson, M. R., 5, 109, 125,
    533, 534, 539
Treybig, H. C., 665
Turnbull, W. J., 507

Ulery, H. H., 49
Ullidtz, P., 55
U.S. Army Corps of Engineers,
    383, 589
U.S. Bureau of Reclamation, 388
U.S. Navy, 2
Uzan, J. T., 453

Van Cauwelaert, F. J., 454
Van der Loo, P. J., 363
Van der Poel, C., 339
Van Til, C. J., 327, 331, 333
Van Wiji, A. J., 466, 601
Verstraeten, J., 348, 539
Vesic, A. S., 57, 211
Von Quintus, H. L., 20, 509

Wambold, J. C., 441
Wang, S. T., 8, 222
Warren, H., 4
Westergaard, H. M., 6, 7, 8,
    176–178, 243
Whitlow, R., 381
Winter, G., 187
Witczak, M. W., 2, 4, 55, 108,
    139, 337. 344
WRI, 189

Yazdani, J. I., 460
Yoder, E. J., 2, 55

Zaniewski, J. P., 435
Zienkiewicz, O. C., 7
Zollinger, D. G., 605

SUBJECT INDEX

AASHTO
    design equations and
        nomographs, 634, 572, 651
    design of overlay, 693
    equivalent axle load factors,
        294, 300
    flexible pavement design, 566
    overlay equations, 696, 707
    rigid pavement design, 632
Airport pavement, 41
Analysis period, 567, 612
Apparent opening size (AOS),
    384
Arlington Experimental Farm, 6,
    173
Asphalt concrete. *See* Hot mix
    asphalt
Asphalt Institute
    equivalent axle load factors,
        296
    failure criteria, 104, 354, 547
    flexible pavement design, 546
    formulas for dynamic
        modulus, 344
    formulas for fatigue of
        bituminous mixtures, 354
    overlay design, 673
Asphalt pavement. *See* flexible
    pavement
Average daily truck traffic, 304,
    549, 628
Axle load distribution, 307, 627

Back-calculation of modulus,
    133, 451, 693
Base courses, 11, 13, 332, 470,
    551, 556, 577
Bates Road Test, 8
Bearing stress, 194, 467
Bending of plate, 169
Benkelman beam, 447, 677
BISAR program, 4, 454

Boltzmann's superposition
    principle, 94
Bonded slabs, 212, 691, 710
British pendulum number, 441
Burgers model, 85, 727

Calibrated mechanistic design
    procedure, 531, 595
California bearing ratio, 2, 328,
    330, 552
Cement-treated base, 257, 333
CHEV program, 4, 454
Climate models, 533, 589
Coefficient of variation, 502, 610
    for performance prediction,
        504, 505
    for traffic prediction, 504
Cohesiometer test, 331
Collector pipe, 400
Collocation method, 89, 92, 726
Combined stresses, 173
Complex modulus. *See* dynamic
    modulus
Composite pavements, 19
Concrete pavement. *See* rigid
    pavement
Contained rock asphalt mat, 12,
    157
Continuous reinforced concrete
    pavement, 18, 607, 645, 647
Conversion factors
    for dual-tandem wheels, 71
    for dual wheels, 69
    for effective thickness, 674,
        676
COPES program, 3, 462, 466,
    600
Corner loading, 175, 269, 609
Covariance, 482
Cracking and seating, 665, 709
Crack relief layer, 21, 666
Creep compliance, 86

Creep test, 89, 363
Critical tensile strain. *See*
    Tensile strain
Curling stresses, 170, 174, 598

Damage analysis, 104, 113, 138,
    235, 298, 746, 750, 758, 761
DAMA program, 4, 121, 138,
    538, 555
Darcy's law, 372
Degree of drainage, 399
Design inflow, 395
Dirichlet series, 88, 92, 94, 726
Discrete element method, 7, 665
Distress, 406, 414, 426, 538
Distress models, 514, 538,
    597–607
Dowel bars
    allowable bearing stress, 193
    bearing stress, 194, 467, 601
    recommended size and length,
        193
    weights and dimensions, 189
Dowel group action, 194, 265
Drainage coefficient, 579, 645
Drainage layer, 21, 375, 396
Duration of moving load, 94, 339
Dynaflect, 448, 682
Dynamic loads, 5
Dynamic modulus
    complex modulus, 334, 731
    dynamic modulus by AI, 344
    dynamic stiffness modulus,
        336
    relationship between dynamic
        stiffness modulus and dy-
        namic modulus, 337
    stiffness modulus by Shell,
        340, 342
    typical ranges, 347
Dynamic test, 320, 336, 362

802

KENTRACK program, 42
Keyed joints, 202, 657

Lane distribution factor, 309
Laplace transform, 82, 722
Layered systems, 60, 73
  multilayer system, 100
  nonlinear layer system, 122,
    747, 749
  three-layer system, 73
  two-layer system, 61
  viscoelastic layer system, 137
Layered theory, 100, 735, 753
LAYERINP program, 111, 117,
  767
Lean concrete subbase, 628
Loading waveform, 317
Load safety factor, 613
Load transfer, 191, 252, 472, 645
Log normal distribution, 490
Longitudinal drain, 375, 473
Low-temperature cracking, 37,
  465, 542

Manning's formula, 401
Marshall test, 331
Maxwell model, 84, 720
Mean monthly air temperature,
  553
Mean texture depth, 441, 443
Mechanical models, 83, 719
Meltwater from ice lenses, 393
Method of successive residuals,
  727
MICH-PAVE program, 5, 37,
  127, 537
Modulus of dowel support, 194
Modulus of rupture, 367, 610,
  644
Modulus of subgrade reaction,
  364, 611, 636
MODULUS program, 453
Moisture warping, 174, 599
Moving loads, 93, 137
Multidepth deflectometer, 460
Multiple axles, 104, 298, 311

Nomographs
  continous reinforced concrete
    pavement, 651
  fatigue of bituminous mix-
    tures, 353
  flexible pavement, 573
  rigid pavement, 635
  stiffness modulus of bitumen,
    340
  stiffness modulus of bi-
    tuminous mixtures, 342

Nondestructive deflection testing, 446, 685, 696
Nonlinear materials
  fine-grained soils, 108
  granular materials, 106, 139,
    551, 556, 577
Normal distribution, 486, 572

Overlays
  design methodology, 670, 705
  flexible overlay on flexible
    pavement, 664, 673, 705
  flexible overlay on rigid pave-
    ment, 664, 681, 707
  minimum thickness, 683, 689
  rigid overlay on flexible pave-
    ment, 668, 713
  rigid overlay on rigid pave-
    ment, 669, 686, 691, 710

Pavement condition survey, 685
Pavement management systems,
  39
Pavement performance, 406, 469
Pavement temperature, 138
PDMAP program, 4, 37, 464
Penetration index, 339, 352
Percent normalized gradient, 441
Performance prediction, 492, 497
Permanent deformation. *See*
  rutting
Permanent deformation parameters, 355
Permeability of some aggregates,
  378
PMRPD program, 521
Poisson ratios for different materials, 366
Portland Cement Association
  failure criteria, 235
  overlay design, 685
  rigid pavement design, 607
  simplified design procedure,
    625
Portland cement concrete:
  elastic modulus, 367, 644, 686
  fatigue, 354
  indirect tensile strength, 367,
    648, 686
  modulus of rupture, 367, 610,
    644, 686
  shrinkage, 648
  thermal coefficient, 648
Portland cement concrete pavement. *See* rigid pavement
Precipitation, 34, 390

Predictive models, 463, 466, 599,
  601, 605
Preformed sealant, 200, 472
Present serviceability index, 39,
  426, 502, 520, 570, 704
Present serviceability rating, 427,
  466
Prestressed concrete pavement,
  18
Prime coat, 10
Probabilistic method, 8, 490,
  505, 509
Profilometer, 429, 435
Pumping, 8, 14, 38, 411, 419,
  599, 609
Punchout, 423, 605

Radius of relative stiffness, 171,
  248
Railroad trackbeds, 42
Reflection cracking, 20, 408
Relative damage, 574, 640
Reliability, 5, 38, 494, 498, 568
Remaining life, 563, 672, 701
Representative rebound deflec-
  tion, 677
Resilient modulus
  asphalt mixtures, 326
  correlations with other tests,
    326
  fine-grained soils, 324
  granular materials, 321
  indirect tension test, 320
  loading waveform
  triaxial compression test, 320
Response-type road roughness
  meter, 436
Riding number, 437
Rigid pavements
  design methods, 595, 607, 632
  effect of design feature on per-
    formance, 469
  types, 16–18
Rigid plate, 54
RISC program, 8
Road rater, 448
Road tests
  AASHO Road Test, 26
  Maryland Road Test, 21
  WASHO Road Test, 23
Rosenblueth method, 505
Roughness, 435
Rut depth, 430, 517, 540
Rutting, 37, 411, 464, 540, 547,
  673
R-value, 327, 330, 552

SCEPTRE program, 461
Seal coat, 9
Sensitivity analysis, 145, 266
Serviceability, 5, 425, 570, 695
Shell nomographs, 340, 342, 353
Shoulder design, 583, 656
Skid number, 438
   recommended minimum skid
     number, 443
   related to concrete finishing
     methods, 445
Skid resistance, 442
SLABSINP program, 229, 239,
   770
Slab-subgrade contact, 223, 643
Slope variance, 429, 519
Spalling, 38, 419, 421, 468
Spring constants of dowel bars,
   218
Stage construction, 563, 581
Standard deviation, 482, 501, 502
Standard normal deviate, 487,
   572
Steel reinforcements
   allowable stress, 189, 651
   weights and dimensions, 189
   welded wire fabric, 190
Stiffness matrix
   foundation, 210, 221
   joint, 216
   multiple slabs, 219
   slab, 215
Stiffness modulus. *See* dynamic
   modulus
Stopping distance number, 439
Stresses due to friction, 185

Stress invariant, 56, 106, 578
Stress modification, 124
Stress point, 110, 452
Structural condition factor, 695,
   701, 703
Structural layer coefficient, 331,
   576, 708, 710
Structural models, 537, 595
Structural number, 492, 576
Subbase, 11, 332, 578
Subdrainage, 373, 473
Subgrade, 11, 327, 550, 554, 574
Superposition of wheel loads,
   101
Surface friction, 437
Surface infiltration, 390
Surface or wearing course, 10
Surface texture, 440
Swell, 16, 411, 421, 569

Tack coat, 10
Taylor's expansion, 484, 507
Temperature adjusting factor,
   677
Tempeature curling, 221, 244,
   254
Temperature differentials, 173,
   684
Tensile strain
   horizontal principal versus
     overall principal, 103
   under dual-tandem wheels, 71
   under dual wheels, 69
   under single wheel, 68
Texas triaxial classification, 329
Thermal fatigue cracking, 37, 542

Tie bars, 191
Tied concrete shoulder, 472, 656
Time-temperature superposition,
   90, 137
Tire contact area, 30
Traffic analysis, 303, 491, 496,
   547, 612, 697
Traffic prediction for shoulder
   design, 586
Transverse slope of highway
   cross section, 374
Truck factor, 304

Unbonded slabs, 213, 686, 710

Variance, 481, 486, 499, 504,
   513–520
Vertical displacement, 51, 64
Vertical interface displacement,
   65
Vertical stress, 49, 62
VESYS program, 5, 37, 121,
   137, 356, 509, 541
Viscoelastic materials, 83, 719
Visual condition factor, 675, 700,
   703

WESDEF program, 454
WESLAYER program, 8
WESLIQID program, 6
Wire fabric, 189

Zero-maintenance or premium
   pavements, 20, 174, 355